Y0-EEV-197

Second Supplements to the 2nd Edition of

RODD'S CHEMISTRY OF CARBON COMPOUNDS

Ref.
547
R61
c.

ELSEVIER SCIENCE PUBLISHERS B.V.
Sara Burgerhartstraat 25
P.O. Box 211, 1000 AE Amsterdam, The Netherlands

Distributors for the United States and Canada:

ELSEVIER SCIENCE PUBLISHING COMPANY INC.
655, Avenue of the Americas
New York, NY 10010, U.S.A.

Library of Congress Card Number 91-16129
ISBN 0-444-88157-3

© Elsevier Science Publishers B.V., 1991

All rights reserved. No part of this publication may be reproduced, stored in a retrieval system or transmitted in any form or by any means, electronic, mechanical, photocopying, recording or otherwise, without the prior written permission of the publisher, Elsevier Science Publishers B.V./Academic Publishing Division, P.O. Box 330, 1000 AH Amsterdam, The Netherlands.

Special regulations for readers in the USA – This publication has been registered with the Copyright Clearance Center Inc. (CCC), Salem, Massachusetts. Information can be obtained from the CCC about conditions under which photocopies of parts of this publication may be made in the USA. All other copyright questions, including photocopying outside of the USA, should be referred to the publisher.

No responsibility is assumed by the Publisher for any injury and/or damage to persons or property as a matter of products liability, negligence or otherwise, or from any use or operation of any methods, products, instructions or ideas contained in the material herein. Because of rapid advances in the medical sciences, the Publisher recommends that independent verification of diagnoses and drug dosages should be made.

This book is printed on acid-free paper.

Printed in The Netherlands

Second Supplements to the 2nd Edition of

RODD'S CHEMISTRY OF CARBON COMPOUNDS

VOLUME I

ALIPHATIC COMPOUNDS
★

VOLUME II

ALICYCLIC COMPOUNDS
★

VOLUME III

AROMATIC COMPOUNDS
★

VOLUME IV

HETEROCYCLIC COMPOUNDS
★

VOLUME V

MISCELLANEOUS
GENERAL INDEX
★

123011

Second Supplements to the 2nd Edition of

RODD'S CHEMISTRY OF CARBON COMPOUNDS

A modern comprehensive treatise

Edited by
MALCOLM SAINSBURY
School of Chemistry, The University of Bath,
Claverton Down, Bath BA2 7AY, England

Second Supplement to

VOLUME I ALIPHATIC COMPOUNDS

Part A: Hydrocarbons (Alkanes, Alkenes, Alkynes, Allenes, Carbenes); Halogen Derivatives
Part B: Monohydric Alcohols, their Ethers and Esters; Sulphur Analogues; Nitrogen Derivatives; Organometallic Compounds

ELSEVIER
Amsterdam — Oxford — New York — Tokyo 1991

CONTRIBUTORS TO THIS VOLUME

R. BOLTON

Department of Chemistry, Royal and New College, The Bourne Laboratory,
Egham Hill, Surrey TW20 0EX

D.W. BROWN

School of Chemistry, University of Bath, Claverton Down, Bath BA2 7AY

D.F. EWING

Department of Chemistry, University of Hull, Hull HU6 7RX

A.J. FLOYD

School of Chemistry, University of Bath, Claverton Down, Bath BA2 7AY

D.T. HURST

Faculty of Science, Kingston Polytechnic, Penrhyrn Road, Kingston upon
Thames, Surrey KT1 2EE

A.M. JONES

Department of Chemical and Life Sciences, Newcastle upon Tyne Polytechnic,
Ellison Building, Newcastle upon Tyne NE1 8ST

M. SAINSBURY

School of Chemistry, University of Bath, Claverton Down, Bath BA2 7AY

S.P. STANFORTH

Department of Chemical and Life Sciences, Newcastle upon Tyne Polytechnic,
Ellison Building, Newcastle upon Tyne NE1 8ST

C.W. THOMAS

Department of Physical and Biological Sciences, Bristol Polytechnic,
Coldharbour Lane, Bristol

Preface to Volume 1A/B

This is the first sub-volume of the second supplement of "Rodd" encompassing the literature since 1970, the date when the first supplement was published. The rate of growth of the subject matter in this time has been so great that a decision was taken to subdivide the original chapters thus to reduce the burden upon individual contributors. The original brief given to authors was that between 40-50 pages of camera ready text should be devoted to each topic. Once the project was underway, however, it was soon appreciated that for certain subject areas this was insufficient space and some chapters in this volume are expanded far beyond their planned length.

Rigid adherence to deadlines is always very difficult for busy chemists and the expansion in the size of some contributions has added considerably to this problem. Despite this I am most grateful to all my co-authors for submitting their chapters on time. Two authors Arthur Floyd and Warner Thomas had an especially arduous task since they were responsible for two chapters each. One problem of timetabling could not be surmounted and it has become necessary to delay the publication of Chapter 6 "Nitrogen derivatives of the Acyclic Hydrocarbons". It will be included in a later sub-volume.

Malcolm Sainsbury

CONTENTS

Aliphatic Compounds; Hydrocarbons (Alkanes, Alkenes, Alkynes, Allenes, Carbenes); Halogen Derivatives; Monohydric Alcohols, Their Ethers and Esters; Sulphur Analogues; Nitrogen Derivatives; Organometallic Compounds.

Preface vii
List of common abbreviations and symbols used xvii

Chapter 1. The Saturated Hydrocarbons - Alkanes
by D.F. EWING

1. Introduction 1
2. Sources of alkanes 1
3. Preparation of alkanes 2
   (a) Reduction of alkenes and alkynes 2
   (b) From alkyl halides and alcohols 4
   (c) From thio- and seleno- compounds 6
   (d) From carbonyl and carboxylic derivatives 6
   (e) Using organometallic coupling reagents 8
   (f) Other methods 9
4. Physical properties of alkanes 9
   (a) Thermochemical properties 9
   (b) Conformation 11
   (c) Bulk properties 12
   (d) Spectroscopic properties 12
   (e) Analysis of alkanes 14
5. Reactions of alkanes 15
   (a) Conversion to other hydrocarbons 15
   (i) Cracking and isomerization, 15 – (ii) Aromatization, 17 – (iii) Hydrogenolysis, 18 – (iv) Oxidative coupling, 18 – (v) Dehydrogenation, 19 –
   (b) Oxidation 20
   (i) Oxidation of methane, 20 – (ii) Alkane oxidation by metal oxo species, M=O, 22 – (iii) The Gif system, 24 – (iv) Other transition metal oxidations, 25 – (v) Other oxidations, 26 –
   (c) Electrochemical reactions 27
   (d) Reactions with transition metal complexes 27
   (e) Reaction with metal ions 29
   (f) Carbonylation and cyanation 29
   (g) Alkanes as alkylating agents 30
   (h) Halogenation 31
   (i) Fluorination, 31 – (ii) Chlorination, 31 – (iii) Bromination, 32 – (iv) Other reactions, 33 –

Chapter 2a. Acyclic Hydrocarbons: Alkenes
by D.T. HURST

1. Structure, properties, theoretical considerations 34
2. Synthesis 35
   (a) Biosynthesis 35
   (b) The Wittig reaction 36
   (c) Alkenes via hydroboration 37
   (d) Alkenes via Grignard reagents and related synthesis 39
   (e) Elimination reactions leading to alkenes 41
   (f) Other methods for the synthesis of alkenes 46
3. Reactions 48
   (a) Isomerisation 48
   (b) Hydrogenation 49
   (c) Homologation, oligomerisation, metathesis, and coupling of alkenes 50
   (d) Hydroboration 51
   (e) Hydroformylation 54
   (f) Addition of alkanes to alkenes 55
   (g) Addition of alkylhalides to alkenes 55
   (h) The Diels-Alder reaction 56
   (i) Addition of hydrogen and hydroxy halides to alkenes 56
   (j) Halogenation of alkenes 58
   (k) Hydrocyanation 62
   (l) Amination and aminomethylation 63
   (m) Oxidation reactions of alkenes 66
      (i) Ozonolysis, 66 – (ii) Epoxidation, 67 – (iii) The "Wacker" process, 69 – (iv) Dihydroxylation and diesterification of alkenes, 72 – (v) Other oxidative reactions of alkenes, 74 – (vi) Miscellaneous reactions of alkenes introducing oxygen functionality, 76 – (vii) Nitration, and related reactions, of alkenes, 80 – (viii) The introduction of sulfur substituents into alkenes, 81 – (ix) Cycloaddition reactions of alkenes with 1,3-dipoles, 82 –

Chapter 2b. Carbenes and Allenes
by M. SAINSBURY

1. Carbenes 84
   (a) Introduction 84
   (b) Structure and reactivity 84
   (c) Reactions of halocarbenes 89
      (i) Cyclopropanation, 89 – (ii) Reactions with imines and enamines, 96 – (iii) Reactions with dienes, 97 – (iv) Insertion, 98 –
   (d) Alkyl and alkylidene carbenes 101
   (e) Arylcarbenes 109
   (f) Carbene reactions mediated by metal ions or organometallic catalysts 109
   (g) Fischer carbene complexes 111
2. Allenes 115
   (a) Introduction 115
   (b) Structure and spectra 115

(c) Synthesis 116
(i) Through elimination reactions of alkenes, 116 – (ii) From propargylic compounds, 116 –
(d) Reactions 128
(i) Metalation, 128 – (ii) Photoaddition reactions, 130 – (iii) Thermal rearrangements, 130 – (iv) Cycloaddition and addition reactions, 131 – (v) Miscellaneous reactions, 139 –
(e) Functionalised allenes 140
(i) Silylallenes, 140 – (ii) Allene oxides, 142 – (iii) Hydroxyallenes, silyloxyallenes, acetoxyallenes, and alkoxyallenes, 142 – (iv) Allenic amides and amines, 151 – (v) Carbonyl derivatives of allenes, 152 – (vi) Allenic nitriles, 153 – (vii) Allenic sulphones, 153 – (viii) Allenic boronates, 155 –

Chapter 2c. Acyclic Hydrocarbons: Alkynes
by A.M. JONES AND S.P. STANFORTH

1. Preparation of alkynes 156
(i) Dehydrohalogenation methods, 157 – (ii) Other eliminations, 159 – (iii) From metal acetylides, 163 – (iv) From allenes, 170 –
2. Reactions of alkynes 171
(i) Addition reactions, 171 – (ii) Metal mediated reactions, 186 – (iii) Heterocyclic synthesis, 195 – (iv) Photochemical reactions, 202 – (v) Pericyclic reactions, 204 – (vi) Oxidation, 210 – (vii) Polymerisation, 211 –

Chapter 3a. Monohalogenoalkanes
by R. BOLTON

1. Toxicity and analysis at low levels 214
(a) Toxicity and environmental effects 214
(b) Analysis 216
2. Monohalogenoalkanes 219
(a) Preparation 219
(i) From alkanes, 219 – (ii) From alkenes, 220 – (iii) From alkyl halides, 221 – (iv) From alcohols, 226 – (v) From amines and ammonium salts, 232 – (vi) From carboxylic acids and derivatives, 233 – (vii) From aldehydes, 234 – (viii) From *N*-halogenoamides, 234 –
(b) Reactions of monohalogenoalkanes 234

Chapter 3b. Polyhalogenoalkanes and Alkenyl Halides
by R. BOLTON

1. Dihalogenoparaffins. Dihalogenoalkanes $C_nH_{2n}X_2$ 245
(a) Preparation 245
(i) With both halogen atoms attached to the same carbon atom, 246 – (ii) With halogen atoms attached to different carbon atoms, 246 –
(b) Reactions of dihalogenoalkanes 249
2. Polyhalogenoalkanes 250
(i) Ion-exchange processes, 250 – (ii) Direct halogenation of methane, 250 – (iii) Halogenation of less fully halogenated methanes, 250 –

3. Polyfluoroalkyl halides (RX where R denotes a perfluoroalkyl residue) 252
(a) Preparation 252
(b) Reactions 252
(i) Ion-exchange, 252 – (ii) Hydrolysis, 252 – (iii) Oxidation, 252 – (iv) Reductive dimerisation, 252 – (v) Addition to alkenes, 253 – (vi) With formal nucleophiles, 254 – (vii) Electrophilic attack, 255 – (viii) Enzymatic attack, 255 – (ix) Coupling with alkyl halides, 255 –
4. Polyfluoroalkenes 255
(a) Preparation 255
(b) Reactions 256
5. Halogeno derivatives of olefins 257
(a) Preparation 257
(i) By addition to alkynes, 257 – (ii) By elimination reactions, 257 – (iii) By displacement reactions, 259 – (iv) From alcohols and alkenyl halides, 260 – (v) From carbonyl compounds, 261 – (vi) Oxychlorination, 262 – (vii) Perfluoroalkenyl halides, 262 –
(b) Reactions of alkenyl halides 262
(i) Ozonolysis, 262 – (ii) With organometallics, 262 –
6. Halogenoalkynes 264
(a) Reactions 264
(i) Organometallic compounds, 264 –
7. Halogenocarbenes (Halogenomethylenes, :CHX and :CX_2) 264
(a) Preparation 265
(b) Reactions of halogenocarbenes 267

Chapter 4a. Monohydric Alcohols and Related Compounds
by C.W. THOMAS

1. Introduction 269
2. Preparation of saturated monohydric alcohols (alkanols) 272
(a) Reduction methods 272
(i) Carboxylic acids, 272 – (ii) Esters, 273 – (iii) Acyl halides, 273 – (iv) Aldehydes and ketones, 273 – (v) Asymmetric reduction of aldehydes and ketones, 275 – (vi) Epoxides, 279 –
(b) Substitution methods 279
(c) Organoborane reactions 280
3. Reactions of saturated monohydric alcohols (alkanols) 284
(a) Oxidation 284
(i) Metal-based oxidants, 284 – (ii) Non-metallic oxidants, 290 –
(b) Substitution 294
(i) Formation of haloalkanes, 294 – (ii) Formation of esters and ethers, 297 –
4. Preparation of unsaturated monohydric alcohols 299
(a) Reduction of aldehydes and ketones 299
(b) Epoxide ring opening 300
(c) From alkenes, alkynes and their derivatives 301
5. Reactions of alkenols 305
(a) Oxidation 305
(b) Epoxidation 306

(c) Substitution . . . 306
(d) Rearrangements . . . 306
6. Preparation and reactions of alkynols . . . 307

Chapter 4b. Monohydric Alcohols and Related Compounds: Acyclic Ethers, Hydroperoxides, Peroxides and Esters of Inorganic Acids
by C.W. THOMAS

1. Preparation of dialkyl ethers . . . 310
(a) Williamson-type substitutions . . . 310
(b) Miscellaneous ether syntheses . . . 312
2. Reactions of dialkylethers . . . 314
(a) Rearrangement . . . 314
(b) Ether fission . . . 314
(c) Oxidation . . . 315
3. Preparation and reactions of alkenic ethers . . . 316
4. Preparation of dialkyl peroxides and alkyl hydroperoxides . . . 317
5. Reactions of dialkyl peroxides and alkyl hydroperoxides . . . 319
(a) Various oxidations . . . 319
(b) Alkene epoxidation . . . 324
6. Preparation of alkyl nitrates . . . 325
7. Reactions of alkyl nitrates . . . 327
8. Preparation of alkyl phosphites . . . 329
9. Reactions of alkyl phosphites . . . 330
10. Preparation of alkyl phosphates . . . 337
11. Reactions of alkyl phosphates . . . 341

Chapter 5. Sulphur Analogues of the Alcohols and their Derivatives
by D.W. BROWN

1. Thiols, mercaptans or hydrosulphides . . . 344
2. Sulphides or thioethers . . . 348
(a) Dialkyl sulphides . . . 348
(b) Alkenyl sulphides . . . 351
(c) Alkynyl sulphides . . . 353
3. Derivatives of alkane sulphenic acids . . . 353
4. Dialkyl disulphides or alkyldithioalkanes . . . 355
5. Polysulphides . . . 357
6. Alkylthiosulphuric acids . . . 358
7. Sulphonium compounds . . . 358
8. Dialkyl sulphoxides or alkylsulphinylalkanes . . . 362
9. Alkanesulphinic acids . . . 365
10. Dialkyl sulphones . . . 368
11. Alkanesulphonic acids and derivatives . . . 374
12. Alkane thiosulphonic acids . . . 376

Chapter 6. Nitrogen Derivatives of the Acyclic Hydrocarbons

This chapter will be included in a later sub-volume

Chapter 7a. Aliphatic Organometallic and Organometalloidal Compounds (Derived from Metals of Groups IA, IIA/B, IIIA, IVA, VA , and VIA)
by A.J. FLOYD

1. Introduction 378
   (a) Bibliography 378
      (i) General articles, 378 – (ii) Methods of preparation, 379 – (iii) Analysis, 380 – (iv) General properties and reactions, 381 –
2. Group IA: Lithium, sodium, potassium, rubidium, caesium 382
   (a) Lithium 382
      (i) Carbon-hydrogen bond formation, 383 – (ii) Carbon-carbon bond formation, 383 – (iii) Carbon-nitrogen bond formation, 388 – (iv) Carbon-sulphur bond formation, 388 –
3. Group IIA: Beryllium, magnesium, calcium, strontium, barium. Group IIB: Zinc, cadmium, mercury 388
   (a) Beryllium 389
   (b) Magnesium 390
      (i) Organomagnesium halides, 390 – (ii) Organomagnesium halides and transition metals, 398 –
   (c) Calcium, strontium, barium 408
   (d) Zinc 409
   (e) Cadmium 414
   (f) Mercury 415
4. Group IIIA: Boron, aluminium, gallium, indium, thallium 429
   (a) Boron 430
   (b) Aluminium 444
   (c) Gallium and indium 451
   (d) Thallium 451
5. Group IVA: Silicon, germanium, tin, and lead 457
   (a) Silicon 457
   (b) Germanium 465
   (c) Tin 466
   (d) Lead 513
6. Group VA: Phosphorus, arsenic antimony, and bismuth 493
   (a) Phosphorus 493
   (b) Arsenic, antimony and bismuth 513
7. Group VIA: Selenium, tellurium, polonium 514
   (a) Selenium 515
   (b) Tellurium, and polonium 536

Chapter 7b. Aliphatic Organometallic and Organometalloid Compounds (Compounds Derived from the Transition Metals)
by A.J. FLOYD

1. Introduction and bibliography 537
      (i) General articles, 538 – (ii) Organo-element chemistry, 538 – (iii) Reactions of transition-metal compounds (including syntheses, catalysis, and hydrogenation), 539 –

2. Compounds with metal bonded to one carbon atom 544
(i) Alkyl and alkenyl copper complexes, 544 – (ii) Alkyl and alkenyl zirconium complexes, 547 – (iii) Alkyl and alkenyl palladium and nickel complexes, 549 – (iv) Transition metal carbonyls, 551 – (v) Transition metal carbenes, 554 –
3. Compounds with metal bonded to two carbon atoms 558
(i) (η^2-Alkene)palladium(II) complexes, 559 – (ii) (η-Alkene)iron(II) complexes, 561 – (iii) Transition-metal alkyne complexes, 563 –
4. Compounds with metal bonded to three carbon atoms 566
(i) (η^3-Allyl)palladium complexes, 566 – (ii) η^3-Allyl complexes of other metals, 570 –
5. Compounds with metal bonded to four or more carbon atoms 572
(i) (η^4-Diene)metal complexes, 572 – (ii) η^5-Dienylmetal complexes, 576

Guide to the Index 579
Index 581

LIST OF COMMON ABBREVIATIONS AND SYMBOLS USED

A	acid
Å	Ångström units
Ac	acetyl
a	axial
as, asymm.	asymmetrical
at.	atmosphere
B	base
Bu	butyl
b.p.	boiling point
c, C	concentration
CD	circular dichroism
conc.	concentrated
D	Debye unit, 1×10^{-18} e.s.u.
D	dissociation energy
D	dextro-rotatory; dextro configuration
d	density
dec. or decomp	with decomposition
deriv.	derivative
E	energy; extinction; electromeric effect
*E*1, *E*2	uni- and bi-molecular elimination mechanisms
E1cB	unimolecular elimination in conjugate base
ESR	electron spin resonance
Et	ethyl
e	nuclear charge; equatorial
f.p.	freezing point
G	free energy
GLC	gas liquid chromatography
g	spectroscopic splitting factor, 2.0023
H	applied magnetic field; heat content
h	Planck's constant
Hz	hertz
I	spin quantum number; intensity; inductive effect
IR	infrared
J	coupling constant in NMR spectra
J	Joule
K	dissociation constant
k	Boltzmann constant; velocity constant
kcal	kilocalories
M	molecular weight; molar; mesomeric effect
Me	methyl
m	mass; mole; molecule; *meta-*
m.p.	melting point
Ms	mesyl (methanesulphonyl)

[M]	molecular rotation
N	Avogadro number; normal
NMR	nuclear magnetic resonance
NOE	Nuclear Overhauser Effect
n	normal; refractive index; principal quantum number
o	*ortho-*
ORD	optical rotatory dispersion
P	polarisation; probability; orbital state
Pr	propyl
Ph	phenyl
p	*para-*; orbital
PMR	proton magnetic resonance
R	clockwise configuration
S	counterclockwise config.; entropy; net spin of incompleted electronic shells; orbital state
S_N1, S_N2	uni- and bi-molecular nucleophilic substitution mechanisms
S_Ni	internal nucleophilic substitution mechanisms
s	symmetrical; orbital
sec	secondary
soln.	solution
symm.	symmetrical
T	absolute temperature
Tosyl	*p*-toluenesulphonyl
Trityl	triphenylmethyl
t	time
temp.	temperature (in degrees centrigrade)
tert	tertiary
UV	ultraviolet
α	optical rotation (in water unless otherwise stated)
[α]	specific optical rotation
ϵ	dielectric constant; extinction coefficient
μ	dipole moment; magnetic moment
μ_B	Bohr magneton
μg	microgram
μm	micrometer
λ	wavelength
ν	frequency; wave number
χ, χ_d, χ_μ	magnetic; diamagnetic and paramagnetic susceptibilities
(+)	dextrorotatory
(−)	laevorotatory
−	negative charge
+	positive charge

Second Supplements to the 2nd Edition of Rodd's Chemistry of Carbon Compounds, Vol. IA/B, edited by M. Sainsbury
© 1991 Elsevier Science Publishers B.V., Amsterdam

Chapter 1

THE SATURATED HYDROCARBONS - ALKANES

D. F. EWING

1. Introduction

This survey of alkane chemistry covers the period from 1972 when a review of this subject last appeared in Rodd. This period began with the oil crisis of 1973 which had a significant impact on the petrochemical industry. Subsequent political factors of a similar type have kept the oil and petrochemical industries in a state of flux right up to the present time and this has permeated back to fundamental chemistry with the result that interest in hydrocarbon chemistry has never been greater. In particular the study of alkane functionalization has come from the academic backwaters to become a mainstream activity, the focus of the interest of many prominent academic and industrial laboratories. Many thousands of papers have appeared mostly in the last ten years. Although all important topics are dealt with here the coverage is necessarily selective and this chapter has a substantially changed emphasis when compared with the preceding edition.

2. Sources of alkanes

Oil and gas are obviously still the major sources of hydrocarbons and since the discovery of new reserves has kept pace approximately with consumption this situation will still obtain for many decades to come. However gas reserves have increased faster than oil reserves and hence there is a move towards increased use of a C_1 feedstock for the chemical and petrochemical industries. Ideally this requires the conversion of methane directly to higher hydrocarbons, a process which is not yet commercially viable. The chemistry of methane dimerization and oligomerization is discussed below. An alternative is to convert methane to methanol *via* syngas and then synthesise higher hydrocarbons from that C_1 source (C.D.Chang, "Hydrocarbons from Methanol", M.Dekker, New York, 1983; "Catalytic Conversion of Synthesis Gas and Alcohols to Chemicals", R.G.Herman (Ed.), Plenum, New York, 1984). Petrol obtained by this route accounts for one third of the domestic market in New Zealand (C.J.Maiden, "Methane Conversion", D.M.Bibby *et al.* (Eds), Elsevier, Amsterdam, 1988, p1). This reaction is catalysed by zeolites and by suitable choice of process conditions the conversion can be tailored to produce ethene, light alkenes or petrol (gasoline) (S.A.Tabak and S.Yurchak, Catal. Today, 1990, **6**, 307).

The utilization of coal as a hydrocarbon source is accomplished by gasification to form syngas then application of the Fischer-Tropsch reaction to afford a range of alkane fractions and other chemicals (R.B.Anderson, "The Fischer-Tropsch Synthesis", Academic Press, 1984; M.E.Dry, Catal. Today, 1990, **6**, 183). This process has provided petrol, diesel and chemicals for decades in South Africa.

More than 95% of organic chemicals are derived from these primary hydrocarbon sources and thus the chemistry of the alkanes is of crucial importance to the world chemical industry.

The use of vegetable oil as a source of hydrocarbons has been of interest for many years but has been little more than a curiosity until recently. It has now been shown that a diesel fraction of hydrocarbons can be readily obtained from soybean oil by hydrocracking at 250 - 350°C with high pressure hydrogen using a catalyst which induces hydrogenation and hydrogenolysis (J.Gusmao *et al.*, Catal. Today, 1989, **5**, 533).

3. Preparation of alkanes

(a) Reduction of alkenes and alkynes

Synthetic applications of hydrogenation, including the use of supported catalysts, have been dealt with in several recent books (M.Freifelder, "Catalytic Hydrogenation in Organic Synthesis: Procedures and Commentary", Wiley, New York, 1978; P.N.Rylander, "Hydrogenation Methods", Academic Press, New York, 1985; F.R.Hartley, "Supported Metal Complexes, A New Generation of Catalysts", D.Reidel Publishing Co., 1985). The work described here is restricted to the direct formation of alkanes. Hydrogenation of an unsaturated linkage in functionalised compounds is not included although there is much overlap in methodology.

Homogeneous catalysis of stereospecific hydrogenation by transition metal catalysts has been reviewed (R.E.Harmon, S.K.Gupta and J.W.Scott, Chem. Rev., 1973, **73**, 21; see also D.Valentine and J.W.Scott, Synthesis, 1978, 329). Applications to simple alkene hydrogenation are few but using a rhodium catalyst with chiral ligands 2-cyclohexylbutane was obtained in 60% enantiomeric excess (T.Hayashi, M.Tanaka and I.Ogata, Tetrahedron Lett., 1977, 295). An interesting water soluble catalyst is obtained from $RhCl_3$ and the phosphine $(m\text{-}NaSO_3\text{-}C_6H_4)_3P$. The exact structure of the active species is unknown but it is air stable and effective for quantitative hydrogenation in a biphasic system (*ie* aqueous solution of catalyst and liquid alkene) using a standard low pressure hydrogenator (C.Larpent, R.Dabard and H.Patin, Tetrahedron Lett., 1987, **28**, 2507).

Reduction of cobalt or nickel salts with $NaBH_4$ gives the corresponding metal borides which are efficient heterogeneous hydrogenation catalysts (reviewed by R.C.Wade *et al.*, Catal. Rev. Sci. Eng., 1976, **14**, 211). Nickel boride is often more effective than the traditional Raney nickel (J.A.Schreifels, P.C.Maybury and W.E.Swartz, J. Org. Chem., 1981, **46**, 1263) and cobalt boride is selective for mono- and di-substituted alkenes and alkynes (J.O.Osby, S.W.Heinzmann and B.Ganem, J. Am. Chem. Soc., 1986, **108**, 67). E.C.Ashby and J.J.Lin (J. Org. Chem., 1978, **43**, 2567) have shown that the chlorides of cobalt, nickel or titanium catalyse the quantitative reduction of alkenes and alkynes with $LiAlH_4$. With $TiCl_4$ greater selectivity is shown for unhindered alkenes (P.W.Chum and S.E.Wilson, Tetrahedron Lett., 1976, 15) and simple lithium hydride with VCl_3 is specific for the reduction of terminal alkenes (E.C.Ashby and S.A.Noding, J. Org. Chem., 1980, **45**, 1041). A heterogeneous catalyst system of similar function is $NaH/Ni(OAc)_2/EtONa$ (or other alcoholate salts). The actual catalytic species have unknown stoichiometry but are stable and easily prepared and are not hazardous (J.-J.Brumet, P.Gallois and P.Caubere, J. Org. Chem., 1980, **45**, 1937).

Ionic hydrogenation of a double bond requires a proton donor and a hydride donor. Since a carbocation is formed initially this reduction is selective for

$$R_2C{=}CR_2 \xrightarrow{CF_3COOH} R_2CH\text{-}CR_2^+ \xrightarrow{Et_3SiH} R_2CH\text{-}CHR_2$$

substituted alkenes. This reaction has been reviewed (D.N.Kursanov, Z.N.Parnes and N.M.Loim, Synthesis, 1974, 633).

A convenient route to alkanes of specific length has been described by M.C.Whiting and coworkers. It starts with the readily available bromoaldehyde (1). This aldehyde is reacted with the Wittig reagent made from the acetal of (1) to give a 24-carbon bromoacetal (2). Further analogous chain extension steps and final reductive removal of double bonds and functional groups has afforded a range of alkanes up to $C_{390}H_{782}$ (I.Bidd *et al.*, J. Chem. Soc., Perkin Trans. 1, 1983, 1369; J. Chem. Soc., Chem. Commun., 1985, 543).

$Br(CH_2)_{11}CHO$ (1)

$Br(CH_2)_{11}CH{=}CH(CH_2)_{10}CH(OEt)_2$ (2)

Further studies of this synthetic strategy (E.Igner *et al.*, J. Chem. Soc., Perkin Trans. 1, 1987, 2447; I.Bidd, D.W.Holdup and M.C.Whiting, J. Chem. Soc., Perkin Trans. 1, 1987, 2455) have resulted in some modifications which give improved yields and minimise the contamination with compounds of shorter chain length. Another modification (E.A.Adegoke *et al.*, J. Chem. Soc., Perkin Trans. 1, 1987, 2465) allows the introduction of lateral alkyl substituents. The properties of many of the compounds longer than C_{50} indicate folding of the chain.

(b) From alkyl halides and alcohols

The substitution of a halogen atom by a hydrogen atom is a reaction of great importance and can be effected by a bewildering variety of reagents. An excellent review of this area (A.R.Pinder, Synthesis, 1980, 425) covers catalytic hydrogenation, hydride substitution, metal/acid and metal/ ammonia reductions, halogen radical abstraction and sundry other reactions.

S.Krishnamurthy and H.C.Brown (J. Org. Chem., 1976, **41**, 3064; 1980, **45**, 849) have made an interesting comparison of some of the hydride reagents used for S_N2 substitution of Cl, Br, I or OTs. These reagents are essentially based on the tetrahydroborate or tetrahydroaluminate anions with many substituted variants (see also R.C.Wade, J. Mol. Catal., 1983, **18**, 273). The best of these hydrides is undoubtedly $LiBHEt_3$ which has much better nucleophilicity than $LiAlH_4$ (S.Krishnamurthy and H.C.Brown, J. Org. Chem., 1983, **48**, 3085) and readily reduces hindered alkyl halides and mesylates (R.W.Holder and M.G.Matturro, J. Org. Chem., 1977, **42**, 2166). This reaction is also effective with LiH and only a catalytic amount of Et_3B.

The use of polar solvents such as DMSO or sulpholane increases the effectiveness of $NaBH_4$ as a reducing agent for tosylates (R.O.Hutchings *et al.*, J. Org. Chem., 1978, **43**, 2259) and $(n\text{-}Bu_4N)BH_3CN$ is a particularly effective reducing agent for derivatives of primary alcohols (R.O.Hutchings *et al.*, J. Org. Chem., 1977, **42**, 82). Tertiary and benzylic alcohols have been reduced directly with $NaBH_3CN/ZnI_2$ by what appears to be a radical process (C.K.Lau *et al.*, J. Org. Chem., 1986, **51**, 3038).

Other new hydride reagents include $KBHPh_3$ (for primary bromides and iodides only, N.M.Yoon and K.E.Kim, J. Org. Chem., 1987, **52**, 5564) and $LiAlH(i\text{-}Bu)_2(n\text{-}Bu)$ which is required in stoichiometric amount only rather than the usual excess (S.Kim and K.H.Ahn, J. Org. Chem., 1984, **49**, 1717). Zinc borohydride shows high selectivity for hydride substitution in tertiary chlorides without any competition from elimination (S.Kim, C.Y.Hong and S.Yang, Angew. Chem., Int. Ed. Engl., 1983, **22**, 562). This is the reverse of the normal S_N2 selectivity for which primary iodides are the most reactive. The presence of a catalytic amount of a transition metal salt can enhance the activity of $LiAlH_4$ towards the reduction of alkyl halides and $CoCl_2$ and $NiCl_2$ are particularly effective (J.O.Osby, S.W.Heinzmann and B.Ganem, J. Am. Chem. Soc., 1986, **108**, 67). E.C.Ashby and coworkers (J. Org. Chem., 1978, **43**, 183) have taken this approach one step further and investigated a series of lithium copper hydrides. Quantitative reduction of Cl, Br, I and OTs at primary sites is achieved with Li_4CuH_5 in THF.

In situ formation of the iodide from primary and secondary alcohols with the reagent $Me_3SiCl/NaI/MeCN$ followed by treatment with zinc and acetic acid is a very convenient one-pot process for the reduction of alcohols (T.Morita, Y.Okamoto and H.Sakurai, Synthesis, 1981, 32). This procedure will also reduce simple ethers. Benzylic alcohols are converted to the hydrocarbon without the reducing agent (T.Sakai et.al., Tetrahedron Lett., 1987, **28**, 3817). Using a similar strategy the reduction of tosylates by zinc and *t*-butanol can be achieved directly in refluxing glyme if excess NaI is added (Y.Fujimoto and T.Tatsumo, Tetrahedron Lett., 1976, 3325).

Hydride tranfer from Et_3SiH to the stable carbocation generated *via* the protonation of a tertiary alcohol is a useful route to some hydrocarbons (reviewed by D.N.Kursanov, Z.N.Parnes and N.M.Loim, Synthesis, 1974, 633). When the carbocation is unstable (with respect to rearrangement or elimination) the protic acid can be replaced with advantage by BF_3 (M.G.Adlington, M.Orfanopoulos and J.L.Fry, Tetrahedron Lett., 1976, 2955).

Deoxygenation of secondary alcohols is often particularly difficult by procedures involving nucleophilic substitution and a radical process can be a convenient alternative. The alcohol is suitably derivatised and then decomposed in presence of a radical activator and hydrogen donor.

$$ROH \longrightarrow \underset{(3)}{ROX} \longrightarrow R^{\cdot} \longrightarrow RH$$

This type of reaction has been thoroughly reviewed by W.Hartwig (Tetrahedron, 1983, **39**, 2069) and since it is rarely applied to the formation of hydrocarbons *per se* it will only be discussed briefly. The derivative (3) usually contains a thiocarbonyl group since the sulphur atom is particularly sensitive to radical attack and many different types have been investigated. The radical reducing agent is usually $(n\text{-}Bu)_3SnH$. The additional presence of triethylborane promotes the reduction at 20°C (K.Nozaki, K.Oshima and K.Utimoto, Tetrahedron Lett., 1988, **29**, 6125) thus providing exceptionally mild conditions for deoxygenation. The readily accessible acetates of secondary alcohols are reduced by Ph_3SiH with *t*-butylperoxide as initiator (H.Sano, M.Ogata and T.Migita, Chem. Lett., 1986, 77). The silane (4) is found to be even more effective since high yields are also obtained with primary and tertiary acetates (H.Sano, T.Takeda and T.Migita, Chem. Lett., 1988, 119).

$Ph_2HSi\text{-}C_6H_4\text{-}SiHPh_2$ (4)

$ROC(=NC_6H_{11})NHC_6H_{11}$ (5)

An analogous catalytic reduction (Pd on carbon) of an N,N'-dicyclohexylisourea (5) gives the hydrocarbon RH quantitatively (E.Vowinkel and I.Büthe, Chem. Ber., 1974, **107**, 1353). Hydrogenation of alkyl halides with Na-K alloy as catalyst is enhanced by the use of a phase transfer agent such as a crown ether or tris(3,6-dioxaheptyl)amine (A.K.Bose and P.Mangiaracina, Tetrahedron Lett., 1987, **28**, 2503).

$$RO_2CCO_2N\text{(pyridine-2-thione)} \xrightarrow[\text{Refluxing } C_6H_6]{R'SH} RH$$

(6)

A new procedure particularly suitable for tertiary alcohols involves formation of an oxalate hydroxamic ester of type (6). Treatment with a suitable thiol results in homolysis of the hydroxamic bond followed by expulsion of two moles of CO_2 (D.H.R.Barton and D.Crich, J. Chem. Soc., Perkin Trans. 1, 1986, 1603). The best thiol was Et_3CSH which has a very low nucleophilicity. Hindered secondary and tertiary acetates are best reduced with Li/$EtNH_2$ or K/18-crown-6/*t*-$BuNH_2$ (D.H.R.Barton and coworkers, J. Chem. Soc., Chem. Commun., 1978, 68).

(c) From thio- and seleno- compounds

The removal of sulphur from alkyl sulphides, sulphoxides or sulphones has long been achieved by hydrogenation with Raney nickel. Recently $Mo(CO)_6$ in acetic acid has been shown to be effective for desulphurization of thiols (H.Alper and C.Blais, J. Chem. Soc., Chem. Commun., 1980, 169) and catalysis by molybdenum compounds is likely to be increasingly important in this area. Other long established but more specific methods of removing the sulphur atom involve Li/$EtNH_2$ at low temperature or sodium amalgum in MeOH. The activity of sodium amalgum is improved by buffering with Na_2HPO_4 (B.M.Trost *et al.*, Tetrahedron Lett., 1976, 3477). With $Pd(Ph_3P)_4$ as catalyst $LiBHEt_3$ is effective for the removal of S, SO_2 and Se from unreactive sites (R.O.Hutchins and K.Learn, J. Org. Chem., 1982, **47**, 4380) and reductive deselenization of phenyl selenides occurs quantitatively with $NaBH_4/NiCl_2$ (T.G.Back *et al.*, J. Org. Chem., 1988, **53**, 3815) or with Ph_3SnH in refluxing toluene (D.L.J.Clive *et al.*, J. Am. Chem. Soc., 1980, **102**, 4438).

(d) From carbonyl and carboxylic derivatives

The traditional methods of converting aldehydes and ketones to the hydrocarbon require strongly acidic (Clemmenson reduction, reviewed by E.Vedejs, Org. Reactions, 1975, **22**, 401) or strongly basic conditions (Wolff-Kishner reduction). These relatively harsh conditions are unsuitable to much of modern synthetic methodology and much milder procedures have been developed based on the reduction of tosylhydrazones using hydride reagents such as $LiAlH_4$, $NaBH_4$, and borane (L.Caglioti and coworkers, Org. Syn., 1972, **52**, 122; Bull. Chem. Soc. Jpn., 1974, **47**, 2323). The use of $NaBH_3CN$ in DMF in presence of *p*-toluenesulphonic acid is applicable to a wide variety of ketones (R.O.Hutchins, C.A.Milewski and B.E.Maryanoff, J. Am. Chem. Soc., 1973, **95**, 3662). In the presence of two moles of $ZnCl_2$ the reagent $NaBH_3CN$ is modified to a species of unknown structure. This new reagent is an effective reductant for hydrazones formed *in situ* (S.Kim *et al.*, J. Org. Chem., 1985, **50**, 1927).

$$\mathrm{RCHO} \xrightarrow[\text{Refluxing MeOH}]{\mathrm{TsNHNH_2/ZnCl_2/NaBH_3CN}} \mathrm{RH}$$

These reactions are particularly useful for the introduction of deuterium or for the synthesis of hydrocarbons without loss of chiral integrity. G.W.Kabalka and J.D.Baker (J. Org. Chem., 1975, **40**, 1834; 1981, **46**, 1217) have found that hydrazones are efficiently reduced with catecholborane or bis(benzyloxy)-borane, only one hydride equivalent being required in contrast to other hydride reagents which are used in large excess.

Decarbonylation of aldehydes can be achieved under neutral conditions using rhodium complexes such as $RhCl(PPh_3)_3$ at 20°C (reviewed by J.Tsuji and K.Ohno, Synthesis, 1969, 157). Very little alkene is formed in contrast to the corresponding reaction of an acid chloride where the alkene predominates. An interesting development of this reaction is the use of the rhodium complex $Rh[Ph_2P(CH_2)_3PPh_2]_2Cl$ at a catalytic level for aldehyde decarbonylation (D.H.Doughty and L.H.Pignolet, J. Am. Chem. Soc., 1978, **100**, 7083). For an example of this catalyst prepared *in situ* see M.D.Meyer and L.I.Kruse, J. Org. Chem., 1984, **49**, 3195.

A different approach to this type of transformation has been descibed by N.C.Billingham, R.A.Jackson and F.Malek, J. Chem. Soc., Perkin Trans. 1, 1979, 1137. Treatment of RCOCl (R is primary or secondary) with Pr_3SiH at 140°C in presence of a radical initiator results in a chain reaction which affords the

$$\mathrm{RCOCl} \xrightarrow{\mathrm{Pr_3SiH}} \mathrm{RCO} \longrightarrow \mathrm{R^{\cdot}} \xrightarrow{\mathrm{Pr_3SiH}} \mathrm{RH}$$

alkane in 50-60% yield. Analogous radical decomposition occurs when $(n\text{-Bu})_3SnH$ is reacted with esters of N-hydroxypyridine-2-thione [analogous to (5)] (D.H.R.Barton, D.Crich and W.B.Motherwell, J. Chem. Soc., Chem. Commun., 1983, 939; Tetrahedron, 1985, **41**, 3901) or with phenylseleno derivatives RCOSePh (J.Pfenninger, C.Heuberger and W.Graf, Helv. Chim.

Acta, 1980, **63**, 2328) although relatively high temperatures are required to minimise formation of the aldehyde in the latter case. A radical mechanism is also indicated in the oxidative decarboxylation of acids with $Ag^+/S_2O_8^{2-}$ (W.E.Fristad, M.A.Fry and J.A.Klang, J. Org. Chem., 1983, **48**, 3575).

An interesting application of the Kolbe reaction is the electrosynthesis of hydrocarbons with multiple quaternary sites, compounds such as 3,3,6,6,9,9-hexaethylundecane (N.Rabjohn and G.W.Flasch, J. Org. Chem., 1981, **46**, 4082).

(e) Using organometallic coupling reagents

Organocopper reagents (7) have been widely applied for coupling of primary, secondary or tertiary alkyl fragments to primary alkyl halides R'X (J.F.Normant, Synthesis, 1972, 63; G.H.Posner, "An Introduction to Synthesis using Organocopper Reagents", Wiley, New York, 1980; see also G.H.Posner, Org. Reactions, 1975, **22**, 253). A practical improvement to this reaction is to use a

$$\underset{(7)}{R_2CuLi} + R'X \longrightarrow R\text{-}R'$$

polymer-bound copper species which reacts with RLi to give an alkylating agent. After the coupling step the 'reagent' is then easily removed from the reaction (R.H.Schwartz and J.San Fillipo, J. Org. Chem., 1979, **44**, 2705). Further work by B.H.Lipshutz, R.S.Wilhelm and D.M.Floyd, (J. Am. Chem. Soc., 1981, **103**, 7672) has shown that a reagent of the type $R_2CuCNLi_2$ is very effective for coupling with secondary iodides.

Alkylation of a tertiary halide or tosylate is more difficult but methylation has been achieved with reagents such as $AlMe_3$ (J.P.Kennedy, J. Org. Chem., 1970, **35**, 532), $ZnMe_2$ (M.T.Reetz *et al.*, J. Chem. Soc., Chem. Commun., 1980, 1202), $MeTiCl_3$ or Me_2TiCl_2 (M.T.Reetz, J.Westermann and R.Steinbach, Angew. Chem., Int. Ed. Engl., 1980, **19**, 900). These alkyl titanium chloride reagents will

$$(C_3H_7)_2CO \xrightarrow[-40^\circ C]{Me_2TiCl_2} (C_3H_7)_2CMe_2$$

also effect dimethylation of ketones to give quaternary hydrocarbons directly (M.T.Reetz, J.Westermann and S.-H.Kyung, Chem. Ber., 1985, **118**, 1050). Z.N.Parnes and G.I.Bolestova have surveyed the use of $Me_4Si/AlCl_3$ to achieve direct methylation of alkenes, alkyl halides or even suitable hydrocarbons (Synthesis, 1984, 991). This reaction probably has undeveloped potential.

Even more sterically demanding is the coupling of two tertiary sites and S.H.Brown and R.H.Crabtree (J. Chem. Soc., Chem. Commun., 1987, 970; Tetrahedron Lett., 1987, **28**, 5599) have demonstrated that this is possible in high

$$RCHMe_2 \xrightarrow{\text{Hg, h}\nu} RCMe_2\text{-}CMe_2R$$

yield on a preparative scale by mercury photosensitised dimerization of suitable alkanes in the vapour phase. The photochemical efficiency is about a thousand times greater in the vapour phase relative to solution. More traditional methods such as the Wurtz reaction and the photolysis of azomethanes have been used to synthesise hindered alkanes such as 1,1,2,2-tetra(*t*-butyl)ethane (W.Bernlöhr *et al.*, Chem. Ber., 1988, **119**, 1911) and *meso*- and D,L-1,2-diadamantyl-1,2-di-(*t*-butyl)ethane (M.A.Flamm-Ter Meer *et al.*, Chem. Ber., 1988, **119**, 1492).

(f) Other Methods

A large number of other functional groups can be replaced by a hydrogen atom. Whilst most of these reactions are not of direct interest for the formation of alkanes they may nonetheless find indirect application. A very comprehensive coverage of these reactions and indeed of the full spectrum of reductive transformations in organic chemistry is given by M.Hudlicky, "Reductions in Organic Chemistry", Ellis Horwood Ltd., Chichester, 1984.

Finally we note a procedure for the removal of minor isoalkane impurities from *n*-alkanes by oxidation with iodine tris(trifluoroacetate) (H.Plettenburg, B.Gosciniak and J.Buddrus, Chem. Ber., 1982, **115**, 2377). By this method isoalkane levels are reduced to less than 0.001%.

4. Physical properties of alkanes

(a) Thermochemical properties

Experimental determination of thermochemical data for alkanes (and most other organic compounds) can be achieved with good accuracy (for a recent compilation see J.B.Pedley, R.D.Naylor and S.P.Kirby, "Thermochemical data of organic compounds", Chapman and Hall, London, 1986). Many attempts have been made to relate experimental standard enthalpies of formation for alkanes (ΔH°_f) to their intrinsic structural features in terms of additivity of particular bond energy terms (with allowance for steric effects in some cases). The early work in this area has been critically discussed by J.D. Cox and G.Pilcher ("Thermochemistry of organic and organometallic compounds", Academic Press, New York, 1970). A recent improvement to the bond energy approach has been to make allowance for the contribution of *gauche* forms in the *n*-alkanes

(J.T.Edwards, Can. J. Chem., 1981, **59**, 3192). The g_+ and g_- conformers constitute about 36% of the total conformer population in *n*-butane at ambient temperature. The most comprehensive attempt at the prediction of ΔH°_f is that described by Pedley *et al.* in their recent book (see above). This work is based on group enthalpies with corrections for steric effects, the parameters being derived from selected experimental enthalpies.

A different approach to the analysis of ΔH°_f has been demonstrated by H.Henry, S.Fliszár and A.Julg (Can. J. Chem., 1976, **54**, 2085). Only C-C bond contributions are considered and these are related to the effective charges at each carbon atom which are in turn related to the corresponding experimental carbon chemical shifts. Thus from the assigned spectra of over 40 branched alkanes excellent agreement was obtained between calculated and experimental ΔH°_f values.

A further development away from simple additivity of empirical energy terms has been the advent of force field (molecular mechanics) methods. These procedures attempt to model molecular properties in alkanes in terms of the energy of fundamental classical functions such as bond stretching, bond bending, bond rotation and steric attraction/repulsion (U.Burkett and N.L.Allinger, "Molecular Mechanics", ACS Monograph 177, American Chemical Society, Washington, 1982). Many different calculational procedures have been developed (see F.Osawa and H.Musso, Top. Stereochem., 1982, **13**, 117) but the best known is probably the MM2 package originated by N.L.Allinger and coworkers (J. Am. Chem. Soc., 1977, **99**, 8127; Adv. Phys. Org. Chem., 1976, **13**, 1. The latest version is MM3, N.L.Allinger and J.H.Lii, J. Comput. Chem., 1987, **8**, 1146; see also a partial reparameterization of MM2 by C.Jaime and E.Osawa, Tetrahedron, 1983, **39**, 2769). The best force field methods are able to take account of zeropoint and thermal energy contributions and can reproduce alkane ΔH°_f values to within experimental error in most cases.

The cleavage of C-C bonds in alkanes involves a wide range of activation energies, about thirty orders of magnitude. The importance of the contribution of strain in the ground state to this range of thermal stabilities for C-C bonds is demonstrated by an excellent correlation, equation (1), of experimental activation energies with strain energies calculated by a force field method

$$\Delta G^{+} = -0.60E_S + 65.6 \qquad (1)$$

(C.Rüchardt *et al.*, Angew. Chem., Int. Ed. Engl., 1977, **16**, 875). It is remarkable that the linear relation of activation energy to strain energy is independent of the type of carbon atoms (secondary,tertiary or quaternary) involved.

(b) Conformation

Development of *ab initio* molecular orbital methods has invariably involved the calculation of the energies of alkanes. This is a large field which we can only deal with very briefly. Although sophisticated methods using large basis sets can give excellent estimates of molecular geometry, molecular energy and net atom charges in alkanes, the relatively simple STO-3G method is just as satisfactory, provided all variational parameters are properly optimised (G.Kean and S.Fliszár, Can. J. Chem., 1974, **52**, 2772). Central to understanding the conformational behaviour of alkanes is the nature of the energy barrier in butane and the difference in energy between the *gauche* and *trans* forms. Recently K.B.Wiberg and M.A.Murcko (J. Am. Chem. Soc., 1988, **110**, 8029) have described the application of a range of *ab initio* methods to this problem, illustrating the effect of variation in the basis set, variation in the method of geometry optimization and the inclusion of electron correlation. The butane rotational barrier is 26.54 kJ mol^{-1} (6.34 kcal mol^{-1}) and the *trans/gauche* energy difference is 3.60 kJ mol^{-1} (0.86 kcal mol^{-1}). This paper is a convenient entry into the literature relating to the most recent *ab initio* methods.

C.Rüchardt and his group have investigated many strained alkanes by NMR, X-ray analysis, thermochemical and theoretical methods. They have recently shown that D,L-1,2-diadamantyl-1,2-di(*t*-butyl)ethane exists as a pair of non-interconverting rotamers which can easily be separated, the first example of this type of stereoisomerism in alkanes. The rotational barrier is estimated by calculation as *ca*.170 kJ mol^{-1} for the least hindered rotation (M.A.Flamm-Ter Meer *et al.*, Chem. Ber., 1988, **119**, 1492 and references therein for earlier work). Similar studies are reported for the analogous compound tetra(*t*-butyl)ethane (E.Osawa, H.Shirahama and T.Matsumoto, J. Am. Chem. Soc., 1979, **101**, 4824).

Recent experimental investigation of the *trans/gauche* ratio in alkanes has included the IR study of *n*-tridecane-7,7-d_2 (H.L.Casal, P.W.Yang and H.H.Mantsch, Can. J. Chem., 1986, **64**, 1544) and the NMR study of *meso-* and D,L-butane-2,3-d_2 (G.Schrumpf, Angew. Chem., Int. Ed. Engl., 1982, **21**, 146). An IR study of the variation of the concentration of *gauche* bonds with temperature shows an abrupt onset of disorder at about 10 - 40°C below the melting point of alkanes in the C_{15} to C_{36} range (Y.Kim, H.L.Strauss and R.G.Snyder, J. Phys. Chem., 1989, **93**, 7520 and references to earlier work). F.M.Menger and L.L.D'Angelo have made doubly ^{13}C-labelled undecane and 8-methylhexa-decane (J. Am. Chem. Soc., 1988, **110**, 8241). Examination of the *J*(CC) values reveals that the average conformation of the chain is insensitive to the solvent. A comparison of ^{13}C chemical shifts of *n*-alkanes (C_3 to C_{12}) in the liquid state and in CCl_4 or MeOH solution suggests that there is little conformational variation and hence that coiling of long alkyl chains is minimal in the condensed phase (L.J.M. van de Ven, J.W. de Haan and A.Bucinská, J. Phys. Chem., 1982, **86**, 2516).

(c) Bulk properties

There is strong interest in the structure of the liquid state of alkanes and studies of orientational order have been reviewed (A.Heintz and R.N.Lichtenthaler, Angew. Chem., Int. Ed. Engl., 1982, **21**, 184). One approach to understanding the relationship of bulk properties to chemical structure has been the attempt to model the experimental data in terms of structure desciptors. J.T.Edwards has examined trends in partial molar volume and enthalpy of vaporization of alkanes in terms of the number and type of carbon atom and the mole fraction of *gauche* conformations (Can. J. Chem., 1980, **58**, 1897; 1982, **60**, 480). A more comprehensive investigation (D.E.Needham, I.-C.Wei and P.G.Seybold, J. Am. Chem. Soc., 1988, **110**, 4186) covers properties such as melting and boiling points, molar volume and refractivity, surface tension, critical temperature and pressure and enthalpy of vaporization. Using correlation and factor analysis, data for 74 alkanes were investigated for relationships with structure indices (see L.B.Kier and L.D.Hall, "Molecular connectivity in chemistry and drug research", Academic Press, New York, 1976), information indices, volume/area values and various *ad hoc* indices. Most properties are well modelled by a 'molecular bulk' factor with a minor contribution from a 'shape' factor. Notably melting point is the property least well accounted for by this approach suggesting that the solid state is complex. Further exploration of the relationship of 24 thermodynamic properties to four topological indices is reported for alkanes in the range C_6 - C_9 by H.Narumi and H.Hosoya (Bull. Chem. Soc. Jpn., 1985, **58**, 1778). These studies are extended to the C_{10} - C_{12} range including 624 isomers but only limited experimental data are available (Y.Gao and H.Hosoya, Bull. Chem. Soc. Jpn., 1988, **61**,.2649).

There are many reports of new physical data but these are mostly of specialist interest. Two compilations of general value are the viscosities of the C_{10} to C_{18} *n*-alkanes in the 20 - 100°C and 1 - 1000 bar ranges (D.Ducoulombier *et al.*, J. Phys. Chem., 1986, **90**, 1692), and a critical collection of solubility data for hydrocarbons in water ("Hydrocarbons with water and seawater, Part I, C_5 - C_7 Hydrocarbons; Part II, C_8 - C_{36} Hydrocarbons", D.G.Shaw (Ed.), Pergamon Press,New York, 1988).

(d) Spectroscopic Properties

The dispersion of proton chemical shifts is too small in alkanes to make ^1H NMR of much value in structural studies. In contrast the ^{13}C chemical shift is easily distinguished for a wide variety of environments even in quite complex hydrocarbons and has proved to be a parameter of great importance in structural and theoretical studies.

The chemical shift of a carbon atom in an alkane (or cycloalkane) decreases when an extra carbon is introduced in a synclinal orientation at a β site. This so-called γ-*gauche* effect has been widely applied to the investigation of conformational effects such as the nature of the order in the chain of atactic polypropylene (A.E.Tonelli and E.E.Schilling, Acc. Chem. Res., 1981, **14**, 233) or the non-equivalence of isopropyl methyl groups in alkanes. (A.E.Tonelli, E.C.Schilling and F.A.Bovey, J. Am. Chem. Soc., 1984, **106**, 1157). Several attempts have been made to devise additivity schemes for the carbon chemical shifts in alkanes both as a means of assisting the identification of unknown species (in fossil fuels for example) and as part of the attempt to understand the general relationship of chemical shift to structure. The early work of E.G.Paul and D.M.Grant (J. Am. Chem. Soc., 1963, **85**, 1701; 1964, **86**, 2984) afforded equation (2) which was based on the idea that the shift of a given carbon $\delta(C)_i$

$$\delta(C)_i = -2.3 + A_k + \Sigma S_{ik} \tag{2}$$

is the sum of the effect of all other carbons (A_k terms are required for sites up to five atoms removed). Correction terms S_{ik} are necessary to account for steric effects. A total of 13 parameters were required although this was reduced to 8 in cyclic compounds due to the introduction of rigidity (D.K.Dalling, D.M.Grant and E.G.Paul, J. Am. Chem. Soc., 1973, **95**, 3718).

L.P.Lindemann and J.Q.Adams (Anal. Chem., 1971, **43**, 1245) modified and extended this approach to 59 normal and branched alkanes by evaluating, separately, parameters appropriate for primary, secondary, tertiary and quaternary (p,s,t,q) carbons, 22 parameters in all. If the type of observed carbon (p,s,t,q) is itself a parameter then relatively few secondary parameters are required, particularly if the conformer populations in complex alkanes are taken into account (H.Beierbeck and J.K.Saunders, Can. J. Chem., 1975, **53**, 1307; 1976, **54**, 2985; 1977, **55**, 771). A simpler method for analysing $\delta(C)$ is obtained if the diamagnetic shifts (for p,s,t,q types of carbon atom) are first subtracted from the experimental data (J.Mason, J. Chem. Soc. (A), 1971, 1038; J. Chem. Soc., Perkin Trans. 2, 1976, 1671). The resulting paramagnetic shifts are accounted for by only five parameters (for the effect of other carbons in α to ε sites).

An alternative rationalization of carbon chemical shifts is obtained with equation (3) which combines a factor B_i, characterising the carbon type, with the

$$\delta(C)_i = -1.94 + B_i + \Sigma H_k \tag{3}$$

sum of the effect of hydrogen atoms at other carbons (H_k are defined for up to five sites removed). The rational for this approach is that $\delta(C)$ depends on the p-character of carbon orbitals and this is effected by the presence of hydrogen atoms (Y.A.Shahab and H.A.Al-Wahab, J. Chem. Soc., Perkin Trans. 2, 1988, 255). There is an approximate correlation of $\delta(C)$ in simple alkanes with local Van der Waals interactions as estimated by force field methods (S.Li and D.B.Chesnut, Magn. Reson. Chem., 1985, **23**, 625).

The usefulness of additivity parameters is illustrated by the assignment of the ^{13}C spectra of branched alkanes of the type $C_2H_5(CHMe)_nR$, n = 1 to 4 (P.Lachance, S.Brownstein and A.E.Eastham, Can. J. Chem., 1979, **57**, 367) and of four isoprenoid alkanes with up to 30 carbons (D.K.Dalling *et al.*, Magn. Reson. Chem., 1986, **24**, 191). With the availability of reference data of this type some structural assignment becomes possible for compounds present in heavy oil fractions.

The relation of ^{13}C chemical shifts to *ab initio* atomic charges (and the implication this has for classical inductive effects) has been examined by S.Fliszár and coworkers (Can. J. Chem, 1974, **96**, 4358; 1976, **54**, 2839). According to their analysis the C-H bond is polarised such as to make the carbon atom positively charged and the chemical shift dependence on charge is shown in equation (4). This assessment of C-H bond polarity implies that the

$$\delta(C)_i = 242.6 - 237.1q_i \qquad (4)$$

donor effect of a methyl group is actually a reflection of a diminished acceptor capacity relative to a hydrogen atom.

The three-bond coupling $^3J(HH)$ in a hydrocarbon moiety has a well established dependence on stereochemistry (the Karplus relation). The effect of substituent groups also has a stereochemical dependence and this has been incorporated in generalised forms of the Karplus relation (C.A.G.Hasnoot, F.A.A.M.de Leeuw and C.Altona, Tetrahedron, 1980, **36**, 2783; W.J.Colucci, S.J.Jungk and R.D.Gandour, Magn. Reson. Chem., 1985, **23**, 335). A Karplus relation has been proposed for $^3J(CH)$ (R.Aydin, J.-P.Loux and H.Günther, Angew. Chem., Int. Ed. Engl., 1982, **21**, 449) and for $^3J(CC)$ (S.Berger, Org. Magn. Reson., 1980, **14**, 65) although non-bonded effects are likely to be important in the latter case.

One-bond couplings are sensitive to the magnitude of the valence angles around the coupling atoms and quantitative relations have been derived for the alkane couplings $^1J(CH)$ (G.Szalontai, Tetrahedron, 1983, **39**, 1783) and $^1J(CC)$ (M.Pomerantz and S.Bittner, Tetrahedron Lett., 1983, **24**, 7).

Mass spectrometry of alkanes up to C_{26} has been reviewed in terms of the rearrangements which occur in fragment ions (J.T.Bursey, M.M.Bursey and D.G.I.Kingston, Chem. Rev., 1973, **73**, 191).

(e) Analysis of alkanes

Established procedures such as gas and liquid chromatography are widely used for the routine quantitative analysis of alkane mixtures. Applications of these and other methods are reviewed biennally and will not be further discussed here. The most recent survey is by J.R.McManus, Anal. Chem., 1989, **61**, 165R.

5. Reactions of Alkanes

The chemistry of alkanes is one of the fastest growing areas in synthetic chemistry and since about 1970, many completely new reactions of alkanes have been developed. The alkanes are the least reactive class of organic compound but since Nature has bestowed on the world a bountiful supply of this type of compound it was inevitable that eventually the industry would turn from its preoccupation with hydrocarbons as a source of energy to the much more challenging task of converting these compounds into useful chemicals. Driven, like so much else in the modern world, by the economic pressures and political realities surrounding the oil industry, interest in alkane functionalization has grown at a phenomenal rate. Within the restricted compass of this chapter only limited reference can be made to original work and the reader is often directed to the review literature. For a excellent survey of the 'new' chemistry of alkanes see "Activation and Functionalization of Alkanes", C.L.Hill (Ed.), Wiley-Interscience, New York, 1989.

(a) Conversion to other hydrocarbons

The catalytic conversion of hydrocarbons with a particular range of molecular size and shape to different hydrocarbons with a more useful range of molecular size and shape is one of the most important industrial chemical reactions. It includes most of the petroleum refining process and the production of feedstocks for chemical synthesis from oil and natural gas. The chemistry of these processes is extremely complex involving the heterogeneous conversion of alkanes by various reactions, including hydrogenolysis, dehydrogenation, fragmentation, isomerization, oligomerization, cyclization and aromatization. There is a vast literature dealing with this area particularly with the development of catalysts and only a selection of aspects are dealt with here.

(i) Cracking and Isomerization

The conversion of long alkane molecules to shorter normal and branched alkanes has been an essential part of petroleum refining for a long time. The catalysts developed originally for hydrocarbon cracking were amorphous aluminosilicates but these have been superseded almost entirely by synthetic zeolites of known structure which have high activity at temperatures in the range 250 - 500°C (reviewed by J.Scherzer, Catal. Rev. Sci. Eng., 1989, **31**, 315). These materials are characterised by a three dimensional lattice which has pores and cages of precisely defined size which permit the sorption of molecules only of a particular size range. The internal surface of the zeolite pores contains sodium ions (paired with negatively charged aluminium) which can be exchanged for protons to create acidic sites. These sites can induce the formation of carbocations from sorbed molecules with subsequent isomerization and fragmentation. Many different zeolites have been investigated as shape-selective catalysts for alkane cracking with structures having a range of pore size but the most intensively studied is the zeolite ZSM-5. This material has medium sized pores (5.1 - 5.5 Å) which allow sorption of *n*-alkanes and *iso*-alkanes and has been widely used in commercial cracking catalysts applied to the enhancement of the octane number of a variety of hydrocarbon feedstocks.

The chemistry of cracking on zeolites has been exhaustively reviewed (A.Corma and B.W.Wojciechowski, Catal. Rev. Sci. Eng., 1985, **27**, 29; N.Y.Chen and W.E.Garwood, Catal. Rev. Sci. Eng., 1986, **28**, 185) and only recent work will be discussed here. Product distribution in the cracking of typical alkanes such as *n*-heptane or *n*-dodecane on ZSM-5 and other zeolites depends on the nature and number of acidic (Bronsted) sites in the catalyst and hence on the extent to which aluminium replaces silicon in the matrix (C.Mirodatus, A.Biloul and D.Barthomeuf, J. Chem. Soc., Chem. Commun., 1987, 149; A.Corma and A.V.Orchillés, J. Catal., 1989, **115**, 551; J.Abbot and B.W.Wojciechowski, J. Catal., 1989, **115**, 521). The detailed mechanism of alkane cracking by zeolites is quite well established (J.Abbot and B.W.Wojciechowski, J. Catal., 1989, **115**, 1) and involves classical and/or non-classical carbocation formation depending on the catalyst structure. After initiation by the zeolite catalyst a chain mechanism

$$R^1\text{-}CH_2CH_2\text{-}R^2 \longrightarrow \overset{\text{Z-H}}{R^1\text{-}CH_2CH_2CH_2\text{-}R^2} \longrightarrow \overset{Z^-}{R^1\text{-}CH_2CH_2^+} + CH_3R^2$$

$$R^1\text{-}CH_2CH_2R^2 + \overset{Z^-}{R^1\text{-}CH_2CH_2^+} \longrightarrow \overset{Z^-}{R^1\text{-}CH_2CH^+CH_2\text{-}R^2} + R^1\text{-}CH_2CH_3$$

can operate in which hydride transfer occurs between feed molecules and cations associated with the surface (Z). Vapour phase cations undergo β-scission followed by hydride shifts to generate alkanes of lower molecular mass. Exacting shape requirements for entry to the pores in ZSM-5 means that only one or two *gauche* contributions can be tolerated in long molecules and many chain conformations must uncoil to enter the catalyst (D.Fârcasu, J.Hutchinson and L.Li, J. Catal., 1990, **122**, 34).

Heptane in presence of the strong acid SbF_5/HF undergoes a mixture of cracking and isomerization, the balance between the two reactions being sensitive to conditions. In contrast pentane and hexane isomerise without fragmentation under similar conditions (M.Bassir, B.Torck and M.Hellin, Bull. Soc. Chim. Fr., 1987, 760; Appl. Catal., 1988, **38**, 211). The general behaviour of alkanes in presence of very strong acids has been widely investigated at temperatures ranging from −78°C to about 40°C. The principal reaction is isomerization of *n*-alkanes to branched alkanes *via* the formation of a carbocation; other reactions such as hydride transfer and alkylation can occur. This area is covered by several reviews (D.M.Brouwer and H.Hogeveen, Prog. Phys. Org. Chem., 1972, **9**, 179; P.-L.Fabre, J.Devynck and B.Trémillon, Chem.

Rev., 1982, **82**; see also G.A.Olah, Acc. Chem. Res., 1987, **20**, 422 for reactions of methane).

Radical reactions in alkanes induced thermally or radiolytically are discussed in recent books ("Thermal Hydrocarbon Chemistry", A.G.Oblad, H.G.Davis and R.T.Eddinger (Eds.), American Chemical Society, Washington, 1979; "Radiation Chemistry of the Hydrocarbons", G.Földiák (Ed.), Elsevier, New York, 1981). P.J.L.Gouverneur has reveiwed the chemistry of the dimerization of hydrocarbon radicals (Ind. Chim. Belg., 1974, **39**, 329 and 1974, **39**, 427). Definitive rate expressions for the unimolecular decomposition and radical combination reactions of methane, ethane, propane and *iso*-butane are presented by W.Tsang (Combust. Flame, 1989, **78**, 71). Thermal cracking (pyrolysis at 800 - 1000°C) has been a practical means of obtaining low molecular mass alkanes from crude oil for many years. Extremely complex radical chain reactions are involved and a satisfactory kinetic model has been difficult to design. Using the parameters for the 23 reactions required to describe thermal cracking of butane at low conversion, a successful computer model of this reaction at high conversion above 600°C has been described by J.H.Purnell in "Frontiers of Free Radical Chemistry", W.A.Pryor (Ed.), Chapman and Hall, London, 1985.

A technique which gives results similar to high temperature thermolysis is sonication and K.S.Suslick *et al.*, (J. Phys. Chem., 1983, **87**, 2299) have investigated the sonochemistry of decane and related alkanes.

(ii) Aromatization

One reaction which occurs during cracking which is of special interest is aromatization (dehydrocyclization). The formation of aromatics from alkanes provides a convenient octane enhancer and a valuable chemical feedstock, and much effort has been directed to optimising the yield of aromatics by the development of suitable catalysts. Several commercial processes (reforming) are in use, typically operating in the range 500 - 525°C at pressures of 10 - 40 atmospheres.

Hexane or higher homologues are readily converted to aromatics using bifunctional catalysts such as Pt/Al_2O_3 (S.Sivasanker and S.R.Padalkar, Appl. Catal., 1988, **39**, 123). Notionally the transition metal effects dehydrogenation and acid sites on the oxide support catalyse the cyclization, although such a naïve separation of catalytic function is probably inappropriate (C.N.Satterfield, "Hetergeneous Catalysis in Practice", McGraw Hill, New York, 1979). This type of catalyst functions poorly for the aromatization of propane and butane. Inclusion of another metal in the catalyst such as Re, Ir, Sn, Ge (J.Beltramini and D.L.Trim, Appl. Catal., 1987, **32**, 71) Sb or Te (C.H.Cheng, K.M.Dooley and G.L.Price, J. Catal., 1989, **116**, 325) may increase the selectivity for aromatics.

The development of zeolite catalysts has greatly increased the scope for aromatization of C_3 and C_4 hydrocarbons (for a review see D.Seddon, Catal. Today, 1990, **6**, 351). ZSM-5 doped with Zn or Ga can convert propane with about 50% selectivity, the other main products being methane and ethane. No effective catalyst is yet available for C_1 and C_2 alkanes although addition of a little oxygen to the feed produces a small enhancement in aromatization of ethane over ZSM-5 (G.Centi and G.Golinelli, J. Catal., 1989, **115**, 452).

(iii) Hydrogenolysis

Intensive investigation of alkane hydrogenolysis has been carried out over many years and two reviews cover the earlier work (J.H.Sinfelt, Catal. Rev., 1972, **3**, 175; Z.Paál and P.Tétényi, Catalysis, Specialist Periodical Rep., 1982, **5**, 80). This reaction is analogous to cracking, involving C-C bond scission on a metal surface. Many transition metal catalysts supported on SiO_2 or Al_2O_3 have been studied and the mechanism is understood in general terms (G.C.Bond, J. Catal., 1989, **115**, 286). Recent work is concerned with detailed mechanistic considerations as modelled by ethane hydrogenolysis over single metal (S.A.Goddard *et al.*, J. Catal., 1989, **117**, 155) or mixed metal catalysts (D.J.Godbey, F.Garin and G.A.Somorjai, J. Catal., 1989, **117**, 144) or with the nature of the surface features of the metal at which sorption and reaction occurs (*eg*. J.R.Engstrom, D.W.Goodman and W.H.Weinberg, J. Am. Chem. Soc., 1988, **110**, 8305; B.Coq *et al.*, J. Phys. Chem., 1989, **93**, 4904).

(iv) Oxidative coupling

Catalytic activation of methane can result in two important classes of reaction, oxidative coupling to give C_2 hydro-carbons or selective oxidation to

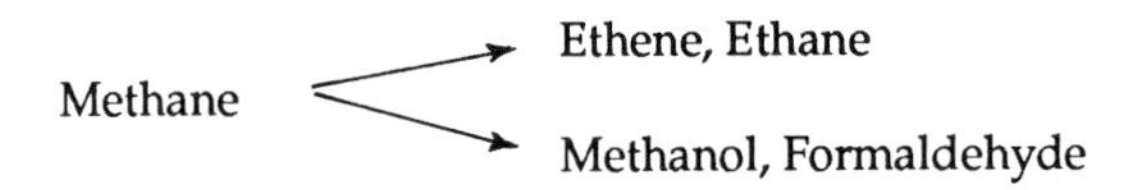

functionalised C_1 species. The economic importance of these reactions arises from the need to develop procedures for the efficient conversion of natural gas at remote field sites into transportable liquid products for use as fuel or chemical feedstock. Since about 1984 the oxidative dimerization of methane to ethene or more especially ethane has been intensively investigated as the first step in the direct conversion of methane to gasoline. Some aspects of this

$$CH_4 \longrightarrow C_2H_6 \longrightarrow C_2H_4 \longrightarrow \text{oligomers}$$

area have been reviewed (J.S.Lee and S.T.Oyama, Catal. Rev. Sci. Eng., 1988, **30**, 249; J.H.Lunsford, Catal. Today, 1990, **6**, 235) and it has been the focus of three symposia ("Methane Conversion", D.M.Bibby *et al.* (Eds.), Elsevier, Amsterdam,

1988; M.Baerns, J.R.H.Ross and K.van der Wiele (Eds.), Catal. Today, 1989, **4**, 271; 1990, **6**, 373).

Oxidative coupling is usually a mixed heterogeneous-homogeneous reaction, methyl radicals being formed by hydrogen atom abstraction at an O^- site in the lattice of a metal oxide catalyst followed by combination in the gas phase to produce ethane. The catalyst may be derived from a main group metal (CaO, MgO), a transition metal (ZnO) or a lanthanide (Sm_2O_3) and is often doped with an alkali salt as promoter (e.g. Li_2CO_3, probably present as Li_2O). Many other oxide/promoter combinations have been reported usually using O_2 as oxidant although CO_2 and N_2O have been found effective. The reaction is highly exothermic overall and in the range 750 - 1000°C methane conversion has a C_2 selectivity approaching 100% at very low conversion rates over some catalysts. More commonly the C_2 selectivity is about 30% for conversions in the 20% - 40% range. Interesting recent variants of the catalyst include Gd_2O_3 promoted with $Na_4P_2O_7$ (R.V.Siriwardane, J. Catal., 1990, **123**, 496), supported palladium oxide at 400°C without gaseous oxidant (K.R.Thampi, J.Kiwi and M.Gratzel, Catal. Lett., 1990, **4**, 49), and the provision of an intrinsic promoter by use of $LiYO_2$ or $LiNO_3$ as sole catalytic species (X.Zhang, R.K.Ungar and R.M.Lambert, J. Chem. Soc., Chem. Commun., 1989, 473). A fuel cell with a solid electrolyte (yttria/zirconia) has been designed which uses the methane coupling reaction to generate ethane and electricity (K.Otsuka, K.Suga and I.Yamanaka, Catal. Today., 1990, **6**, 587). Another unorthodox approach to this reaction involves catalysis by oxygen atoms obtained by the diffusion of O_2 through a silver membrane (A.G.Anshits *et al.*, Catal. Today, 1990, **6**, 593).

Direct photochemical dimerization of C_1, C_2 and C_3 alkanes is promoted by silver atoms and clusters supported on zeolites, selectivity for dimerization being 90% at 1% conversion (G.A.Ozin and F.Hugues, J. Phys. Chem., 1982, **80**, 5174).

Oxidative coupling of methane is also possible under strongly acidic conditions which permit the formation of cationic species. The coupling reaction and other related electrophilic reactions are surveyed by G.A.Olah, Acc. Chem. Res., 1987, **20**, 422.

(v) Dehydrogenation

An intrinsic part of aromatization of an *n*-alkane is the initial dehydrogenation catalysed heterogeneously by a transition metal such as Pt. An analogous photochemical reaction is achieved using the novel homogeneous catalyst, the heteropolytungstic acid ($H_3PW_{12}O_{40}$.10.7 H_2O) in acetonitrile solution (R.F.Renneke and C.L.Hill, New J. Chem., 1987, **11**, 763; J. Am. Chem. Soc., 1988, **110**, 5461). This reaction is about 90% selective for alkenes from branched alkanes but linear alkanes largely form oxygenated products. The mechanism involves the tranfer of an electron from the alkane to an excited state of the catalyst. The resulting radical cation eliminates hydrogen ions regiospecifically to afford the most heavily substituted and hence the most stable alkenes. If Pt(0) is present it catalyses the reoxidation of the heteropolytungstate anion and markedly improves the turnover rate, but is not involved in the actual dehydrogenation step.

(b) Oxidation

Study of the conversion of alkanes to oxygenated products is an area which has undergone enormous expansion in recent years. Oxidation is thermodynamically favourable overall (*ca.* 105 kJ mol^{-1} for each mole of oxygen consumed) but each partial oxidation step, as shown schematically below for methane, is thermodynamically more favourable than the previous one and hence controlled partial oxidation is difficult to achieve in high yield.

$$CH_4 \longrightarrow CH_3\text{-}OH \longrightarrow CH_2{=}O \longrightarrow CHO_2H \longrightarrow CO + CO_2$$

Some of the early industrial processes are reviewed by D.J.Hucknall, "Selective Oxidation of Hydrocarbons", Academic Press, New York, 1974. Many new methods have been developed for the oxygenation of the more reactive alkanes. Since methane is of particular industrial interest it is dealt with separately.

Complete oxidation of hydrocarbons by oxygen (autooxidation) in both the liquid phase and the gas phase has been investigated intensively for many years. An enormous amount of detail concerning the mechanisms has been elucidated from very sophisticated kinetic studies and these radical chain reactions are now quite well understood. In the liquid phase the mechanism is relatively simple but the complexity of the gas phase reaction is illustrated by a study of some aspects of propane oxidation (169 reactions in the mechanistic scheme, S.Refael and E.Sher, Combust. Flame, 1989, **78**, 326) and butane oxidation (244 reactions with a further 82 reactions for a submechanism required in the later stage of the reaction, W.J.Pitz and C.K.Westbrook, Combust. Flame, 1986, **63**, 113). A very thorough review of mechanism appears in volumes 16 (Liquid phase oxidation) and 17 (Gas phase oxidation) of "Comprehensive Chemical Kinetics", C.H.Bamford and C.F.H.Tipper (Eds.), Elsevier, Amsterdam, 1980. The theory of combustion, experimental procedures and computer methods for modelling chain reactions are discussed by D.J.Hucknall, "Chemistry of Hydrocarbon Combustion", Chapman and Hall, London, 1985.

(i) Oxidation of methane

The conversion of methane to methanol *via* syngas is a well established process using a highly evolved technology. As the schematic equation below

$$CH_4 + H_2O \xrightarrow[\text{NiO-Al}_2\text{O}_3]{\textit{ca.}\ 750^\circ\text{C, 2.5 MPa}} CO + CO_2 + H_2$$

$$CO + CO_2 + H_2 \xrightarrow[\text{Cu-ZnO-Al}_2\text{O}_3]{\textit{ca.}\ 200^\circ\text{C, 4.5 MPa}} CH_3OH + H_2O$$

indicates, the two steps require very different conditions.. A number of variations to both steps are possible and current developments in this area aimed at improving efficiency and reducing costs are reviewed by D.L.Trim and M.S.Wainwright (Catal. Today, 1990, **6**, 261). Selectivity for methanol can be virtually 100% but the proportion of higher oxygenates (methyl formate, ethanol etc.) can be enhanced by some newer catalysts such Cs/MoS_2.

The direct oxidation of methane to methanol or formaldehyde is not yet an industrially viable process. Although both homogeneous and heterogeneous reactions have been investigated these are largely of academic interest since no process can approach the syngas reaction in terms of economics. A critical comparison of different catalysts has been made by R.Pitchai and K.Klier (Catal. Rev. Sci. Eng., 1986, **28**, 13). Using Mo or MoO_3 on silica as catalysts and N_2O as oxidant selectivity for methanol is 96% at 400°C but only when the conversion is limited to 0.1% (M.M.Khan and G.A.Somorjai, J. Catal., 1985, **91**, 263: H.F.Liu *et al.*, J. Am. Chem. Soc., 1984, **106**, 4117). At higher temperatures and higher conversion rates (3%) formaldehyde is the main product and over the catalyst Bi_2O_3/SnO_2 selectivity for formaldehyde is 92% (F.Solymosi, I.Tombác and G.Kutsán, J. Chem. Soc., Chem. Commun., 1985, 1455).

Catalysis of methane oxidation by supported metals such as Pd/Al_2O_3 results in formation of CO_2. However the addition of CH_2Cl_2 inhibits the reaction and formaldehyde is formed with a selectivity of 40% at a conversion of 8% (R.S.Mann and M.K.Dosi, J. Chem. Technol. Biotechnol., 1979, **29**, 467). S.Kasztelan and J.B.Moffat (J. Catal., 1987, **106**, 512) have investigated the heteropolyoxometalates as methane oxidation catalysts but the selectivity for CH_2O and CH_3OH is poor (ca 10%) in contrast to their performance as alkane hydroxylation catalysts (see below). The superconducting oxide $YBa_2Cu_3O_{6+X}$ has an unusual selectivity/conversion profile (I.Lee and K.Y.S.Ng, Catal. Lett., 1989, **2**, 403). Methanol is not formed at all and formaldehyde production is maximum at about 5% conversion. Selectivity is increased if the catalyst oxygen atoms are replaced with fluorine.

(ii) Alkane oxidation by metal oxo species, M=O

Biochemical hydroxylation of saturated hydrocarbons is achieved efficiently by the monooxygenase group of enzymes. This selective oxidation of unactivated alkanes by enzymes *in vivo* has no parallel in synthetic chemistry. The oxygenation catalyst is the Fe(III) porphyrin protein, cytochrome P-450, which binds, activates and reduces O_2 to form an active Fe(V) oxo derivative, designated {P}Fe=O. Hydroxylation of a C-H bond plays a central role in the biological response of many organisms to the presence of exogenous compounds including alkanes and facilitates their elimination ("Cytochrome P-450 : Structure, Mechanism and Biochemistry", P.R.Ortiz de Montellano (Ed.), Plenum Press, New York, 1986). The active part of P-450 is a heme moiety *ie* an iron atom ligated with a porphrin molecule. This forms an iron-oxo species which induces homolytic cleavage of a C-H bond in a saturated hydrocarbon followed by transfer of a hydroxyl radical from Fe to C as shown below. This

$$\{P\}Fe^{V}{=}O \ + \ R_3C\text{-}H \longrightarrow \{P\}Fe^{IV}\text{-}OH \ + \ R_3C^{\cdot} \longrightarrow \{P\}Fe^{III} \ + \ R_3C\text{-}OH$$

type of mechanism is now thought to be involved in many chemical oxidations based on transition metal ions or complexes. For example the oxidation of alkanes by chromic acid or chromyl chloride has long been known as a radical reaction and it probably involves a chromium-oxo species. A similar mechanism operates for acid permanganate (R.Stewart and U.A.Spitzer, Can. J. Chem., 1978, **56**, 1273) with a tertiary/secondary selectivity of about 30.

Intensive efforts have been made in the last few years to mimic the bio-oxidation of saturated hydrocarbons using simple iron or manganese complexes of tetraphenylporphyrin (TPP) with a reactive oxidant such as PhIO, peracid or peroxide, operating at 20°C (for a review see B.Meunier, Bull. Soc. Chim. Fr., 1986, 578). An Fe or Mn oxo complex is the active intermediate and the radical mechanism is confirmed by the observed selectivity for tertiary C-H (J.T.Groves and T.E.Nemo, J. Am. Chem. Soc., 1983, **105**, 6243). For the catalysts Fe(TPP)Cl and Mn(TPP)Cl the turnover number is usually low (< 10) but this can be increased greatly by the introduction of halogen substituents into the phenyl groups (T.G.Traylor and S.Tsuchiya, Inorg.Chem., 1987, **26**, 1338; P.E.Ellis and J.E.Lyons, Catal. Lett., 1989, **3**, 389). The combined electronic and steric effects suppress the formation of the unreactive μ-oxo dimer and reduce degradation of the catalyst. For example oxidation of *iso*-butane by O_2 (in air) has a turnover of 12,000 for the fluorinated complex $Fe(TPPF_{20})OH$.

Altering the electrophilicity of the porphyrin ring by the introduction of nitro groups increases the catalytic reactivity so much that even methane can be hydroxylated albeit *ca.* 400 times more slowly than cyclohexane (V.S.Belova, A.M.Khenkin and A.E.Shilov, Kinet. Katal., 1987, **28**, 1011). Steric hindrance of the FeO site by phenyl substituents [*e.g.*in tetra-(2,4,6-triphenylphenyl)-

porphyrin] also alters the selectivity in favour of primary C-H bonds by placing the active site on the flat porphyrin ring at the bottom of a deep pocket (B.R.Cook, T.J.Reinhart and K.S.Suslick, J. Am. Chem. Soc., 1986, **108**, 7281). The shape selectivity of these modified TPP complexes can approach that of some P-450 isozymes.

Oxidation of isobutane has been demonstrated with a triphasic system consisting of alkane/Mn(TPP)X/solvent and NaX/H_2O in presence of solid PhIO. Both ROH and RX (X = Cl, Br, NCO) are formed (C.L.Hill, J.A.Smegal and T.J.Henly, J. Org. Chem., 1983, **48**, 3277). Hydrogen peroxide can be employed as oxidant if the catalyst Mn(TPP)Cl is activated by the presence of imidazole. This heteocyclic base presumably coordinates to Mn and promotes the formation of the Mn=O group. Conversion to oxygenated products is *ca*. 50% including a small amount of ketone (P.Battioni *et al*., J. Chem. Soc., Chem. Commun., 1986, 341). An interesting application is the use of Mn(TPP)Cl/imid-azole as the catalyst in the electrochemical oxidation of alkanes by O_2 with CH_3COOH as proton source. Yields are low (<10%) but the usual radical selectivity is observed (P.Leduc *et al*., Tetrahedron Lett., 1988, **29**, 205). Another example of dioxygen activation is the conversion of heptane to the isomeric heptanones with Mn(TPP)Cl/Ascorbate/O_2 in a biphasic system at 20°C (D.Mansuy, M.Fontecave and J.F.Bartoli, J. Chem. Soc., Chem. Commun., 1983, 253). It is interesting that the use of ascorbate as reducing agent leads exclusively to ketone formation. In this and similar reactions requiring reductive activation of O_2 the yield of oxidised hydrocarbon is low since there is competition between hydrocarbon and reductant for the active oxidising species (M.Fontecave and D.Mansuy, Tetrahedron, 1984, **40**, 4297).

As discussed above the polyoxotungstate anion (8) is an active photochemical catalyst for the dehydrogenation of alkanes. Replacement of a WO group with Fe^{II}, Mn^{II}, Cu^{II}, or Co^{II} to give a species of type (9) results in a novel oxo catalyst which contains no oxidisable organic moieties and is hence intrinsically more stable than the the porphyrin complexes (M.Faraj and C.L.Hill, J. Chem. Soc.,

$PW_{12}O_{40}^{5-}$ (8)

$FePW_{11}O_{39}^{3-}$ (9)

Chem. Commun., 1987, 1487; C.L.Hill *et al*., in "The Role of Oxygen in Chemistry and Biochemistry", W.Ando and Y.Moro-oka (Eds.), Elsevier, Amsterdam, 1988). With iodosylbenzene or alkylhydroperoxide as oxidant for the hydroxylation of alkanes, compounds of type (8) can survive over ten thousand catalytic cycles.

N.Herron and C.A.Tolman (J. Am. Chem. Soc., 1987, **109**, 2837) have described an intriguingly different approach to the selective oxidation of alkanes which involves the doping of a zeolite catalyst with ca 1% Fe(II) and 1% Pd(0).

Hydrogen peroxide is formed *in situ* from O_2 and H_2 (catalysed by Pd) and oxidation of *n*-alkanes occurs at the Fe sites. The shape-selectivity conferred by the 5Å pore size means that cyclic alkanes do not react within the pores and that the oxidation products do not diffuse out and hence the catalyst is rapidly poisoned. However the selectivity of the in-pore hydroxylation reaction (primary/secondary = 0.6) is similar to that of the natural enzymes and this type of catalyst has significant potential.

Complexes of ruthenium involving a bipyridyl (bipy) ligand have some similarity to the porphyrin complexes and the catalytic activity of (10) and related species has been examined for the oxidation of hexane and heptane using *t*-butylhydroperoxide (T.-C.Lau *et al.*, J. Chem. Soc., Chem. Commun., 1988, 1406). Both alcoholic and ketonic products are formed from the secondary sites at 25°C with turnover numbers of *ca.* 1000. PhIO is a much less effective

$Ru[(6,6\text{-diClbipy})_2(OH_2)_2](CF_3SO_3)_2$ (10)

$[RuCl_2(dpp)_2]$ (11)

oxidant. A marked reduction in oxidising power is observed for ruthenium complexes with soft ligands such as diphenylphospinomethane (dpp). Using a biphasic system with LiClO in water as the oxidant, decane was unreactive to oxidation by complex (11) although cyclcohexane reacted normally. The active catalyst may be a ruthenium oxo species (M.Bressan and A.Morvillo, J. Chem. Soc., Chem. Commun., 1989, 421).

(iii) The Gif system

D.H.R.Barton and his group have developed a procedure for the selective oxidation of hydrocarbons to ketones without the intermediate formation of an alcohol which has some analogy with the metal oxo systems described above. This procedure has evolved several variants but essentially requires an iron catalyst, pyridine, acetic acid, a reductant (iron or zinc dust) and an oxidant (O_2 or H_2O_2). The original version (designated Gif-III in its final form since the work was carried out at Gif-sur-Yvette) required a stoichiometric amount of iron to act as catalyst and reductant and gave ketones with nearly 100% selectivity at about 20% conversion (D.H.R.Barton *et al.*, J. Chem. Soc., Perkin Trans. 1, 1986, 947). Higher conversions reduced the selectivity. Alternative use of zinc dust as reductant still required iron as catalyst (D.H.R.Barton *et al.*, New J. Chem., 1986, **10**, 387) and this was also the case in an electrochemical version where a cathode provided the electron input (G.Balavoine *et al.*, Tetrahedron, 1988, **44**, 1091). Detailed study of the mechanism has been required to account for the unusual selectivities of this reaction when compared to hydroxylation by P-450 and the synthetic analogues discussed above (D.H.R.Barton *et al.*, New J. Chem., 1989,

13, 177). A radical mechanism is ruled out by the predominance of products arising from attack at secondary sites (secondary/tertiary ratio is about 100 times greater than than would be expected for the mechanism shown above for the iron-oxo porphyrins). Nonetheless an iron-oxo species is thought to be involved but this does not develop radical character (which seems to require

$$Fe^{V}{=}O + R_2CH_2 \longrightarrow Fe^{V}(OH){-}CHR_2 \longrightarrow Fe^{V}{=}CR_2 \longrightarrow Fe^{IV} + O{=}CR_2$$

the delocalization provided by the porphyrin ligand). Insertion into the C-H bond is followed by formation of a carbene which is further oxidised by superoxide to a ketone. Most of the work on the Gif system has been carried out with cylcoalkanes but the reaction is probably general for any alkane.

A similar type of oxidation of 2-methylbutane has been achieved with H_2O_2, pyridine, acetic acid and copper or iron perchlorate (Yu.V.Geletii, V.V.Lavrushko and G.V.Lubinova, J. Chem. Soc., Chem.Commun., 1988,936). Other transition metal salts are effective substitutes for iron dust provided that the other ingredients are present (Yu.V.Geletii *et al.*, Dokl. Akad. Nauk SSSR, 1986, **228**, 139).

(iv) Other transition metal oxidations

Vanadium forms complexes of type (12) where pic = picolinic acid. Some activation of alkanes to oxidation (to alcohols and ketones) is observed with these complexes but the active species is thought to involve the V-O-O radical rather than a vanadium oxo species (H.Mimoun *et al.*, J. Am. Chem. Soc., 1983, **105**, 3101). Mixed vanadium and phosphorus oxides such as (13) have been

$VO(O_2)(pic).2H_2O$ (12) $\qquad$ $(VO)_2P_2O_7$ (13)

used as catalysts in the conversion of butane to maleic anhydride on an industrial scale. B.K.Hodnett (Catal. Rev. Sci. Eng., 1985, **27**, 373) has surveyed the important factors in this reaction, usually carried out with a feed of *ca.*1% butane in air. A recent study (H.Bosch, A.A.Bruggink and J.R.H.Ross, Appl. Catal., 1987, **31**, 323) indicates that the optimal selectivity for maleic anhydride (49%) is obtained with 15% butane in air. The potential of this catalyst is illustrated by the oxidation of *n*-pentane which, surprisingly, gives 15% maleic anhydride and 23% phthalic anhydride at 80% conversion (G.Centi, M.Burrattini and F.Trifuro, Appl. Catal., 1987, **32**, 353). Clearly there are aspects to this important catalytic system which require further investigation.

Another novel transition metal oxidation is the electrophilic attack of Pd(II) ions in trifluoracetic acid solution at 80°C. As shown below even methane reacts to give the methyl ester in 60% yield (E.Gretz, T.F.Oliver and A.Sen, J. Am. Chem. Soc., 1987, **109**, 8109).

$$Pd^{2+} + CH_4 \longrightarrow Pd\text{-}CH_3^+ \xrightarrow{CF_3CO_2H} Pd(0) + CH_3OOCF_3$$

Alkane oxidation with Co^{III} in acetic or trifluoracetic acid has been examined by several groups (see J.R.Chipperfield and D.E.Webster in "Chemistry of the Carbon-Metal Bond", Vol.4, F.R.Hartley (Ed.), Wiley, New York, 1987). This a complex reaction which is not fully characterised as yet.

(v) Other oxidations

In super acid medium ($HSO_3F\text{-}SbF_5\text{-}SO_2ClF$) oxidation of alkanes is characterised by electrophilic attack to form a penta-coordinate cation (G.A.Olah, N.Yoneda and D.G.Parker, J. Am. Chem. Soc., 1976, **98**, 5261; 1977, **99**, 483). The relatively high basicity of the sigma bond between hydrogen and a tertiary carbon induces preferential attack of the electrophile at this site, probably involving a two electron three centre bond. With protonated hydrogen peroxide ($H_3O_2^+$) or protonated ozone (HO_3^+) as electrophiles oxygenated species are formed leading to the corresponding alcohol or ketone. The products derived from *n*-alkanes are characteristic of attack by an incipient OH^+ species but the reaction is more complex with branched alkanes. A detailed review of this chemistry is given by the above authors in Angew. Chem., Int. Ed. Engl., 1978, **17**, 909.

The electrophilic reagent iodine tri(trifluoroacetate) readily converts tertiary sites in alkanes to the corresponding esters (J.Buddrus and H.Plettenberg, Angew. Chem., Int. Ed. Engl., 1976, **15**, 436). Further oxidation to give

$$Me_3CH + I(OCOCF_3)_3 \longrightarrow Me_3COCOCF_3 + Me_2C(OCOCF_3)CH_2OCOCF_3$$

diesters involves the normally unreactive primary C-H bonds. The selectivity for tertiary relative to secondary sites in this oxidative reaction is the basis of a procedure for the removal of branched alkanes from *n*-alkanes (H.Plettenberg, B.Gosciniak and J.Buddrus, Chem. Ber., 1982, **115**, 2377).

Direct reaction of alkanes absorbed on silica with ozone (at a level of 3% in an oxygen stream) at −40 °C results predominantly in C-C bond cleavage to give ketones (50-80%) (D.Tal, E.Keinan and Y.Mazur, J. Am. Chem. Soc., 1979, **101**, 502). Although no solvent is present the mechanism is thought to be similar to normal solution ozonization *ie* electrophilic attack mainly at tertiary C-H bonds.

Hydroxylation of branched alkanes with 4-nitroperoxybenzoic acid (H.-J.Schneider and W.Müller, Angew. Chem., Int. Ed. Engl., 1982, **21**, 146) shows typical radical characteristics, selectivity for the tertiary alcohol being ca 95%. Similar results are found for reaction with ground state oxygen atoms, 2,3-dimethylbutane giving mainly the tertiary alcohol (E.Zadok and Y.Mazur, Angew. Chem., Int. Ed. Engl., 1982, **21**, 303). A recent synthesis of dimethyldioxirane (14) has allowed R.W.Murray, R.Jeyaraman and L.Mohan to study the oxygen insertion reactions of this compound (J. Am. Chem. Soc., 1986, **108**, 2470). The initial products are characteristic of a radical reaction but in the case of decane further oxidation occurs to give decan-2-one and decan-3-one.

Me O
Me O

(14)

$$Me_3C^+ \longrightarrow Me_2C{=}CH\text{-}COMe$$

(c) Electrochemical reactions

Electrochemical functionalization of alkanes is readily achieved with a Pt electrode in a strongly acidic medium such as HSO_3F (J.Bertram *et al.*, J. Chem. Soc., Perkin Trans. 2, 1973, 373). Evidence shows that the alkane (RH) is protonated to RH_2^+, electro-oxidised to RH_2^{2+} and then converted to a carbocation R^+, a two electron oxidation overall. If the medium contains acetic acid the carbocation reacts to form an α,β-unsaturated ketone as shown above for the *t*-butyl cation. In weakly acidic medium (MeCN/TFA) the alkane is anodically oxidised directly to a carbocation which reacts with the solvent to form the corresponding trifluoracetoxy derivatives (D.B.Clark, M.Fleischmann and D.Pletcher, J. Chem. Soc., Perkin Trans. 2, 1973, 1578). A non-statistical distribution of isomeric products is observed with *n*-alkanes reflecting increased stability for cationic centres remote from the end of the chain (H.P.Fritz and T.Würminghauzen, J. Chem. Soc., Perkin Trans. 1, 1976, 610).

(d) Reactions with transition metal complexes

Many complexes are known in which a transition metal can react intramolecularly with a C-H bond in an alkyl group. These cyclometallation reactions are possible because the C-H bond is activated by the transition metal, by other ligands or by steric effects. In contrast, intermolecular reaction with unactivated alkanes is much more difficult to achieve and only a few examples are known. It has been suggested (W.D.Jones and F.J.Feher, Acc. Chem. Res., 1989, **22**, 91) that such a reaction may actually be more common than is realised simply because the alkyl complex is unstable at the temperature employed in many cases. The subject of C-H bond activation by and toward transition metals is well served by reviews (A.E.Shilov, "Activation of Saturated Hydrocarbons by Transition Metals", Reidel, Dordrecht, 1984; R.H.Crabtree, Chem. Rev., 1985, **85**, 245; M.Ephritikhine, New J. Chem., 1986, **10**, 9; J.R.Chipperfield and D.E.Webster, Chapter 13 in "Chemistry of the Metal-Carbon Bond, Vol.4, F.R.Hartley (Ed.), Wiley, New York, 1987).

Reactivity towards alkanes at room temperature was first observed for complexes (15, M = Ir or Rh) which are activated by photochemical dehydrogenation to form a high energy coordinatively unsaturated species (16).

$$\underset{(15)}{Cp^*(Me_3P)MH_2} \xrightarrow{h\nu} \underset{(16)}{Cp^*(Me_3P)M} \xrightarrow{RH} \underset{(17)}{Cp^*(Me_3P)M\begin{smallmatrix} H \\ R \end{smallmatrix}}$$

This undergoes oxidative addition to an alkane to form complex (17) (A.H.Janowicz and R.G.Bergmann, J. Am. Chem. Soc., 1982, **104**, 352; R.A.Periana and R.G.Bergmann, Organometallics, 1984, **3**, 508). Study of a range of substrates shows that complexes (15) have a high selectivity for primary C-H bonds including methane. The detailed chemistry of these reactions and of analogous complexes such as $CpRe(PMe_3)_3$, $Cp^*Ir(CO)_2$, Cp^*_2LuH and Cp^*Th(cycloalkyl) is critically surveyed by Chipperfield and Webster (see above).

Complex (18) can photoeliminate hydrogen and then add to the C-H bond in an alkane as expected. However the η^2-carboxylate ligand can become η^1-bonded which results in β-elimination in the alkyl group (M.J.Burk and R.H.Crabtree, J. Am. Chem. Soc., 1987, **109**, 8025). The complexed alkene is

$$\underset{(18)}{(PH_3P)_2(CF_3CO_2)IrH_2} \qquad \underset{(19)}{Fe(Me_2PCH_2CH_2PMe_2)_2H_2}$$

expelled to regenerate species (18). Thus complexes of type (18) are catalytic for alkane deydrogenation. Two recent reports illustrate the high reactivity which transition metal complexes may achieve in oxidative addition. The iron phosphine hydride (19) reacts photochemically with pentane at −90°C (M.V.Baker and L.D.Field, J. Am. Chem. Soc., 1987, 109, 2825) and complex

$CpIr(CO)H_2$ experiences photoaddition with the matrix in solid methane at −261 °C (12 K) (P.E.Bloyce *et al.*, J. Chem. Soc., Chem. Commun., 1988, 846).

Probably all of the complexes discussed above have a similar mechanism of addition to alkanes. A detailed theoretical study (J.-Y.Saillard and R.Hoffmann, J. Am. Chem. Soc., 1984, **106**, 2006) of the orbital changes involved in the reaction of methane with species (16) suggests that the initial interaction requires perpendicular approach of a C-H bond to the two unoccupied sites on the metal. Electrons are tranferred between orbitals on the metal and the alkane resulting in concerted formation of M-H and M-C bonds. In spite of the reactivity of these complexes toward unactivated alkanes this chemistry is of little value for direct alkane functionalization on a preparative scale since both reactants and products are difficult to handle in most cases. However there is an obvious potential for catalytic application and many interesting developments may appear in the future.

Alkanes react with platinum(IV) in the presence of platinum(II) to form an alkyl hydride complex by oxidative addition. Reductive elimination may give an alkyl chloride or oxygenated species. This is a complex reaction which is discussed at length in the reviews noted above.

(e) Reaction with metal ions

Many coordinatively unsaturated metal ions react exothermically with alkanes in the gas phase. Such reactions can only be studied by mass spectrometry and related techniques and are used to probe fundamental orbital interactions between metal and C-C and C-H bonds. Most of the work in this area has come from the groups of J.L.Beauchamp and B.S.Freiser and has appeared in the last ten years (see J.B.Schilling and J.L.Beauchamp, Organometallics, 1988, **7**, 194 and Y.Huang and B.S.Freiser, J. Am. Chem. Soc., 1988, **110**, 387 and references therein to earlier work). The most thoroughly investigated metal ions are Fe^+, Co^+ and Ni^+ but other ions include Cr^+, Mn^+, Ru^+, Rh^+, Pd^+, some lanthanides and clusters such as Co_2^+ and $LaFe^+$. The effect of partial ligation of the metal to give ML^+, L = H, CH_3, OH, CO, and Cl, depends upon the metal ion involved. Both C-C and C-H bonds in simple alkanes are activated to reaction with metal ions and dehydrogenation and fragmentation may be observed. This chemistry is complex however, the reaction path depending on the strength of the M-H and M-C bonds and on the nature of the orbital configuration of the metal.

(f) Carbonylation and cyanation

T.Sakakura and M.Tanaka (J. Chem. Soc., Chem. Commun., 1987, 758) have photochemically carbonylated pentane to give hexanal (1 atmosphere CO at 20°) using $RhCl(CO)(PMe_2)$ as catalyst. An very high selectivity (50:1) is observed for the primary relative to secondary C-H bond suggesting that oxidative addition is involved, as discussed above.

Carbon monoxide gas is absorbed by a solution of CuO in FSO_3H/SbF_5 forming a mixture of the complex ions $Cu(CO)^+$ and $Cu(CO)_4{}^+$. Carbonylation of alkanes can be achieved with this ion mixture but isomerization of the intermediate cations gives rise to several products in most cases, identified as carboxylic acids after treatment with water. Hexane for example gave five products including cyclopentane carboxylic acid (Y.Souma and H.Sano, Bull. Chem. Soc. Jpn., 1976, **49**, 3335). A more extensive investigation by N.Yoneda *et al.*, (Chem. Lett., 1983, 17) of the reaction of alkanes with CO in HF/SbF_5 solution has confirmed that this potentially useful reaction is dominated by cation rearrangements which results in a mixture of secondary and tertiary carboxylic acids in most cases. For *n*-alkanes C-C bond cleavage is a further complication.

Facile insertion of the cyano group into an alkane would be an attractive prospect and methyl cyanoformate has been investigated for its reactivity toward alkanes under radical conditions. If alkyl radicals are generated thermally with benzoyl peroxide these attack the cyanoformate only at the cyano carbon to give an iminyl radical (D.D.Tanner and P.M.Rahimi, J. Org.

$$CH_3O_2C\text{-}CN + R^{\cdot} \longrightarrow CH_3O_2C\text{-}\underset{\displaystyle R}{\underset{|}{C}}{=}N^{\cdot} \longrightarrow CH_3^{\cdot} + CO_2 + RCN$$

Chem., 1979, **44**, 1674). The decomposition of this iminyl radical affords the alkyl cyanide in some cases (2,3-dimethylbutane is converted to the 2-cyano derivative in 77% yield) but intramolecular cyclisation to give a nitrogen hetercycle predominates in other cases.

Gas phase reactions of the cyano radical have been studied by several groups (W.P.Hess, J.L.Durant and F.P.Tully, J. Phys. Chem., 1989, **93**, 6402 and references therein). This most recent work includes rate coefficients for the reaction with ethane and propane with cyanide radical generated by laser photolysis of $(CN)_2$.

(g) Alkanes as alkylating agents

Addition of methane to alkenes is calculated to be an exothermic process at 300°C and this interesting reaction can be achieved in practice using a methane-treated nickel catalyst. A similar reaction occurs with benzene (I.D.Löffler *et al.*, J. Chem. Soc., Chem. Commun., 1984, 1177). Although the amount of addition is limited this reaction may have interesting potential.

Alkylation of benzene is normally achieved with an alkyl halide using $AlCl_3$ as catalyst to create an incipient carbocation. In presence of $CuCl_2/AlCl_3$ the alkane 2-methylbutane will produce the cation Me_2EtC^+ by oxidative electron transfer. Reaction with benzene gives a mixture of the isomeric pentylbenzenes in 40% yield at 21°C (L.Schmerling and J.A.Vesely, J. Org. Chem., 1978, **38**, 312).

(h) Halogenation

(i) Fluorination

If an undiluted liquid alkane is mixed with chlorine trifluoride an explosive reaction occurs. However if the alkane is present in dilute solution then fluorination occurs readily at - 75°C (K.R.Brower, J. Org. Chem., 1987, **52**, 798).

$$RH + ClF_3 \longrightarrow RF + HF + ClF$$

This substitution shows high selectivity for tertiary sites, primary sites being unreactive. A hydride abstraction mechanism is implicated.

Electrophilic fluorination is reported using a stream of 2% F_2 in nitrogen (C.Gal and S.Rozen, Tetrahedron Lett., 1984, **25**, 449). Thus monsubstitution at the tertiary hydrogen in 3-methylnonane is achieved, a much better control than is usually observed under radical conditions.

(ii) Chlorination

Interaction of an alkane with chlorine atoms is perhaps the archetypal radical chain reaction and as such continues to attract attention. For liquid phase reaction the nature of the solvent has a substantial effect on the selectivity for secondary over primary hydrogen atom abstraction in butane (A.Potter and J.M.Tedder, J. Chem. Soc., Perkin Trans. 2, 1982, 1689). In CCl_4 the secondary/primary selectivity is 3.0, similar to that for the gas phase reaction (3.9) but in a solvent such as CS_2 which can 'complex' with chlorine atoms this selectivity ratio increases to 9.3. A similar effect is observed for chlorination in benzene although there has been some dispute about the nature of the complex in this case (K.U.Ingold, J.Lusztyk and K.D.Raner, Acc.Chem.Res, 1990, **23**, 219). An interesting catalytic effect has been reported by G.A.Olah and coworkers (J. Org. Chem., 1974, **39**, 3472) who found that butane reacts with chlorine at 25°C in the absence of light but in the presence of PCl_5. This Lewis acid catalyst does not alter the secondary/primary selectivity.

Chlorination with $SOCl_2$ shows a normal tertiary/primary selectivity of about 3.2 for substitution in 2,3-dimethylbutane (J.M.Krasniewski and M.W.Mosher, J. Org. Chem., 1974, **39**, 1303) indicative of hydrogen atom abstraction by chlorine radical. This is a rather complex reaction in which the mechanism is different for thermal and photochemical initiation. N-chlorophthalimide is also a mechanistically complex chlorinating reagent since the selectivity varies markedly with the extent of reaction (M.W.Mosher and G.W.Estes, J. Am. Chem. Soc., 1977, **99**, 6928). In contrast reaction with $ClSO_2NCO$ is well behaved giving the alkyl chloride as 90% of products (M.W.Mosher, J. Org. Chem., 1982, **47**, 1875). A chain length of about 20 and a tertiary/primary selectivity of 115 for the photochemical reaction are in keeping with initiation by isocyanate radical. Similar results are obtained for thermal initiation.

N.J.Turro *et al.*, have described an new type of heterogeneous methodology involving the photochemical reaction of chlorine with *n*-alkanes adsorbed on a pentasil zeolite (J. Org. Chem., 1988, **53**, 3731). The selectivity depends upon factors such as the Si/Al ratio and the water content of the catalyst, or the substrate loading but in all circumstances the terminal methyl sites are favoured. For reaction in the pores of the zeolite primary/secondary selectivity is about 6 and there is also some selectivity between secondary sites. It is not clear whether the photolysis of Cl_2 occurs inside or outside of the catalyst pores but this intriguing process must have considerable potential for the development of synthetic procedures with selectivities unattainable in solution.

Several catalysts which promote the reaction of Cl_2 as an electrophile have been surveyed by G.A.Olah and coworkers (J. Am. Chem. Soc., 1973, **95**, 7686). With SbF_5 in SO_2ClF reaction with alkanes occurs even at –78 °C but there is substantial scission of C-C bonds in addition to substitution. The less powerful Lewis acid $AlCl_3$ restricts the reaction to substitution at C-H but subsequent ionisation and elimination/addition reactions give rise to many products. With $AgSbF_5$ alkanes react only at tertiary sites to give the corresponding alkyl chlorides in low yield. None of these reactions are synthetically useful.

(iii) Bromination

In presence of mercuric oxide the reactivity of bromine is increased substantially and bromination of alkanes can readily be achieved in a useful yield (72% for tetramethylbutane) (N.J.Bunce, Can. J. Chem., 1972, **50**, 3109). As might be expected the selectivity of this reagent is reduced relative to bromine as indicated by a tertiary/secondary/primary ratio of 400/28/1. Although some bromine radical is probably present the principal active species is thought to be the bromoxy radical, generated from Br_2O.

$$Br_2 + HgO \longrightarrow HgBr_2 + Br_2O$$

Electrophilic bromination with Br_2 has been investigated with $AgSbF_6$ as catalyst (G.A.Olah and P.Schilling, J. Am. Chem. Soc., 1973, **95**, 7680). Substitution occurs at tertiary or secondary sites without any accompanying scission of C-C bonds. However rearrangement is prevalent.

(iv) Other reactions

Homolytic cleavage of the sulphur chlorine bond in sulphenyl chlorides RSCl gives chlorine and thiyl radicals. A detailed study (D.D.Tanner, N.Wada and B.G.Brownlee, Can. J. Chem., 1973, **51**, 1870) of the reaction of butane and isobutane with C_6Cl_5SCl reveals that H atoms are abstracted mainly by chlorine radicals and that the chain propagation is achieved by ArS radical. The

$$ArS^{\cdot} + ArSCl \longrightarrow ArSSAr + Cl^{\cdot}$$

$$R^{\cdot} + ArSSAr \longrightarrow ArSR + ArS^{\cdot}$$

tertiary/secondary/primary selectivity of this radical is 25/8/1. The products from the reaction of alkanes with a variety of sulphenyl chlorides such as (20) and (21) show that formation of alkyl thioethers is about ten times more likely than formation of alkyl chlorides (J.F.Harris, J. Org. Chem., 1978, **43**, 1319; 1979, **44**, 563). This product selectivity and the secondary/primary ratio are dependent on the structure of the thiyl radical. Other sulphur containing

C_6F_6SCl (20

$(CF_3)_3CSCl$ (21)

$CH_3C{-}C{-}S{-}C{-}C(=S)$ (ring) (22)

products are also formed in many cases and this reaction has no preparative value. Sulphur dioxide will promote the dehydrogenation of *n*-butane and isobutane to butenes at 350 - 400° over palladium on Al_2O_3. The conversion rate is low and further reaction occurs to give cyclic species of type (22) (F.M.Ashmany, J. Catal., 1977,**46**,424).

Salts of the nitronium ion ($NO_2^+.X^-$ where X^- is PF_6^-, SbF_6^- or BF_4^-) in methylene chloride/sulpholane solution will react with alkanes at 20°C (G.A.Olah and H.C.Lin, J. Am. Chem. Soc., 1971, **93**, 1259). However this reaction is of little value as a synthetic procedure since the substitution reaction to form nitroalkanes in low yield is accompanied by extensive isomerization and cleavage of C-C bonds. Similarly nitration of isobutane in HNO_3/H_2SO_4 gives all possible C_1 to C_4 nitroalkanes and other products.

The photochemical reaction of nitrosyl chloride with alkanes is a route to the corresponding oximes. The major products are ketoximes resulting from reaction of the nitrosyl radical with secondary alkyl radicals but a surprisingly large amount of aldoximes are formed (E.Müller and A.E.Böttcher, Chem. Ber., 1975, **108**, 1475). The secondary/primary selectivity is about 5. No products are detected deriving from nitrosyl reaction at tertiary sites and hence 2,3-dimethylbutane is converted to the aldoxime in quantitative yield.

Second Supplements to the 2nd Edition of Rodd's Chemistry of Carbon Compounds, Vol. IA/B, edited by M. Sainsbury
© 1991 Elsevier Science Publishers B.V., Amsterdam

Chapter 2a

ACYCLIC HYDROCARBONS: ALKENES

DEREK T. HURST

1. Structure, properties, theoretical considerations

The description of the carbon-carbon double bond as a "banana", i.e. bent equivalent hybrid, bond system first proposed by Pauling in 1931 has been shown using Valence-bond calculations, without the imposition of any orthogonality constraints, to be more stable than the description of Hückel of a one σ-bond and one π-bond system (W.E.Palke, J.Am.Chem.Soc., 1986,108,6543).

The isodemic enthalpies of hydrogenation of the lower alkenes have been shown to be almost the same regardless of chain branching and to be about -21.5 kcal mol^{-1} (R.Fuchs, J.Chem.Ed., 1984,61,133; D.W.Rogers Tetrahedron Lett., 1987,28,1967).

The enthalpies of hydration of a number of alkenes have also been measured and have been shown to be about -11.9 kcal $mol^{-1}$with slight differences according to the alkene (K.B.Wiberg and D.J. Wasserman, J.Am.Chem.Soc., 1981,103, 6563; K.B. Wiberg, D.J. Wasserman and E.Martin, J.Phys.Chem., 1984,88,3684).

A review with 211 references on the *cis trans* isomerisation of alkenes was published in 1980 (J.Saltiel and J.L. Charlton in "Rearrangements in Ground and Excited States", Ed.P.de Mayo, Academic Press, N.Y., 1980,Vol.3,p.25).
The rotational barriers of strained alkenes have also been studied and can be described by a torsional potential which is independent of the degree of substitution (W.v.E.Doering, *et al.*, Chem.Ber., 1989,122,1263).

A dynamic NMR study has investigated the rotation about alkyl-alkene bonds using substituted isopropylethenes (J.E. Anderson, *et al.*, J.Chem.Res.Synop., 1986,420).

In further studies of sterically hindered alkenes a number of attempts have been made to synthesise tetra-*tert*-butylethene (1). However the tetraaldehyde (3), obtained by metathesis of the alkene (2) is the precursor closest to (1) that has yet been obtained (A.Krebs *et al.*, Tetrahedron, 1986,42,1693).

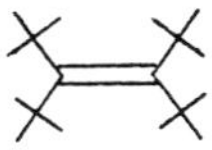

OHC CHO OHC CHO CO_2H NH_2

(1) 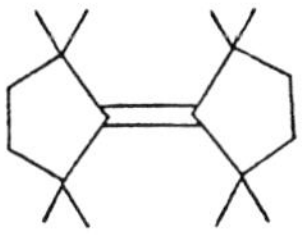 (2) 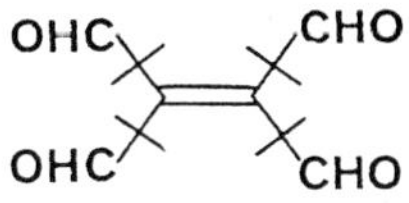 (3) 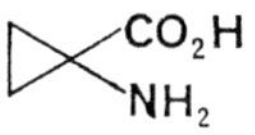 (4)

A theoretical study of the structure and stability of protonated alkenes in both the gas phase and in solution has shown that there is a distinction between alkenes which have an equal number of alkyl substituents on both sides of the double bond and those which have a different number. In the latter case no bridged carbenium ions could be defined. However the effects of solvents have a profound effect on the stability of the cationic structures and change the gas phase data remarkably (H.J. Rauscher *et al.*, Theor.Chim. Acta, 1980,57,255).

2. Synthesis

(a) Biosynthesis

Ethene is known to be a plant growth regulator and its biosynthesis from methionine *via* 1-aminocyclopropane-1-carboxylate (4) has been proposed (D.O. Adams and S.F.Yang, Proc.Natl.Acad.Sci.USA, 1979,76,170; K.Lürssen, K.Naumann and R.Schroder, Z.Pflanzenphysiol. Bd., 1979,92.s.285). It has now been shown that $[^2H_4]$-aminocyclopropane carboxylate is converted by apple slices into $[^2H_4]$ethene without loss of deuterium and the proposed biosynthetic pathway has been confirmed (R.M. Adlington *et al.*, J.Chem. Soc.Chem.Commun., 1982,1086).

(b) The Wittig Reaction

The Wittig reaction continues to be a widely used and well studied method for the synthesis of alkenes. Of recent reviews of this reaction one covers the olefination of compounds other than carbonyls (P.J. Murphy and J.Brennan, Chem.Soc.Rev., 1988,17,1), and another is all-embracing having 558 references principally since 1978 (B.E.Maryanoff and A.B. Reitz, Chem.Rev.1989,89,863). Another recent review on phosphorus addition at sp^2 carbon centres gives 230 references (R.Engel, Org.React., 1988,36,175).

A number of modifications to the earlier experimental techniques have been proposed. The use of potassium carbonate as catalyst reduces aldolisation of the carbonyl compound and introduces the effect of a phase-transfer process to a solid-liquid heterogeneous medium. It gives greatly improved yields (Y. Le Bigot, M.Delmas, and A,Gaset, Synth.Commun.,1982,12,107). A variety of other bases have also been used in a number of instances. The use of phase-transfer techniques has also proved to be beneficial for alkene synthesis using the Wittig reaction. Examples include gas-liquid phase-transfer catalysts (E. Angeletti, P. Tundo, and P. Venturello, J.Chem.Soc. Chem.Commun., 1983,269) and liquid-liquid transfer (J.Villieras and M. Rambaud, Synthesis, 1982,924;1983, 300; A.Kh.Khusid and B.G. Kovalev, Zh.Org.Khim.,1987,23, 71; Chem.Abstr.,1988,107,175154). The Wittig reaction in heterogeneous media has been reviewed (Y. Le Bigot, M. Delmas, and A. Gaset, Inf.Chim., 1987,286,217).

The mechansim, stereochemistry, and substituent effects of the Wittig reaction have been extensively investigated by Vedejs and co-workers (E.Vedejs, C.F. Marth, and R.Ruggeri, J.Am.Chem.Soc.,1988,110,3940; E.Vedejs and C.F. Marth, J.Am.Chem.Soc., 1988,110,3948) who have found that the variation in stereochemistry is attributed to dominant kinetic control. Formation of the intermediate *cis* or *trans* oxaphosphetanes is the decisive step. Stabilisation of the phosphorus ylide and steric crowding of the ylide or of the aldehyde influences E or Z stereoselectivity. (R.Tamura *et al.*, J.Org.Chem., 1988,53,2723).

The use of selenium in Wittig alkene synthesis has been independently reported by two groups (K. Okuma *et al.*, Tetrahedron Lett., 1987,28,6649; G.Erker, R. Hock, and R.Noble, J.Am.Chem.Soc., 1988,110,624). The reaction proceeds through selenocarbonylintermediates to give *trans* alkenes (Scheme 1).

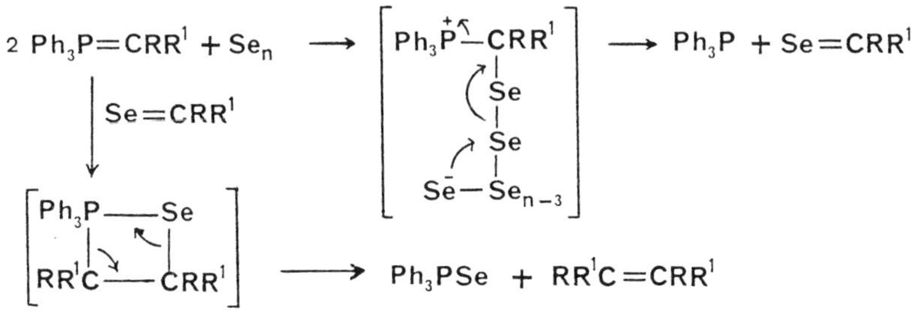

$$2\ Ph_3P{=}CRR^1 + Se_n \longrightarrow \left[Ph_3\overset{+}{P}{-}CRR^1{-}Se{-}Se{-}Se_{n-3}{-}\overset{-}{Se}\right] \longrightarrow Ph_3P + Se{=}CRR^1$$

$$2\ Ph_3P{=}CRR^1 \xrightarrow{Se=CRR^1} \left[\begin{array}{cc} Ph_3P & Se \\ RR^1C & CRR^1 \end{array}\right] \longrightarrow Ph_3PSe + RR^1C{=}CRR^1$$

Scheme 1

A similar process has been described for the reaction of Wittig reagents with episulfides or sulfur (K. Okuma *et al.*, Chem.Lett., 1987,357; Bull.Chem.Soc., Japan, 1980,61,4323).

(c) Alkenes via Hydroboration

The application of organoboranes for the stereospecific synthesis of alkenes has been extensively studied, particularly by H.C. Brown and co-workers. The stereospecific synthesis of trisubstituted alkenes is conveniently carried out by the hydridation of dialkylhaloboranes followed by hydroboration and iodination of an internal alkene. This is exemplified for (Z)-3-methyl-2-pentene (5) in Scheme 2. The reaction proceeds *via trans* addition, followed by *trans* elimination resulting in the *trans* stereochemistry of the two alkyl groups from the alkyne. (H.C.Brown,

D. Basavaiah, and S.U. Kulkarni, J.Org.Chem., 1982,47,171).

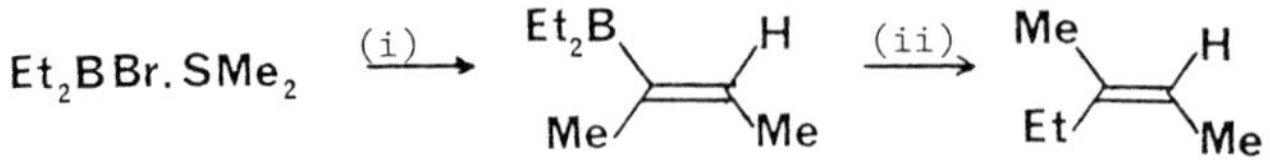

$Et_2BBr.SMe_2$ →(i) $Et_2B(Me)C=C(H)Me$ →(ii) $Me(Et)C=C(H)Me$

Reagents: (i) MeC≡CMe, $LiAlH_4$; (ii) NaOMe,I_2, -78°C (5)

Scheme 2

These workers have used related methodologies using bromoalkynes, alkylhaloboranes, and variations in procedure to develop general stereospecific syntheses of trisubstituted alkenes and disubstituted alkenes (H.C. Brown *et al.*, J.Org.Chem., 1988,53,239, and papers cited therein).

Hydroboration has now been extended to that of enamines which provides an easy route to either [Z]-or[E]- alkenes. The synthesis of pure [Z]- or [E]- alkenes is readily achieved from the same acyclic ketone enamine by modification of the hydroboration-elimination procedure. The two approaches are: hydroboration of the enamine by 9-borabicyclo[3.3.1]nonane (9-BBN), followed by methanolysis, or hydroboration by borane methyl sulfide (BMS), followed by methanolysis and hydrogen peroxide oxidation. The ready availability of stereospecific enamines thus makes the first general synthesis of the two stereoisomers from a single intermediate (B. Singaram *et al.*, J.Am.Chem.Soc., 1989,111,384). The method is indicated for [E]-4-pyrrolidinyl-3-heptene (6) in Scheme 3.

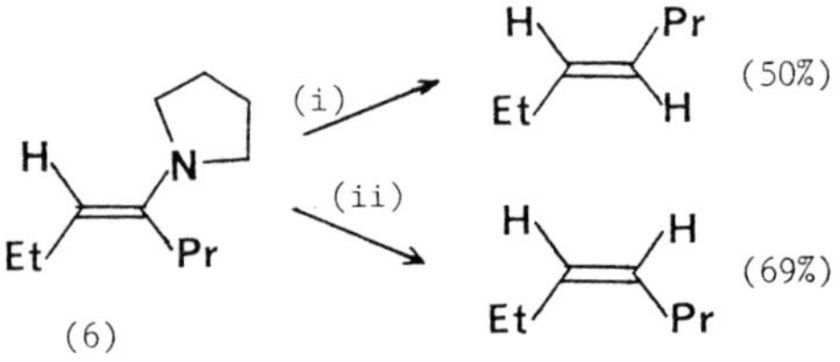

(6) H(Et)C=C(Pr)(N-pyrrolidinyl) →(i) H(Et)C=C(Pr)H (50%)

(6) →(ii) H(Et)C=C(H)Pr (69%)

Reagents: (i) BMS,MeOH,[O]; (ii) 9-BBN, MeOH

Scheme 3

[E]-2-Bromoethenyl)dibromoborane has also been shown to be a useful precursor for the synthesis of [E]-olefins by the stepwise cross-coupling reaction with organic chlorides then with organic halides in the presence of base (catalysed by Pd complexes) (S.Hyuga *et al.*, Chem.Lett., 1988,809; S.Hyuga *et al.*, Chem.Lett., 1987,1757).

(d) Alkenes via Grignard reagents and related synthesis

The substitution of halogen, alkoxy, or alkylthio groups on alkenes by alkyl groups in reactions with Grignard reagents has led to a general method for the stereospecific synthesis of alkenes (E. Wenkert and T.W. Ferreira, J.Chem. Soc.Chem.Commun., 1982,840; V. Ratovelomanana and G. Linstrumelle, Synth.Commun., 1984,14,179). For example, the reaction of the readily available [Z]-1-bromo-2-phenylthioethene (7) with butylmagnesium bromide in the presence of nickel (II) or palladium (II) complexes as catalyst gives [Z]-5-decene (8) (72%) in 98% purity (V.Fiandanese, *et al.*, J.Chem.Soc.Chem.Commun., 1982,647).

Br(H)C=C(H)SPh (7) —(i)→ Bu(H)C=C(H)SPh —(ii)→ Bu(H)C=C(H)Bu (8)

Reagents: (i) BuMgBr, Et_2O, $PdCl_2(PPh_3)_2$, r.t. 1h
(ii) BuMgBr,Et_2O, $NiCl_2(Ph_2PCH_2CH_2PPh_2)$, r.t. 24h

This reaction has also been extended to the stereoselective synthesis of trisubstituted alkenes (V.M. Dzhemilov *et al.*, Izv.Akad.Nauk SSSR, Ser.Khim., 1988,2385; Chem.Abstr., 1989,110,231055).

The stereoselective synthesis of alkenes has also been achieved through the use of the addition of vinyl cuprates to α,β-unsaturated sulfones (G. De Chirico *et al.*, J.Chem. Soc.Chem.Commun., 1981,523) (Scheme 4) or by the sodium dithionite reduction of vinylsulfones (J. Bremner *et al.*, Tetrahedron Lett., 1982,23,3265). The cross coupling of vinylic sulfones with Grignard reagents catalysed by nickel and iron complexes also provides a stereospecific synthesis of alkenes (J.-L. Fabre, M. Julia, and J.-N. Verpeaux, Tetrahedron Lett., 1982,23,2469).

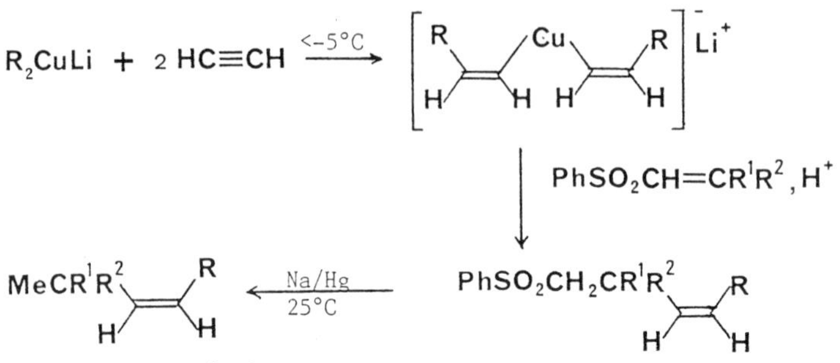

$$R_2CuLi + 2\,HC{\equiv}CH \xrightarrow{<-5^\circ C} [(RCH{=}CH)_2Cu]^- Li^+$$

$$\xrightarrow{PhSO_2CH{=}CR^1R^2,\ H^+} PhSO_2CH_2CR^1R^2CH{=}CHR \xrightarrow[25^\circ C]{Na/Hg} MeCR^1R^2CH{=}CHR$$

(R = n,s,t-Bu; R^1,R^2 = H,Me)

Scheme 4

The reaction of ethyl trimethylsilyl acetate with Grignard reagents, followed by acid treatment, provides a convenient synthesis of symmetrical 1,1-disubstituted ethenes (Scheme 5) (G.L. Larson and D.Hernandez, Tetrahedron Lett., 1982,23,1035).

$$Me_3SiCH_2CO_2Et \xrightarrow{(i),(ii)} CH_2{=}CR_2 \quad (R = \text{various alkyl})$$

Reagents: (i) $2RMgX, Et_2O$; (ii) H_2SO_4, THF, or $BF_3.OEt_2$, CH_2Cl_2

Scheme 5

A one-pot regioselective synthesis of alkenes has been devised using α-chlorocarboxylic acid chlorides *via* a tandem addition of two different nucleophiles, such as Grignard reagents, and further lithiation. Many such compounds have been obtained by the general method shown in Scheme 6 (J.Barluenga *et al.*, J.Org.Chem., 1983,48,609).

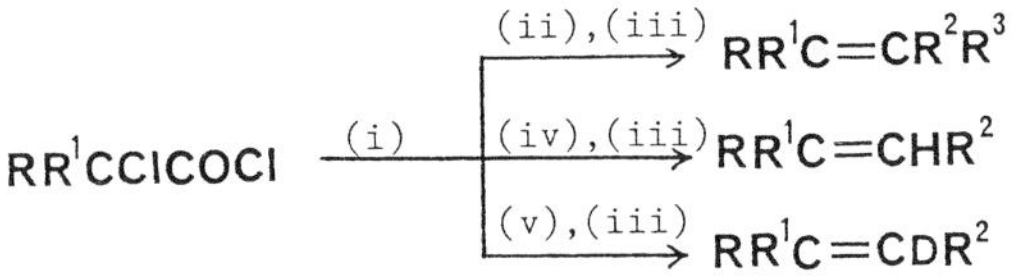

$$RR^1CClCOCl \xrightarrow{(i)} \begin{cases} \xrightarrow{(ii),(iii)} RR^1C{=}CR^2R^3 \\ \xrightarrow{(iv),(iii)} RR^1C{=}CHR^2 \\ \xrightarrow{(v),(iii)} RR^1C{=}CDR^2 \end{cases}$$

Reagents: (i) R^2MgBr, THF, -78°C to -60°C; (ii) R^3MgBr, Et_2O, -60°C; (iii) Li, -60°C to 20°C; (iv) $LiAlH_4$,$AlCl_3$, EtO, -60°C (v) $LiAlD_4$,$AlCl_3$,Et_2O, -60°C

Scheme 6

(e) Elimination reactions leading to alkenes

Elimination reactions are extensively studied and are regularly reviewed (see Org.React.Mech., published annually by J. Wiley).

In recent years some new elimination reactions which produce alkenes have been investigated. *sec*-Alkyl phenyl telluroxides have been shown to decompose under mild conditions to yield alkenes (S. Uemura and S. Fukuzawa, J.Am.Chem.Soc., 1983,105,2748). For example, the treatment of the dibromide (9) with aqueous sodium hydroxide at room temperature gives a mixture of 1-octene and *cis*-and *trans*-2-octenes (80%).

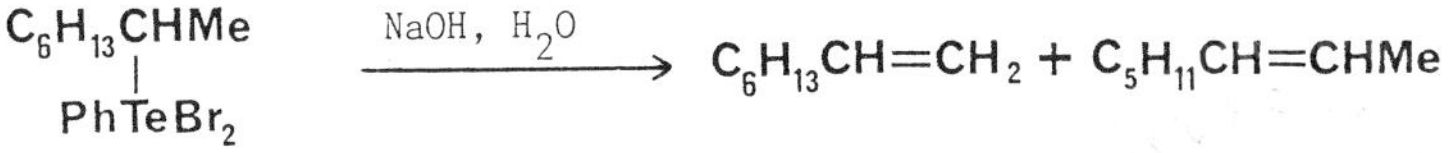

$$\underset{(9)}{C_6H_{13}CH(PhTeBr_2)Me} \xrightarrow{NaOH,\ H_2O} C_6H_{13}CH{=}CH_2 + C_5H_{11}CH{=}CHMe$$

The ratio of terminal to linear alkenes in such an elimination is about 2.5:1 and the elimination seems to occur *via* an internal *syn* process. A development of this reaction is the formation of alkenes *via* the telluroxide elimination by the direct oxidation of *sec*-alkyl phenyl tellurides using m-chloroperoxybenzoic acid in ether (S. Uemura, K. Ohe, and S. Kukuzawa, Tetrahedron Lett., 1985,26,895).

The use of the pyridylselenoxy group as a leaving group for alkene synthesis has been shown to be much preferable to the use of the phenylselenoxy group giving excellent yields under mild conditions. Thus treatment of (pyridin-2-yl)selenyloctadecane with 30% hydrogen peroxide in THF at 32°C for 0.6h gives 82% 1-octadecene (A.Toshimitsu *et al.*, Tetrahedron Lett., 1980,21,5037). Another new synthesis of alkenes uses the elimination of β-tributylstannyl organosulfur compounds (M. Ochiai *et al.*, Chem.Pharm. Bull., 1984,32,1829).

An interesting and useful new synthesis of alkenes utilises α-triflyl-dimethylsulfone ($CF_3SO_2CH_2SO_2CH_3$) which allows successive polyalkylation of the two carbon atoms with regiocontrol. The polyalkylated triflyl-sulfone then undergoes a Ramberg-Bäcklund reaction with loss of triflinate anion and extrusion of sulfur dioxide to yield the alkene (Scheme 7) (J.B. Hendrickson, G.J. Boudreaux, and P.S. Palumbo, Tetrahedron Lett., 1984,25,4617).

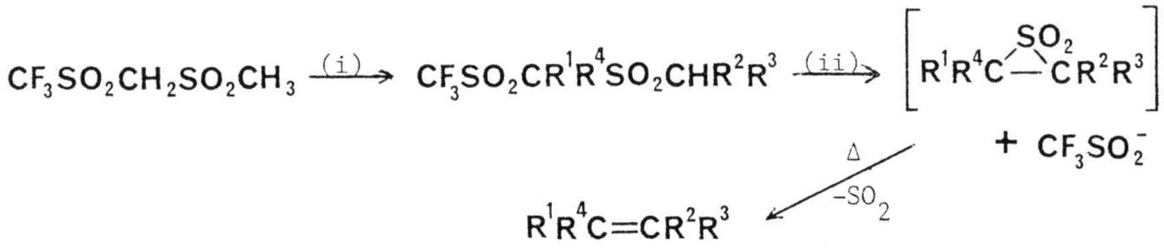

$$CF_3SO_2CH_2SO_2CH_3 \xrightarrow{(i)} CF_3SO_2CR^1R^4SO_2CHR^2R^3 \xrightarrow{(ii)} \left[R^1R^4C\overset{SO_2}{-}CR^2R^3\right] + CF_3SO_2^-$$

$$\xrightarrow[-SO_2]{\Delta} R^1R^4C{=}CR^2R^3$$

Reagents: (i) BuLi,R^xI(or Br),successively, THF,various temp., time;(ii) BuLi,THF

Scheme 7

A highly satisfactory, direct, one-pot synthesis of terminal alkenes has been developed using carboxylic acid chlorides and the method shown in Scheme 8 (J.Barluenga *et al.*, J.Org.Chem., 1983,48,3116).

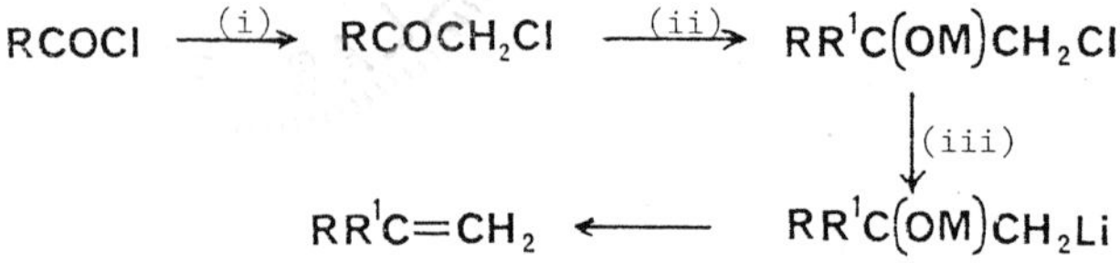

$$RCOCl \xrightarrow{(i)} RCOCH_2Cl \xrightarrow{(ii)} RR^1C(OM)CH_2Cl \xrightarrow{(iii)} RR^1C(OM)CH_2Li \longrightarrow RR^1C{=}CH_2$$

Reagents: (i) CH_2N_2, EtO, then HCl; (ii) R^1MgBr, $MgBr_2$ (or $LiAlH_4$, $AlCl_3$); (iii) Li (various R,R^1; yields up to 99%)

Scheme 8

Another method which gives alkenes under relatively mild conditions is the oxidative decarboxylation of carboxylic acids in aqueous solution using persulfate in the presence of silver (I) and copper (II) ions. The persulfate and silver (II) generates silver (I) which reacts with the carboxylic acid to generate carboxylate radicals which decompose to give alkyl radicals. The alkyl radical is then oxidised via an alkyl copp. (II) species (W.E. Fristad, M.A. Fry and J.A. Klang, J.Org.Chem., 1983,48,3575).

Mechanism and the stereochemistry of eliminations involving alkyltrimethylammonium salts continue to be of interest and to be rationalised. For example a study of the E2 eliminations of $R^1R^2CHCHDNMe_3^+$ shows that *syn* elimination becomes the major reaction path when R^1 and R^2 are both bulky groups, the results being rationalized by simple conformational arguments (Y.-T. Tao and W.H. Saunders, J.Am.Chem.Soc., 1983,105,3183). These studies have continued (B.R. Dohner and W.H. Saunders, J.Am.Chem.Soc., 1986,108, 245; R. Subramanian and W.H. Saunders, J.Am.Chem.Soc., 1984,106,7887).

An alternative conversion of secondary alkyl primary amines into alkenes under very mild conditions (chloroform, room temp.) involves their reaction with the pentacyclic pyrylium salt (10). The reaction proceeds *via* the non-isolated pyridinium intermediate (11) and the corresponding secondary carbenium ion (12). Isomeric mixtures of alkenes are obtained resulting from carbenium ion rearrangements (A.R. Katritzky and J.M. Lloyd, J.Chem.Soc.Perkin Trans.1, 1982,2347).

(10) $\xrightarrow{R^1CH_2CH(R)NH_2}$ (11)

$RCH{=}CHR^1 \leftarrow R\overset{+}{C}HCH_2R^1$ (12)

Interesting syntheses of alkenes from alkyl amines have also resulted from ruthenium-catalysed hydrogenolysis of butylamine (G.C.Bond and L.P.Vousden, J.Chem.Soc.Chem. Commun., 1986,538) and the zirconium oxide catalysed elimination of ammonia from 2-butanamine (A.Satoh, H.Hattori, and K.Tanabe, Chem.Lett., 1983,497).

Alcohols are useful precursors for the synthesis of alkenes. The first example of an *anti*-Saytzeff orientation for the dehydration of an alcohol with an alumina catalyst was reported in 1976 (B. Floris, J.Org.Chem., 1976,41,2774). This reaction has been investigated further and it is concluded that the *anti*-Saytzeff selectivity is a result of kinetic control, the product being the primary reaction product that desorbs to the gas phase. For heterogeneous catalysis the geometry of the catalytic site and electronic factors, as well as steric factors, control the reaction product (B.H. Davis, J.Org.Chem., 1982,47,900; H.A.Dabbagh and B.H. Davis, J.Mol.Catal. 1988,47,123).

Alkene synthesis by deoxygenation of *vic* diols has been reviewed (E.Block, Org.React., 1984,30,457) (189 refs.) and a new, mild, stereospecific conversion of *vic* diols into alkenes has been found involving the intermediacy of 2-methoxy-1,3-dioxolane derivatives which decompose to give alkenes on refluxing in acetic anhydride (M. Ando, H.Ohhara, and K.Takese, Chem.Lett., 1986, 879).

In recent years the conversion of methanol into hydrocarbons has attracted considerable attention and is of great industrial importance. At the mechanistic level the conversion of methanol to ethylene on zeolite catalysts is also of great interest. The favoured mechanism involves the formation of dimethyl ether then methylation of a dimethyloxonium methylide intermediate (Scheme 9) rather than the formation of a zeolite bound carbenoid species (G.A. Olah *et al.*, J.Chem.Soc.Chem.Commun., 1986,9; F.X. Cormerais *et al.*, J.Chem.Res.Synop., 1980,362).

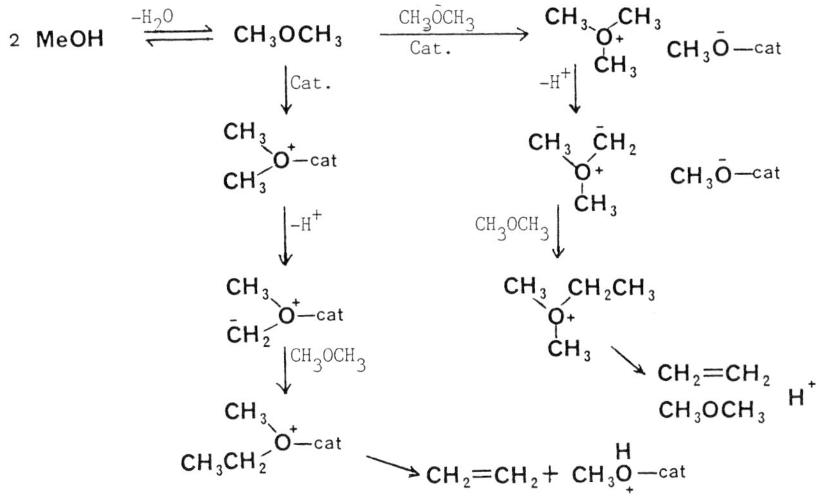

$$2\ MeOH \underset{}{\overset{-H_2O}{\rightleftharpoons}} CH_3OCH_3 \xrightarrow[Cat.]{CH_3\bar{O}CH_3} (CH_3)_3O^+ \quad CH_3\bar{O}\text{—cat}$$

$$CH_3OCH_3 \xrightarrow{Cat.} (CH_3)_2O^+\text{—cat} \xrightarrow{-H^+} (CH_3)(\bar{C}H_2)O^+\text{—cat} \xrightarrow{CH_3OCH_3} (CH_3)(CH_3CH_2)O^+\text{—cat} \rightarrow CH_2{=}CH_2 + CH_3\overset{H}{O^+}\text{—cat}$$

$$(CH_3)_3O^+ \xrightarrow{-H^+} (CH_3)(\bar{C}H_2)(CH_3)O^+ \quad CH_3\bar{O}\text{—cat} \xrightarrow{CH_3OCH_3} (CH_3)(CH_2CH_3)(CH_3)O^+ \rightarrow CH_2{=}CH_2 \quad CH_3OCH_3 \quad H^+$$

Scheme 9

An "in-depth" study has been carried out of the gas-phase, base-induced, elimination reactions of onium intermediates (G. Occhiucci *et al.*, J.Am.Chem.Soc., 1989,111, 7387; G. Angelini, G. Lilla, and M. Speranza, J.Am.Chem. Soc., 1989,111,7393).

The elimination:substitution ratio associated with the attack of amines on diethylmethyloxonium and halonium ions has been studied in the gas phase using Fourier transform ion cyclotron resonance mass spectrometry. Evidence has been gained pointing to the possibility of alkyl radical-amine ionradical recombination and other mechanisms being involved in these processes. The mechanism involved in high pressure, in low pressure, and in solution systems seem to differ

The conversion of methanol to a mixture of ethene, propene and butenes can also be effected in the liquid phase using cobalt catalysts (M. Blanchard *et al.*, J.Chem. Soc.Chem.Commun., 1982, 570).

The dehalogenation of *vic*-dihalides can be a useful reaction. The common debrominating reagent is zinc dust in acetic acid or ethanol. A useful new method is the use of zinc dust with a catalytic amount of titanium (IV) chloride in THF at 0°C (F. Sato, T. Akiyama, K. Iide, and M. Sato, Synthesis 1982,1025).

A study of the zinc-promoted dehalogenation of *vic*-dihalides has shown that the proportion of syn-elimination in the overall reaction varies with the variation of the halogen bearing group in the order I<Br<Cl<F between 3 - 30% in the erythro and 5 - 60% in the threo series. On the basis of the variable transition state theory, the observed pattern of *syn - anti* dichotomy has been correlated with the extent of double bond development in the transition state (M.Pankova *et al.*, Collect.Czech.Chem.Commun., 1983,48,2944).

A new approach to the elimination of hydrogen halide from primary and secondary alkyl bromides and iodides involves the use of nickel (0) complexes in the presence of DBU. This method requires mild conditions and is devoid of substitution reactions (M.C. Henningsen, S. Jeropoulos, and E.H. Smith, J.Org.Chem., 1989,54,3015). The use of DBU and DBN has proved to be valuable in alkene synthesis including a route to 1,1-disubstituted alkenes from primary tosylates and iodides (S. Wolff, M.E. Huecas, and W.C. Agosta, J.Org.Chem., 1982,47,4358).

(f) *Other methods for the synthesis of alkenes*

The Fischer-Tropsch synthesis of hydrocarbons is an old-established process. The use of this method for the production of alkenes is of industrial as well as academic interest. When carbon monoxide and hydrogen are heated with triiron dodecacarbonyl supported on inorganic oxides (magnesium oxide or silicon dioxide) propene is formed with a similar degree of selectivity as the reaction of ethene suggesting that this is an intermediate in the CO/H_2 reaction. A carbene + alkene metallocyclobutane mechanism has been proposed for this reaction (F.Hugues, B.Besson, and J.M. Basset, J.Chem.Soc.Chem.Commun.,1980,719).

A light olefin product (methane and C_2 - C_6 hydrocarbons containing over 75% alkenes) can be obtained from CO/H_2 in a liquid phase synthesis using a cobalt catalyst (from cobalt (II) acetylacetonate and triethylaluminium) in an alkylterphenyl solvent (M. Blanchard *et al.*, J.Chem. Soc.Chem.Commun., 1980,908).

Further alkene syntheses include photocatalytic dehydrogenation of alkenes (K.Nomura and Y.Saito), J.Chem.Soc. Chem.Commun.,1988,161), the selective reduction of dienes via

H transfer in the presence of carbonyl clusters (J.Kaspar, R. Spogliarich, and M. Graziani, J.Organomet.Chem., 1985, 281,299), and the electrochemical reduction of carbon dioxide at a copper electrode at ambient temperature (Y.Hori *et al.*, J.Chem.Soc.Chem.Commun., 1988,17).

Reductive coupling of the t-butyl ketone (13) by $TiCL_3/LiAlH_4$ gives rise to the highly strained E-3,4-diethyl-2,2,5,5,tetramethylhex-3-ene (14). Such compounds are of interest for structural and chemical reasons (D. Lenoir, D. Malwitz, and B.Meyer, Tetrahedron Lett., 1984, 25,2965; D. Lenoir, H.R. Seikaly, and T.T. Tidwell, Tetrahedron Lett., 1982, 23,4987).

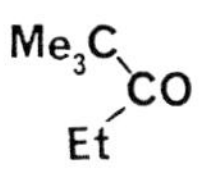

$Me_3C(Et)CO$ (13) $\qquad$ $Me_3C(Et)C{=}C(Et)CMe_3$ (14)

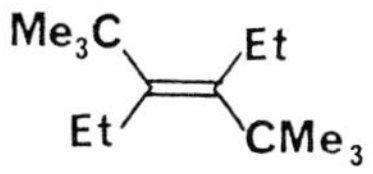

3-Hydroxyalkanoic acids in the presence of oxophilic metal species undergo not only oxidative decarboxylation, but also deoxygenation to yield alkenes, this representing a direct route from aldol compounds to alkenes (Scheme 10) (I.K. Meier and J. Schwartz, J.Am.Chem.Soc., 1989,111,3069). This provides a new route to tri and tetra substituted alkenes.

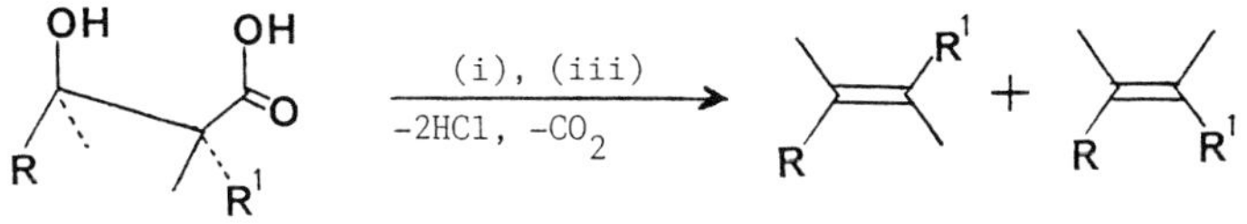

(i), (iii); $-2HCl$, $-CO_2$

Reagents: (i) $Cl_3V^V{=}NC_6H_4Me$; (ii) heat

Scheme 10

3. Reactions

The reactions of alkenes have been regularly reviewed in "Organic Reaction Mechanisms" and in the "Annual Reports of the Royal Society of Chemistry". Photochemical reactions have been surveyed annually in "Photochemistry" and other reviews have appeared in the "RSC Special Publ. Photochemistry and Organic Synthesis (1985)". Reviews of nucleophilic addition to alkenes (C.F. Bernasconi, Tetrahedron, 1989,45,407) and palladium (II) assisted reactions of alkenes (L.S. Hegedus, Tetrahedron, 1984,40,2415) have also appeared.

(a) *Isomerisation*

Alkenes are isomerised by heating with a wide variety of inorganic and organometallic catalysts, the products which are obtained being highly dependent on the catalyst and on the conditions employed. Internal alkenes are isomerised to terminal alkenes by heating over alumina containing sodium oxide (U.S. Pat., 4,229,610 1980; Chem. Abstr., 1981,94,83585), whilst polystyrene bound bis(cyclopentadienyl)titanium dichloride reacts with Grignard reagents to form reactive alkene isomerisation catalysts which convert 1-alkenes primarily into E-2-alkenes at room temperature (D.E. Bergbreiter and G.L. Parsons, J.Organomet. Chem., 1981,208,47). The hydrogen sulfide promoted thermal isomerisation of *cis*-2-butene to 1-butene or *trans*-2-butene is interpreted as a free radical process involving butene-2-yl and thiyl free radicals (C. Richard, A.Boiveaut, and R.Martin, Int.J.Chem.Kinet., 1980,12,921). Silica supported zirconium hydrides show high activity towards alkene isomerisation under mild conditions and terminal alkenes are isomerised to their thermodynamic mixtures very rapidly (J. Schwartz and M.D. Ward, J.Mol.Catal., 1980, 8,465). Butenes can be isomerised by various catalysts including aluminium phosphate (H.Itoh, A. Tada, and H.Hattori, J.Catal., 1982,76,235) and lanthanum modified alumina (M. Stoecker, T. Riis, and H. Hagen, Acta Chem. Scand.Ser.B, 1986,B40,200).

Photochemical isomerisation using 185 nm radiation provides a facile preparation of *cis*-di-*tert*-butylethylene (W.Adam, Z.Naturforsch.,B,Anorg.Chem.Org.Chem.,1981,36B,658).

Photo-assisted isomerisation of alkenes by platinum complexes effects *cis*-*trans* interconversion with high efficiency (P. Courtot, R. Pichon, and J.-Y. Salaun, J. Chem.Soc.Chem.Commun., 1981, 542).

An example of alkene inversion involves the *cis*-chlorotelluration - *trans* - dechlorotelluration process. Thus E-2-butene when treated with tellurium (IV) chloride in acetonitrile °0^{c}, 3 hr) followed by treatment with aqueous sodium sulfide give 2-butene with the E:Z ratio 3:97. The reaction is solvent dependent and may involve an epitelluride intermediate (J.E. Barchvall and L. Engman, Tetrahedron Lett., 1981,22,1919). A highly selective isomerisation of mono-substituted 1-alkenes to E-2-alkenes is mediated by the catalytic action of pentamethylcyclodienyl titanium (II) chloride-sodium naphthenate (M. Akita *et al.*, Bull.Chem.Soc., Jpn., 1983,56,554).

Significant E to Z isomerisation of simple alkenes takes place under flash vacuum pyrolysis conditions above 650°C. The evidence indicates that the reaction proceeds *via* a diradical which can be trapped by intramolecular H-transfer (C.L. Hickson and H. McNab, J.Chem.Res.Synop., 1989,176).

(b) Hydrogenation

A detailed *"ab initio"* MO study of the full catalytic cycle of alkene hydrogenation by the Wilkinson catalyst [$(PR_3)RHCl$] has been made to further the understanding of the intermediate and geometry of the process (C. Daniel, *et al.*, J.Am.Chem.Soc., 1988,110,3773).

A new catalytic hydrogenation of alkenes involves the use of uranium (III) chloride in tetrahydrofuran with LiH or $LiAlH_4$, agents which do not reduce alkenes themselves. A number of simple alkenes can be reduced directly to the corresponding alkene (G. Folcher, J.F. Marechal, and H. Marquet-Ellis, J.Chem.Soc.Chem.Commun., 1982,323). Mono-alkenes are also readily hydrogenated over amorphous nickel-phosphorus or nickel-boron alloys (S. Yoshida *et al.*, J.Chem.Soc.Chem.Commun., 1982, 964), whilst other polymer-supported catalysts which have been found to be effective are ruthenium complexes (C.P. Nicolaides and N.J. Coville, J.Organomet.Chem., 1981,222,285) and palladium (II) complexes (S.D. Mayak, V.Mahadevan, and M. Srinivasan, J.Catal., 1985,92,327).

A very mild catalytic hydrogenation of alkenes (room temperature and atmospheric pressure) can be carried out in biphasic-water systems or without solvent using triphenylphosphine meta-trisulfonate as an efficient reagent in the presence of rhodium salts. High yields of easily isolable products can be obtained (C. Larpent, R. Dabard, and H. Patin (Tetrahedron Lett., 1987,28,2507).

(c) Homologation, oligomerisation, metathesis, and coupling of alkenes

The synthesis of alkenes and alkanes using the Fischer-Tropsch synthesis has been studied for many years as it is a process of considerable industrial interest. The mechanisms of the processes using different catalysts are still being investigated.

Using supported tungsten oxides the results indicate that the C-C bond in the homologation of ethylene may proceed by polymerisation of "carbenes and alkyl groups" on the catalyst surface (T. Yamaguchi, S. Nakamura, and K. Tanabe, J.Chem.Soc.Chem.Commun., 1982,621). The homologation of alkenes can also be carried out, in competition with hydrogenation, by hydrogen over an iron oxide catalyst (R.A. Strehlov and E.C. Douglas, J.Chem.Soc.Chem.Commun., 1983,259).

The dimerisation of the lower alkenes, catalysed by transition-metal complexes, has been extensively studied after the pioneering work of Ziegler and an extensive review is available (567 references) (S.M. Pillai, M. Ravindranathen, and S. Sivaram, Chem.Rev., 1986,86,353).

The palladium, nickel or rhodium catalysed coupling reaction has provided a very useful method for the formation of C-C bond formation. The topic was reviewed in 1982 (E. Negishi, Acc.Chem.Res. 1982,15,340). Coupling of alkenes has proved to be a very useful reaction and can be used to synthesise alkadienes (T. Sakakura *et al.*, Chem. Lett., 1988,885) and a variety of functionalised alkenes. The reaction is carried out under mild conditions and can provide stereoselective synthesis and the introduction of a hydroboration coupling sequence has proved very useful (N. Miyaura *et al.*, J.Am.Chem.Soc., 1989,111,314).

$$C_6H_{13}CH{=}CH_2 + HB\text{(9-BBN)} \longrightarrow C_6H_{13}CH_2CH_2B\text{(9-BBN)} \xrightarrow[\text{(i)}]{Me_2C{=}CHBr} Me_2C{=}CHC_8H_{17}$$

(several other examples)

[HB(9-BBN) = 9-borabicyclo[3.3.1]nonane (9-BBN)]

Reagents: (i) $PdCl_2$(dppf), NaOH,THF,65°C

[dppf = 1,1[1]-bis(diphenylphosphino)ferrocene]

Alkene metathesis is of continuing interest and newer catalysts for this include zero valent tungsten complexes (A. Korda, R. Grieznski, and S. Krycinski, J.Mol.Catal., 1980,9,51), titanium tetrachloride and low-valent lanthanides (H. Imanura *et al.*,Chem.Lett., 1989,171), the Schrock carbene complex (K. Weiss, Angew.Chem., 1986,98,350), and anion promoted ruthenium clusters (J.L. Zuffa, M.L. Blohm, and W.L. Gladfelter, J.Am.Chem.Soc., 1986,108,552). Alkene metathesis has been reviewed (226 refs.) in 1981 (R.L. Banks, Catalysis (London) 1981,4,100).

(d) Hydroboration

The remarkable work of H.C. Brown and coworkers on hydroboration has continued throughout the 1980's (Paper No.82 in the series "Hydroboration" appearing in 1989) and many other papers have also been published.

The reaction of 2,3-dimethyl-2-butene with one equivalent of monochloroborane-methyl sulfide complex gives exclusively the product of monohydroboration, thexylchloroborane-methyl sulfide, which is very stable at room temperature and which hydroborates reactive alkenes quantitatively with high regiospecificity. Subsequent oxidation produces alcohols in nearly quantitative yield with high regiospecificity.

Ketones and long chain alkenes may also be made from this intermediate (Scheme 11) (H.C. Brown *et al.*, J.Org. Chem., 1982,47,863; J.A. Sikorski and H.C. Brown, J.Org. Chem., 1982,47,872). See also H.C. Brown *et al.*, J.Org. Chem., 1980,45,4540; *ibid* p.4542, and other references).

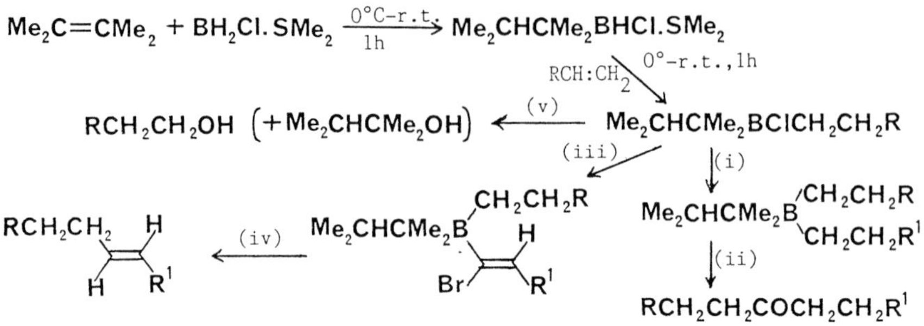

$$Me_2C{=}CMe_2 + BH_2Cl.SMe_2 \xrightarrow[1h]{0°C\text{-}r.t.} Me_2CHCMe_2BHCl.SMe_2$$

$$Me_2CHCMe_2BHCl.SMe_2 \xrightarrow[0°\text{-}r.t.,1h]{RCH:CH_2} Me_2CHCMe_2BClCH_2CH_2R$$

$$RCH_2CH_2OH\ (+Me_2CHCMe_2OH) \xleftarrow{(v)} Me_2CHCMe_2BClCH_2CH_2R$$

$$Me_2CHCMe_2BClCH_2CH_2R \xrightarrow{(iii)} Me_2CHCMe_2B(CH_2CH_2R)C(Br){=}CHR^1 \xrightarrow{(iv)} RCH_2CH_2CH{=}CHR^1$$

$$Me_2CHCMe_2BClCH_2CH_2R \xrightarrow{(i)} Me_2CHCMe_2B(CH_2CH_2R)(CH_2CH_2R^1) \xrightarrow{(ii)} RCH_2CH_2COCH_2CH_2R^1$$

Reagents: (i) H^-,$R^1CH{=}CH_2$; (ii) NaCN,$(CF_3CO)_2O$.[O]; (iii) H^-,$BrC{\equiv}CR^1$; (iv) NaOMe, i-$PrCO_2H$; (v) OH^-, H_2O_2

Scheme 11

Other dihaloborane-methylsulfide complexes ($HBX_2.SMe_2$, X=Cl,Br,I) have been used as hydroborating agents, and ultrasound has been shown to accelerate the reaction (H.C. Brown and U.S. Racherla, Tetrahedron Lett., 1985,26,2187); J.Org.Chem., 1986,51,895; H.C. Brown, N. Ravindran, and S.U. Kulkarni, J.Org.Chem., 1980,45,384).

The preparation of the simple organoboranes methyl- and dimethylborane has been simplified and their characteristics as hydroborating agents have been investigated. These reagents react with good regioselectivity and selectivity with terminal and trisubstituent alkenes and good yields of tertiary alcohols containing one or two methyl groups can be achieved. For example, from methylborane and 1-hexene 7-methyl-7-tridecanol can be obtained in 90% yield (H.C. Brown *et al.*, J.Org.Chem., 1986,51,4925).

The kinetics and reactivities of the hydroboration process have been investigated (K.K. Wang and H.C. Brown, J.Org.Chem., 1980,45,5303); H.C. Brown and J.Chandarasekharan, J.Org.Chem., 1983,48,644; *ibid* p.5080) including the hydroboration of sterically hindered alkenes under high pressure, which may provide a useful variant of the reaction (J.E. Rice and Y. Okamoto, J.Org.Chem., 1982,47,4189). However, in spite of all the studies of this important reaction, precise mechanistic details are still subject to discussion. "*Ab initio*" calculations of the reaction of the

$BH_3.H_2O$ complex with ethene indicate an S_N2 type of displacement of the solvent by the alkene with no involvement by free BH_3 (T. Clark, D. Wilhelm, and P. van R. Schleyer, J.Chem.Soc.Chem.Commun., 1983,606), although this has been suggested by Wang and Brown (K.K. Wang and H.C. Brown, J.Am.Chem.Soc., 1982,104,7148).

A comparison of steric effects in the electrophilic addition reactions of alkenes indicates that the transition states in the rate determining steps of hydroboration and oxymercuration are similar but both are different from that of bromination (D.J. Nelson, P.J. Cooper, and R. Soundararajan, J.Am.Chem.Soc., 1989,111,1414).

The continuing interest and development of this important process, including the use of modified reagents [e.g. cobalt (II) chloride-sodium borohydride (N. Satyanarayana and N. Periasamy, Tetrahedron Lett., 1984,25,2501)], the modification of reactions [e.g. the development of a simple synthesis of dialkylketones from alkenes via hydroboration-carbenoidation (C. Narayana and N. Periasamy, Tetrahedron Lett., 1985,26,6361)], and the development of new processes, e.g. the conversion of alkenes to alkyl bromides via hydroboration (H.C. Brown and C.F. Lane, Tetrahedron, 1988,44,2763), must mean that many papers will continue to be published in this area.

The value of the hydroboration of alkenes as the initial step for further transformation has been well recognised (H.C. Brown *et al.*, "Organic Synthesis via Boranes", Wiley-Interscience, N.Y., 1975). In recent years the uses of organoboranes for the synthesis of alkanes (a non-catalytic conversion of alkenes to alkanes) (H.C. Brown and K.J. Murray, Tetrahedron, 1986,42,5497), for alcohols (H.C. Brown *et al.*, Tetrahedron, 1986,42,5505; H.C. Brown, M.M. Midland, and G.W. Kabalka, *ibid* p.5523), for ketones (H.C. Brown and C.P. Garg, Tetrahedron, 1986, 42,5511), for aldehydes (H.C. Brown *et al.*, Tetrahedron, 1986,42,5515), for alkyl bromides (G.W. Kabalka *et al.*, J.Org.Chem., 1981,46,3113), and for hydroalumination of alkenes (K. Mamoka *et al.*, J.Am.Chem.Soc., 1986,108,6036) have been reported. This represents only a selection of the reactions available.

(e) *Hydroformylation*

The hydroformylation (also carbonylation and carboxylation) reactions of alkenes have recently been reviewed. (J.A. Davies, Chem.Met.-Carbon Bond, 1985,3,361; B.A. Murrer and M.J.H. Russell, Catalysis (London), 1983,6,169). This commercially important reaction continues to be developed and the catalysts improved. Recent developments in this area include the use of rhodium phosphine complexes (T. Okano *et al.*, Bull.Soc.Chem. Japan, 1981,54,3799; A.M. Trzeciak and J.J. Ziolkowski, J.Mol.Catal., 1986,34,213), and such complexes supported on zeolites (E. Rode, M.E. Davis, and B.E. Hanson, J.Chem.Soc.Chem.Commun., 1985,1477), Pt(II)-Sn(II) systems (H.C. Clark and J.A. Davis, J. Organomet.Chem., 1981,213,503; S.C. Tang and L. Kim, J.Mol. Catal., 1982,14,231), ruthenium complexes (M. Bianchi *et al.*, J.Organomet.Chem., 1983,247,89), and ruthenium-cobalt carbonyl clusters (A. Fukuoka *et al.*, Chem.Lett.,1987,941).

Photochemical acceleration of alkene hydroformylation in the presence of a number of cobalt carbonyls has also been studied and has been shown to be advantageous over traditional methods (M.J. Mirbach *et al.*, Angew.Chem., 1981, 93,391; J.Am.Chem.Soc., 1981,103,7590;7594).

Rhodium complexes have also been shown to catalyse the hydroformylation of alkenes by paraformaldehyde thereby constituting a novel C-C bond forming reaction between formaldehyde and alkenes (Scheme 12) (T. Okano *et al.*, Tetrahedron Lett., 1982,23,4967).

$$C_4H_9CH{=}CH_2 \longrightarrow \underset{67\%}{C_6H_{13}CHO} + \underset{4\%}{C_6H_{13}CH_2OH} + \underset{13\%}{C_6H_{13}CO_2Me} + \underset{3\%}{C_6H_{14}}$$

Reagents: $RhH_2(O_2COH)[P(\underline{i}\text{-}Pr)_3]_2$, $(CH_2O)_n$, THF, 120°C, 20h

Scheme 12

Although hydroformylation principally gives aldehydes, by carrying out the reaction in aqueous trifluoracetic acid solution in the presence of a palladium (II) triphenylphosphine complex good yields of dialkyl ketones are obtained under mild conditions (V.N. Zudin *et al.*, J.Chem.Soc.Chem. Commun., 1984,545).

(f) Addition of alkanes to alkenes

The long established Friedel-Crafts alkylation of arenes by alkenes will be considered under "aromatic compounds". However, alkanes and alkenes can react together with addition of the alkane to the alkene either by the thermal production of alkyl radicals (J. Hartmanns and J.O. Metzger, Chem.Ber., 1986,119,500), by the catalytic action of supported nickel catalysts (I.D. Loffler *et al.*, J.Chem.Soc.Chem.Commun., 1984,1177), or by a carbocation process using an hydrogen fluoride-tantalum pentafluoride system (J. Sommer, M. Muller and K.Laali, Nouv.J.Chem., 1982,6,3).

(g) Addition of alkylhalides to alkenes

A number of free-radical generators initiate the addition reactions of polyhalogenoalkenes to alkenes, and such a reaction can also be initiated by a variety of inorganic salts or organometallic complexes [copper catalysed additions of organic polyhalides to olefins were reviewed in 1985 (D. Bellus, Pure Appl.Chem., 1985,57, 1827)]. Palladium salts or complexes also catalyse such a reaction, in the presence of base, under mild conditions. Some reactions of 1-decene are shown in Scheme 13 (J. Tsuji, K. Sato and H. Nagashima, Tetrahedron, 1985,41,393).

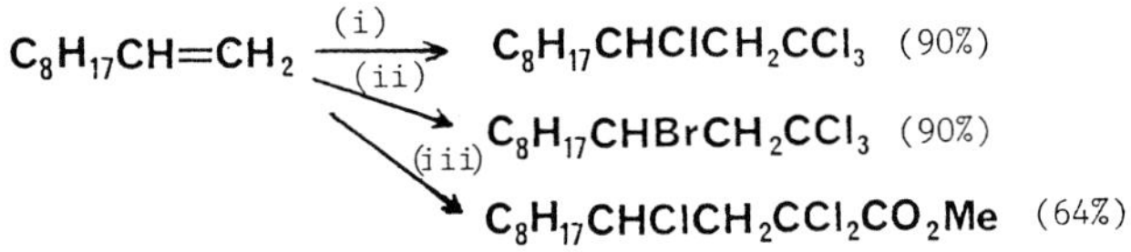

$$C_8H_{17}CH{=}CH_2 \xrightarrow{(i)} C_8H_{17}CHClCH_2CCl_3 \quad (90\%)$$

$$C_8H_{17}CH{=}CH_2 \xrightarrow{(ii)} C_8H_{17}CHBrCH_2CCl_3 \quad (90\%)$$

$$C_8H_{17}CH{=}CH_2 \xrightarrow{(iii)} C_8H_{17}CHClCH_2CCl_2CO_2Me \quad (64\%)$$

Reagents: (i) CCl_4, CO, r.t., 120h; (ii) $BrCCl_3$, Ar, 100°C, 3.5h; (iii) CCl_3CO_2Me, Ar, 100°C, 15h

Scheme 13

Other catalysts which can be used include nickel complexes (D.M. Grove, and G. Van Koben, J.Mol.Catal., 1988,45,169) and copper complexes (M. Hajek and P.Silhavy, Collect.Czech.Chem.Commun., 1983,48,1710). Copper complexes also catalyse the photoaddition of alkyl halides to alkenes (M. Mitani, I. Kato and K. Koyama, J.Am.Chem.Soc., 1983, 105,6719).

Lewis acid catalysis (zinc chloride-ether) can be used to catalyse the addition of alkyl chlorides to alkenes to give linear products with high Markownikow regioselectivity (but no stereospecificity). The products, which still contain an olefinic bond, can go further to yield cyclic products by cycloaddition reactions or give polymeric products (H. Mayr, H. Klein, and G. Kolberg, Chem. Ber., 1984,117,2555).

(h) The Diels-Alder reaction

The Diels-Alder reaction was first discovered 60 years ago (O. Diels and K. Alder, Chem.Ber., 1929,62,554) yet even now it has been quoted that "the mechanism of the simplest example, i.e. the reaction of ethene with butadiene, still remains uncertain" (M.J.S. Dewar, S. Olivella, and J.J.P. Stewart, J.Am.Chem.Soc., 1986,108,5771). These workers infer that the reaction cannot be synchronous (M.J.S. Dewar and A.B. Pierini, J.Am.Chem.Soc., 1984,106, 203; M.J.S. Dewar and E.F. Healy, J.Am.Chem.Soc., 1984, 106,7127) and the evidence suggests that the reactions proceed *via* very unsymmetrical transition states close to biradicals in structure and energy. However, other workers provide experimental details and calculations which indicate that the reaction is a synchronous, concerted, process (K.N. Houk, Y.T. Lin, and F.K. Brown, J.Am. Chem.Soc., 1986,108,554). Another group of workers find that "it appears that both synchronous and non-synchronous paths exist and that the synchronous path is favoured by about 2k cal mol^{-1} (F. Bernardi *et al.*, J.Am.Chem.Soc., 1988,110,3050).

There will, no doubt, be further discussions of this topic in the next decade.

(i) Addition of hydrogen and hydroxy halides to alkenes

Experimental results on the products and kinetics of the addition of hydrogen bromide to alkenes in non-polar media, together with theoretical calculations, indicate a molecular mechanism for the process involving intermediate cyclic structures of 2:1 hydrogen bromide:alkene or vice-versa (G.B. Sergeev *et al.*, Kinet.Katal., 1980,21,1130; Chem.Abstr., 1981,94,64769; Tetrahedron, 1982,38,2585).

The hydrochlorination of 2-methylpropene and the 2-butenes in the gas phase gives 2-chloro-2-methylpropene and 2-chlorobutanes exclusively, and the reaction seems to require surface catalysis, surface adsorption of both the hydrogen chloride and the alkene being required to consummate the reaction (F. Costello, D.R. Dalton, and J.A. Poole, J.Phys.Chem., 1986,90,5352; J. Tierney, F. Costello, and D.R. Dalton, J.Org.Chem., 1986,51,5191).

The geometries of eleven chloronium ion intermediates have been determined using MO calculations. The chloronium ion affinities correlate well with the reported rate constants for the chlorination of alkenes (S. Yamane, T. Tsuji, and K. Hirao, Chem.Phys.Lett., 1988,146,236).

A convenient method for the addition of HI to alkenes has been developed which uses iodine in refluxing petroleum ether (35 - 36°C) in the presence of dehydrated alumina. The iodine reacts with the surface hydroxyl groups on the alumina to give hydrogen iodide which then adds to the alkene. For example, 1-octene is converted to 2-iodooctane (83%) in 2h. The reaction occurs via the ionic Markownikow addition since both 3,3-dimethylbut-1-ene and 2,3-dimethylbut-2-ene give the same product, namely 2-iodo-2,3-dimethylbutane (L.J. Stewart *et al.*, Tetrahedron Lett., 1987,28,4497).

A very easy hydroiodination of alkenes has been developed which uses trimethylchlorosilane and sodium iodide in the presence of 0.5 mole equivalent of water. The reaction gives good yields of alkyl iodides (Scheme 14) (S.Isifune *et al.*, Synthesis, 1988,366).

$$RR^1C{=}CR^2R^3 + Me_3SiCl + NaI + 0{\cdot}5\,H_2O \longrightarrow RR^1CHCIR^2R^3$$

Conditions: MeCN,r.t., 1-3h. (65-96%; R=H, various alkyl)

Scheme 14

Chlorohydroxylation of unfunctionalised alkenes can be carried out quite readily using either t-butyl hydroperoxide or di-*tert*-butyl peroxide in the presence of titanium tetrachloride (Scheme 15). The intermediary of an epoxide is probable which reacts with the titanium (IV) chloride to give the chlorohydrinated product. The stereochemistry of the addition is pure anti and there seems to be some regioselectivity in the ring opening reaction, for example, 1-decene gives 96% 2-chlorodecan-1-ol and 4% 1-chlorodecan-2-ol (overall yield 62%) (J.M. Klunder *et al.*, J.Org. Chem., 1985,50,912).

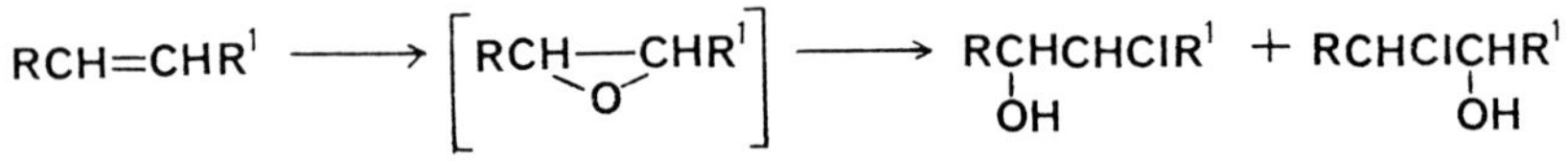

$$RCH{=}CHR^1 \longrightarrow \left[RCH\!\!-\!\!\underset{O}{\diagdown\!\diagup}\!\!-\!\!CHR^1 \right] \longrightarrow \underset{OH}{RCH}CHClR^1 + RCHCl\underset{OH}{CHR^1}$$

Reagents: $TiCl_4$, TBHP (or DTBP), CH_2Cl_2, -78°C, 2h

Scheme 15

Kinetic and product analysis of the chlorohydroxylation of 3-ethyl-2-pentene using gaseous chlorine in water and hypochlorous acid indicates that the rate-determining steps for the two reagents are different whereas the product determining step is the same (E. Buss, A.Rockstuhl, and D. Schnurpfeil, J.Prakt.Chem., 1982,324,197).

Fluoroalkyl iodides add to alkenes under the catalytic influence of several transition metal complexes including copper, palladium, and platinum. However this addition can also be catalysed under mild conditions simply by Raney nickel or by samarium diiodide (Q.-Y.Chen and Z.-Y.Yang, J.Chem.Commun., 1986,498; X.Lu, S.Ma, and J. Zhu, Tetrahedron Lett., 1988, 29,5129).

(j) Halogenation of alkenes

The direct addition of elemental fluorine to simple alkenes can now be achieved using low temperatures (-75°C) in polar solvents (typically a mixture of trichlorofluormethane, chloroform and ethanol) which suppress fluorine radical formation and encourage the required polar processes. Using such a method **7-tetradecene** (15) has been shown to give racemic-7,8-difluorotetradecane (16a) (50%) and meso-7,8-difluorotetradecane (16b) (20%).

The terminal alkenes 1-octene and 1-dodecene give, in addition to the expected 1,2-difluoro products some

1,1,2-trifluoro products. Such products may result from intermediate alkene formation and a second reaction with F_2 (Scheme 16) which is evidence against a concerted four-centre reaction mechanism (S. Rozen and M. Brand, J.Org. Chem., 1986,51,3607).

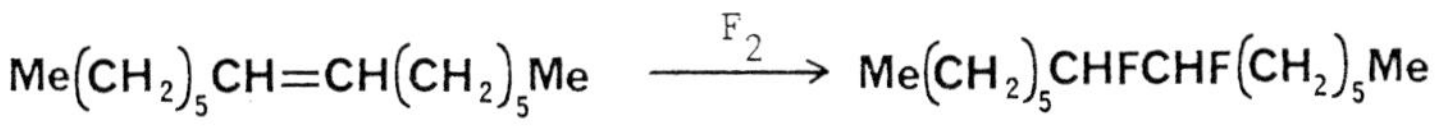

$$Me(CH_2)_5CH{=}CH(CH_2)_5Me \xrightarrow{F_2} Me(CH_2)_5CHFCHF(CH_2)_5Me$$

(15) (16)(a) racemic (b) meso

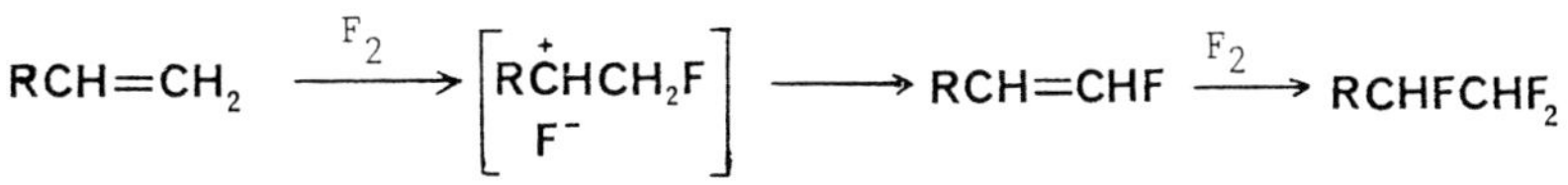

$$RCH{=}CH_2 \xrightarrow{F_2} \left[\begin{matrix} R\overset{+}{C}HCH_2F \\ F^- \end{matrix}\right] \longrightarrow RCH{=}CHF \xrightarrow{F_2} RCHFCHF_2$$

($R = Me(CH_2)_x$, $x = 5,9$)

Scheme 16

N-Haloelectrophiles such as N-chlorodimethylamine etc. react with alkenes in the presence of boron trifluoride etherate to give halofluoroalkanes. In addition to the normal products of addition rearranged products are found which presumably arise from hydride shifts of the intermediate carbocation (Scheme 17) (G.E. Heasley *et al* ., Tetrahedron Lett., 1985,26,1811).

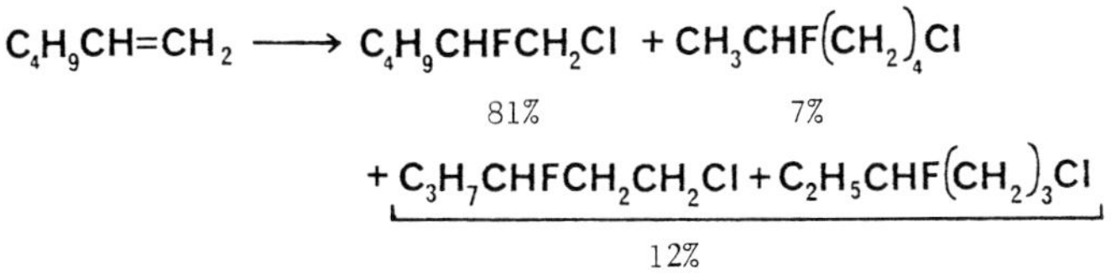

$$C_4H_9CH{=}CH_2 \longrightarrow \underset{81\%}{C_4H_9CHFCH_2Cl} + \underset{7\%}{CH_3CHF(CH_2)_4Cl}$$

$$+ \underbrace{C_3H_7CHFCH_2CH_2Cl + C_2H_5CHF(CH_2)_3Cl}_{12\%}$$

Reagents: $MeNCl_2$, $BF_3.OEt_2$, CH_2Cl_2, r.t.

Scheme 17

The treatment of alkenes with manganese (III) acetate and a chloride salt (calcium chloride or sodium chloride) in acetic acid at reflux temperature results in the synthesis of vicinal dichloroalkanes in good yield. The *trans* addition of chlorine is strongly favoured and the reaction seems to involve oxidative addition across the alkene of chloride ion *via* some magnesium (III) chloride species (Scheme 18) (K.D. Donnelly *et al.*, Tetrahedron Lett., 1984,25,607).

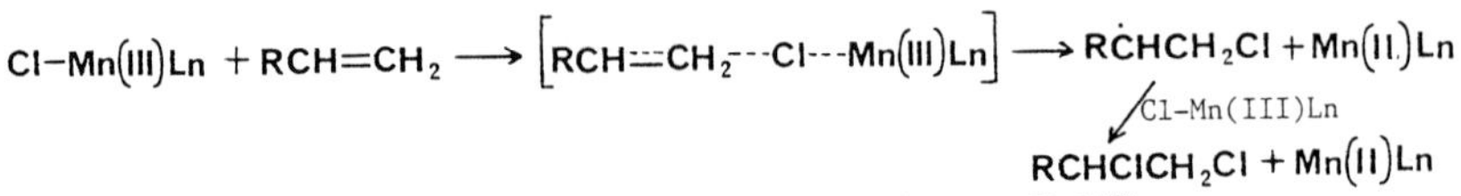

$$Cl{-}Mn(III)Ln + RCH{=}CH_2 \longrightarrow [RCH{\cdots}CH_2{\cdots}Cl{\cdots}Mn(III)Ln] \longrightarrow R\dot{C}HCH_2Cl + Mn(II)Ln$$

$$\xrightarrow{Cl-Mn(III)Ln} RCHClCH_2Cl + Mn(II)Ln$$

R = $Me(CH_2)_3$, $Me(CH_2)_5$; $Me(CH_2)_2CH{=}CH(CH_2)_2Me$ (trans)(yields 79-91%)

Scheme 18

Using phenylselenyl chloride reactions with alkenes give high yields of 1,2-dichloroalkanes having *cis* geometry. The initially formed adducts are oxidised, either by chlorine or by further phenylselenyl chloride, to give adducts which undergo displacement by chloride sources such as tetra-n-butylammonium chloride to give the products (Scheme 19) (A.M. Morella and A.D. Ward, Tetrahedron Lett., 1984,25,1197).

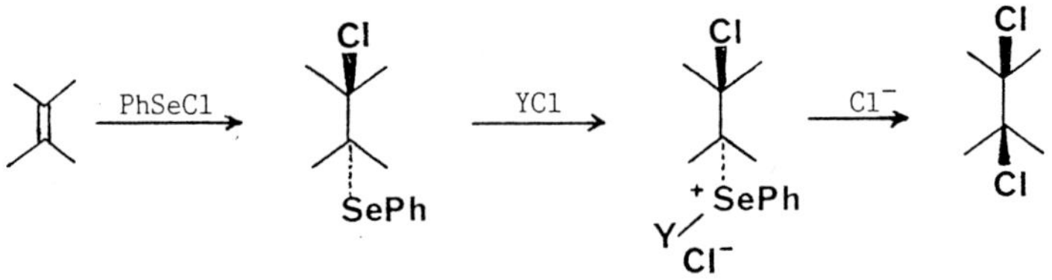

Alkene $\xrightarrow{PhSeCl}$ Cl/SePh adduct $\xrightarrow{YCl}$ Cl/$^{+}$SePh(Y)Cl^- $\xrightarrow{Cl^-}$ Cl/Cl

Reagents: alkene, PhSeCl,Cl_2, $(n\text{-}Bu)_4N^+Cl^-$, MeCN, reflux, 40 mins.
(Alkenes: 1-octene, *cis*-2-octene, *cis*-2-hexene, *trans*-2-hexene)

Scheme 19

Tellurium tetrachloride reacts with alkenes to give β-chloroalkyltellurium trichlorides. A more or less concerted, stereospecific *syn* addition, and a competing radical chain reaction are proposed as the major pathways and mixtures of *syn* and *anti* products are obtained. The addition of 2-naphthyltellurium trichloride gives complete antistereospecificity (Scheme 20) (J.E. Baeckvall,

J. Bergmann, and L. Engman, J.Org.Chem., 1983,48,3918).

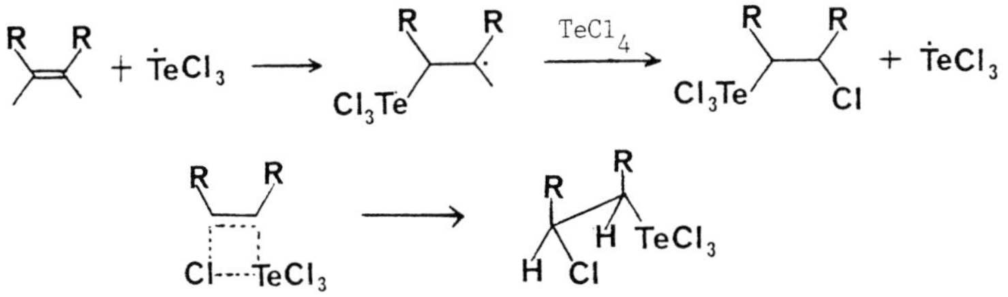

Scheme 20

The reaction of phenyltellurium tribromide with alkenes in alcoholic solution yields β-alkoxyalkylphenyltellurium dibromides. The reaction is *trans* stereospecific and is regiospecific in the case of terminal alkenes with the tellurium species attacking the terminal carbon atom. The size of the phenyltellurium reagent controls the site of attack and the alkanol solvent attacks specifically *anti* to the tellurium atom and at the better carbocationic centre of the molecule (Scheme 21) (S. Uemur, S. Fukuzawa, and A. Toshimitsu, J. Organomet Chem., 1983,250,203).

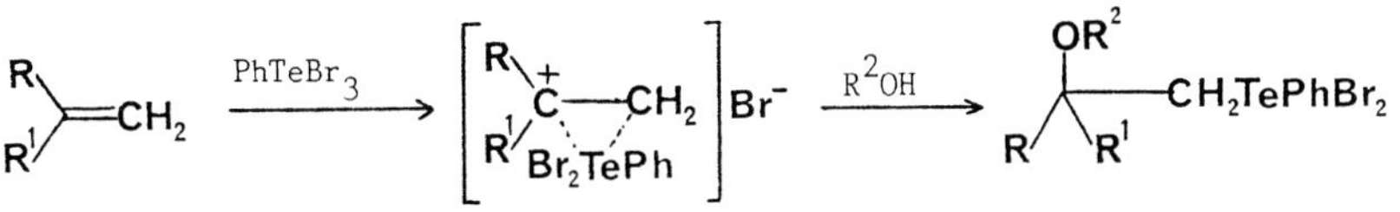

Scheme 21

Kinetic investigations of the chlorination and bromination of alkenes show that change from bromination to chlorination is associated with a large reactivity increase, but a small drop in selectivity. This result can be explained in terms of the charge distribution in the halonium transition states (E. Bienvenne-Goetz *et al.*, Tetrahedron Lett., 1982,23,3273). The solvent has also been shown to play an important role in the ionisation step of bromination (M.-F. Ruasse and B.-L. Zhang, J.Org.Chem., 1984,49,3207).

The use of transition metal salts to catalyse the halogenation of alkenes has proved to be an extremely useful reaction. Bromo and iodo functionalisation of alkenes has now been found to be readily carried out using either mercury (II) salt-halogen or copper (II) iodine combinations to give a range of products (J. Barluenga *et al.*, J.Chem.Res.(S), 1986,274; J.Chem.Soc.Chem.Commun., 1985,1422; *ibid*, 1987,1491)(Scheme 22).

$$RCH{=}CHR^1 \longrightarrow RCHZCHXR^1$$

R = H, n-Bu; Z = I, Br; X = Cl,Br: R^1 = H, n-Bu, t-Bu; Z = I, Br; X = Cl,Br

Reagents: HgX_2, halogen, CH_2Cl_2

Scheme 22

The synthesis of fluorohaloalkanes can be achieved by the boron trifluoride promoted reaction of alkyl hypohalites with alkenes. This reaction is convenient for both fluorochloro as well as fluorobromo products, but the use of reagents such as methyl hypochlorite also gives methoxy products (V.L. Heasley *et al.*, J.Org.Chem., 1983,48,3915). The bromochlorination of alkenes may be carried out using dichlorobromate (1-)ion (T. Negoro and Y. Ikeda, Bull.Chem. Soc., Japan, 1984,57,2116; 1986,59,2547).

In contrast to the addition to the double bond, the chlorination of simple alkenes using *tert*-butylhypochlorite in the presence of silica gel in non-protic solvent is reported to give good yields of alkylic dichlorides regioselectively (W. Sata, N. Ikeda, and H. Yamamoto, Chem.Lett., 1982,141).

(k) Hydrocyanation

The nickel(0)-catalysed reaction of two molecules of hydrogen cyanide to butadiene is the industrial process for the manufacture of adiponitrile, but the mechanism of the reaction has been little understood. The mechanism of the addition of deuterium cyanide to alkenes has been demonstrated to involve cis addition. This is interpreted as being the result of oxidative addition of deuterium cyanide to the nickel to give a complex, coordination of the alkene to the complex followed by σ-bond formation, then reductive elimination to result in the *cis* adduct (J.-E. Baeckvall and O.S. Andell, J.Chem.Soc.Chem.Commun., 1981,1098;Organometallics,1986,5,2350). Homogenous nickel-catalysed alkene hydrocyanation has also been reported by Seidel and Tolman (W.C. Seidal and C.A. Tolman, Ann.N.Y. Acad.Sci., 1983, 415,201).

A convenient method for the laboratory synthesis of alkyl nitriles equivalent to the *anti*-Markownikow hydrocyanation of alkenes has been developed using zirconocene halides to produce a variety of nitriles in good yield (Scheme 23) (S.L. Buchwald and S.J. LaMaire, Tetrahedron Lett., 1987,28,295).

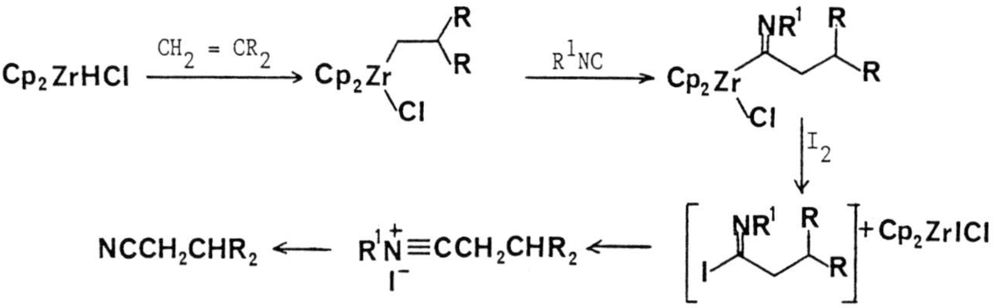

$Cp_2ZrHCl \xrightarrow{CH_2=CR_2} Cp_2Zr(Cl)CH_2CHR_2 \xrightarrow{R^1NC} Cp_2Zr(Cl)C(=NR^1)CH_2CHR_2 \xrightarrow{I_2} [IC(=NR^1)CH_2CHR_2] + Cp_2ZrICl \longrightarrow R^1\overset{+}{N}{\equiv}CCH_2CHR_2\ I^- \longrightarrow NCCH_2CHR_2$

Reagents: t-BuNC, C_6H_6,I_2,r.t., 15h

Scheme 23

(l) Amination and aminomethylation

The amination of alkenes has recently been extensively reviewed (M.B. Gase, A. Lattes, and J.J. Perie, Tetrahedron, 1983,39,703; J.J. Brunet, D. Neibecker, and F. Niedercorn, J.Mol.Catal., 1989,49,235; D. Steinborn and R. Taube, Z. Chem., 1986,26,349).

The traditional methods for ethylamine production involve alkali metal amides, but acidic zeolites can also catalyse the conversion of ammoniacal ethylene to ethylamine and may prove to be a useful alternative method for alkene amination. This method has already been applied to the preparation of *tert*-butylamine (M. Deeba, M.E. Ford, and T.A. Johnson, J.Chem.Soc.Chem.Commun., 1987,562; M.Deeba and M.E. Ford, J.Org.Chem., 1988,53,4594).

The aminomethylation of alkenes initially discovered by Reppe (W. Reppe, Experientia, 1959,5,93) seems to have given only low yields of alkylamines. However the use of rhodium rather than iron pentacarbonyl as a catalyst seems to provide greatly improved yields and to be a useful practical method for the preparation of a number of tertiary amines from simple alkenes (F. Jachimowicz and J.W. Rakiss, J.Org. Chem., 1982,47,445).

A method for the ready synthesis of vicinal diamines from simple alkenes has been developed. This involves reaction of the alkene with cyanamide and NBS to give an alkyl cyanamide which is then treated with ethanol and HCl. The resulting urea is then treated with base to give an imidazoline which is hydrolysed to the diamine (Scheme 24) (H.Kohn and S.-H.Jung, J.Am.Chem.Soc., 1983,105,4106).

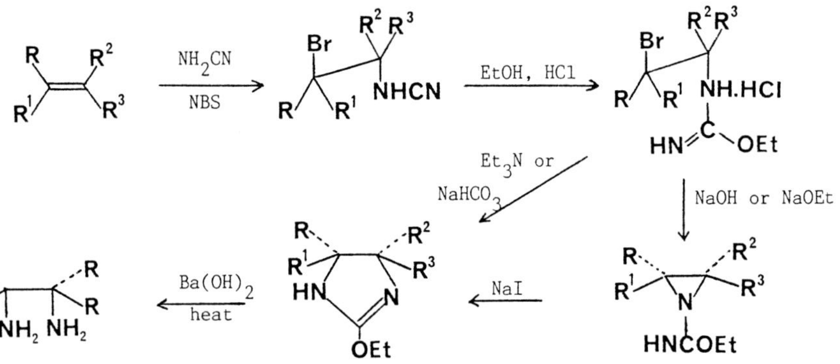

$R = R^3 = Me$, $R^1 = R^2 = H$	Product	racemic	51%
$R = R^2 = Me$, $R^1 = R^3 = H$	"	meso	53
1 - hexene	"		63
trans-4-octene	"	racemic	71

Scheme 24

Although hydrazoic acid is usually unreactive towards alkenes, the use of Lewis acids, in particular titanium (IV) chloride, catalyses its addition to 1,1-disubstituted alkenes to give the alkylazide (Scheme 25) (A. Hassner, R. Fibiger, and D. Audisik, J.Org.Chem., 1984,49,4237).

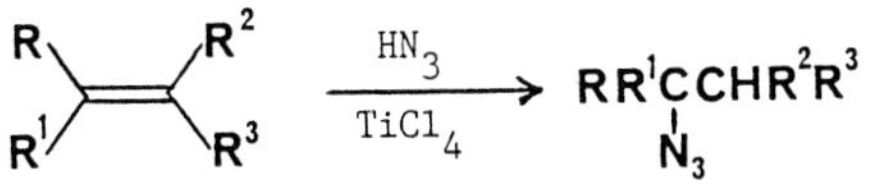

$$\mathrm{RR^1C{=}CR^2R^3} \xrightarrow[\mathrm{TiCl_4}]{\mathrm{HN_3}} \mathrm{RR^1C(N_3)CHR^2R^3}$$

$R = R^1 = R^2 = R^3 = Me$
$R = R^1 = Me,\ R^2 = H,\ R^3 = Et$
$R = R^1 = Me,\ R^2 = H,\ R^3 = Me$

Scheme 25

By the use of Manganese (III) acetate hydrate in the presence of azide ion the conversion of alkenes to 1,2-diazides is effected very easily, presumably through a manganese (III)-azide ligand transfer oxidative process. Reduction of the diazide then leads to the 1,2-diamines (W.E. Fristad *et al*, J.Org.Chem., 1985,50,3647). Alternatively such vicinal diazides may be obtained using hypervalent iodine catalysis (R.M. Moriarty and J.S. Khosrowshahi, Tetrahedron Lett., 1986,27,2809).

The aminomercuration-demercuration of alkenes can be problematic due to the formation of ammonia-mercury (II) complexes instead of the products of addition to the double bond. However, amidomercuration seems to be a relatively facile reaction to provide amides and related compounds such as urethanes, ureas, and sulfonamides (Scheme 26) (J.Barluenga *et al*, J.Chem.Soc.Perkin Trans.1, 1983,591; 1984, 721).

$$\mathrm{RR^1C{=}CHR^2 + R^3CONH_2 \xrightarrow{(i)\ (ii)} R^3CONHCRR^1CH_2R^2} \quad \text{(Rs, various)}$$

Reagents: (i) $HgNO_3$, CH_2Cl_2, reflux, 24h; (ii) $NaBH_4$, aq.NaOH

Scheme 26

Other reactions which involve the introduction of amino functionalisation into alkenes include cyanamidoselenylation, which leads to cyanamides (R. Hernandez *et al.*, J.Chem.Soc.Chem.Commun., 1987,312), the addition of phenylselenium azide, which occurs *via* a 3-membered selenonium ion intermediate (A. Hassner and A.S. Amarasekara, Tetrahedron Lett., 1987,28,5185), or phenylselenyl chloride and carbamates (C.G. Francisco *et al.*, Tetrahedron Lett., 1986,27,2513).

Aminotellurinylation of alkenes provides a useful synthesis of 2-oxazolidinones, and amidotellinurinylation provides a novel synthesis of 4,5-dihydrooxazoles from alkenes (Scheme 27) (N.X. Hu *et al.*, J.Org.Chem., 1989,54, 4398; J.Chem.Soc.Perkin Trans.1, 1984,1775).

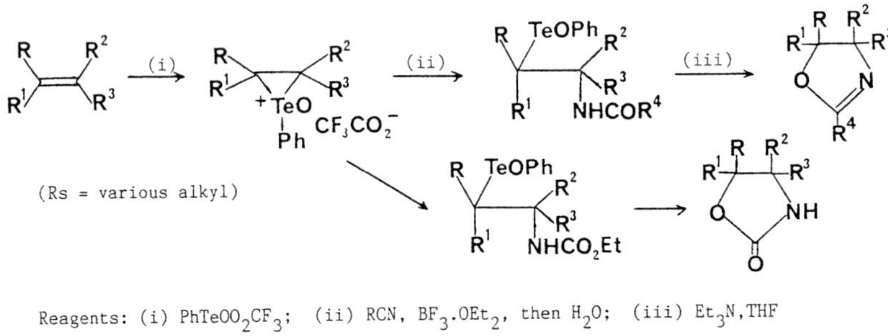

(Rs = various alkyl)

Reagents: (i) $PhTeOO_2CF_3$; (ii) RCN, $BF_3.OEt_2$, then H_2O; (iii) Et_3N,THF

Scheme 27

There is also evidence for the reversible aminopalladation of alkenes (J.E. Baeckvall and E.E. Bjoerkmann, Acta Chem.Scand.Ser.B, 1984,B38,91).

(m) Oxidation reactions of alkenes

(i) Ozonolysis

The formation and structure of ozonides has been reviewed in 1983 (R.L. Kuczkowski, Acc.Chem.Res., 1983,16, 42).

A theoretical discussion of the ozonolysis of alkenes has indicated that the stereochemistry of the primary ozonide conversion to carbonyl compound and carbonyl oxide is determined by the electronic properties of the carbonyl oxide for lower alkenes but by the conformation of the primary ozonide for higher alkenes (D. Cremer, Angew.Chem., 1981,93,935).

However a study of the mechanism of ozonolysis of propene in the liquid phase has indicated that the picture is not simple and that a variety of factors control ozonide stereochemistry including solvent environment (J.-I.Choes, M.Srinivasan, and R.L. Kuczkowski, J.Am.Chem.Soc., 1983,105, 4703). The mechanisms of ozone-alkene reactions have also been studied in the gas phase using stopped-flow techniques (R.I. Martinez and J.T. Herron, J.Phys.Chem., 1988,92,4644). The ozonolysis of alkenes adsorbed on to silica gel also gives different proportions of products depending on the nature of the alkene and on the state of the silica gel (I.A. Den Besten and T.H. Kinstle, J.Am.Chem.Soc., 1980, 102,5968); C. Aronovitch, D. Tal, and Y. Mazur, Tetrahedron Lett., 1982,23,3623).

A useful development for the preparation of aldehydes from alkenes by ozonolysis in methanol is to use thiourea to reduce the hydroperoxide products initially formed (D. Gupta, R. Soman, and S. Dev, Tetrahedron, 1982,38,3013).

Further discussions of the mechanism of the ozonolysis of alkenes have also centred on the epoxidation which occurs in competition with the intended reaction. In general the greater the steric crowding around the double bond the more is epoxidation favoured, which is in agreement with the initial formation of a π-complex rather than an open σ-complex (P.S. Bailey, H.H. Hwang, and C.-Y.Chiang, J.Org. Chem., 1985,50,231). A theoretical study indicates that the carbonyl oxide can act as an oxygen transfer agent leading to the epoxide and aldehyde (D. Cremer and C.W. Bock, J.Am. Chem.Soc., 1986,108,3375). Free radicals have also been observed in the reaction of ozone with simple alkenes using BHT(2,6-di-*tert*-butyl-4-cresol) as a radical trap. In the case of 3,3-dimethylbutene the adduct can be rationalised as arising from *tert*-butylperoxyl radicals (W.A. Pryor, J.G. Gu, and D.F. Church, J.Org.Chem., 1985,50,185; W.A. Pryor, D. Giamalva, and D.F. Church, J.Am.Chem.Soc., 1985,107,2793).

(ii) Epoxidation

The mechanism of ethene epoxidation was reviewed in 1981 (W.M.H. Sachfler, C. Backx, and R.A. Van Santen, Cat. Rev.Sci.Eng., 1918,23,127) and the reaction has continued to attract much attention. The catalytic epoxidation of alkenes with organic hydroperoxides was also reviewed in 1981 (197 references) (J.Sobezak and J.J. Ziolkowski, J. Mol.Catal., 1981,63,575).

Hydrogen peroxide in aqueous solution can be used to form alkene epoxides by the use of molybdenum (VI) and tungsten (VI) catalysis using a two-phase system and phase-transfer technology (O. Bortolini *et al.*, J.Org.Chem.,

1985,50,2688); Y. Ishi *et al.*, J.Org.Chem., 1988,53 3587). High yields of the epoxide products are obtained even with the use of dilute hydrogen peroxide under relatively mild conditions, the key intermediate apparently being a molybdenum or tungsten peroxo species (H. Mimoun, Angew.Chem.Intl.Edn.Engl., 1982,21,734). A palladium superoxo-complex has also been found to specifically oxidise linear alkenes to epoxides (E.P. Talso *et al.*, J.Chem.Soc. Chem.Commun., 1985,1768), as has an hydroxoplatinum (II) complex (G. Strukul and R.A. Michelin, J.Chem.Soc.Chem. Commun., 1984,1538) and rhodium (III) (M. Faraj *et al.*, J.Organomet.Chem., 1984,276,C23).

A number of workers have been studying the epoxidation reactions catalysed by cytochrome P-450 models (for example, see D. Ostovic and T.C. Bruice, J.Am.Chem.Soc., 1989,111, 5611, and refs. therein). The evidence seems to suggest that proposed metallaoxetane intermediates are not involved, but hypervalent porphyrin metal-oxo species participate in the rate-determining process.

The mechanism of ethene epoxidation using oxygen over silver catalysts seems to be in dispute. Some workers suggest that it involves direct epoxide formation with dioxygen (R.A. Van Santen and C.P.N. De Groot, J.Catal., 1986,98,530).
However, others suggest that dioxygen plays no direct role but that chemisorbed atomic oxygen is the crucial surface species (R.B. Grant and R.M. Lambert, J.Chem.Soc. Chem.Commun., 1983,662).

Cobalt-catalysis of alkene epoxidation seems to show dual pathways for the oxygen transfer to the alkene (J.D. Koola and J.K. Kochi, J.Org.Chem., 1987,52,4545).

An interesting epoxidation reaction involves the use of dimethyldioxirane which seems to proceed *via* a spiro transition state (Scheme 28) (A.L. Baumstark and C.J. Mc-Closky, Tetrahedron Lett., 1987,28,3311).

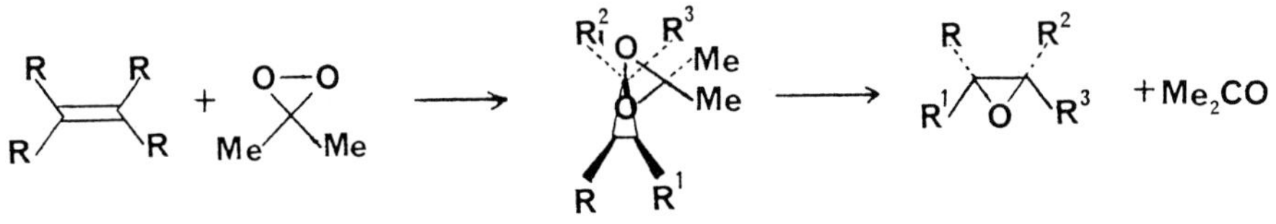

Scheme 28

It has been found that iodosobenzene in the presence of various iron, manganese or chromium porphyrins gives, not only epoxides, but also allylic alcohols and aldehydes these not being derived from the corresponding epoxides (D.Mansuy *et al.*, Tetrahedron, 1984,40,2847).

Another interesting epoxidation is that of propene by thermal or photochemical reaction in acetonitrile in the presence of sulfur dioxide. The reaction can be carried out at 0°C and may proceed through formation of some sulfinyl oxide species. Other alkenes may also undergo epoxidation under these conditions, the success of the reaction depending on the stability of the peoxide under the reaction conditions (T. Sasaki, J.Am.Chem.Soc., 1981,103, 3882).

The epoxy group can act as an alkene protection group since it can be readily deoxygenated using diazomalonate and rhodium (II) acetate catalysis (as shown below) (M.G. Martin, and B. Ganem, Tetrahedron Lett., 1984,25,251).

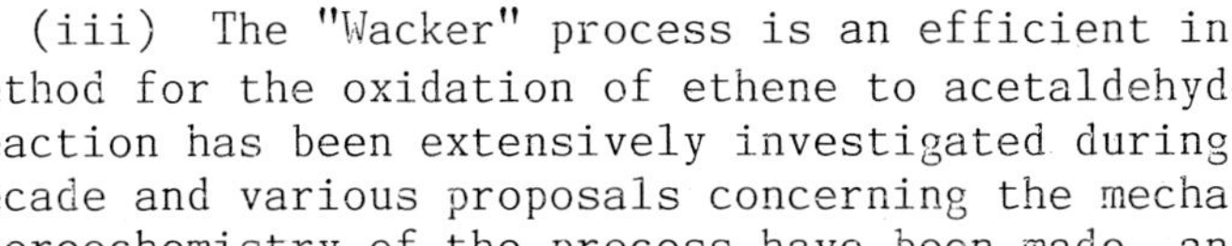

Reagents: $Rh_2(OAc)_4$, C_6H_6(or C_6H_5Me), reflux, various times

(iii) The "Wacker" process is an efficient industrial method for the oxidation of ethene to acetaldehyde. The reaction has been extensively investigated during the last decade and various proposals concerning the mechanism and stereochemistry of the process have been made, and various modifications have been proposed. The reaction involves the palladium (II)-catalysed oxidation of ethene in water in the presence of cupric chloride and chloride ion. It has been suggested that the mechanism involves the reversible formation of a hydrated palladium species (17) followed by a rate-determining dissociation of chloride to give (18) (Scheme 29) (J.E. Baeckvall, B. Akermark, and S.O.Ljunggren, J.Am.Chem.Soc., 1979,101,2411).

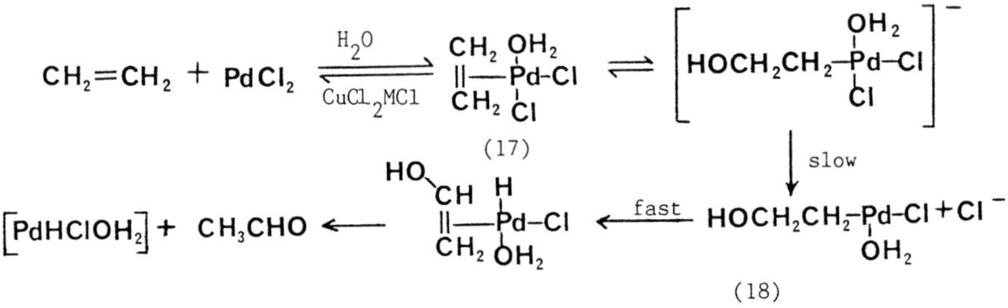

$CH_2{=}CH_2 + PdCl_2 \xrightleftharpoons[CuCl_2,MCl]{H_2O}$ $CH_2{=}CH_2$–$Pd(OH_2)(Cl)$–Cl (17) $\rightleftharpoons$ $[HOCH_2CH_2{-}Pd(OH_2)(Cl){-}Cl]^-$

$\downarrow$ slow

$HOCH_2CH_2{-}Pd(OH_2){-}Cl + Cl^-$ (18) $\xrightarrow{fast}$ $HOCH{=}CH_2$ / $H{-}Pd(OH_2){-}Cl$ $\longrightarrow$ $CH_3CHO + [PdHClOH_2]$

Scheme 29

Other workers have suggested that a mechanism utilising a rate-determining hydroxy-palladation step involving an unsymmetrical π-coordinated ethene molecule is involved (Y. Saito and S. Shinoda, J.Mol.Catal., 1980,9,461).

The palladium-catalysed oxidation of terminal alkenes to methyl ketones is conveniently effected by hydrogen peroxide, which seems to involve a pseudo hydroperoxy-palladation of the coordinated alkene (M. Roussell and H. Mimoun, J.Org.Chem., 1980,45,5387).

The synthetic applications of the palladium catalysed oxidation of alkenes to ketones has been reviewed (J.Tsuji, Synthesis, 1984,369).

Cobalt and cobalt-nitrocomplexes (B.S. Tovrog, F. Mares, and S.E. Diamond, J.Am.Chem.Soc., 1980,102,6616; A. Zombreck, D.E. Hamilton and R.S. Drago, J.Am.Chem.Soc., 1982,104,6782), palladium (II)-nitro complexes (M.A. Andrews and K.P. Kelly, J.Am.Chem.Soc., 1981,103,2894; J.E. Baeckvall and A.Henmann, J.Am.Chem.Soc., 1986,108,7107, and refs. therein), rhodium complexes (R.S. Drago *et al.*, J.Am. Chem.Soc., 1985,107,2893) manganese (III) complexes

(W.E. Fristad *et al.*, Tetrahedron, 1986,42,3429), and also ruthenium (III) complexes (M.M.T. Khan and A.P. Rao, J.Mol.Catal., 1988,44,95) have all been used in modifications of the original Wacker process. The reaction can also be carried out in chloride-free media, using p-benzoquinone as oxidant, under very mild conditions (J.E. Baeckvall and R.B. Hopkins, Tetrahedron Lett., 1988,29, 2885). Phase-transfer catalysis can also be applied to the palladium-catalysed oxidation of terminal and internal alkenes (K. Januszkiewicz and H. Alper, Tetrahedron Lett., 1983,25,2519, H. Alper, K. Januszkiewicz, and D.J.H. Smith, Tetrahedron Lett., 1985,26,2263) and photo-activated palladium-catalysed molecular oxygen oxidation of alkenes to ketones has also been observed (J. Muzart, P. Pale, and J.-P. Pete, Tetrahedron Lett., 1982,23,3577).

The cobalt complex [bis(salicylidene-γ-iminopropyl)-methylamino]cobalt(II), CoSMDPT (19), has been found to catalyse the oxidation of alkenes in the presence of dioxygen or hydrogen peroxide to give, from terminal alkenes, methyl ketones and secondary alcohols. However, this does not seem to be a Wacker-type of reaction and it is proposed that a cobalt hydroperoxide is formed which adds to the alkene to give a hydroperoxide which is subsequently decomposed by CoSMDPT to give the observed products (D.E. Hamilton, R.S. Drago, and A. Zombeck, J.Am.Chem.Soc., 1987,109,374).

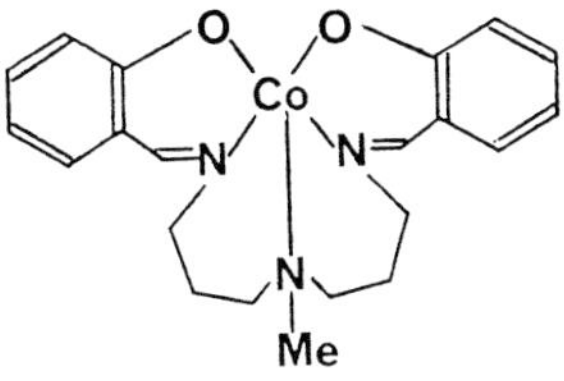

(19)

When the Wacker reaction is carried out in the presence of acids esters are also produced and, in the case of propene in the presence of acetic acid, the catalytic oxidation in the presence of palladium catalysts gives alkyl acetate and carbon dioxide as the sole products (A.K. Zaidi, Zh.Obsch.Khim., 1981,51,1422); Chem.Abstr., 1981, 95,149597).

A study of the acetoxylation of ethylene using PdX_4^{2-} (X = halogen) catalysts has shown that the reactions proceeds mainly by a binuclear π-complex between C_2H_4 and Pd, then isomerisation to a σ-complex and heterolysis of the Pd-C bond with the participation of water (A.V. Devekki, Yu.N. Koshlev, and D.V. Mushenko, Zh.Org.Khim., 1980,16,2518; Chem.Abstr., 1981,94,102411).

The reaction of alkenes with CO, H_2O, O_2 and HCl in the presence of palladium and copper chlorides gives good yields of branched chain carboxylic acids under very mild conditions (THF at r.t. and atmospheric pressure), this being an important development in this reaction which has considerable industrial importance (H. Alper *et al.*, J.Chem. Soc.Chem.Commun., 1983,1270; B. Despeyroux and H. Alper, Ann.N.Y.Acad.Sci., 1983,415,148).

In another modification of the Wacker process the palladium acetate catalysts have been bound to an insoluble aromatic oligoamide to give "heterogenised" homogenous catalysts with good activity but with a different product distribution than usual. In the presence of oxygen and perchloric acid in ethanol-water there is direct conversion to alkynes under mild conditions and in high yields, but ketonization occurs with these same reagents in dioxane-water (G. Cum *et al.*, J.Chem.Soc.Chem.Commun., 1985,1571).

(iv) Dihydroxylation and diesterification of alkenes

Mono and disubstituted alkenes can be converted to *trans*-vicinal diacetates in good yield by heating with iron (II), persulfate, acetic acid. The mechanism of the process seems to involve initial addition of the sulfate radical ion to the alkene, followed by solvolysis, addition, hydrolysis and acetylation (Scheme 30) (W.E. Fristad and J.R. Peterson, Tetrahedron Lett., 1983,24,4547; Tetrahedron, 1984,40,1469; C. Arnoldi, A. Citterio, and F. Minisci, J.Chem.Soc.Perkin Trans.II, 1983,531).

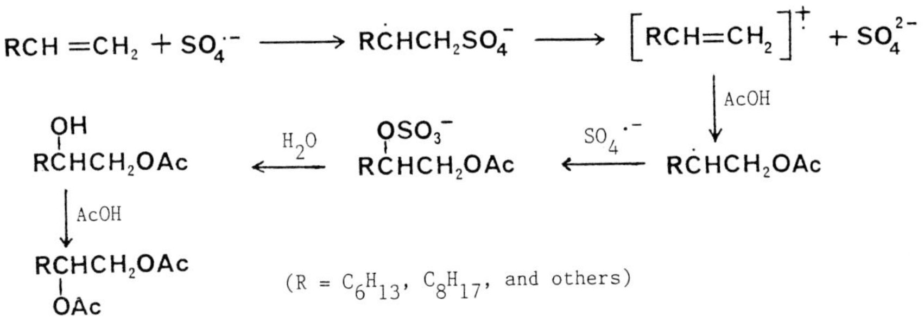

$$RCH{=}CH_2 + SO_4^{\cdot -} \longrightarrow R\dot{C}HCH_2SO_4^- \longrightarrow [RCH{=}CH_2]^{\dot{+}} + SO_4^{2-}$$

$$[RCH{=}CH_2]^{\dot{+}} \xrightarrow{AcOH} R\dot{C}HCH_2OAc \xrightarrow{SO_4^{\cdot -}} RCH(OSO_3^-)CH_2OAc \xrightarrow{H_2O} RCH(OH)CH_2OAc \xrightarrow{AcOH} RCH(OAc)CH_2OAc$$

(R = C_6H_{13}, C_8H_{17}, and others)

Scheme 30

Studies of the oxidation of alkenes by potassium permanganate have confirmed that the reaction proceeds *via* a cyclic ester of type (20) and that the pathway leading to the α-ketol product proceeds *via* an acyclic manganese (IV) ester (21) (S. Wolfe, C.F. Ingold, and R.U. Lemieux, J.Am.Chem.Soc., 1981,103,938).

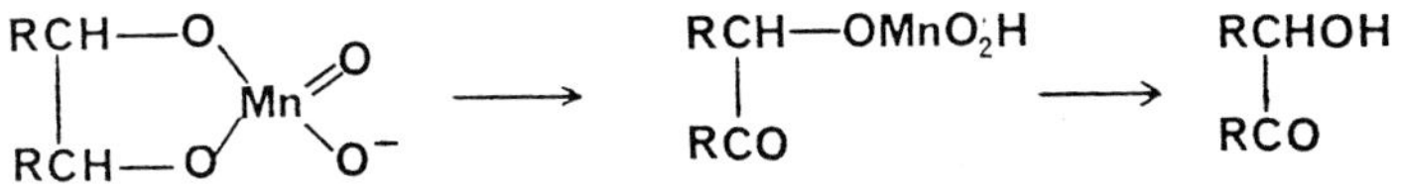

$$\text{(20) cyclic } RCH{-}O{-}Mn(=O)(O^-){-}O{-}CHR \longrightarrow \text{(21) } RCH(OMnO_2H){-}RCO \longrightarrow RCHOH{-}RCO$$

(20) (21)

A simple and mild method for the *cis* hydroxylation of alkenes utilises the reagent cetyltrimethylammonium permanganate (V. Bhushan, R. Rathmore, and S. Chandrasekaran, Synthesis, 1984,431).

An interesting new reaction to introduce two oxygen functional groups into alkenes involves an oxidative carboxylation of alkenes using palladium (II)chloride, copper (II)-chloride, and sodium chloride as catalyst. In acetic acid/acetic anhydride acetoxy anhydrides are formed, and in ethanol/triethyl orthoformate ethoxy ethyl esters are formed (Scheme 31) (H. Urata, A. Fujita, and T. Fuchikami, Tetrahedron Lett., 1988,29,4435).

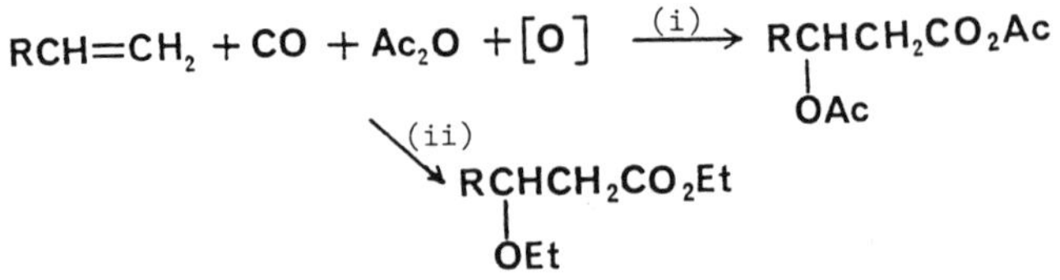

$$RCH{=}CH_2 + CO + Ac_2O + [O] \xrightarrow{(i)} RCH(OAc)CH_2CO_2Ac$$

$$\xrightarrow{(ii)} RCH(OEt)CH_2CO_2Et$$

Reagents: (i) $PdCl_2$, $CuCl_2$, NaCl, AcOH, 80°C, 24h
(ii) $PdCl_2$, $CuCl_2$, NaCl, CO + EtOH,O_2, $HC(OEt)_3$, 50°C, 24h

Scheme 31

A simple procedure for the stereospecific synthesis of vicinal dicarboxylic acids has been devised utilising the addition of dichloroketene to alkenes to give α,α-dichlorocyclobutanones, followed by treatment with lithium dimethylcopper, then sodium metaperiodate-ruthenium dioxide (Scheme 32) (J.-P. Depres, F. Coelho, and A.E. Green, J.Org.Chem., 1985,50,1972).

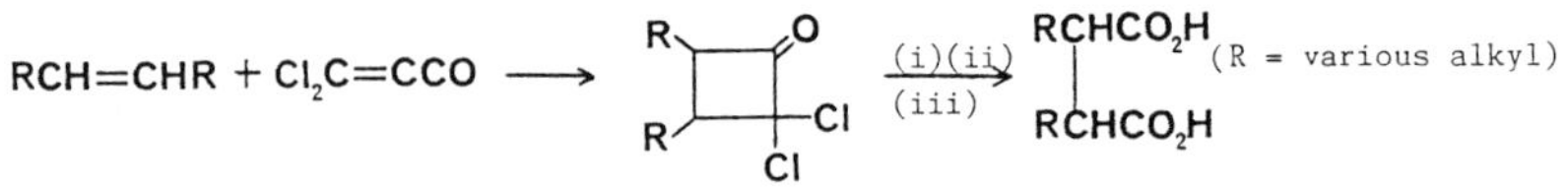

$$RCH{=}CHR + Cl_2C{=}CCO \longrightarrow \text{(2,3-dialkyl-4,4-dichlorocyclobutanone)} \xrightarrow[(iii)]{(i)(ii)} RCHCO_2H{-}RCHCO_2H$$ (R = various alkyl)

Reagents: (i) $LiMe_2Cu$, Ac_2O, hexane, -78°C to r.t.; (ii) $NaIO_4$, RuO_2, MeCN,CCl_4, H_2O, r.t., 6h; (iii) NaOH, H_2O, then HCl

Scheme 32

(v) Other oxidative reactions of alkenes

Oxidative cleavage of alkenes into carboxylic acids can be achieved by using hydrogen peroxide in the presence of catalytic amounts of tungstic acid (T. Oguchi *et al.*, Chem.Lett., 1989,857).

The thiol-olefin cooxidation (TOCO) reaction was first described in 1951 (M.S. Kharasch, W. Nudenberg, and G.J. Mantell, J.Org.Chem., 1951,16,524) and has since been extensively investigated, particularly by Szmant and coworkers. The reaction between a thiol, an alkene, and molecular oxygen results in the formation of a β-hydroxy sulfoxide (22). However, the precise mechanism of the reaction is still actively under investigation and attempts have been made to explain the dependence of product distribution on the concentration of thiol and alkene and on the reaction temperature, the variation in reaction rate with alkene structure, and other factors (H.H. Szmant and coworkers, J.Org.Chem., 1980,45,4902; 1987,52,1720,1725, 1729, 1741).

The ene reaction of singlet oxygen with simple alkenes gives allylic hydroperoxides (23).

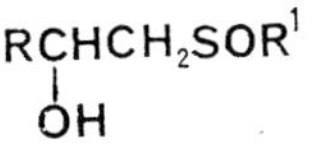

$RCH(OH)CH_2SOR^1$ (22)

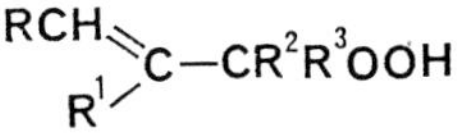

$RCH{=}C(R^1)-CR^2R^3OOH$ (23)

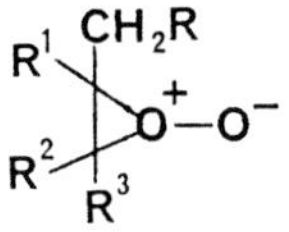

(24)

This reaction has been studied for over twenty years, an early suggestion being that it involves a peroxide intermediate (24). Evidence for the existence of such an intermediate has now been demonstrated (J.R. Hurat, S.L. Wilson, and G.B. Schuster, Tetrahedron, 1985,41,2191).

The dicyanoanthracene sensitised photooxygenation of alkenes also gives products similar to those obtained in the singlet oxygen ene reaction for simple alkenes, but arylalkenes undergo a different reaction leading to cleavage resulting in the formation of alkenes and carbonyl compounds (C.S. Foote, Tetrahedron, 1985,41,2121).

The metal induced formation of an allylic C-O bond has been reviewed (J. Muzart, Bull.Soc.Chim.France, 1986,65).

The stereochemistry of the autoxidation and photosensitised oxidation which produces allylic hydroperoxides has been investigated both with and without photosensitisation. There are significant product differences and, in the case of autoxidation, the relative proportions of isomeric hydroperoxides are explicable in terms of the conformations of the parent alkenes and the delocalised radicals formed on hydrogen abstraction. In the case of photosensitised oxidation the isomeric distribution is that expected from the concerted ene reaction of singlet oxygen with the alkene orientated in favour of the *cis* cyclic process (E.N. Frankel *et al.*, J.Chem.Soc.Perkin Trans.I, 1982,2707,2715).

(vi) Miscellaneous reactions of alkenes introducing oxygen functionality

The ene reaction, mentioned as an example of alkene oxidation above, is a general reaction of alkenes containing an allylic hydrogen which has been extensively studied. It is a thermally allowed ($2n_\pi + 2n_\pi + 2\sigma$) process related to the Diels-Alder reaction and a 1,5-sigmatropic shift. The ene reactions of alkenes with chloral and bromal in the presence of Lewis acid catalysts have been investigated the products being principally trihalomethylalkenols (25) and dihalomethyl halogeno ketones (26). Anhydrous aluminium (III)chloride seems to be the most effective catalyst and reaction mechanisms, reactivities, and conditions for product optimisation have been established in some cases (J.P. Benner *et al.*, J.Chem.Soc.Perkin Trans I, 1984, 291,215,331).

$$RCH_2CH{=}CH_2 + X_3CCHO \longrightarrow RCH{=}CHCH(OH)CX_3 + RCH_2CHXCH_2COCHX_2$$

(25) (26)

A ^{13}Cn.m.r. study of the reaction between maleic anhydride and a number of alkenes has provided detailed information concerning the stereochemistry of the reaction and has led to a mechanistic picture of the process and product formation. The general pattern to emerge is that trans alkenes result in the formation of *erythro* diastereomers and *cis* alkenes give *threo* diastereomers and that the diastereomer preference is essentially independent of chain length.

Regioselectivity is strongly dependent on the size of the groups flanking the double bond in the reactants whereas *cis/trans* selectivity and diastereoselectivity are not, (S.H. Nahm and H.N. Cheng, J.Org.Chem., 1985,51,5093).

$$RCH{=}CHCH_2R^1 \xrightarrow{\text{maleic anhydride}} \text{succinic anhydride-}CH(R)CH{=}CHR^1$$

An interesting new conversion of alkenes to methyl esters involves the addition of methyl formate to alkenes catalysed by the complex tetra[triphenylphosphinyl]ruthenium (II)hydride. The proposed reaction is that shown in Scheme 33 (W. Uede *et al.*, J.Mol.Catal., 1988,44,197).

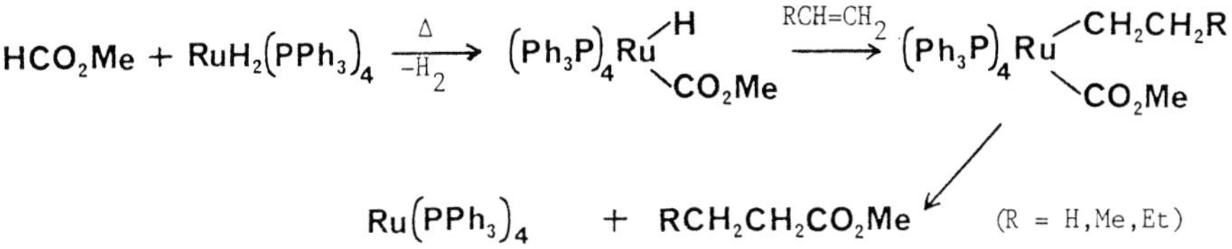

$$HCO_2Me + RuH_2(PPh_3)_4 \xrightarrow[-H_2]{\Delta} (Ph_3P)_4Ru\begin{matrix}H \\ CO_2Me\end{matrix} \xrightarrow{RCH=CH_2} (Ph_3P)_4Ru\begin{matrix}CH_2CH_2R \\ CO_2Me\end{matrix}$$

$$\longrightarrow Ru(PPh_3)_4 + RCH_2CH_2CO_2Me \quad (R = H,Me,Et)$$

Scheme 33

Other esterifications of alkenes can produce keto esters, for example, the acylation of isobutene with acetyl fluoroborate in the presence of acetic anhydride yields (27) (A.V. Shastin and E.S. Balenkova, Zh.Org.Khim., 1984,20,956; Chem.Abstr., 1984,101,229959), whilst hydroxyphenyliodoso mesylate (28) leads to dimesyl products, e.g. (29) (N.S. Zefirov *et al.*, Zh.Org.Khim., 1985,21,2461; Chem.Abstr., 1986,105,225736). Lithium perchlorate can react with alkenes to give covalently bonded organic perchlorates (N.S. Zefirov *et al.*, J.Org.Chem., 1982,47,3679).

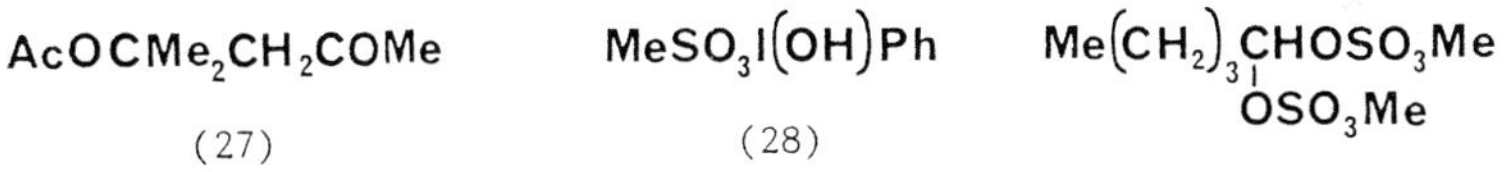

$AcOCMe_2CH_2COMe$ (27)

$MeSO_3I(OH)Ph$ (28)

$Me(CH_2)_3CH(OSO_3Me)OSO_3Me$ (29)

Caesium fluoroxysulfate undergoes addition to alkenes to give the new vicinal fluoralkyl sulfates. The regioselectivity is solvent dependent, but the predominance of *cis* product formation is consistent with a concerted mechanism for the addition (Scheme 34) (N.S. Zefirov *et al.*, Tetrahedron, 1988,44,6505).

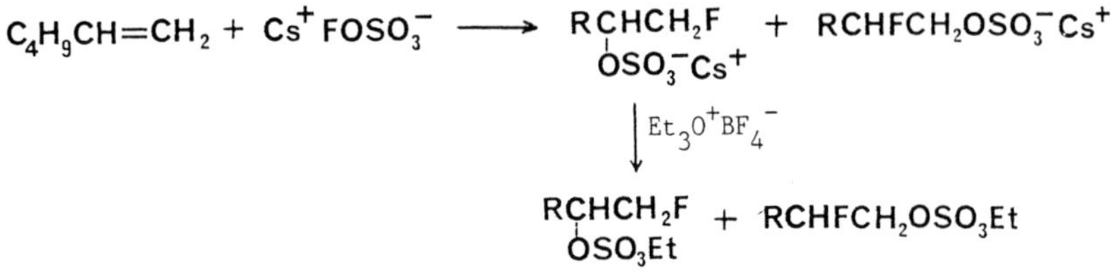

$$C_4H_9CH{=}CH_2 + Cs^+FOSO_3^- \longrightarrow \underset{OSO_3^-Cs^+}{RCHCH_2F} + RCHFCH_2OSO_3^-Cs^+$$

$$\downarrow Et_3O^+BF_4^-$$

$$\underset{OSO_3Et}{RCHCH_2F} + RCHFCH_2OSO_3Et$$

Scheme 34

A series of new reagents (e.g. $R^1I(OX)_2$) containing hypervalent iodine has also been developed which react with alkenes to give vicinal diesters such as vicinal ditriflates (Scheme 35) (N.S. Zefirov *et al.*, Tetrahedron Lett., 1986,27,3971); R.T. Hembre, C.P. Scott, and J.R. Norton, J.Org.Chem., 1987,52,3650, and refs. therein).

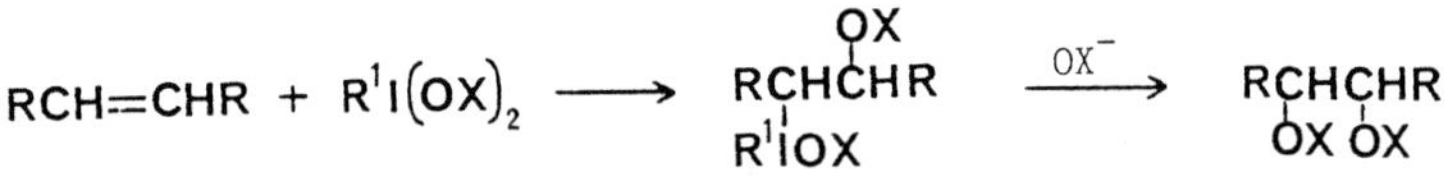

$$RCH{=}CHR + R^1I(OX)_2 \longrightarrow \underset{R^1IOX}{RCHCHR(OX)} \xrightarrow{OX^-} \underset{OX\ OX}{RCHCHR}$$

($R{=}H$, $n\text{-}C_4H_9$; $R^1 = Ph$; Ms, Ts, $COCF_3$)

Scheme 35

The reagent chlorine chlorosulfate (30) undergoes addition to alkenes to give generally good yields of β-chloroalkyl chlorosulfates (31), but both Markownikov and *anti*-Markownikov products are obtained (N.S. Zefirov, A.S. Koz'min, and V.D. Sorokin, J.Org.Chem., 1984,49,4086).

$ClOSO_2Cl$ (30) $\quad$ $RCHClCH_2OSO_2Cl$ (31)

The Hydroacylation of alkenes has received much attention. It has been shown that aldehydes and alkenes react to give ketones with some ruthenium complexes as catalyst, although in modest yields (P. Isnard *et al.*,

J.Organomet.Chem., 1982,240,285), and a convenient method for the homolytic alkylation of ketones by alkenes is to use the silver ion catalysed decomposition of peroxydisulfate as the initiation process (A. Citterio, F. Ferrario, and S.De Bernardinis, J.Chem.Res.Synop., 1983,310).

Catalytic α-alkylation of aldehydes by alkenes using transition metal acetate catalysts involves interaction of the metal enolate complex with the alkene. The best yields are obtained using manganese (III) (M.G. Vinogradov, I.P. Kovalev, and G.I. Nikishin, Izv.Akad.Nauk SSSR Ser.Khim., 1981,1569; Chem.Abstr., 1981,95,149596). Other metal-catalysed aldehyde-alkene reactions have also been investigated as alternative approaches to Wittig-type reactions.

A convenient synthesis of α-bromo ketones has been reported which involves treating alkenes with sodium bromite in aqueous acetic acid at room temperature (T. Kageyama *et al.*, Chem.Lett., 1983,1481).

An ether synthesis from alkenes having a wide generality involves the solvomercuration-demercuration reaction in alcohol solvents in the presence of mercuric trifluoroacetate. The product is principally that of Markownikov addition (i.e. the 2-ether) and yields of up to 90% can be obtained in favourable cases (Scheme 36) (H.C. Brown *et al.*, J.Org.Chem., 1984,49,2551; 1985,50, 1171); H.C. Brown, P.J. Geoghan, and J.T. Kurek, J.Org. Chem., 1981,46,3810).

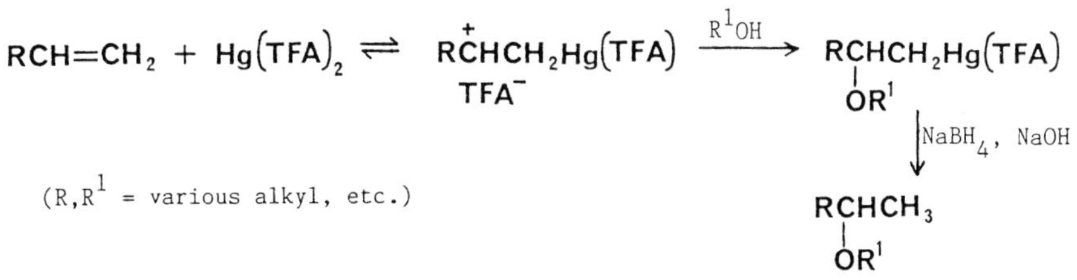

$$RCH{=}CH_2 + Hg(TFA)_2 \rightleftharpoons \underset{TFA^-}{R\overset{+}{C}HCH_2Hg(TFA)} \xrightarrow{R^1OH} \underset{\displaystyle OR^1}{RCHCH_2Hg(TFA)} \xrightarrow{NaBH_4,\ NaOH} \underset{\displaystyle OR^1}{RCHCH_3}$$

(R,R^1 = various alkyl, etc.)

Scheme 36

Other mercuration reactions which can be carried out include peroxymercuration and bromodemercuration, which leads to stereoselective conversion of non-terminal alkenes into β-bromoalkyl peroxides (A.J. Bloodworth and J.L. Courtneidge, J.Chem.Soc.Perkin Trans.I, 1981,3258), and peroxymercuration followed by reduction using tributyl tin hydride which converts nonterminal alkenes into secondary *tert*-butyl peroxides (A.J. Bloodworth and J.L. Courtneidge, J.Chem.Soc.Chem.Commun., 1981,1117).

The hydration of alkenes to give alcohols is an important industrial process in which a very useful improvement has been made with the use of proton-exchanged ferrierite-type zeolites as efficient catalysts at relatively low temperatures (473°K (M. Iwamoto, M. Tajima and S. Kagawa, J.Chem.Soc.Chem.Commun., 1985,228). Recent studies in the acid-catalysed hydration of butene show that the reaction does not proceed through the same carbocation intermediate as for the oxygen exchange into 2-butanol (P.E. Dietze and W.P. Jencks, J.Am.Chem.Soc., 1987,109,2057).

(vii) Nitration, and related reactions, of alkenes

The reactions of nitrogen oxides with alkenes is of potential importance in the chemistry of atmospheric pollution as well as for synthetic possibilities. There seem to be differences in the reactions of NO_2 and N_2O_4 in the gas and liquid phases. A study of this reaction shows that N_2O_4 seems to undergo addition to alkenes to give dinitro products whilst NO_2 reactions principally proceed via allylic abstraction of hydrogen to give products which include allyl nitro compounds (D.H. Giamalva *et al.*, J.Am. Chem.Soc., 1987,109,7059).

New types of nitrogen containing compounds - nitroperoxypropyl nitrate (32), and other products of this type, have been identified from reactions of propene with nitroxy radicals. The reaction is initiated by the NO_3 radical adding to propene but the intermediates are unstable and the final product is propylene glycerol-1,2-dinitrate (33) (H. Bandow, M. Okuda, and H. Akimoto, J.Phys.Chem., 1980, 84,3604).

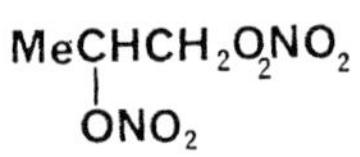

$MeCH(ONO_2)CH_2O_2NO_2$ (32)

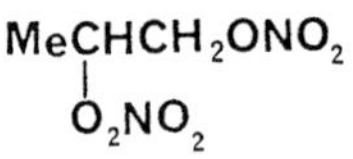

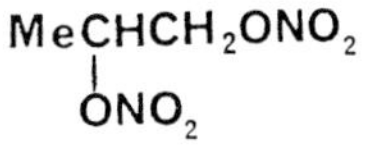

$MeCH(O_2NO_2)CH_2ONO_2$

$MeCH(ONO_2)CH_2ONO_2$ (33)

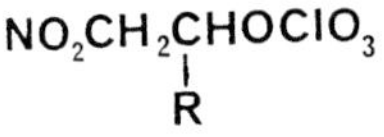

$NO_2CH_2CH(R)OClO_3$ (34)

Nitroperchlorates (34) are formed by the reaction of alkenes with nitronium tetrafluoroborate and lithium perchlorate (N.S. Zefirov *et al.*, Zh.Org.Khim., 1981,17,195; Chem.Abstr., 1981,95,24353), and alkenes react reversibly with mercury (II) nitrate to form β-nitratoalkyl mercury (II) nitrates (35). These products readily react with bromine at 0°C to give β-bromoalkyl nitrates (36) (A.J. Bloodworth and P.N. Cooper, J.Chem.Soc.Chem.Commun., 1986,709).

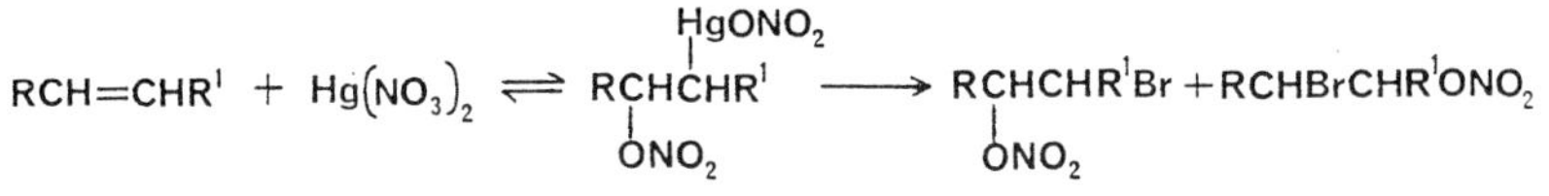

$$RCH{=}CHR^1 + Hg(NO_3)_2 \rightleftharpoons RCH(ONO_2)CHR^1(HgONO_2) \longrightarrow RCH(ONO_2)CHR^1Br + RCHBrCHR^1ONO_2$$

(R = alkyl; R^1 = H, alkyl) (35) (36)

The nitration of alkenes can also be effected using palladium nitro complexes (M.A. Andres *et al.*, Organometallics, 1984,3,1479).

The ene reaction of nitroso compounds with alkenes, which can lead to allyl hydroxyamines, has been shown to proceed in a two-step process via an aziridine N-oxide (Scheme 37) (C.A. Seymour and F.D. Green, J.Org.Chem., 1982,47,5226).

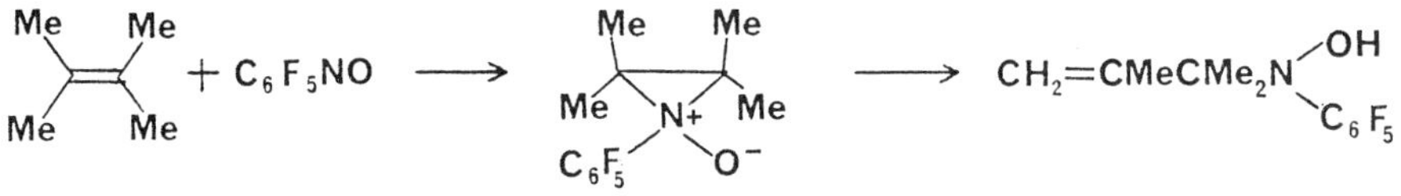

$$Me_2C{=}CMe_2 + C_6F_5NO \longrightarrow \text{(tetramethylaziridine } N\text{-oxide, } N^+(C_6F_5)O^-) \longrightarrow CH_2{=}CMeCMe_2N(OH)C_6F_5$$

Scheme 37

(viii) The introduction of sulfur substituents into alkenes

Sulfenyl compounds RSX where X is an electron-attracting group add readily to alkenes and the reaction is well known. However, where X is not electron-attracting then the reagents do not normally add to alkenes unless a catalyst is present. The use of boron trifluoride etherate catalyses the addition of such compounds to alkenes, for example, sulfenamides and methylthiodimethylsulfonium salts (Scheme 38) (M.C. Caserio, C.L. Fisher, and J.K. Kim, J.Org. Chem., 1985,50,4390; J.Am.Chem.Soc., 1982,104,3231; B.M.

Trost and T. Shibata, J.Am.Chem.Soc., 1982,104,3225).

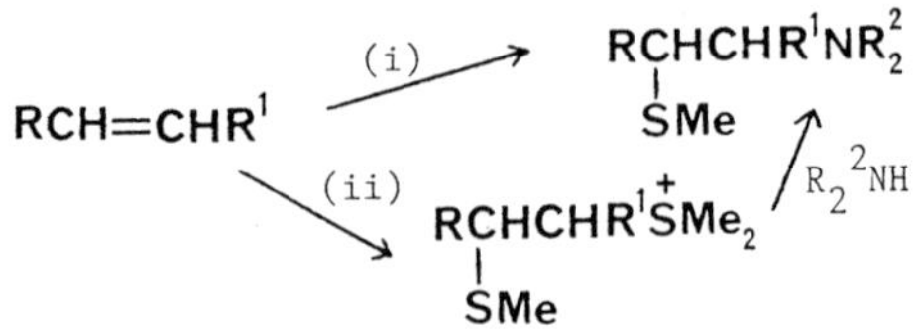

$$RCH{=}CHR^1 \xrightarrow{(i)} RCH(SMe)CHR^1NR^2_2$$

$$RCH{=}CHR^1 \xrightarrow{(ii)} RCH(SMe)CHR^1\overset{+}{S}Me_2 \xrightarrow{R^2_2NH} RCH(SMe)CHR^1NR^2_2$$

Reagents: (i) $MeSNH_2$, $BF_3.OEt_2$; (ii) $Me_2S^+SMe.BF_4^-$, CH_2Cl_2, $MeNO_2$, 0-25°C

Scheme 38

Phenylsulfenamides also react with alkenes in the presence of trifluoromethane sulfonic acid to give phenylthioamidines or amines (P. Brownbridge, Tetrahedron Lett., 1984,25,3759), and anodic or lead (IV) oxidation of disulfides in the presence of alkenes and acetonitrile gives β-thioacetamides (A. Bewick *et al.*, J.Chem.Soc.Perkin Trans.I, 1985,1033,1039).

A potentially useful conversion of alkenes into thiiranes involves the reaction of carbon disulfide with tertiary N-oxides in the presence of alkenes. The results indicate the formation of a species such as (37) or (38) which acts as the reagent for episulfidation of the alkene. (M.F. Zipplies, M.-J.De Vos, and T.C. Bruice, J.Org.Chem., 1985,50,3228).

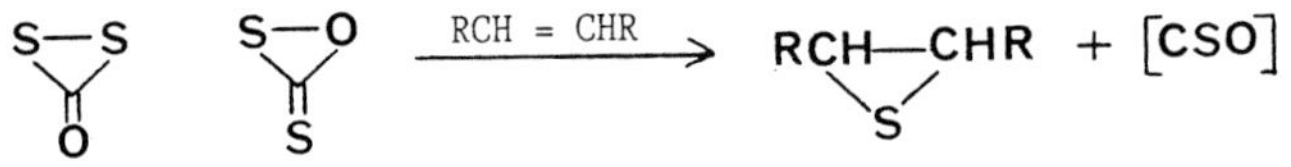

$$\text{(37) or (38)} \xrightarrow{RCH\,=\,CHR} RCH{-}CHR\ (\text{bridged by } S) + [CSO]$$

(37) (38)

Selenosulfonation of alkenes can be carried out using boron trifluoride catalysis giving a novel regio- and stereo-controlled synthesis of vinylsulfones (T.G. Back, and S. Collins), J.Org.Chem., 1981,46,3249).

(ix) Cycloaddition reactions of alkenes with 1,3-dipoles

Alkene cyclisation processes to form C-C bonds via the Dields-Alder reaction have been mentioned above. The addition of 1,3-dipoles to alkenes to form heterocycles is also a widely used reaction.

The 1,3-dipolar cycloaddition of diazomethane with alkenes has been shown to be in accord with a diradical mechanism and the regiochemistry of the reaction has been explained (R.A. Firestone, Tetrahedron Lett., 1980,21, 2209). A quantitative analysis of the charge-transfer perturbation model for the process has also been made (M. Burdisso *et al.*, Tetrahedron, 1987,43,159).

The 1,3-dipolar cycloadditions of nitrile oxides with Z-disubstituted alkenes gives a high degree of regioselectivity to give the isoxazolines (39). However, the E-isomers give mixtures of the isomeric products (40). This regioselectivity is explicable by considering the energy of the transition states in the process (S.F. Martin and B. Dupre, Tetrahedron Lett., 1983,24,1337).

$R\overset{+}{C}N\text{-}\bar{O}$ + alkene (R^1, R^2, Me) ⟶ (39) + (40)

(R^2 = alkyl, R^3 = H; R^2 = H,R^3 = alkyl) (39) (40)

Second Supplements to the 2nd Edition of Rodd's Chemistry of Carbon Compounds, Vol. IA/B, edited by M. Sainsbury
© 1991 Elsevier Science Publishers B.V., Amsterdam

Chapter 2b

CARBENES AND ALLENES

M. SAINSBURY

1. Carbenes

(a) *Introduction*

A brief historical overview of the origins of carbene chemistry can be found in an introduction to a Tetrahedron Symposium-in-print (P.S.Skell, *Tetrahedron* 1985, **41**, 1427), and a summary of carbenic activity is provided by R.A.Moss (*Acc.Chem.Research*, 1989, **22**, 15). For a more general reviews, see K.G.Taylor, *Tetrahedron*, 1982, **38**, 2751; G.B.Schuster, *Adv.Phys.Org.Chem.*, 1986, **22**, 311, and R.A.Aitken, Organic Reaction Mechanisms 1986, J.Wiley and Sons Ltd., New York, 1988.

(b) *Structure and Reactivity*

Molecular orbital calculations have provided much information about the geometry and electronic structures of simple carbenes, and singlet-triplet energy gaps have been predicted (P.J.MacDougall and R.F.W.Bader, *Can.J.Chem.*, 1986, **64**, 1496; G.E.Scuseria *et al.*, *J.Amer.Chem.Soc.*, 1986, **108**, 3248; D.A.Dixon, *J.Phys.Chem.*, 1986, **90**, 54).

The singlet-triplet interconversion of diphenylmethylene, and the reactivities of its different spin states have also been studied (K.B.Eisenthal *et al.*, *Tetrahedron*, 1985, **41**, 1543). Similarly, by comparing the chemical behaviour and the reaction kinetics of a series of arylcarbenes it is possible to order the compounds in terms of their singlet-triplet energy separations (B.B.Wright and M.S.Platz, *J.Amer.Chem.Soc.*, 1984, **106**, 4175). Much related work has been done: thus the chemical reactions of diphenylcarbene and fluorenylidene (1) with cyclohexane have been analysed to prove that diphenylcarbene reacts with cyclohexane predominantly, if not exclusively, as its triplet. On the otherhand, fluorenylidene exhibits substantial chemistry from its low lying singlet state, as well as showing some triplet chemistry. These results agree with predictions derived from laser flash photolysis experiments which indicate that the singlet-triplet splitting of fluorenylidene is smaller than that of diphenylcarbene (T.G.Savino, V.P.Senthilnathan and M.S.Platz, *Tetrahedron*, 1986, **46**, 2167). Like fluorenylidene, 1,8-diazafluorenylidene (2) exists as the triplet, but reacts in the singlet form (Y.-Z.Li and G.B.Schuster, *J.Org.Chem.*, 1986, **51**, 3804).

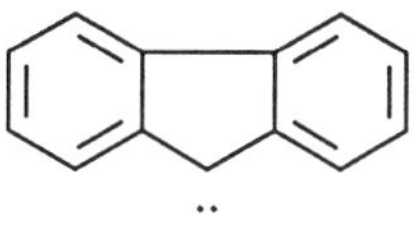

(1)

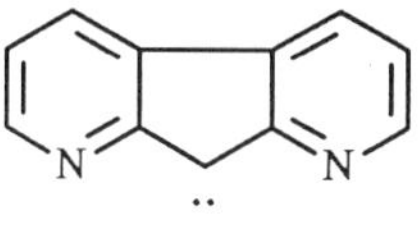

(2)

Indeed most arylcarbenes have a triplet ground state, but when they are generated photochemically they tend to react in the singlet form. This is due to the fact that the rate of reaction of the singlet carbene is faster than that of the competitive intersystems crossing. The influence of substituents upon intersystems crossing efficiency is of minor importance, but the reactivity of the singlet carbene does depend upon the nature of the aryl substituents (H.Tomioka *et al.*, *Tetrahedron,* 1985, **41**, 1435).

The central C-C-C angle within the structures of diaryl carbenes has an influence upon the energy separation between triplet and singlet states (this subject is widely discussed in papers compiled in Tetrahedron Symposium-in-print No. 19; *Tetrahedron* 1985, **4**, 1423). In the case of 9,9-dianthrylcarbene the central angle is 180° and the carbene shows a high degree of allenic character. Laser flash photolysis, conventional flash photolysis, and optical modulation spectroscopy indicate that the carbene has a structure in which the unpaired electrons are highly delocalised. In the singlet state dianthrylcarbene fails to react with methanol and rapidly self quenches. It is also slow to combine with oxygen (D.J.Astles *et al.*, *J.Org.Chem.*, 1988, **53**, 6053).

Triplet arylcarbenes (ArCH:) react intramolecularly with their unsaturated *ortho* substituents, when these are present; while the corresponding singlet arylcarbenes react preferentially with solvent methanol. Rotation about the exocyclic C-CH bond is necessary in order for cycloaddition to occur from the singlet states and quantitative calculations show that this process has a relatively high energy barrier. In the triplet states, however, little or no rotation is needed (G.Hoemberger, *J.Amer.Chem.Soc.*, 1989, **111**, 475). The effect of aryl substituents upon arylcarbene reactivity has been analysed and for appropriate substrates the influence which the electronic properties of these substituents play in the determination of C-H *versus* O-H bond insertion selectivity has been quantified (H.Tomioka *et al.*, *Tetrahedron*, 1985, **41**, 1435).

It was once believed that planar singlet carbenes of cyclopropene, pentadiene, heptatriene, and nonatetrene could achieve stable aromatic structures by the inclusion of the carbenic electron pair in the sigma or pi-system in order to give $(4n+2)\pi$ electrons. MNDO calculations now indicate that this concept is untenable, since such extensive resonance delocalisations would result in extensive charge separations, at least for the species (4), (5) and (6). This does not apply to the carbene (3) of cyclopropene, which because of its small ring size can take advantage of the aromatic stabilisation without extensive separation of charge. Thus only this species is likely to exist as a planar ground state singlet. Planar versions of the others are unstable and nonplanar singlet carbenes or their allenic counterparts are favoured (M.Kassaee *et al.*, *Tetrahedron*, 1985, **41**, 1579).

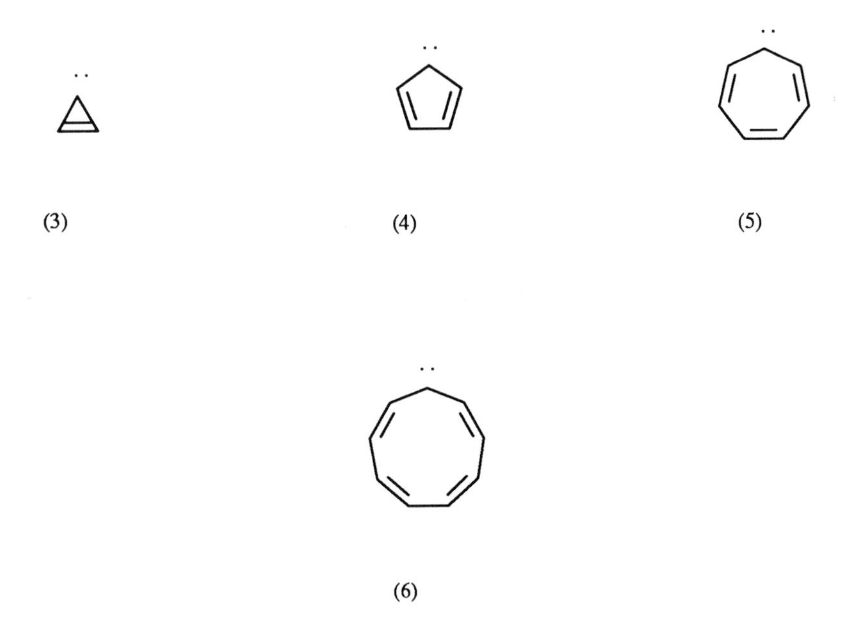

The product of flash vacuum pyrolyses of both 1- (7) and 2-naphthyl- (11) diazomethanes is cyclobutano[*d,e*]naphthalene (12). Thus it appears that while there is an equilibration between the carbenes (8),(9), and (10) the irreversibility of the insertion reaction leading from the benzoheptadienocarbenoid (8) to the tricyclic product outweighs all other factors (J.Becker and C.Wentrup, *Chem. Commun.*, 1980, 190).

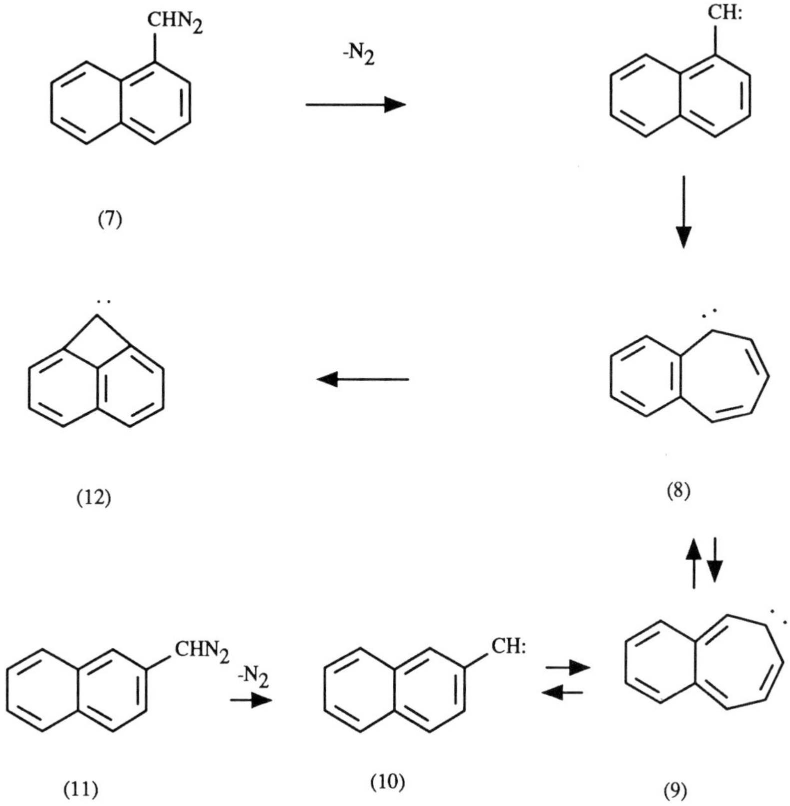

CHN_2 $-N_2$ CH:

(7)

(12)

(8)

CHN_2 $-N_2$ CH:

(11)

(10)

(9)

The complex nature of the rearrangements of arylcarbenes has been reviewed and interpreted (P.P.Gaspar *et al.*, *Tetrahedron*, 1985, **41**, 1479), and the use of frozen organic matrices to study their chemistry has been discussed. The matrix chemistry of arylcarbenes can be explained in terms of a thermal singlet-triplet equilibration in which the relative amounts of singlet and triplet chemistry is determined (a) by the energy gap between the two states and (b) by the rate constants of the reactions from each state. Insertion into O-H bonds and stereoselective cyclopropanation are typical reactions of the singlet state, whereas H-atom abstraction and non-stereoselective cyclopropanation are common reactions of triplets. There is evidence that the effects of the rigid matrix environment can alter the reactivity of the abstraction process, yet the extent of these changes is uncertain (B.B.Wright, *Tetrahedron*, 1985, **41**, 1517).

Steric effects in the cycloaddition reactions of alkenes and singlet carbenes prevent the establishment of a general free energy relationship, but it has been noted that rate retarding effects do manifest themselves when the alkenes bear bulky groups (B.Giese, W.B.Lee, and C.Stiehl, *Tetrahedron Letters*, 1983, **24**, 881).

The mechanism of the addition reactions between alkenes and carbenes is a subject of both historical and current interest and the Skell-Woodworth rules predict that for triplet arylcarbenes addition to an alkene should proceed through a stepwise pathway involving prior complex or biradical formation. Direct evidence for a biradical intermediate (14) has been obtained for the reaction of diarylcarbenes with alkenes (13), where it is noted that the first bond forming reaction *en route* to the cyclopropanes (15) is reversible (L.M.Tolbert and M.Ali, *J.Amer.Chem.Soc.*, 1985, **107**, 4589).

$Ar_2\ddot{C}$ + $MeOCOCH{=}CHCO_2R$ ⇄

(13)

Ar, Ar, CO_2Me, H, H, CO_2R → Ar, Ar, CO_2Me, H, H, CO_2R

(14) (15)

In the last decade, or so, laser flash photolysis has provided quantitative data on the behaviour of carbenes towards typical substrates. For example, the rates at which singlet arylhalocarbenes react with alkenes to form cyclopropanes, and the mechanisms of such reactions have been studied (K.N.Houk, N.G.Rondan, and J.Marena, *Tetrahedron*, 1985, **14**, 1555). Thus the reactions of phenylchlorocarbene with electron rich alkenes such as tri- or tetra-methylethene give rise to curved Arrhenius plots, which reveal negative activation energies even at temperatures as low as 230°K. A possible explanation is that no intermediate complexes are involved, so that the temperature dependent free energy barrier is solely entropic in origin. This concept is supported by calculations which best represent conditions in the gaseous phase; however, it is uncertain how this proposal relates to reactions which occur in polar solvents (J.E.Jackson *et al., Tetrahedron Letters*, 1989, **30**, 1335). Although values determined for the absolute rate constants of cycloadditions of arylhalocarbenes to alkenes rule out the possibility of *late* stage two-bond transition states and also bipolar single-bond transition states, these results are consistent with reversible carbene-alkene complexes or *early* one or two bond transition states (N.J.Turro *et al., J.Amer.Chem.Soc.*, 1987, **109**, 4973).

Diazirines are commonly employed as carbene precursors, and both thermal and photochemical methods are used to initiate the extrusion of dinitrogen from them. The characteristics of the carbenes so generated are of much interest and there has been some discussion regarding the preferential reactivity with nucleophilic alkenes (E.Schmitz, *Adv. Heterocyclic Chem.*, 1979, **24**, 63; M.T.H.Liu and D.H.T.Chien, *J.Chem.Soc. Perkin Trans., 2*, 1974, 937; R.A.Moss, *Acc. Chem. Res.*, 1980, **13**, 58; *idem. ibid.*, 1989, **22**, 15). Ambident behaviour is apparent in some cases and methoxychlorocarbene, for example, reacts with electron rich alkenes as an electrophile and as a nucleophile with electron poor alkenes (R.A.Moss and L.A.Perez, *Tetrahedron Letters*, 1983, **24**, 2719). The geometric isomerism of aryl-, vinyl- and carbonyl- carbenes has been evaluated (H.D.Roth and R.S.Hutton, *Tetrahedron Letters*, 1988, **29**, 1567). Simple frontier orbital calculations indicate that dimethylvinylcarbene should show ambiphilic reactivity, but this conclusion is not confirmed by more rigorous treatments which take orbital overlap into account. These computations predict that the molecule is electrophilic in nature, a view borne out by experimental results (Y.Cao *et al., Chem.Abs.*, 1986, **105**, 171480x).

(c) *Reactions of halocarbenes*

(i) *Cyclopropanation*

The reaction of fluorotrichloromethane with titanium tetrachloride and lithium aluminium hydride in tetrahydrofuran solution at 0°C gives rise to chlorofluorocarbene, which adds to alkenes, such as the vinyl ether (16) to yield 1-chloro-1-fluorocyclopropanes (17) (W.R.Dolbier and C.R.Burkholder, *Tetrahedron Letters*, 1988, **29**, 6749).

Dichlorocarbene reacts with the dichloroalkene (18) to afford the tetrachlorocyclopropane (19), this molecule on treatment with methyl lithium undergoes 1,2-dechlorination to give the cyclopropene (20), which is in equilibrium with the acyclic carbene (21). This carbene can be trapped with alkenes to form adducts (22). If dibromocarbene is used in the first reaction the dibromodichlorocyclopropane (23) is formed. Treatment of this product with two molecular equivalents of methyl lithium leads to the allenic carbene (24) which combines with 2,3-dimethylbutene to afford the cyclopropanoallene (25) (J.AlDulayymi and M.S.Baird, *Tetrahedron Letters*, 1988, **29**, 6147; AlDulayymi, Baird, and W.Clegg, *ibid.*, 6149).

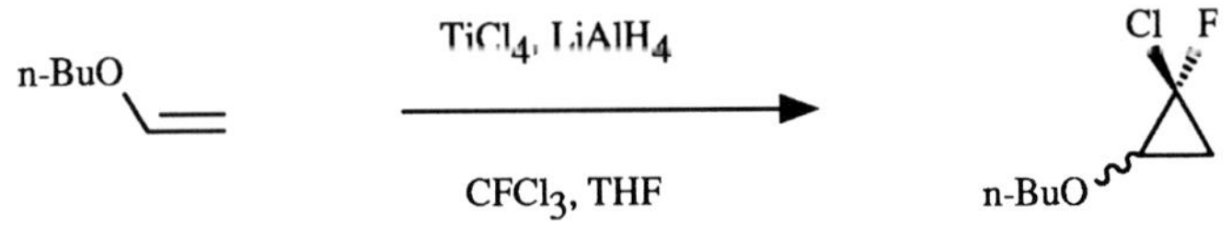

n-BuO

$TiCl_4$, $LiAlH_4$

$CFCl_3$, THF

Cl F

n-BuO

(16)

(17)

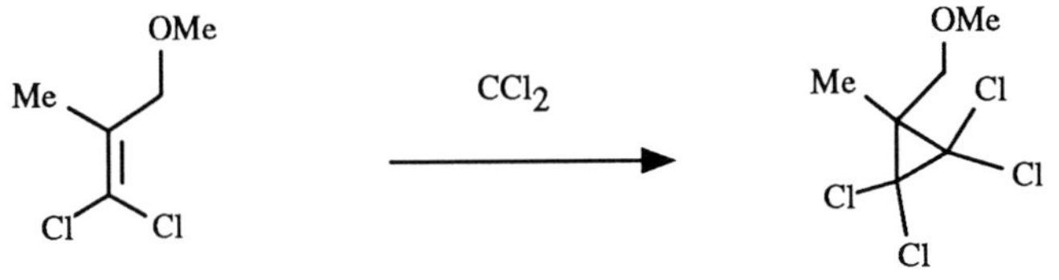

OMe

Me

Cl Cl

CCl_2

OMe

Me Cl

Cl

Cl

Cl

(18)

(19)

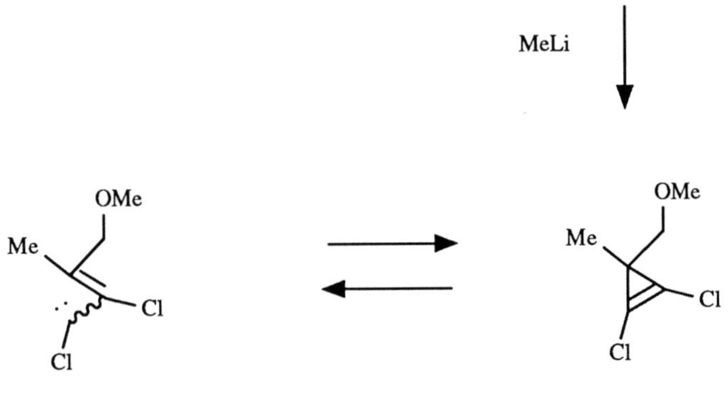

MeLi

OMe

Me

Cl

Cl

OMe

Me

Cl

Cl

(21)

(20)

(21) $\xrightarrow{R_2C{=}CR_2}$ (22)

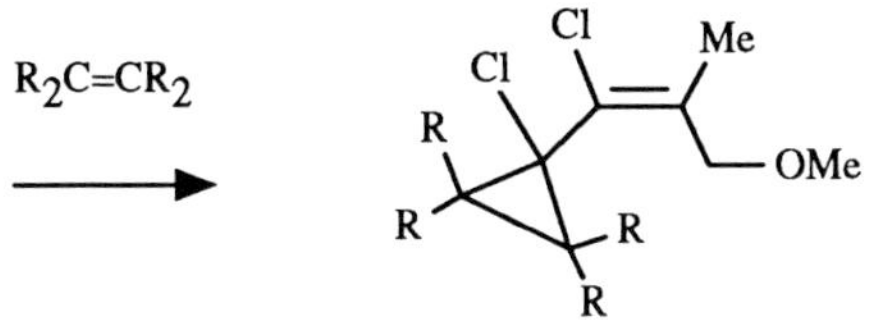

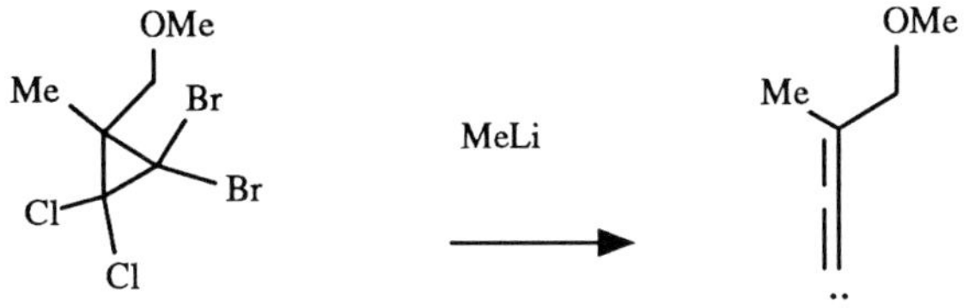

(23) $\xrightarrow{MeLi}$ (24)

$\longrightarrow$ (25)

More conventionally, difluorocarbene adds to perfluoro-3-methylindene (26) to yield the cycloadduct (27); this compound is thermally stable up to 270°C when its reforms the starting materials. Similarly the spiro compound (29) is formed from the reaction between methyleneindane (28) and difluorocarbene. This last product is rather less stable and degrades to its starting materials at 250°C (I.P.Chuikv *et al.*, *Izv. Akad. Nauk. S.S.S.R. Ser. Khim.*, 1988, 1839).

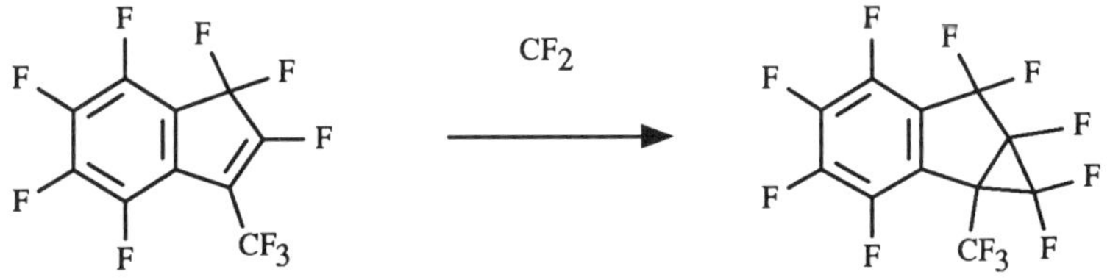

CF_2

(26) (27)

CF_2

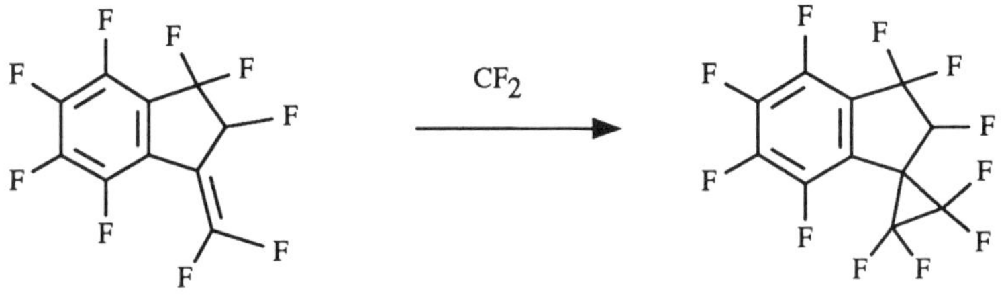

(28) (29)

Chlorophenylcarbene may be regarded as a typically electrophilic carbene, and as such should be relatively unreactive towards electron poor alkenes. In practice, however, this is not so and when generated by heating the diarizine (30) with diethyl maleate at 80°C in the absence of solvent it gives three cyclopropanes (31), (32) and (33). While the first two adducts retain the *cis*-stereochemistry of the alkene, the last does not, suggesting that two molecules of the alkene are involved in the reaction, leading to the participation of a Zwitterionic intermediate (34, where B = *cis*-$EtO_2CH{=}CHCO_2Et$). This allows a time interval for bond rotation to occur, prior to the elimination of a molecule of diethyl maleate and the formation of the cyclopropanes (M.P.Doyle, J.W.Terpstra, and C.H.Winter, *Tetrahedron Letters*, 1984, **25**, 901).

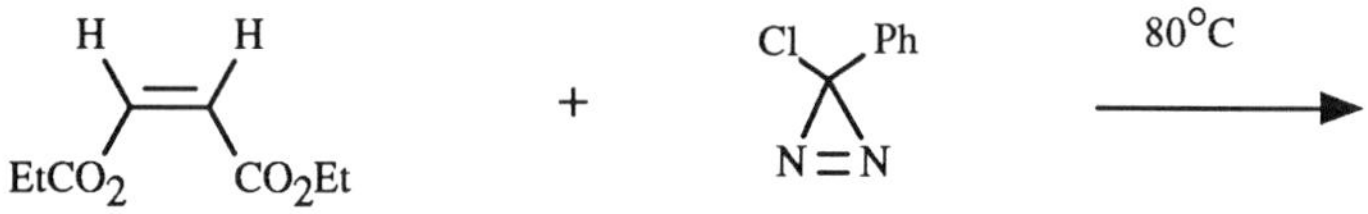

H H EtCO2 CO2Et + Cl Ph N=N $\xrightarrow{80^{\circ}C}$

(30)

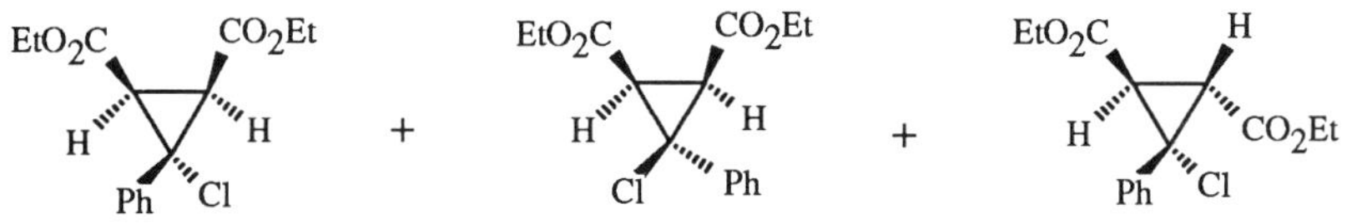

EtO2C CO2Et H H Ph Cl + EtO2C CO2Et H H Cl Ph + EtO2C H H CO2Et Ph Cl

(31) (32) (33)

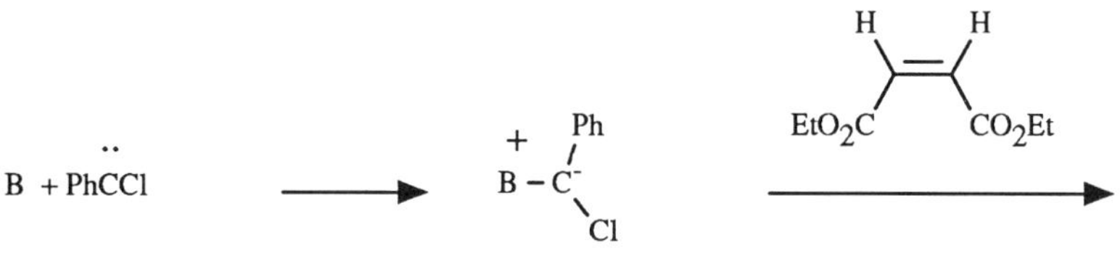

$\ddot{B}$ + PhCCl ⟶ $\overset{+}{B}-C^{-}$(Ph)(Cl) $\xrightarrow{\text{EtO}_2\text{C(H)C=C(H)CO}_2\text{Et}}$

^{+}B Cl Ph H - H EtO2C CO2Et $\xrightarrow{-B}$ (31-33)

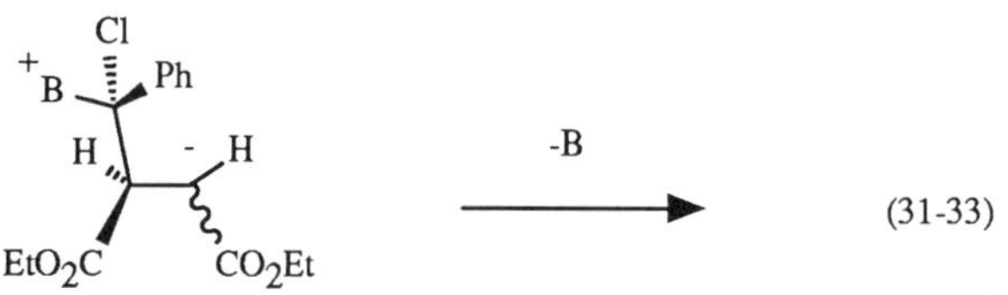

(34)

Whereas an ylide is only assumed to be involved in the above reaction, laser flash photolysis of diazirines (35, R= 4-Cl, or 4-CF_3) yields the corresponding carbenes which react with acetone to produce ylides (36) which are easily detected, before they cyclise to epoxides (37) and then rearrange to chloroketones (38) (N.Soundarajan *et al.*, *Tetrahedron Letters*, 1988, **29**, 3419).

Similarly the carbonyl ylide (40, R= 4-C_6H_4), generated from 3-chloro -3-(4-nitrophenyl)diazirine (39, R= 4-C_6H_4) by reaction with acetone in acetone-diethyl ether solution, enters into 1,3-dipolar addition reactions with benzaldehydes (ArCHO; Ar= 3-NO_2-, 4-NO_2-, 4-CN-, 4-Cl-, 4-Me-, or 4-OMe-C_6H_4-) to form 4-aryl-2,2-dimethyl-5-(4-nitrophenyl)- dioxoles (41). Alternatively, the ylide may also undergo intramolecular cyclisation and rearrangement to give the chloroketones (42) (T.Ibata, J.Toyoda and M.T.H.Liu, *Chem. Letters*, 1987, 2135).

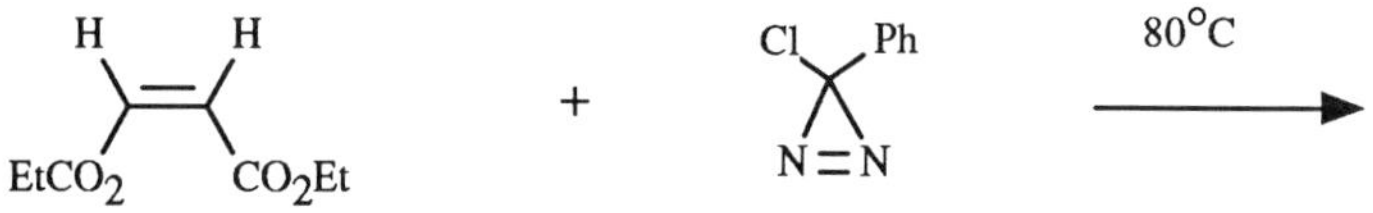

80°C

(30)

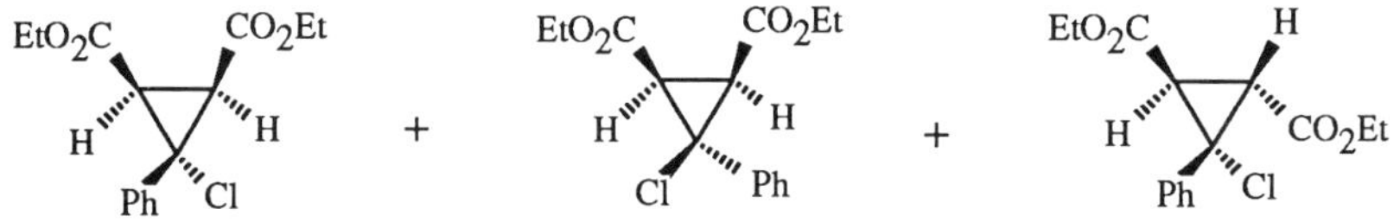

(31) (32) (33)

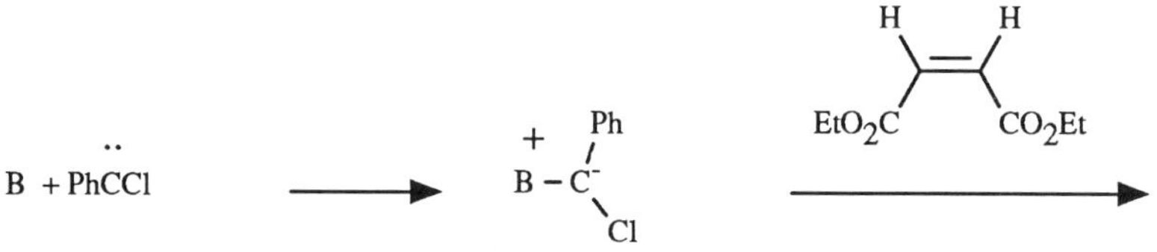

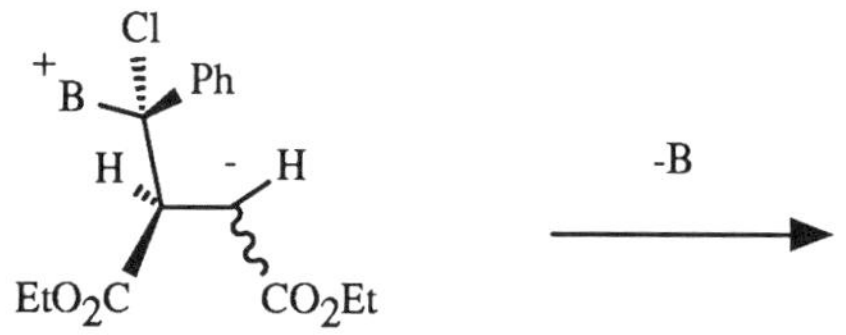

-B

(31-33)

(34)

Whereas an ylide is only assumed to be involved in the above reaction, laser flash photolysis of diazirines (35, R= 4-Cl, or 4-CF_3) yields the corresponding carbenes which react with acetone to produce ylides (36) which are easily detected, before they cyclise to epoxides (37) and then rearrange to chloroketones (38) (N.Soundarajan *et al.*, *Tetrahedron Letters*, 1988, **29**, 3419).

$R\text{-}C_6H_4$ N N Cl (35) $\xrightarrow{CH_3COCH_3}$ Cl C⁻ Ar O⁺ (36)

$\longrightarrow$ Cl Ar O (37) $\longrightarrow$ O Cl Ar (38)

Similarly the carbonyl ylide (40, R= 4-C_6H_4), generated from 3-chloro -3-(4-nitrophenyl)diazirine (39, R= 4-C_6H_4) by reaction with acetone in acetone-diethyl ether solution, enters into 1,3-dipolar addition reactions with benzaldehydes (ArCHO; Ar= 3-NO_2-, 4-NO_2-, 4-CN-, 4-Cl-, 4-Me-, or 4-OMe-C_6H_4-) to form 4-aryl-2,2-dimethyl-5-(4-nitrophenyl)- dioxoles (41). Alternatively, the ylide may also undergo intramolecular cyclisation and rearrangement to give the chloroketones (42) (T.Ibata, J.Toyoda and M.T.H.Liu, *Chem. Letters*, 1987, 2135).

Cl R N=N (39) $\xrightarrow{\text{Acetone}}$ R Cl $\overset{+}{O}-\overset{-}{C}Me_2$ (40) $\xrightarrow{\text{ArCHO}}$ R Ar O O Me Me (41)

(40) ⟶ (42)

Non-stabilised alkylidenetriphenylphosphoranes (43) react with chlorodifluoromethane to give difluoromethylene alkenes (45). This is a useful alternative to the Wittig synthesis of such compounds. Both primary and secondary ylides react provided they do not contain electron withdrawing groups, and it is thought that the reaction proceeds by initial deprotonation of the chlorodifluoromethane by the ylide to form difluorocarbene. This molecule is then trapped by a second molecule of the ylide, now acting as a nucleophile. The synthesis is complicated by the formation of 1*H*-1-fluoro-1-alkenes (47), which is explained by equilibration of the acyclic difluorovinyphosphonium intermediates (44) with the cyclic phosphonium salts (46). The latter on hydrolysis form the unrequired by-products (G.A.Wheaton and D.J.Burton, *J.Org.Chem.*, 1983, **48**, 917).

$$\overset{+}{Ph_3P}\text{-}\overset{-}{C}R^1R^2 \;(43) + HCClF_2 \longrightarrow [\overset{+}{Ph_3P}\text{-}CHR^1R^2]\overset{-}{X} + CF_2$$

$$CF_2 \xrightarrow{\overset{+}{Ph_3P}\text{-}\overset{-}{C}R^1R^2} \overset{+}{Ph_3P}\text{-}C(R_1)(R_2)\text{-}\overset{-}{CF_2} \;(44)$$

$$(44) \xrightarrow{-\,Ph_3P} F_2C{=}CR^1R^2 \;(45)$$

(44) ⇄ (46) $\xrightarrow{H_2O}$ [$Ph_3-P(OH)-CR^1R^2-CF_2H$] $\xrightarrow[-HF]{-Ph_3PO}$ (47)

(ii) *Reactions with imines and enamines*

A Zwitterionic species (49) is presumed to be involved in the reaction between difluorocarbene and the imine (48). Here the initial intermediate is considered to rearrange to give a second Zwitterion (50), which then cyclises to the aziridine (51) (J.R.McCarthy *et al., Chem. Commun.*, 1987, 469).

Evidence has also been presented for the existence of a Zwitterionic intermediate (53) in the reaction of 2,3-dimethylindole (52) and phenyl(trichloromethyl)mercury(II)-sodium iodide. This intermediate then gives rise to 3-dichloromethyl-2,3-dimethylindole (54) and 3-chloroquinoline (55) (M.Botta, F.De Angelis and A.Gambocorta, *Tetrahedron*, 1982, **38**, 2315).

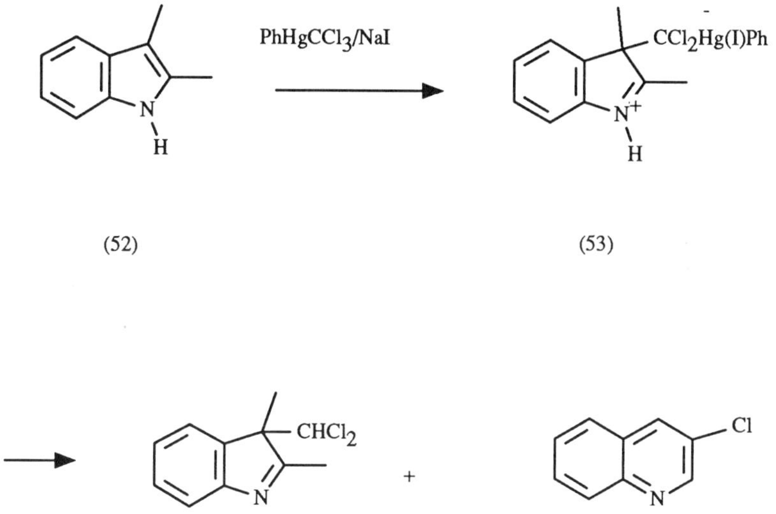

$PhHgCCl_3/NaI$

(52) (53)

(54) + (55)

(iii) *Reactions with dienes*

1,2-Addition of dihalocarbenes to 1,3-dienes is commonplace, but until recently 1,4-addition was unknown. In the first example of the latter reaction type it has been shown that various dihalocarbenes CXY (where X and Y can be fluorine, chlorine and, or bromine) react with the dienes (56) to give mixtures of the expected 1,2-adducts (57) and the unexpected 1,4-adducts (58) (the proportions of the products depending upon the halogens and the ring size) (L.A.M.Turkenburg, W.H.De Wolf, and F.Bickelhaupt, *Tetrahedron Letters*, 1982, **23**, 769; L.W.Jenneskens, De Wolf, and Bickelhaupt, *Angew. Chem. Internat. Ed. Engl.*, 1985, **24**, 585).

In a further illustration of this reaction type 1,1,2,2,3,3-hexa methyl-4,5-bis(methylene) cyclopentane (59) reacts with either dichlorocarbene or with dibromocarbene to give the adducts (60) and (61). The former being the more abundant, typically in the ratio of 7:3 (L.M.Tolbert and M.B.Ali, *J.Amer.Chem.Soc.*, 1985, **107**, 4589).

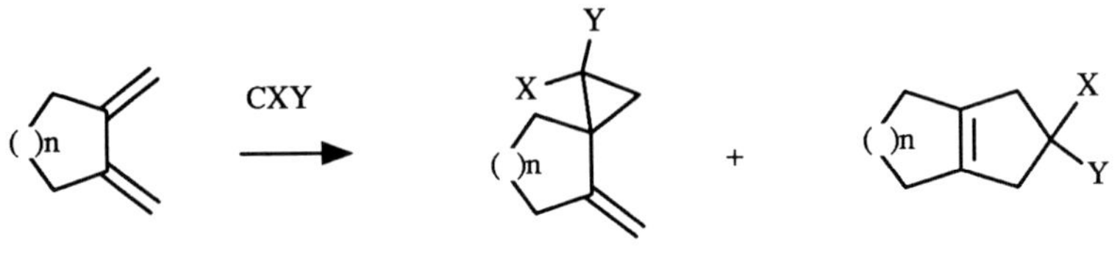

Y

X

CXY

( )n ( )n + ( )n X

Y

(56) (57) (58)

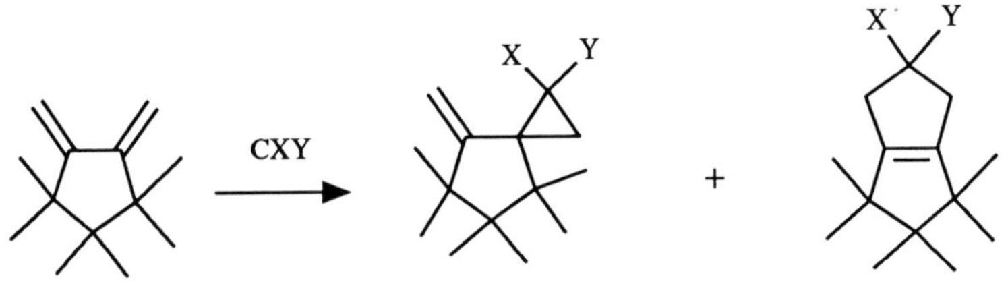

X Y X Y

CXY +

(59) (60) (61)

(iv) *Insertion*
Success in achieving intramolecular carbenoid sigma-bond insertion and thence ring expansion depends upon the ability of the reaction centre to eclipse a neighbouring sigma bond. When through constraints of molecular geometry this is not possible only substitution and reduction products are formed. Thus, for example, treatment of the dichloro compound (62, R=Cl) with phenyl lithium leads only to the diphenyl derivative (62, R=Ph). On the otherhand, when the molecular architecture is favourable similar reactions work well and this approach has been used to synthesise cyclopropano derivatives (64) of dichloromethylated spherical hydrocarbons (63) (L.A.Paquette, T.Kobayashi, and J.C.Gallucci, *J.Amer.Chem.Soc.*, 1988, **110**, 1305).

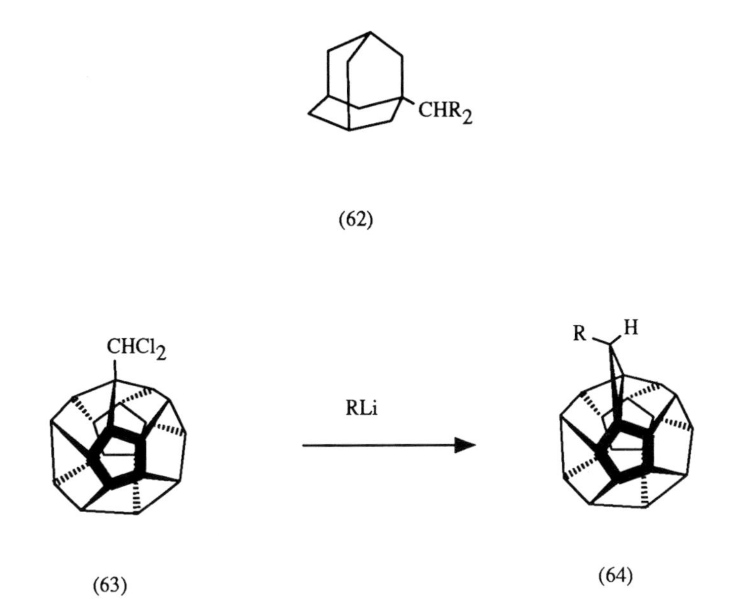

While there are many examples of carbene insertion reactions into C-H bonds, similar insertions into C-C bonds are less common. However, it has been demonstrated that difluorocarbene combines with quadricyclane (65) to give *endo*-6-(2,2-difluorovinyl)bicyclo[3.1.0]hex-2-ene (66). The precise mechanism for this reaction is uncertain, but it is possible that the product is formed as a result of a concerted reaction of the carbene with the bicyclo[2.1.0] system of quadricyclane (U.Misslitz, M.Jones Jr., and A.de Meijere, *Tetrahedron Letters*, 1985, **26**, 5403). There are a number of minor products formed in this reaction and the nature and significance of these compounds has caused some controversy. For example, there is a discrepancy between the results obtained by Jones' group and the members of C.W.Jefford's team (Jefford, J.C.E.Gehret and V.de los Heros, *Bull. Soc. Chim. Belg.,* 1979, **88**, 973; Jefford, J.Roussilhe, and M.Papadopoulos, *Helv.,* 1985, **68**, 1557). It is suggested that the purity of the quadricyclane used may lie at the centre of this problem especially since this substrate is difficult to obtain free from norbornadiene.

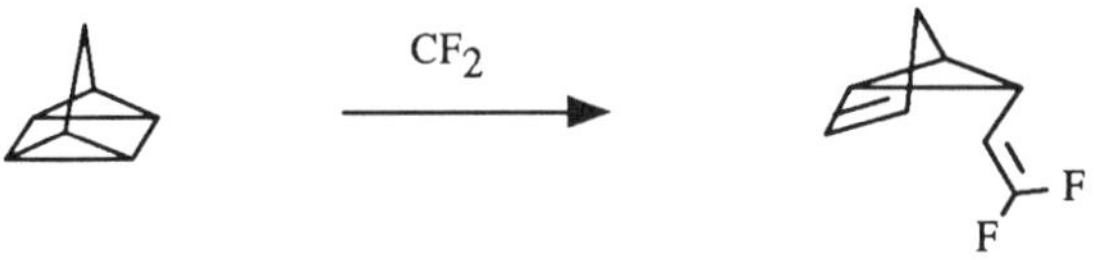

CF_2

F F

(65) (66)

Certainly, it is known that difluorocarbene reacts with 2-methoxy- and 2-methoxycarbonyl- 7,7-dimethylnorbornadiene (67, R=OMe) and 67, R=MeO_2C), respectively, to give homo-1,4-adducts such as (68) and (69), compounds analogous to some of those formed as by-products in the previous reaction (Jefford and P.T.Huy, *Tetrahedron Letters*, 1982, **23**, 391).

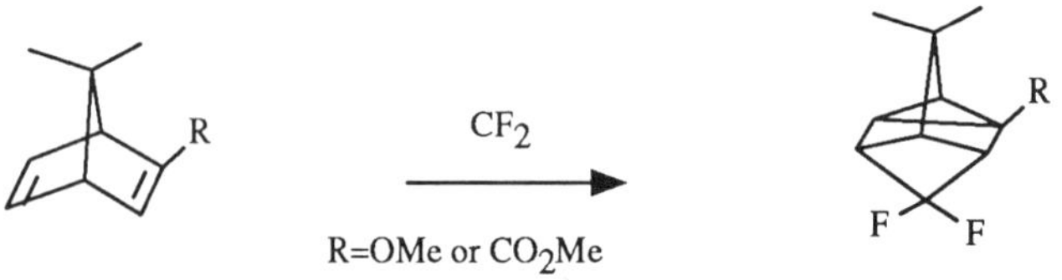

R

CF_2

R=OMe or CO_2Me

R

F F

(67) (68)

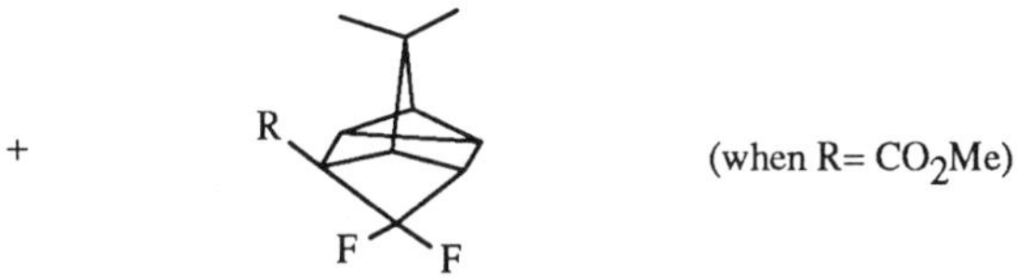

\+

R

F F

(69)

(when R= CO_2Me)

Dimethoxy- and dichloro- carbene add to 1,2,2,-trimethylbicyclo-[1.1.0]butane (70) to give derivatives of 2,3,3-trimethylpenta-1,4-diene, and it is suggested that this reaction occurs through the agency of the tertiary radical (71). This species is favoured since, although attack of the carbene could also take place at the 'a-b' side of the substrate, the 'c-d' face is less sterically encumbered. Cleavage of either bond c or d then affords the observed products (72) or (73), respectively (G.B.Mock and M.Jones Jr., *Tetrahedron Letters*, 1981, **22**, 3819).

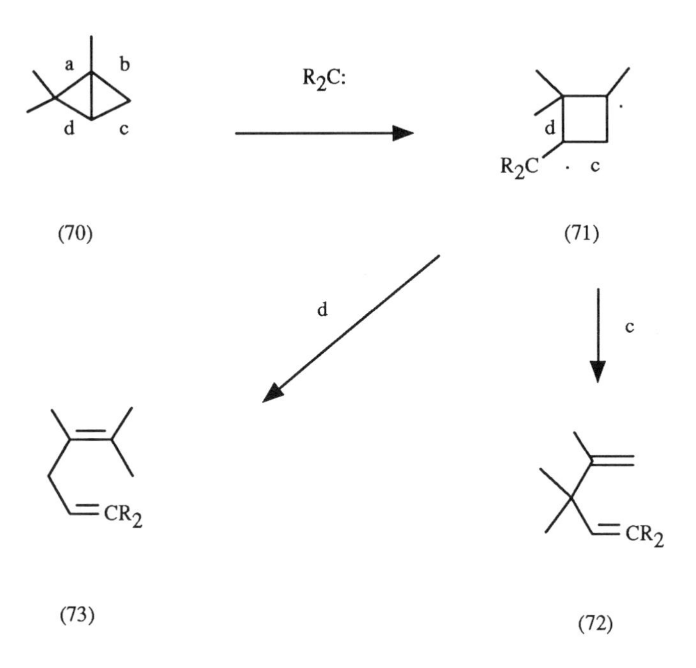

(d) *Alkyl and Alkylidene carbenes*

Insertion reactions

Phenylthiocarbene inserts into the α-CH bond of alkoxides to give the corresponding phenylsulphides. The reaction is stereoselective and in the case of the cyclohexane derivative (74, M=K) the major product is the β-orientated sulphide (76). It is suggested that the oxyanionic substituent of the substrate causes a decrease in the α-CH bond energy, and also helps to stabilise the polar transition state (75) by electron donation. Together these two effects facilitate the insertion process (T.Harada and A.Oku, *J.Amer.Chem.Soc.*, 1981, **103**, 5965).

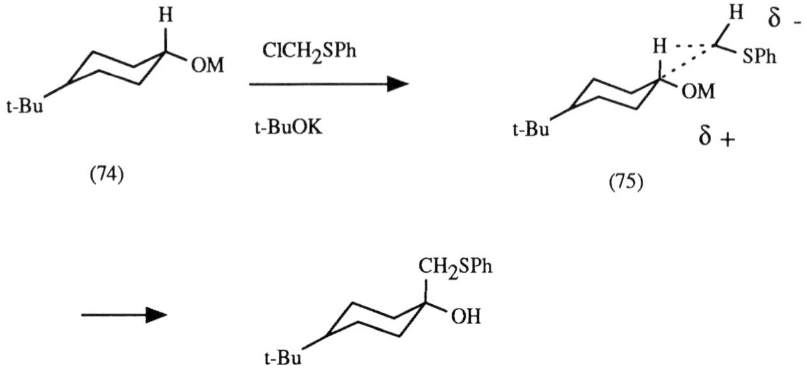

$ClCH_2SPh$

t-BuOK

(74) (75)

(76)

When alkylidenecarbenes (77, R= Me or [t]Bu) react with chiral α-naphthylphenylmethylsilane (78) the corresponding allenes (79) are formed in diastereomeric excess, which demonstrates the transfer of central chirality *via* the silicon based auxiliary (P.J.Stang and A.E. Learned, *J.Org.Chem.*, 1989, **54,** 1779).

$R\ddot{C}CH{=}CHR$ + Ph(Me)(naphth-1-yl)SiH

(77) (78)

⟶ Ph(Me)(naphth-1-yl)SiCH=C=CHR

(79)

Primary alkoxides (82, M = K or Li) and haloalkenes (80) together when treated with butyl lithium react to give allylic alcohols (83), and it is considered that these reactions involve the insertion of the alkylidenemethylene carbenoids (81) into the α-C-H bond of the alkoxide, perhaps through a hydride ion abstraction-recombination mechanism. Secondary alkoxides under similar reaction conditions afford tertiary alcohols through addition reactions, as well as products from insertion processes (A.Oku *et al.*, *J.Org.Chem.*, 1988, **53**, 3089).

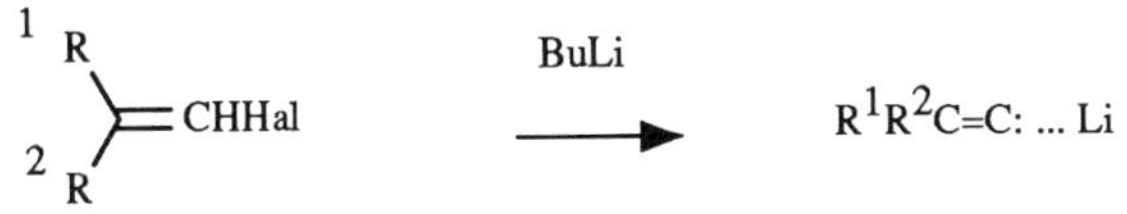

$R^1R^2C{=}CHHal$ $\xrightarrow{BuLi}$ $R^1R^2C{=}C: ... Li$

(80) (81)

$\xrightarrow{RCH_2OM\ (82)}$ $RCH(OH)CH{=}CR^1R^2$

(83)

Isopropylidene (84), from either vinyl triflate and potassium ^{t}butoxide, or from silylvinyl triflate and benzyltrimethylammonium fluoride, reacts with azobenzene to give 2-phenyl-3-isopropylindazole (86). Evidence has been presented to show that the most likely course for this reaction is that the singlet carbene initially forms the Zwitterionic ylide (85), which then cyclises and rearomatises to the bicyclic product (K.Krageloh, G.H.Anderson, and P.J.Stang, *J.Amer.Chem.Soc.*, 1984, **106**, 6015).

Alkylidene and allylidene cyclopropanones are prepared by treating the intermediates formed on addition of aldehydes to lithio(cyclopropyl)silanes with potassium ^{t}butoxide at low temperature, followed by a period of heating. The lithiated reagents are themselves obtained from the corresponding sulphides through reaction with lithium 1-(dimethylamino)naphthalenide (LDMAN). Thus, for example, the sulphides (87) when reacted with LDMAN afford the lithio derivatives (88), which react with aldehydes to form the corresponding adducts (89). These may then be decomposed to yield the cyclopropanes (90), the whole procedure being carried out in one pot (T.Cohen *et al.*, *Tetrahedron Letters*, 1988, **29**, 25).

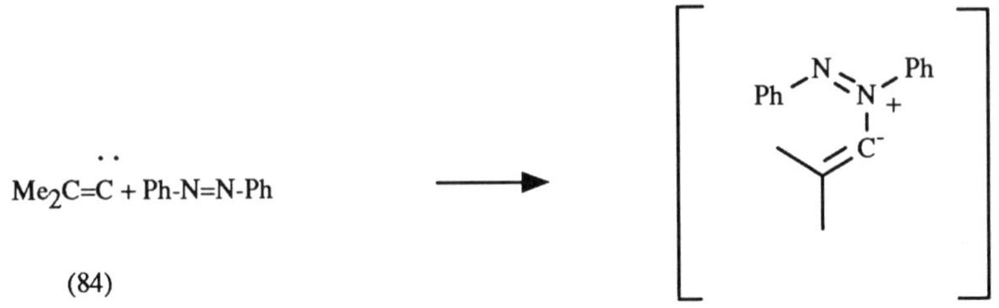

$Me_2C{=}\ddot{C}$ + Ph-N=N-Ph ⟶ [Ph–N=N⁺(Ph)–C⁻=CMe₂]

(84) (85)

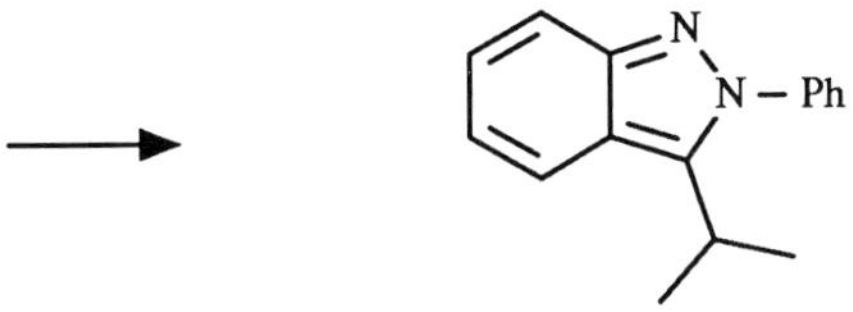

⟶ 2-phenyl-3-isopropyl-2H-indazole

(86)

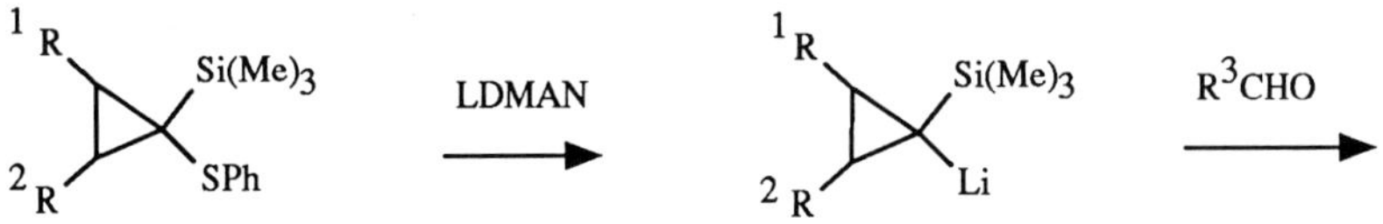

(87) + LDMAN ⟶ (88) + R^3CHO ⟶

(87): ^1R, ^2R, Si(Me)$_3$, SPh

(88): ^1R, ^2R, Si(Me)$_3$, Li

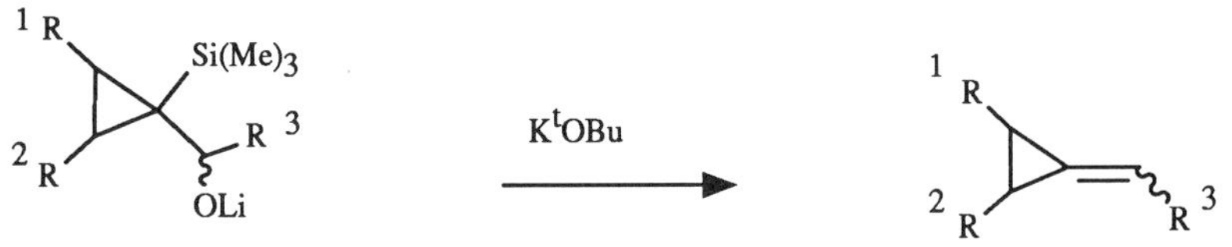

(89) (90)

Cyclobutylidene (91) tends to undergo intramolecular reactions; but when it is generated from 1,1-dibromocyclobutane, by treatment with methyl lithium at -70°C, intermolecular additions with alkenes occur to give spirocyclic products (94). This may in fact become the dominant reaction, and then only small amounts of methylenecyclopropane (92) and cyclobutene (93) are formed. However, if the carbene is obtained *via* the diazo compound (95) the same products may also form, but now, in addition, 1,3-cycloaddition between the alkene and the diazo affords the diazolidine (96). In this case (96) may be the major product (V.H.Brinker and M.Boxberger, *Angew.Chem.Internat.Ed.Engl.*, 1984, **23**, 974).

When the nitrosourea (97) is treated with methanol in the presence of sodium bicarbonate the 2-vinylcyclopropanediazonium ion (98) is formed. In this weakly basic medium, apart from deprotonation, the only reaction observed is ring opening to give pentadienyl cations which ultimately afford the corresponding isomeric methoxy derivatives (99), (100), and (101). No cyclopentenyl products are produced, but if sodium methoxide is present the more basic conditions favour loss of nitrogen, deprotonation of the resultant cation, and the formation of the vinylcyclopropylidene (102). This last compound may rearrange to 3-cyclopentenylidene (103), which is efficiently trapped by methanol to give 4-methoxycyclopentene (104) (W.Kirmse *et al.*, *Tetrahedron*, 1985, **41**, 1441).

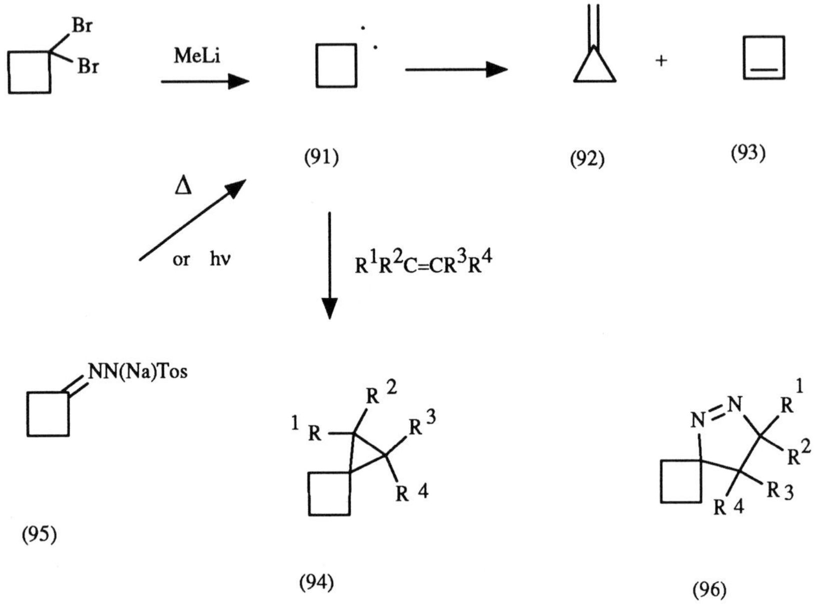

Br
Br
MeLi
(91)
(92)
(93)
Δ
or hv
$R^1R^2C{=}CR^3R^4$
NN(Na)Tos
(95)
(94)
(96)

The reaction of diazomethane and 1,3-bis(trimethylsilyl)cyclopropanone (105) at room temperature leads to the cyclobutanone (106, R=Me) in 75% yield. If trimethylsilyldiazomethane is employed as the reagent the tris(trimethylsilyl)butanone (106, R=Me_3Si) is obtained (G.S.Zaitseva, *Zh.Obsch.Khim.*, 1988, **58**, 1677).

1,5-Dipentalene (113) is formed as the main product when *trans*-1,2-bis(2,2-dibromocyclopropyl)ethene (107) is treated with methyl lithium at -40°C. In addition, smaller amounts of 1- and 2-propadienylcyclopentadienes (116 and (117) and 1,2,4,6,7-octapentaene (114) are produced. It seems likely that the first equivalent of methyl lithium generates the carbene (108) and this rearranges to the isomer (109). A 1,2-H shift then produces the cyclopentadiene derivative (110) which gives rise to the carbene (111) when it is reacted with a equivalent of the base. This carbene is then a precursor for a second carbene-carbene rearrangement to give the fused carbene (112) and, after another 1,2-H shift, 1,5-dihydropentalene. The other products result from the various carbenes through ring-opening of the cyclopropyl units perhaps as indicated (U.H.Brinker and I.Fleischauser, *Tetrahedron*, 1981, **37**, 4495).

O

MeOH-NaHCO$_3$

N NH$_2$

NO

+

N_2

MeONa

(97)

(98)

(102)

OMe

OMe

(99)

(100)

(103)

OMe

(101)

MeOH

OMe

(104)

O

$RCHN_2$

R O

Me_3Si $SiMe_3$

Me_3Si $SiMe_3$

(105)

(106)

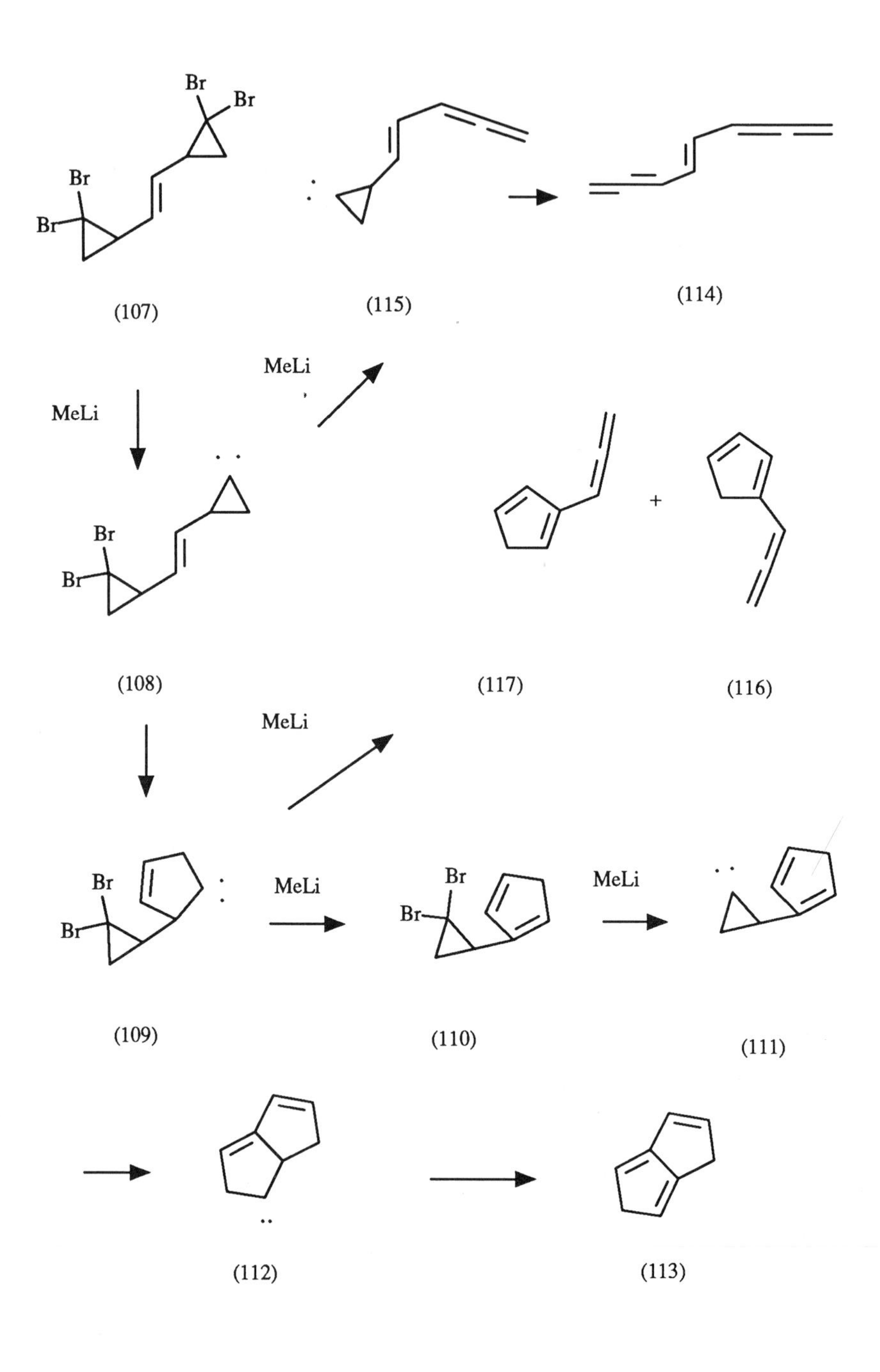
Br
Br
Br
Br
(107)
(115)
(114)
MeLi
MeLi
Br
Br
(108)
(117)
(116)
MeLi
Br
Br
MeLi
Br
Br
MeLi
(109)
(110)
(111)
(112)
(113)

(e) *Arylcarbenes*

3-Methoxylphenylcarbene, generated from 3-methoxy-3-phenyldiazirine (118), reacts with 2,3-dimethylbut-2-ene to give the expected cycloadduct (119), but the yield is reduced if the alkene is not freshly purified. For example, the presence of the peroxide (120, n=2), or the related alcohol (120, n=1), leads to acyclic compounds (121, n=2) and (121, n=1) respectively, the products of insertion into the O-H bonds (R.A.Moss and J.Wlostowska, *Tetrahedron Letters*, 1988, **29**, 2559).

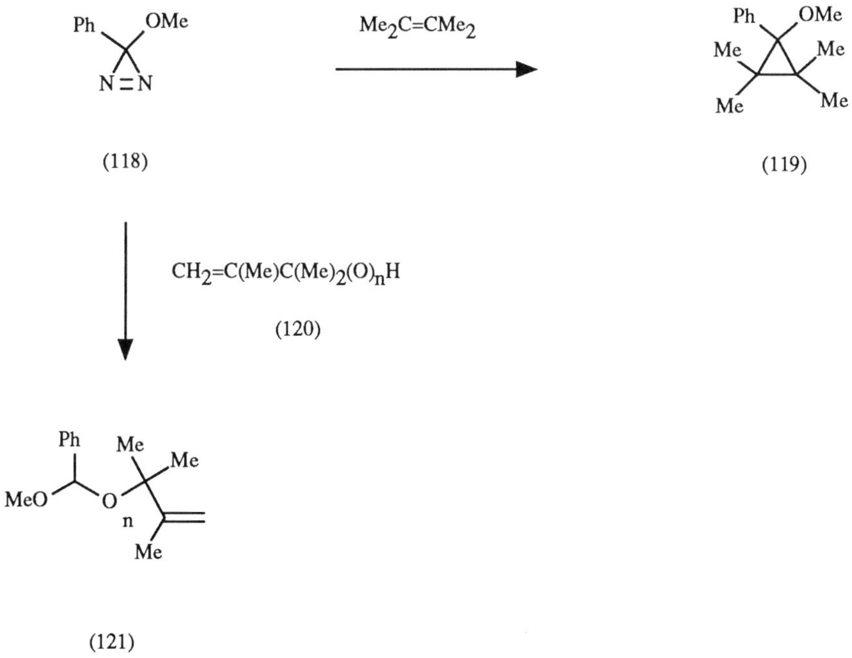

Ph OMe N=N

(118)

$Me_2C=CMe_2$

Ph OMe Me Me Me Me

(119)

$CH_2=C(Me)C(Me)_2(O)_nH$

(120)

Ph Me Me MeO O n Me

(121)

(f) *Carbene reactions mediated by metal ions or organometallic catalysts*

The synthetic value of organic diazo compounds as precursors of carbenes has been revitalised by the application of transition metal catalysts to the generation reactions (Chapter 7b). Indeed these catalytic methods are usually superior to thermal and photochemical techniques, and can be applied to a wide range of typical carbene reactions such as insertion and cyclopropanation (for a review see M.P.Doyle, *Chem.Reviews*, 1986, **86**, 919).

α-Diazo-β-ketoalkylphosphonates [122, R=P(EtO)$_2$O], or phosphine oxides [122, R=P(O)Ph$_2$], undergo intramolecular carbenoid cyclisations in dichloromethane solution containing rhodium (II) acetate catalyst to give diethoxyphosphono- [123, R=P(EtO)$_2$O],) or diphenylphosphono-cyclopentan-2-ones [123, R=P(O)Ph$_2$]) respectively (B.Corbel *et al.*, *Tetrahedron Letters*, 1987, **28**, 6605). Unfortunately the yields of these products are reduced owning to competitive Wolff rearrangement of the carbene, or carbenoid intermediates, to ketenes (124). If ethanol is added to the reaction mixtures the ketenes then react to form the appropriate phosphonoacetate esters (125).

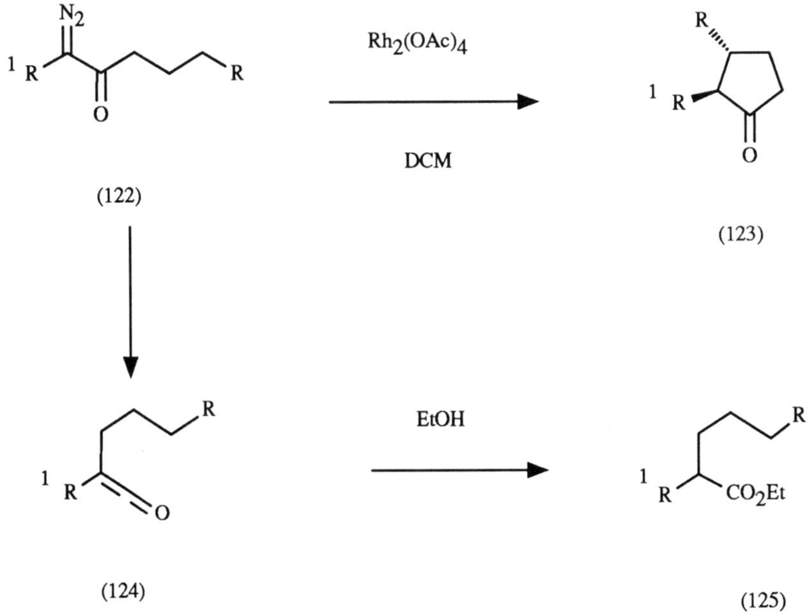

N_2 R^1 R O Rh$_2$(OAc)$_4$ DCM R R^1 O

(122) (123)

R^1 R O EtOH R^1 R CO_2Et

(124) (125)

Bis(methoxycarbonyl)carbene, formed from the copper(II) catalysed fragmentation of 2,5-dichlorothiophenium-1-bis(methoxylcarbonyl) methylide (126), undergoes a rapid and efficient insertion reaction into the N-H bond of primary and secondary aliphatic amines to afford *N*-substituted aminomalonate esters (127) in good yields. (S.Husinec *et al.*, *Synthesis*, 1988, 721).

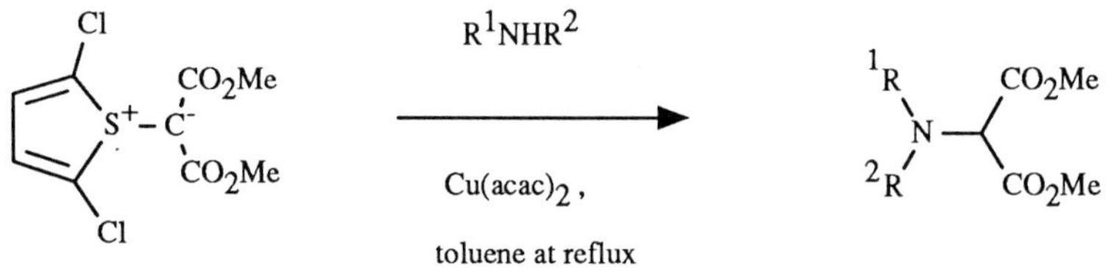

Cl
CO_2Me
S^+-C^-
CO_2Me
Cl

R^1NHR^2

$Cu(acac)_2$,

toluene at reflux

^1R CO_2Me
N
^2R CO_2Me

(126) (127)

(g) *Fischer carbene complexes*

The reactions of Fischer carbene complexes of group VI metals with alkynes has proved to be of great value in organic synthesis (for an account see W.D.Wulff in Advances in Metal-Organic Chemistry, ed. L.S.Liebskind, J.A.I. Press Inc., Greenwich, C.T., 1987, Vol. 1).

Similarly, the synthesis of cyclopropanes from the reactions of transition-metal-carbene complexes with alkenes has been reviewed by M.Brookhart and W.B.Studabaker (*Chem. Reviews*, 1987, **87**, 411).

Some recent examples of the utility of these types of reactions include the demonstration that the phenylchromium carbene complex (128) reacts with 1-hexyne in the presence of acetic anhydride and triethylamine to give the naphthol derivative (129, R=Ac) in 56% yield, plus the vinyl ether [129, R=COCH(Bu)CH=C(OMe)Ph] in 18% yield. Repetition of the reaction, but without acetic anhydride and triethylamine, followed by acetylation of the products, gives the naphthol (129, R=Ac) in 56% yield, together with smaller amounts of the cycloadducts (130) and (131) (A.Yamashita, J.M.Timko and W.Watt, *Tetrahedron Letters*, 1988, **29**, 2513).

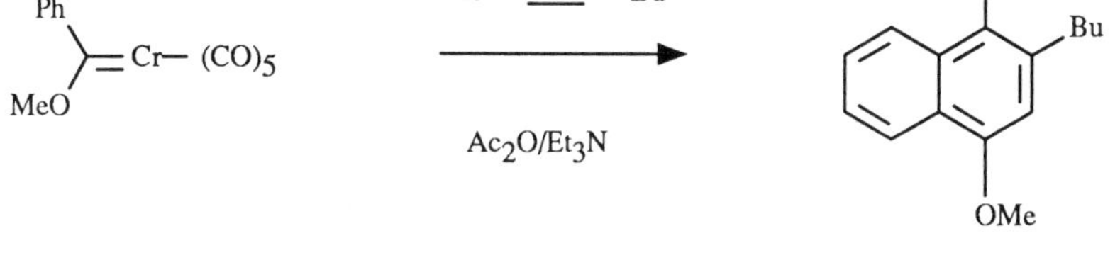

Ph
$=Cr-(CO)_5$
MeO

H ≡ Bu

Ac_2O/Et_3N

OR
Bu
OMe

(128) (129)

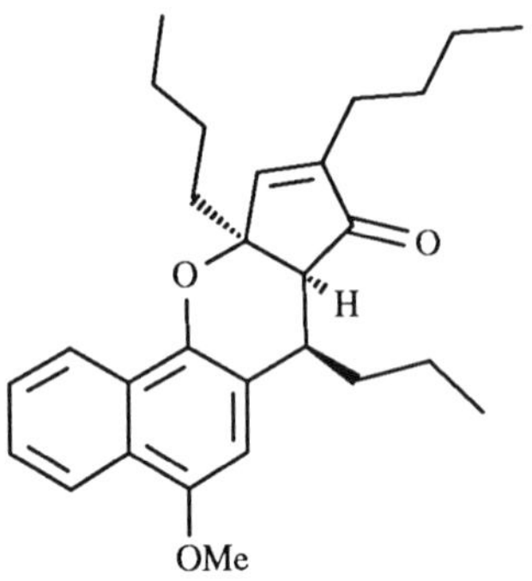

OMe MeO OAc O H OMe

(130)

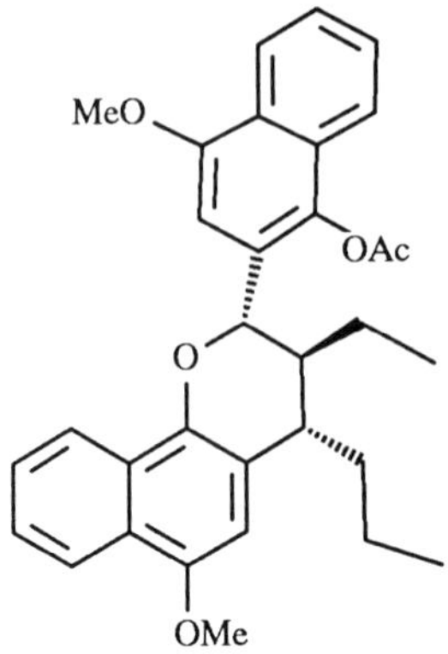

(131)

The same chromium carbene complex (128) reacts with morpholine, or pyrrolidine, to give the appropriate amino complexes (132, NR^1R^2= morpholino), or (132, NR^1R^2=pyrrolidino) respectively, which can be reacted with alkynes to form the indenes (136) and (137). It appears reasonable that initial extrusion of carbon monoxide from the amino complexes affords metallocyclobutenones (133), which rearrange to the complexes (134). These in a series of steps eventually give the intermediates (135). Alternative sigmatropic shifts then allow for the production of the indenes (136) and (137) (A.Yamashita, *Tetrahedron Letters*, 1986, **27**, 5915).

(128) $\xrightarrow{R^1NHR^2}$ Ph, $^1R-N(R^2)$, $=Cr(CO)_5$ $\xrightarrow[-CO]{R^3C\equiv CR^4}$

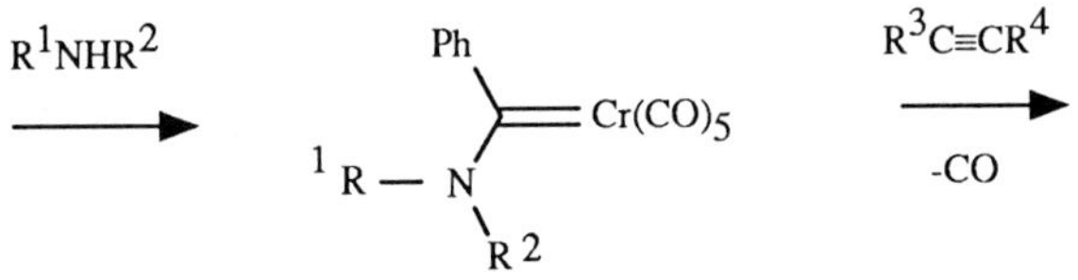

(132)

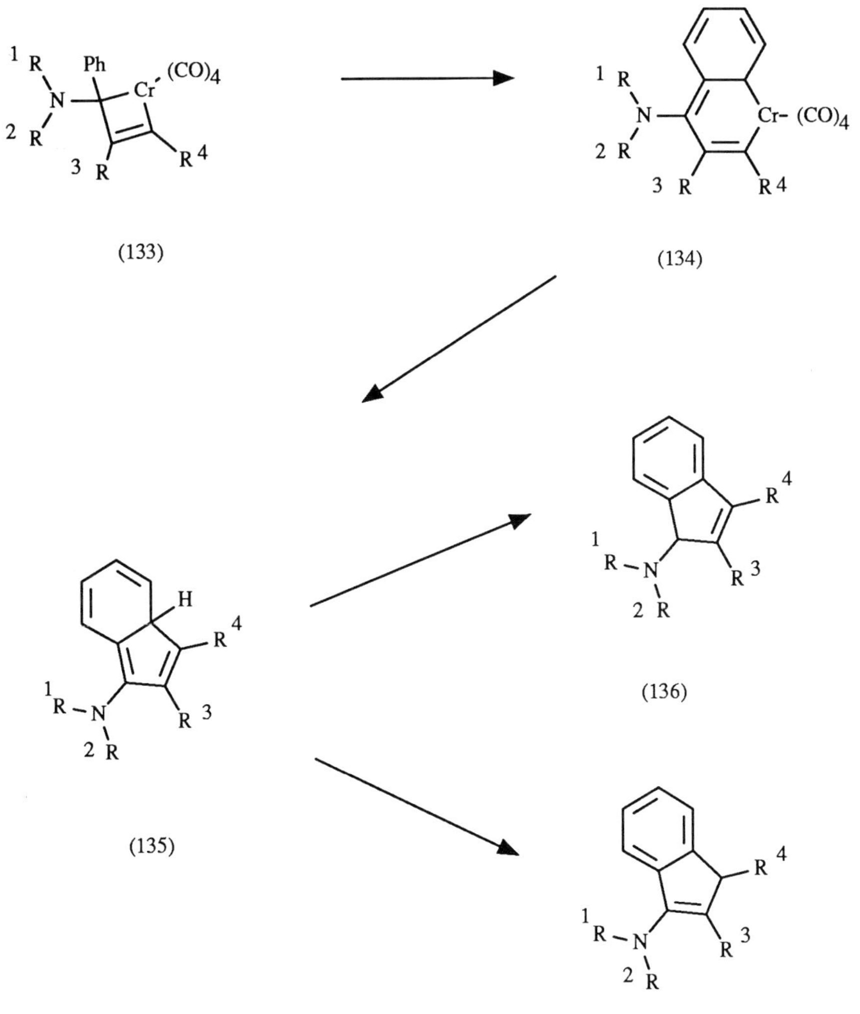

(133)

(134)

(135)

(136)

(137)

The complex (138) enters into a thermal reaction with diphenylacetylene to afford enaminoketones (139) which react further with imines e.g. (140) to give bicyclo[3.1.0]lactams (141), rather than cycloaddition products (L.S.Hegedus and D.B.Miller, *J.Org.Chem.*, 1989, **54**, 1241).

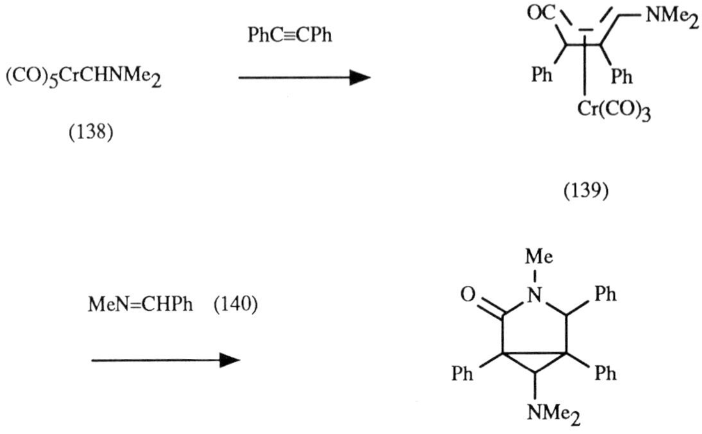

PhC≡CPh

(CO)$_5$CrCHNMe$_2$ ⟶ OC, NMe$_2$, Ph, Ph, Cr(CO)$_3$

(138) (139)

MeN=CHPh (140) ⟶ Me, O, N, Ph, Ph, Ph, NMe$_2$

(141)

Nitriles (142) insert into the metal-carbon bond of the heteroatom stabilised Fischer carbene complex (128) to give a variety of imino complexes. However, this reaction may be reversible since in the case of the complex (143), formed from the nitrile (142, R=^{t}Bu) and (128), further reaction with 1-pentyne affords the naphthalene (144). Significantly this last compound is known to be a product from the reaction of the starting complex (128) with 1-pentyne (D.Yang *et al., J.Amer.Chem.Soc.*, 1988, **110**, 307; see p.111).

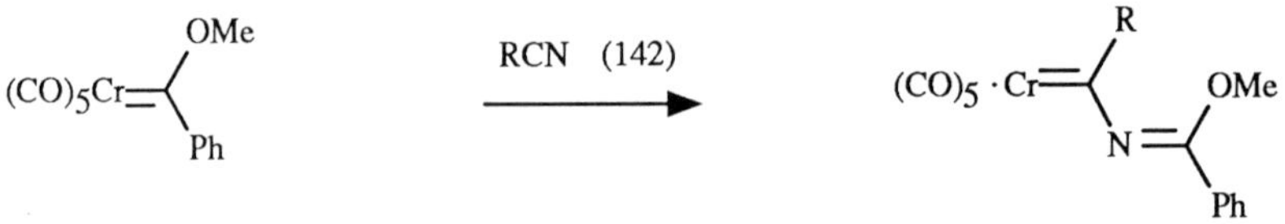

(CO)$_5$Cr=C(OMe)Ph —RCN (142)⟶ (CO)$_5$·Cr=C(R)–N=C(OMe)Ph

(128) (143)

(143,R=^{t}Bu) $\xrightarrow[\text{(1) THF, 70°C; (2) Air}]{HC\equiv CPr}$ (144)

OH

OMe

(144)

2. Allenes

(a) *Introduction*

The chemistry and synthesis of allenes are under constant review. Three particularly useful publications are the Synthesis of Acetylenes, Allenes, and Cumulenes, by L.Brandsma and H.D.Verkruijsse, Elsevier, Amsterdam, 1981; The Chemistry of Allenes, ed. S.R.Landor, Vols.1-3, Academic Press, New York, 1982, and The Chemistry of Allenes, G.M.Cappola and G.B.Schuster, J.Wiley and Sons, New York, 1984.

(b) *Structure and spectra*

An empirical method for the calculation of ^{13}C NMR shifts for allenes has been deduced (H.A.M. Jansen, J.J.Ch.Lousberg, and M.J.A.de Bie, *Rec.Trav.Chim.*, 1981, **100**, 950), these data correlate well with experimentally derived shifts (see, for example, the values quoted for substituted allenes by J.W.Munson in 'The Chemistry of Ketenes, Allenes and Related Compounds, ed., S.Patai, Wiley-Interscience, New York, 1980, 176).

The geometries and heats of formations of allenes can be calculated by means of an extended version of the MM2 approximation (N.L.Allinger and A.Pathiaseril, *J.Comput.Chem.*, 1987, **8**, 1225).

Whereas the constitution of 1,2-cyclohexadiene is predicted from INDO-MO calculations to be an allyl cation with a negative charge located at the central carbon atom in the in-plane sp orbital (P.W.Dillon and G.R.Underwood, *J.Amer.Chem.Soc.*, 1974, **96**, 779), more recent evidence from Diels Alder reaction studies with this compound suggest that it exists as a twisted bent allene (M.Balci and W.M.Jones, *ibid.*, 1980, **102**, 7607).

The stretching band at *ca* 1950cm^{-1}in the vibrational circular dichroism spectra of optically active substituted allenes is shown to correlate directly with the absolute configuration of the molecule, being positive for allenes with the *S*-configuration (U.Narayanan and T.A.Keiderling, *J.Amer.Chem.Soc.*, 1988, **110**, 4133).

(c) *Synthesis*

(i) *Through elimination reactions of alkenes*

Terminal allenes are often synthesised by elimination reactions. Thus Posner has shown that when 2-alkenylaryl sulphoxides (1) are treated with lithium tetramethylpiperidide (LTMP) the corresponding allenes (2) are formed (G.H.Posner, P.Tang, and J.P.Mallamo, *Tetrahedron Letters*, 1978, 3995).

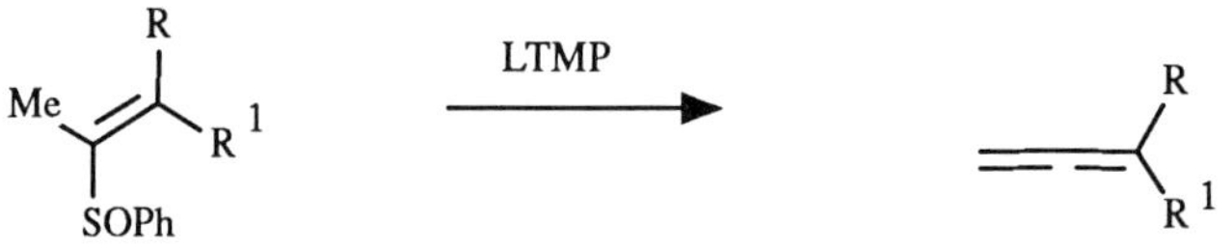

(1) (2)

Vinylsilanes (3) containing a good leaving group X when treated with fluoride ion form fluorosilanes and terminal allenes (4). The rate of the reaction is influenced by the nature of X and also by the substituents on the silyl unit. For example, triphenylsilyl fluoride is eliminated more readily than trimethylsilyl fluoride, and chlorine ion is removed faster than the triflate anion (T.H.Chan *et al., Org.Chem.*, 1978, **43**, 1526).

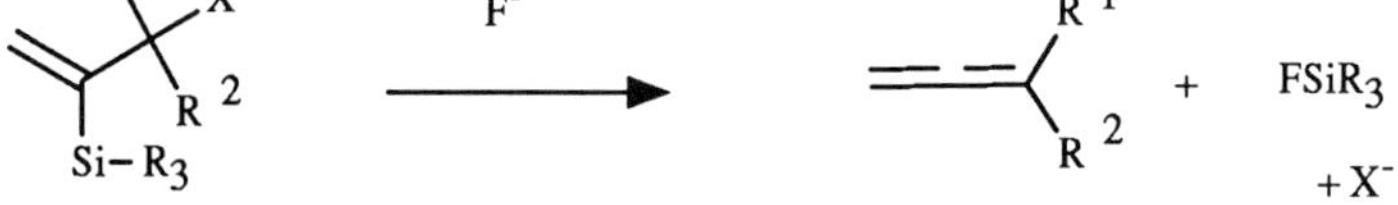

(3) (4)

(ii) *From propargylic compounds*

(1) Base and organometallic catalysed syntheses

The most important source of allenes is propargylic compounds and a common approach is *via* the elimination reactions of substituted propargylic derivatives. For instance, secondary propargylic halides are readily converted into allenes by treatment with base. Often the products are contaminated with the corresponding 1-alkyne, but if a high boiling ether, such as triethyleneglycoldimethyl ether, is used as solvent the allene can be distilled directly out of the reaction mixture in almost quantitative yield (J.Meijer *et al., Synthesis*, 1981, 551.)

R
H — C(Br) — C≡C — H
Br

$LiAlH_4$ →

(R= lower alkyl)

R(H)C=C=CH_2

Grignard reagents are frequently used as bases in such reactions, but when various transition metals are added prior to the addition of the organomagnesium halide the reactions are greatly accelerated and the amounts of alkynes in the reaction mixture are diminished. The efficiency of the metallic catalyst decreases in the order Fe > Co > Ni > Cu, and, for example, in the presence of ferric chloride as catalyst both terminal and secondary propargylic halides react with Grignard reagents to form allenes in excellent yields (D.J..Pasto *et al., J.Org.Chem.*, 1978, **43**, 1385). Palladium generated *in situ*, by the reduction of palladium chloride through treatment with diisobutylaluminium hydride in the presence of triphenylphosphine, is also an effective catalyst for these types of reactions (T.J.-Luong and G.Linstrumelle, *Tetrahedron Letters*, 1980, **21**, 5019). Thus 3-chlorobutyne reacts with octylmagnesium chloride under these conditions to give a 62% yield of the allene (5) and the alkyne (6) in a ratio of 99:1.

Me(Cl)CH—C≡C—H

Pd →
OctylMgBr

$MeCH{=}C{=}CH(CH_2)_7Me$

(5)

+ $Me(CH_2)_7CHMeC{\equiv}CH$

(6)

Much work has evolved around the use of organocopper and cuprate reagents. They react with propargylic functionalised derivatives (7), where X is a good leaving group, to give allenes (10). In general homocuprates (R_2CuMg Hal) tend to favour the formation of alkynes, whereas alkylheterocuprates [(RCuBr)MgHal], or organocopper complexes

(RCu.LiBr.MgBrI) lead to allenes (T.L.Macdonald, D.R.Reagan, and R.S.Brinkmeyer, *J.Org.Chem.*, 1980, *45*, 4740). When the substrates are chiral their asymmetry may be transferred to the products (G.Tadema *et al.*, *Tetrahedron Letters* 1978, 3935; L.Olsson and A.Claesson, *Acta Chem. Scand.Ser. B*, 1979, **33**, 679). Normally the stereochemical outcome of reactions of this type indicates that an overall *anti* S_Ni process is involved probably *via* the participation of a copper (III) intermediate (8), although π-complexes (9) are also possible participants in the reaction (D.J.Pasto *et al.*, *J.Org.Chem.*, 1978, **43**, 1389).

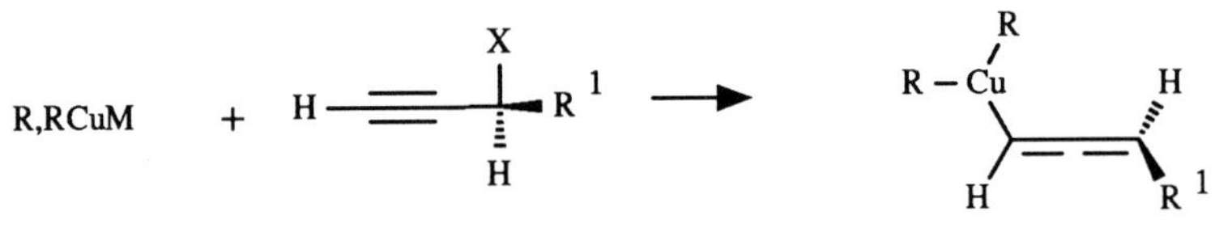

(7) (8)

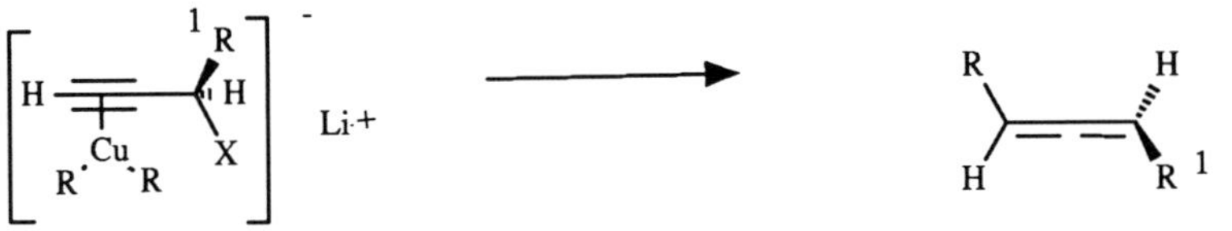

(9) (10)

The copper catalysed reaction of butyl magnesium bromide with the asymmetric propargylic ether (11) gives the chiral allene (13), in high optical yield. Here the leaving group is a poor one, and in this case an alternative mechanism may operate, proceeding first through *syn* addition of the reagent to the triple bond forming a covalently bonded copper complex (12). The addition is followed by a *syn* β-elimination step. For other Grignard reagents both *syn* and *anti* elimination modes may occur (*I.Marek et al., Tetrahedron Letters*, 1986, **27**, 5499).

Chiral alkoxyallenes are synthesised *via* the reactions of chiral acetylenic acetals with Grignard reagents (R=alkyl or phenyl) in the presence of a catalytic amount of copper(I) bromide. The crucial step in the process seems to be a highly diastereoselective β-elimination from within a transient alkenyl organometallic intermediate. Thus when the acetal (14) is treated with Grignard reagents RMgX (R = Me, Bu, Me_3C, or Ph) in diethyl ether in the presence of 5% copper bromide the allenes (15) are formed in 46-100% diastereoselective excess (A.Alexakis, P.Mangeney, and J.F.Normant, *Tetrahedron Letters*, 1985, **26**, 4197).

BuCu, MgBr, $(EtO)_3P$, Et_2O

(11) (12)

(13)

RMgX, CuBr

(14) (15)

A widely applicable synthesis of silylallenes (17) involves the 1,3-addition of organoheterocuprates to the *O*-mesylates (16) of propargylic alcohols already bearing a silyl group at the terminal position. Yields are high in these conversions and only when R^2=Ph or ^{t}butyl are any alkynes produced (R.L.Danheiser, D.J.Carini, and A.Basak, *J.Amer.Chem.Soc.*, 1981, **103**, 1604).

[RCuBr]MgHal

(16) (17)

The racemic propargylic bromide (18) can be converted into the (*R*)(-)-allene (20) if chromium (II) chloride is used as the reagent in the presence of (-)-menthol as a chiral auxiliary (Au*). The reaction requires three molar equivalents (with respect to the reductant) of HMPA. In the absence of HMPA both *R*- and *S*- isomers are formed in equal amounts, and less than three equivalents of this reagent leads to a poorer enantiomeric excess. If (-)-borneol is used as the auxiliary instead of (-)-menthol the (*R*)-allene is formed, but in only 22.5% *ee*. The reaction is considered to involve the intermediacy of chromium complex (19), in which both the auxiliary and HMPA, and possibly solvent THF, are bound to the metal atom (B.Cazes, C.Vernier and J.Gore, *Tetrahedron Letters*, 1982, **23**, 3501).

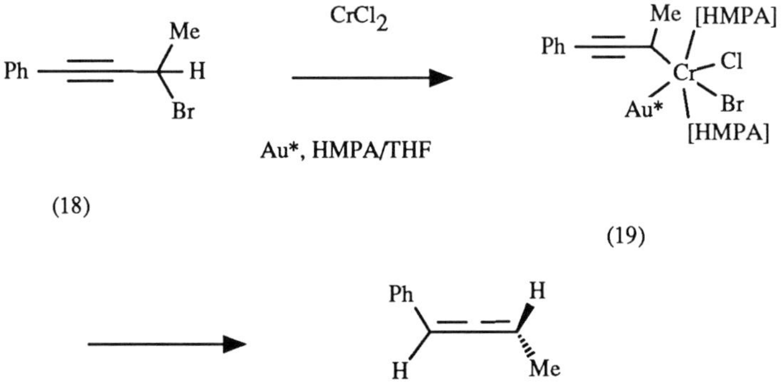

Ph—≡—C(Me)(H)(Br) (18) —[$CrCl_2$; Au*, HMPA/THF]→ complex (19) [Ph—≡—C(Me)—Cr(Cl)(Br)(Au*)[HMPA]$_2$] → allene (20) [Ph(H)C=C=C(H)(Me)]

(18) (19) (20)

This type of reaction, when performed in the presence of an enone (21), instead of a chiral auxiliary gives rise to racemic allenic alcohols (22). Again the presence of HMPA is essential if the reaction is to be successful (P.Place, C.Vernier, and J.Gore, *Tetrahedron* 1981, **37**, 1359).

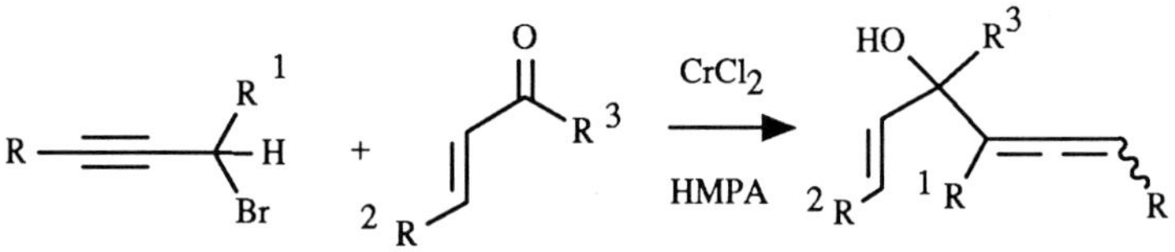

R—≡—C(R^1)(H)(Br) + R^2CH=CH–CO–R^3 (21) —[$CrCl_2$; HMPA]→ R^2CH=CH–C(OH)(R^3)–C(R^1)=C=CH–R (22)

(21) (22)

Propargylic acetates or tosylates (23) under the influence of copper(I) ions react with the trimethylsilylmethylmagnesium bromide to give the corresponding allenes (24). These products are useful intermediates and react further with boron trifluoride-acetic acid to afford 1,3-dienes (25). The geometries of the dienes depend upon the substitution pattern of the allene and the temperature of the reaction (M.Montury, B.Psaume, and J.Gore, *Tetrahedron Letters*, 1980, **21**, 163; *idem.*, *Syn.Commun.*, 1982, **12**, 409).

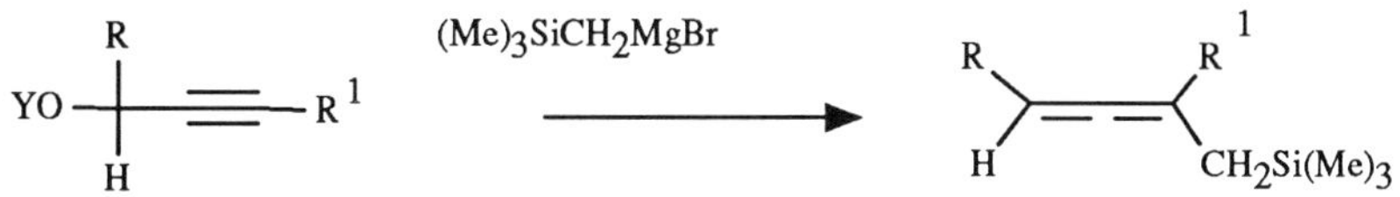

$(Me)_3SiCH_2MgBr$

(23)

Y=Ac or 4-$MeC_6H_4SO_2$

(24)

BF_3 HOAc

(25)

(2) From silylalkynes

Propargylic trimethylsilanes (26), when protonated with trifluoroacetic acid, generate cations which then react with the trifluroacetate anion to give the corresponding allenes (27) (T.Flood and P.Peterson, *J.Org.Chem.*, 1980, **45**, 5006).

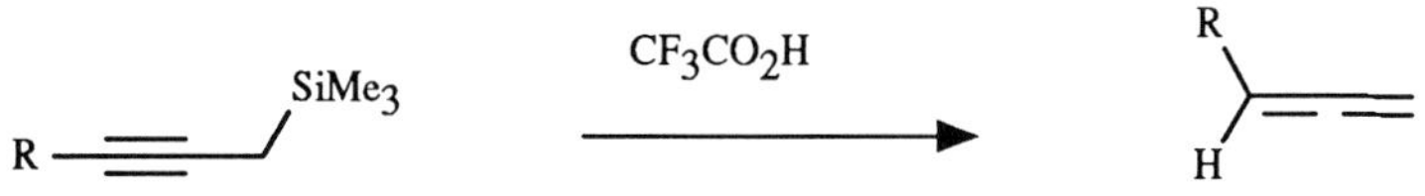

CF_3CO_2H

(26)

(27)

(3) Pyrolysis of silylalkynes

Trimethylsilylation of methoxyalkynes (28, R=H) gives the corresponding silylated derivatives (28, R=$SiMe_3$), which when pyrolysed at low pressure in a flow apparatus at 650°C afford good yields (79-99%) of the allenes (29) (H.Hopf and E.Naujoks, *Tetrahedron Letters*, 1988, **29**, 609).

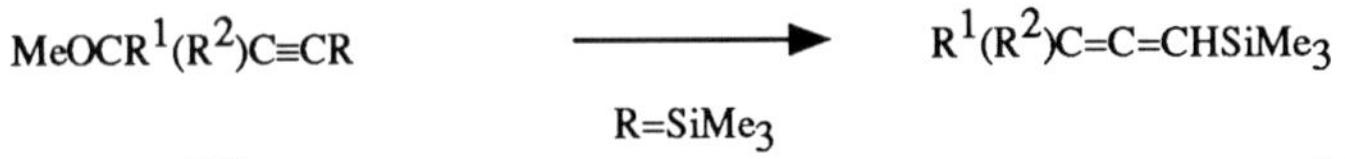

$$MeOCR^1(R^2)C{\equiv}CR \xrightarrow[R=SiMe_3]{} R^1(R^2)C{=}C{=}CHSiMe_3$$

(28) (29)

R^1=H, R^2=Me

R^1=R^2=Me

R^1=Ph, R^2=H

R^1R^2=$(CH)_n$ [n= 4 or 5]

(4) Through coupling reactions

Triorganoalanates or triorganoboronates (30) react with allyl halides (31) to yield the corresponding allylic allenes (32), only minor amounts of alkynes are formed as side products (N.R.Pearson, G.Hahn, and G.Zweifel, *J.Org.Chem.*, 1982, **47**, 3364).

Bu–C≡C–MLi + R^1R^2C=CH–CH$_2$–Hal

(30) (31)

M=$Al(i\text{-}C_4H_9)_3$

M=$B(s\text{-}C_4H_9)_3$

Hal=Cl or Br

R^1,R^2=H or alkyl

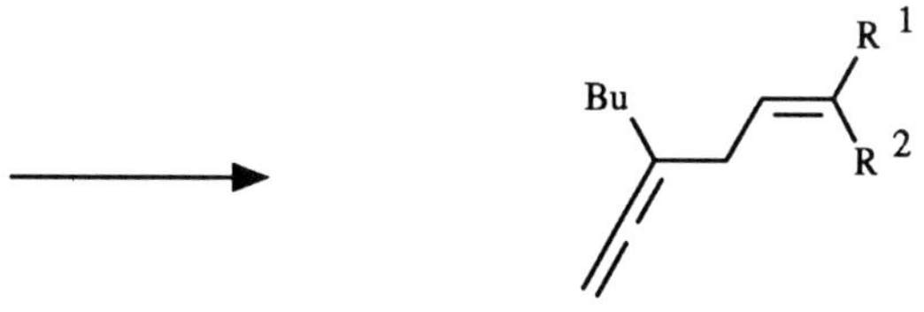

(32)

Yields 76-84%

Copper (I)-1-alkenyltrimethylborates (33) cross-couple with propargylic bromide to provide access to 1,2,7-trieneoct-5-ynes (34), probably through intermediate hydroboration reactions (N.Miyaura, T.Yano, and A.Suzuki, *Bull.Chem.Soc.Jap.*, 1980, **53**, 1471).

$[RCH{=}C(R^1)B(Me)_3]Cu$ + $H{-}C{\equiv}C{-}CH_2Br$

(33)

(34)

R, R^1=H,alkyl   Yields 40-70%

Alkenylation of aliphatic ketones (R^1R^2CO) with carbon tetrabromide affords dibromides ($R^1R^2C{=}CBr_2$) which can be lithiated with butyl lithium and condensed with aliphatic or aromatic aldehydes (RCHO) to give bromoallyl alcohols [$R^1R^2C{=}C(Br)CH(R)OH$]. These when reacted firstly with trimethylsilyl chloride and hexamethylsilazane, and then with ^{t}butyl lithium produce the allenes ($R^1R^2C{=}C{=}CHR$) (R.Hässig, D.Seebach, and H.Siegel, *Chem.Ber.*, 1984, **117**, 1877).

Alkynes can be homologated to allenes (37) by reaction with formaldehyde and di(^{i}propyl)amine in the presence of copper(I) bromide. The reaction involves the formation of a 1:1 Mannich base-CuBr complex (35), within which the π-complexed copper atom first abstracts a hydrogen atom from one of the *N*-alkyl groups forming a hydridocopper(III) species (36), and then transfers it to the acetylenic carbon atom. This concept adds sophistication to earlier proposals for the mechanism of this type of process, which simply invoked [1,5] hydrogen shifts in Mannich type intermediates and did not explain the function of the metal atom (S.Searles *et al.*, *J.Chem.Soc.Perkin Trans 1*, 1984, 747).

R—≡—H → (HCHO, ^{i}Pr$_2$NH; CuBr) → (35)

(35): R—C≡C—CH$_2$—N(CHMe$_2$)—CHMe$_2$ with Br—Cu π-complexed to the triple bond

(35) → (36): R—C≡C—CH$_2$—N^{+}(CHMe$_2$)=CMe$_2$ with Br—Cu$^-$—H

(36) → (37) + $Me_2{-}N^+{=}CMe_2$ Br-

(37): RHC=C=CH$_2$

N-(1-Chloro-2-methylpropenyl)pyrrolidine (38) is an efficient and regioselective reagent for the coupling of benzylmagnesium bromide and propargylic alcohols leading to allenes (39) as the final products (T.Fijisawa, S.Iida, and T.Sato, *Tetrahedron Letters*, 1984, **25**, 4007).

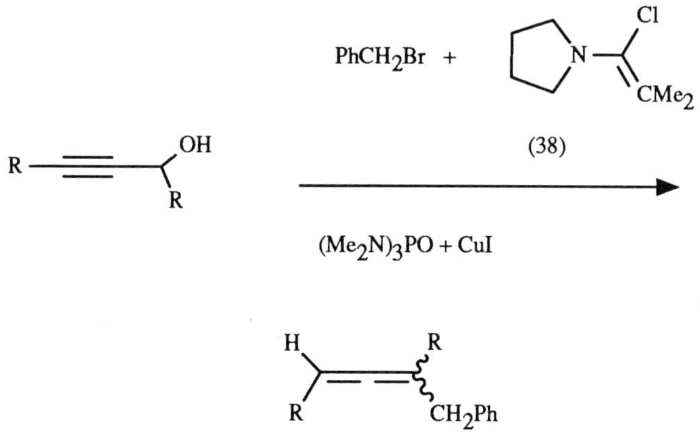

$PhCH_2Br$ + (38)

$(Me_2N)_3PO$ + CuI

(39)

(5) Rearrangement reactions

Acetoxyallenes (41) are obtained through the silver (I) or copper (I) catalysed rearrangement of propargylic acetates (40) (R.C.Cookson, M.C.Cramp, and P.J.Parsons, *Chem.Commun.*, 1980, 197; R.E.Banks *et al.*, *J.Chem.Soc.Perkin Trans.* 1, 1981, 1096).

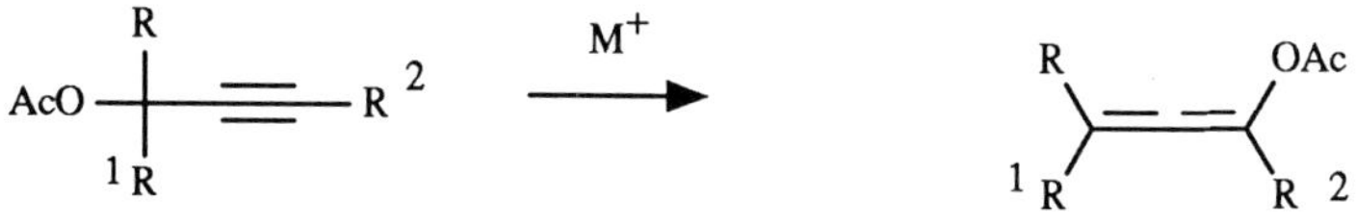

(40) $\xrightarrow{M^+}$ (41)

M= Cu, or Ag

In a similar manner allenic sulphoxides are synthesised from propargylic alcohols by reaction with sulphenyl chlorides in the presence of a tertiary amine. The initial step is the formation of the sulphenic esters (42) which then undergo [2,3] sigmatropic rearrangement to the allenic products (43) (H.Altenbach and H.Soicke, *Liebigs Ann.*, 1982, 1096; E.M.G.A. van Kruchten, and W.H.Okamura, *Tetrahedron Letters*, 1982, **23**, 1019; A.J.Bridges and J.W.Fischer, *Chem.Commun.*, 1982, 665).

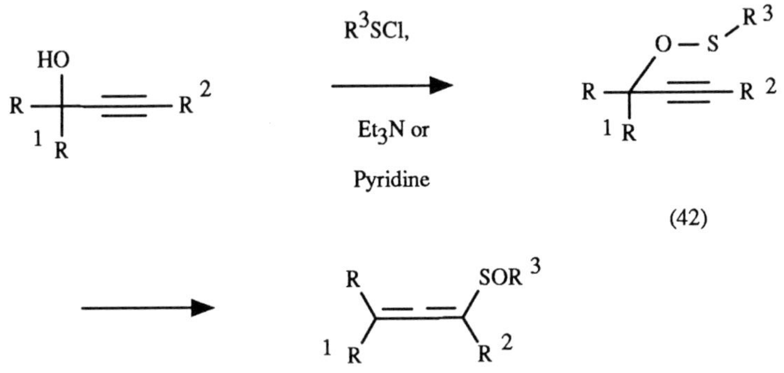

R^3SCl, Et_3N or Pyridine

(42)

(43)

R,R^1,R^2 = H, or alkyl; R^3 = Ar, or CCl_3

In the case of the acetylenic diol (44) the product is the disulphoxide (45) formed through two successive sigmatropic shifts (S.Jeganathan and W.H.Okamura, *Tetrahedron Letters*, 1982, **23**, 4763).

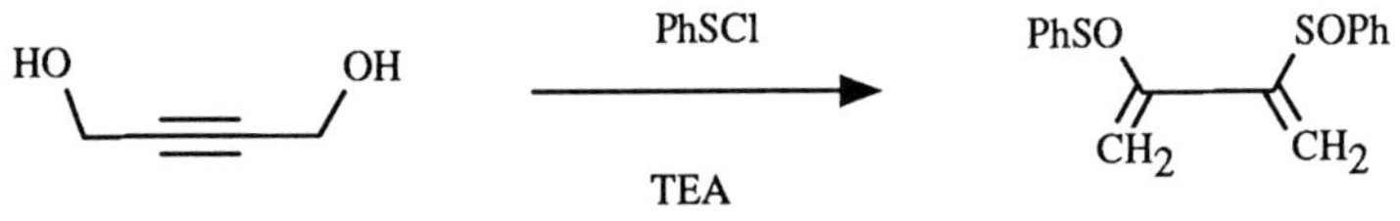

PhSCl, TEA

(44)

(45)

Haloallenes are available through the S_Ni rearrangement of propargylic chlorosulphinates, formed *in situ* by the action of propargylic alcohols and thionyl chloride in the presence of triethylamine (T.Toda *et al., Chem. Letters*, 1983, 523). Alternatively propargylic tosylates may converted into bromoallenes under mild reaction conditions simply by treatment with copper(I) bromide in the presence of lithium bromide with THF as solvent (M.Montury and J.Gore, *Syn. Commun.*, 1980, **10**, 873).

(6) Miscellaneous syntheses

If propargylic alcohols are reacted with sulphur dichloride the mixed allenic acetylenic sulphinates (46) are formed, but if either bromine, or methanesulphenyl chloride, are added as electrophiles to these products then cyclisation to the oxathioles (47) occurs (S.Braverman and Y.Duar, *J.Amer.Chem.Soc.*, 1983, **105**, 1061).

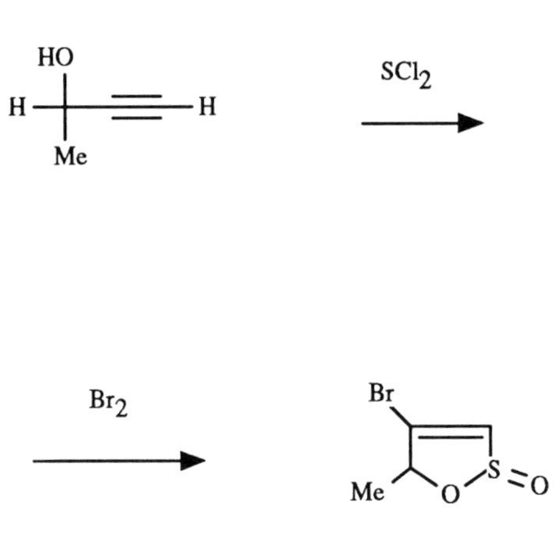

(46)

(47)

(d) *Reactions*

(i) *Metalation*

Treatment of 1,1,3,3-tetraphenylallene (48) with lithium at low temperature in ethereal solvents leads to the formation of a dilithium allenide which apparently consists of two equilibrating 2-lithioallyl lithium species. A similar, but less stable, disodium allenide is produced if sodium is used as the reductant, whereas with potassium an analogous dipotassium allenide cannot be detected. Now protonation by the solvent affords 1,1,3,3-tetraphenylallyl potassium (49), which deprotonates to the trianion (50). If the temperature is raised to -30°C the trianion abstracts a proton from the solvent to yield the dianion (51). Evidence for the presence of the tri- and dianions comes from trapping experiments with deuteriomethanol. These give either 1,2,3-trideuterio-1,1,3,3-tetraphenylpropane or 1,3-dideuterio-1,1,3,3-tetraphenylpropane depending upon the temperature at which the deuterated alcohol is added (A.Rajca and L.M.Tolbert, *J.Amer.Chem.Soc.*, 1987, **109**, 1782).

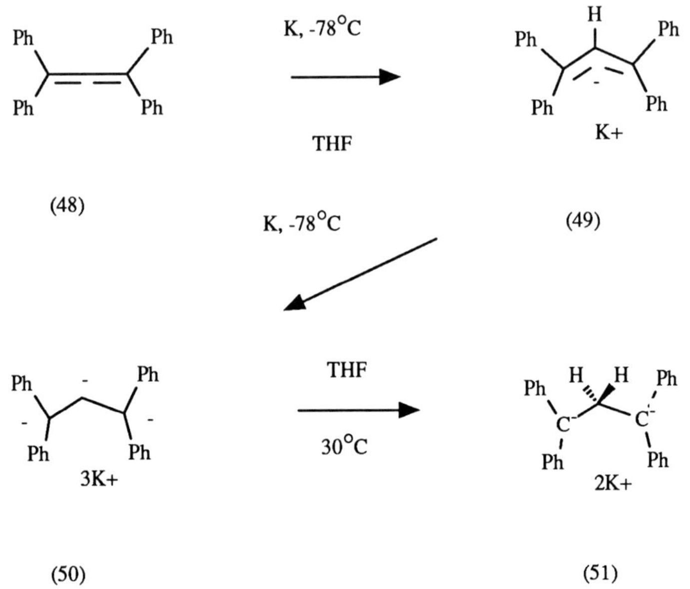

(48) —K, -78°C / THF→ (49) K+

(49) —K, -78°C→ (50) 3K+

(50) —THF / 30°C→ (51) 2K+

Lithiation and reaction of monosubstituted allenes (52, R = hexyl or cyclohexyl) with ^{t}butyl lithium and then reaction of the resultant 1-lithio-1,2-dienes (53) with triisobutylalane affords allenic alanates (54). Although these possess two potential sites for electrophilic attack, with aliphatic aldehydes and ketones (R^1R^2CO), for example, they give the corresponding homopropargylic alcohols (55) in 79-99% yields. In similar reactions using carbon dioxide as the electrophile the corresponding homopropargylic acids are also the major products, with only minor amounts of the isomeric allenic acids as by-products (G.Hahn and G.Zweifel, *Synthesis*, 1983, 883).

R, H (52) —^{t}BuLi→ R, H, CHLi (53) —$(^{i}Bu)_3Al$→

R, H, $\overset{-}{C}HAl(CHCH_2Me_2)_3$ Li+ (54) —R^1R^2CO→ R, R^1, R^2, OH (55)

The 1,1-dimetallated alkene (56), obtained from 1-heptyndimethylalane, by the addition of trimethylaluminium and Cl_2TiCp_2, reacts with aldehydes at -30°C and ketones, at 0°C, to give the corresponding allenes (57) in good yields (T.Yoshida and E.-i.Negishi, *J.Amer.Chem.Soc.*, 1981, **103**, 1276).

C_3H_7, Me, $Al(Me)_2$, $TiCp_2Cl$ (56) —R^1COR^2→ C_3H_7, Me, R^1, R^2 (57)

Direct silylcupration of allene with bis(dimethylphenylsilyl-copper)lithiate, followed by treatment of the intermediate silylallyl-copper species (58) with electrophiles such as methyl iodide, acetyl chloride, or cyclohexanone, gives rise to the corresponding vinylsilanes (59). With iodine as electrophile the product is the iodo compound (59, E=I) (P.Cuadrado *et al., Tetrahedron Letters*, 1988, **29**, 1825).

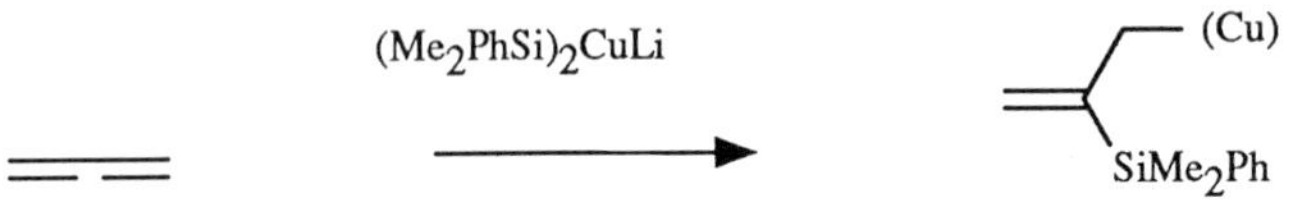

$(Me_2PhSi)_2CuLi$

(Cu)

$SiMe_2Ph$

(58)

EX

E

$SiMe_2Ph$

EX = MeI, AcCl, cyclohexanone, or I_2

(59)

(ii) *Photoaddition reactions*

Ultraviolet irradiation of di-, tri-, and tetra-phenylallenes in methanol solution gives mixtures of allyl methyl ethers. Molecular orbital calculations suggest that such products arise through the participation of an excited allene state initially protonated at C-2. This reacts with a second solvent molecule to form the observed products (M.W.Klett and R.P.Johnson, *Tetrahedron Letters*, 1983, **24**, 1107).

(iii) *Thermal rearrangements*

The vinylallene (60) undergoes thermally induced [1,5] sigmatropic rearrangements to give the trienes (61) and (62); whereas the dienic allene (63) affords the tetraenes (64) and (65), through [1,7] sigmatropic changes (J.A.Palenzuela, H.Y.Elnagar, and W.H.Okamura, *J.Amer.Chem.Soc.*, 1989, **111**, 1770).

(60)

(61)

+

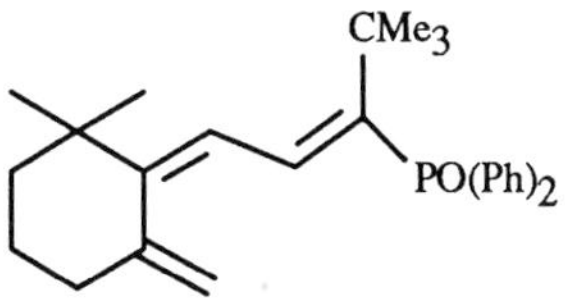

(62)

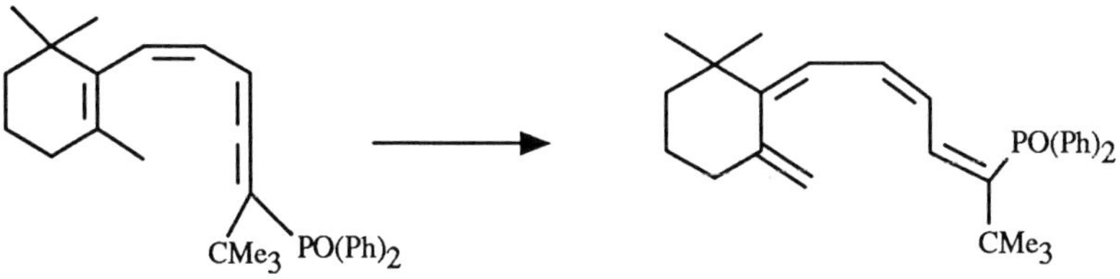

(63)

(64)

+

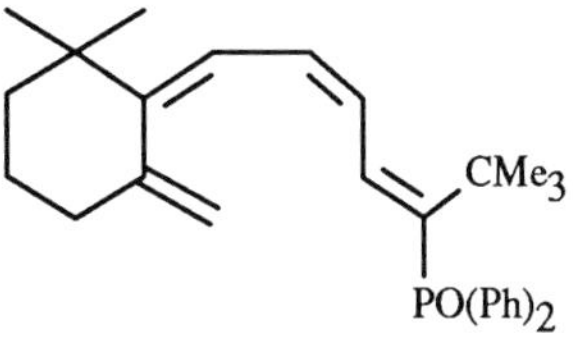

(65)

(iv) *Cycloaddition and addition reactions*

The cycloaddition of alkenes and alkynes with unsymmetrically *C*-substituted allenes normally occurs preferentially at the less substituted double bond of the allene. Thus 1,1-dicyclopropylallene reacts principally at the unsubstituted double bond with alkenes such as methyl acrylate to give the corresponding [2+2] adducts at reasonable reaction rates (Z.Komiya and S.Nishida, *J.Org.Chem.*, 1983, **48**, 1500). In the cycloadditions between 1,1-dimethylallene (66) and propiolate esters the major products are the butenes (67), together with very low yields of the ene-products (68) (D.J.Pasto and W.Kong, *J.Org.Chem.*, 1988, **53**, 4807).

The mechanisms of these reactions has provoked some debate and there is evidence that, at least in some cases, the reactions proceed through two step diradical intermediates (see D.J.Pasto and P.F.Heid, *J.Org.Chem.*, 1982, **47**, 2204; Pasto and S.E.Warren, *J.Amer.Chem.Soc.*, 1982, **104**, 3670). The cycloaddition reactions are also mediated by the addition of Lewis acids, especially ethylaluminium dichloride. In this case the arguments best fit the concept of a stereospecific $[\pi^2{}_s + \pi^2{}_a]$ addition mode and, for example, a reaction between methoxycarbonylallene (69) and 2,3-dimethylbutene (70) in the presence of this catalyst affords mainly the (*E*)-cyclobutane (71) and only minor amounts of the alternative isomer (72) (B.B.Snider and D.K.Spindell, *J.Org.Chem.*, 1980, **45**, 5017).

Phenyldimethylsilyl cuprates combine with α,β-unsaturated carbonyl compounds, alkynes, and tertiary and secondary allylic acetates so that the silyl cuprates behave as silicon nucleophiles. They thus react in the same way as noted for alkyl cuprates (D.J.Ager *et al., J.Chem.Soc.Perkin Trans* 1, 1981, 2520). In the case of allenes, however, bis(phenyldimethyl)cuprate enters into *syn* additions and the silyl group and the copper become individually bonded to the carbon atoms of one of the double bonds of the allene. The reactions are selective, and, after the reaction mixtures are quenched with ammonium chloride in aqueous methanol, either vinylsilanes or allylsilanes are obtained. Allene and phenylallene give the appropriate vinylsilanes, whereas simple di- and tri- alkylallenes yield allylsilanes (I.Fleming and F.J.Pulido, *Chem.Commun.*, 1986, 1010).

Thiophenol adds in a regioselective manner to the allene (73) to afford a mixture of the *cis*- and *trans*- sulphides (74). Whereas the *cis*- compound then undergoes a spontaneous Cope rearrangement to the cycloheptadiene (75), the *trans*-isomer has to be heated to 160°C for three hours before it yields this product (P.M.Cairns, L.Crombie, and G.Pattenden, *Tetrahedron Letters*, 1982, **23**, 1405).

Methoxyallene enters into Ad_E type reactions and with 4-methylphenylsulphenylchloride at -70°C in dichloromethane solution, for example, it yields the two adducts (76) and (77).

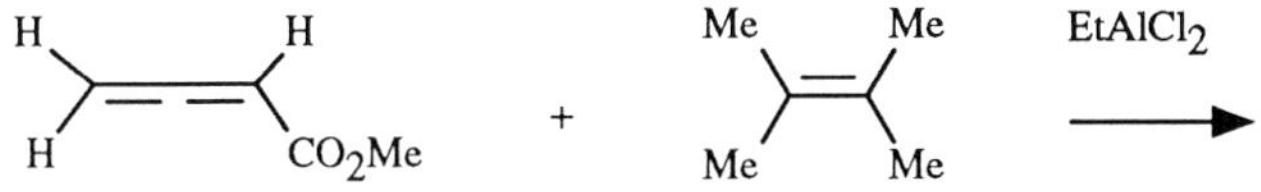

H H H CO_2Me + Me Me Me Me $EtAlCl_2$

(69) (70)

Me Me Me Me CO_2Me + Me Me Me Me CO_2Me

(71) (72)

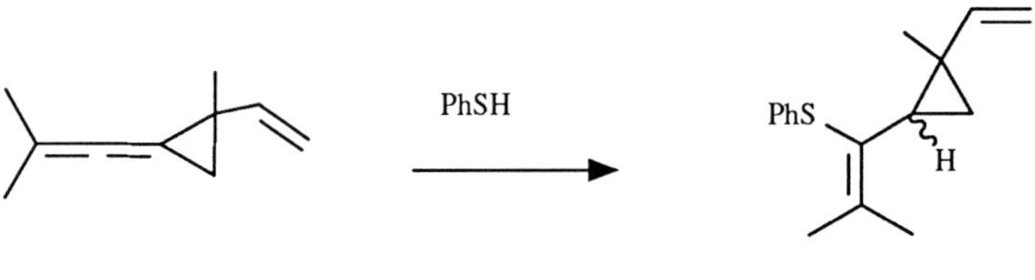

PhSH PhS H

(73) (74)

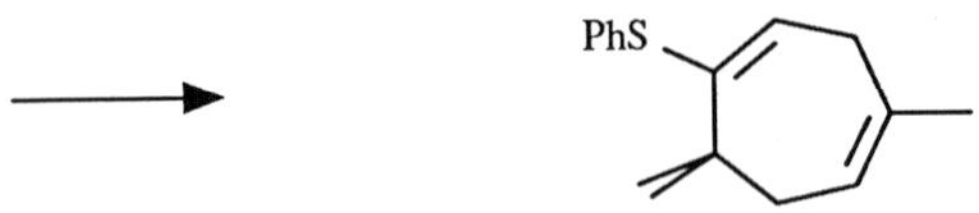

PhS

(75)

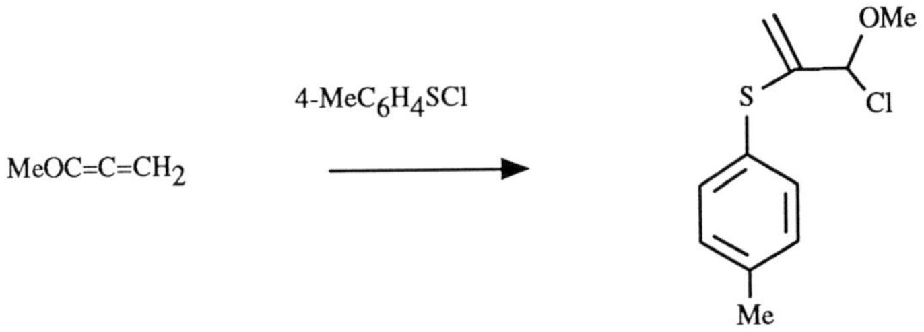

4-MeC_6H_4SCl

$MeOC{=}C{=}CH_2$

S OMe Cl Me

(76)

Cl OMe S Me

+

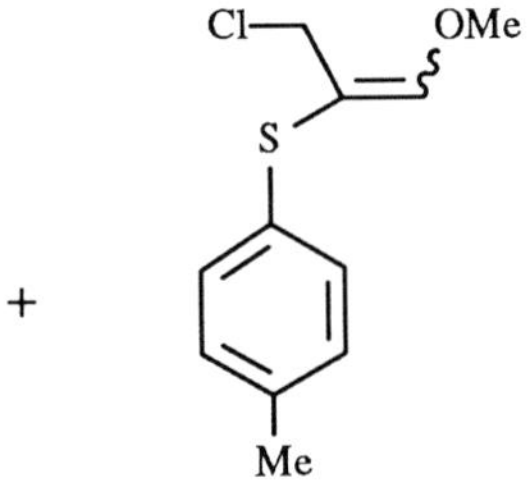

(77)

On the otherhand, although phenylselenyl chloride reacts with allenes (78, $R{=}R^1{=}$ H or Me; R=H, $R^1{=}$ Me, or Et) to give a mixture of the addition products (79) and (80), with phenylselenyl bromide in dichloromethane solution at -78°C, allene itself gives only the vinylselenide (81). This product when allowed to stand in solvents such as nitromethane or dichloromethane-trifluoroacetic acid rearranges to the isomeric allyl selenides (82). It appears that the products obtained in reactions of this type depend upon, not only the halogen atom of the phenylselenide and the nature of the substitution pattern about the allene, but also on the reaction conditions (S.Halazy and L.Hevesi, *Tetrahedron Letters*, 1983, **24**, 2687).

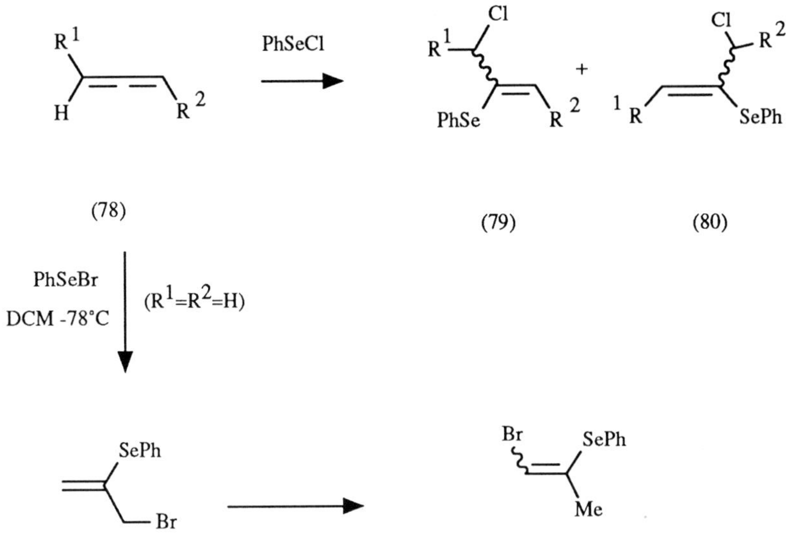

R^1 R^2 H PhSeCl → R^1 Cl PhSe R^2 + R^1 Cl R^2 SePh

(78) (79) (80)

PhSeBr DCM -78°C (R^1=R^2=H)

SePh Br → Br SePh Me

(81) (82)

Alcohols (ROH) add to allenic ketones (83) to afford the vinylethers (84) which readily isomerise to their more conjugated forms (85) (H.J.Cristau, J.Viala, and H.Christol, *Tetrahedron Letters*, 1982, **23**, 1569). In similar manner, hydroxylamine adds to the same substrates to yield isoxazoles (86) and (87) (C.Santelli *ibid*., 1980, **21**, 2893). 1,2-Oxaphospholenes (88) are produced when trivalent phosphorus compounds are the addends (G.Bouno and J.R.Llinas, *J.Amer.Chem.Soc*., 1981, **103**, 4532).

Vinyl halides enter into palladium mediated addition reactions with allenes (M.Ahmar *et al*., *Tetrahedron*, 1987, **43**, 513), and Gore and his colleagues have demonstrated that vinyl bromides (89; R_1,R_2= H or alkyl, or R_1= $SiMe_3$, R_2= H) are palladated by a mixture of bis(dibenzylideneacetone)palladium and 1,2-bis(diphenylphosphino)ethane to give complexes (90), which react with allenes to afford vinylallenic intermediates (91). These can then be trapped with enolate anions to produce the corresponding 1,3-dienes (92) B.Cazes, V.Colovray, and J.Gore, *Tetrahedron Letters,* 1988, **29**, 627; see also B.Friess, B.Cazes, and J.Gore, *ibid*., p.4089).

RO O
RCH R
(85)

^{1}R H
ROH
O
^{2}R
(83)

RO O
^{1}RCH R^{2}
(84)

RP(OMe)$_3$
(R=O-alkyl, O-aryl, S-Me,
Me, Ph, CN, or vinyl)

NH_2OH

R^{2}
N
^{1}RCH$_2$ O
(86)

+

R^{2}
O
^{1}RCH$_2$ N
(87)

R^{2}
O
^{1}RCH P
R OMe
OMe
(88)

$Pd(0)L_n$ → $RCH{=}C{=}CH_2$ →

(89) (90)

$\overset{-}{R^3CH}\,CO_2Me$ →

$R^3{=}COMe$ or CO_2Me

(91) (92)

Methoxyallenes (93) act as components in Diels Alder reactions with inverse electron demand requirements. Thus, for example, they react with oximes (94) in the presence of base to form the exocyclic methyleneoxazolines (95). The latter easily isomerise to the corresponding 6*H*-1,2-oxazines (96) (R.Zimmer and H.U.Reissig, *Angew.Chem.*, 1988, **100**, 1576).

$HON{=}C(R)CH_2R^1$ (94) →

(93) (95)

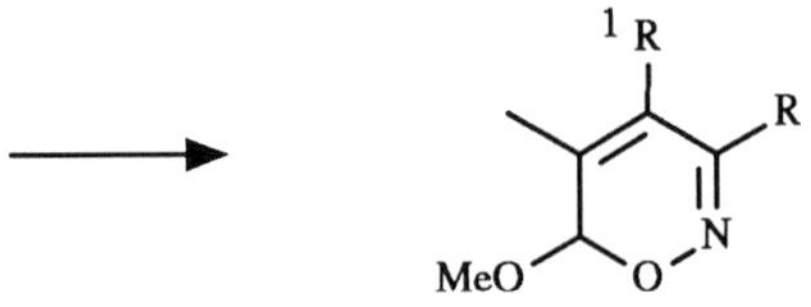

(96)

R=Ph, CF_3 or CO_2Et

R^1=Cl or Br

Molecular orbital calculations suggest that allenes will combine with 1,3-dipoles in a highly selective manner and afford adducts with the more activated double bond. This is observed in practice and with diazopropane, for example, the phenylsulphonylallene (97) reacts to form the pyrazoline (98) which isomerises to the pyrazole (99) (A.Padwa *et al., J.Org.Chem.*, 1988, **53**, 2232).

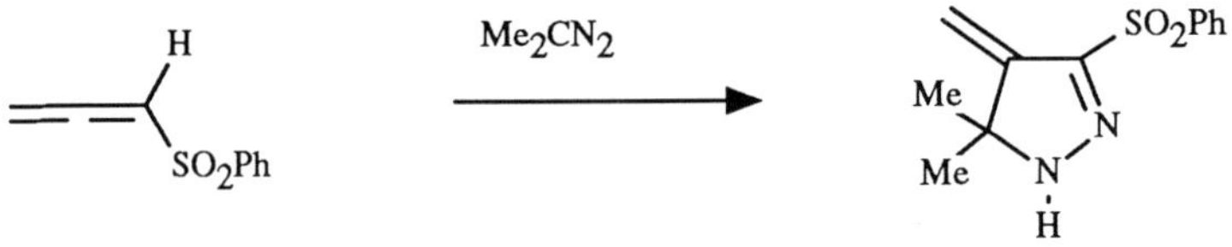

Me_2CN_2

(97)

(98)

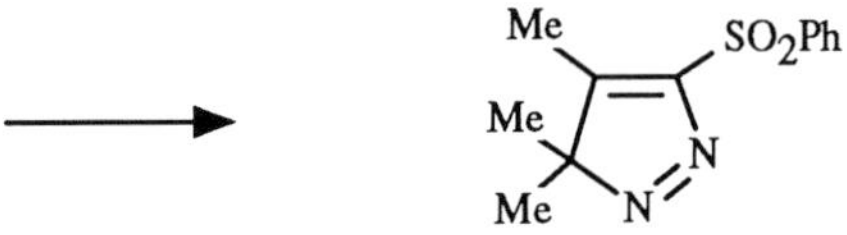

(99)

The allene (97) has a markedly low LUMO energy compared to allene itself and it also reacts with bis(phenylsulphonyl)methane in the presence of a trace of sodium hydride to give 2,4,4-(trisphenylsulphonyl)but-1-ene in high yield (Padwa and P.E.Yeske, *J.Amer.Chem.Soc.*, 1988, **110**, 1617). Similarly electron deficient allenes react with *C*-phenyl-*N*-alkylnitrones with complete specificity to give 5-methylene substituted isoxazolidines.

The orientation observed follows the prediction that maximum orbital overlap should be established between the nitrone HOMO and the LUMO of the allene. In some cases when stereochemical factors compete with electronic considerations diastereomeric mixtures of isoxazolidines are produced and now the reactions may operate through 'two-plane' intermediates. The transition state which predominates will thus depend upon the nature of the substituents attached to the *N*-atom of the nitrone and to the double bond of the dipolarophile (Padwa *et al., J.Org.Chem.*, 1987, **52**, 3909; *ibid.*, 1989, **54**, 810).

(v) *Miscellaneous reactions*

Acetylcobalt tetracarbonyl reacts with allenes under phase transfer conditions to form hydroxyenones (101). The reaction is thought to proceed *via* a 2-acetyl-π-allylcobalt tricarbonyl complex (100), which is selectively attacked by hydroxyl ion at the more substituted terminus (S.Gambarotta and H.Alper, *J.Org.Chem.*, 1981, **46**, 2142).

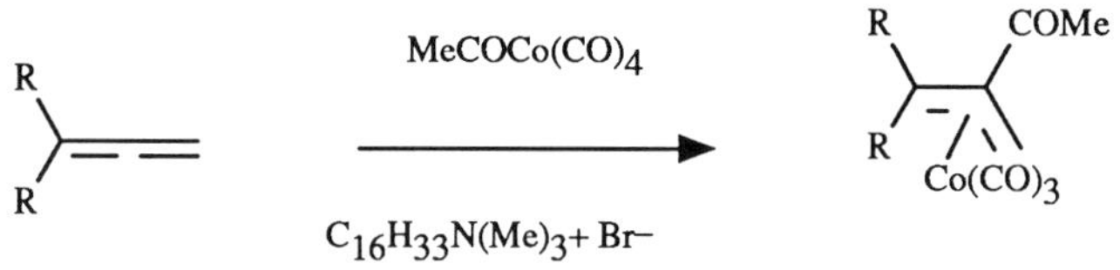

$MeCOCo(CO)_4$

$C_{16}H_{33}N(Me)_3^+ Br^-$

R COMe R $Co(CO)_3$

(100)

H_2O

R COMe HO R

(101)

Tetramethylallene when dissolved in liquid sulphur dioxide undergoes an ene reaction to afford the sulphinic acid (102) recalling similar reactions known to occur, for example, with diethyl acetylenedicarboxylate [where the product is (102, R=N(CO_2Et)NHCO_2Et)] (G.Capozzi *et al., Tetrahedron Letters*, 1980, 3289).

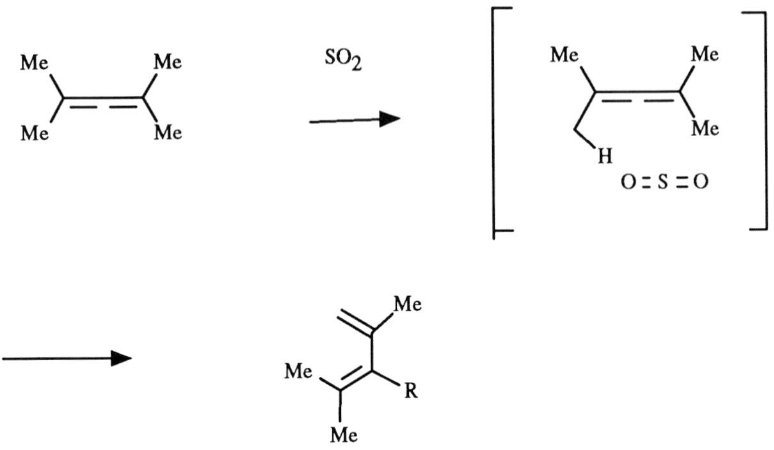

(102, R=SO_2H)

Photochemical irradiation of the propenylallene (103) affords the housane (104) in 70% yield (W.Kudrawcew *et al., Heterocycles*, 1982, **17**, 139).

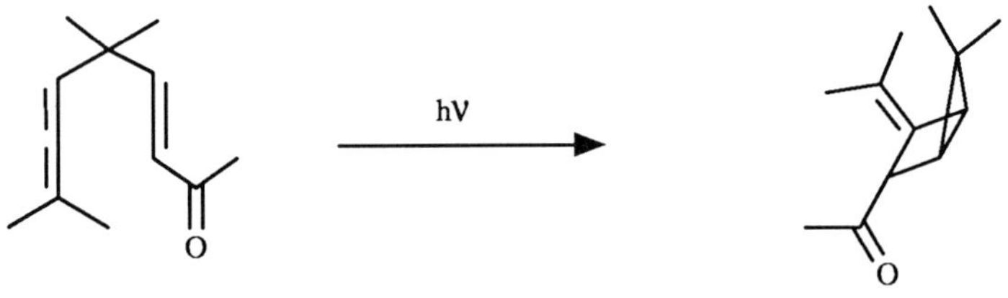

(103)

(104)

(e) *Functionalised allenes*

(i) *Silylallenes*

Trimethylsilylallenes (105) react regioselectively with aldehydes and ketones in the presence of titanium tetrachloride to generate homopropargylic alcohols (106), when R^1=H the products are accompanied by small amounts of (trimethylsilyl)vinyl halides (R.L.Danheiser

D.J.Carini, and A.Basak, *J.Amer.Chem.Soc.*, 1981, **103**, 1604). In a similar study the reactions between trimethylsilylalkynes (107), ^{t}butyl lithium and tetra(isopropoxy)titanium were shown to give titanylallenes (108). These compounds combine with aldehydes in a highly stereoselective manner to afford *threo*-β-hydroxyalkylacetylenes (109) (M.Ishiguro, N.Ikeda, and H.Yamamoto, *J.Org.Chem.*, 1982, **47**, 2225). The influence of the titanium is essential for other metal derivatives are much less effective, and it is probable that the oxygen atom of the aldehyde is bonded to the titanium atom of the titanylallene in the transition state so as to afford the least steric interaction between the substituent groups.

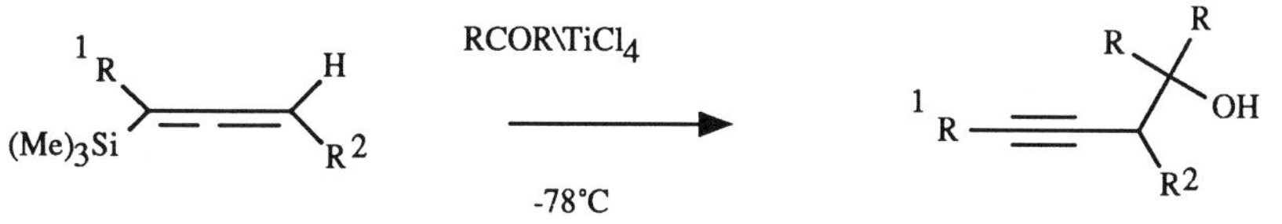

^1R, $(Me)_3Si$, H, R^2 — RCOR\TiCl$_4$, -78°C → R, R, OH, ^1R, R^2

(105) (106)

$(Me)_3Si$, H, R^1 — ^{t}BuLi, Ti(OR)$_4$ → $(Me)_3Si$, $(RO)_3Ti$, R^1, H — RCHO, below -78°C →

(107) (108)

[$(Me)_3Si$, $(RO)_3Ti$, O, H, R^1, H, R] → $(Me)_3Si$, H, R, H, ^1R, OH

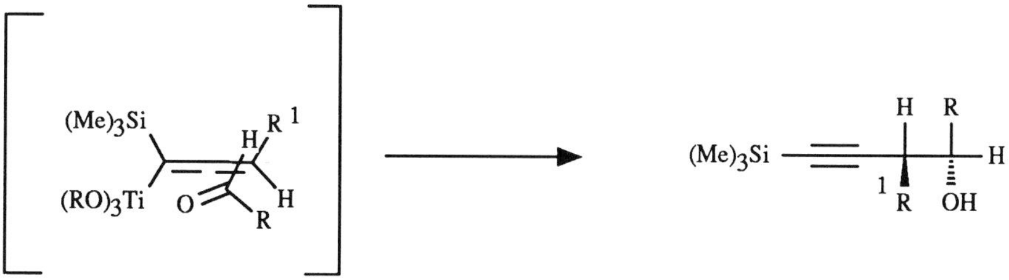

(109)

It is interesting that *(E)*-alkenynes (111) are formed as the major products when the lithiated allenes (110) are reacted with aldehydes in the presence of five molecular equivalents of HMPA; whereas the *(Z)*-isomers (112) predominate when the magnesium or titanium analogues are the substrates (Y.Yamakado *et al., J.Amer.Chem.Soc.*, 1981, **103**, 5568; E.J.Corey and

C.Rucker, *Tetrahedron Letters*, 1982, **23**, 719).

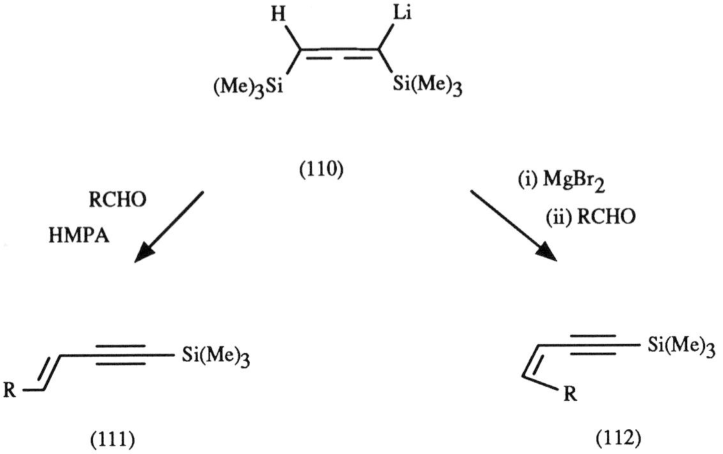

H Li

(Me)3Si Si(Me)3

(110)

RCHO
HMPA

(i) $MgBr_2$
(ii) RCHO

R Si(Me)3

(111)

Si(Me)3
R

(112)

(ii) *Allene oxides*

Allene oxides (113, R=alkyl) are versatile synthons and react, for example, with methyl acrylate to give cyclopentanonyl esters (114), and with cyclopentadiene to form cyclopentadienyl ketones (115) (T.H.Chan and B.S.Ong, *Tetrahedron*, 1980, **36**, 2285; 2286).

(iii) *Hydroxyallenes, silyloxyallenes, acetoxyallenes, and alkoxyallenes*

The unsymmetrical butyne-1,4-diol (116, R=H) can be monoacetylated at the less hindered hydroxyl group by treatment with acetic anhydride in pyridine. The product (116, R=Ac), if reacted with silver perchlorate, is considered to undergo a silver(I) catalysed isomerisation to the allene (117) *en route* to the dihydrofuran (118) (H.Saimoto, T.Hiyama, and H.Nozaki, *J.Amer.Chem.Soc.*, 1981, **103**, 4975). Ring closures of this type are also noted if the alcohols (119) are treated with phenylselenyl chloride The rates of the formation of the dihydrofurans (120) are accelerated for terminal allenes by the addition of triethylamine (P.L.Beauieu, V.M.Morisset, and D.G.Garratt, *Tetrahedron Letters*, 1980, **21**, 129).

The trimethylsilyl derivative (121) when treated with hydrogen peroxide in benzonitrile solution rearranges to the enones (122) (M.Bertrand, J.P.Dulcere, and G.Gil, *ibid.*, 4271) and in the case of the acetate (123) similar treatment gives the dienone (124).

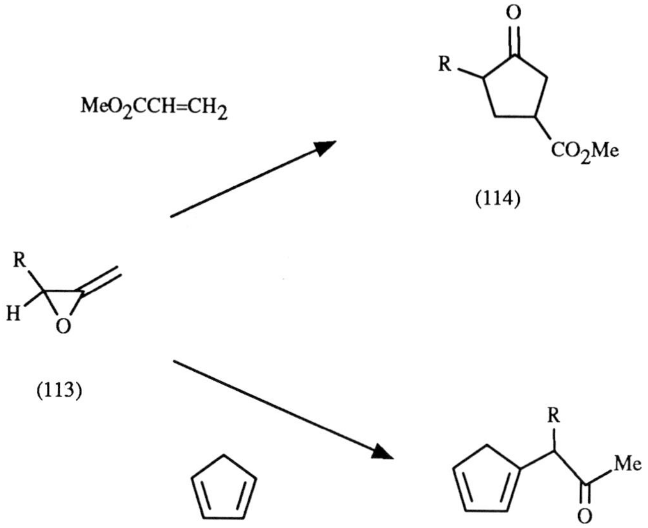

$MeO_2CCH{=}CH_2$

R, O, CO_2Me

(114)

R, H, O

(113)

R, Me, O

(115)

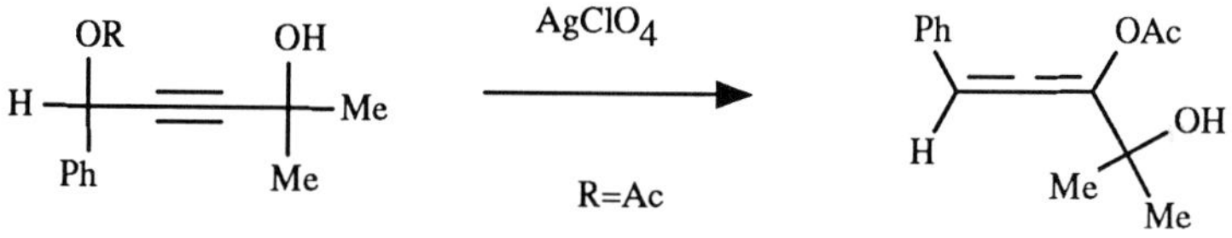

OR, OH, H, Me, Ph, Me

(116)

$AgClO_4$

R=Ac

Ph, OAc, H, OH, Me, Me

(117)

OAc, Me, Ph, O, Me

(118)

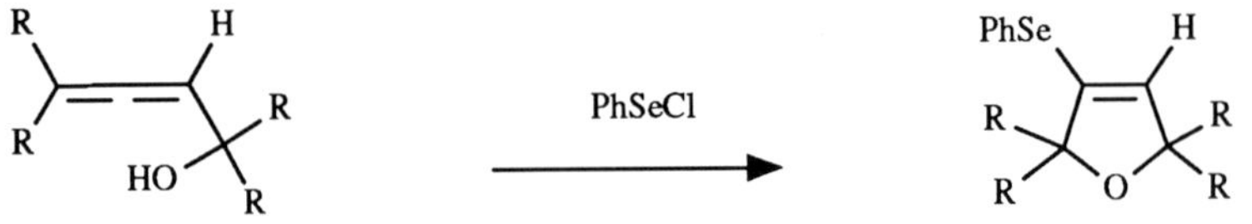

PhSeCl

(119) (120)

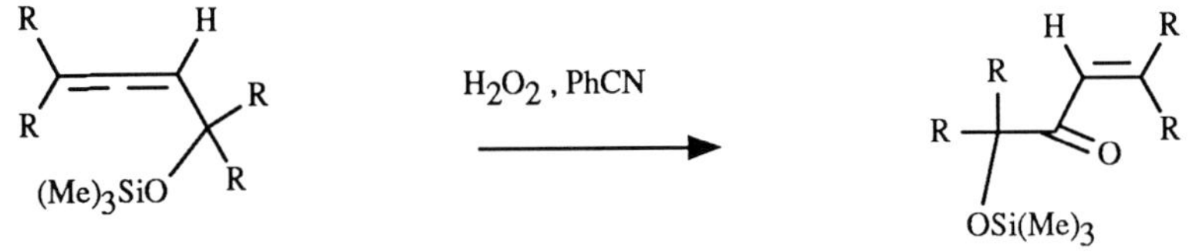

H_2O_2, PhCN

(121) (122)

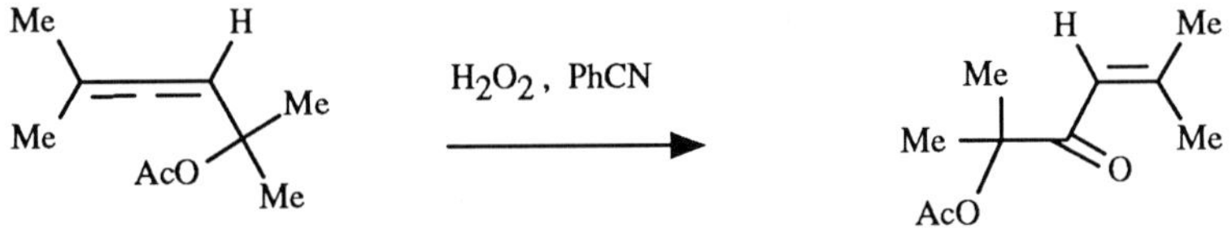

H_2O_2, PhCN

(123)

-HOAc

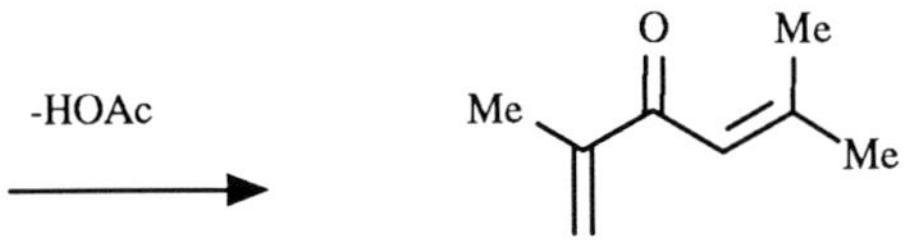

(124)

The silyl analogues (125, R=H or alkyl) react with hydrogen peroxide in the presence of benzonitrile and methanol to form δ-lactones (128). A reasonable mechanism for these reactions is that cyclopropanones (126) are formed first. These then suffer intramolecular attack by the lone pair electrons of the hydroxyl group at the carbonyl function to give anions

(127), which are quenched by proton abstraction from methanol. Importantly, if the starting materials are chiral this asymmetry is transferred to the products (M.Bertrand, J.P.Dulcere, and G.Gil., *ibid.*, 1980, **21**, 1945).

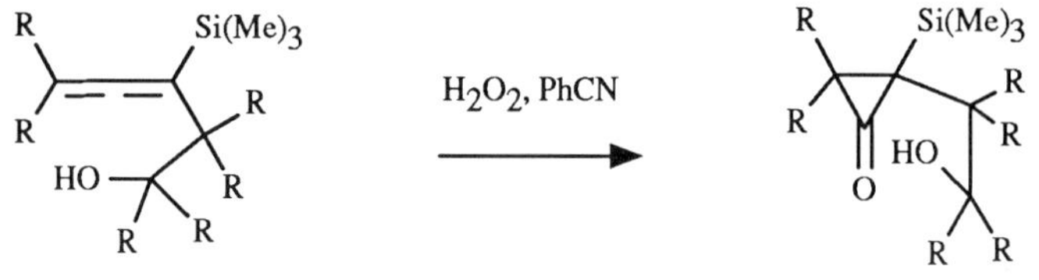

R Si(Me)3 R R HO R R R H_2O_2, PhCN R Si(Me)3 R R R HO O R R

(125) (126)

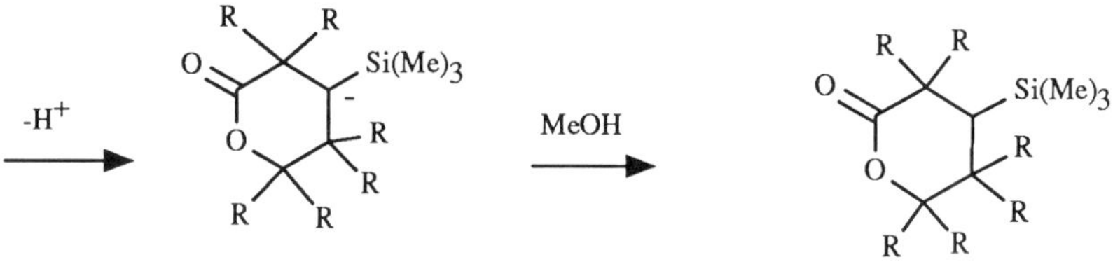

$-H^+$ R R O Si(Me)3 O R R R R MeOH R R O Si(Me)3 O R R R R

(127) (128)

Alkylallenes (129, n=1) bearing a γ-hydroxyl group in the presence of a catalyst such as aqueous silver nitrate undergo cyclisation to vinyltetrahydrofurans (130, n=1) (A.Audin, *et al., Bull.Soc.Chim.Fr.*, Part II, 1981, 313; A.Audin, A.Doutheau, and J.Gore, *Tetrahedron Letters*, 1982, **23,** 4337); whereas their homologues (129, n=2) give the corresponding pyrans (130, n=2), when reacted with either silver (I) or mercury (II) salts.

Methoxyallenes can be lithiated at the α-position and the anions so formed then reacted with a range of electrophiles. On *O*-demethylation, the latent carbonyl groups of the products are unmasked and the methoxyallenes thus serve as acyl anion equivalents. For example, the lithio derivative (131) reacts with phenyl iodide in the presence of tetrakis(triphenylphosphine)palladium to give the corresponding aryl derivative (132, R=Ph), which on treatment with sulphuric acid affords the phenylpropenone (133, R=Ph) (T.J.-Luong and G.Linstrumelle, *Synthesis*, 1982, 738).

This reaction becomes more generally applicable if the original lithiated allene is transmetallated by treatment with zinc chloride prior to reaction with the electrophile. Now alkenyl halides may also be used in place of aryl halides (C.E.Russell and L.S.Hegedus, *J.Amer.Chem.Soc.*, 1983, **105**, 943). The lithiated allene (131) may be silylated by reaction with trimethylsilyl chloride to afford the corresponding *C*-silylallenes (134) (J.C.Clinet and G.Linstrumelle, *Synthesis*, 1981, 875; P.Pappalardo, E.Ehlinger, and P.Magnus, *Tetrahedron Letters*,1982, **23**, 309).

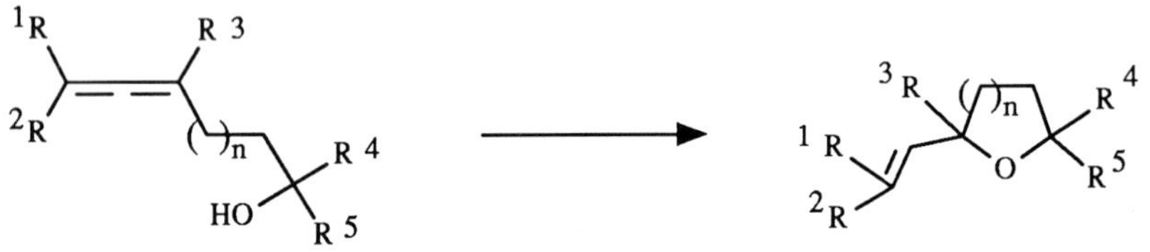

(129) (130)

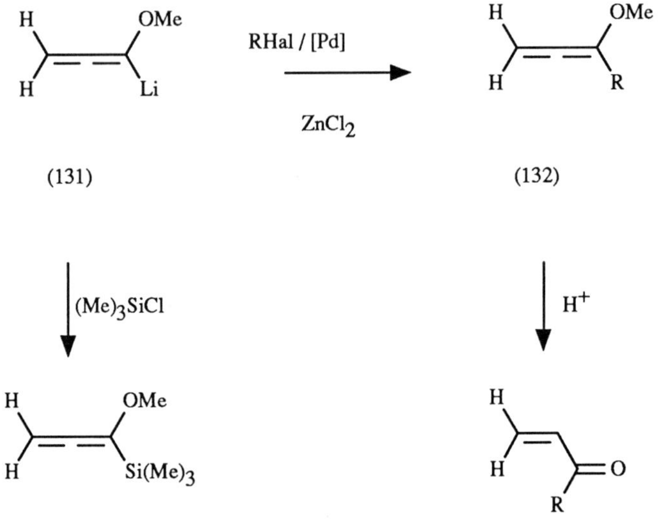

RHal / [Pd]

$ZnCl_2$

(131) (132)

$(Me)_3SiCl$

H^+

(134) (133)

Similarly treatment of the alkoxyallene (135) with butyllithium and trimethylsilyl chloride yields the silyl derivative (136). This compound may then be reacted with (a) sulphuryl chloride to give the chlorovinyl ketone

(137); (b) with 3-chloroperbenzoic acid (MCPBA) to afford diketone (138); and (c) with phenylselenyl chloride to yield the phenylselenide (139) (H.Reich and M.J.Kelly, *J.Amer.Chem.Soc.*, 1982, **104**, 119).

When the lithiated alkoxyallene (131) is reacted with *B*-alkylborabicyclononanes (140) salts of the type (141) are formed, which when acidified undergo cyclisation to cyclopropanes (142) in such a way that the alkyl group of the original boron reagent is transferred to the product (N.Miyaura *et al.*, *Tetrahedron Letters*, 1980, **21**, 537).

γ-Lithioalkoxyallenes (143) are potential homoenolate equivalents and they can be synthesised from alkoxyallenes by direct lithiation, provided the alkoxy group is bulky enough to direct attack away from the α-proton. If not, the α-site is lithiated and the anion formed may then be silylated with trimethylsilyl chloride. Should the product be reacted further with an alkyl lithium the desired γ-lithiated alkoxyallenes are obtained. These may then be combined with electrophiles and the silyl protecting groups removed in the final step (P.Pappalardo, E.Ehlinger, and P.Magnus, *Tetrahedron Letters*, 1982, **23**, 309; J.C.Clinet and G.Linstrumelle, *ibid.*, p. 3987). Electrophilic attack at the γ-position of alkoxylated allenes can also be achieved through first reacting propargylic ethers with ^{t}butyl lithium, and then treating the lithiated products (144) with aryl, alkyl, or silyl halides. If diethylaluminium chloride is used as electrophile the aluminium allenes (145) are formed, which then react with aldehydes to provide the corresponding allenic alcohols (146). The last compounds are readily converted into furans (147) by hydrolysis with aqueous acid (M.Ishiguro, N.Ikeda, and H.Yamamoto, *Chem.Letters,* 1982, 1029).

The choice of lithium alkyl in these reactions is important, for conversion of 1-trimethylsilyl-1-methoxyallene (148) into the lithiated alkyne (149) occurs if lithium diisopropylamide is used. Should the lithiated alkyne be treated with aldehydes or ketones the corresponding propargylic alcohols (150) are formed. With dry dimethylsulphoxide alone these products decompose into a mixture of the isomeric alkenylalkynes (151), but if potassium hydride is present then cyclisation to the dihydrofurans (152) takes place (I.Kuwajima, S.Sugahara, and J.Enda, *Tetrahedron Letters*, 1983, **24**, 1061).

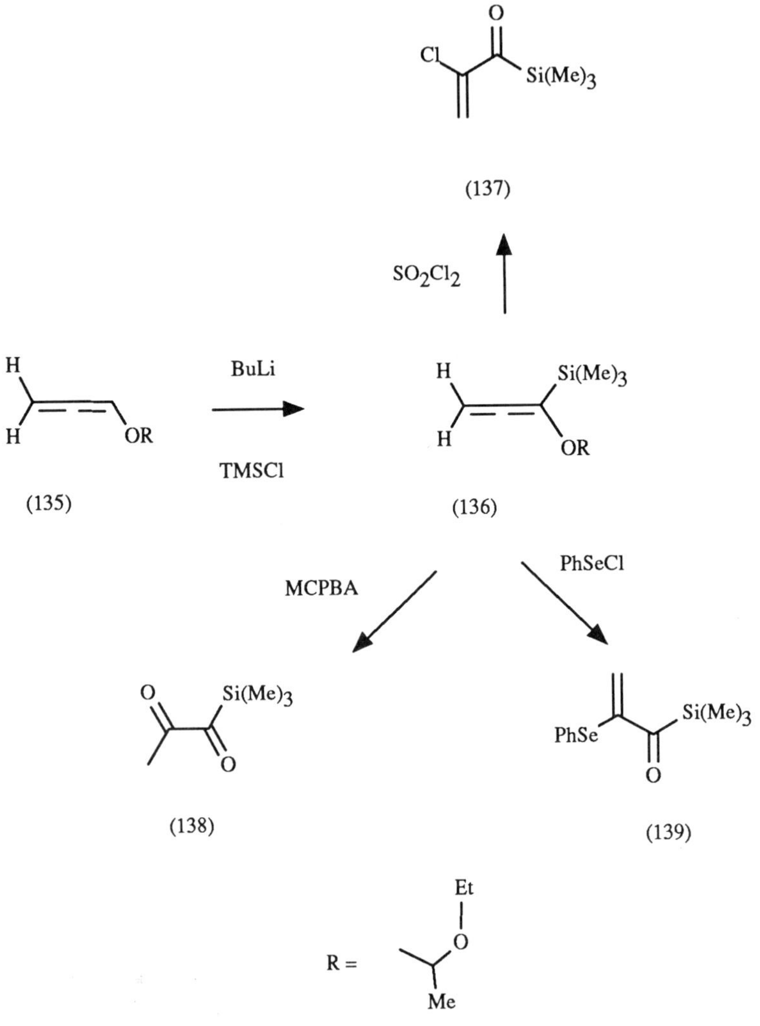

O
Cl
$Si(Me)_3$

(137)

SO_2Cl_2

H
H
OR

BuLi
TMSCl

H
H
$Si(Me)_3$
OR

(135)

(136)

MCPBA

PhSeCl

O
$Si(Me)_3$
O

PhSe
$Si(Me)_3$
O

(138)

(139)

Et
O
R =
Me

H OMe H Li + R-B

(131) (140)

H OMe H B⁻ R Li^+ HOAc OMe R

(141) (142)

H OR H H R^1Li H OR Li H E^+ H OR E H

(143)

R^1Li TMSCl F^-

H OMe H $Si(Me)_3$ R^1Li H OMe Li $Si(Me)_3$ E^+ H OMe E $Si(Me)_3$

R — ≡ — OMe

BuLi

R, Li, OMe

R^1Hal

R, ^1R, OMe

(144)

$(Et)_2AlCl$

R, Et – Al, Et, OMe

R^2CHO

R, ^2R, O – H, OMe

H^+ / H_2O

R, ^2R, O

(145)

(146)

(147)

H, H, OMe, $Si(Me)_3$

LDA

Li, OMe, H, $Si(Me)_3$

(148)

(149)

RCOR

HO, R, R, OMe, H, $Si(Me)_3$

(150)

DMSO

(if one R=H)

KH / DMSO

R, OMe

R, R, O, OMe

(151)

(152)

(iv) *Allenic amides and amines*

Allenic amides and amines [153, R=CH_2Ph; R=$SO_2(4\text{-}CH_3C_6H_4)$; or R=$CO_2Me$; n=1 or 2] have been prepared and cyclised to the corresponding pyrrolidines or piperidines (154, n=1 or 2) respectively by the action of palladium (II) chloride (0.1 equiv.) and cupric chloride (3 equiv.) under an atmosphere of carbon monoxide (D.Lathbury, P.Vernon, and T.C.Gallagher, *Tetrahedron Letters*, 1986, **27**, 6009).

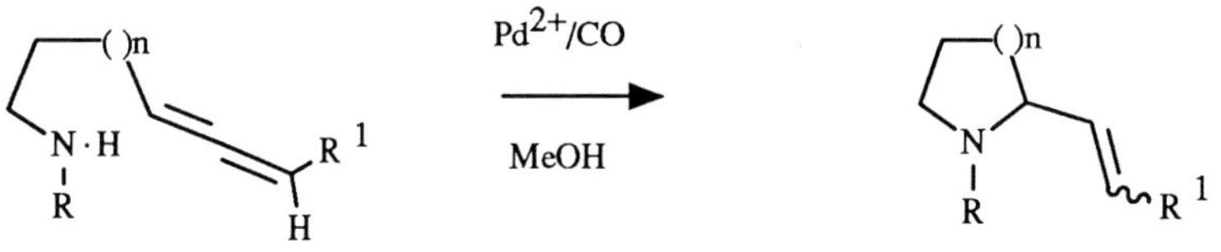

Pd^{2+}/CO, MeOH

(153) (154)

The reaction can be made stereoselective if silver (I) fluoroborate (0.1-1 equiv.) in dichloromethane is used as the cyclisation reagent. Thus, for example, the sulphonamide [155, R=$SO_2(4\text{-}CH_3C_6H_4)$] is cyclised exclusively to the *cis*-pyrrolidine (156). The corresponding amine (155, R=H), however, gives a mixture of both *cis*- and *trans*- isomers (R.G.Kinsman *et al., Chem. Commun.*, 1987, 243). This valuable reaction, which clearly involves a cyclic transition state involving the π-bonded metal atom and the *N*-sulphonamido group, has been used in the stereoselective syntheses of several alkaloids including (R)(-)-coniine and racemic anatoxin-a (D.Lathbury and T.C.Gallagher, *Chem Commun.*, 1986, 114; P.Vernon and T.C.Gallagher, *ibid.*, 1987, 245).

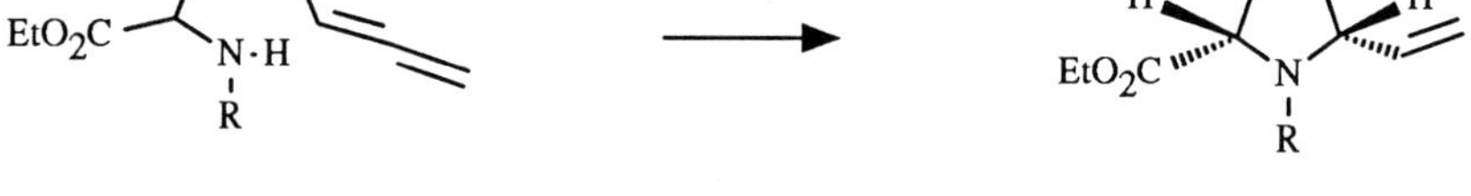

$AgBF_4$

(155) (156)

(v) *Carbonyl derivatives of allenes*

Trimethylsilyloxyenones (157) on flash vacuum pyrolysis give the allenyl ketones (158), but if the contact time is extended beyond 10^{-3} seconds at a temperature of 800°C then intramolecular cyclisation and rearrangement afford the corresponding furans (159) in modest yields (36-50%) (J.Jullien *et al.*, *Tetrahedron*, 1982, **38**, 1413).

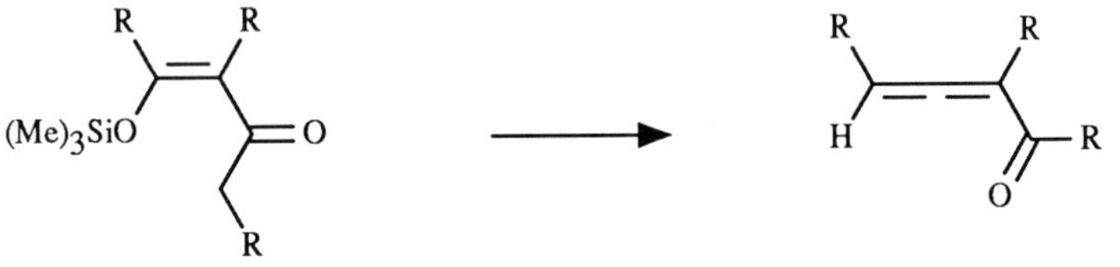

(157) (158)

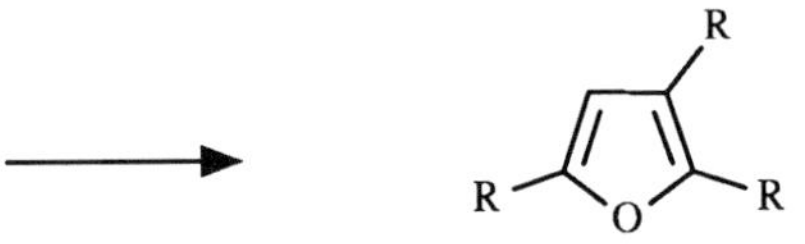

(159)

The chiral aldehyde (160) when heated undergoes an intramolecular hetero-ene reaction to form the cyclopentenol (161) with retention of chirality (M.Bertrand, M.L.Roumesttant, and P.S.-Panthet, *Tetrahedron Letters*, 1981, **22**, 3589).

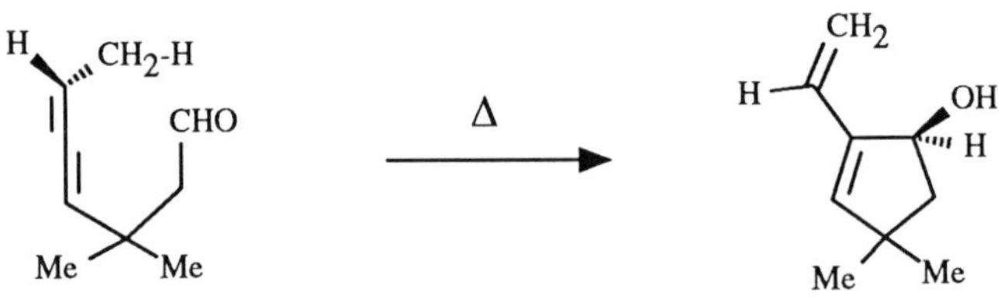

(160) (161)

Isomerism to the 2*E*,4*Z*-dienes (164) occurs when alkyl penta-3,4-dienoates (162) are heated with alumina. It is considered that the carbonyl oxygen atom of the allenic ester is coordinated to the surface of the alumina catalyst through an enolate species (163), and that a proton from a hydroxyl function bonded to the surface is then transferred on to central carbon atom *via* less hindered face of the allene. Finally release from the catalyst allows the generation of the *trans* geometry of the α,β-double bond of the product (S.Tsuboi, T.Masuda, and A.Takeda, *J.Org.Chem.*, 1982, **47**, 4478).

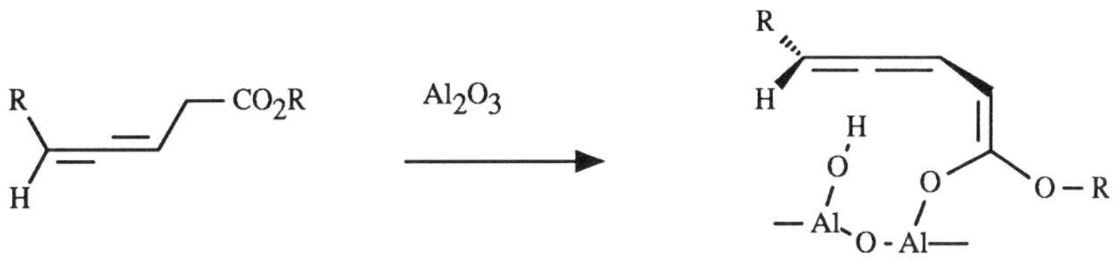

Al_2O_3

(162)

(163)

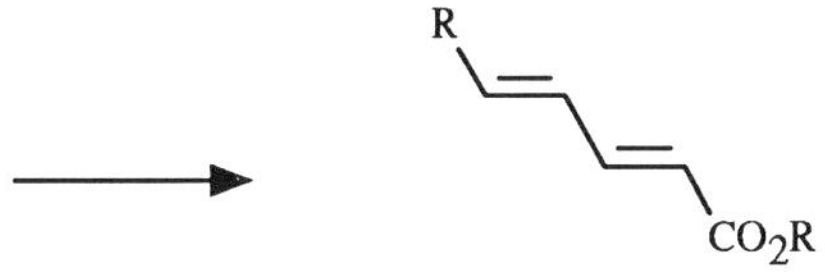

(164)

(vi) *Allenic nitriles*

Hydroxylamine reacts with cyanoallenes (165) to give aminoisoxazoles (166); while hydrazines afford pyrazoles (167) (Z.T.Fromm *et al.*, *J.Chem.Soc. Perkin Trans* I, 1981, 2997).

(vii) *Allenic sulphones*

A general synthesis of allenic sulphones from alkynes (168) may be achieved through selenosulphonation using arylsulphonylphenyl selenide as the reagent in the presence of AIBN. This free radical process yields β-(phenylseleno)vinylsulphones (169) which when treated with base (normally triethylamine, but sodium hydride or LDA in resistant cases) give rise to the anions (170). These can be protonated and the selenides (171, E=H) which are formed may then be oxidised to afford the allenes (172, E=H). Alternatively the anions can be reacted with other electrophiles, such as alkyl or allyl iodides, or trimethylsilyl chloride, prior to oxidation to provide the corresponding derivatives (172; E=alkyl, allyl, or trimethylsilyl) (T.G.Black, M.V.Krishna and K.R.Muralidharan, *Tetrahedron Letters*, 1987, **28**, 1737).

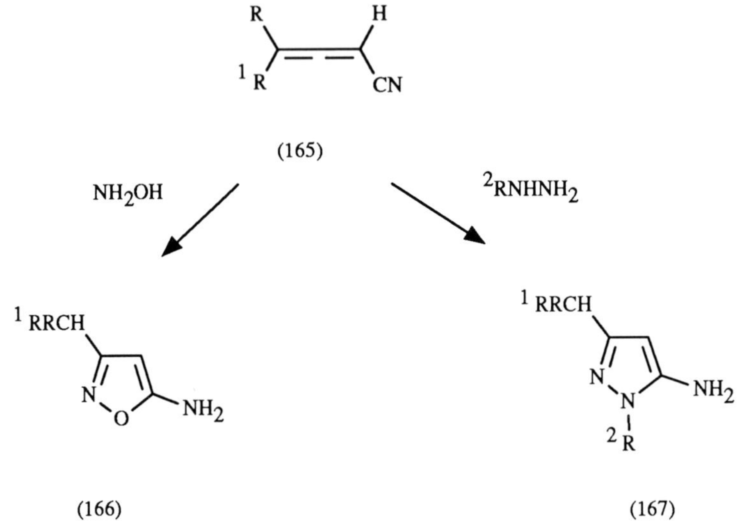

(165)

NH_2OH

$^2RNHNH_2$

(166)

(167)

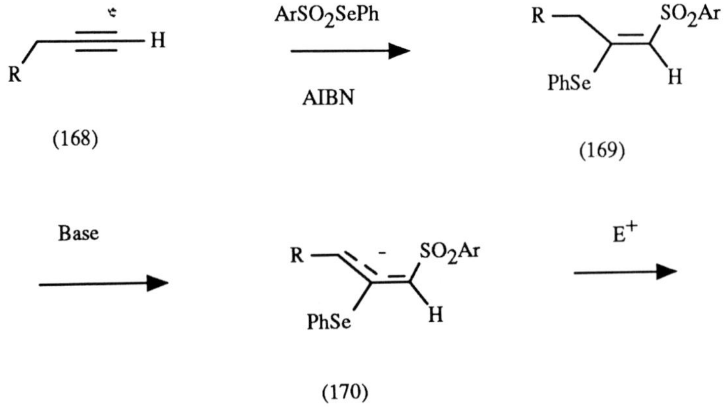

$ArSO_2SePh$

AIBN

(168)

(169)

Base

E^+

(170)

(171) $\xrightarrow{[O]}$ (172)

(viii) *Allenic boronates*

Allenylboronates (174) are formed by the reaction of propargylic bromides (173), firstly with aluminium, then with trimethyl borate and finally with water. Further reaction of these compounds with aliphatic aldehydes in the presence of dialkyl tartrates yields homopropargylic alcohols (176), the stereochemistry of the product depending upon which enantiomer of the tartrate is used. It is certain that a highly ordered transition state such as that shown (175) is involved, wherein the tartrate unit is bonded *via* the boron atom in such a way that steric interactions are minimised (R.Haruta *et al., J.Amer.Chem.Soc.*, 1982, **104**, 7667).

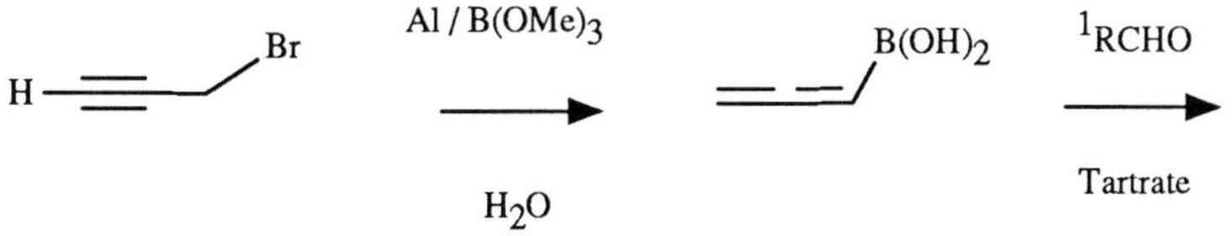

(173) $\xrightarrow[H_2O]{Al / B(OMe)_3}$ (174) $\xrightarrow[Tartrate]{^1RCHO}$

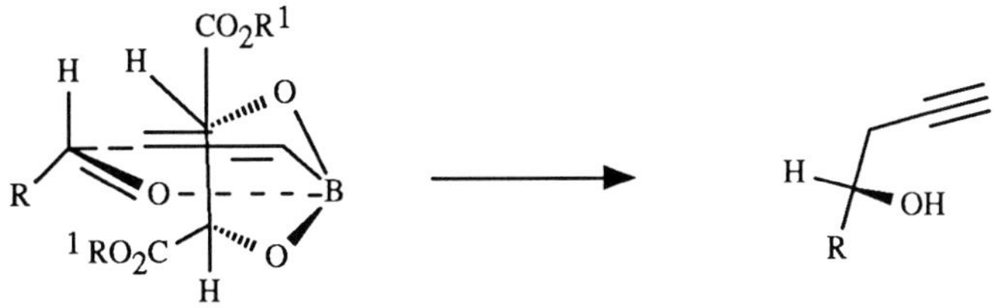

(175) $\longrightarrow$ (176)

Second Supplements to the 2nd Edition of Rodd's Chemistry of Carbon Compounds, Vol. IA/B, edited by M. Sainsbury
© 1991 Elsevier Science Publishers B.V., Amsterdam

Chapter 2c

ACYCLIC HYDROCARBONS: ALKYNES

A.M. JONES and S.P. STANFORTH

Major publications concerned with alkynes include "Synthesis of Acetylenes, Allenes and Cumulenes: A Laboratory Manual" ed. L. Brandsma and H.D. Verkruijsse, Elsevier Scientific, Amsterdam, 1981; "Acetylene Based Chemicals from Coal and other Natural Sources", R.J. Tedeschi, Marcel Dekker, New York, 1982; "New Reactions and Chemicals Based on Sulphur and Acetylene", B.A. Trofimov, Harwood Academic, London, 1983; "Chemistry of Functional Groups, Supplement C: The Chemistry of Triple-Bonded Functional Groups", ed. S. Patai and Z. Rappoport, Wiley-Interscience, Chichester, 1983; "Preparative Acetylene Chemistry" 2nd Edn., L. Brandsma, Elsevier-Scientific, Amsterdam, 1988 and "Chemistry and Biology of Natural Occuring Acetylenes and Related Compounds: Proceedings of a Conference", ed. J. Lam, H. Breteler, T. Arnason and L. Hansen, Elsevier Scientific, Amsterdam, 1989.

(1) Preparation of Alkynes

Reviews of the preparations of alkynes include the texts cited above together with "Organic Functional Group Preparations", 2nd Edn., Vol.1, S.R. Sandler and W. Karo, Academic Press, London, 1983 and "Compendium of Organic Synthetic Methods" Vols. IV, V and VI, Wiley-Interscience, New York, 1980, 1984 and 1988. Annual compilations of acetylene syntheses appear in "Annual Reports on the Progress of Chemistry, Section B", The Royal Society of Chemistry, London, 1980-88; "Annual Reports in Organic Synthesis", Academic Press, Orlando, 1980-87 and "General and Synthetic Methods, Specialist Periodical Report", Vols. 5-11, ed. G. Pattenden, The Royal Society of Chemistry, London, 1980-88.

(i) Dehydrohalogenation Methods

Base induced elimination from vicinal dihalides, accessible from alkenes, has continued to attract attention and the methodology refined *via* the use of crown ethers (E.V. Dehmlow and M. Lissel, *Liebigs Ann. Chem.*, 1980, 1), phase transfer catalysis (A. Le Coq and A. Gorgues, *Org. Synth.*, 1980, **59**, 10; E.V. Dehmlow and M. Lissel, *Tetrahedron*, 1981, **37**, 1653) and tertiary diols to enhance reactivity (E.V. Dehmlow *et al.*, *ibid.*, 1986, **42**, 3569). Functionalised alkynes are available from ethyne (M.A. Pericas, F. Serratosa and E. Valenti, *ibid.*, 1987, **43**, 2311) and the use

$$\text{H-C}\equiv\text{C-H} \xrightarrow[\text{Br}_2]{\text{Bu}^t\text{OK/Bu}^t\text{OH}} \text{H-C}\equiv\text{C-OBu}^t$$

of heterocycles as masking groups yields acetylenic carboxylic acids (P. Sarti-Fatoni, *J. Heterocycl. Chem.*, 1983, **20**, 105; A.Svensonn, J.O. Karlsson and A. Hallberg, *ibid.*, 1983, **20**, 729).

$$\text{Ar-CH=CH-(3,4-dinitroisoxazol-5-yl)} \xrightarrow[\text{H}_3\text{O}^{\oplus}]{\text{Br}_2/\text{OH}^{\ominus}} \text{Ar-C}\equiv\text{C-COOH}$$

1,1-Dihaloalkenes, from modified Wittig reactions (S. Hayashi, T. Nakai and N. Ishikawa, *Chem. Lett.*, 1980, 935; F. Marcacci, G. Giacomelli and R. Mencagli, *Gazz. Chim. Ital.*, 1980, **110**, 195), 1,1,2-trihaloalkenes [from 1,1-dichloro-2,2-difluoro-ethene and organometallic compounds (K. Okuhara, *Bull. Chem. Soc. Jpn.*, 1981, **54**, 2045) or tribromoethanal (R.H. Smithers, *Synthesis*, 1985, 556)], yield alkynes when treated with organolithium compounds. Variations in methodology allow functionalisation of the termini of the triple bond.

$$\text{R-C}_3\text{H}_3\text{(Cl)-C(Cl)=CCl}_2 \xrightarrow[\text{E}^{\oplus}]{\text{BuLi}} \text{R-C}_3\text{H}_3\text{(Cl)-C}\equiv\text{C-E}$$

(A. de Meijere *et al.*, *Angew. Chem., Int. Ed. Engl.*, 1982, **21**, 65)

$$F_2C{=}CH_2 \xrightarrow[CH_3CHO]{Bu^tLi} Bu^t\text{-}C{\equiv}C\text{-}CH(OH)CH_3$$

(J.F. Normant *et al. Tetrahedron Lett.*, 1982, **23**, 4325; K. Oshima, *Bull. Chem. Soc. Jpn.*, 1982, **55**, 3941)

$$R^1R^2N\text{-}C{\equiv}C\text{-}C(Cl){=}CCl_2 \xrightarrow[R^3R^4NH]{BuLi/BrCN} R^1R^2N\text{-}C{\equiv}C\text{-}C{\equiv}C\text{-}NR^3R^4$$

(G. Himbert *et al.*, *Synthesis*, 1987, 73; A. Greene *et al.*, *J. Org. Chem.*, 1987, **52**, 3641)

1,1,2-Trichlorostyrene is converted smoothly to phenylacetylene by zinc or copper (M. Ballester *et al.*, *J. Org. Chem.*, 1986, **51**, 1100, 1413) and ring opening of substituted trichlorocyclopropanes gives acetylenic ketones in good yield (K. Musigmann, H. Mayr and A. de Meijere, *Tetrahedron Lett.*, 1987, **28**, 4517; M.R. Detty *et al.*, *J. Org. Chem*, 1987, **52**, 3662). 1-Bromoalkynes are available from geminal dibromides using polyethylene glycols and sodium hydroxide (P. Li and H. Alper, *ibid.*, 1986, **51**, 4354). Eliminations from mono-haloalkenes have been accomplished with a variety of bases selected for their tolerance to other functional groups present in the alkyne.

$$HC(Br){=}CH(OSiR_3) \xrightarrow{LDA} H\text{-}C{\equiv}C\text{-}OSiR_3$$

(R.L. Danheiser *et al.*, *Tetrahedron Lett.*, 1988, **29**, 4917)

$$ArOCH_2CH{=}C(Br)CH_2OH \xrightarrow{K_2CO_3} ArOCH_2C{\equiv}CCH_2OH$$

(P.F. Schuda and M.R. Heimann, *J. Org. Chem.*, 1982, **47**, 2484)

R-C≡CH —(Pd°cat., (Z)-ClCH=CHCl)→ R-C≡C(H)C=C(H)Cl —(Bu_4^nNF)→ R-C≡C-C≡CH

(A.S. Kende and C.A. Smith, *ibid.*, 1988, **53**, 2655)

(ii) Other Eliminations

Metal amide induced elimination from α-haloacetals leads to alkoxyacetylides which can be functionalised with electrophiles (S. Raucher and B.L. Bray, *J. Org. Chem.*, 1987, **52**, 2332; W.M. Stalick, R.N. Hazlett and R.E. Morris, *Synthesis*, 1988, 287) and a related procedure demonstrates the retention of stereochemical integrity in the synthesis of alkynols from an enantiomerically pure precursor (J.S. Yadav, M.C. Chander and B.V. Joshi, *Tetrahedron Lett.*, 1988, **29**, 2737).

(+)-Tartaric acid → Cl–(acetonide)–O–Ar → (OH)–O–Ar

Ketones afford alkynes (J.B. Hendrickson and Md. S. Hussoin, *Synthesis*, 1989, 217).

$R^1COCH_2R^2$ —($(Ph_3P^+)_2O\,(^-OTf)_2$/NEt_3/$ClCH_2CH_2Cl$, reflux)→ R^1-C≡C-R^2

Methyl ketones may be transformed into terminal acetylenes *via* the enol phosphonates (E.-i. Negishi, A.O. King and W.L. Klima, *J. Org. Chem.*, 1980, **45**, 2526; D.H. Hua, *J. Am. Chem. Soc.*, 1986, **108**, 3835; C.H. Heathcock and S.J. Hecker, *ibid.*, 1986, **108**, 4586) and the conversion of enol triflates into conjugated triynes has been refined.

$$Bu^tCH{=}C(OSO_2CF_3){-}C{\equiv}C{-}C{\equiv}C{-}SiMe_3 \xrightarrow{ArOK/diglyme/50^\circ C} Bu^tC{\equiv}C{-}C{\equiv}C{-}C{\equiv}CH$$

(M. Ladika and P.J. Stang, *J. Chem. Soc., Chem. Commun.*, 1981, 459)

The synthetic utility of phosphinyldiazomethane for the preparation of non-terminal alkynes has been reviewed (J.C. Gilbert and U. Weerasooriya, *J. Org. Chem.*, 1982, **47**, 1837) and the quantitative conversion of benzaldehyde into the alkynyl phosphoric diamide reported (G. Bertrand *et al.*, *J. Am. Chem. Soc.*, 1987, **109**, 4711).

$$(Pr^i_2N)_2P(Cl){=}C{=}\overset{\oplus}{N}{=}\overset{\ominus}{N} \xrightarrow{PhCHO} (Pr^i_2N)_2P(O){-}C{\equiv}C{-}Ph$$

Base induced fragmentation of 3-substituted cyclohexa-2-enones (R. Friary and V. Seidl, *J. Org. Chem.*, 1986, **51**, 3214) or the derived alcohol leads to γ-alkynyl carbonyl compounds (T. Shimizu, R. Ando and I. Kuwajima, *ibid.*, 1981, **46**, 5246).

HO, R^2, R^1, SO_2Ph (cyclohexene) $\xrightarrow{NaH/THF}$ O, R^2, R^1 (alkynyl ketone)

1,3-Dicarbonyl compounds afford, *via* pyrazolinone intermediates, acetylenic acids (F.M. Simmross and P. Weyerstall, *Synthesis*, 1981, 72) or the corresponding esters (R.M. Moriarty *et al., Tetrahedron,* 1989, **45**, 1605) using bromine and base and phenyliodonium acetate respectively. The action of potassium fluoride and fluorinated tertiary amines on β-diketones produces alkynones in reasonable yield (T. Kitazume and N. Ishikawa, *Chem. Lett.*, 1980, 1327).

$$R-\overset{O}{\overset{||}{C}}-CH_2-\overset{O}{\overset{||}{C}}-R^1 \xrightarrow{KF/Et_2NCF_2CHFX} R-C{\equiv}C-\overset{O}{\overset{||}{C}}-R^1$$

Decomposition of 1,2,3-selenodiazoles, obtained from ketones *via* the semicarbazones and selenium dioxide, results in the formation of β-alkynylselenoketones in excellent yield (I. Lalezari *et al.*, *J. Heterocycl. Chem.*, 1986, **23**, 893).

$$\text{4-Ar-1,2,3-selenadiazole} \xrightarrow{Bu^tOK \quad XCH_2COR} Ar-C{\equiv}C-SeCH_2\overset{O}{\overset{||}{C}}R$$

Flash vacuum pyrolysis of acylphosphoranes has been shown to yield alkynes uncontaminated by allenes and 3,3,3-trifluoro-1-aryl-propynes have been produced by a related thermolytic procedure.

$$R^1-\underset{\underset{O}{||}}{C}-\underset{\underset{PPh_3}{||}}{C}-R^2 \longrightarrow R^1-C{\equiv}C-R^2$$

(R.A. Aitken and J.I. Atherton, *J. Chem. Soc., Chem. Commun.*, 1985, 1140)

$$R-C_6H_4-CH_2X \xrightarrow[TFAA \quad \Delta]{Ph_3P \quad BuLi} R-C_6H_4-C{\equiv}C-CF_3$$

(Y. Kobayashi *et al.*, *Tetrahedron Lett.*, 1982, **23**, 343)

α-Acyl-α-(arylseleno)phosphoranes, from α-acylphosphoranes and arylselenyl bromides, thermolyse to arylselenoalkynes which may then be functionalised *via* the corresponding alkynide anion (A.L. Braga, J.V. Comasseto and N. Petragnani, *Synthesis*, 1984, 240). Thermal decomposition of appropriately substituted cyclopropenones results in diarylalkynes (D.H. Wadsworth and B.A. Donatelli, *ibid.*, 1981, 285) and bis-dialkyaminoalkynes (C. Wilcox and R. Breslow, *Tetrahedron Lett.*, 1980, **21**, 3241). Aldehydes can be converted into β-acetoxysulphones, which undergo a double elimination in the presence of *t*-butoxide to give alkynes in excellent yield (J. Otera *et al.*, *J. Am. Chem. Soc.*, 1984, **106**, 3670; *J. Org, Chem.*, 1986, **51**, 3830).

$$R^1CH_2SO_2Ph \xrightarrow[Ac_2O\ Py\ PTSA]{BuLi\ R^2CHO} R^1CH(SO_2Ph)CH(OAc)R^2 \xrightarrow[Bu^tOK\ THF]{Bu^tOK\ Bu^tOH} R^1\text{-}C{\equiv}C\text{-}R^2$$

Hydrogen peroxide induced elimination from phenylseleno-alkenylsulphones results in alkynylsulphones (T. Miura and M. Kobayashi, *J. Chem. Soc., Chem. Commun.*, 1982, 438). Peroxy promoted coupling of nitrostyrenes (P.G. Karmarkar *et al., Tetrahedron Lett.*, 1981, 22, 2301) and selenoalkynes (J.V. Comasseto *et al., J. Chem. Soc., Chem. Commun.*, 1986, 1067) to diarylalkynes and conjugated diynes respectively has been reported. Palladium acetate has been found to catalyse the oxidation of alkenes to alkynes (G. Cum *et al., ibid.*, 1985, 1571).

$$RCH{=}CHR \xrightarrow[EtOH\ H_2O]{Pd(OAc)_2\ O_2} RC{\equiv}CR$$
$$RCH_2CH{=}CH_2 \longrightarrow RCH_2C{\equiv}CH$$

Novel eliminations include a retro Diels-Alder scheme

BuLi R^1Br; R^2MgX PTSA; Δ → $R^1\text{-}C{\equiv}C\text{-}R^2$

(B. Tarnchampoo, Y. Thebtaranonth and S. Utampanaya, *Chem. Lett.*, 1981, 1214) and the transformation of 1,1-dimethyl-alkenes into alkynes under mild conditions (S. Abidi, *Tetrahedron Lett.*, 1986, **27**, 267). A mechanistic study (E.J.

Corey *et al.*, *ibid.*, 1987, **28**, 4921) has demonstrated selectivity to the isopropyl moiety.

OH $\xrightarrow[H_2O]{NaNO_2/AcOH}$ OH

The observation of phenylethynol has been reported (A.J. Kresge *et al.*, *J. Am. Chem. Soc.*, 1989, **111**, 2355) and the photochemical generation of ethynol described (R. Hochstrasser and J. Wirz, *Angew. Chem., Int. Ed. Engl.*, 1989, **28**, 181).

(iii) From Metal Acetylides

The interaction of alkynyl anions with electrophilic species is the most versatile synthetic route to functionalised alkynes. The search for chemoselectivity, regio- and stereo- control has resulted in the development of an array of anionic reagents modified by differing counter cations, complexing agents and catalytic systems, which, together with variations in solvent systems and reaction conditions, provide powerful tools in synthesis.

Lithium alkynides, from terminal alkynes and organolithium compounds or LDA , are the usual source of acetylides and also precursors of acetylides of other metals. Nucleophilic displacement of halogen is facilitated for monolithium acetylide-EDA complex by the use of polar aprotic solvents such as HMPA (W.J. De Jarlais and E.A. Emken, *Synth. Commun.*, 1980, **10**, 653) and DMSO (J. Salaum, J.-P. Barnier and B. Karkour, *J. Chem. Soc., Chem. Commun.*, 1985, 1269; S.K. Chiu and P.E. Peterson, *Tetrahedron Lett.*, 1980, **21**, 4047). Iodoalkynes are available from lithium acetylides (A. Ricci *et al.*, *Synthesis*, 1989, 461; J. Barluenga *et al.*, *ibid.*, 1987, 661). Lithium alkynides afford chiral α-amino-α,β-ynones in good yield by nucleophilic displacement of the heterocyclic moiety in isooxazolidines of *N*-protected amino acids (T.L. Cupps, R.H. Boutin and H. Rapoport, *J. Org. Chem.*, 1985, **50**, 3972).

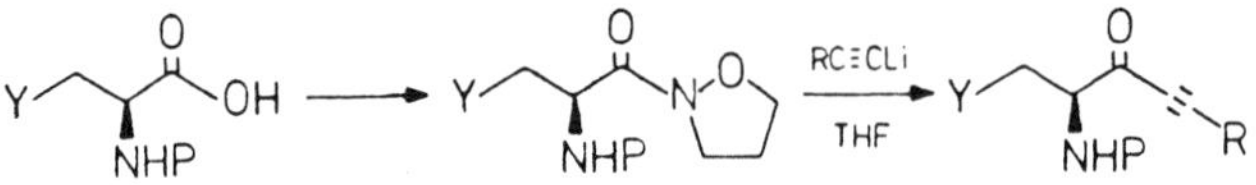

Metallation at C-1 of propargyl chloride has been achieved using methyl lithium at low temperature (M. Olomucki, J.-Y. Le Gall and I. Barrand, *J. Chem. Soc., Chem. Commun*, 1982, 1290). Stereoselection in the addition of lithium alkynides to aldehydes has been reported using chiral complexing agents (T. Mukaiyama and K. Suzuki, *Chem. Lett.*, 1980, 255) and the application of the Felkin model for non-chelated transition states to chiral aldehydes with bulky substituents has been described.

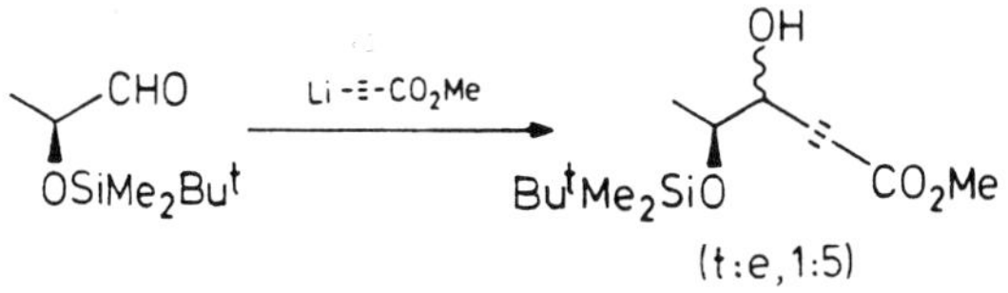

(t:e,1:5)

(M. Hirama *et al., J. Chem. Soc., Chem. Commun.*, 1986, 393)

Boron trifluoride etherate has been shown to enhance the reactivity of lithium acetylides with oxygen functions. Ring opening of oxetanes to novel γ-hydroxyalkynes (M. Yamaguchi, Y. Nobayashi and I. Hirao, *Tetrahedron*, 1984, **40**, 4261), and the conversion of (E)-allylic alcohols to erythro acetylenic diols *via* attack at C-1 of the epoxide are of note (M. Yamaguchi, I. Hirao, *J. Chem. Soc., Chem. Commun*, 1984, 202).

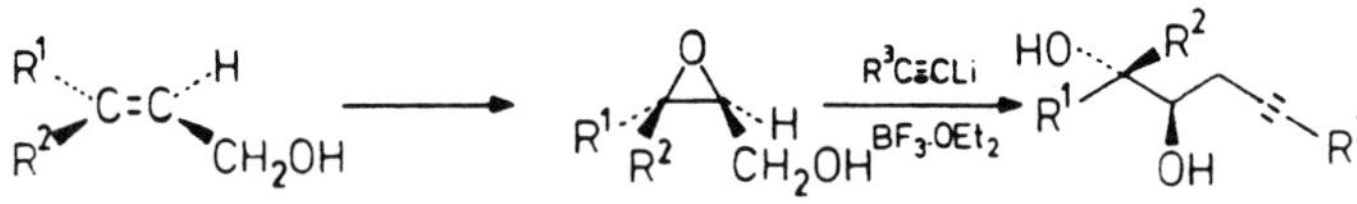

Conjugated ynones are available from lithium acetylides with dimethylamides (M. Yamaguchi, T. Waseda and I. Hirao, *Chem. Lett.*, 1983, 35; K. Suzuki, T. Ohkhuma and J. Tsuchihashi, *J. Org. Chem.*, 1987, **52**, 2929; H.D.

Verkruijsse, Y.A. Heus-Kloos and L. Brandsma, *J. Organomet. Chem.*, 1988, **388**, 298) and novel 2-alkynyl-5-methoxy-1,4-benzoquinones from 4,5-dimethoxy-1,2-benzoquinones (K.F. West and H.W. Moore, *J. Org. Chem.*, 1982, **47**, 3591).

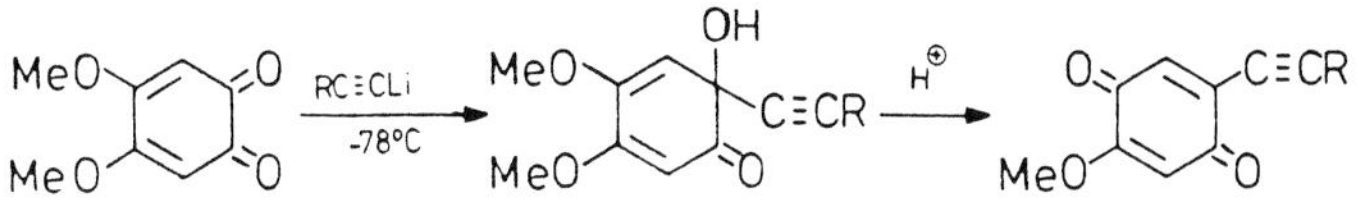

Conjugated enynes have been prepared from lithium acetylides and phosphonium salts (Y. Shen and W. Qiu, *J. Chem. Soc., Chem. Commun.*, 1987, 703).

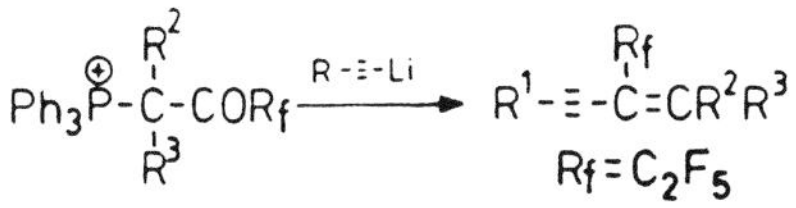

Lithium acetylides undergo addition to iminium salts derived from cyclic thioamides in a synthesis of nitrogen heterocycles (H. Takahata *et al.*, *J. Chem. Soc. Perkin Trans. I*, 1989, 1211).

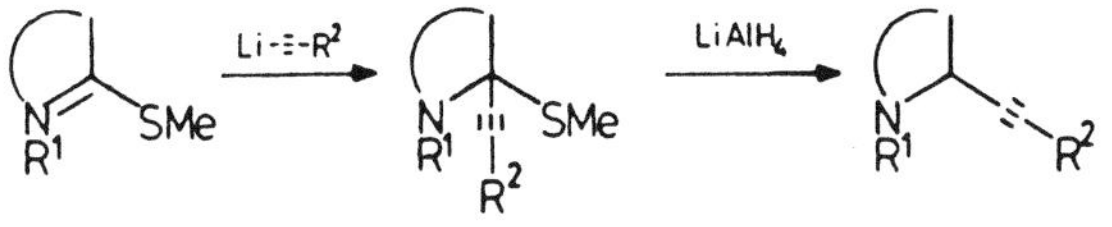

Propargyl amine derivatives have been prepared from lithium acetylides (A.R. Katritzky, J.K. Gallos and K. Yannakapolou, *Synthesis*, 1989, 31; J. Barluenga, P.J. Campose and G. Canal, *ibid.*, 1989, 33).

Copper acetylides displace halogen from alkynyl halides (J.A. Miller and G. Zweifel, *Synthesis*, 1983, 128), vinyl halides (G. Struve and S. Seltzer, *J. Org. Chem.*, 1982, **47**, 2109), propargylic halides (W. Boland and K. Mertes, *Synthesis*, 1985, 705) and alkyl halides (J.C. Miller and E.W. Underhill, *Can. J. Chem.*, 1986, **64**, 2427) to yield coupled products. Alkynyl cuprates complexed with dimethylsulphide readily couple with propargylic tosylates (R. Baker, M.J. O'Mahoney and C.J. Swain, *J. Chem. Res. (S)*,

1984, 190).

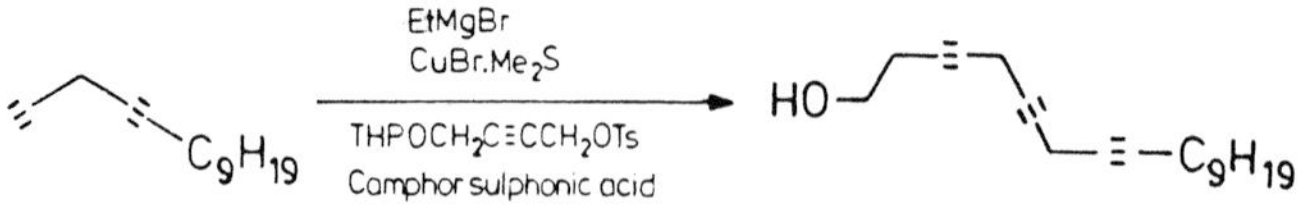

Treatment of 2-chloromethyltetrahydrofuran with lithium in ammonia affords the dianion of pent-4-yne-1-ol which has been reacted with epoxides (J.S. Yadav, P.R. Krishna and M.K. Gurjar, *Tetrahedron*, 1989, **45**, 6263).

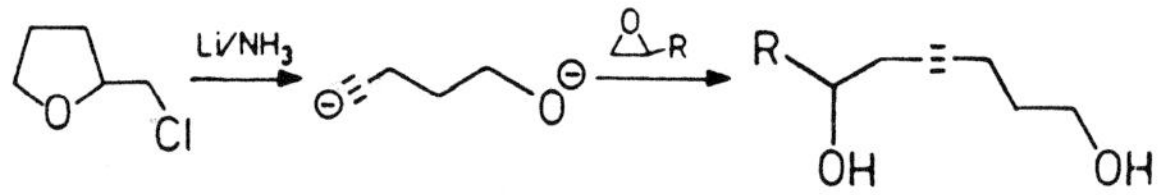

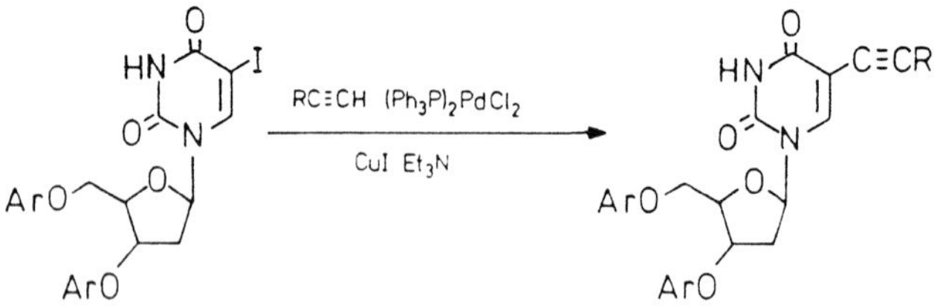

The use of palladium triphenylphosphine complexes in the presence of cuprous iodide and amines allows the direct alkylation of terminal alkynes with vinyl and aryl halides under mild conditions. Mono-alkynation of 1,1-dichloroethene (G. Linstrumelle *et al., Tetrahedron Lett.*, 1987, **27**, 5857) and di-alkynation of (Z)-1,2-dichloroethene (K.P.C. Vollhardt and L.S. Winn, *ibid.*, 1985, **26**, 709) have been achieved and the sequential alkynation of 2,5-dibromopyridine reported (J.W. Tilley and S. Zawoiski, *J. Org. Chem.*, 1988, **53**, 386). The utility of this approach is the tolerance of other functional groups *i.e.* hydroxyl (A. Bumagin and I.P. Beletskaya, *Synthesis*, 1984, 728), carbonyl (H. Yamanaka, *Heterocycles*, 1986, **24**, 31), lactone (A. Krantz *et al., J. Am. Chem. Soc.*, 1984, **106**, 6849) and deoxyribonucleoside derivatives (M.J. Robins and P. J. Barr, *Tetrahedron Lett.*, 1981, **22**, 421) in either the alkyne or alkylating function.

The chemoselectivity of zinc alkynides in coupling has been used to advantage in the preparation of conjugated diynes (E.-i. Negishi *et al., J. Org. Chem.*, 1984, **49**, 2629).

$$RC{\equiv}CH \xrightarrow[(E)\text{-}ClCH{=}CHCl,\ Pd(PPh_3)_4]{BuLi,\ ZnCl_2} RC{\equiv}CCH{=}CHCl \xrightarrow{NaNH_2} RC{\equiv}CC{\equiv}C^{\ominus}Na^{\oplus} \xrightarrow{E^{\oplus}} RC{\equiv}CC{\equiv}CE$$

Alkynylstannanes yield alkynylketones with acid chlorides in a Pd(0) catalysed reaction (J.K. Stille, *Angew. Chem., Int. Ed. Engl.*, 1986, **25**, 508).

$$R^1\text{-}{\equiv}\text{-}SnBu^n_3 \xrightarrow{R^2COCl/Pd(0)} R^1\text{-}{\equiv}\text{-}\overset{O}{\overset{\|}{C}}\text{-}R^2$$

The intermediacy of alkynylboron compounds has been reviewed extensively (M.M. Midland and D.C. McDowell, *Prepr. Div. Pet. Chem., Am. Chem. Soc.*, 1979, **24**, 176; H.C. Brown and J.B. Campbell Jr., *ibid.*, 1979, **24**, 185; H.C. Brown and M. Zaidlewicz, *Chem. Stosow.*, 1982, **26**, 155; E.-i. Negishi and M.J. Idacavage, *Org. Reactions*, 1985, **33**, 1). Treatment of alkynylborate complexes with iodine or iodine/oxidant is a versatile route to mono-and di-substituted alkynes, 1-iodoalkynes and alkynols (H.C. Brown *et al., J. Org. Chem.*, 1986, **51**, 4507, 4514, 4518, 4521).

$$R^1\text{-}C{\equiv}CLi \xrightarrow{R_3B} R^1\text{-}C{\equiv}C\text{-}\overset{\ominus}{B}R_3Li^{\oplus} \xrightarrow{I_2} R^1\text{-}C{\equiv}C\text{-}R$$

Silylated alkynes when reacted with acyl halides and Lewis acid catalysts provide access to mono- and di-tertiary alkylated acetylenes

$$R\text{-}C{\equiv}C\text{-}SiMe_3 \xrightarrow[AlCl_3]{R^1_3CHal} R\text{-}C{\equiv}C\text{-}CR^1_3$$

(G. Capozzi *et al.*, *J. Chem. Soc., Chem. Commun.*, 1982, 959; *Gazz. Chim. Ital.*, 1985, **115**, 311) and, with acetals, yield ethers in excellent yield (T. Mukaiyama, *Chem. Lett.*, 1987, 1975) with favourable stereoselectivity.

$$\xrightarrow[TiCl_4/CH_2Cl_2/-78°C]{R^2\text{-}C{\equiv}C\text{-}SiMe_3}$$

(W.S. Johnson, R. Elliott and J.D. Elliott, *J. Am. Chem. Soc.*, 1983, **105**, 2904)

2-Methyl-3-butyn-2-ol behaves as an equivalent of acetylene in the synthesis of aryl (E.T. Sabourin and A. Onopchenko, *J. Org. Chem.*, 1983, **48**, 5135) and diaryl (A. Carpita, A. Lessi and R. Rossi, *Synthesis*, 1984, 571) acetylene derivatives.

$$Me_2C(OH)\text{-}{\equiv} \xrightarrow{ArX/Pd(PPh_3)_4/CuI} Ar\text{-}{\equiv}\text{-}C(OH)Me_2 \xrightarrow{Ar^1X} Ar\text{-}{\equiv}\text{-}Ar^1$$

Silylated 1,3-enynes and 1,5-diyne-3-enes have been synthesised (R. Rossi *et al.*, *Tetrahedron*, 1989, **45**, 5621). 1,3-Enynes are prepared from stannanes (J.K. Stille and J.H. Simpson, *J. Am. Chem. Soc.*, 1987, **109**, 2138).

$$Me_3Si\text{-}{\equiv}\text{-}SnMe_3 \xrightarrow{X\text{-}CH{=}CH\text{-}Bu/Pd(PPh_3)_4} Me_3Si\text{-}{\equiv}\text{-}CH{=}CH\text{-}Bu$$

Propargyl acetate may be readily converted to non-conjugated dialkynes *via* its complex with cobalt hexacarbonyl and a trialkynylalane (S. Padnambhan and K.M. Nicholas, *Tetrahedron Lett.*, 1983, **24**, 2239) and β-ketoalkynes are formed in the nickel catalysed addition of monoalkynylalanes to α,β-unsaturated carbonyl compounds (J. Schwartz *et al.*, *J. Org. Chem.*, 1980, **45**, 3053). Functionalised alkynylstannanes are precursors to conjugated ynones by treatment with acyl halides (G. Himbert and L. Henn, *Tetrahedron Lett.*, 1981, **22**, 2637; M. Logue and K. Teng, *J. Organometal. Chem.*, 1982, **184**, 317).

New methods include the intermediacy of vanadium trichloride in the acylation of alkynes (T. Hirao, D. Misu and T. Agawa, *Tetrahedron Lett.*, 1986, **27**, 933), regioselective opening of substituted oxiranes *via* titanium alkynides (N. Krause and D. Seebach, *Chem. Ber.*, 1988, **121**, 1315) and the chemoselective addition of alkyne moieties to dicarbonyl compounds in the presence of chromium(II) (K. Takai *et al.*, *Tetrahedron Lett.*, 1985, **26**, 5585).

H O O BuC≡CI $CrCl_2$/DMF Bu OH O

Lead acetylides react with β-ketoesters (M.G. Maloney, T. Pinhey and E.G. Roche, *ibid.*, 1986, **27**, 5025).

O CO_2Et R-≡-Pb(OAc)$_3$ O R CO_2Et

Acetylenic esters are prepared from terminal alkynes, carbon monoxide and alcohols in the presence of palladium dichloride (J. Tsuji, M. Takahashi and T. Takahashi, *ibid.*, 1980, **21**, 849).

R-≡ —CO/MeOH/$PdCl_2$/Cu(II)→ R-≡-CO_2Me

A review of acylsilanes (A. Ricci and A. Degl'Innocenti, *Synthesis*, 1989, 647) includes acylsilane derivatives of acetylenic acids.

Selenoalkynes are available from terminal alkynes (T. Hayama *et al.*, *Chem. Lett.*, 1982, 1249; S. Tomoda, Y.

Takeuchi and Y. Nomura, *ibid.*, 1982, 252).

$$R\text{-}\equiv \xrightarrow[\text{or PhSeBr/AgNO}_2]{\text{PhSeCNO/CuCN/Et}_3\text{N}} R\text{-}\equiv\text{-SePh}$$

Alkynyl dialkylphosphates have been prepared from alkynyl(phenyl)iodinium phosphates (P.J. Stang *et al., J. Am. Chem. Soc.*, 1989, **111**, 2225).

(iv) From Allenes

Prototropic rearrangement of allenes has provided a fruitful route to alkynes. Allenes with organometallic reagents produce functionalised terminal alkynes (G. Hahn and G. Zweifel, *Synthesis*, 1983, 883; J. Hooz *et al., Tetrahedron Lett.*, 1985, **26**, 271) and internal alkynes (J.M. Oostven *et al., J. Org. Chem.*, 1980, **45**, 1158).

$$CH_2{=}C{=}C(OMe)(I) \xrightarrow[\text{THF}/-85°\text{C}]{\text{Bu}^t\text{CuBr.MgCl}} Bu^tCH_2\text{-}C{\equiv}C\text{-}OMe$$

Incorporation of the allene function in the organometallic moiety as a borane allows the synthesis of β-hydroxyalkynes from carbonyl compounds with excellent stereocontrol (H. Yamamoto *et al., J. Am. Chem. Soc.*, 1982, **104**, 7667; 1986, **108**, 483; E.J. Corey *et al., ibid.*, 1984, **106**, 3875).

THPO ... CHO $\xrightarrow{CH_2{=}C{=}C(Et)(BEt_2)}$ THPO ... OH

Isomerisation of alkynes to terminal alkynes *via* the "acetylene zipper" continues to demonstrate synthetic utility (M. Midland *et al., Tetrahedron Lett.*, 1981, **22**, 4171; S.A. Abrams, *Can. J. Chem.*, 1984, **62**, 1333; A.V. Rao, *Tetrahedron*, 1986, **42**, 4523).

Triphenylstannylallenes and α,β-unsaturated ketones provide alkynes in the presence of titanium(IV) chloride (J.-i. Haruta *et al.*, *J. Chem. Soc., Chem. Commun.*, 1989, 1065).

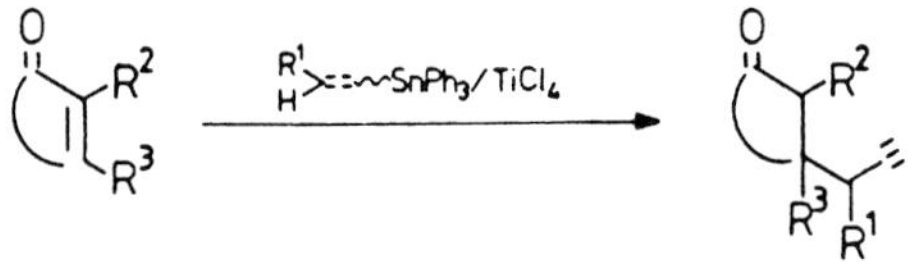

(2) Reactions of Alkynes

(i) Addition Reactions

The addition reactions of alkynes giving alkenes has continued to attract considerable attention. Numerous methods for the formal addition of the components X-Y across the alkynyl bond have been developed.

Addition reactions to alkynes to yield heterocycles or adducts which have subsequently been transformed into heterocycles are considered in Section (iii).

The reduction of terminal and internal alkynes affording alkenes has been described by several groups of workers and reviewed "Best Synthetic Methods-Hydrogenation Methods", P.N. Rylander, Academic Press, Orlando, 1985. Many methods of reduction have been employed including using zinc-copper in boiling methanol (B.L. Sondengam, G. Charles and T.M. Akam, *Tetrahedron Lett.*, 1980, **21**, 1069), hydrogen in the presence of Ni-graphite (A. Umani-Ronchi *et al.*, *J. Org. Chem.*, 1981, **46**, 5344), hydrogen in the presence of Pd-graphite (D. Savoia *et al.*, *J. Chem. Soc., Chem. Commun.*, 1981, 540), niobium(V) chloride/ sodium aluminium hydride (M. Sato and K. Oshima, *Chem. Lett.*, 1982, 157), modified palladium catalyst (R.A.W. Johnstone and A.H. Wilby, *Tetrahedron*, 1981, **37**, 3667), modified Lindlar catalyst (S.

Dev *et al.*, *ibid.*, 1983, **39**, 2315), palladium dichloride/ sodium borohydride (N. Suzuki *et al.*, *J. Chem. Soc., Chem. Commun.*, 1983, 515; *Tetrahedron*, 1985, **41**, 2387), the action of lithium aluminium hydride on 4-(trimethylsilyl)-3-butyn-2-ol (M.L. Mancini and J.F. Honek, *Tetrahedron Lett.*, 1983, **24**, 4295) and propargyl alcohols (J.W. Blunt *et al.*, *Aust. J. Chem.*, 1983, **36**, 581, 1387), calcium and methylamine (R.A. Benkeser and F.G. Belmonte, *J. Org. Chem.*, 1984, **49**, 1662), hydrogen in the presence of a complex formed from palladium(II) acetate (J.-J. Brunet and P. Caubere, *ibid.*, 1984, **49**, 4058), hydrogen in the presence of nickel(II) acetate, sodium hydride and alkoxide (J.J. Brunet, P. Gallois and P. Caubere, *ibid.*, 1980, **45**, 1937, 1947), reduction at the 3,4-position of 1,3-diyne-5-ol derivatives (R.E. Doolittle, *Synthesis*, 1984, 730), reduction at the 3,4-position of terminal 1,3-diynes (M.H.P.J. Aerssens and L. Brandsma, *J. Chem. Soc., Chem. Commun.*, 1984, 735), reduction of 1-trimethylsilyl-2-ynes with bis(1,2-dimethylpropyl)borane or diisobutylaluminium hydride followed by protonolysis (S. Rajagopalan and G. Zweifel, *Synthesis*, 1984, 113), the action of diisobutylaluminium hydride on ynones (T. Tsuda *et al.*, *J. Org. Chem.*, 1987, **52**, 1624), reduction with hydrogen in the presence of epitaxial palladium (J.G. Ulan, W.F. Maier and D.A. Smith, *ibid.*, 1987, **52**, 3132) and reduction by interlamellar montmorillonite-diphenylphosphine Pd(II) complex (B.M. Choudary *et al.*, *ibid.*, 1989, **54**, 2997). Organometallic complexes of titanium yield alkenes from alkynes and water (B. Demerseman and P.H. Dixneuf, *J. Chem. Soc., Chem. Commun.*, 1981, 665).

Conjugated diynes can be reduced *via* hydroalumination giving 1,3-dienes or 1,3-enynes (J.A. Miller and G. Zweifel, *J. Amer. Chem. Soc.*, 1983, **105**, 1383). Reductive dimerisation of terminal alkynes also yield 1,3-dienes (N. Satyanarayana and M. Periasamy, *Tetrahedron Lett.*, 1986, **27**, 6253; S.A. Rao and M. Periasamy, *J. Chem. Soc., Chem. Commun.*, 1987, 495).

1-Bromo-1,3-dienes are available from alkynes *via* organoboron intermediates (A. Suzuki *et al.*, *Tetrahedron Lett.*, 1986, **27**, 977; *Chem. Lett.*, 1986, 459).

A mixture of lithium aluminium hydride and titanium(III)

chloride affords a substituted cumulene (H.M. Walborsky and H.H. Wust, *J. Am. Chem. Soc.*, 1982, **104**, 5807) and a similar reaction gave a highly substituted cyclobutane (M. Iyoda, Y. Kuwatani and M. Oda, *ibid.*, 1989, **111**, 3761).

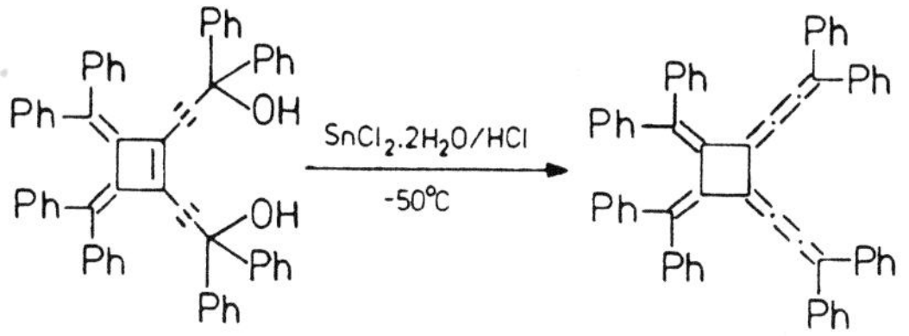

$SnCl_2.2H_2O/HCl$, -50°C

Halogenation of alkynes has been achieved with sulphuryl chloride (S. Uemura *et al.*, *Bull. Chem. Soc. Jpn*, 1981, **54**, 2843), tetrabutylammonium tribromide (J. Berthelot and M. Fournier, *Can. J. Chem.*, 1986, **16**, 1357), amberlyst A26 (perbromide form) (A. Bongini *et al.*, *Synthesis*, 1980, 143) and iodine in the presence of aluminium oxide (G.W. Kabalka *et al.*, *Tetrahedron Lett.*, 1988, **29**, 35). Tetrabutylammonium iodide and aldehydes yield iodoalkene derivatives from acetylenic ketones (M. Taniguchi and T. Hino, *Tetrahedron Lett.*, 1986, **27**, 4767).

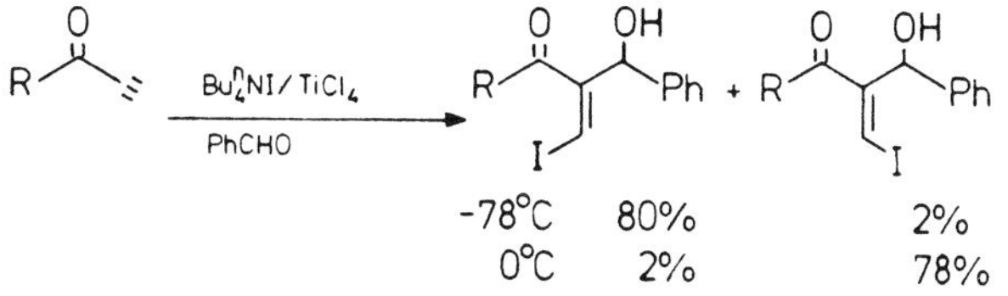

$Bu^n_4NI/TiCl_4$, PhCHO

-78°C	80%	2%
0°C	2%	78%

Interhalogens (FI, FBr) add to alkynes (S. Rozen and M. Brand, *J. Org. Chem.*, 1986, **51**, 222). Addition of hydrogen bromide to terminal alkynes has been achieved in the presence of tetraethylammonium bromide (J. Cousseau, *Synthesis*, 1980, 805). Nitrosyl chloride adds to acetylenic esters (M.M. Siddiqui, F. Ahamad and S.M. Osman, *J. Chem. Res. (S)*, 1984, 186) as does methylmercury cyanate (F. De Sarlo *et al.*, *J. Organometal. Chem.*, 1984, **269**, 115). Fluorinated alkene derivatives have been prepared from alkynes (T. Kitazume and T. Ikeya, *J. Org. Chem.*, 1988, **53**, 2350).

$$R^1\text{-C}\equiv\text{C-}R^2 \xrightarrow{RfCF_2I/\text{enzyme}} R^1CH{=}C(R^2)CF_2Rf$$

Addition of hydrogen cyanide to alkynes yielding α,β-unsaturated nitriles occurs in the presence of a nickel catalyst (W.R. Jackson *et al.*, *Synthesis*, 1987, 1032).

$$R^1\text{-C}\equiv\text{C-}SiR^2_3 \xrightarrow{Ni(0)/HCN} R^1(NC)C{=}CH(SiR^2_3)\ (\text{major}) + R^1HC{=}C(SiR^2_3)CN$$

Alkyldinitriles have been prepared by the formal addition of hydrogen cyanide and hydrogen to diynes and vinyl cyanides are formed from the nickel promoted addition of hydrogen cyanide to alkynylsilanes (T. Funabiki, Y. Sato and S. Yoshida, *Bull. Chem. Soc. Jpn.* 1983, **56**, 2863).

Trimethylsilylcyanide addition to arylalkynes is achieved in the presence of palladium or nickel catalysts.

$$R^1\text{-C}\equiv\text{C-}R^2 \xrightarrow{Me_3SiCN/PdCl_2/Py} R^1(NC)C{=}C(R^2)SiMe_3$$

(N. Chatani and T. Hanafusa, *J. Chem. Soc., Chem. Commun.*, 1985, 838; P. Perlmutter *et al.*, *ibid.*, 1985, 4; P. Moreau and A. Commeyras, *ibid*, 1985, 817; N. Chatani *et al.*, *J. Org. Chem.*, 1988, **53**, 3539).

Hydration of alkynes to methyl ketones has been reported using tellurium (N.X. Hu *et al.*, *Tetrahedron Lett.*, 1986, 27, 6099) and mercury chemistry (V. Janout and S.L. Regen, *J. Org. Chem.*, 1982, **47**, 3331). Hydration of alkynes has also been achieved in a palladium and ultrasound mediated reaction (K. Imi, K. Imai and K. Utimoto, *Tetrahedron Lett.*,

1987, **28**, 3127). Hydration of arylacetylenes with sodium sulphide under aqueous acidic conditions yields arylketones (M.J Chapdelaine, P.J. Warwick and A. Shaw, *J. Org. Chem.*, 1989, **54**, 1218). 1,3-Diketones have also been prepared by hydration of diynes (M.G. Constantino, P.M. Donate and N. Petragnani, *Tetrahedron Lett.*, 1982, **23**, 1501).

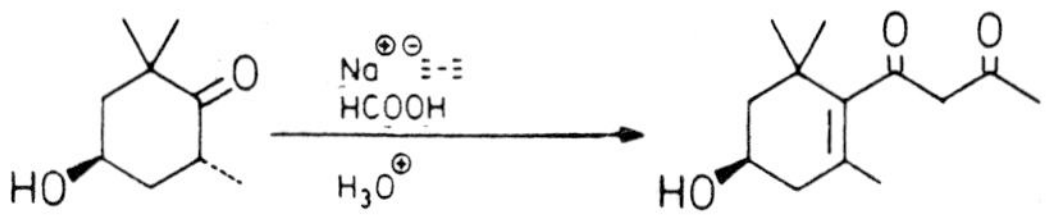

Esters of 2-hydroxyketones are available from alkynyliodonium salts (M. Ochiai *et al.*, *J. Org. Chem.*, 1989, **54**, 4038).

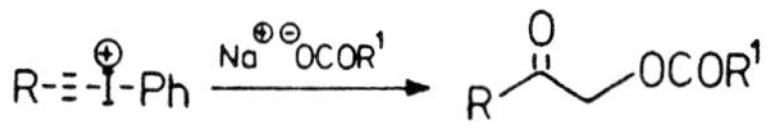

The preparation of the *t*-butyl ester of 2,3-dioxo-pent-3-ynoic acid has been achieved. This alkyne reacts with nucleophiles at both the 2- and 5-positions (H.H. Wasserman *et al.*, *J. Org. Chem.*, 1989, **54**, 6012).

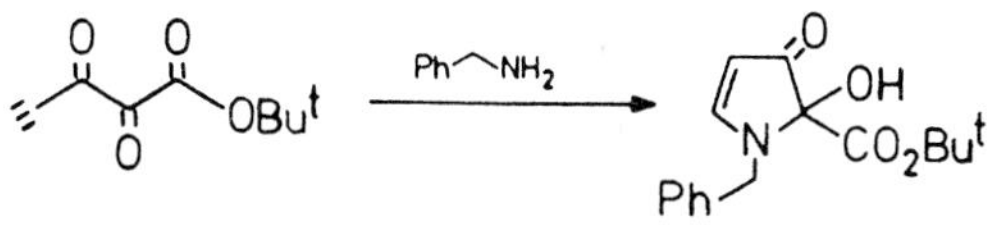

The base promoted intramolecular addition of an anion derived from a β-keto-ester to an alkyne provides a route to tricyclic ketone derivatives (J.-F. Lavallee and P. Deslongschamps, *Tetrahedron Lett.*, 1987, **28**, 3457).

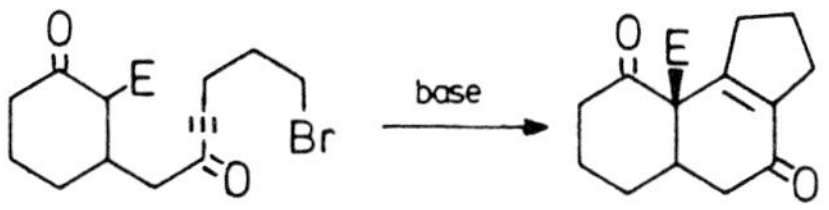

Esters of maleic and fumaric acids have been prepared from acetylene. Mono-substituted alkynes behave similarly and di-substituted alkynes afford only mono-esters (H. Alper, B. Despeyroux and J.B. Woell, *ibid.*, 1983, **24**, 5691).

$$H\text{-}C{\equiv}C\text{-}H \xrightarrow{ROH/CO/PdCl_2/O_2/CuCl_2} (RO_2C)HC{=}CH(CO_2R)$$

Addition of chlorofluorocarbene to alkynes under aqueous conditions yields cyclopropenones (E.V. Dehmlow and A. Winterfeldt, *Tetrahedron*, 1989, **45**, 2925).

$$R^1\text{-}{\equiv}\text{-}R^2 \xrightarrow{Cl_2FCH/NaOH/PTC} \text{cyclopropenone } (R^1, R^2)$$

The preparation of di- and tri-substituted alkenes from the corresponding mono- and di-substituted alkynes *via* organoborane chemistry (E.-i. Negishi and M.J. Idacavage, *Org. Reactions*, 1985, **33**, 1) has continued to attract attention (H.C. Brown *et al.*, *J. Org. Chem.*, 1982, **47**, 171, 754, 1792, 3806, 3808, 5407; *ibid.*, 1986, **51**, 5270, 5282; H.C. Brown, D. Basavaiah and S.U. Kulkarni, *J. Organometal. Chem.*, 1982, **225**, 63; A. Latters *et al.*, *Tetrahedron*, 1982, **38**, 2355; *Tetrahedron Lett.*, 1982, **23**, 2785; H.C Brown and T. Imai, *Organometallics*, 1984, **3**, 1392; B. Singaram, T.E. Cole and H.C. Brown, *ibid.*, **3**, 1520; *Synthesis*, 1984, 303, 919, 920; K.K. Wang *et al.*, *Tetrahedron Lett.*, 1987, **28**, 1003, 1007).

Boron chemistry provides a route to vinyl halides (H.C. Brown, N.G. Bhat and S. Rajogopalan, *Synthesis*, 1986, 480; A. Suzuki *et al.*, *Tetrahedron Lett.*, 1983, **24**, 731, 735; A. Suzuki *et al.*, *J. Am. Chem. Soc.*, 1985, **107**, 5225; H.C. Brown *et al.*, *J. Org. Chem.*, 1989, **54**, 6064, 6068) and vinyl mercury compounds (R.C. Larock and K. Narayanan, *J. Org. Chem.*, 1984, **49**, 3411).

Boron derivatives of alkynes have provided α,β-unsaturated aldehydes and α,β-unsaturated carboxylic acids.

$$R^1\text{-}{\equiv}\text{-}Li \xrightarrow{R^2_3B} R^1\text{-}{\equiv}\text{-}\overset{\ominus}{B}R^2_3 \xrightarrow[AcOH]{CO_2} (R^2)HC{=}C(R^1)CO_2H$$

(D. Min-zhi, T. Yong-ti and X. Wei-hua, *Tetrahedron Lett.*, 1984, **25**, 1797)

$$R^1\text{-}\equiv\text{-}\overset{\ominus}{B}R^2_3 \xrightarrow[\text{Pr}^i\text{CO}_2\text{H};\ \text{HgO/BF}_3]{\text{benzodithiolium}^{\oplus}\ \text{BF}_4^{\ominus}} R^1(H)C{=}C(R^2)CHO$$

(A. Pelter, R. Rupanie and P. Stewart, *J. Chem. Soc., Chem. Commun.*, 1981, 164)

(E)-Vinylsulphides have been prepared from iodoalkynes *via* organoboron intermediates (M. Hoshi, Y. Masuda and A. Arase, *J. Chem. Soc., Chem. Commun.*, 1985, 1068). Mixtures of (Z)- and (E)-vinylsulphides were prepared from terminal alkynes (K. Oshima *et al.*, *Chem. Lett.*, 1987, 1647) and substituted vinylsulphides are available from diethylaminoprop-1-yne (V.H.M. Elferink, R.G. Visser and J.T. Bos, *Rec. Trav. Chim.*, 1981, **100**, 414).

$$Et_2N\text{-}\equiv\text{-}Me \xrightarrow{ArCS_2R/MeCN/\Delta} Et_2NC(O)(Me)C{=}C(SR)Ar$$

Vinylsulphides obtained by addition of thiophenol to propargyl alcohols are transformed into α,β- unsaturated aldehyde derivatives when treated with acid (M. Julia and C. Lefebvre, *Tetrahedron Lett.*, 1984, **25**, 189).

$$RR^1C(OH)\text{-}\equiv \xrightarrow{PhSH} RR^1C(OH)CH{=}CHSPh \xrightarrow{H_3O^{\oplus}} RR^1C{=}CH\text{-}CHO$$

Addition of 4′-nitrobenzenesulphenanilide to alkynes in the presence of boron trifluoride and acetic acid yields products of acetoxysulphenylation *via* thiiranium intermediates.

$$R^1\text{-}\equiv\text{-}R^2 \xrightarrow[BF_3\cdot Et_2O/AcOH]{4\text{-}NO_2PhNHSPh/} \left[\text{thiirenium}(S^{\oplus}Ph;\ R^1, R^2)\right] \longrightarrow (PhS)R^1C{=}C(OAc)R^2$$

(L. Benati *et al.*, *J. Chem. Soc. Perkin Trans. I*, 1989, 1113)

$$R^1\text{-}\equiv\text{-}R^2 \xrightarrow[\text{hydrol.}]{ArNR^3SPh/BF_3.Et_2O/MeCN} (PhS)R^1C{=}C(R^2)N{=}C(Me)NR^3Ar$$

(L. Benati, P.C. Montevecchi and P. Spagnolo, *ibid.*, 1989, 1105)

Vinylselenides are available from reduction of the corresponding alkynylselenides with lithium aluminium hydride or by addition of phenylselenol to alkynes (J.V. Comasseto, J.T.B. Ferreira and N. Petragnani, *J. Organometal. Chem.*, 1981, **216**, 287).

Vinylstannanes have been prepared from acetylenes α,β-acetylenic *NN*-dimethylamides and acetylenic ketones.

$$H\text{-}C{\equiv}C\text{-}H \xrightarrow[RX]{Ph_3SnCu} (H)(Ph_3Sn)C{=}C(H)(R)$$

(P. Vermeer *et al.*, *Tetrahedron Lett.*, 1982, **23**, 2797)

$$R^1\text{-}\equiv\text{-}\overset{O}{\overset{\|}{C}}NMe_2 \xrightarrow[R^2X]{Me_3SnCu.Me_2S} (R^1)(Me_3Sn)C{=}C(CONMe_2)(R^2)$$

(E. Piers, J.M. Chong and B.A. Keay, *ibid.*, 1985, **26**, 6265)

$$R^1\text{-}\equiv\text{-}CO_2R^2 \xrightarrow[\text{or }[Me_3SnCuY]Li]{Me_3SnCu.Me_2S} (Me_3Sn)(R^1)C{=}C(H)(CO_2R^2) + (Me_3Sn)(R^1)C{=}C(CO_2R^2)(H)$$

(E. Piers, J.M. Chong and H.E. Morton, *Tetrahedron*, 1989, **45**, 363)

$$R-\equiv \xrightarrow[E^{\oplus}]{Bu_3SnAlEt_2/Cu(I)\ or\ Pd(0)} (Z)\text{-}R(Bu_3Sn)C{=}CH(E) + (E)R\text{-}CH{=}\ldots$$

(S. Sharma and A.C. Oehlschlager, *J. Org. Chem.*, 1989, **54**, 5064)

$$R^1-\equiv-C(=O)-R^2 \xrightarrow{(Me_3Sn)_2/(Ph_3P)_4Pd} R^1(Me_3Sn)C{=}CH\text{-}C(=O)R^2$$

(E. Piers and R.D. Tillyer, *J. Chem. Soc. Perkin Trans. I*, 1989, 2124)

$$R^1-\equiv \xrightarrow[Pd(PPh_3)_4]{R^2_3SnSiMe_2Ph/} R^1(R^2_3Sn)C{=}CH(SiMe_2Ph) \xrightarrow{Bu^n_4NF} R^1(R^2_3Sn)C{=}CH_2$$

(K. Ritter, *Synthesis*, 1989, 218)

$$R-\equiv-SiMe_3 \xrightarrow[Pd(PPh_3)_4]{Bu_3SnH/} R(Bu_3Sn)C{=}CH(SiMe_3) \xrightarrow{MeCOCl/AlCl_3} R(Bu_3Sn)C{=}CH(COMe);\ \xrightarrow{I_2} R(I)C{=}CH\text{-}SiMe_3$$

(B.L. Chenard and C.M. Van Zyl, *J. Org. Chem.*, 1986, **51**, 3561)

The addition of organometallic silane reagents to alkynes followed by treatment with electrophiles provides a versatile route to substituted vinylsilanes.

$$R-\equiv \xrightarrow[E^{\oplus}]{(Me_2PhSi)_2CuLi.LiCN} R(E)C{=}CH\text{-}SiMe_2Ph$$

(I. Fleming, T.W. Newton and F. Roessler, *J. Chem. Soc. Perkin Trans. I*, 1981, 2527)

$$R\text{-}\equiv \xrightarrow[E^{\oplus}]{PhMe_2SiMgMe/Pt} R(E)C=CH\text{-}SiMe_2Ph$$

(F. Sato *et al.*, *Tetrahedron Lett.*, 1983, **24**, 1041; K. Oshima *et al.*, *J. Am. Chem. Soc.*, 1983, **105**, 4491)

Vinylsilanes are available from 1-trimethylsilylalkynes.

$$R^1\text{-}\equiv\text{-}SiMe_3 \xrightarrow[CuI/P(OEt)_3,\ R^2X]{HAlBu^i_2,\ MeLi} R^1(H)C=C(SiMe_3)R^2$$

(F.E. Ziegler and K. Mikami, *Tetrahedron Lett.*, 1984, **25**, 131; K. Oshima *et al.*, *ibid.*, 1984, **25**, 3217, 3221)

$$R\text{-}\equiv\text{-}SiMe_3 \xrightarrow{ArI/Pd(OAc)_2(PPh_3)_2} Ar(R)C=C(H)SiMe_3$$

(A. Arcadi, S. Cacchi and F. Marinelli, *ibid.*, 1986, **27**, 6397)

Vinyl acetates and related compounds are prepared from silver carboxylates (Y. Ishino *et al.*, *Chem. Lett.*, 1981, 641).

$$R^1\text{-}\equiv\text{-}R^2 \xrightarrow{Ag^{\oplus}\ {}^{\ominus}OCOR^3} R^1C(OCOR^3)=CHR^2$$

A variety of substituted alkenes and α,β-unsaturated compounds are available from alkynes using organocopper additions [J.F. Normant *et al.*, *Pure Appl. Chem.*, 1983, **55**, 1759; "An Introduction to Synthesis Using Organocopper Reagents", G. Posner, Wiley-Interscience, New York, 1980; *Tetrahedron*, 1989, **45**(2)].

MeO_2C-≡-CO_2Me $\xrightarrow[NH_4Cl\ aq.]{RCu.Me_2S}$ (R)(H)C=C(MeO_2C)(CO_2Me)

(H. Nishiyama, M. Sasaki and K. Itoh, *Chem. Lett.*, 1981, 905; M.T. Crimmins, S.W. Mascarella and J.A. DeLoach, *J. Org. Chem.*, 1984, **49**, 3033)

R^1-≡ $\xrightarrow[R^3COX/(Ph_3P)_4Pd]{R^2CuMgX_2}$ R^2C(R^1)=CH–CO–R^3

(J.P. Marino and R.J. Linderman, *ibid.*, 1983, **48**, 4621)

R^1-≡ $\xrightarrow{R_3CuMet.(LiBr)n/H^{\oplus}}$ $R^1R^2C=CH_2$

(P. Vermeer *et al., Rec. Trav. Chim.*, 1981, **100**, 98)

R^1-≡ $\xrightarrow[ClCH_2NMeCHO]{R_2CuLi}$ (H)(R^2)C=C(R^1)CH_2N(Me)CHO

(C. Germon, A. Alexakis and J.F. Normant, *Synthesis*, 1984, 40, 43)

H-≡-H $\xrightarrow{R_2CuLi/R^1X}$ R–CH=CH–CH=CH–R^1

(J.P. Foulon, M. Bourgain-Commercon and J. F. Normant, *Tetrahedron*, 1986, **42**, 1389, 1399; M. Furber, R.J.K. Taylor and S.C. Burford, *J. Chem. Soc. Perkin Trans. I*, 1986, 1809)

Terminal alkynes have been converted into α,β-unsaturated ketones (N. Jabri, A. Alexakis and J.F. Normant, *ibid.*, 1983, **24**, 5081; Y. Tamaru, H. Ochiai and Z.-i. Yoshida, *ibid.*, 1984, **25**, 3861).

$$Bu^n\text{-}\equiv \xrightarrow[H_2O]{CO/Pd(PPh_3)_2Cl_2/ZnCu/Cp_2TiCl_2} Bu^nCH{=}CHCOAr$$

α,β-Unsaturated nitriles are available from alkynylnitriles and either organo-silver or copper reagents (P. Vermeer *et al., J. Organometal. Chem.*, 1981, **206**, 257).

$$R^1\text{-}\equiv\text{-}CN \xrightarrow[H_3O^{\oplus}]{R_2MetMgCl} R^1(R^2)C{=}C(CN)H$$

Hydroalumination and carboalumination of alkynes has been reviewed and synthetic uses of the resulting organo-aluminium compounds described (G. Zweifel and J. A. Millar, *Org. Reactions*, 1984, **32**, 375). Carbometallation reactions of alkynes as a method for the stereospecific synthesis of alkenyl derivatives has been reviewed (J.F. Normant and A. Alexakis, *Synthesis*, 1981, 841) as have carbometallation catalysts (E.-i. Negishi, *Pure Appl. Chem.*, 1981, **53**, 2333). Carbotitanation of is achieved with titanium(IV) chloride and trimethylaluminium (M.D. Schiavelli, J.J. Plunkett and D.W. Thompson, *J. Org. Chem.*, 1981, **46**, 807).

$$\text{(alkynol)} \xrightarrow[MeOH/H_2SO_4]{Me_3Al,\ TiCl_4} \text{(alkenol)}$$

Hydromagnesation of alkynes provides substituted alkenes (F. Sato, H. Ishikawa and M. Sato, *Tetrahedron Lett.*, 1981, **22**, 85).

Trialkylaluminium compounds and zirconium reagents yield alkene derivatives from terminal alkynes (E.-i. Negishi and H. Matsushita, *J. Am. Chem. Soc.*, 1981, **103**, 2882, T. Yoshida and E.-i. Negishi, *ibid.*, 1981, **103**, 4985; E.-i. Negishi, D.E. Van Horn and T. Yoshida, *ibid.*, 1985, **107**, 6639).

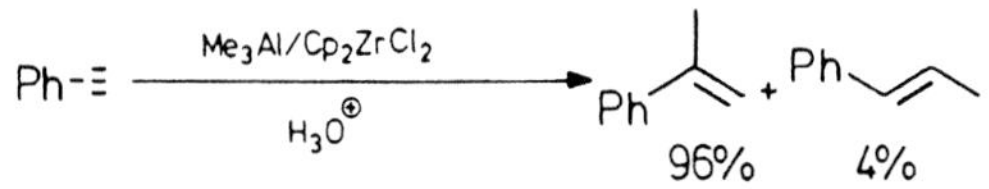

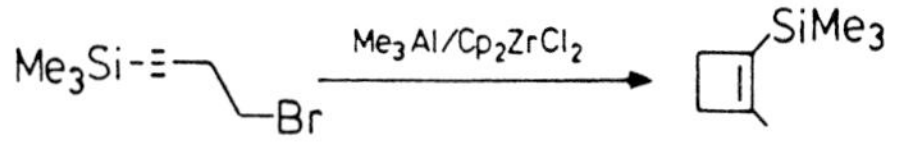

(T.J. Zitzelberger, M.D. Schiavelli and D.W. Thompson, *J. Org. Chem.*, 1983, **48**, 4781)

Spiroketals, such as milbemycin derivatives, are often synthesised *via* acetylenic intermediates (F. Perron and K.F. Albizati, *Chem. Rev.*, 1989, **89**, 1617).

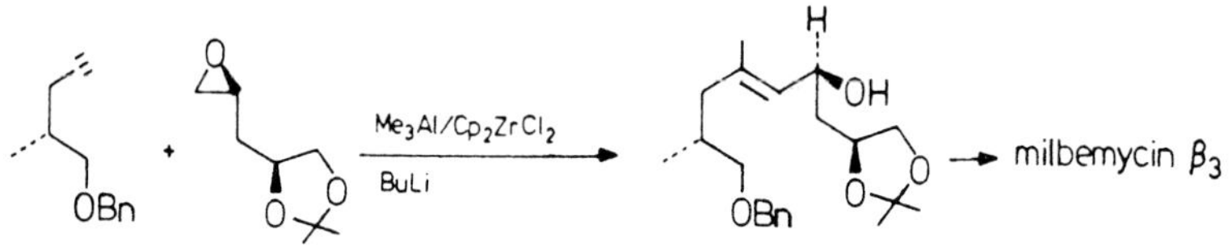

(P.J. Kocienski *et al., J. Chem. Soc. Perkin Trans. I*, 1987, 2189)

Vinyl iodides are available by the application of aluminium chemistry (E.-i. Negishi and J.A. Millar, *Tetrahedron Lett.*, 1984, **25**, 5863; E.-i. Negishi, D.E. Van Horn and T. Yoshida, *J. Am. Chem. Soc.*, 1985, **107**, 6639).

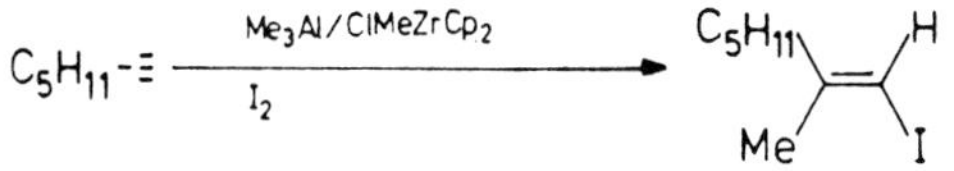

Addition of nucleophiles to a but-2-yne complex of iron yields a number of alkenic products (D.L. Reger and P.J. McElligott, *J. Am. Chem. Soc.*, 1980, **102**, 5923).

$$[Fe]^{\oplus}\text{—}(MeC{\equiv}CMe)\ BF_4^{\ominus} \xrightarrow{Nu^{\ominus}} [Fe](Me)C{=}C(Me)Nu \xrightarrow{X_2} X(Me)C{=}C(Me)Nu$$

$[Fe]$ = $(Cp)Fe(CO)(PPh_3)$

Allylic bromides add to alkynes in the presence of zinc yielding a great variety of highly functionalised 1,4-dienes (P. Knochel and J.F. Normant, *J. Organomet. Chem.*, 1986, **309**, 1; *Tetrahedron Lett.*, 1984, **25**, 1475, 4383).

$$Me_3Si\text{-}C{\equiv}C\text{-}CH_2CH_2Br \xrightarrow{CH_2{=}CHCH_2ZnBr} \text{(allyl)cyclobutenyl-}SiMe_3$$

Propargyl bromide and ketones yield allenic alcohols in the presence of tin(II) (M. Iyoda *et al.*, *Bull. Chem. Soc. Jpn*, 1989, **62**, 3380).

$$R^1R^2C{=}O + BrCH_2C{\equiv}CH \xrightarrow{SnCl_2.2H_2O} \underset{\text{major}}{R^1R^2C(OH)CH{=}C{=}CH_2} + R^1R^2C(OH)CH_2C{\equiv}CH$$

The addition of mercury(II) salts to alkynes and subsequent transformation of the adduct provides routes to a variety of products.

$$HC{\equiv}C\text{-}OEt \xrightarrow[\text{Zn, RCHO, } H_2O]{\text{pyridine N-oxide}/HgCl_2} RCH(OH)CH(Cl)CO_2Et$$

(T. Mukaiyama and M. Murakami, *Chem. Lett.*, 1981, 1129)

$$R^1\text{-}Y\text{-}CH_2C{\equiv}CH + R^2R^3NH \xrightarrow{HgX_2/K_2CO_3} R^1\text{-}YCH{=}C(NR^2R^3)Me \xrightarrow{\text{hydrol.}} R^1YCH_2COMe$$

(J. Barluenga, F. Aznar and R. Liz, *Synthesis*, 1984, 304)

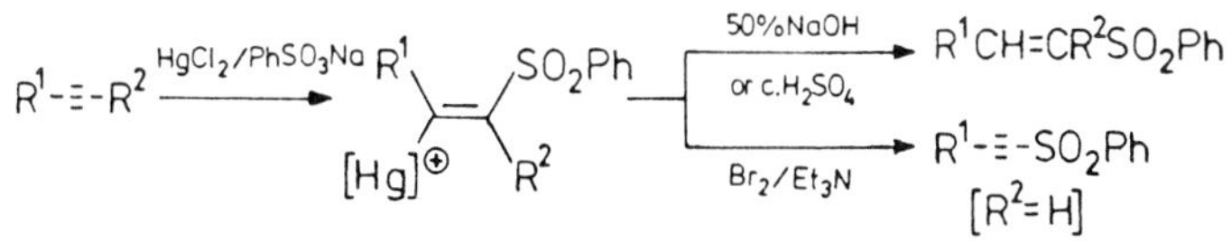

(P. Rajakumar and A. Kannan, *J. Chem. Soc., Chem. Commun.*, 1989, 154)

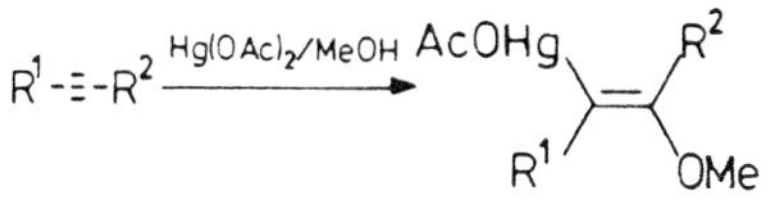

(M. Bassetti, B. Floris and G. Spadafora, *J. Org. Chem.*, 1989, **54**, 5934)

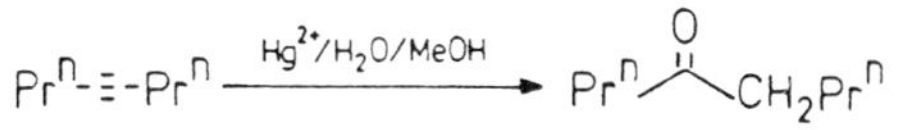

(M. Bassetti and B. Floris, *Gazz. Chim. Ital.*, 1986, **116**, 595)

Phenylselenoalkenylsulphones have been prepared from alkynes (T.G. Back, S. Collins, R.G. Kerr, *J. Org. Chem.*, 1983, **48**, 3077; T.G. Back *et al.*, *ibid.*, 1983, **48**, 4776; T.G. Back, M.V. Krishna and K.R. Muralidharan, *ibid.*, 1989, **54**, 4146).

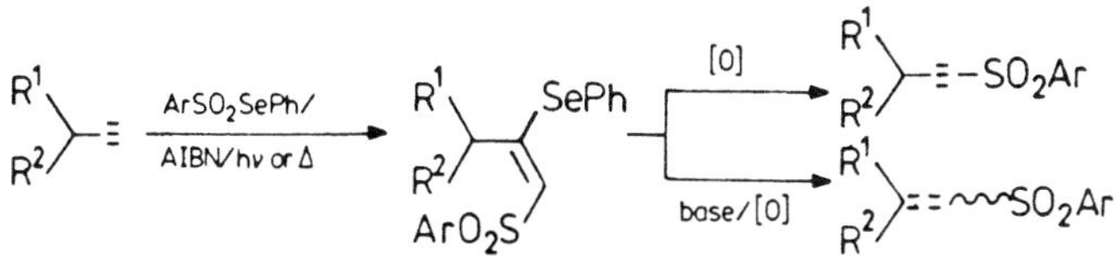

Tin, and other non-carbon radicals (E. Nakamura, D. Machii and T. Inubushi, *J. Am. Chem. Soc.*, 1989, **111**, 6849; K. Nozaki *et al.*, *ibid.*, 1987, **109**, 2547; H. Pak, J.K. Dickson Jr. and B. Fraser-Reid, *J. Org. Chem.*, 1989, **54**, 5357; W.P. Newmann, *Synthesis*, 1987, 665) add to alkynes giving substituted alkene derivatives.

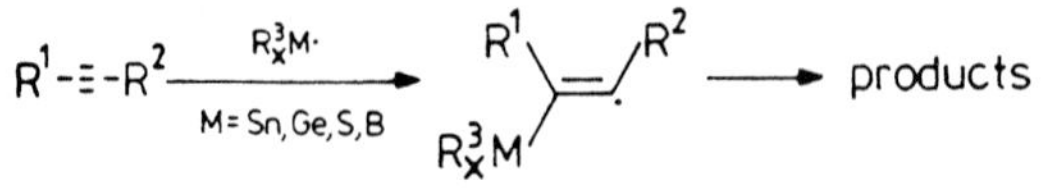

$$R^1\text{-}\equiv\text{-}R^2 \xrightarrow[M=Sn,Ge,S,B]{R^3_xM\cdot} R^1(R^3_xM)C{=}\dot{C}R^2 \longrightarrow \text{products}$$

The intramolecular addition of carbon radicals to alkynes continues to attract interest as a method for preparing carbocycles and intermolecular additions comprise an important method for carbon-carbon bond forming reactions ("Radicals in Organic Synthesis: Formation of Carbon-Carbon Bonds", B. Giese, Pergamon, Oxford, 1986; D.P. Curran, M.-H. Chen and D. Kim, *J. Am. Chem. Soc.*, 1989, **111**, 6265; K. Last and H.M.R. Hoffman, *Synthesis* 1989, 901). The addition of tin radicals to alkynes has been developed into a useful method for carbocyclic syntheses (G. Stork and R. Mook, *J. Am. Chem. Soc.*, 1987, **109**, 2829).

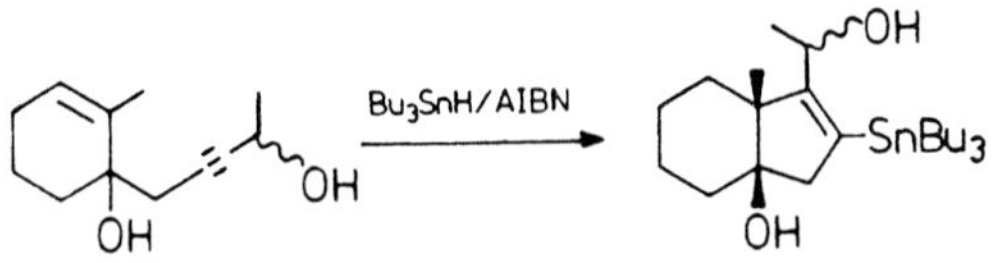

$\xrightarrow{Bu_3SnH/AIBN}$ OH, OH, $SnBu_3$

Methylenecyclopentannulation reactions have been achieved using free-radical reactions (D.P. Curran, M.-H. Chen and D. Kim, *ibid.*, 1986, **108**, 2489; D.P. Curran *et al.*, *ibid.*, 1989, **111**, 8872).

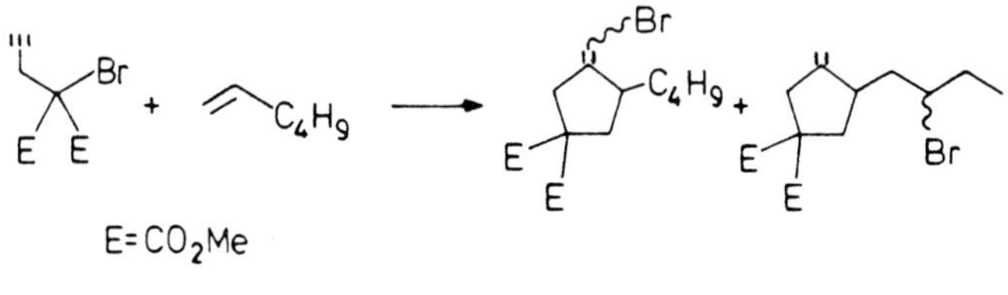

Br + C_4H_9 ⟶ Br, C_4H_9 + Br

E=CO_2Me

(ii) Metal Mediated Reactions

Metal mediated reactions which result in the formal addition of the X-Y across an alkyne giving alkene derivatives are considered in Section (i) above.

Para-quinone derivatives and related compounds have been prepared from metalcarbene complexes (K.H. Dotz and I. Pruskil, *J. Organomet. Chem.*, 1981, **209**, C4; W.D. Wulff, P.-C. Tang and J.S. McCallum, *J. Am. Chem. Soc.*, 1981, **103**, 7677; W.D. Wulff *et al.*, *Tetrahedron Lett.*, 1987, **28**, 1381; K.H. Dotz and M. Popall, *Tetrahedron*, 1985, **41**, 5797; M.F. Semmelhack *et al.*, *ibid*, 1985, **41**, 5803; W.D. Wulff *et al.*, *ibid.*, 1985, **41**, 5813; L.S. Liebeskind *et al.*, *ibid.*, 1985, **41**, 5839; A. Yamashita and T.A. Scahill, *J. Org. Chem.*, 1989, **54**, 3625; K.H. Dotz, *Angew. Chem., Int. Ed. Engl.*, 1984, **23**, 587) and also from metallocycles (L.S. Liebeskind, S.L. Baysdon and M.S. South, *J. Am. Chem. Soc.*, 1980, **102**, 7397; S.L. Baysdon and L.S. Liebeskind, *Organometallics*, 1982, **1**, 771; L.S. Liebeskind *et al.*, *ibid.*, 1986, **5**, 1086).

Ph(MeO)C=Cr(CO)$_5$ $\xrightarrow{R^1-\equiv-R^2}$ [naphthalene-1,4-diol, 2-R^1, 3-R^2]–Cr(CO)$_5$ $\xrightarrow{Ag_2O}$ 2-R^1-3-R^2-1,4-naphthoquinone

[phthaloyl metallocycle] $\xrightarrow{R^1-\equiv-R^2}$ 2-R^1-3-R^2-1,4-naphthoquinone

[M] = $Fe(CO)_4$; $CoCl(PPh_3)_2$

1-Naphthol derivatives have been synthesised from alkynes and substituted phthaldialdehydes (J.B. Hartung Jr. and S.F. Pedersen, *J. Am. Chem. Soc.*, 1989, **111**, 5468).

4,5-X$_2$-phthaldialdehyde $\xrightarrow[KOH]{R^1-\equiv-R^2/Nb(III)}$ 6,7-X$_2$-2-R^1-3-R^2-1-naphthol

Thioether carbene complexes yield substituted naphthalene derivatives (A. Yamashita *et al.*, *J. Org. Chem.*, 1989, **54**, 4481).

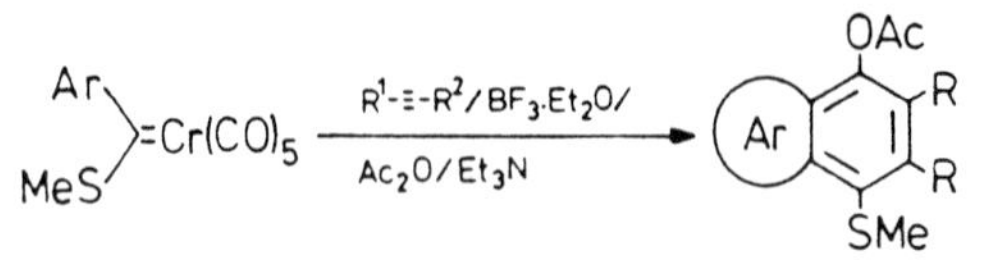

Carbene complexes offer a route to cyclopropylium cation derivatives (M. Brookhardt *et al.*, *J. Am. Chem. Soc.*, 1980, **102**, 7802).

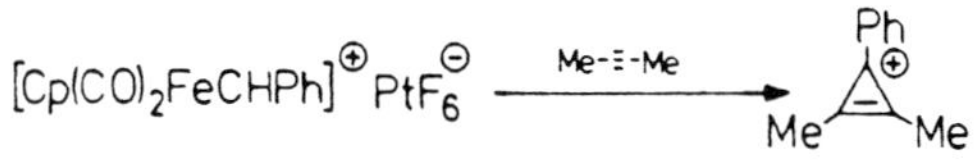

An amino-carbene complex and diethylacetylene affords an aminoindene derivative (K.H. Dotz, D. Grotjahn and K. Harms, *Angew. Chem., Int. Ed. Engl.*, 1989, **28**, 1381).

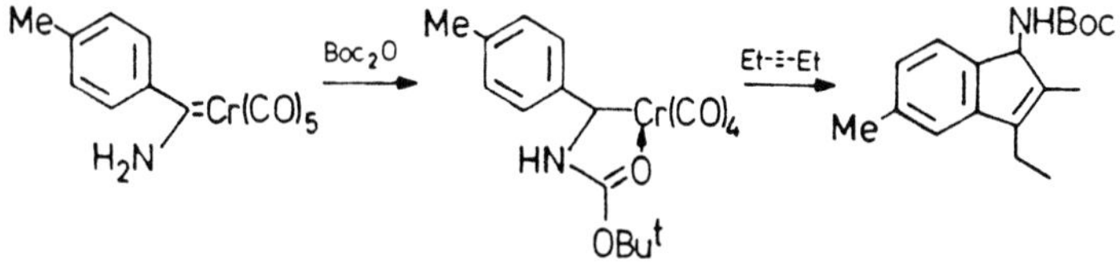

Several products are isolated from the reaction of acetylenic carbene complexes and alkynes (Y.-C. Xu *et al.*, *J. Am. Chem. Soc.*, 1989, **111**, 7269).

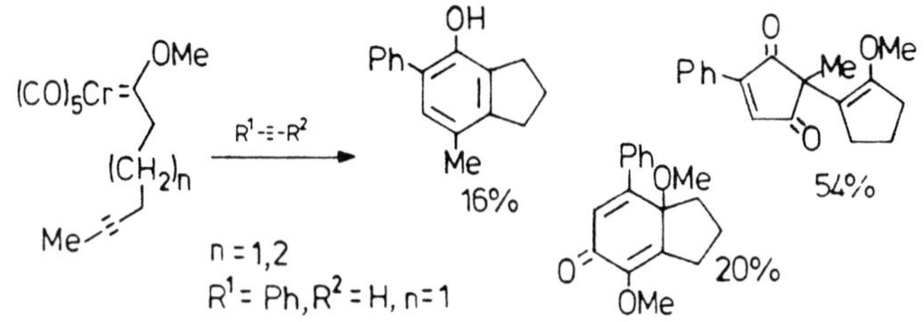

Numerous reactions have been reported in which non-conjugated di-ynes are converted into benzene derivatives (E.R.F. Gesing, J.A. Sinclair and K.P.C. Vollhardt, *J. Chem. Soc., Chem. Commun.*, 1980, 286; R.L. Funk and K.P.C. Vollhardt, *J. Am. Chem. Soc.*, 1980, **102**, 5245, 5253; R.L. Halterman, N.H. Hguyen and K.P.C. Vollhardt, *ibid.*, 1985, **107**, 1379; R. Grigg, R. Scott and P. Stevenson, *Tetrahedron Lett.*, 1982, **23**, 2691), phenols (W.D. Wulff *et al.*, *J. Am. Chem. Soc.*, 1985, **107**, 1060; T.M. Sivavec and T.J. Katz, *Tetrahedron Lett.*, 1985, **26**, 2159), pyridine derivatives (C.A. Parnell and K.P.C. Vollhardt, *Tetrahedron*, 1985, **41**, 5791) and 2-pyrone derivatives (T. Tsuda *et al.*, *J. Org. Chem.*, 1988, **53**, 3140). Pyridine derivatives are also available from two molecules of an alkyne and one molecule of a nitrile in a similar reaction (Y. Wakatsuki and H. Yamazaki, *Bull. Chem. Soc. Jpn.*, 1985, **58**, 2715).

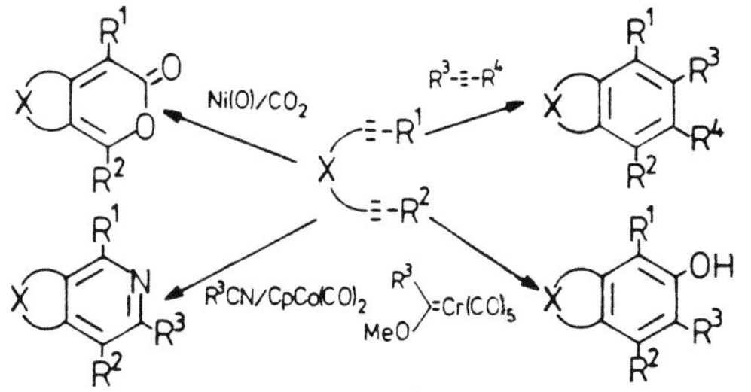

1,3-Dienes have been prepared from both alkynes and non-conjugated diynes (W.A. Nugent and J.C. Calabrese, *J. Am. Chem. Soc.*, 1984, **106**, 6422; T. Takahashi, D.R. Swanson and E.-i. Negishi, *Chem. Lett.*, 1987, 623; K. Tamao, K. Kobayashi and Y. Ito, *J. Am. Chem. Soc.*, 1989, **111**, 6478; T. Tsuda *et al.*, *ibid.*, 1988, **110**, 8570).

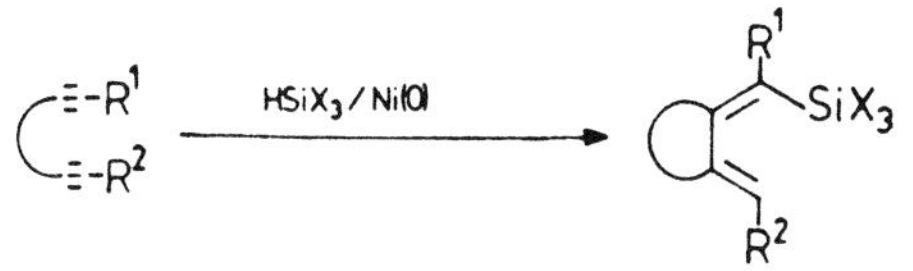

Imines of cyclopentadienone derivatives are available in a related reaction (K. Tamao, K. Kobayashi and Y. Ito, *J. Org. Chem.*, 1989, **54**, 3517) and

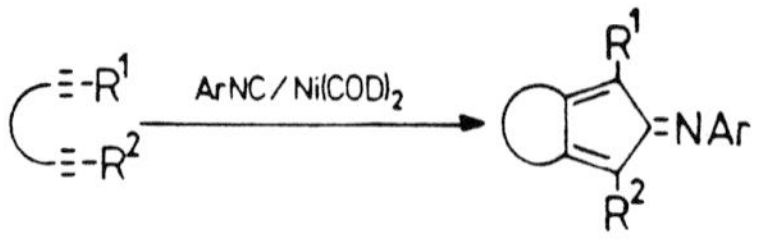

di-*t*-butoxyacetylene has been used in the synthesis of dimethoxycyclopentene trione (A. Bou, M.A. Pericas and F. Serratosa, *Tetrahedron Lett.*, 1982, **23**, 361).

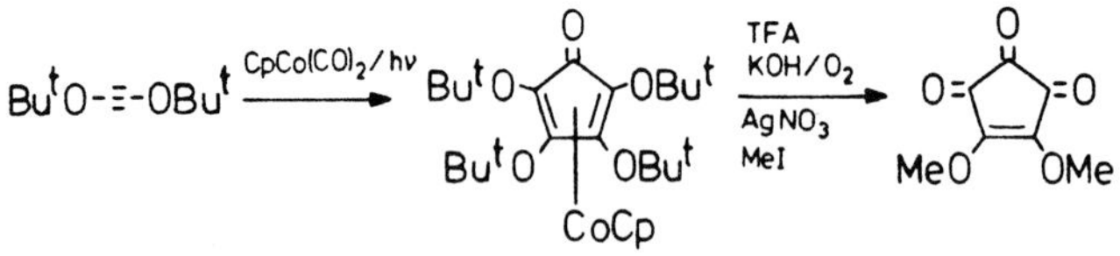

Cyclic enynes have been prepared from non-conjugated diynes (B.M. Trost, S. Matsubara and J.J. Caringi, *J. Am. Chem. Soc.*, 1989, **111**, 8745).

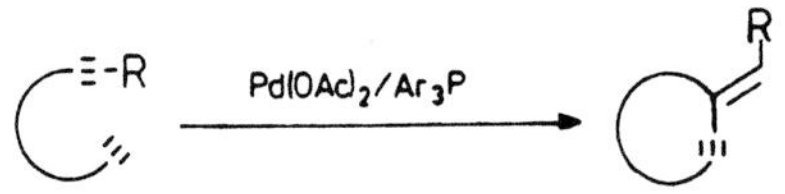

Steroid derivatives (K.P.C. Vollhardt *et al.*, *J. Am. Chem. Soc.*, 1986, **108**, 856; J.-C. Clinet, E. Dunach and K.P.C. Vollhardt, *ibid.*, 1983, **105**, 6710; E.D. Sternberg and K.P.C. Vollhardt, *J. Org. Chem.*, 1982, **47**, 3447; H. Butenschon, M. Winkler and K.P.C. Vollhardt, *J. Chem. Soc., Chem. Commun.*, 1986, 388) have been prepared as well as other carbocycles (E.D. Sternberg and K.P.C. Vollhardt, *J. Org. Chem.*, 1984, **49**, 1564).

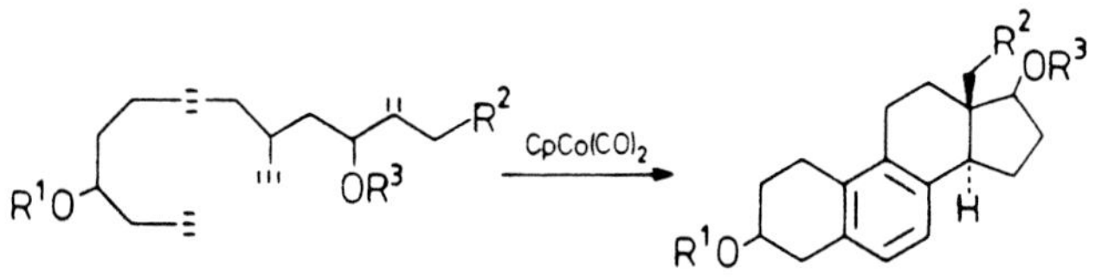

The reaction of non-conjugated eneynes with organometallic reagents gives intermediate metallocycles which subsequently afford cyclopentenones (N.E. Shore and M.J. Knuden, *J. Org. Chem.*, 1987, **52**, 569; E.-i. Negishi *et al.*, *J. Am. Chem. Soc.*, 1985, **107**, 2568; *ibid.*, 1989, **111**, 3336; P. Magnus and L.M. Principe, *Tetrahedron Lett.*, 1985, **26**, 4851; P. Magnus, C. Exon and P. Albaugh-Robertson, *Tetrahedron*, 1985, **41**, 5861) or alkenes when treated with acid (E.-i. Negishi *et al.*, *Tetrahedron Lett.*, 1987, **28**, 917; E.C. Lund and T. Livinghouse, *J. Org. Chem.*, 1989, **54**, 4487).

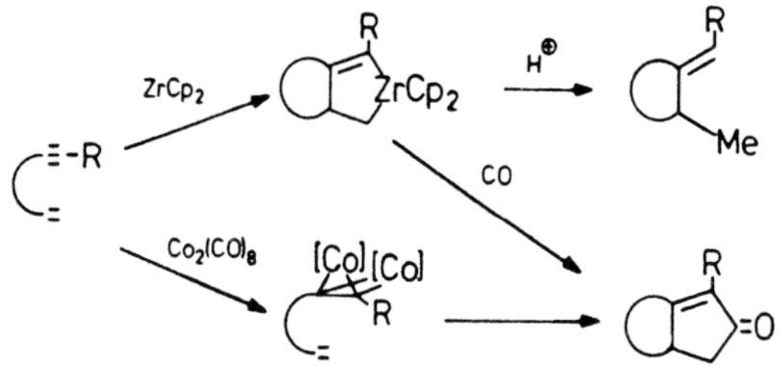

Non-conjugated enynes afforded 1,3-dienes under mild conditions in a palladium mediated reaction (B.M. Trost, E. Edstrom and M.B. Carter-Petillo, *J. Org. Chem.*, 1989, **54**, 4489).

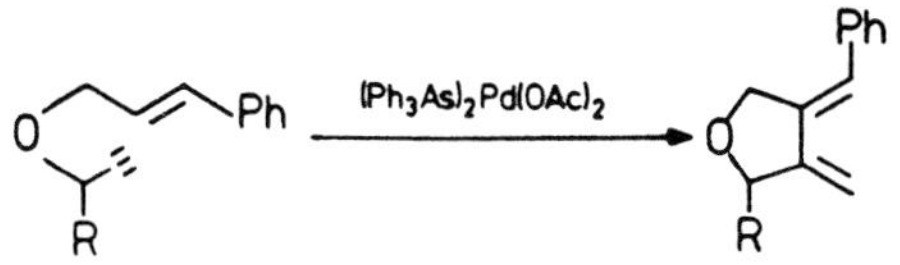

A variation on this theme also yields cyclopentenones (T. Tamao, K. Kobayashi and Y. Ito, *J. Am. Chem. Soc.*, 1988, **110**, 1286).

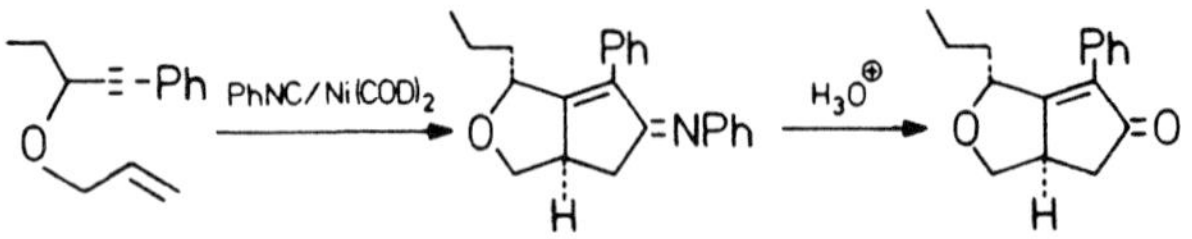

Intermolecular variations of these routes (the Khand reaction) continue to attract interest (P.L. Pauson, *Tetrahedron*, 1985, **41**, 5855; M.E. Krafft, *J. Am. Chem. Soc.*, 1988, **110**, 968; M.E. Price and N.E. Schore, *J. Org. Chem.*, 1989, **54**, 5662).

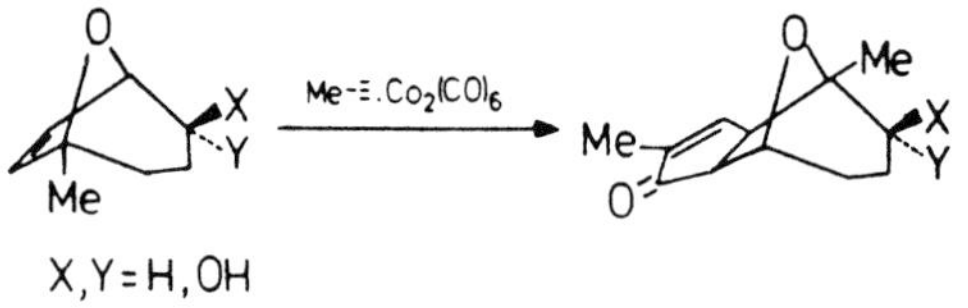

X,Y = H,OH

Cyclopentenones are available from metal carbenes.

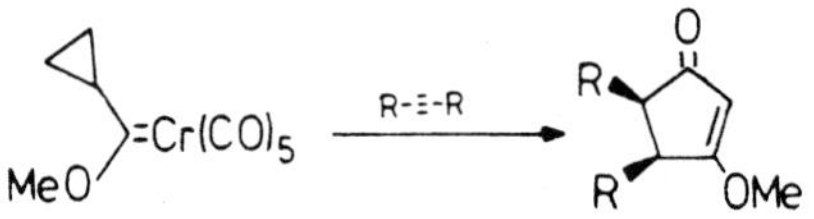

(J.W. Herndon, S.U. Turner and W.K.F. Schnatter, *J. Am. Chem. Soc.*, 1988, **110**, 3334; J.W. Herndon and L.A. McMullen, *ibid.*, 1989, **111**, 6854)

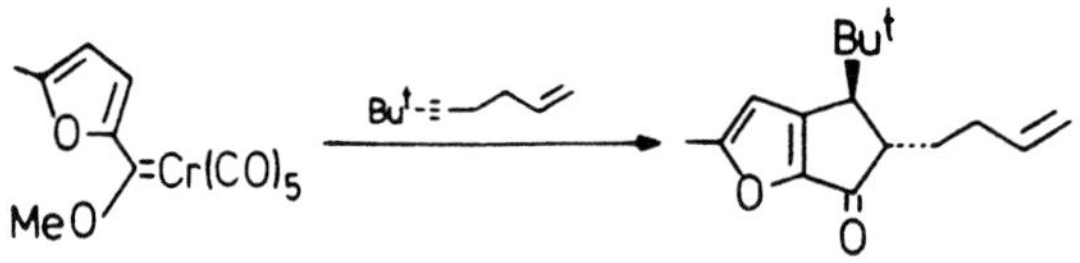

(A. Yamashita *et al.*, *Tetrahedron Lett.*, 1988, **29**, 3403)

Iminium salts and alkynes have also provided a route to cyclopentenones (L. Ghosez *et al.*, *ibid.*, 1984, **25**, 5043).

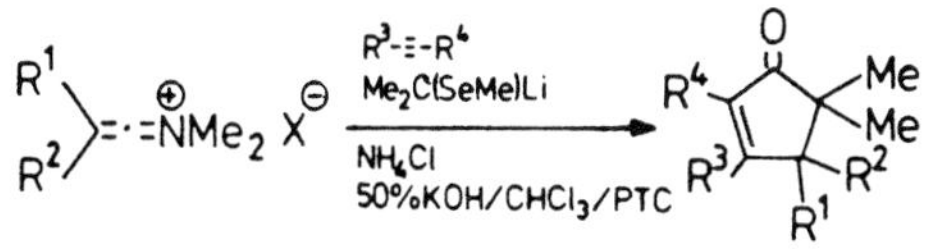

Unsaturated ester derivatives of indane have been prepared by an intramolecular Heck reaction (Y. Zhang and E.-i. Negishi, *J. Am. Chem. Soc.*, 1989, **111**, 3454).

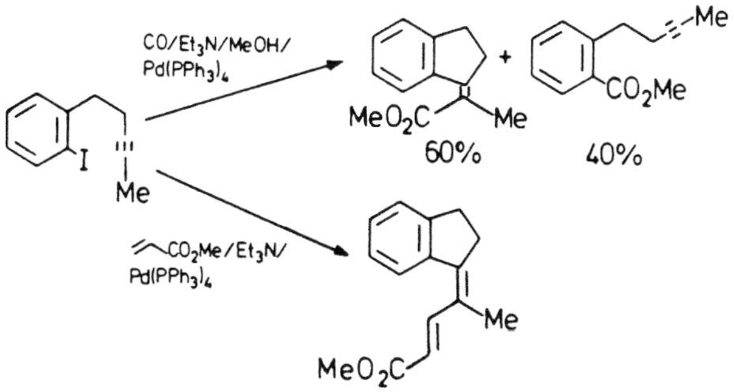

2,3-Diarylindan-1-ones (W.R. Cullen *et al.*, *J. Chem. Soc., Chem. Commun.*, 1987, 439) and tetraphenyl-cyclopentadienone (J.J. Eisch, A.A. Aradi and K.I. Han, *Tetrahedron Lett.*, 1983, **24**, 2073) have been prepared from diphenylacetylene.

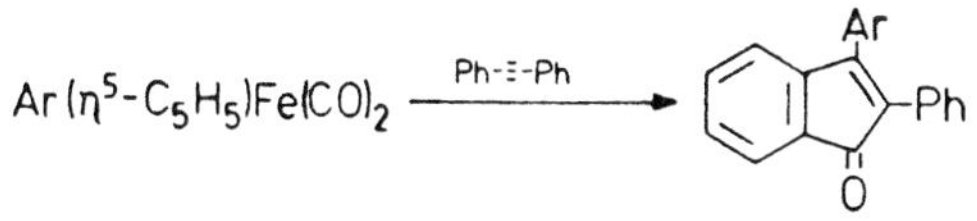

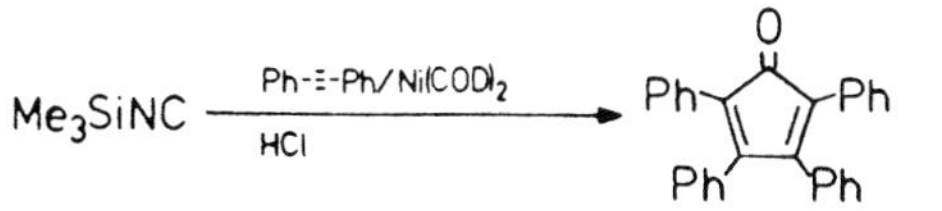

2-Substituted-5-oxo-hexanoic acids have been synthesised from terminal alkynes (H. Alper and J.-F. Petragnani, *J. Chem. Soc., Chem. Commun.*, 1983, 1154).

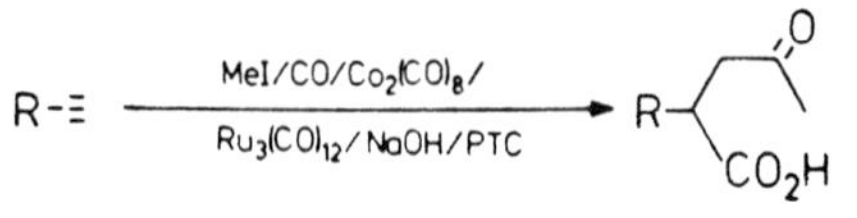

Allenes have been prepared from alkynes and organozinc compounds and in titanium(IV) chloride mediated intramolecular cyclisation reactions.

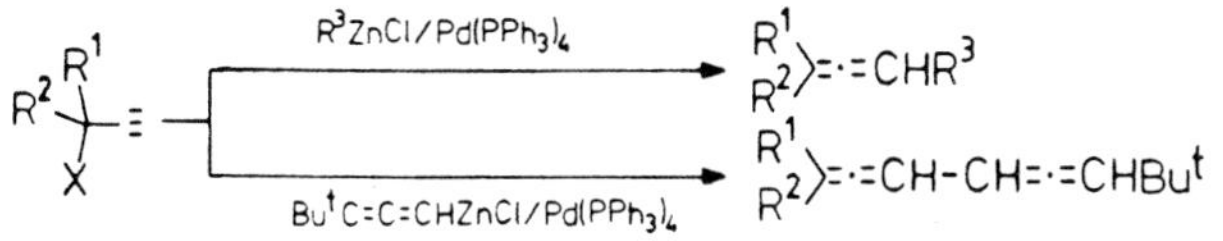

(P. Vermeer *et al.*, *Tetrahedron Lett.*, 1981, **22**, 1451)

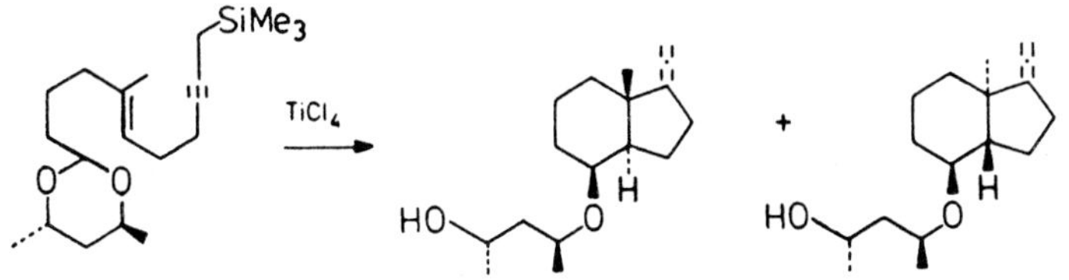

(W.S. Johnson, J.D. Elliott and G.J. Hanson, *J. Am. Chem. Soc.*, 1984, **106**, 1138)

NN-Diethyl β-ketoamides are available from propargylic alcohols, carbon dioxide and diethylamine (Y. Sasaki and P.H. Dixneuf, *J. Org. Chem.*, 1987, **52**, 314, 4389). Propargyl alcohols, carbon monoxide, sulphur and triethylamine give heterocycles (T. Mizuno *et al.*, *Synthesis*, 1989, 770) and alkynes, carbon dioxide and diethylamine yield carbamates

(P. H. Dixneuf *et al.*, *J. Org. Chem.*, 1989, **54**, 1518).

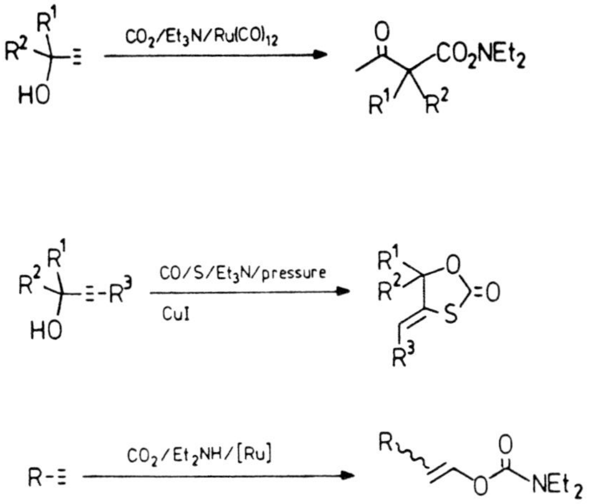

Cyclopentadiene, iron and alkynes yield substituted ferrocenes (R.D. Cantrell and P.B. Shevlin, *J. Am. Chem. Soc.*, 1989, **111**, 2348).

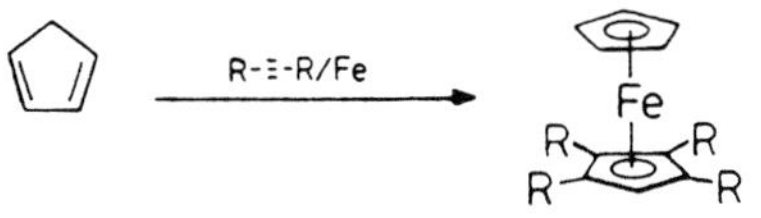

(iii) Heterocyclic Synthesis

A wide variety of substituted furans have been prepared from alkynes. Pyrazole and thiophene derivatives are also available [(a) H. Ishibashi *et al.*, *Tetrahedron Lett.*, 1983, **24**, 3877; (b) K. Itoh and H. Nishiyama, *Heterocycles*, 1984, **22**, 449; (c) J. Koshino, T. Sugawara and A. Suzuki, *ibid.*, 1984, **22**, 489; (d) L. Ghosez *et al.*, *Tetrahedron Lett.*, 1984, **25**, 5043; (e) G. Himbert, S. Kosack and G. Mass, *Angew. Chem., Int. Ed. Engl.*, 1984, **23**, 321; (f) F. Sato, H. Kanbara and Y. Tanaka, *Tetrahedron Lett.*, 1984, **25**, 5063; (g) H. Sheng, S. Lin and Y. Huang, *ibid*, 1986, **27**, 4893; (h) H. Abdallah and R. Gree, *ibid.*, 1980, **21**, 2239; (i) G. Le. Guillanton, Q.T. Do and J. Simonet, *ibid.*, 1986, **27**, 2261].

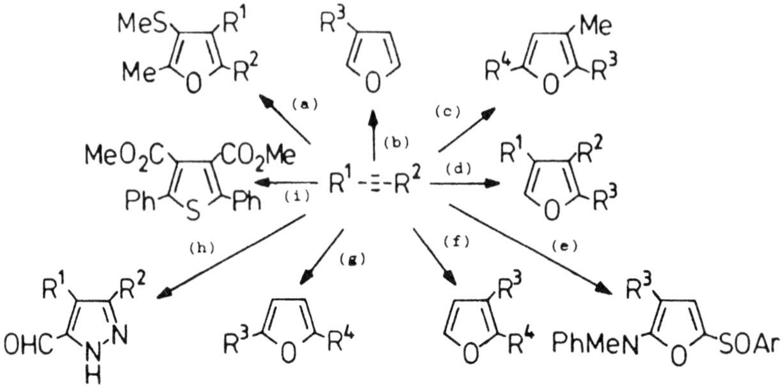

Reagents:

(a) i, MeCOCHClSMe/$SnCl_4$; (b) R^1, R^2=CO_2Me: i, $R^3Cu(SMe_2)$.$MgBr_2$; ii, aq. NH_4Cl; iii, $LiAlBu^nBu^iH$; iv, PCC;

(c) R^1, R^2=CH_2OMe: i, 2BuLi/ (catechol boronate BR^4) ; ii, AcOH; iii,

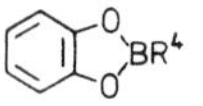

R^3CHO; iv, ^-OH; (d) i, $R^3R^4C=C=\overset{\oplus}{N}Me_2X^{\ominus}$; ii, aq. NaOH; iii,

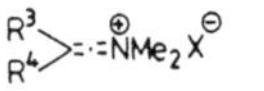

MCPA/$NaHCO_3$; iv, DIBAL; (e) R^1=H, R^2=SO_2Ar: i, R^3–≡–NMePh; (f) R^1=$SiMe_3$, R^2=CH_2OH: i, Bu^iMgBr/Cp_2TiCl_2; ii, R^3COR^4; iii, $BF_3.OEt_2$; iv, MCPA; v, aq.H_2SO_4; (g) R^1=COR^3, R^2=CH_2R^4: i, $Pd(dba)_2/PPh_3$; (h) R^1=H,CO_2Me, R^2=CO_2Me,COMe,COPh: i, $N_2CHCH(OMe)_2$; (i) R^1=Ph, R^2=CO_2Me: i, e^-/DMF

Alkynyl epoxide derivatives yield furans when treated with mercury(II) sulphate and sulphuric acid (K. Eichinger *et al.*, *J. Chem. Res. (S)*, 1983, 167) and with β-keto-esters, 3-alkylidene furans were formed in a palladium catalysed reaction (I. Minami, M Yuhara and J. Tsuji, *Tetrahedron Lett.*, 1987, **28**, 629).

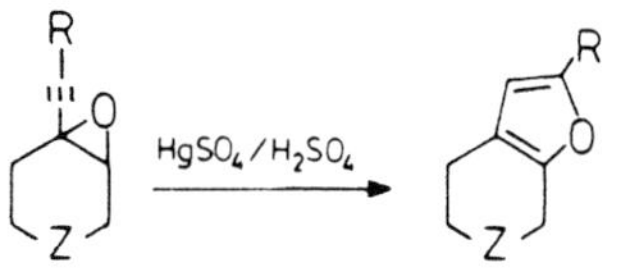

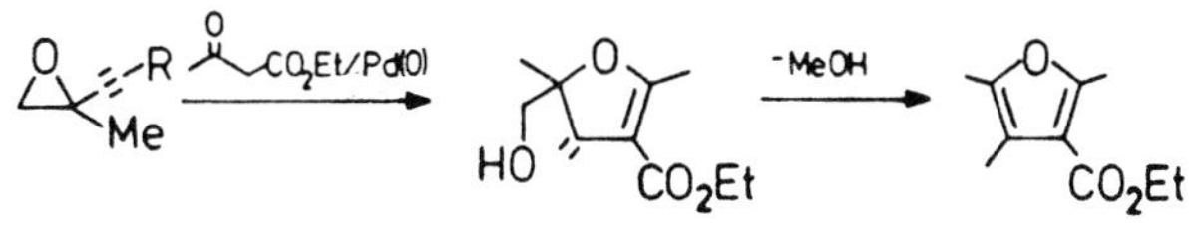

Metal carbene complexes and alkynes provide a route to furan derivatives (W.D. Wulff, S.R. Gilbertson and J.P. Springer, *J. Am. Chem. Soc.*, 1986, **108**, 520; M.F. Semmelhack and J. Perk, *Organometallics*, 1986, 2550) and pyran-2-ones (J.M. Moreto *et al.*, *J. Chem. Soc., Chem. Commun.*, 1989, 1560).

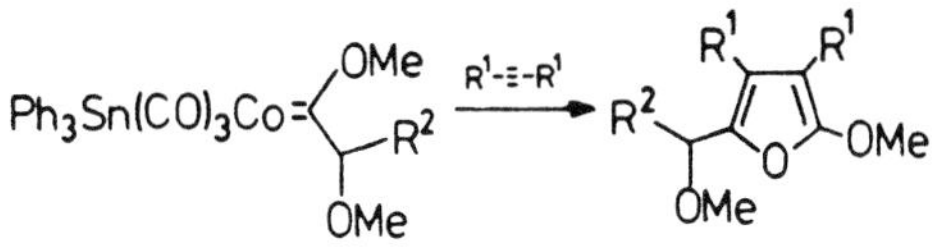

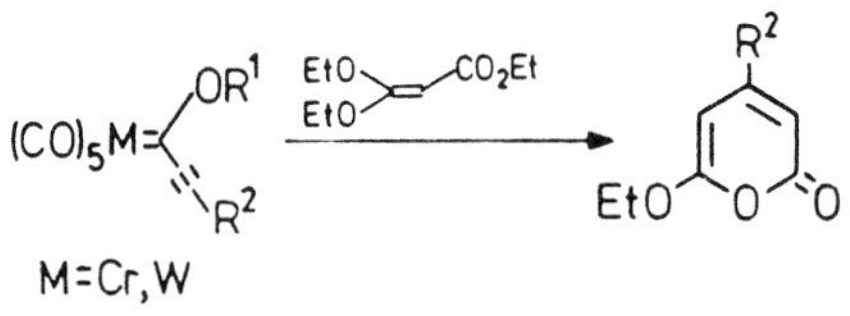

Benzofuran derivatives have been prepared by demethylation of 2-alkynylanisoles (D.R. Buckle and C.J.M. Rockell, *J. Chem. Soc. Perkin Trans. I*, 1985, 2443). Benzothiophene derivatives are similarly made from thioanisoles (A. Tundo *et al.*, *J. Chem. Soc., Chem. Commun.*, 1985, 1390). A related reaction yields indoles (T. Sakamoto, Y. Kondo and H. Yamakana, *Heterocycles*, 1986, **24**, 31, 1845).

R; X–Me; demethylate; R; X; EtO⁻; R; $NHCO_2Et$

X=O,S   X=O,S,NH

Palladium mediated cyclisation reactions of 2-alkynyl-aniline derivatives continues to attract attention as a method for the formation of the indole ring (Y. Kondo, T. Sakamoto and H. Yamanaka, *ibid.*, 1989, **29**, 1011).

R^1; NHR^2; $PdCl_2/CO/CuCl_2/MeOH/$ $NaOAc/K_2CO_3$; CO_2Me; R^1; N; R^2

Furan-2-ones are available from alkynes and metal carbonyls (T. Mise, P. Hong and H. Yamazaki, *J. Org. Chem.*, 1983, **48**, 238; J.-X. Wang and H. Alper, *ibid.*, 1986, **51**, 273), but-3-ynoic acids and Pd(II) in the presence of triethylamine (C. Lambert, K. Utimoto and H. Nozaki, *Tetrahedron Lett.*, 1984, **25**, 5323) and protected but-3-ynol derivatives (W.R. Jackson, P. Perlmutter and A.J. Smallridge, *J. Chem. Soc., Chem. Commun.*, 1985, 1509). Substituted propargyl alcohols have been converted to furan-3-one derivatives (L. Gomez, P. Calas and A. Commeyras, *ibid.*, 1985, 1509).

Lactones have been prepared from non-conjugated ynoic acids (D.M.T. Chan *et al.*, *J. Am. Chem. Soc.*, 1987, **109**, 6385).

O=; OH; n; R; $(LRhCl)_2$; $L=(C_6H_{11})_2PCH_2CH_2P(C_6H_{11})_2$; O=; O; n; R

The Diels-Alder reaction of 1,3-oxazoles with alkynes yield furan derivatives by elimination of a nitrile from the intermediate cycloadduct (K. Konig, F. Grof and V. Werberndorfer, *Liebigs Ann. Chem.*, 1981, 668; D. Liotta, M. Saindane and W. Ott, *Tetrahedron Lett.*, 1983, **24**, 2473; M.S. Ho and H.N.C. Wong, *J. Chem. Soc., Chem. Commun.*, 1989, 1238).

The Diels-Alder reaction of 6-membered heterocycles which

can behave as latent hetero-dienes has attracted considerable attention (D.L. Boger and R.S. Coleman, *J. Org. Chem.*, 1984, **49**, 2240; H.C. van der Plas *et al.*, *Tetrahedron*, 1989, **45**, 5151, 5611, 6211, 6499, 6511, 6519, 6891; E.C. Taylor and J.E. Macor, *J. Org. Chem.*, 1989, **54**, 4984): the formation of dihydropyrrolo[2,3-b]pyridines is a representative example (E.C. Taylor and J.L. Pont, *Tetrahedron Lett.*, 1987, **28**, 379).

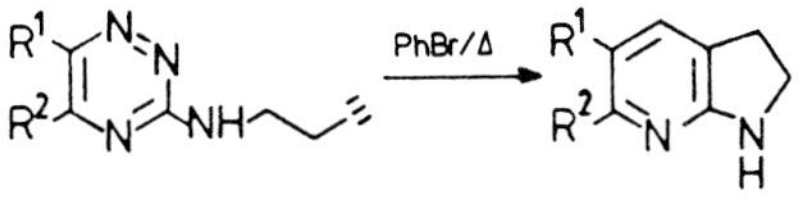

Phosphor us-containing heterocycles have been synthesised from dimethyl acetylenedicarboxylate.

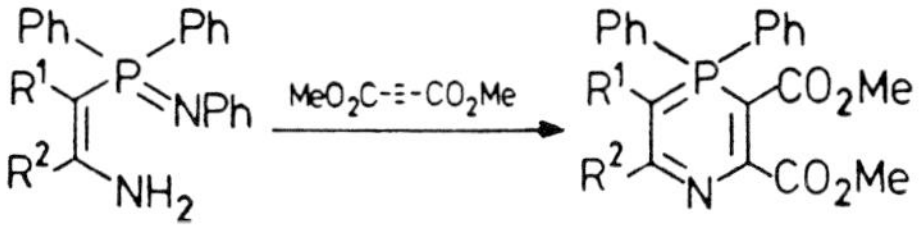

(J. Barluenga, F. Lopez and F. Palacios, *J. Chem. Soc., Chem. Commun.*, 1985, 1681)

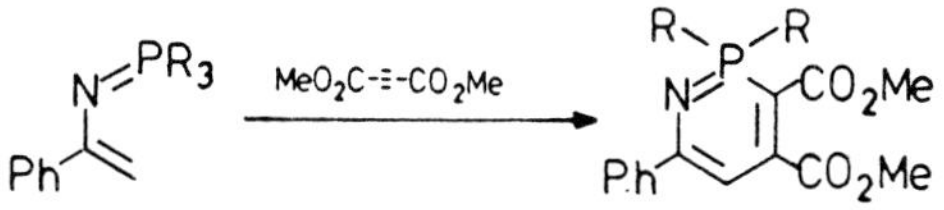

(T. Kobayashi and M. Nitta, *Chem. Lett.*, 1985, 1459)

Sulphur-nitrogen containing heterocycles have been synthesised (P.J. Dunn and C.W. Rees, *J. Chem. Soc., Chem. Commun.*, 1987, 59).

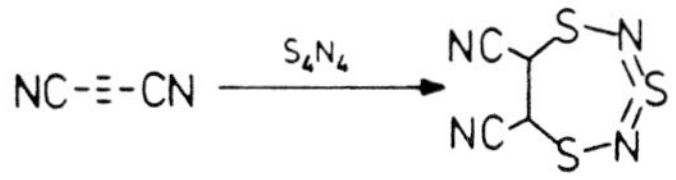

The base-promoted isomerisation of alkynes to allenes with subsequent intramolecular cyclisation provides routes to a variety of heterocycles.

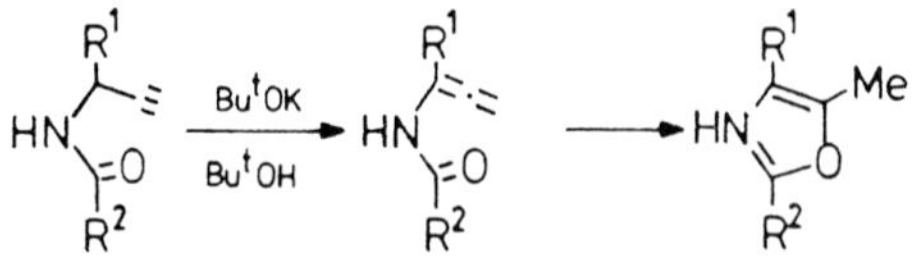

(B.M. Nilsson and U. Hacksell, *J. Heterocycl. Chem.*, 1989, **26**, 269)

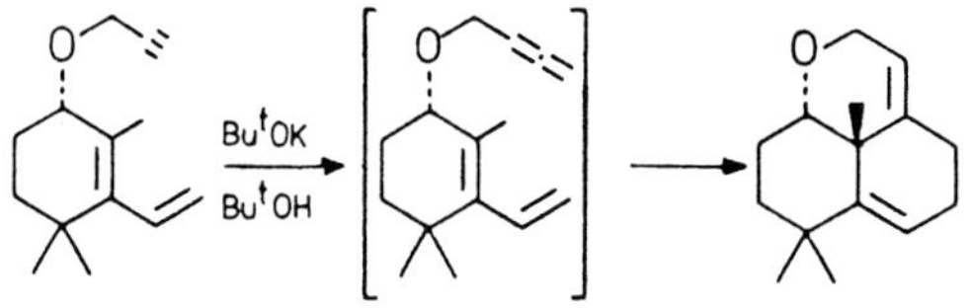

(K. Kanematsu and S. Nagashima, *J. Chem. Soc., Chem. Commun.*, 1989, 1028; K. Kanematsu *et al., J. Am. Chem. Soc.*, 1989, **111**, 5312)

Diethyl ketenedicarboxylate and aminoalkyne derivatives provide a route to pyran-4-ones (G. Himbert and L. Henn, *Liebigs Ann. Chem.*, 1987, 381).

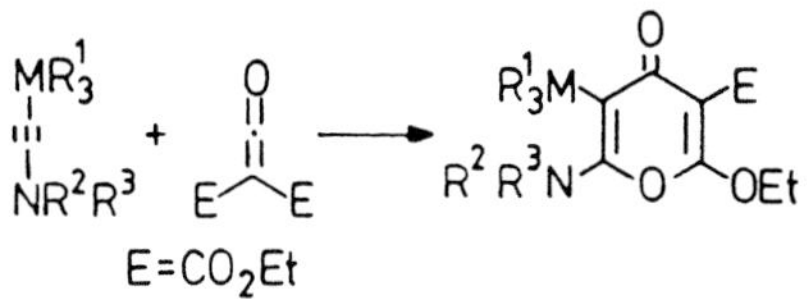

$E=CO_2Et$

Pyrid-2-one derivatives are available from alkynes (J.S. Swenton *et al., J. Org. Chem.*, 1983, **48**, 2337; L. Ghosez, *J. Am. Chem. Soc.*, 1982, **104**, 1428).

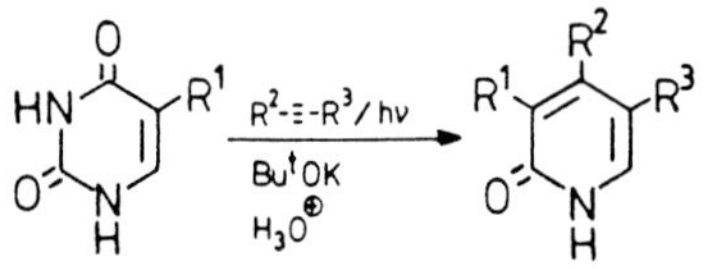

Pyrid-2-ones and 2-methoxypyridine derivatives react with cyanoalkynes (J.M. Rao *et al.*, *Tetrahedron*, 1989, **45**, 7093).

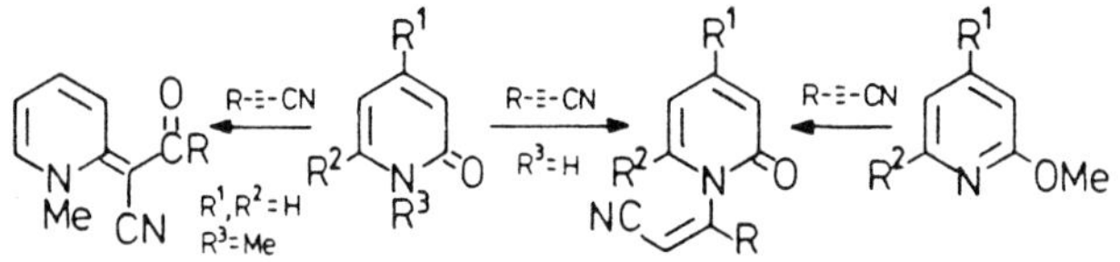

Organometallic chemistry has facilitated the preparation of numerous types of heterocycles from non-conjugated diynes [see Section (ii)]. Cyclisation of *N*-propargyl-2-vinylpyrrolidine and related compounds affords a nitrogen-containing heterocycles (B.M. Trost and S.F. Chan, *J. Am. Chem. Soc.*, 1986, **108**, 6053).

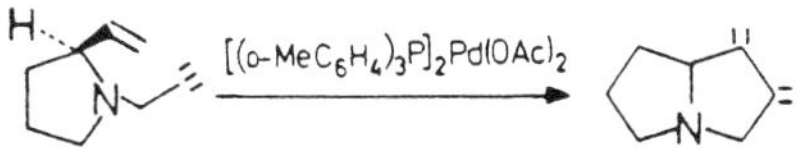

Isoquinolinium salts are prepared from aldimines (R.F Heck *et al.*, *J. Org. Chem.*, 1988, **53**, 3238).

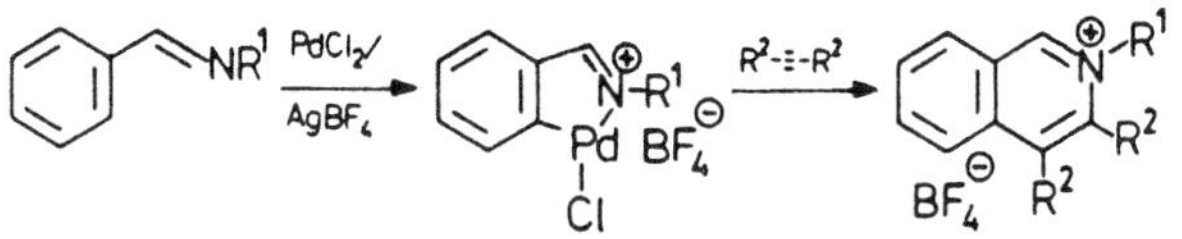

Acetylenic ketones and *N*-stannyltetrazole derivatives yield imidazoles in a two step synthesis (C.W. Rees *et al.*, *J. Chem. Soc. Perkin Trans. I*, 1985, 741).

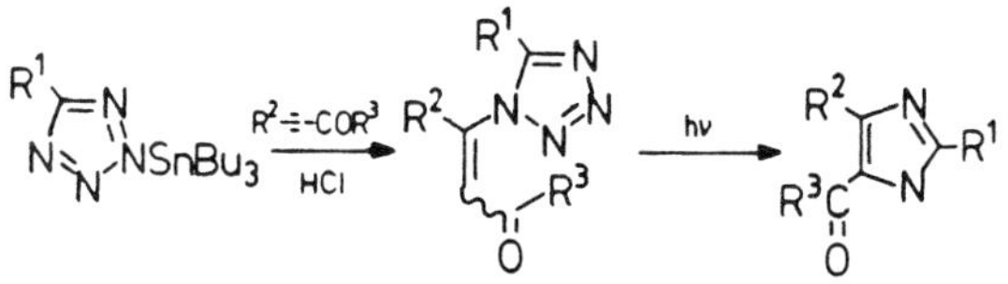

Inter- and intramolecular reactions of free-radicals [cf. references for free-radical additions to alkynes, Section (i)] continue to provide a versatile route to a variety of heterocycles.

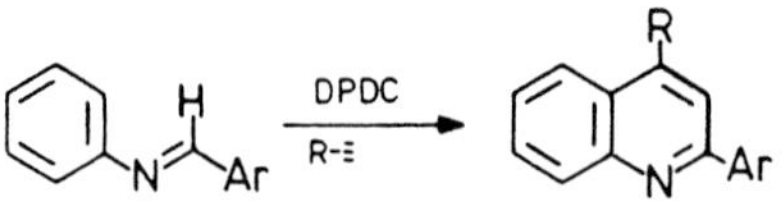

(A. Tundo *et al.*, *J. Chem. Soc., Chem. Commun.*, 1984, 1320)

Flash vacuum pyrolysis has been used to prepare a 1,1-dimethylsilaindene (T.J. Barton and B.L. Groh, *Organometallics*, 1985, **4**, 575).

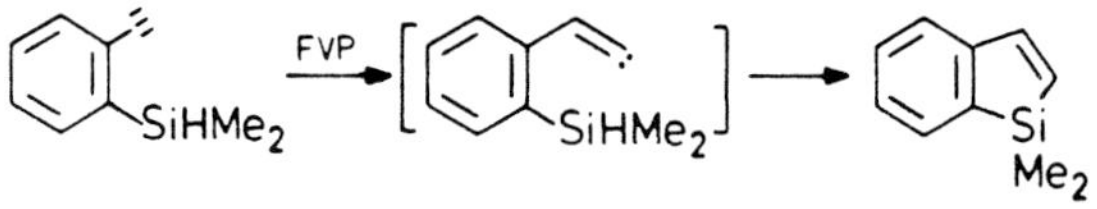

(iv) Photochemical Reactions

The photochemistry of alkynes has been reviewed extensively (J.D. Coyle in " Organic Photochemistry", Vol. 7, ed. A. Padwa, Marcel Dekker, New York, 1985, pp. 1-73, Y. Inoue and T. Hakushi, *Kagaku to Kogyo,* 1982, **56**, 165 and E.P. Serebryakov, *Izv. Akad. Nauk. S.S.S.R., Ser. Khim.*, 1984, 136).

Photolysis of organomercury compounds in the presence of acetylenes leads to alkylated alkenes (G.A. Russell *et al.*, *J. Org. Chem.*, 1986, **51**, 1986) whereas iodoalkynes yield substitution products (G.A. Russell, and P. Ngoviwatchi, *Tetrahedron Lett.*, 1986, **27**, 3479). Alkynes are reduced stereoselectively *via* platinum-titanium dioxide catalysis (H. Yamanaka *et al.*, *J. Chem. Soc., Chem. Commun.*, 1985, 788) and reductive cleavage of terminal alkynes in the presence of ruthenium catalysts has been reported (Y. Degami and I. Willner, *ibid.*, 1985, 648).

Photoinduced intramolecular cycloaddition reactions are a valuable source of bicyclic systems. Indole derivatives with electron-deficient alkynes yield benzoazepines (P.D. Davies, D.C. Neckers and J.R. Blount, *J. Org. Chem.*, 1980, **45**, 462), whereas the interaction of benzoisothiazoles and benzothiazoles with electron-rich alkynes results in 1,4- and 1,5-benzothiazepines respectively.

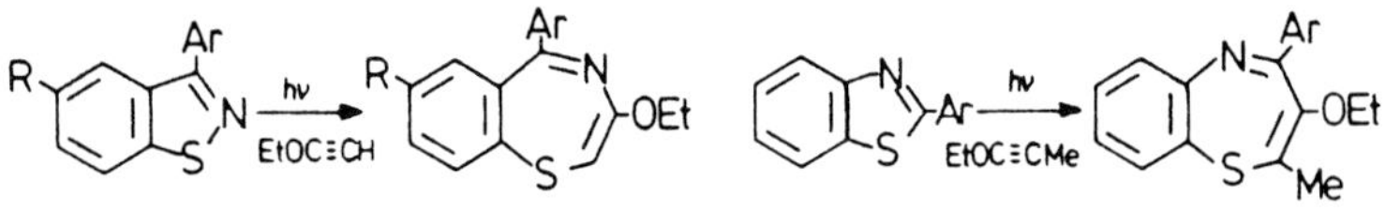

(M. Sindler-Kulyk and D. Neckers, *Tetrahedron Lett.*, 1981, **22**, 529; *J. Org. Chem.*, 1982, **47**, 4914; *ibid.*, 1983, **48**, 1983). A diradical mechanism is invoked to explain these processes. Protected 3-hydroxy-5-aryl-1-pentynes produce cyclopentane annulated cyclooctatetrenes in reasonable yield (M.C. Pirrung, *ibid.*, 1987, **52**, 1635)

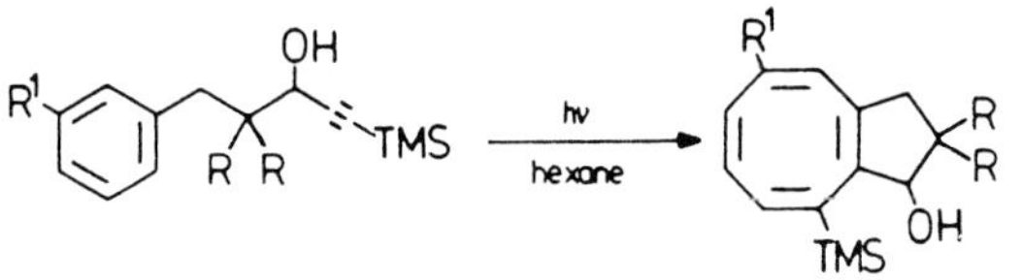

and photocyclisation of conjugated (V.B. Rao, S. Wolff and W.C. Agosta, *J. Am. Chem. Soc.*, 1985, **107**, 521)

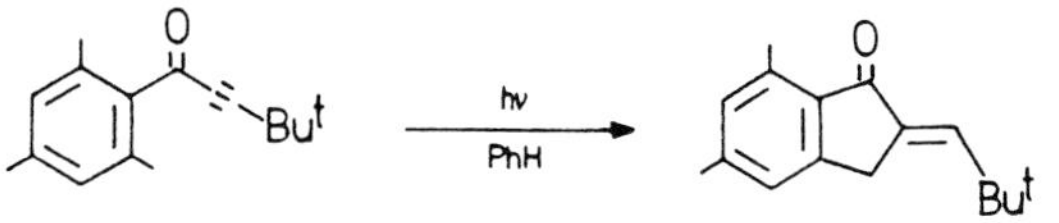

and non-conjugated ynones (J. Cossy *et al.*, *J. Org. Chem.*, 1986, **51**, 4196) yields exocyclic alkenes.

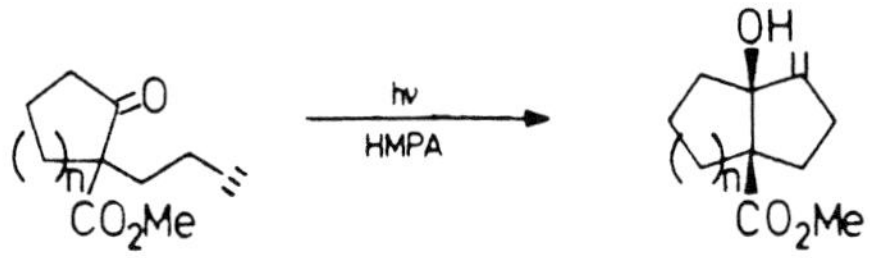

A novel cyclisation of dimethylacetylene dicarboxylate with a non-conjugated diene has been reported (T. Zaima, Y. Matsunaga and K. Mitsuhashi, *J. Heterocycl. Chem.*, 1983, **20**, 1).

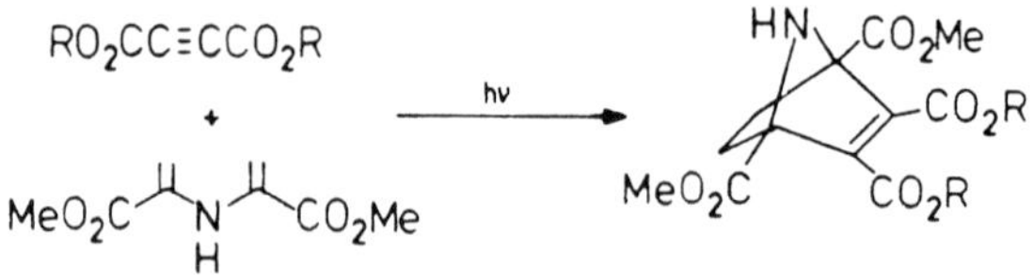

RO$_2$CC≡CCO$_2$R + MeO$_2$C–C(=)–N(H)–C(=)–CO$_2$Me $\xrightarrow{h\nu}$ [bicyclic product: HN, CO$_2$Me, CO$_2$R, MeO$_2$C, CO$_2$R]

(v) Pericyclic Reactions

The intramolecular Diels-Alder reaction (D. Craig, *Chem. Soc. Rev.*, 1987, **16**, 187; E. Ciganek, *Organic Reactions*, 1984, **32**, 1) and the transition metal mediated cycloaddition reactions of alkynes have been reviewed (N.E. Schore, *Chem. Rev.*, 1988, **88**, 1081). Intramolecular cyclisation reactions of acetylenic ketones giving cyclopentenone derivatives have been reported (A.S. Dreiding *et al.*, *Helv. Chim. Acta.*, 1982, **65**, 13, 2413, 2517; J. Hugnet, M. Karpf and A.S. Dreiding, *Tetrahedron Lett.*, 1983, **24**, 4177).

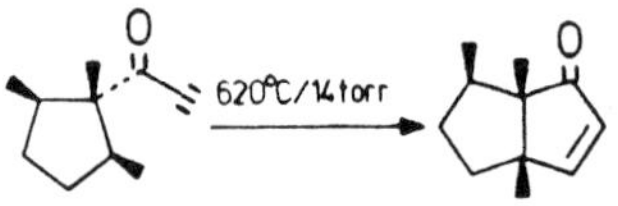

620°C/14 torr

Unactivated alkynes participate in the intramolecular Diels Alder reaction both in the absence (P. Helquist *et al.*, *Tetrahedron Lett.*, 1985, **26**, 5393) and presence (I. Matsuda *et al.*, *ibid.*, 1987, **28**, 3361; P.A. Wender and T. Jenkins, *J. Am. Chem. Soc.*, 1989, **111**, 6432) of metal catalysts .

Alkynylstannanes react with 1,3-dienes giving cyclohexa-1,4-diene derivatives (B. Jousseaume and P. Villeneuve, *Tetrahedron*, 1989, **45**, 1145).

The thermally produced dimer of cyanoacetylene gave an adduct with [2.2]paracyclophane (H. Hopf *et al.*, *Angew. Chem., Int. Ed. Engl.*, 1989, **28**, 1279).

Alkynes and dichloroketene yield cyclobutenones (R.L. Danheiser and H. Sard, *Tetrahedron Lett*, 1983, **24**, 23; C. J.

Kowalski, *J. Am. Chem. Soc.*, 1988, **110**, 3693).
Cyclobutenones and electron-rich alkynes yield phenols (R.L. Danheiser and S.K. Gee, *J. Org. Chem.*, 1984, **49**, 1672).

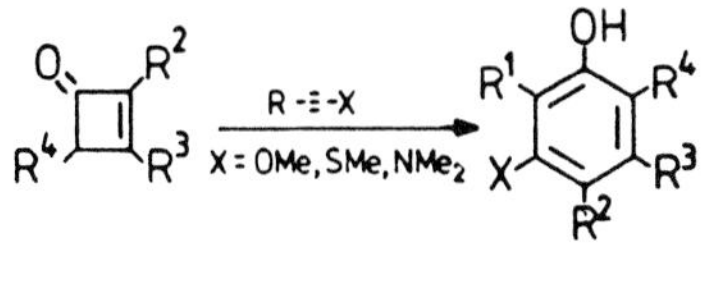

R -≡-X
X = OMe, SMe, NMe_2

Acetylenic ketenes and alkynes afford 2-alkynylcyclobut-3-enones (D.J. Pollart and H.W. Moore, *ibid*, 1989, **54**, 5444) which undergo a thermal rearrangement to 1,4-dibenzoquinones and 2-alkylidene-1,3-cyclopentenediones (H.W. Moore *et al.*, *J. Am. Chem. Soc.*, 1989, **111**, 975).

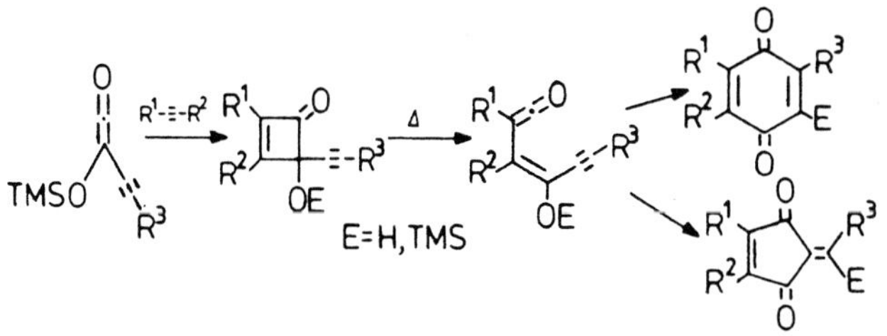

E=H,TMS

Recent interest in the mechanism of action of natural products possessing the enediyne moiety, such as calicheamicin and esperamicin derivatives (J.N. Haseltine, S.J. Danishefsky and G. Schulte, *ibid.*, 1989, **111**, 7638) has

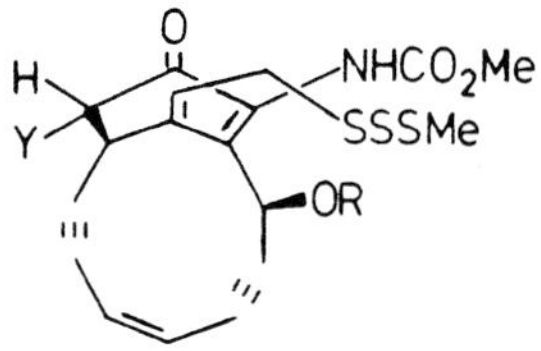

Y= H, R= carbohydrate : calicheamicin
Y= O-carbohydrate, R= carbohydrate : esperamicin

generated interest in the cyclisation reactions of conjugated alkynyl-ene-allenes (A.G. Meyers, E.Y. Kuo and N.S. Finney, *ibid.*, 1989, **111**, 8057; A.G. Meyers and P.S. Dragovitch, *ibid.*, 1989, **111**, 9130) and 1,4-diynes (J.P. Synder, *ibid.*, 1989, **111**, 7630).

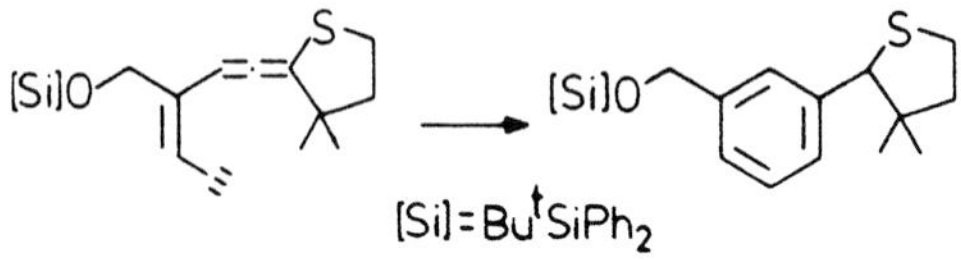

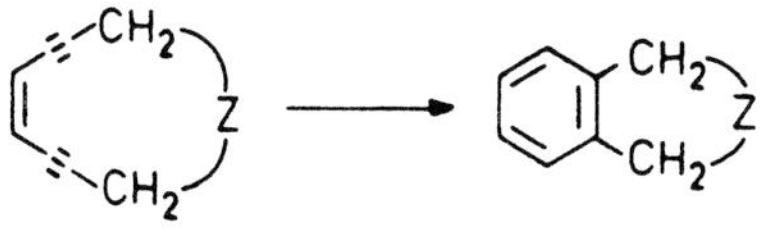

[Si]O [Si]O
[Si] = Bu^{t}SiPh$_2$

CH$_2$ Z CH$_2$ → CH$_2$ Z CH$_2$

Cyclopropenones and aminoacetylene derivatives yield cyclopentenones (T. Eicher and M. Urban, *Chem. Ber.*, 1980, **113**, 408). With cyclohexenone, *NN*-diethylaminoprop-1-yne gives a [2+2] adduct in the absence of magnesium bromide but yields an ethylenic amide in the presence of magnesium bromide (J. Ficini *et al.*, *Tetrahedron Lett.*, 1981, **22**, 725) and with pyranones aniline derivatives are formed (P. Martin *et al.*, *ibid.*, 1985, **26**, 3947).

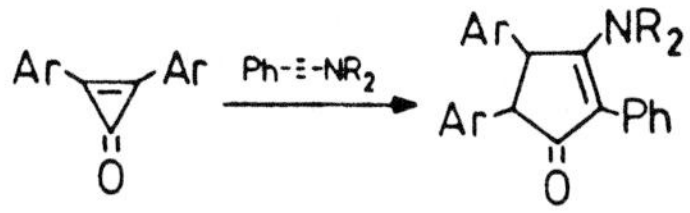

Ar Ar O Ph-≡-NR$_2$ → Ar NR$_2$ Ar Ph O

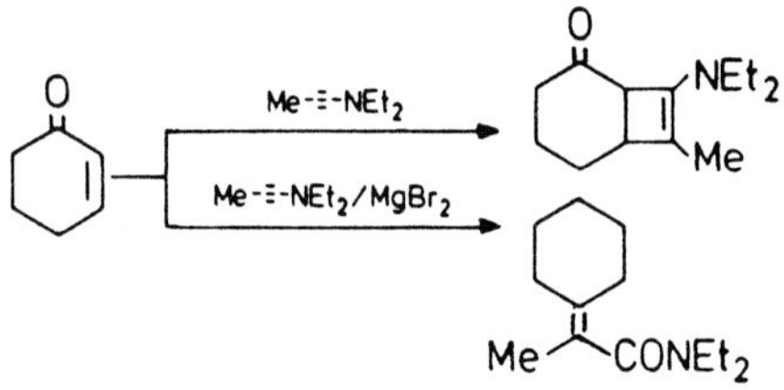

O Me-≡-NEt$_2$ → O NEt$_2$ Me
Me-≡-NEt$_2$/MgBr$_2$ → Me CONEt$_2$

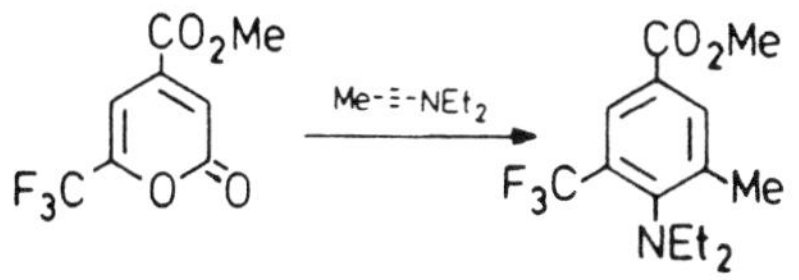

Aniline derivatives are also formed from the reaction of vinylketimines and acetylenic esters (L. Ghosez *et al.*, *ibid.*, 1987, **28**, 397).

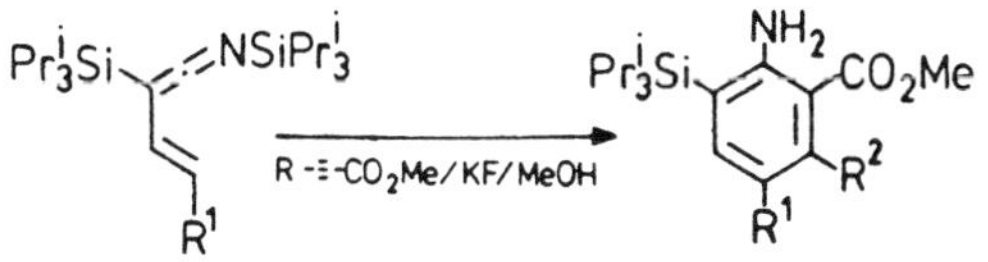

Dienamines and methyl propiolate or dimethyl acetylenedicarboxylate afford spiro-compounds (V. Nair and T.S. Jahnke, *ibid.*, 1984, **25**, 3547).

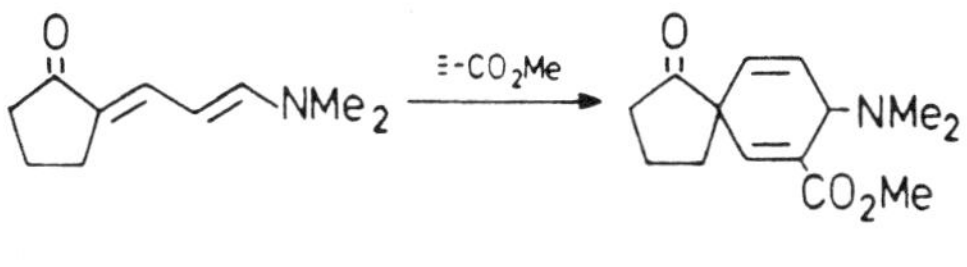

Acetylene derivatives of metal carbene complexes react with 1,3-dienes giving cycloadducts which can be transformed into bicyclic compounds (W.D. Wulff and C.D. Jung, *J. Am. Chem. Soc.*, 1984, **106**, 7565).

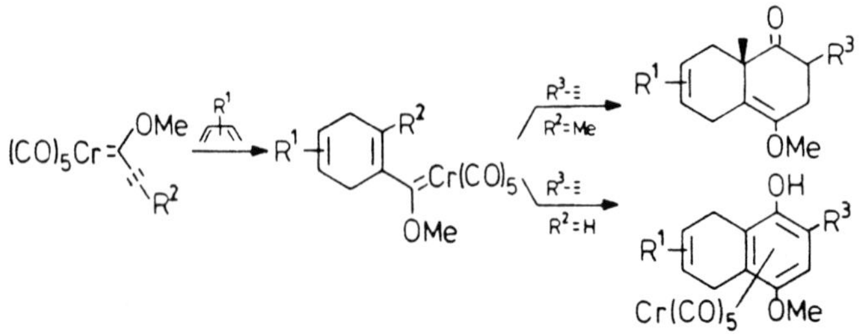

$(CO)_5Cr=C(OMe)$–$C\equiv C$–R^2 + R^1-diene → R^1-cyclohexadienyl–R^2, $=Cr(CO)_5$, OMe; R^3–≡, R^2=Me → R^1-bicyclic ketone (O, R^3, OMe); R^3–≡, R^2=H → R^1-bicyclic arene (OH, R^3, OMe, $Cr(CO)_5$)

Cyclobutenes have been prepared from an acetylenic iron complexes (M. Rosenblum and D. Scheck, *Organometallics*, 1982, **1**, 397).

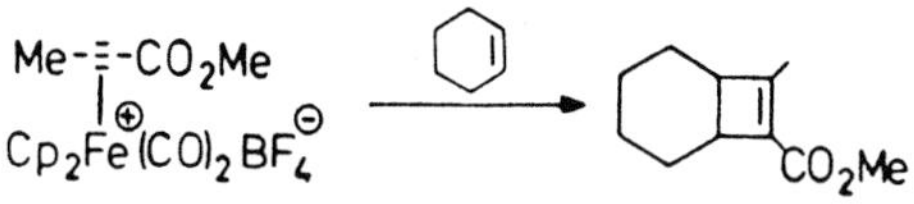

Me–≡–CO_2Me / $Cp_2Fe^{\oplus}(CO)_2 BF_4^{\ominus}$ + cyclohexene → bicyclo[4.2.0]octene, Me, CO_2Me

1,3-Dienes and alkynes, in the presence of iron complexes, yield cyclohexa-1,4-dienes (H. tom Dieck and R. Diercks, *Angew. Chem., Int. Ed. Engl.*, 1983, **22**, 778).

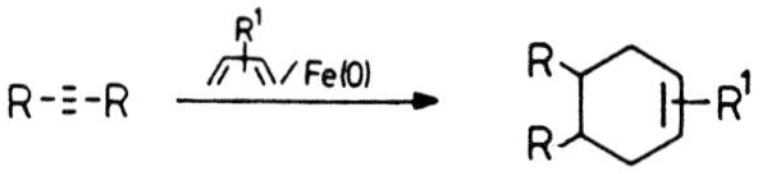

R–≡–R + R^1-diene / Fe(0) → R, R-cyclohexadiene–R^1

Nickeloles behave as 1,3-dienes with alkynes (J.J. Eisch *et al.*, *J. Organomet. Chem.*, 1986, **312**, 399).

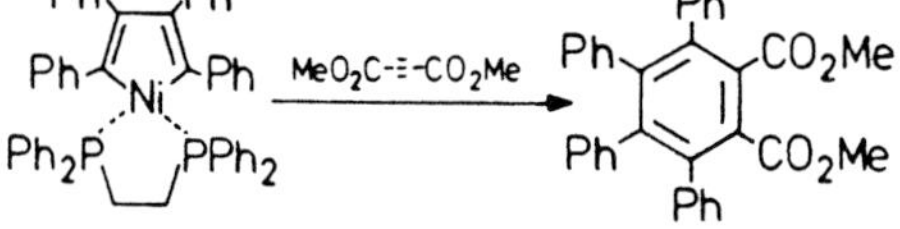

Ph, Ph, Ph, Ph-nickelole (Ni, Ph_2P, PPh_2) + MeO_2C–≡–CO_2Me → Ph, Ph, Ph, Ph, CO_2Me, CO_2Me-benzene

A cyclophane derivative has been prepared from dimethylacetylenedicarboxylate (H. Hopf, I. Bohm and J. Kleinschroth, *Org. Synth.*, 1981, **60**, 41).

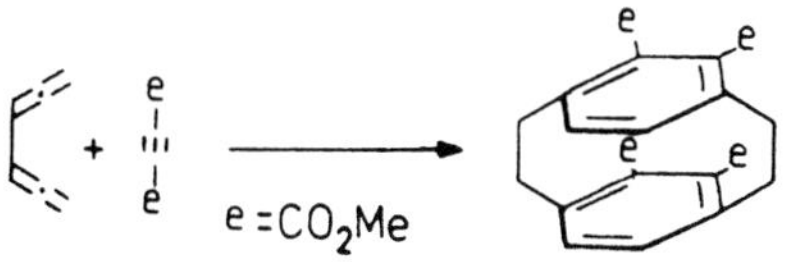

Routes to biphenylene derivatives have been developed which involve a metal mediated transformation of a 1,4-diyne moiety to a 1,3-diene which then cycloadds to an alkyne (G.H. Hovakeemian and K.P.C. Vollhardt, *J. Chem. Soc., Chem. Commun.*, 1983, 502; K.P.C. Vollhardt *et al.*, *J. Am. Chem. Soc.*, 1985, **107**, 5670; *ibid.*, 1986, **108**, 2481; K.P.C. Vollhardt *et al.*, *Angew. Chem., Int. Ed. Engl.*, 1986, **25**, 266).

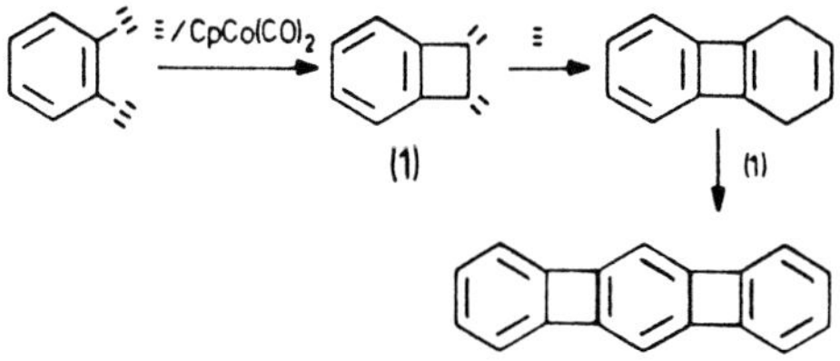

Methyl propiolate participates in an ene reaction in the presence of ethylaluminium dichloride (A.D. Batcho, D.E. Berger and M.R. Uskokovic, *J. Am. Chem. Soc.*, 1981, **103**, 1293) or diethylaluminium chloride (W.G. Dauben and T. Brookhardt, *J. Org. Chem.*, 1982, **47**, 3921).

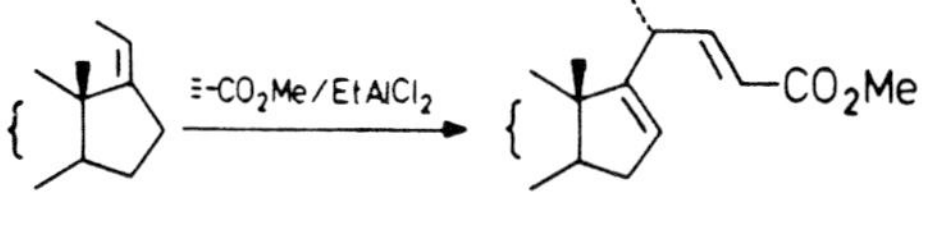

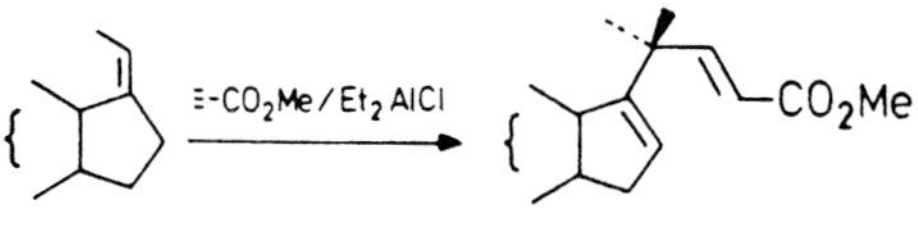

Allylic chlorides and alkynes afford cyclopentene derivatives (J.A. Miller and M. Moore, *Tetrahedron Lett.*, 1980, **21**, 577; J.A. Miller *et al., ibid.*, 1987, **28**, 689).

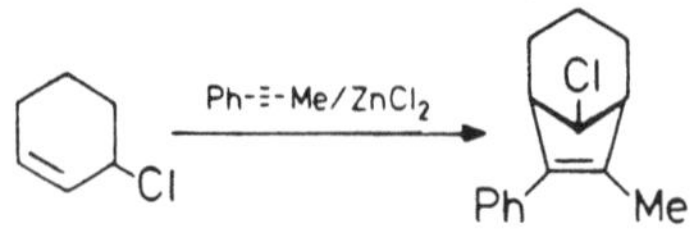

$Ph\text{-}\equiv\text{-}Me/ZnCl_2$

(vi) Oxidation

Several methods are available for preparing 1,2-diketones from disubstituted alkynes (F.P. Ballistreri *et al., Tetrahedron Lett.*, 1986, **27**, 5139; S. Toru, T. Inokuchi and Y. Hirata, *Synthesis*, 1987, 377; R. Zibuck and D. Seebach, *Helv. Chim. Acta.*, 1988, **71**, 237). When one of the substituents is either an ether or disubstituted amino group then the corresponding α-ketoesters or α-ketoamides are formed (P. Muller and J. Godoy, *Tetrahedron Lett.*, 1982, **23**, 3661). α-Ketoesters are also available from trimethylsilyl derivatives of terminal alkynes (P.C.B. Page and S. Rosenthal, *ibid.*, 1986, **27**, 1947).

An oxidative rearrangement of disubstituted alkynes has been reported (R.M. Moriarty *et al., ibid.*, 1987, **28**, 2845).

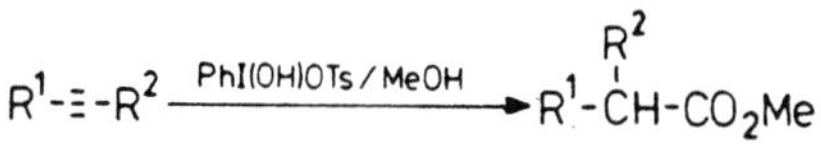

$$R^1\text{-}\equiv\text{-}R^2 \xrightarrow{PhI(OH)OTs/MeOH} R^1\text{-}CH(R^2)\text{-}CO_2Me$$

Oxidation of terminal acetylenes can yield α-ketoaldehydes (F.P. Ballistreri, S. Failla and G.A. Tomaselli, *J. Org. Chem.*, 1988, **53**, 830), carboxylic acids (G.A. Tomaselli *et al., ibid.*, 1989, **54**, 947; K. Tamao, M. Kamuda and K. Maeda, *Tetrahedron Lett.*, 1984, **25**, 321), α-ketoaldehydes (T. Satoh *et al., Synthesis*, 1985, 406) or α-hydroxyketones (Y. Tamura *et al., Tetrahedron Lett.*, 1985, **26**, 3837).

(vii) Polymerisation

Over the last decade a major research programme concerning the preparation, characterisation and applications of alkyne and di-alkyne polymers has been developed. The electronic and non-linear optical properties of these systems resulting from their delocalised electronic stuctures has been the focus of attention. A comprehensive treatment is beyond the scope of this review, but the following comprise an overview of this extensive subject: "Polyacetylene-Chemistry, Physics and Materials Science", J.C.W. Chien, Academic Press, Orlando, 1984; "Nonlinear Optical and Electroactive Polymers", ed. P.N. Prasad and D.R. Ulrich, Plenum Press, New York, 1988; "Handbook of Conducting Polymers", Vols. 1 & 2, ed. T.A. Skotheim, Marcel Dekker, New York, 1989; P. Hegenrother, in "Encyclopaedia of Polymer Science and Engineering", Vol. 1, ed. J.I. Kroschwitz, John Wiley and Sons, New York, 1985, pp. 61-86; H.W. Gibson and J.M. Pouchan, *ibid.*, Vol. 1, pp. 87-130; R.R. Chance, *ibid.*, Vol. 4, pp. 767-769; R.R. Chance and J.E. Fommer, *ibid.*, Vol. 5, pp. 462-507; C. Krohnke and G. Wegner in Houben-Weyl´s "Methoden der Organischen Chemie", Vol. E20, ed. K.H. Buchel and J. Falbe, Thieme-Verlag, Stuttgart, 1987, p. 1312; M. Schott and G. Wegner in "Nonlinear Optical Properties of Organic Molecules and Crystals", Vol. 2, ed. D.S. Chemla and J. Zyss, Academic Press, Orlando, 1987, pp. 3-49; W.D. Huntsman in "The Chemistry of Functional Groups, Supplement C: The Chemistry of Triple-Bonded Functional Groups", ed. S. Patai and Z. Rappoport, Wiley-Interscience, Chichester, 1983, pp. 917-980; A.G. MacDairmid and M. Maxfield in "Electrochemical Science and Technology of Polymers-1", ed. R.G. Linford, Elsevier Applied Science, London, 1987, pp. 67-98.

The main synthetic routes to polyacetylenes include the interaction of alkynes with Ziegler-type catalysts (G. Wegner, *Angew. Chem., Int. Ed. Engl.*, 1981, **20**, 361),

$$H\text{-}C\equiv C\text{-}H \xrightarrow{cat.} [\ \ldots\ ]_n \xrightarrow{\Delta} [\ \ldots\ ]_n$$

the cycloaddition of hexafluorobut-2-yne to cyclooctatetrene followed by ring-opening metathesis polymerisation (ROMP) of the adduct, the "Durham route" to polyacetylene (J.H. Edwards and W.J. Feast, *Polymer*, 1980, **21**, 595; D.C. Bott, J.H. Edwards and W.J. Feast, *ibid.*, 1984, **25**, 395; D.C. Bott *et al.*, *Synth. Met.*, 1986, **14**, 245, *Polymer*, 1987, **28**, 601)

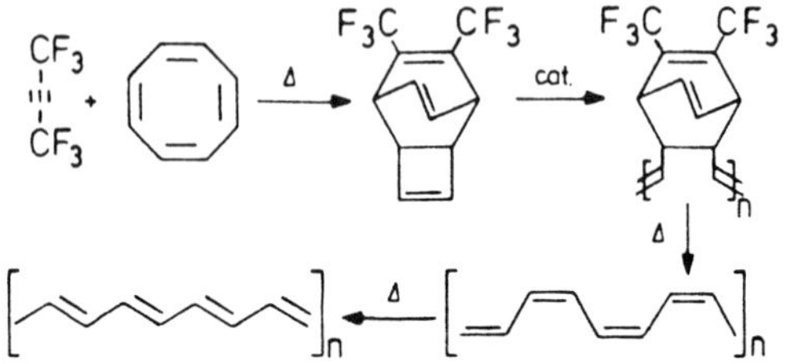

and the ROMP of cyclooctatetrenes (F.L. Klavetter and R.H. Grubbs, *J. Am. Chem. Soc.*, 1988, **110**, 7807; R.H. Grubbs *et al.*, *Angew. Chem., Int. Ed. Engl.*, 1989, **28**, 1571).

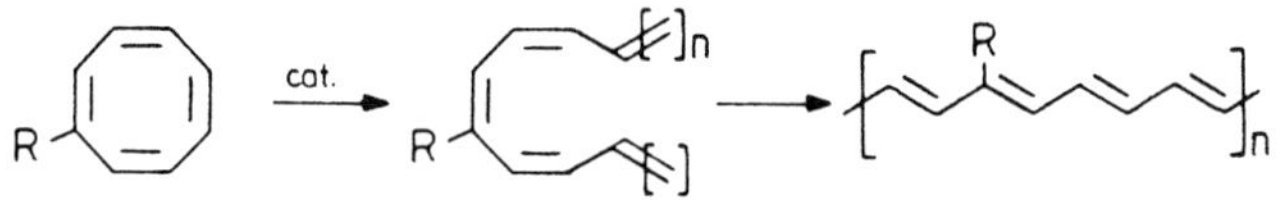

Other references to the developments in the preparation, properties and uses of polyacetylenes include: C.R. Fincher Jr. *et al.*, *Synth. Met.*, 1983, **6**, 243; T. Nagotomo *et al.*, *J. Electrochem. Soc.*, 1985, **132**, 1380, 1987, **134**, 305; W.J. Feast, *Chem. Ind.(London)*, 1985, 623; V.M. Misin and M.I. Cherkashin, *Usp. Khim.*, 1985, **54**, 956; A.J. Heeger, D. Moses and M. Sinclair, *A.C.S. Symp. Ser.*, 1987, **346**, 372; S.G. Grigoryan and A.A. Matnishyan, *Arm. Khim. Zh.*, 1987, **40**, 498; V.A. Lopyrev *et al.*, *Vysokomol. Soedin., Ser. A*, 1988, **30**, 2019; A. Montaner *et al.*, *Polymer*, 1988, **29**, 1101; J. Kunzler and V. Perec, *Polym. Prep., Am. Chem. Soc. Div. Polym. Chem.*, 1988, 1; A. Stowell *et al.*, *Polymer*, 1989, **30**, 195.

1,4-Disubstituted but-1,3-diynes are polymerised to poly(diacetylenes) under the influence of heat, radiation (visible, ultraviolet, X-ray and γ-ray) and mechanical stress.

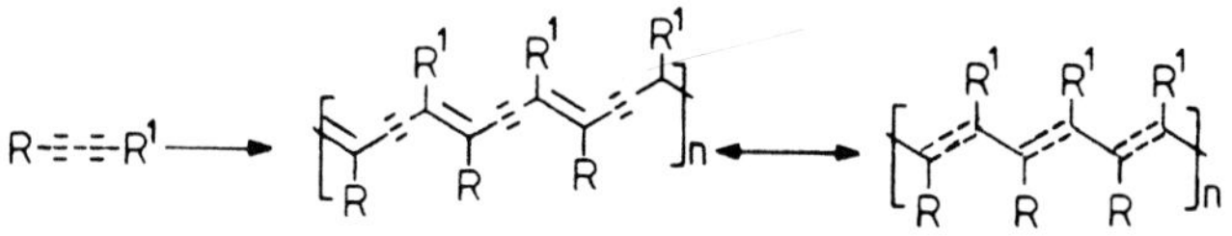

Progress in the field of synthesis, structural investigation and application of numerous poly(diacetylenes) is exemplified in: R.R. Chance, *A.C.S. Symp. Ser.*, 1983, **233**, 235; "Proceedings of N.A.T.O. Advanced Study Institute on Quantum Chemistry of Polymers: Solid State Aspects", ed. D. Bloor, J. Ladik and J. Andre, Reidel, The Netherlands, 1984, p. 191; D. Bloor, *Synth. Met.*, 1987, **21**, 71; D. Bloor, *Phys. Scr.*, 1987, **T19A**, 266; H. Matsuda and H. Nakanishi, *Nippon Gomu Kyokaishi*, 1988, **61**, 637; H. Matsuda, H. Nakanishi and S. Okada, *Kino Zairyo*, 1988, **8**, 5; K. Araya, *Hyomen*, 1988, **26**, 113; G.H.W. Milburn *et al.* in "Organic Materials for Nonlinear Optics", ed. R.A. Haan and D. Bloor, Special Publication No. 69, The Royal Society of Chemistry, London, 1989, p. 196; S. Etemad and G.L. Baker, *Synth. Met.*, 1989, **28**, D159.

Second Supplements to the 2nd Edition of Rodd's Chemistry of Carbon Compounds, Vol. IA/B, edited by M. Sainsbury
© 1991 Elsevier Science Publishers B.V., Amsterdam

Chapter 3a

MONOHALOGENOALKANES

R. BOLTON

This Review will follow the same guidelines as the previous Supplements, with the exceptions that (i) there will be less specific detail of individual compounds, and (ii) a short preliminary section addresses the environmental problems caused by alkyl halides and the analytical implications. Chemical Abstracts citations have been included for the less common sources.

1. Toxicity and analysis at low levels

(a) Toxicity and environmental effects

The toxicity of alkylating agents has been well documented, and their ability to initiate or predispose towards cancer is reflected in the growing number of restrictions in their general use. This property of alkyl halides has been reviewed ((*a*) L. Fishbein, Sci. Total Environ., 1978, 11(3), 223; Chem. Abs., 90, R198386. (*b*) K. Fukuda, Hen'igen to Dokusei, 1981, 4(6), 36; Chem. Abs., 97, R139656. (*c*) M. W. Anders and L. R. Pohl, Bioact. Foreign Compd., 1982, 283. Ed. M.W. Anders, Academic Press, Orlando, Florida.; (*d*) D. Henschler, *idem.*, 317) with specialist reviews on chlorinated methanes and ethanes (P. F. Infante and T. Tsongas, Environ. Sci. Res., 1982, 25; Chem. Abs. 97, 34189). Toxic limits in the workplace in France, USA and USSR have been proposed for polyhalogenoalkanes ((*a*) Anon., Cah.

Notes Doc., 1983, 110, 53. (*b*) Anon., Fed. Registr. 12 Oct 1977, 42 (197) 55026-80; Chem. Abs., 88, 46123) and for alkyl fluorides (G. N. Bakhishev, Farmakol. Toksikol. (Kiev) 1973, No. 8, 137; Chem. Abs., 80, 116844); such determinations are quantified in a 'vapour hazard index' (M. J. Pitt, Chem. Ind. (London), 1982, (20), 804).

The carcinogenicity of halogenoalkanes ((*a*) Y. T. Woo *et al.*, "Chemical Induction of Cancer: Structural Bases and Biological Mechanisms." Vol IIIB. Aliphatic and Polyhalogenated Carcinogens; Chem. Abs., 103, 83370) and the mutagenicity of pesticides (Y. Shirosu *et al.*, Envir. Sci. Res., 1984, 31, 617; Chem. Abs. 103, 180566) have implications upon the environmental incidence of these substances (U. Bauer, Zentralbl. Bakteriol, Mikrobiol. Hyg. Abt.1, Orig.B, 1981, 174(6), 556; Chem. Abs., 96, 211935) which points the relevance of studies of the mechanism of biotransformation and metabolism of these substances ((*a*) M. W. Anders, Trends Pharmacol. Sci., 1982, 3(9), 356. (*b*) S. M. Naqvi and M. C. Blois, Report 1980, EPA560/ 13-79-018. Order No. PB81-232811. Gov. Rep. Announce. Index (U. S.) 1981, 81(24), 5094; Chem. Abs., 96, R47030. (*c*) O. T. Love *et al.*, Report 1983, EPA-600/8-83-019; Order No., PB83-239434. From Gov. Rep. Announce Index (U. S.) 1983, 83(22) 5392; Chem. Abs., 100, R12190).

The carcinogenicity of drinking water has been linked to the presence of alkyl halides (R. Koch, Toxikol. Anal. Probl. Loesungsmittelexpo. [Vortr. Mini-symp] 1979 (Publ. 1980) 17; Chem. Abs., 96, R46982) and especially of haloforms (H. Arai *et al.*, Report 1983, JAERI-H-83-149. From INIS Atomindex 1985, 16(9), Abstr. No. 16:029630s; Chem. Abs., 103, 165719). These arise principally through the halogenation of organic contaminants, and may be minimised by prior ozonisation of the water ((*a*) Japan Atomic Energy Research Institute. Jpn. Kokai Tokkyo Koho JP 59,150,594; Chem. Abs., 102, P12153. (*b*) H. Arai *et al.*, Water Res., 1986, 20(7), 885; Chem. Abs., 105, 48748. (*c*) M. Schalekamp, Aqua (London) 1983, (3), 89; Chem. Abs., 99, 93429).

Despite the ubiquitous argument on the influence of chloro-fluorocarbons upon the atmosphere ((*a*) J. Russow, Nachr. Chem. Tech. Lab. 1977, 25(9), 507, 511; Chem. Abs. 88, R65225 (*b*) H. B. Singh *et al.*, Atmos. Environ. 1981, 15(4),

601; Chem. Abs. 95, 48234 (*c*) B. P. Block and J. R. Soulen, Stratos. Ozone Man 1982, 2, 149. Ed. F. A. Bower and R. B. Ward, CRC, Boca Raton, Fla.. (*d*) C. R. Patrick, J. Fluorine Chem., 1984, 25(1), 7. (*e*) P. L. Hanst, Chemistry, 1978, 51(2), 6; Chem. Abs., 88, 125532. (*f*) G. K. Moortgat, U. S. Fed. Aviat. Adm. Off. Environ. Energy [Tech. Rep.] FAA-EE 1980, FA-EE-80-20, Proc. NATO Adv. Study Inst. Atmos. Ozone: Var. Hum. Influences; AD A088889, 599-646; Chem. Abs., 94, 160300. (*g*) M. de Bortoli and E. Pecchio, Atmos. Environ., 1976, 10(11), 921; Chem. Abs., 86, 95259) the replacement of $CFCl_3$, CCl_2F_2, and $C_2Cl_2F_2$ by CH_3CClF_2 or by $CHClF_2$ is recommended because of the lack of toxicity, carcinogenicity and mutagenicity of the last two compounds (M. C. Renand and M. Vivet, Parfums, Cosmet., Aromes, 1985, 61, 61). The relative effects of these CFC upon the ozone layer do not seem to have been considered.

(b) Analysis

This awareness of the toxicity has encouraged studies of the analysis of halogenoalkanes in the environment at low levels in air, water, and soil. A review of the sources and pathways by which alkyl halides contribute to environmental pollution is provided by Radding (S. B. Radding *et al.*, U.S. NTIS, PB Rep. 1977 PB-267121. Gov. Rep. Announce. Index (US) 1977, 77(17), 101; Chem. Abs., 88, R11108). The disinfecting or purification of water by chlorine causes the formation of halogenoalkanes from dissolved or suspended organic material, and their presence has been shown in the air and the water in indoor swimming pools (U. Lahl *et al.*, Water Res., 1981, 15(7), 803; Chem. Abs., 95, 103078), in cooling water from nuclear reactors ((*a*) R. M. Bean, C. L. Gibson and D. R. Anderson, Report 1981, PNL-3271; Order No. NUREG/CR-1300. Gov. Rep. Announce. Index (U.S.) 1981, 81(21), 4494; Chem. Abs., 96, 11404 (*b*) R. M. Bean, Report, 1983, NUREG/CR-3408, PNL-4788: Order No. DE84004274. From Energy Res. Abs. 1984, 9(5), Abs. No. 8549; Chem. Abs., 101, 59828. (*c*) W. G. Haltrich *et al.*, Vom Wasser 1983, 61, 305; Chem. Abs. 100 73665) and in other water in which the natural presence of bromide ion contributes to the range of organic halides formed (A. Eisenberg *et al.*, Dev. Arid Zone Ecol. Environ. Qual. Proc. Int. Meet., Sci. Conf. Isr. Ecol. Soc. 12th, 1981, 263; Chem. Abs., 97, 97959).

The methods for their determination in air are relatively less well discussed. K. Kadlec *et al.*, (Czech CS 203,567, 15 August 1983; Chem. Abs., 100, P144398) describe a means to monitor alkyl halide levels continuously; R. W. Coutant (Report 1982, EPA-600/4-83-014, Order No. PB 83-194464. From Gov. Rep. Announce Index (U.S.) 1983, 83(16), 4018; Chem. Abs., 99, 127615) describes a passive dosimeter and other authors ((*a*) H. B. Singh *et al.*, Atmos. Environ. 1981, 15(4), 601; Chem. Abs., 95 48234. (*b*) I. G. Zenkevich, Deposited Doc. 1980, VINITI 2552-80, 23; Chem. Abs., 95 191541) describe other techniques to measure atmospheric concentrations of these materials. Many reports deal with such determinations in aqueous media ((*a*) U. Bauer, Zentralbl. Bakteriol. Mikrobiol., Hyg. Abt. 1, Org. B., 1981, 174(1-2), 39; Chem. Abs., 96, R192616. (*b*) J. K. Reichert and J. Lochtmann, Gewaesserschutz, Wasser, Abwasser 1982, 57, 42; Chem. Abs., 98, R95235 (*c*) *ibid.*, Environ. Technol. Lett., 1983, 4(1) 15; Chem. Abs., 98, 113417. (*d*) Mitsubishi Chem. Co., Jpn. Kokei Tokkyo Koho JP 58,153,163; Chem. Abs., 100, P131851. (*e*) K. Ballschmiter, Environ. Specimen Banking Minot. Relat. Banking, Prac. Int. Workshop 1982 (Publ. 1984) 264; Chem. Abs., 100, R114261. (*f*) G. E. Carlsberg and A. Kringstad, Commun. Eur. Communities [Rep] EUR 1984, EUR 8518, Anal. Org. Micropollut. Water, 276; Chem. Abs., 101, 97377).

Two stages are required in the measurement of alkyl halides at such levels. In the first, the alkyl halide must be accumulated until its quantity is accurately measurable. The generally available techniques for this include liquid-liquid extraction ((*a*) W. C. Glaze and C. C. Lin, Report 1983, EPA-600/4-83-052. Order No. PB84-112937. Gov. Rep. Announce. Index (U.S.) 1984, 84(3), 69; Chem. Abs., 100, 144718w. (*b*) P. Belouschik, V. Neitzel and I. Belouschek, GWF, Gas-Wasserfach: Wasser Abwasser 1986, 127(1), 25; Chem. Abs., 104, 212579m) and absorption and concentration on charcoal ((*a*) M. Schnitzler, Schriftenr. Ver. Wasser.- Boden-Lufthyg., 1983, 56 (Zellstoffabwasser Umwalt) 185; Chem. Abs., 101, 136675. (*b*) M. Schnitzler *et al.*, Vom Wasser, 1983, 61, 263; Chem. Abs., 100, 90979. (*c*) R. S. Summers and P. V. Roberts, Adv. Chem. Ser., 1983, 202 (Treat. Water Granular Act. Carbon), 503; Chem. Abs., 98, 166700) or on other materials such as XAD-resin, which is preferable to both charcoal and liquid-liquid extraction for the determination of volatile halides, but is inferior in the estimation of non-volatile halides (J. Alberti, Gewaesserschutz, Wasser, Abwasser, 1984, 65, 371; Chem. Abs., 100, 179698).

The second step involves the identification and quantitative analysis of the alkyl halide, which may be done by gas chromatography (E. Fogelqvist and M. Larsson, J. Chromatog, 1983, 279, 297) using negative ion CIMS with methane as reagent gas (S. Daishima, J. Iida and T. Kajiku, Nippon Kagaku Kaishi, 1984, (7), 1146; Chem. Abs., 101, 97393) or using electron capture detectors ((*a*) K. Hasegawa, S. Itoh and S. Naito, Kanagawa-ken Eisei Ken- kyusho Kenkyu Hokoku, 1983, (13) 19; Chem. Abs., 101, 197769. (*b*) B. Kolb, M. Auer and P. Pospisil, J. Chromatog., 1983, 279, 341). Alternatively, microwave-induced plasma atomic emission methods may be used (H. Raimund and H. J. Hoffmann, Vom Wasser, 1984, 63, 225; Chem. Abs., 102 R154335) or pyrolysis of the absorbed material and micro- coulometry (F. J. Springer, Gewaesserschutz, Wasser, Abwasser, 1982, 57, 88; Chem. Abs., 98, R77712). Such methods have been applied to the analysis of halogenoalkanes in wastewater ((*a*) A. Grots, Gewaes- serschutz, Wasser, Abwasser, 1982, 57, 70; Chem. Abs., 98, R77711. (*b*) M. Schnitzler, LaborPraxis 1986, 10(3), 214; Chem. Abs., 104, R229996), sea-water (B. Josefsson, J. Chromatog., 1983, 279, 119), rivers (J. Alberti, Gewaesserschutz, Wasser, Abwasser 1982, 57, 1; Chem. Abs., 98, 113381), sewage and especially in drinking water ((*a*) Y. Takahashi, R. T. Moore and R. J. Joyce, Chem. Water Reuse, 1981, 2, 127; Chem. Abs., 97, 43981. (*b*) *ibid.*, N.A.T.O. Comm. Challenges Mod. Soc. [Tech. Rep.] NATO/CCMS 1984, CCMS 12, Absorpt. Tech. Drinking Water Treat., 521; Chem. Abs., 102, R137305 (*c*) A. A. Stevens *et al.*, J. Am. Water Works Assoc. 1985, 77(4) 146; Chem. Abs., 103, R26813).

Conditions for the simultaneous analysis of bromo-, chloro-, and iodo-alkanes in admixture after combustion have been reported (A. Campiglio, Mikrochim. Acta 1982, 2(5-6), 347; Chem. Abs. 98, 10897). Interestingly, the selectivity of halide determination is improved by titrating with I^+ generated at a platinum anode in acetic acid solution 0.1-0.5M in perchloric acid and 0.002-0.08M in alkyl iodide (A. I. Kostromin *et al.*, USSR SU 1,154,608. Otkrytiya, Izobret. 1985, (1), 154; Chem. Abs., 103, P226689).

2. Monohalogenoalkanes

(a) Preparation

(i) *From alkanes*

(1) *By replacement of hydrogen.* Homolytic substitution of hydrogen by halogen in alkanes has been brought about using molecular chlorine ((*a*) M. Sano, Japan, 72 05,922; Chem. Abs., 77, 4854.(*b*) L. Bayer and A. L. Coldiron, U.S. 3,655, 801; Chem. Abs., 76, P112657. (*c*) Kh. Karagozov, E. Prodanov and A. Dinov, Chem. Prom., 1984, 34(3), 141; Chem. Abs., 101, 38089)

$$RH + X_2 \longrightarrow RX + HX \qquad (1)$$

Alkanes appear to undergo regioselective photochemical chlorination when sulphuric acid solutions are employed; the success of the method relies upon the instability of the secondary chlorides towards sulphate ester formation, so that primary chlorides predominate in the low yields (4-18%) of chloro-alkane produced (B. Wang *et al.*, Kexue Tongbao (Foreign Lang. Ed.) 1985, 30(6), 755; Chem. Abs., 105, 60278). A mixture of Cu(I) and Cu(II) chlorides is used to chlorinate alkanes in the presence of oxygen (U. Tsao, Ger. Offen. 2,706, 409; Chem. Abs., 87, P200764), conditions reminiscent of the Deacon reactor earlier reported (L. Bayer and A. L. Coldiron, *loc. cit.*). N-Chloroamines (N. C. Deno, E. J. Gladfelter, and D. E. Pohl, J. Org. Chem., 1979, 44(21), 3728) have been used as sources of homolytic chlorine in the presence of Fe(II). Some selectivity of attack could be obtained using the steric effects associated with the N-chloroamines; this encouraged the formation of secondary alkyl chlorides.

The photochemical reaction of BrCl upon alkanes has been reported (C. Walling, G. M. El-Taliawi and A. Sopchik, J. Org. Chem., 1986, 51(5), 736) using N-chlorosuccinimide and molecular bromine. 2,2-Dimethylpropane (CMe_4) gives only $Me_3C.CH_2Br$ in dichloromethane in a process in which the photo-decomposition of the N-chloroamide is catalysed by HCl but decelerated by other species.

Radical halogenation reactions have been reviewed (A. Alberti and C. Chatgilialoglu, Org. React. Mech. 1981-2 (Publ. 1984) 73). They are exemplified by the reaction of t-butyl hypochlorite with cyclohexane (H. J. Schneider and K. Philippi, J. Chem. Res., Symp. 1984 (4) 104) and the well-established electrochemical halogenation processes are also the subject of a review (S. H. Langer, J. C. Card and M. J. Foral, Pure Appl. Chem., 1986, 58(6), 895).

(ii) *From alkenes*

(1). *Direct addition of HX to alkenes* The addition of hydrogen halide to alkenes

$$>C{=}C< + HX \longrightarrow >CH{-}CX< \qquad (2)$$

is still regarded as an important route to halogenoalkanes. Hydrogen chloride, either alone ((*a*) J. K. Boggs, U.S. 3,852,368; Chem. Abs., 82, P170004; (*b*) A. V. Patrin *et al.*, U. S. S. R. SU 1,168,547; Chem. Abs., 104, P88104) or in the presence of catalysts (J. K. Boggs, U.S. 4,049,730; Chem. Abs., 87 P200766) gives alkyl chlorides. Hydrogen fluoride reacts analogously, so that but-2-ene provides 2-fluorobutane (W. P. Krause and T. Hutson, U.S. 4,049,728; Chem. Abs., 87, P200767). Alkyl bromides were made similarly by the interaction of sodium bromide, alkenes, and sulphuric acid at temperatures between 25° and 100° (S. Kobori and J. Kiro, Japan Kokai 75 05,306; Chem. Abs., 83, P27534).

(2) *Indirect addition of HX to alkenes.* Trialkyl boranes, which may be prepared by adding compounds such as di-(cyclohexyl)borane across the double bond of an alkene, form alkyl chlorides upon reaction with a number of reagents of which NCl_3 is reported (H. C. Brown and N. R. DeLuc, Tetrahedron, 1988, 44, 2875) to be the most effective. Similarly, alkyl iodides may be obtained by the iodination of tri-alkyl boranes. In this way alkyl halides may be prepared whose orientation of addition (anti-Markownikow) is contrary to that found when hydrogen halide adds across the unsaturated system in an alkene. For example, 1-iodohexane is obtained from hex-1-ene by the addition of dialkylborane to the alkene in tetrahydrofuran followed by iodinolysis with iodine monochloride and sodium acetate (G. W. Kabalka and E. E.

Gooch, Synth. Commun., 1981, 11(7), 521; Chem. Abs., 96, 6109) or with iodine (G. W. Kabalka, K. A. R. Sastry and K. U. Sastry, Synth. Commun., 1982, 12(2), 101; Chem. Abs., 96, 162067).

$$C_4H_9CH{=}CH_2 + R_2BH \longrightarrow C_4H_9CH_2CH_2BR_2 \longrightarrow C_4H_9CH_2CH_2I \qquad (3)$$

and the method may be used to insert radiolabelled iodine as ^{123}I (G. W. Kabalka, *et al.*, Int. J. Appl. Radiat. Isot., 1982, 33(3), 223) or as ^{125}I (G. W. Kabalka and E. E. Gooch, J. Chem. Soc., Chem. Commun., 1981, 1011).

(iii) *From alkyl halides*

(1) *Halogen exchange* The Finkelstein reaction involves the exchange of halogen by halide ion in a classical nucleophilic displacement;

$$RX + X'^{-} \longrightarrow RX' + X^{-} \qquad (4)$$

Eberson has used this exemplary SN2 process in discussing the application of Marcus theory towards understanding other types of substitution process (L. Eberson, Acta Chem. Scand., Ser. B 1982, B36(8), 533). Other workers (Y. Kondo *et al.*, Bull. Chem. Soc. Jpn., 1979, 52(8), 2329) have used the reaction of bromide ion with ethyl iodide to determine the role of anion solvation by various solvent mixtures, and from there to deduce the energetics of solvation of the transition state.

The process can be extended (i) by using other displacable groups such as *p*-toluenesulphonate and (ii) by bringing about the exchange by other mechanisms. An example is the fluidized bed process with an onium phase-transfer catalyst to encourage the exchange reaction; both kinetic studies and a proposed mechanism have been reported for such a system (P. Tundo, F. Trotta and G. Moraglio, J. Chem. Soc., Perkin Trans. 2, 1988, (9), 1709).

Among the other mechanisms available for such reactions, alkyl iodides may yield alkyl halides by processes in which the iodine is removed by oxidative assistance, encouraging the attack of the halide ion nucleophile.(R. J. Davidson and P. J. Kropp, J. Org. Chem., 1982, 47(10), 1904).

(i) Alkyl fluorides Alkyl fluorides are often made by such techniques. Mixtures of potassium fluoride-calcium fluoride (1:2 or 1:4 ratio, ground together and dried *in vacuo* at 150°) give good results when the exchange is carried out in acetonitrile; thus, benzyl fluoride is obtained in 81% yield from benzyl bromide after 15 hours boiling (J. Ichihara *et al.*, J. Chem. Soc. Chem. Commun. 1986, (10) 793). Silver fluoride supported on calcium fluoride also shows improved nucleophilicity as a source of F^- (T. Ando *et al.*, Chem. Lett., 1988, (11), 1877).

$$1\text{-}C_8H_{17}Br \xrightarrow[\text{MeCN, }75^\circ]{\text{AgF-CaF}_2\text{, 30 min}} 1\text{-}C_8H_{17}F \quad (41\%) \qquad (5)$$

Prolonged drying (40-45°/2 days/high vacuum) of commercial tetra-n-butylammonium fluoride is recommended as a means of lowering the necessary reaction temperature and improving the yields of alkyl fluorides made by ion-exchange (D. P. Cox, J. Terpinski and W. Lawrynowicz, J. Org. Chem., 1984, 49(17), 3216). The high heat of hydration of fluoride ion is probably significant, as the energetics of the exchange reaction depend critically upon the availability of the nucleophilic halide ion and the removal of the displaced group. The use of *aqueous* fluoride with a phosphonium salt and either an alkyl halide (F. Montanari, D. Landini and G. Lazzerini. Swiss 605, 492; Chem. Abs., 90 P5892) or an alkyl tosylate (*ibid.*, Swiss 605,473; Chem. Abs., 89, 214889) is therefore remarkable. Conversely, the report (B. Escoula, I. Rico and A. Lattes, Mol. Cryst. Liq. Cryst., 1987 (Publ. 1988), 161(Part B), 4887) that the reaction between KF and alkyl chlorides, which is catalysed by ammonium salts, is also improved when formamide is used in place of water supports the thesis that solvation, and especially hydration, is a critical factor in determining the success of such ion-exchange reactions (see, for example, D. Landini, A. Maia and A. Rampoli, J. Org. Chem., 1989, 54(2), 328). In this context, the inertness of methyl chloride towards SN2 attack by $^{37}Cl^-$ in the gas phase is relevant and supportive (J. M. Van Doren, C. H. DePuy amd V. M. Bierbaum, J. Phys. Chem., 1989, 93(3), 1130). Polyethylene glycols (specifically, PEG-400) are recommended as promoters of the reaction between alkyl sulphonates and potassium fluoride; simple alkyl fluorides are obtained in near quantitative yields after 8-168 hours at 50-100°.

$$R\text{-}SO_2OR' + KF \longrightarrow R\text{-}SO_2O^-K^+ + R'F \quad (6)$$

A solvent-free set of reaction conditions, in which nearly every other beneficial additive (CsF, $Bu_4N^+F^-$, phase-transfer agent) is present has been advocated to bring about the exchange preparatively (G. Bram, A. Loupy and P. Pigeon, Synth. Commun., 1988, 18(14), 1661). Halide ion exchange between Bu_3PFMe and alkyl chloride, bromides, or tosylates is carried out in THF to give 30-100% yields of alkyl fluoride (J. Bensoem *et al.*, Tetrahedron Lett., 1979, (4), 353). Eight cycloalkyl or tertiary alkyl halides were fluorinated in 75-100% yield by Cu_2O-HF in THF or diethyl ether (N. Yoneda *et al.*, Chem. Lett 1985, (11), 1693) Thus bromocyclohexane in THF-tetrachloromethane at 20° gave 95% yield of fluorocyclohexane.

A comparable reaction between alkyl bromides and BrF_3 gives alkyl fluorides. The attack of $BrCH_2.CHBr.CH_2Cl$ occurs predominantly at the secondary carbon atom, implying a carbocationic transition state, but the relative extents of displacement of bromine (2°:1°, 3:1) argue little selectivity and possibly a free-radical character to the process (N. N. Churatkin *et al.*, Zh. Org. Khim., 1987, 23(2), 267).

Radio-labelled alkyl fluorides have been obtained from alkyl iodides by a silver oxide induced nucleophilic displacement in MeCN at 120°. $H^{18}F$ was obtained from irradiated lithium carbonate followed by ion-exchange, and was neutralised with tetraethylammonium hydroxide to give labelled $NEt_4^+F^-$ (S. J. Gatley *et al.*, Int. J. Appl. Radiat. Isot. 1981, 32(4), 211; for an assessment of the sources of loss from such a method, see *ibid.*, Int. J. Appl. Radiat. Isot. 1982, 33(4), 255). Amino-polyethers are also advocated as catalysts in the formation of labelled alkyl fluorides by the attack of $^{18}F^-$ upon alkyl bromides (D. Block *et al.*, J. Labelled Compd. Radiopharm. 1986, 23(5), 467).

(ii) Alkyl chlorides Transesterification allows acetate esters to give alkyl chlorides by reaction with allyl chloride (L. Kh. Rakhmatullina, M. V. Lukashevich, and V. S. Markevich, Neftapererab. Neftekhim. (Moscow) 1982, (2), 31; Chem. Abs., 96, 217206p), and alkyl borates give the

corresponding chloride on treatment with aqueous HCl ((*a*) V. P. Savel'yanov *et al.*, Zh. Org. Khim., 1977, 13(3), 659 (*b*) V. P. Savel'yanov and R. T. Savel'yanova, U.S.S.R. 546,599; Chem. Abs., 87, P22349u).

Radiolabelled alkyl chlorides have been obtained through the reaction of $Ag^{39}Cl$ (itself obtained from the reaction of labelled molecular chlorine upon silver wool) with alkyl bromides or iodides. The labelled alkyl chloride was prepared in the gas-phase, and was separated and identified by gas chromatography ((*a*) M. Yagi, G. Izawa and Y. Murano Kakuriken Kenkyu Hokoku (Tokoku Daigaku) 1981, 14(2), 181; Chem. Abs., 96, 196002n (*b*) *ibid.*, Int. J. Appl. Radiat. Isot., 1985, 36(1), 69).

(iii) Alkyl bromides Similarly, alkyl bromides may be obtained from the corresponding chlorides by ion-exchange with calcium bromide in the presence of tetra-n-hexylammonium bromide (M. Yonovich and Y. Sasson, Synthesis, 1984, (1), 34); for example, 89% yields of 1-bromooctane are obtained after 24 h. at 110°.

The position of the equilibrium set up during such trans-halogenation reactions may be greatly affected by the presence of water. Sasson (Y. Sasson *et al.*, J. Chem. Soc., Chem. Commun., 1986, (16), 1250) has reported conditions for the preparation of *either* RCl *or* RBr by such reactions at 85-110° in the presence of Aliquat 336, and has described (Y. Sasson, M. Weiss and G. Barak, Bromine Compounds - Chemistry and Applications. Eds. D. Price, B. Iddon and B. J. Wakefield, Elsevier (1988) 272) a range of studies with various alkali salts and alkyl halides. Lithium bromide in the presence of methyltri(octyl)ammonium chloride is also reported as a means to convert other alkyl halides to the bromide (A. Loupy and C. Pardo, Synth. Commun., 1988, 18(11), 1275) The best conditions required to prepare octyl halides use a three-phase system comprising an organic solvent, water, and a divinylbenzene polymer cross-linked with poly-(4-vinylpyridine) residues which have been quarternised with C_2-C_{16} alkyl halides; the best catalytic effect is seen with long-chain alkyl residues on the heterocyclic nitrogen atom, and using protic solvents (H. Serita *et al*, Kobunshi Ronbunshu 1979, 36(8), 527; Chem. Abs., 91, 192388).

(iv) Alkyl iodides Ion-exchange processes between more readily commercially available alkyl derivatives and sources of nucleophilic iodine have been extensively reported as sources of alkyl iodides; formally,

$$RX + I^- \longrightarrow RI + X^- \quad (7)$$

The driving force in many cases is either the strength of the new bond being formed in such a process or the sparing solubility of the inorganic product. Thus, alkyl halides gives the organic iodides with potassium iodide in the presence of 18-crown-6-ether (N. A. Tsarenko *et al.*, Dokl. Akad. Nauk SSSR 1981, 258(2), 366), and with sodium iodide the process has been advanced as a source of ^{13}C-labelled alkyl iodides (A. Heusler, P. Ganz and T. Gaermann, J. Labelled Compd., 1975, 11(1) 37). Similarly, alkali iodides supported on zeolite achieve this exchange with alkyl bromides (T. Tatsumi, H. Ohta and H. Tominaga, Nippon Kagaku Kaishi, 1988, (10), 1758; Chem. Abs., 110, 191884). With anion-exchange resins (G. Cainelli, and F. Manescalchi, Synthesis, 1976, (7), 4723) ion-exchange occurs preparatively to give all four alkyl halides; a similar range of applications is reported with tetra-n-butylammonium salts as the source of reagent (I. Bidd and M. C. Whiting, Tetrahedron Lett., 1984, 25(51), 5949), and the corresponding process in the vapour phase, using SiO_2 - KI - $n\text{-}Bu_4N^+I^-$ (P. Tundo and P. Venturello, Synthesis, 1979, (12), 952) gives good conversions of alkyl chlorides to alkyl iodides (56-83%); it is amenable for industrial application. A copolymer of β-cyclodextrin with epichlorohydrin is advocated as the most suitable phase-transfer catalyst to achieve attack by sodium iodide upon alkyl bromides (N. Tanaka *et al.*, Chem. Lett., 1987, (4), 715). A Russian patent (N.A. Tsarenko, V. V. Yakshin and N. G. Zhukova, USSR SU 895,976. Otkrytiya, Izobret., Prom. Obraztsy, Tovarnye Znaki 1982, (1), 114; Chem. Abs., 96, P180759) reports the use of an alkali metal iodide with 1-5 mol% of 18-crown-6 ether or its dicyclohexyl analogue at 75-110° with counter-current feeding of the reagents to convert RCl to RI.

The same exchange may be achieved by treating alkyl chlorides with aluminium iodide in CS_2 at 20° (F. J. Arneiz and J. M. Bustillo, An. Quim., Ser. C, 1986, 82(3), 270; Chem. Abs., 107, 197464).

Alkyl oleates react with alkali or alkali-earth iodides analogously (H. Shiina and T. Hashimoto, Yukagaku, 1980, 29(4), 243; Chem.Abs., 93, 149744y) and trimethylsilyl iodide similarly converts alkyl chlorides or fluorides to the appropriate iodide (G. A. Olah, S. C. Narang and L. D. Field, J. Org. Chem., 1981, 46(18), 3727).

(iv) *From alcohols*

(1) *Alkyl fluorides* The pyridine-HF mixture has been thoroughly studied in a range of reactions in which carbocationic intermediates are attacked by fluoride ion (G. A. Olah *et al.*, J. Org. Chem. 1979, 44(22), 3872), but mixtures of HF with 14-23% melamine are advocated (T. Fukuhara *et al.*, Nippon Kagaku Kaishi, 1985, (10), 1951; Chem. Abs., 104, 206662s) as superior in the synthesis of alkyl fluorides from alkenes or from alcohols. Pentane or tetrachloromethane may be used as diluents in this versatile reaction. Interestingly, S. Matsubara and his colleagues failed to prepare cyclododecyl fluoride from the alcohol directly; they were obliged to dehydrate this to the alkene, add BrF across the multiple bond, and then reduce with stannane (Tetrahedron, 1988, 44(10), 2855).

1,1,2,3,3,3-Hexafluoropropyldiethylamine gives alkyl fluorides (RF) with alcohols; CF_3.CHF.CO.OR derivatives occur in considerable quantity (*ca.* 30% total yield in the case of dodecanol. (*a*) S. Watanabe *et al.*, J. Amer. Oil. Chem. Soc., 1983, 60(9), 1678. (*b*) *ibid.*, Yakagaku, 1984, 33(1), 58; Chem. Abs., 101, 7445).

Alcohols and their silylated derivatives may be selectively fluorinated by an alkyl or aryl sulphonyl fluoride in the presence of $R_4\overset{+}{N}\overset{-}{F}$ in THF; in this way 1-fluoro-tetradecane was prepared from the alcohol in 83% yield (M. Shimizu, Y. Nakahara and H. Yoshioka, Tetrahedron Lett., 1985, 26(35), 4207).

(2) *Alkyl chlorides* Alkyl chlorides are still prepared commercially by the action of alcohols upon hydrogen chloride (Mitsui Toatsu Chemicals Inc., Jpn. Kokai Tokkyo Koho JP 58,144,328; Chem. Abs. 100, P5828u).

$$ROH + HCl \longrightarrow RCl + H_2O \quad (8)$$

Charcoal, activated by absorbed metal chlorides, also encourages the formation of alkyl chlorides from alcohols with HCl (W. Linke *et al.*, Ger(East) DD 237,065; Chem. Abs. 107, P96318), and catalytic amounts of cetyltrimethylammonium salts promote good yields (80-97%) of C_4-C_{16} alkyl halides from the corresponding alcohols (B. Jursic, Synthesis, 1988, 868). Using HMPT as a solvent for the process encouraged milder reaction conditions (30-45 min., 70-85°) in which neopentyl alcohol did not suffer rearrangement (R. Fuchs and L. L. Coles, Can. J. Chem., 1975, 53(23), 3620). A continuous flow reactor using $ZnCl_2$ as Lewis acid catalyst gave a mixture of isomeric chlorides, unreacted (and unisomerised) alcohol and water (S. Yamagiwa and A. Takabe, Jpn. Kokai Tokkyo Koho JP 61,200,933; Chem. Abs., 106, 49597). In the presence of tertiary amines, such as pyridine or triethylamine, tertiary alcohols and their related alkenes underwent addition of hydrogen halide without the usual concomitant polymer- isation of the organic material (J. K. Boggs, US 3,852,368; Chem. Abs., 82, P170004q). Consistent with this is the use of amine hydrochlorides and hydrobromides as reagents to convert primary alcohols into alkyl chlorides or bromides, though here the presence of triphenylphosphine and the necessary addition of diethyl azodicarboxylate suggests a somewhat different mechanism (M. Alpegiani, A. Bedeschi and E. Perrone, Gazz. Chim. Ital., 1985, 115(7), 393). The reaction of alcohols with diphenyl(chlorophenylmethylene)iminium chloride ($PhC(Cl){:}N^+Ph_2Cl^-$), a route to alkyl chlorides in which optically active alcohols suffer complete inversion in forming the halide, may be included in this section (T. Fujisawa, S. Iida and T. Sato, Chem. Lett., 1984, (7), 1173).

A recently described process converts the alcohol, through reaction with trifluoroacetic anhydride, to the triacetate, displacing this group using lithium halide in a mixture of THF and HMPT to give the alkyl halide in 70-98% yield. Double bonds are not affected, and nor are acetate residues, which may therefore be used to protect a hydroxyl function, as in the synthesis of $AcO.CH_2.CH{:}CH.CH_2.Br$ (F. Camps, V. Gasoh and A. Guerrano, Synthesis, 1987, (5), 511).

Esters of chloroformic acid, obtained from the reaction of

the alcohol with phosgene, may be de- carboxylated to give alkyl chlorides

$$ROCOCl \longrightarrow RCl + CO_2 \quad (9)$$

by heating with polyvinylpyridine (100-175°; 0.5-10 h.: R. G. Briody and H. C. Stevens, US 4,622,431; Chem. Abs., 106, 49598) or in the presence of NH_4^+ or PH_4^+ as in a recent industrial patent (N. Grief and K. Oppenlaender, Ger. Offen. DE 3,611,419; Chem. Abs., 109, P6089b).

Phosphorus halides are popular reagents, especially when the recovery of phosphorous acids makes the process economically feasible (e.g. R. Yokoyama and T. Kojima, Japan 74 19,246; S. Kojima and M. Azuma, Japan 74, 19,247; Chem. Abs., 82, P30943-4). The treatment of long-chain primary alcohols with PCl_3 in three steps is reported to give excellent yields (97-99%) of the alkyl chloride (J. Eriksson, Z. A. Dadekian, Ger. Offen. 2,635,350; Chem. Abs., 86 P189173).

Similarly, the use of selenium(IV) or tellurium (IV) chlorides to convert alcohols into alkyl chlorides (T. Yamauchi *et al.*, Bull. Chem. Soc. Jpn, 1986, 59(11), 3617) is complicated, in the case of $TeCl_4$, by the incidence of further Friedel-Crafts alkylation of the aromatic solvents used as diluents and by the racemisation which accompanies the reaction with optically active Ph.CH(OH).Me. Alcohols react with trimethylsilyl chloride in the presence of selenium dioxide to give alkyl chlorides; in the course of the reaction selenium oxychloride is made also (J. G. Gun and K. K. Kang, J. Org. Chem., 1988, 53(15), 3634).

Analogously, Si-O bond formation probably explains the use of silicon tetrachloride to make alkyl chlorides from alcohols in *ca.* 80% yield (S. Yuki and Y. Koizumi, Japan. Kokai 78 50,107; Chem. Abs., 89, P75240z). The catalytic use of zinc chloride suggests a cleavage by HCl of the initially formed silicate ester, reminiscent of the formation of tertiary alkyl chlorides from alcohols and acetyl chloride. Secondary alcohols similarly give alkyl chlorides with *p*-toluenesulphonyl chloride in pyridine at 100° (N. P. Volynskii, L. L. Perepalitchenko and L. A. Zegel'man, USSR SU 1,296,558. Otkrytiya, Izobret., 1987, (10), 111; Chem. Abs., 107, P58494).

Copper(II) halides attack alcohols to give the alkyl halide in the presence of triphenylphosphine (S. Miyano *et al.*, Nippon Kagaku Kaishi, 1978, (1), 138; Chem. Abs., 88, 120543b). The process is held to involve nucleophilic attack by halide ion upon an alkoxyphosphonium intermediate, with net inversion of configuration; thus, (-)-R-2-octanol gives (+)-S-2 chlorooctane in 82% yield, but dehydration can diminish the yield in other instances, and no product is formed with phenol. Tetrachloromethane also provides halogen in such processes, supported by triphenylphosphine (L. A. Jones *et al.*, J. Org. Chem., 1978, 43(14), 2821). The rate of formation and decomposition of the phosphorylated intermediates lies in the order primary > secondary > neopentyl; the neopentyl adduct was sufficiently stable to be isolated. $Me_3C.CD_2OH$ showed a slight positive isotope effect and a number of plausible mechanisms were discussed. Com- parable processes involving tetrachloromethane and various phosphinic polymers have been studied ((*a*) C. R. Harrison *et al.*, J. Org. Chem. 1983, 48(21), 3721; (*b*) D. E. Bergbeiter and J. R. Blaton, J. Chem. Soc. Chem. Commun., 1985 (6) 337). In each case, the formation of strong P-O bonds is the probable driving force.

(3) *Alkyl bromides* Alkyl bromides are still reported by classical methods using alcohols, sodium bromide and sulphuric acid (R. Kozlowski, Z. Kubica and B. Rzeszotarska, Org. Prep. Proc. Int., 1988, 20(1-2), 177; Chem. Abs., 109, 37478g). Lebedev and his colleagues report in a series of patents (O. V. Lebedev *et al.*, Chem. Abs., 81, P169092-5. Okrytiya, Izobret., Prom. Obraztsy, Tovarnye Znaki 1974, 51(33), 74) the use of a number of unexpected catalysts in the reaction of alcohols with bromine and sulphur to provide alkyl bromides. These agents include cobalt or cerium oxides (USSR 430,624), vanadium pentoxide (USSR 330,830), chromium trioxide (USSR 392,682) and manganese dioxide (USSR 406, 448). Alcohols or cyclic ethers may also be converted into the alkyl bromides using concentrated aqueous HBr with tetrachloromethane, the water being removed by distillation with the solvent (N. V. Kuznetsov and R. A. Myrsina, USSR 435, 223. Otkrytiya, Izobret., Prom. Obraztsy, Tovarnye Znaki 1974, 51(25), 78; Chem. Abs., 81, P91037d).

Ethers are also cleaved by PBr_3-HBr to give alkyl bromides, as well as by HBr (D. Landini, F. Montanari and F. Rolla,

Synthesis, 1978, (10), 771) and also by trimethylsilyl bromide. The last reaction is most rapid with oxiranes, while primary acyclic ethers require 7 days at reflux temperatures to provide 40-50% yields of alkyl bromide (H. R. Kricheldorf, G. Moerber and W. Regel, Synthesis, 1981, (5), 383).

Correspondingly, trimethylsilyl bromide makes bromo-alkanes in $CHCl_3$ (M. E. Jung and G. L. Hatfield, Tetrahedron Lett., 1978, (46), 4483) with (-)-2-heptanol providing (+)-2-bromoheptane (88% optical purity; 91% yield). Preformed trimethylsilyl ethers of alcohols are cleaved by the bromine-triphenylphosphine reagent (J. M. Aizpurua, F. P. Cossio and C. Palomo, J. Org. Chem., 1986, 51(25), 4941) in a process which is catalysed by small amounts of zinc bromide, and the cleavage of t-butyldimethylsilyl ethers of alcohols by BBr_3 is claimed to show high regiospecificity of attack. For instance, 2-bromo-octane is obtained in 90% yield and with 71% retention of configuration (S. Kim and J. H. Park, J. Org. Chem., 1988, 53(13), 3111). Polymer-bound phosphorus tribromide or trichloride is reported to give alkyl halides (G. Cainelli, Synthesis, 1983, (4), 306), and the mixture of bromoform and phosphorus tribromide is also recommended in the conversion of alcohols into their bromides (Q. Huang, S. Tang and Z. Wong, Zhongshan Daxue Xuebao, Ziran Kexueban 1982, (4) 70; Chem. Abs., 98, 215138k). Bromine in an excess of phosphorus trichloride gives alkyl bromides from alcohols; alkenyl and alkynyl bromides are formed from the corresponding hydroxy-compounds (N. K. Bliznyuk, Ot. Izobret., 1987, (30), 106; Chem. Abs., 108, 204209).

Added lithium or calcium halides show a remarkable effect upon the purity and yield of alkyl bromides and iodides obtained from tertiary alcohols by the action of the hydrogen halide; the incidence of dehydration and rearrangement is greatly diminished and often removed (H. Masada and Y. Murotani, Bull. Chem. Soc. Jpn., 1980, 53(4), 1181).

Mixtures of $Na^{77}Br$, alcohol, triphenylphosphine and tetrachloromethane give labelled alkyl bromides after 5-30 minutes at 0-5° (M. R. Kilbourn and M. J. Welch, Int. J. Appl. Radiat. Isot., 1982, 33(12), 1479); if tetrabromomethane is present, some dilution of the label is caused by the formation of alkyl bromides from this substrate. Iodine-exchange (with ^{131}I) is catalysed by polymer-supported phosphonates ((*a*) M. Maeda *et al.*, Int. J. Appl. Radiat. Isot., 1979, 30(4), 255. (*b*) *ibid.*, *idem.*, 1979, 30(11), 713).

Labelled alkyl bromides may be obtained (M. R. Kilbourn, K. D. McElvany and M. J. Welch, Int. J. Appl. Radiat. Isot., 1982, 33(5), 391) from primary or secondary alcohols using $Na^{77}Br$ and trimethylsilyl chloride in acetonitrile, and the degree of isotopic labelling is limited only by the purity of the sodium bromide.

(4) *Alkyl iodides* Alcohols may be converted into their iodides using cyanuric chloride and sodium iodide in a process giving variable yields (30-76%; S. R. Sandler, Chem. Ind. (London) 1971, 133(49), 1416). 2-Chloroethyl chloroformate together with sodium iodide also brings about this reaction (J. J. Brunet, H. Laurent and P. Caubere, Tetrahedron Lett., 1985, 26(44), 5445; Chem. Abs., 104, 167680z), and a mixture of sodium iodide and trimethylsilyl pyrophosphate is particularly effective (T. Imamoto *et al.*, Synthesis, 1983, (6), 460; Chem. Abs., 99, 53034).

Trimethylsilyl iodide, either prepared *in situ* or added as such, allows alcohols to be converted into their corresponding iodides under mild reaction conditions (ambient temperature, short time). The reaction appears to be general (M. E. Jung and P. L. Ornstein, Tetrahedron Lett., 1977, (30), 2659; T. Morita *et al.*, Synthesis, 1979, (5), 379; G. A. Olah *et al.*, J. Org. Chem., 1983, 48(21), 3667).

Iodides may also be made by the cleavage of ethers, a recent variation being the use of phenyldichlorophosphate with sodium iodide in boiling xylene (H. J. Liu and V. Wiszniewski, Synth. Commun., 1988, 18(2), 1189)

A major difficulty in the use of more vigorous reagents is the competition between dehydration and the formation of alkyl halide. The use of P_2I_4, either in ether or benzene in the presence of a tertiary amine, has been studied. In ether at room temperature secondary alcohols of the structure Ar.CHOH.R provide alkyl iodides, whereas in benzene at the boiling point they give the alkenes; both *cis*- and *trans*-1,4-cyclohexanediols give mixtures of the appropriate di-iodides and 4-iodo-cyclohexene (H. Suzuki and T. Fuchita,

Nippon Kagaku Kaishi, 1977 (11) 1679; Chem. Abs., 88, 62081). In carbon disulphide at 20°, P_2I_4 gives good yields of alkyl iodides from alcohols; thus, 1-iodo-octane is obtained in 80-88% yield from 1-octanol (M. Lauwers, *et al.*, Tetrahedron Lett., 1979, (20), 1801).

(v) *From amines and ammonium salts*

Katritzky has combined the ease of preparation of pyrylium ions and the thermal decomposition of quaternary ammonium salts such as N-alkylpyridinium halides (which reverses the usual Menschutkin reaction between tertiary amines and alkyl halide) to provide a general route to halogenoalkanes among other species.

2,4,6-Triphenylpyrylium ion readily undergoes nucleophilic attack by aliphatic primary amines to generate N-alkyl-2,4,6-triphenylpyrylium ions. Pyrolysis of the appropriate salts gave alkyl chlorides (A. R. Katritzky, K. Horvath and B. Plau, Synthesis 1979, (6), 437-8), alkyl fluorides ((*a*) A. R. Katritzky, A. Chermprapai and R. C. Patel, J Chem. Soc. Chem. Commun., 1979, (5), 238. (*b*) *ibid.*, J. Chem. Soc., Perkin Trans. 1, 1980, (12), 2901) and alkyl iodides (A. R. Katritzky, N. F. Eweiss and P.-L. Nie, J. Chem. Soc., Perkin Trans. 1, 1979 (2), 433)

$$\text{2,4,6-Ph}_3\text{-pyrylium } (O^+) + RNH_2 \longrightarrow \text{2,4,6-Ph}_3\text{-pyridinium } (N^+\text{-}R)\ Hal^- \xrightarrow{\Delta} \text{2,4,6-Ph}_3\text{-pyridine} + RHal \qquad (10)$$

Alkylamines ($R.CH_2.CH_2.NH_2$) may also provide carbocationic intermediates which, in the presence of sources of incipient halide ion, may provide alkyl halides. Thus, with peroxodisulphate ion in the presence of copper(II) chloride, alkylamines yield alkyl chlorides though the major product is the nitrile $R.CH_2.CN$ along with $R.CCl_2.CHO$ and $R.CH_2.CO_2H$ (E. Troyanskii *et al.*, Izv. Akad. Nauk SSSR, Ser. Khim. 1983, (11), 2554). In principle, the diazotisation method which allowed the preparation of a number of aryl bromides by treatment of the corresponding amine with a mixture of alkyl nitrite and copper(II) bromide may be applied to aliphatic amines with similar success, but this has not been reported recently.

N-Alkylpiperidines have also been reported (R. A. Olofson and D. E. Abbott, J. Org. Chem., 1984, 49(15), 2795) to react with 1-chloroethyl chloroformate ($ClCO.O.CHCl.CH_3$) to give alkyl chlorides (90-97% yield) in a way superficially similar to that already described (Brunet, Laurent and Caubere, *loc.cit.*) for alcohols.

(vi) *From carboxylic acids and derivatives*

Despite the expense of the reagents the Hunsdiecker synthesis of alkyl bromides by the reaction of molecular halogen upon the silver or mercury(II) salts of carboxylic acids is still employed. A modern variant (R. C. Cambie *et al.*, J. Chem. Soc., Perkin Trans. 1, 1981, 2608) relies upon the formation of thallium(III) carboxylate dibromides from Tl(I) carboxylate and bromine; subsequent decomposition of this provides the alkyl bromide, carbon dioxide, and TlBr. The mechanism of this reaction , which is useful only for the synthesis of primary alkyl bromides, is probably closer to those of organic iodine(III) derivatives rather than the Hunsdiecker process.

$$RCO_2^{-}M^{+} + Br_2 \longrightarrow MBr + RBr + CO_2 \qquad (11)$$

Alkyl chlorides may be obtained from carboxylic acids by the formation of acyl derivatives (esters) of benzophenone oxime ($Ph_2C{=}NOCOR$). Photolysis of such esters in tetrachloromethane may give good yields (up to 87%) of alkyl chlorides (M. Hasebe and T. Tsuchiya, Tetrahedron Lett., 1988, 29, 6278) in a reaction which reflects the well-documented instability of benzophenone oximes. Esters also react with iodine under an atmosphere of hydrogen and (optionally) carbon monoxide under pressure, and in the presence of a platinum catalyst and a quaternary nitrogen or phosphorus compound as promoter, to give alkyl iodides.(H. K. Kuebbeler *et al.*, E. P. Appl. EP 46,870; Chem. Abs., 97, P5788)

Alkyl derivatives of acetyl chloride, $R_2CH.COCl$ and $R_3C.COCl$, react with hydroxypyridinethione in the presence of polyhalogenomethanes XCY_3 to give secondary (R_2CHX) or tertiary

(R_3CX) alkyl halides (R. E. Stafer and C. Lion, Bull. soc. chim. Belg., 1987, 96(8), 623).

(vii) *From aldehydes* Reductive halogenation could be brought about by treating aliphatic aldehydes (RCHO) with 1,1,3,3-tetramethyldisiloxane and trimethylsilyl halide, when RCH_2X was obtained in good yield. In the presence of trimethylsilyl trifluoroacetate, however, only the dialkyl ethers $(R.CH_2)O$ were made (J. M. Aizpurua, B. Lecca and C. Palomo, Can. J. Chem., 1986, 64(12), 2342).

(viii) *From N-halogenoamides* N-Chloro- and -bromo-amides have long been used to initiate radical chains by which alkyl, and particularly allyl, halides may be prepared. N-Fluoro-sulphonamides ($R.SO_2.NFR'$) are obtained by direct fluorination of $R.SO_2.NHR'$, and act as fluorinating agents to carbanionic species, so that indene anion gives 1-fluoro-indene, and alkyl Grignard reagents provide the corresponding alkyl fluorides (W. F. Barnette, J. Amer. Chem. Soc., 1984, 106, 452). The latter reaction may be regarded as an alternative to the Finkelstein reaction, but its general applicability has not yet been assessed.

(*b*) *Reactions of monohalogenoalkanes*

Reviews of the reactions and properties of alkyl halides are plentiful ((*a*) E. W. Colvin, Annu. Rep. Prog. Chem., Sect. B, 1975, 72, 199. (*b*) G. M. Brooke, Int. Rev. Sci.: Org. Chem., Ser. Two 1975, 2, 83. (*c*) R. C. F. Jones, Gen. Synth. Methods 1978, 1, 156. See also previous sections of this report, and particularly C. A. Buehler and D. E. Pearson, Survey of Organic Synthesis, Wiley-Interscience NY 1970, 329; Vol. II, 1977, 345) and their reactions are numerous. They may be divided into processes involving displacement of the halogen, and those in which the alkane structure is attacked; within each of these subsections further division may be made, depending upon the nature of the reagent and the mechanism of the process. There have been a number of attempts to rationalise displacement reactions; a conventional study of solvent effects upon the rates of reaction of ethyl iodide with some substituted pyridines (S. M. M. Elshafic, M. M. Sayed and F. A. Fouli, Egypt J. Chem., 1986, 29(6), 647; Chem. Abs., 110, 134446) and a detailed dissection of the contributions of the initial state and the transition state in the solvolysis of t-butyl bromide and chloride in 20-30 systems (M. H. Abrahams *et al.*, J. Chem. Soc., Perkin Trans. 2, 1988, (9), 1717) serve as examples.

Three mechanisms of displacement of halogen from alkyl halides seem to prevail. The classical Ingold mechanisms of unimolecular or bimolecular nucleophilic attack may be encouraged by the presence of metal ions which are reported ((*a*) V. V. Zamashchikov, T. V. Bezbozhnaya and N. A. Kovach, React. Kinet. Catal. Lett., 1986, 32(1), 233; Chem. Abs., 107, 153774. (*b*) V. V. Zamashchikov, E. S. Rudakov and T. V. Bezbozhnaya, React. Kinet. Catal. Lett., 1984, 24(12), 65; Chem. Abs., 101, 71911) to act both as charged particles affecting selectivity in hydrolysis and as Lewis acid promoting the heterolysis of the C-X bond.

Also, halogen may be lost under conditions in which single-electron transfer (SET) permits a range of reactions with a free-radical character. For example, the reaction of some lithium alkoxides with alkyl iodides gives the expected ethers, but the reaction of the same lithium alkoxides with perylene (1) provide the well-authenticated perylene radical anion

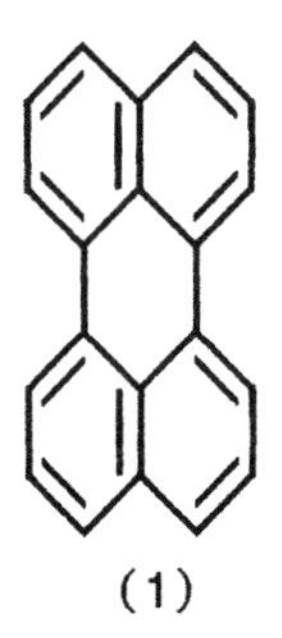

(1)

which is apparently derived from a single-electron transfer process. Lithium isopropoxide with 2,2-dimethyl-1-iodo-5-hexene also gives cyclisation products, indicative of an intermediate with sufficient free-radical character to attack the alkene system (E. C. Ashby *et al.*, Tetrahedron Lett. 1984, 25(45), 5107). Theoretical considerations also support such a mechanism, making a clear distinction between cases where the classical mechanism is expected (as in the Finkelstein reaction) and those in which single electron transfer is encouraged (L. Eberson, Acta Chem. Scand., Ser. B. 1982, B36(8), 533). In addition to Eberson's studies, T. Lund and H. Lund (Acta Chem. Scand., Ser. B, 1988, B42(5), 269) have examined a VCBM model for nucleophilic displacement upon alkyl halides. Bordwell has also made careful studies of the conditions under which electron transfer becomes important in the displacement of halogen by nucleophiles such as substituted fluorene anions (2) (*e.g.* F. G. Bordwell and C. A. Wilson, J. Amer. Chem. Soc., 1987, 109(18), 5470; F. G. Bordwell and J. A. Harrelson, J. Amer. Chem. Soc., 1989, 111(3), 1052) and analogous systems derived from phenothiazine, carbazole, and similar structures.

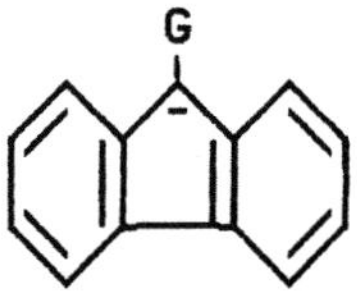

G

(2)

Steric effects are clearly recognised as important in determining the occurrence of electron transfer in preference to classical nucleophilic displacement (F. G. Bordwell and J. A. Harrelson, J. Org. Chem., 1989, 54(20), 4893).

The oxidative displacement of iodine from alkyl iodides has also been associated with unexpected consequences. Zefirov *et al.* (Zh. Org. Khim., 1982, 18(12), 2608) have pointed out that the oxidative displacement of iodine from alkyl iodides under the influence of chlorine, *m*-chloro-perbenzoic acid, or periodic acid encourages the formation of products in which very weak nucleophiles have been incorporated to the exclusion of relatively more potent species such as water or chloride ion. In such conditions, 20-50% yields of alkyl tosylates, methanesulphonates, and trifluoromethanesulphonates are obtained in aqueous media. In aprotic organic media, even alkyl perchlorates may be obtained in the presence of lithium or tetra-alkylammonium perchlorate (N. S. Zefirov, V. V. Zhdankin and A. S. Koz'min, Izv. Akad. Nauk SSSR, Ser. Khim., 1982, (7), 1676; Chem. Abs., 97, 126970). Macdonald and Naresimham (J. Org. Chem., 1985, 50(24) 5000) similarly found that alkyl iodides could be encouraged to give fluorides, chlorides, bromides, acetates, trifluoroacetates and tosylates under similar mildly oxidising conditions and in the presence of the appropriate nucleophiles.

Although these reactions have also been described as involving single-electron transfer processes, there is some confusion in the literature; for example, Zefirov (N. S. Zefirov *et al.*, Chem. Scr., 1983, 22(4), 195; Chem. Abs., 100, 50767) includes the oxidation of alkyl iodides among a number of classically electrophilic processes (*e.g.* addition to alkenes) in which such "super weak" nucleophiles become incorporated into the aliphatic system. Davidson and Kropp (J. Org. Chem., 1982, 47(10), 1904) suggest that the oxidation of alkyl iodides by *m*-chloroperbenzoic acid to give RIO_n encourages nucleophilic attack by unusually weak reagents, and in this respect the observation that alkyl iodides react with aryliodine(III) derivatives such as $ArI(OCOCF_3)_2$ and $(ArI.ONO_2)_2O$ at 25° in $CHCl_3$ or CH_2Cl_2 to give products such as $R.OCOCF_3$, $R.ONO_2$ and $R.OClO_3$ becomes significant (N. S. Zefirov, V. V. Zhdankin and A. S. Kos'min, Izv. Akad. Nauk SSSR, Ser. Khim., 1983, (7), 1682). The analogous reaction of alkyl iodides with $ArI(OCOR')_2$ to give esters $(R'CO_2R)$ in $CHCl_3$ at room temperature over 2-8 hours (J. Gallos, A. Varvoglis, J. Chem. Soc., Perkin Trans. 1, 1983, (9), 1999) has preparative use also.

At the mechanistic extreme is the homolysis of the carbon-halogen bond induced in the attack of alkyl bromides or iodides by triphenylpropargyltin ($HC{\equiv}CCH_2SnPh_3$) to give terminal allenes, a reaction carried out under the gradual addition of AIBN as a radical source (J. E. Baldwin, R. M. Adlington and A. Basak, J. Chem. Soc., Chem. Commun., 1984 (19) 1284).

1. *Electrochemical Reduction*

The electrochemical reduction of alkyl halides has been extensively studied. ((*a*) G. M. McNamee, Diss. Abs. Int. B. 1977, 37(8), 3917; Chem. Abs., 86, 154735f. (*b*)T. Iwasaki and K. Harada., J. Chem. Soc.., Chem. Commun., 1974, (9), 338. (*c*) D. M. La Perriere, Diss. Abs. Int. B, 1979, 39(7), 3334; Chem. Abs., 90, 120715). In contrast to the corresponding hydrogenodehalogenation of aryl halides, where the $[ArX]^{\cdot-}$ is an intermediate, reduction of alkyl halides at inert electrodes is a concerted process which takes place to give alkyl radicals directly. The ease of C-X bond stretching, as well as the C-X bond dissociation energy, determines the $RX/(R^{\cdot}X^{\cdot})$ standard potential ((*a*) C. P. Andrieux *et al.*, J. Amer. Chem. Soc., 1986, 108(4), 638). (*b*) C. P. Andrieux, J. M. Saveant and K. B. Su, J. Phys. Chem., 1986, 90(16), 3815). The electrocatalysed reduction of alkyl chlorides or bromides in THF-HMPT (2:3) is quantitative in the presence of ethylenebis(salicylideniminato)cobalt (J. M. Duprilot *et al.*, J. Organomet. Chem., 1985, 286(1), 77). β-Hydroxyalkyl bromides may be reduced electrochemically; the hydroxyl function is protected as the tetra- hydropyranyl ether, and the resulting alkyl bromide is electrolysed in DMF solution in the presence of chromium(II), ethylene diamine, and butane-1-thiol when bromine is replaced by hydrogen (J. Wellmann and E. Steckhan, Angew. Chem., Int. Ed. Engl., 1980, 19, 46). Both dehydration and dehydrohalogenation of the parent material are minimised by this variant of the Barton reagent.

The reduction of alkyl halide by chemical means is also widely reported. Thus, alkyl halides and mesylates are reduced by $[CNBH_3]^-$ (R. O. Hutchins *et al.*, J. Org. Chem., 1977, 42(1), 82) (*b*) G. W. Kabalka, J. D. Baker, and G. W. Neal, J. Org. Chem., 1977, 42(3), 512. (*c*) R. J. Kinney, W. D. Jones and R. G. Bergmann, J. Amer. Chem. Soc., 1978, 100(2), 635). Lithium aluminium hydride has been used, either alone (S. Krishnamurthy and H. C. Brown, J. Org. Chem., 1980, 45, 849) or in the presence of metal chlorides such as $CoCl_2$ and $NiCl_2$ (E. C. Ashby and J. J. Lim, Tetrahedron Lett., 1977, (51), 4481) when catalytic amounts are needed to reduce primary halides, but stoichiometric quantities are required to attack secondary halides.. Lithium triethylborohydride is superior to LAH in the reduction of alkyl halides (Krishnamurthy and Brown, *loc. cit.*); interestingly, bromine (iodine) or tosylate groups may each be selectively displaced by hydrogen using LAH. In diglyme alkyl iodides were reduced, but alkyl tosylates were not; the opposite preference is seen in diethyl ether solutions. The situation is exemplified by the reduction of 11-bromoundecyl tosylate, in which either halogen or sulphonate group may be removed according to the nature of the reaction solvent (S Krishnamurthy, J. Org. Chem., 1980, 45, 2550). The reduction of alkyl bromides to alkanes by AlH_3 or BH_3 seems to involve conventional nucleophilic displacement by H^- rather than some single-electron transfer process. No evidence of radical products were found (S. U. Park, S. K. Chung, and M. Newcomb, J. Org. Chem., 1987, 52(15), 3275).

Lithium copper hydrides are involved in the reduction of alkyl halides to alkanes by the product of reaction of methyl-lithium, copper(I) iodide and then LAH in ether. The heavier halogens are cleaved quantitatively, and even fluorine is displaced to the extent of 10% under the reaction conditions (E.C. Ashby, A. B. Goel and J. J. Lin, Tetrahedron Lett., 1977, (42), 3695). The same reduction has been reported using sodium hydride, a tertiary alkoxide ($EtCMe_2ONa$) and a transition metal halide as catalyst in either THF or dimethoxyethane (B. Loubinoux *et al.*, Tetrahedron Lett., 1977, (45), 3951). Sodium naphthalene also brings about reduction; the process clearly involves radical intermediates, since the products of $R^{\cdot}$ and RH recombination in the reactions of EtBr, Me_2CHCl, Me_2CHBr and $Me_2CH.CH_2I$ show a pronounced CIDNP effect (M. A. Komba *et al.*, Tetrahedron Lett., 1977, (15), 1513).

In addition, a range of anionic (nucleophilic) metal hydride derivatives have been studied such as the vanadium system $[(Ph_3P)_2N]^+[\eta^5\text{-}C_5H_5V(CO)_3H]^-$ which in THF or MeCN at room temperature rapidly forms RH from RX, a process in which the alkyl radical may be identified (R. J. Kinney, W. D. Jones, and R. G. Bergmann, J. Amer. Chem. Soc., 1978, 100(2), 635). Alkyl iodides are also reported to be reduced by sodium iodide and perchloric acid in the presence of Pt(II). Oxidative addition of RI gives a species RPt(IV), which is firstly reduced by iodide ion to RPt(II) and then suffers protolysis to give RH ((*a*) V. V. Zamashchikov *et al.*, Zh. Obshch. Khim., 1982, 52(10), 2356. (*b*) *ibid.*, React. Kinet. Catal. Lett., 1982, 21(1-2), 141; Chem. Abs., 98, 88627).

2. Sulphur nucleophiles Sulphur nucleophiles have been widely used in preparative chemistry involving alkylation. The most widely reported synthesis is the alkylation of alkali metal sulphides or polysulphides to form thiols or dialkyl disulphides.

$$RX + R'S_n^- \longrightarrow RS_nR' + X^- \quad (R' = H, \text{alkyl}; n = 1, 2) \qquad (12)$$

For example, A. B. Kuliev *et al* (Zh. Org. Khim., 1978, 14(3), 661. See also V. I. Koshutin *et al.*, USSR 380,645; Chem. Abs., 79, 52784) describe a classical synthesis of sulphides and disulphides by the interaction of alkyl halides, sodium thiolates, and (where appropriate) sulphur in ethanolic solution. The process may be carried out in aqueous media (*e.g.* R. Kolta, K. Mihalszky and I. Czeka, Hung. Teljes 9,289; Chem. Abs., 82, 170041). Trialkyl phosphites in small amount (0.1-0.5%) are recommended catalysts in the synthesis of C_1 to C_4 thiols from SH^- and RX (S. V. Golubkov *et al.*, USSR 537,998. Otkrytiya, Izobret., Prom. Obraztsy, Tovarnye Znaki 1976, 53(45), 92; Chem. Abs., 86, 139389) although A. M. Vasil'tov, B. A. Trefimov and S. V. Amosova (Zh. Org. Khim., 1983, 19(6), 1339) claim that C_2-C_{12} straight-chain alkyl bromides give satisfactory yields of the thiol upon treatment at room temperature with 1-1.5 mole of NaSH in DMSO. The reaction of S_8 with THF solutions of $LiEt_3BH$ to give Li_2S or Li_2S_2 allowed the formation of R_2S or R_2S_2 in 70-80% yields from alkyl halides (J. A. Gladysz, V. K. Womg and B. S. Jick, J. Chem. Soc., Chem. Commun., 1978, (19), 858 and Tetrahedron, 1979, 35(20), 2329).

Other published methods for the preparation of thiols rely upon the use of phase-transfer catalysts such as $C_{16}H_{33}\overset{+}{P}Bu_3\ Br^-$ to encourage the reaction of S^{2-} or of thiolate anion with RX (D. Landini and F. Rolla, Org. Synth., 1978, 58, 143) or the use of labile intermediates. These include the preparation of RS^tBu from the attack of thio-t-butoxide upon alkyl halides (T. Yamamoto *et al.*, Bull. Chem. Soc. Jpn., 1983, 56(4), 1249) and the use of $Me_3COCS_2^-K^+$ whose attack upon RX provides intermediates which decompose at 75- 80° to give RSH (I. Degani, R. Fuchi and M. Santi, Synthesis, 1977, (12) 873), a process which is much quicker with primary alkyl bromides than with either the corresponding chlorides or the analogous secondary bromides. Ion-exchange columns may also be used; thus, the hydroxide form of Amberlyst A26 may be treated first with thiols and then with alkyl halides to give RSR' and in the disulphide form yields R_2S_2 (G. Cainelli *et al.*, Gazz. Chim. Ital., 1982, 112(11-12), 461).

The reaction of KCNS with alkyl halides proceeds to give the expected substitution products, although the yields are improved with tetraalkylammonium salts as catalysts or phase-transfer agents acting between the aqueous salts solutions and the organic reactant (thus, n-BuBr gives n-BuSCN in 70% yield over 18 h) yields are further increased either by stirring or by sonication, when 97% yields are obtained (W. P. Reeves, J. V. McClucky, Tetrahedron Lett., 1983, 24(15), 1585). The alkylation of mercury(II) thiocyanate gives different products according to the reaction conditions. N-alkylation predominates in hexane, di-butyl ether or 1,2-dichloroethane (N. Watanabe, S. Uemura and M. Okano, Bull. Chem. Soc. Jpn., 1974, 47(11), 2745), when a kinetic study suggests the intermediacy of R^+ and $[XHg(SCN)_2]^-$, but in tetrahydrofuran the products are $R[O(CH_2)_4]_nNCS$ and $R[O(CH_2)_4]_nCNS$ in which one or more molecules of the solvent appear to have been intruded between the initially formed ions. Ions such as (3) are suggested as intermediates (*ibid., idem.*, 1975, 48(11), 3205).

(3)

Sulphones may be prepared by alkylation of sulphinates,

$$ArSO_2^-M^+ + RX \longrightarrow ArSO_2R + M^+X^- \quad (13)$$

a process in which O-alkylation is clearly unimportant. The reaction may be brought about in good yield by using preformed Bu_4N^+ *p*-toluenesulphinate (G. E. Vannstra and B. Zwanenburg, Synthesis, 1975, 519) in preference to the commercially available sodium salt, or also by using the benzenesulphinate form of a polystyrene anion-exchange resin (F. Manescalchi, M. Orena and D. Savoia, Synthesis, 1979, 445). The advantage of the tetrabutylammonium system could be realised by using sodium *p*-toluenesulphinate in 1,2-dimethoxyethane in the presence of tetrabutylammonium bromide (J. Wildeman and A. M. van Leusen, Synthesis, 1979, 733) and more recently conditions were described in which catalytic amounts of this salt are used to bring about good yields of sulphones from alkylation of sodium *p*-toluene- sulphinate in water-acetone-benzene mixtures (J. K. Crandell and C. Predat, J. Org. Chem., 1985, 50(8), 1327). Catalysis was also reported when poly-ethylene glycols, either alone or in the presence of methanol, were added to the mixture of sodium salt and alkyl halide (K. Sukata, Bull. Chem. Soc. Jpn., 1984, 57(2), 613).

3. *Halide nucleophiles* While the preparative aspects of the Finkelstein reaction have been detailed in the earlier section, alkyl bromides and iodides have been used as sources of halogen in redistribution processes. An example is the formation of di- and tri-butyltin halides (Bu_3SnX or Bu_2SnX_2 where X=Br or I) from the corresponding chlorides using tetra-n-butylammonium halide as the catalyst and RBr or RI as the source of the entrant halogen (E. C. Friedrich C. B. Abma and G. Delucca, J. Organomet. Chem., 1982, 228(3), 217). Trimethylsilyl chloride, however, was not affected under these conditions, although the earlier reports of the use of sodium iodide as a promoter of the reaction of this chlorosilane are most readily understood in terms of such a halogen exchange process.

4. *Oxidation of alkyl halides.* Among less usual reagents, dimethyl sulphoxide with sodium iodide converts alkyl chlorides or bromides to the corresponding aldehydes

$$\mathrm{RCl} \longrightarrow \mathrm{RCHO} \tag{14}$$

and this may be used as a one-step synthetic route (P. Dave, H. S. Byun, and R. Engel, Synth. Commun., 1986, 16(11), 1343; Chem. Abs., 107, 22941).

5. *Carbon nucleophiles*

The alkylation of malonic ester derivatives was achieved in a new way when the dimethyl ester of trifluoromethylmalonic acid was deprotonated by the electrochemically-generated pyrrolidone anion in DMF; subsequent treatment of the malonic ester anion with alkyl halide (R = Me, Bu) then gave the product $R.(CF_3)C(CO_2Me)_2$ (T. Fuchigami and Y. Nakagawa, J. Org. Chem., 1987, 52(23), 5276).

Alkyl iodides react with ketones in the presence of SmI_2 to provide tertiary alcohols in a Grignard-type process which is catalysed by $FeCl_3$. This suggests electron-transfer to RI from the samarium iodide, but no products of coupling of the putative intermediate alkyl radical are found (H. B. Kagan, J. L. Namy and P. Girard, Tetrahedron Suppl. 1981, (9), 175).

Electrochemical alkylation of monosubstituted acetylenes in $(Me_2N)_3PO$ with $Bu_4\overset{+}{N}\overset{-}{I}$ gives 80-100% PhC:CR via $Bu_4\overset{+}{N}\overset{-}{C}{\equiv}CPh$. (M. Tokuda, J. Chem. Soc., Chem. Commun., 1976, (15), 606). The reaction of sodio derivatives of terminal acetylenes (e.g. PhC≡CNa) with alkyl iodides (RI; R = Me, Et, Pr) may be brought about at 60-80° in benzene, and accelerated using 2 mol% dibenzo-18-crown-6-ether (45-72% after 5 h.). The crown ether effect is less for the longer R groups (O. A. Tarasova, S. V. Amosova and B. A. Trofimov, Zh. Org. Khim, 1983, 19(8), 1765). The alkylation of benzyl cyanide or of methyl benzyl ketone by methyl, ethyl, or butyl iodides is catalysed by 2,6-di-sulphinylpyridine derivatives (N. Furukawa *et al.*, Zh. Org. Khim., 1982, 18(12), 2608).

The corresponding reaction of alkyl bromides or iodides with $PhS.CH(Li).SiMe_3$ in hexane (0°/1 h; 45-99%) gives an alkylated analogue which provides, ultimately, the corresponding aldehyde (D. J. Agar, J. Chem. Soc., Perkin Trans. 1, 1983, (6), 1131)

$$PhS{-}CHLi{-}SiMe_3 \xrightarrow{RX} PhS{-}CHR{-}SiMe_3 \xrightarrow{[O]}$$

$$PhS{-}CHR{-}OSiMe_3 \xrightarrow{H_2O} RCHO \qquad (15)$$

Ketones (R.CO.R') may be obtained in good to excellent yields from esters of 2-hydroxypyridine, or of the corresponding thiol, $(R.CO.X.C_5H_3N.Z$; X = S, O) which reacts with alkyl iodides (R'I) in the presence of nickel(II) chloride and zinc powder at 50° over 5 h. (Mitsubishi Chem. Ind. Co. Ltd., Jpn Kokai Tokkyo Koho JP 57,165,336; Chem. Abs., 98, P71490).

6. Miscellaneous

Alkyl iodides may also undergo reductive carbonylation by treatment with a precious-metal catalyst (e.g. $PtCl_2(PPh_3)_2$) in the presence of hydrogen and carbon monoxide under pressure. In the presence of methanol, the methyl ester of the corresponding carboxylic acid is formed (R. Takeuchi *et al.*, J. Chem. Soc., Chem. Commun., 1986, (4), 351). If $Co_2(CO)_8$ and sodium acetate are used under high pressure of H_2 and CO, even alkyl chlorides undergo the carbonylation process to give aldehydes (P. Leconte, Eur. Pat. Appl., EP 165,881; Chem. Abs., 105 P45225). In a variant of such processes, an acyl chloride (R.COCl) and an alkyl iodide (R'I) react in the presence of a Pd complex to provide the ketone R.CO.R' (Y. Tamaru *et al.*, Tetrahedron Lett., 1985, 26(45), 5529).

Alkyl iodides also react with protonated heterocyclic bases in the presence of benzoyl peroxide, when the alkyl group is introduced at sites *ortho* and *para* to the hetero-atom (G. Castaldi *et al.*, Tetrahedron Lett., 1984, 25(35), 3897).

"Reactive" alkyl halides, such as methyl iodide and allyl bromide, act as sources of halogen in the reaction between alcohols and N,N'-carbonyldi-imidazole to give RBr or RI (T. Kamijo, H. Harada and K. Iizuka, Chem. Pharm. Bull., 1983, 31(11), 4189).

Second Supplements to the 2nd Edition of Rodd's Chemistry of Carbon Compounds, Vol. IA/B, edited by M. Sainsbury
© 1991 Elsevier Science Publishers B.V., Amsterdam

Chapter 3b

POLYHALOGENOALKANES AND ALKENYL HALIDES

R. BOLTON

Much of the chemistry and synthesis of polyhalogenoalkanes and some of the chemistry of unsaturated halides is similar to that of the monohalogenoalkanes, and the reader is advised to refer to the previous Chapter, which should be read in conjunction with the present review.

1. Dihalogenoparaffins. Dihalogenoalkanes $C_nH_{2n}X_2$

(a) Preparation.

The methods of synthesis of polyhalogenoalkanes may be subdivided into general groups. The first are simple extensions of techniques already described in the earlier Chapter; thus, 1,6-dibromohexane may be readily prepared by treating hexane-1,6-diol with $HBr-H_2SO_4$ mixture. The second type of method uses the peculiarities of the system in order to bring about the reaction; thus, 1,1-di-fluoroalkanes can be prepared by the reaction of an aldehyde with phenyl sulphur trifluoride and the replacement of a carbonyl system by halogen has not been earlier discussed.

(i) With both halogen atoms attached to the same carbon atom

1,1-Difluorides may be obtained from other *gem*-dihalides through the combined effect of HgO and pyridine-HF mixtures (30:70; G. A. Olah *et al.*, J. Org. Chem., 1979, 44(2), 3872).

1,1-Di-iodides may be prepared from 1,1-dibromides using CuI in *ca.* 70% yields (H. Suzuki *et al.*, Synthesis, 1988, 236). An interesting variant upon the cleavage of ethers by HI comes in the synthesis of *gem*-di-iodides from 2,2-dialkylated benzdiazoles or benz-oxathioles. Cleavage is achieved by acetyl chloride - sodium iodide mixtures (L. Corda *et al.*, J. Heterocycl. Chem., 1988, 25(1), 311).

(ii) With halogen atoms attached to different carbon atoms

The addition of halogen to alkenes provides 1,2-di-halides by one of the oldest reactions in organic chemistry (G. B. Sergeev and V. V. Smirnov, "Molekulyarnoe Galogenirovanie Olefins" MGU, Moscow, USSR (1985); Chem. Abs. 104, B167967). The selection of electrophilic reagents available has been discussed in the application of zeolites and other supports towards controlling the orientation and selectivity of halogen addition (K. Smith *et al.*, Brit UK Pat. Appl. GB 2,155,009; Chem. Abs., 104, P128604) and similar discussions have been concerned with the attack of perfluoroalkenes (Yu. V. Zeifman *et al.*, Izv. Akad. Nauk SSSR, Ser. Khim., 1984, (5), 1116. (*b*) L. S. German and G. I. Savicheva, *idem.*, 1984 (2) 478. (*c*) A. V. Fokin *et al.*, *idem.*, 1985, (10), 2298). Perfluoroalkenes are attacked by halogen in fluorosulphonic acid, either alone or with added SbF_5; under these conditions both the nature of the halogenating agent and the possible fates of intermediate carbocations are not readily predictible. The isolation of various products from tetrafluoroethene and heptafluoropropene show the complexity of the process, although some of the products are easily explicable (X=Hal; R=CF_3 or F)

$$F_2C{=}CFR \longrightarrow FO_2S{-}O{-}CF_2{-}CFR + X{-}CF_2{-}CFR{-}X \qquad (1)$$

When iodine is present the initially formed alkyl iodide may then suffer oxidative displacement of halogen to give ultimately a fluorine substituent.

Mixed halogen addition may occur either by using the inter-halogen (*e.g.,* BrF) or by diverting the halogeno-carbocationic intermediate in the addition process by providing another halide ion.

$$>C{=}C< + X_2 \xrightarrow{-X^-} >CX{-}\overset{+}{C}< \xrightarrow{+Y^-} >CX{-}CY< \qquad (2)$$

In this way XY may be formally added to an alkene (*e.g.* (*a*) N. S. Zefirov *et al.*, J. Org. Chem., 1982, 47(19), 3679. (*b*) *ibid.*, Zh. Org. Khim., 1984, 20(2), 233). From the preparative viewpoint it is immaterial whether the sequence above is followed, or whether the inter-halogen is first formed and then adds

$$>C{=}C< + XY \xrightarrow{-Y^-} >CX{-}\overset{+}{C}< \xrightarrow{+Y^-} >CX{-}CY< \qquad (3)$$

or whether a trihalide ion (*e.g.* $BrCl_2^-$) is involved. The mechanistic distinctions are considerable, however, and have been argued extensively (see, for example, P. B. D. de la Mare and R. Bolton, "Electrophilic Addition to Unsaturated Systems", 2nd Edn., Elsevier, Amsterdam (1982)).

The preparative addition of BrF to dichloroethylenes (N. B. Mel'nikova *et al.*, Izv. Vyssh. Uchebn. Zaved., Khim. Khim. Tekhnol., 1981, 24(9), 1070; Chem Abs., 96, 19523b) and to unsaturated steroids (C. D. Ksiezny *et al.*, Rom. RO 70,443; Chem. Abs., 99, P176162n) have been reported, and the prior synthesis of BrF and IF by union of their elements has prepared reagents which attack more rapidly and which show Markownikow orientation of attack of alkenes (S Rozen and M. Brand, J. Org. Chem., 1985, 50(18), 3342). Polymer-supported systems (A. Gregorcic and M. Zupan, J. Fluorine Chem., 1984, 24(3), 291) have been used successfully, and a detailed study has been made of the N-bromosuccinimide - HF system where a combination of solvent effects and the possible protonation of various reactants causes great changes in the stereoselectivity of addition, although always with Markownikow orientation (S. Hamman and C. G. Beguin, J. Fluorine Chem., 1983, 23(6), 515).

Apart from these methods, the addition of BrCl may also be

brought about using the dichlorobromide anion. The attack of alkenes is reported (T. Negoro and Y Ikada, Bull. Chem. Soc. Jpn., 1984, 57(8) 2111) and a mechanism proposed for the complex processes observed both here (*ibid., idem.*, 2116) and in the attack of conjugated dienes, when both Markownikow and anti-Markownikow addition occurs; thus butadiene gives 1-bromo-2-chlorobut-3-ene (84%) and 1-chloro-2-bromobut-3-ene (16%) (*ibid., idem.*, 1985, 58(12), 3655).

ClF adds to halogenoalkenes (N. Amirtha, S. Viswanathan and R. Ganesan, Bull. Chem. Soc. Jpn., 1983, 56(1), 314) and a range of alkenes have been treated with ClX (Chem. Abs., 98, 125095g. A. K. Singh and S. M. Verma, Indian J. Chem., Sect. B, 1984, 23B(7), 635); the analogous reactions of crotonaldehyde have also been reported. The cyclodextrin complex of cinnamic acid ester is used to bring about such additions, relying upon the encapsulating effect of this reagent (Y. Tanaka, H. Sakuraba and Y. Oka, J. Inclusion Phenom., 1984, 2(3-4), 841).

Iodine monofluoride (S. Rozen and M. Brand, *loc. cit.*) may also be prepared by using mixtures of iodine and IF_5 (Daikin Kogyo Co. Ltd., JP 59 51,225 [84 51,225]; Chem. Abs., 101, P54544c). 'Iodofluoride' adducts are also obtained when $MeIF_2$ reacts with *cis-* or *trans*-1- phenylalkenes. The addition is stereospecifically *anti-*, and follows Markownikow orientation to give, respectively, (±)-*threo-* and (±)-*erythro*-1-fluoro-2-iodo-1-phenylalkanes. Indene correspondingly gives *trans*-1-fluoro-2-iodoindane. (M. Zupan and A. Pollak, J. Chem. Soc., Perkin Trans. 1, 1976, (16), 1745). The addition of other iodine halides (M. Adinolfi *et al.*, Tetrahedron, 1984, 40(11), 2183) is discussed, and evidence is presented for the intermediacy of complexes between ICl and alkenes in their heterolytic reaction (G. H. Schmidt and J. W. Gordon, Can. J. Chem., 1984, 62(11), 2526).

Manganese(III) chloride, from $Mn(OAc)_3$-$CuCl_2$ mixtures in acetic acid, chlorinates alkenes (M. Okano *et al.*, Nippon Kagaku Kaishi, 1978, (4), 578). Xenon difluoride gives 1,2-difluoro adducts with 1,1-diarylethenes in HF or trifluoroacetic acid; $Ar_2C{:}CHF$ however incorporates CF_3CO_2-residues by nucleophilic diversion of the cation intermediate (M. Zupan and A. Pollak, J. Org. Chem., 1976,

41(25), 4002).

Enzymatic methods of bringing about the addition of halogen (S. L. Neidleman and J. Geigert, Biochem. Soc. Symp., 1983, 48 (Biotechnology) 39; Chem Abs., 101 R168922e) have been reviewed. Thus, the bromochlorination of alkenes may be brought about using bromoperoxidase in sea water (J. Geigert and S. L. Neidleman, Phytochemistry 1984, 23(2) 287). The corresponding reaction of chloroperoxidase shows evidence of neighbouring group participation in the addition to allyl chloride, when rearranged products are formed (T. D. Lee *et al.*, Biochem. Biophys. Res. Commun. 1983, 110(3), 880). This reflects the similarity in mechanism between such processes and the classic electrophilic addition mechanisms, in which such rearrangements have been carefully detailed (P. B. D. de la Mare *et al* J. Chem. Soc., (1954) 3990; (1962) 443; (1963) 3429).

The addition of bromine has been advocated as a selective means of protecting one unit in an unsymmetrical and non-conjugated diene; the more electrophilically sensitive unit is saturated first, and halogen may later be removed by a number of reductive processes (U. Husstedt and H. J. Schaefer, Synthesis, 1979, (12), 966).

(b) Reactions of dihalogenoalkanes.

Where the reactions of polyhalogenoalkanes are identical in type to those of the monohalogenoalkanes they are reported under this heading. Correspondingly, the elimination of hydrogen halide from dihalogenoalkanes is discussed under the synthesis of alkenyl halides. Some ring-closure reactions of appropriately substituted dihalogenoalkanes have been reported in the literature, such as the formation of cyclic amines, ethers, or sulphides, but the chemistry of such processes has already been well documented and such new products have been ignored in this Review when their syntheses proceed by such classical methods.

We include, however, the reaction of α,ω-diiodides to give excellent yields of three to five membered cycloalkanes by their reaction with Me_3CLi in pentane-diethyl ether at -23° (W. F. Bailey and R. P. Gagnier, Tetrahedron Lett., 1982, 23(49), 5123).

2. Polyhalogenoalkanes

Tetrahalogenomethanes have been usually made by three methods:

(i) Ion-exchange processes, such as the reaction of aluminium iodide with carbon tetrachloride to give carbon tetraiodide (*e.g.* P. Pouret, Comp.rend., 1900, 130, 1191).

$$3CCl_4 + 4AlI_3 \longrightarrow 3CI_4 + 4AlCl_3 \qquad (4)$$

CBr_3I may be obtained from CBr_4 and NaI in acetone (J. R. Kennedy *et al.*, J. Chem. Eng. Data, 1979, 24(3), 251; Chem. Abs., 91, 56256).

(ii) Direct halogenation of methane or its derivatives, as exemplified by the synthesis of $BrCF_3$ from CHF_3 on a mixed $Cr/\gamma\text{-}Al_2O_3$ catalyst with bromine (N. Koketsu *et al.*, JP 78 34,705; Chem. Abs., 89, 42399); and

(iii) Halogenation of less fully halogenated methanes, notably the haloforms. W. M. Dehn (J. Amer. Chem. Soc., 1909, 31, 1220) used alkaline solutions of the appropriate halogen to make a range of tetrahalogenomethanes with more than one halogen attached (*e.g.* CXY_3), and the process has been advocated more recently to obtain high yields (76%) of labelled tetrabromomethane from ^{13}C-labelled bromoform (H. Siegel and D. Seebach, J. Labelled Comp. Radiopharm., 1980, 17, 279) . The process is wasteful of halogen; for example, the preparation of tetraiodomethane by iodination of iodoform, as reported by French workers ((*a*) J. F. Durand, Bull. chem. soc. Fr., 1927, 4 41, 1252; Chem. Zentralblatt, 1927 II 2662 (*b*) M. Lantenois, Compt. Rend., 1913, 156, 1385), provides less than 10% yields of product based upon iodine used, in contrast with the exchange process which is recommended as a synthetic method. None the less, $CHCl_2I$ and $ICCl_3$ have both been made by the reaction of chloroform with NaOI, and $CHBr_2I$ has been prepared analogously from $CHBr_3$ (J. R. Kennedy *et al.*, J. Chem. Eng. Data, 1979, 24(3), 251; Chem. Abs., 91, 56256)

Mixed halogenated methanes (*e.g.* $CHBr_2I$) are generally prepared either by direct halogenodeprotonation by radical methods, or by exchange reactions in which halogen is scrambled between different methane derivatives. Thus, both iododibromomethane and bromodiiodomethane may be obtained from bromoform. Iron pentacarbonyl assists the exchange between iodoform and alkyl halides, a route to alkyl iodides, and also encourages the scrambling of halogen between various haloforms (R. A. Amriev, F. K. Velichko and R. Kh. Friedlina, C1 Mol. Chem., 1985, 1(4), 319; Chem. Abs., 104, 224531).

Polyhalogenomethanes themselves are sources of higher polyhalogenoalkanes through the Prins reaction

$$XCCl_3 + {>}C{=}C{<} \longrightarrow Cl_3C\text{-}C\text{-}CX{<} \qquad (5)$$

Originally the process was catalysed by Lewis acids, usually $AlCl_3$, and a more modern equivalent is found in the reaction between either 1,1- or 1,2-dichloroethane and tetrafluoroethene. In mixtures of $HF-SbF_5$ at 20° these give $CH_3.CHCl.CF_2.CF_3$ (66-80%) along with the unexchanged and presumably primary product ($CH_3.CHCl.CF_2.CF_2Cl$) and the isomeric $Cl.CH_2.CH_2.CF_2.CF_2Cl$ (G. G. Belen'kii, V. A. Petrov and L. S. German, Izv. Akad. Nauk SSSR, Ser. Khim., 1980 (5) 1099). Free- radical mechanisms are also available for this addition, and have been very popular. Copper(I) chloride and ethanolamine together encourage the addition of $Cl_3C.CF_3$ to isoprene which, in t-butanol, gives 80% yield of the 1:1-adduct $Me_2CH.CHCl.CH_2.CCl_2.CF_3$ (D. Holland, D. J. Milner and R. K. Huff, Ger. Offen. 2,920,536; Chem. Abs., 92, 128384), whereas the corresponding addition of CCl_4 to 1-hexene in benzene at the boil for 5 hours gives good yields of the adduct only in the presence of a phosphino ruthenium hydride such as $RuHCl(PPh_3)_3$ (H. Kono, Y. Nagai and H. Matsumoto, JP 78 23,907; Chem. Abs., 89, 42397).

H. Mayr (Angew. Chem., 1981, 93(2), 202) has pointed out that addition of RX across an alkene may give either the 1:1-adduct or induce polymerisation depending upon whether the product halide dissociates more readily than RX (when polymerisation is observed) or less. Solvolysis studies were sources of information about the relative extents of dissociation of the various alkyl halides. The analogous argument has been applied to the free-radical addition mechanism, when the relative stabilities of the various radicals were important in determining whether or not polymerisation accompanied the 1:1-addition.

3. Polyfluoroalkyl halides ($\mathcal{R}$X where $\mathcal{R}$ denotes a perfluoroalkyl residue)

(a) Preparation.

Ulm (K. Ulm, Spec. Chem. 1988, 8(5), 418; Chem. Abs., 110, 59882) has reviewed the synthesis and preparation of short-chain polyfluoroalkyl iodides ($\mathcal{R}$I).

Fluorosulphonates such as RR'C:CR'.CF_2.O_3SF (R, R'= Cl; R=$\mathcal{R}$, R'=F) are attacked by KI in sulpholan to give RR'C:CR'.CF_2I in 70-88% yield and without evidence of allyl rearrangement (V. F. Cherstkov *et al.*, Izv. Akad. Nauk SSSR, Ser. Khim., 1985 (1) 220)

(b) Reactions.

(i) *Ion-exchange.* Polyfluoroalkyl iodides may be obtained by the action of alkali metal iodides upon the corresponding perfluoroalkyl bromide (K. Nakumura, M. Suwa and A. enomoto, JP 74 48288; Chem. Abs., 82 170013)

(ii) Hydrolysis. In fuming sulphuric acid of different strengths (15-45% SO_3) perfluoroalkyl primary iodides ($\mathcal{R}$.CF_2I) hydrolyse to give perfluoroalkyl carboxylic acids

$$\mathcal{R}.CF_2I \longrightarrow \mathcal{R}.CO_2H \qquad (6)$$

(K. von Werner and A. Gisser, Eur. Pat. Appl. EP 52, 285; Chem. Abs., 97, P162372).

(iii) Oxidation. At 0°, $(CF_3)_3$C.O.Cl reacts with perfluoroalkyl iodides to give $\mathcal{R}$.I[OC$(CF_3)_3$]$_2$ (J. A. M Cornish *et al.*, Inorg. Chem., 1986, 25(17), 3030).

(iv) Reductive dimerisation. Zinc metal in dioxan brings about the conversion of perfluoroalkyl iodides to the bialkyl system (V. F. Cherstkov *et al.*, *loc. cit.*).

$$2\mathcal{R}I \longrightarrow (\mathcal{R})_2 \qquad (7)$$

(v) Addition to alkenes. Perfluoroalkyl iodides add across the multiple bond of alkenes and alkynes under a range of reaction conditions most of which may be understood in terms of single electron transfer processes.

$$\mathcal{R}I + {>}C{=}C{<} \longrightarrow {>}C(\mathcal{R}).C(I){<} \quad (8)$$

Iron metal (Q. Y. Chen, Y. B. Ho, and Z. Y. Yang, J. Fluorine Chem., 1986, 34(2), 255), Raney nickel (Q. Y. Chem and Z. Y. Yang, J. Chem. Soc., Chem. Commun., 1986, (7), 498), tetrakis(triphenylphosphine)palladium (T. Ishihara, M. Kuroboshi, and Y. Okada, Chem. Lett., 1986, (11), 1895), and sodium dithionite (W, Huang, W. Wong and B. Huang, Huaxue Xuebao, 1986, 44(5), 488; Chem. Abs., 106, 83935) are all reported catalysts to the addition.

e.g. $Cl(CF_2)_4I + C_4H_9.CH{:}CH_2 \longrightarrow$

$$Cl(CF_2)_4.CH_2.CH(I).C_4H_9 \quad (9)$$

Both PbO_2 and $Pb(OAc)_4$ promote this addition which was apparently free-radical in character, since di-allyl ether gave tetrahydrofuranyl derivatives through ring-closure (Q. Chen and J. Chen., Huaxue Xuebao, 1988, 46(3), 301; Chem. Abs., 110, 74759).

Functional groups may be present in the alkene, so that allyl alcohols ($R'.CH{:}CH.CH_2OH$) may be converted into the perfluoroalkyl derivative $R'(\mathcal{R})CH.(CH_2)_2.OH$; allyl esters behave correspondingly (N. O. Brace, J. Fluorine Chem., 1982, 20(3), 313).

Polar effects are also apparent in the light-catalysed addition of 1,1,2,3,3-pentafluoropropyl iodide to ethene in which only the 1:1-adduct is formed

$$CF_2{=}CFCF_2I + CH_2{=}CH_2 \longrightarrow CF_2{=}CFCF_2CH_2CH_2I \quad (10)$$

with no evidence of attack upon the polyfluoroalkenyl iodide by the intermediate radical (Daikin Kogyo Co. Ltd., JP 82 85 327; Chem. Abs., 97, 144328).

Such adducts ($\mathcal{R}.CH_2.CH_2.I$) undergo hydrolysis in aqueous

DMSO to give the corresponding primary alcohol (68% yield). Aeration of the reaction mixture and the presence of copper markedly improves both the yield (98%) and the conversion (97%) of the alcohol (T. Yoshida and O. Yamamoto, JP 63,270,633; Chem. Abs., 110, 212128).

(vi) With formal nucleophiles. Perfluoroalkyl iodides react with sulphite ion in the presence of tetra-n-butylammonium bromide as a phase-transfer agent (24h, 85°. I Tabuse and N. Morioka, JP 60,228,453; Chem. Abs., 105, P81150) a reaction which also takes place in aqueous dioxan at similar temperatures (B. Huang, W. Huang and C. Hu, Huaxue Xuebao 1981, 39(5), 481; Chem. Abs., 96, 142214).

The ease of homolysis of the C-I bond in perfluoroalkyl iodides (*ca.* 230 kJ/mol; S. I. Ahonkai and E. Whittle, Int. J. Chem. Kinet., 1984, 16(5), 543) supports both homolytic and heterolytic processes. Attack upon nucleophiles may be encouraged by radiolysis or by u.v. radiation. Imidazoles are trifluoromethylated in methanolic solution by CF_3I under these free-radical conditions to give a mixture of 2- and 4-substituted products (H. Kimoto, S. Fuji and L. A. Cohen, J. Org. Chem., 1982, 47(15), 2867).

The perfluoroalkylation of arenes (heterocyclics and PhX, where X = H, Me, MeO, Cl, CO_2Me) may be brought about by $\mathcal{R}$I in the presence of a base (*e.g.* potassium carbonate) and a 2% Ru/C catalyst under pressure at 170° for 30 hours. Since the process is dissuaded neither by electron-withdrawing or electron-donating substituents in the benzene ring (K. von Werner, Ger. Offen DE 3,247,728; Chem. Abs., 102, 5889) it apparently involves either a homolytic rate-determining step or at least a free-radical attack upon the aromatic system.

The reaction of perfluoroalkyl iodides and $[(EtO)_2P]_2O$ is overtly homolytic in character, being carried out at 120° in 1,1,2-trichlorotrifluoroethane and in the presence of di-*tert*-butyl peroxide. The reaction product, $\mathcal{R}.P(OEt)_2$, may be oxidised to the alkylphosphate ester by *tert*-butyl hydroperoxide at -5° to -10° (M. Kato, K. Akiyama and M. Yamabe, Asahi Garasu Kenkyu Hokoku, 1982, 32(2), 117; Chem. Abs., 99, 53873). The reaction of the corresponding alkyl iodides with ethylene carbonate is also free radical in character, since it requires the presence of the Zn-Cr couple. At 80°-90° the main product (60-80%) is the half ester of ethylene glycol and a perfluoroalkanoic acid; at higher temperatures (>150°) the parent acid is the major product (60-85%. S. Benefice, H. Blancou and A. Commeyres, Tetrahedron, 1984, 40(9), 1541).

(vii) Electrophilic attack. A. V. Fokin *et al.* (Izv. Akad. Nauk SSSR, Ser. Khim., 1986 (12) 2734) describe the substitution of halogen in alkyl iodides upon treatment with halogeno fluorosulphates, or peroxydisulphuryl difluoride as "oxidative electrophilic substitution"

(viii) Enzymatic attack. Perfluoroalkyl iodides undergo attack by enzymes in the present of alkynes under very mild conditions (T. Kitsatsume, JP 63,222,695). This preparation of perfluoroalkanoic acids (C_1-C_7) is exemplified by the formation of trifluoroacetic acid in 29% yield from C_2F_5I in a phosphate buffer containing catalase and PhC≡CH over 1-4 weeks.

(ix) Coupling with alkyl halides. Trifluoromethyl copper complexes obtained from CF_3I by conventional methods (Cu in HMPA) replace halogen in alkyl and alkenyl halides by the $-CF_3$ group (13-81% yield).

e.g. $Bu_2C{=}CHI \longrightarrow Bu_2C{=}CHCF_3$ (11)

Halogen exchange may intervene to give the corresponding alkyl fluoride (A. Takaoka, H. Iwakiri and N. Ishikawa, Bull. Chem. Soc. Jpn., 1979, 52(11), 3377).

4. *Polyfluoroalkenes*

(a) Preparation

(i) Polyfluoroalkenes may be obtained from trifluorochloroethene by lithiation and silylation to give *Z*-R.CF:CF.$SiMe_3$ which may then react either with KF to provide *E*-RCF:CHF or with an alkyl halide R'X to give *E*-R'CF:C(R)F (S. Martin, R. Sauvetre and J. F. Normand, J. Organomet. Chem., 1984, 264(1-2), 155)

(b) Reactions.

(i) Addition.

(1) Electrophilic: In fluorosulphuric acid perfluoroalkenes undergo addition of halogen (Cl, Br). Antimony(V) fluoride may be used as a catalyst. The process appears to be mechanistically similar to the more widely studied addition of halogen to simple alkenes; adducts of the structure $\mathcal{R}$.CF(X).CF(X).$\mathcal{R}'$ are formed, and also alkyl fluorosulphonates are obtained, evidently through the solvolysis of an intermediate carbocationic species (A. V. Fokin *et al.*, Izv. Akad. Nauk SSSR, Ser. Khim., 1985, (10), 2298).

(2) Nucleophilic: Perfluoropropene reacts with dialkylamines to give a mixture of α,α-difluoro-monoalkylamines and α-fluoro-enamines. Such mixtures are useful fluorinating agents which replace -OH by -F in both alcohols and carboxylic acids. They are superior to the corresponding adduct of diethylamine with chlorotrifluoroethylene in both stabil- ity and ease of preparation (A. Takaoka, H. Iwakiri and N. Ishikawa, Bull. Chem. Soc. Jpn., 1979, 52(11), 3377). Watanabe and his colleagues (S. Watanabe *et al.*, J. Fluorine Chem., 1986, 31(2), 135) have investigated the application of 1,1,2,3,3,3-hexafluoropropyldiethylamine as such a reagent. Although halogeno-alcohols give the corresponding alkyl fluoride,

$$e.g.\ C_2H_5.CHBr.CH_2OH \longrightarrow C_2H_5.CHBr.CH_2F\ (52\%) \qquad (12)$$

α-halogenocyclic alcohols give the corresponding 2,3,3,3-tetrafluoropropionate esters.

Perfluoro-1-heptene reacts with methoxide ion to give 1-methoxyperfluoro-2-heptene (U. Gross and W. Storek, J. Fluorine Chem., 1984, 26(4), 457) and analogously with sodium sulphite (U. Gross and G. Engler, *idem.*, 1985, 29(4), 425) in which

$$C_3F_7.CF_2.CF{:}CF_2 + Y^- \longrightarrow C_3F_7.CF{:}CF.CF_2Y + F^- \qquad (13)$$

an allylic rearrangement has clearly occurred.

5. Halogeno derivatives of olefins

Under this heading may be logically included (i) allyl halides, with the C=C–C–X structure, (ii) vinylic halides, with the halogen attached to a sp^2 carbon atom (C=C–X), and (iii) those structures in which the unsaturated carbon atoms and the halogen substituents are too far separated to interact. Many of the reactions of types (i) and (iii) are already described in the Monohalogenoalkane sections, and the following section is biassed towards vinyl halide systems the formation and reactions of which have not yet been reviewed.

(a) Preparation

(i) By addition to alkynes Mixtures of alkenes, 1,2-alkadienes and alkynes (*e.g.* $CH_3CH{:}CH_2$, $CH_2{:}C{:}CH_2$ and $CH_3C{\equiv}CH$) add hydrogen halide in the presence of alumina to give alkenyl halides in a process which does not consume the mono-alkene but attacks the alkadiene and alkyne with 95% selectivity (R. T. Klun, C. B. Murchison and D. A. Hucel US 4,480,121; Chem. Abs., 102, 46415).

In the presence of a source of hydrogen, iron(III) chloride attacks alkynes RC≡CR'to give the isomeric chloroalkenes RC(Cl)=CHR' and RCH=C(Cl)R' as well as polychlorinated material (D. Cassagne, M. Gillet and J. Braun, Bull. soc. chim. Fr., 1979, (3-4, Pt. 2), 140). Cycloaddition and rearrangement are also said (K. Griesbaum *et al.*, Liebig's Ann. Chem., 1979, (8), 1137) to accompany the addition of hydrogen bromide to alkynes, though the conditions used encourage a range of available mechanisms of addition.

Halogen or mixed halogen addition to alkynes provides 1,2-dihalogenoalkenes, and $MeIF_2$ in MeI reacts with PhC≡CR to give *Z*-Ph.CF=CRI (65-80%); when R is Me, isomerisation to the *E*-isomer (60%) may be induced by HF in MeI (M. Zupan, Synthesis, 1976, (7), 473).

(ii) By elimination reactions.

1. *From polyhalogenoalkanes.* Elimination of HX from polyhalogenoalkanes remains a popular route to halogeno olefins. The use of polyethylene glycols (n = 1 to 10) to

facilitate the alkaline elimination of HCl from derivatives of 1,1,1-trichloroethane (R.R'CH.CCl_3) is the subject of a patent dealing with the formation of 1,1-dichloroalkenes (M. Fischer, Eur. Pat. Appl., EP 53,687; Chem. Abs., 97, P144326). Lithium chloride is an excellent base in DMF, providing 78% yield of Me_2C:CH.CH:C(Cl).CF_3 from the adduct of Cl_3CCF_3 and isoprene. Correspondingly, 1,2-dihalides provide a mixture of *cis*- and *trans*-1-halogeno-1-alkenes, 2-halogeno-1-alkenes, alkenes and alkadienes when tertiary amines are used as bases (J. Beger and E. Meerbote, J. Prakt. Chem., 1984, 326(1), 12).

Base-catalysed elimination from 5-fluorononane gives a mixture of *cis*- and *trans*-olefins, the distribution of which, like the rate of the reaction, is critically dependent upon the base used. NaOMe-MeOH gives a *cis*-/*trans*-ratio of about 1:3, but the much more rapid reaction with lithium diisopropylamide forms the *trans*-isomer almost exclusively. Similar observations were found in the corresponding reactions of cyclododecyl fluoride (S. Matsubara *et al.*, Tetrahedron 1988, 44(10), 2855); with *meso* and of (±)-6,7-difluorododecane KOBu in THF at 75° reacts rapidly and stereospecifically to give exclusively *E*- or *Z*-6-fluoro-6-dodecane by antiperiplanar elimination. Lithium isopropylamide is a remarkably potent base in hexane, but loses its effect almost completely in ethers such as Et_2O and THF. This was held to imply assistance by lithium, presumably through the high lat- tice energy of LiF (H. Matsuda *et al.*, Tetrahedron 1988, 44, 2865 and 2875).

Catalytic dehydrohalogenation (Fe/C) of 1,2-dichloropropane gives a mixture of *cis*- (6.1%) and *trans*-1-chloropropene (59.8%) (H. Marschner *et al.*, (*a*) Ger (East) DD 227,129; Chem. Abs., 105, 116969 (*b*) Ger (East) DD 235,631; Chem. Abs., 105, 174815). In this context, the Fe-catalysed dehydrobromination of 1,2-dibromo-1,1-diphenylethane in CCl_4 has been shown to involve metal-insertion intermediates (A. R. Suarez, M. R. Mazzieri and A. G. Suarez, J. Amer. Chem. Soc., 1989, 111(2), 763).

Dehydrohalogenation accompanies many displacement reactions; for example, 1,2-dibromoalkanes give the alkyne (33-80%) on treatment with Ph_2XLi (X = P, As) buta-2,3-dichlorobuta-1,3-diene gives $Ph_2PCH_2CH=C=CH_2$ and 2-chlorobutadiene gives the SN2' product, $Ph_2PCH_2C{\equiv}CCH_2PPh_2$ (D. L. Gillespie and B. J. Walker, J. Chem. Soc., Perkin Trans. 1, 1983, (8), 1689).

(iii) By displacement reactions.

(1) *From boron derivatives.* *Z*-Vinyl boronate esters are attacked by solutions of iodine in aqueous sodium hydroxide to give *Z*-iodo-alkenes; reaction with bromine in dichloromethane, followed by treatment with methanolic sodium methoxide, gives the *E*-bromo-alkene (H. C. Brown and V. Somayaji, Synthesis 1984, (11), 919). 1-Alkynes may also be attacked by halogenoboranes (XBH_2; X = Br, I) to give *cis*-adducts which, on protonolysis in acetic acid give 2-halogeno-1-alkenes. 1-Octyne, for example, gives 99% yields of *Z*-2-bromo-1-octene of *ca.* 96% stereochemical purity. Further, treating the intermediate borane with lithium acetylides offer a route to *Z*-1-alkynyl-2-halogeno-1-alkenes (S. Hara *et al.*, Tetrahedron Lett., 1983, 24(7), 731 and 735).

As with the alkyl iodides, radiolabelling with ^{125}I has been brought about by this technique, using NaI as the source of radiolabel in THF under oxidising conditions (G. W. Kabalka, K. A. R. Sastry and K. Muralidhur, J. Labelled Compd. Radiopharm., 1982, 19(6), 785).

Hydroboration of 1-chloro-1-alkanes by R_2BH gives 1-chloro-1-alken-1-ylboranes the borinate ester derivatives of which are then brominated to provide *Z*-1-bromo-1-chloroalk-1-enes. Bromodesilicobromination of *E*-1-chloro-1-alkenylsilanes or chlorodesilicochlorination of *Z*-1-bromo-1-alkenylsilanes provides *E*-1-bromo-1-chloro-1-alkenes (R. P. Fisher, H. P. On, J. T. Snow and G. Zweifel, Synthesis, 1982, (2) 127).

(2) *From organosilanes.* The analogous reaction of trimethylvinylsilanes with bromine in dichloromethane at -78° provides the vinyl bromide of opposite orientation; thus, the *E*-isomer gives the *Z*-bromide in 65-87% yield. Cyanogen bromide with aluminium chloride in the same solvent at 0° gives the isomeric *E*-bromide (53-73%), and iodine at room temperature also provides the iodide (60-80% yield) with retention of configuration about the double bond (T. H. Chan, P. W. K. Lau and W. Mychajlowskij, Tetrahedron Lett., 1977 (38) 3317)

(3) *From organoselenium derivatives.* The addition of PhSeBr across the double bond of $RCH:CH_2$ proceeds to give both Markownikow and anti-Markownikow oriented products. The former product, on ozonolysis in CCl_4 at -20° and then boiling with di-isopropylamine gives 2-bromo-1-alkenes; for example, the formation of $C_{14}H_{29}C(Br):CH_2$ in 87% yield. The anti-Markownikow product, oxidised by 30% H_2O_2 with one equivalent of pyridine, provides allyl or vinyl bromides, so that $C_6H_{13}.CH(SePh).CH_2Br$ gives 74% $C_5H_{11}CH:CH.CH_2Br$ whereas $Me_3C.CH(SePh).CH_2Br$ gives *E*-$Me_3C.CH:CH.Br$ in 51% yield (S. Raucher, Tetrahedron Lett., 1977, (44), 3909).

(*iv*) *From alcohols and alkenyl halides.* The reaction of allylic alcohols or their alkyne analogues with many reagents often gives rise to rearrangements. For example, acetylenic alcohols of the formula R.CH(OH).C≡CH react with thionyl chloride in boiling dioxan to give *ca.* 50% yields of the cumulene system R.CH=C=CHCl (R = Me_2CH or Me) together with small amounts of the expected chloride R.CHCl.C≡CH and, in the former case, of the alkenyne Me_2C=CH.C≡CH. The more extended system $Me_2CH.CH(OH).C≡C.C≡CH$ gives only the unrearranged chloride and the corresponding dehydration product, however (A. N. Patel, Zh. Org. Khim., 1977, 13, 2226). High yields and stereospecificity of attack are claimed for the hexachloroacetone-PPh_3 system, so that *E*-2-buten-1-ol (crotyl alcohol; 0.05 mol) with $(CCl_3)_2CO$ (0.11 mol) and PPh_3 (0.052 mol) for 20 min. at 10° gives 99.3% *E*-1-chloro-2-butene (R. M. Magid, O. S. Fruchey and W. L. Johnson, Tetrahedron Lett., 1977, (35), 2999).

Alkenyl iodides may be converted to the corresponding fluorides (71-88%) through the vinyl-lithium by trapping with $Ph.SO.N(F).CMe_3$ (S. H. Lee and J. Schwartz, J. Amer. Chem. Soc., 1986, 108(9), 2445). Allyl chlorides may be made from the corresponding dialkyl-allylamines by reaction with ethyl chloroformate with K_2CO_3 in benzene, when the stereochemistry about the double bond system is not altered (R. Mornet and L. Gouin, Synthesis, 1977, (11), 786). Correspondingly, the reaction of alcohols with Br_2/PPh_3 also allows alkenyl bromides to be made (N. K. Bliznyuk *et al*, SU 1 330 121. Ot. Izobret., 1987, (30), 106; Chem. Abs., 108, 204209). Tribromoacetate in the presence of PPh_3 also converts such alcohols into their bromides. In preparation of alkenyl chlorides *E*-$MeCH=CHCH_2OH$ gave only the *E*-chloride (E. D. Matreeva *et al.*, Vestn. Mosk. Univ., Ser. 2: Khim. 1986, 27(6), 603; Chem. Abs., 108, 37163).

Vinyl bromides exchange with copper(I) iodide in the presence of HMPT or of N-methylpyrrolidone. DMF or diglyme are less effective solvents, and the exchange does not proceed with potassium iodide so that the copper halide is clearly an essential reagent (H. Suzuki *et al.*, Synthesis, 1988, 236).

Alkyl halides (R'X) also react with allyl-lithiums of the structure $R(Cl)C{:}CH.CH_2Li$ to provide the isomeric products $CH_2{:}CH.C(Cl)RR'$(1) and $R.C(Cl){:}CH.CH_2R'$(2), the latter mainly in the *Z*-configuration. Trimethylsilyl chloride with $ClHC{:}CH.CH_2Li$ gives only the product of allylic rearrangement (1), whereas the methallyl analogue (R = Me) gives both (1) and (2) (B. Mauze, P. Ongoka and L. Miginiac, J. Organomet. Chem., 1984, 264(1-2), 1).

1-Fluoroalkenes are formed by the alkylation ($R.CH_2I$) in THF at 20° for 3 h. of Ph.SO.CHFLi, giving $Ph.SO.CHF.CH_2.R$ (60-85%) which on heating to 180° for 30 min. gives R.CH=CHF (73-95% yield. V. Reutrakul and V. Rukachaisirikul, Tetrahedron Lett., 1983, 24(7), 725). 1,1-Diphenylethene reacts with $CsSO_4F$ in CH_2Cl_2 (r.t., 24 h.) to give $Ph_2C=CHF$ in 70% yield; 2-norbornene, under these conditions, provides 7-fluoronortricyclene (22%) and 7-*syn*-fluoronorborn-2-ene (31%; S. Stavber and M. Zupan, J. Chem. Soc., Chem. Commun., 1981, 795).

(v) From carbonyl compounds

Chloroiodomethane gives (chloromethyl)triphenylphosphonium ion with Ph_3P; the ion loses a proton to KO^tBu and the resulting ylide reacts with aldehydes or ketones to give the vinyl chloride; *e.g.*

$$PhCOMe \longrightarrow PhC(Me){=}CHCl \qquad (14)$$

in which the stereoisomers are nearly evenly distributed (*E*/*Z*, 56:44. S. Miyamo, Y. Izumi and H. Hashimoto, J. Chem. Soc., Chem. Commun., 1978, (10), 446).

The great tendency to form P-O bonds is again seen in the reaction between tetraiodomethane and aldehydes (RCHO) in the presence of PPh_3 to give 1,1-di-iodoalkenes ($R.CH:CI_2$), a reaction which takes place under mild conditions (CH_2Cl_2, 25°. F. Gavina *et al.*, J. Chem. Soc., Chem. Commun., 1985, (5), 296).

Vinyl fluoride derivatives are side-products in the reaction between perfluoroalkylacetylene ($\mathcal{R}.CF_2C{\equiv}CH$) and ethyl chloroformate. The acetylide anion is formed by the reaction of butyl-lithium with the alkyne at -60°; at -90° $ClCO_2Et$ reacts with this anion to give a mixture of the expected ester $\mathcal{R}.CF_2.C{\equiv}C.CO_2Et$ (84-89% of the isolated product) and small amounts (11-16%) of the rearranged product $\mathcal{R}.C(F):CH.C_5H_{11}$ (A. Chauvin *et al.*, J. Fluorine Chem., 1984, 25(2), 259). Vinyl fluorides ($\mathcal{R}.C(F):CHP(O)(OEt)_2$) are also obtained by the reaction of perfluoroalkanoyl chlorides ($\mathcal{R}.CF_2.COCl$) with triethyl phosphite; with $Bu_4N^+F^-$ at 0-20° these species provide $\mathcal{R}C{\equiv}CH$.

(*vi*) *Oxychlorination* is exemplified by the formation of tri- and tetra-chloroethene from the action of mixtures of oxygen and chlorine upon 1,2-dichloroethane (L. M. Kartashov, I. N. Prokhorova and Yu. A. Tregev, Zh. Prikl. Khim. (Leningrad) 1983, 56(8), 1026; Chem. Abs., 99, 212100).

(*vii*) *Perfluoroalkenyl halides*

Nucleophilic attack of perfluoroalkenes may be brought about by RLi, so that perfluoro-3,4-dimethylhex-3-ene gives $MeC(CF_3)(C_2F_5)C(CF_3):CF.CF_3$, $MeC(CF_3)(C_2F_5)C(C_2F_5):CF_2$ and $MeC(CF_3)(C_2F_5)C(CF_3): C(Me).CF_3$ (13%, 23% and 15% yields respectively) with MeLi (R. D. Chambers *et al.*, J. Chem. Soc., Perkin Trans. 1, 1983, (6), 1235 and 1239).

(b) Reactions of alkenyl halides

(i) The *ozonolysis* of halogeno-alkenes has been reviewed (C. W. Gillies and R. L. Kuszkowski, Isr. J. Chem., 1983, 23(4), 446).

(ii) With organometallics. *E*-1-Alkenylzirconium derivatives may be made from the interaction between 1-alkynes and $HZrCp_2Cl$ (Cp = C_5H_5) and react with alkenyl halides under

palladium catalysis to give conjugated dienes. An example is the synthesis of *E,E*-hexadecan-5,7-diene from 1-iodobut-1-ene and *E*-C_5H_{11}.CH=CH.$ZrCp_2Cl$ in 91% yield (N. Okukado *et al.*, Tetrahedron Lett., 1978 (12) 1027).

Analogously, vinyl copper derivatives couple with 1-halogeno-1-alkenes in the presence of $Pd(PPh_3)_4$ to give high yields of conjugated dienes in a stereospecific process. Thus, *Z*-Me_3C.CH:CH.Cu and *E*-I.CH:CH.$(CH_2)_4$.Me in THF at -15° over 30 min followed by alkaline hydrolysis gave 55% yield of *Z,E*–$Me_3C.(CH:CH)_2.(CH_2)_4$.Me of 98.9% isomer purity (N. Jabri, A. Alexakis and J. F. Normand, Tetrahedron Lett., 1982, 23(15), 1589). In the presence of potassium acetate and the same palladium catalyst, 1-bromo-octene adds to norbornene to give (3) through an insertion reaction followed by reductive elimination. The reaction is evidently mechanistically complex, and may involve a series of processes operating synchronously (M. Catellani and G. P. Chiurschi, J. Organomet. Chem., 1982, 232(2), C21).

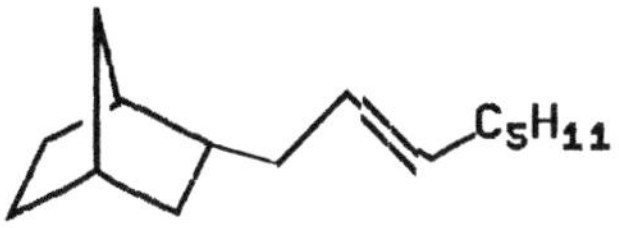

C_5H_{11}

(3)

Similarly, a mixture of geometrical isomers of crotyl chloride (Me.CH:CH.CH_2Cl; *trans- :cis-*, 85:15) reacted in ether at -76° with the product of interaction of butyllithium and copper(I) iodide/dimethyl sulphide to provide 2-octene in the same ratio of geometrical isomers (G. L. Van Mourik and H. J. J. Pabon, Tetrahedron Lett., 1978 (30), 2705).

Unsaturated nitriles may be obtained from the reaction of various bromostyrenes with $[Co(CN)_5]^{3-}$-KOH-H_2O under hydrogen at 45° (T. Funabiki, S. Yoshida and K. Tarama, J. Chem. Soc., Chem. Commun., 1978, (23) 1059); the process may be general to vinyl halides, since it proceeds with both α- and β-bromostyrenes.

Y. N. Usov (Deposited Doc. 1981, SPSTL 691 Khp-D81; Chem. Abs., 98, 88850) reports that the dehydrohalogenation of alkenyl halides yields cyclopropane derivatives, whereas the phase-transfer catalysed dehydrochlorination of some vinyl chlorides gives alkynes (A. E. Kalaidzhyan *et al.*, Arm. Khim. Zh., 1982, 35(6), 402; Chem. Abs., 97, 215462) *e.g.*

$$MeC(Cl)=CHCH_2OR \longrightarrow MeC{\equiv}CCH_2OR. \qquad (15)$$

Alkenyl bromides and iodides are reduced by $CrCl_2$ in DMF at 25°. The resulting alkenyl chromium compound reacts selectively with aldehydes; ketones and nitriles are inert to the intermediate (K. Takai *et al.*, Tetrahedron Lett., 1983, 24(47), 5281).

6. Halogenoalkynes

(a) *Reactions*

(i) Organometallic compounds. 1-Iodohept-1-yne reacts with one molecular proportion of BuLi to provide undec-5-yne in 61% yield in HMPT

$$R.C{\equiv}C.I + BuLi \longrightarrow R.C{\equiv}C.Bu \qquad (16)$$

but with 2 molecular proportions of BuLi reductive coupling occurs. Octane (49%) and 1-heptyne (43%) are formed. With 1-bromohept-1-yne under the same conditions butylation takes place at C3, presumably through a form of allylic rearrangement (S. Bhanu, E. A. Khan and F. Scheinmann, J. Chem. Soc., Perkin Trans. 1, 1976, (15), 1609).

7. Halogenocarbenes (Halogenomethylenes, :CHX and $:CX_2$)

Both monohalogenocarbenes and dihalogenocarbenes have enjoyed extensive study, including M.O. calculations of their heats of formation and allied parameters ((*a*) :CHF, :CHCl, :CHBr, G. E. Scuseria, M. Duran and R. G. A. R. Maclagen, J. Amer. Chem. Soc., 1986, 108(12), 3248. (*b*) :CHCl M. J. S. Dewar and H. S. Rzepa, J. Comput. Chem., 1983, 4(2), 158 (*c*):CFCl, $:CCl_2$ S. G. Lias, Z. Karpas and J. F. Liebman, J. Amer. Chem. Soc., 1985, 107(21), 6089. (*d*) :CHCl, $:CCl_2$ E. Goldstein *et al.*, THEOCHEM 1983, 14(3-4), 315; Chem. Abs., 100, 138245. (*e*) $:CX_2$, R. C. Binning, Diss Abstr. Int. B., 1980, 40(7), 3183 (*f*) :CHF, M. J. S. Dewar and H. S. Rzepa, J. Amer. Chem. Soc., 1978, 100(1), 58) and upon their physical properties (Dewar and Rzepa, *loc.cit.*). Reviews include Burton's definitive discussion of difluorocarbene and carbenoid systems (D. J. Burton and J. L. Hahnfeld, Fluorine Chem. Rev., 1977, 8, 119) which deals with the literature until June 1976, a specialist review of the formation and reactions of dichlorocarbene by the action of concentrated NaOH upon $CHCl_3$ in the presence of tetraalkylammonium ions or of crown ethers (M. Fedorynski, Wiad. Chem., 1976, 30(9), 575; Chem. Abs., 86, R42567), and a text devoted to the addition of dichlorocarbene to alkenes (N. S. Zefirov, I. V. Kazimirchik and K. A. Lukin, "Tsikloprisoedinenie Dikhlorkarbene k Olefinam", Nauka (Moscow) 1985; Chem. Abs., 104, B224619).

The cyclisation ("cyclopropanation") which addition of carbenes such as $:CCl_2$ and $:CF_2$ to alkenes produces has also been the subject of M.O. calculation (K. N. Houk, N. G. Rondan and J. Mareda,(*a*) J. Amer. Chem. Soc., 1984, 106(15), 4291. (*b*) *ibid.*, Tetrahedron, 1985, 41(8), 1555).

Halocarbenes have been indicted as the agents responsible for the toxicity of halogenoalkanes (V. Ullrich *et al.*, Adv. Pharmacol. Ther., Proc. Int. Congr. Pharmacol., 7th 1978 (Publ. 1979), 9, 131. Ed. Y. Cohen, Pergamon, Oxford, England; Chem. Abs., 91, R134884), and dichlorocarbene has been claimed as a product of the metabolism of CCl_4 (L. R. Pohl, J. W. George, Biochem. Biophys. Res. Commun., 1983, 117(2), 367). Evidence ofthis came from the identification of 1,1-dichloro-2,2,3,3-tetramethylcyclopropane in the presence of 2,3-dimethyl-2-butene, which reflects the common addition reaction of carbenes to alkenes (see below).

(a) Preparation

(1) From trihalogenomethanes. The early and definitive work of J. Hine (J. Amer. Chem. Soc., 1950, 72, 2438) upon the mechanism of hydrolysis of chloroform led to the postulation of dichlorocarbene as a necessary intermediate to explain the formation of the range of one-carbon products including both formate ion and carbon monoxide. Mixtures of strong bases with $CHCl_3$ are still used to obtain dichlorocarbene; the further uses of ultrasound (S. L. Regen and A. Singh, J. Org. Chem., 1982, 47(8), 1587), of phase-transfer catalysts

such as $R_4N^+Cl^-$ and $R_3NMe^+Cl^-$ ((*a*) Yu. Sh. Goldberg and M. V. Shimanskaya, Zh. Org. Khim.,, 1984, 20(6), 1332. (*b*) B. Sain *et al.*, Heterocycles 1984, 22(3), 449) and polyethylene glycols and their mono-ethers (F. Tao and J. Xu, Gaodeng Xuexiao Huaxue Xuebao 1981, 2(4), 460; Chem. Abs., 97, 38121) or of non-aqueous media such as superoxides in benzene ((*a*) S. T. Purrington and G. B. Kenion, J. Chem. Soc., Chem. Commun., 1982, (13), 731. (*b*) J. L. Roberts, T. S. Calderwood and D. T. Sawyer, J. Amer. Chem. Soc., 1983, 105(26), 7691) are proposed as means of improving the yield of the desired reaction product and decreasing the loss of carbene through hydrolysis or other side-reactions.

Other techniques are also used to fragment tri- or tetra-halogenomethanes. Laser photo-dissociation of $CHClF_2$ causes the formation both of chlorocarbene and of chloro-fluorocarbene (R. I. Martinez and J. T. Herron, Chem. Phys. Lett., 1981, 84(1). 180) and these may react with chlorine or oxygen molecules present to provide $CFCl_3$ and COXF (X = F or Cl) in a simulation of atmospheric processes (J. Shi, X. Shen, J. Song and J. Yao, Youji Huaxue, 1985, (2), 124; Chem. Abs., 103, 159871). Deuterochloroform similarly forms dichlorocarbene, and the details of the laser-induced process have been reported (R. D. McAlpine, J. M. Goodale and D. K. Evans, Can. J. Chem., 1985, 63(11), 2995) Other polyhalogenomethanes are analogously decomposed and the intermediate carbenes have been identified and studied (R. D. Kenner, H. K Haak and F. Stuhl, J. Chem. Phys., 1986, 85(4), 1915). For example, the mechanism of formation of difluorocarbene by the photolysis of CCl_2F_2 has been reported (R. J. S. Morrison and E. R. Grant, J. Chem. Phys., 1979, 71(8), 3537), and so has the formation of various halogenocarbenes by the thermolysis of $CBrCl_3$ and of CCl_2F_2.

(2) *From trihalogenoethanoic acid derivatives.*
Dichlorocarbene is also obtained by the thermolysis of sodium trichloroacetate

$$CCl_3\text{-}CO_2^- \; Na^+ \longrightarrow NaCl + CO_2 + \; :CCl_2 \qquad (17)$$

and this affords the intermediate under conditions much less alkaline that those often used to generate the trihalogenomethyl carbanion. More modern alternatives include the use of $Ph_4Sb^+ \; CCl_3CO_2^-$ (N. A. Nesmeyanov *et al.*,

Dokl. Akad. Nauk SSSR 1981, 261(5) 1152 [Chem]), diaryltellurium bis(trichloroacetate) (I. D. Sadekov, B. B. Rivkin, and A. A. Maksimenko, Izv. Sev.-Kavk. Nauchn, Tsentra Vyssh. Shk., Estestv., Nauki 1982, (2) 54; Chem. Abs., 98, 71583) and trimethylsilyl trichloroacetate (S. V. Volovik *et al.*, Dopov. Akad. Nauk Ukr. RSR, Ser. B: Geol., Khim. Biol. Nauki 1982, (12) 36; Chem. Abs., 98, 142714) as sources of $:CCl_2$ under virtually non-basic conditions.

(3) *From organometallics* Various organometallics decompose to give reactive intermediates such as the carbenes. The reaction of bis(trifluoromethyl)cadmium with acyl chlorides in glyme gives not only the acyl fluoride (RCOF) but also $:CF_2$ which may be trapped at sub-zero temperatures by its stereospecific addition to, for example, *cis*-2-butene (L. J. Krause and J. A. Morrison, J. Amer. Chem. Soc., 1981, 103(11), 2995).

(4) *From organosilicon derivatives* The formation of strong silicon-halogen bonds leads to the use of a number of polyhalogenoalkylsilanes as precursors to carbenes which may be generated by thermolysis. An example is the generation of 1,2,2-tri-fluoroethylidene ($CHF_2.CF:$) by heating the silane $CHF_2.CF_2.SiF_3$ to 150°. This carbene inserts into C-H bonds of alkanes and cycloalkanes, the order of attack being 3° > 2° > 1°(R. N. Haszeldine *et al.*, J. Chem. Soc., Perkin Trans 1, 1979, (8), 1943).

(b) Reactions of halogenocarbenes

By far the most widely reported use of these intermediates is in their 1,2-addition to alkenes:

$$e.g.\ \rangle C{=}C\langle\ +\ :CCl_2 \longrightarrow \text{(1,1-dichlorocyclopropane)} \qquad (18)$$

and more than 100 reports of such processes occur in the recent abstracted literature.

In addition to studies of the effects of substituents (*e.g.* (*a*) E. G. Larson, Diss. Abs. Int. B. 1985, 46(4), 1175. (*b*) W. W. Schoeller, N. Aktekin and H. Friege, Angew. Chem., 1982, 94(12), 930 (*c*) B. Giese, W. B. Lee and C. Neumann, *idem.*, 1982, 94(4), 320) upon such processes, and the formation of new and unusual structures as a result of such additions (*e.g.* reactions with (*a*) quadricyclane: C. W. Jefford, J. Roussilhe and M. Papadopoulos, Helv. Chem. Acta 1985, 68(6) 1557-68. (*b*) norbornadienes: (*i*) C. W. Jefford *et al.*, Helv. Chim. Acta 1982, 65(5), 1467; (*ii*) C. W. Jefford and Phan Thanh Huy, Tetrahedron Lett., 1982, 23(4), 391) the carbenes can also undergo 1,4-addition to suitably substituted 1,3-dienes. These include hexamethyldimethylenecyclopentane (H. Mayr and V. Heigl, Angew. Chem., 1985, 97(7) 567) among other such dimethylenecycloalkanes (L. W. Jenneskens, W. H. de Wolf and F. Bickelhaupt, Angew. Chem., 1985, 97(7), 568).

Second Supplements to the 2nd Edition of Rodd's Chemistry of Carbon Compounds, Vol. IA/B, edited by M. Sainsbury

© 1991 Elsevier Science Publishers B.V., Amsterdam

Chapter 4a

MONOHYDRIC ALCOHOLS AND RELATED COMPOUNDS

C.W.THOMAS

1. *Introduction*

The chemistry of alcohols was covered in the corresponding chapter in the Main Second Edition which includes some references to the 1972 literature. A major review by S.G. Wilkinson ("Comprehensive Organic Chemistry", ed. J.F. Stoddart, Vol. 1, Chap. 4.1, Pergamon Press, 1978) has appeared. Interest in the synthesis and reactions of alcohols and their derivatives continues unabated and coverage in this review is, of necessity, selective. Progress in this area is reviewed annually in "General and Synthetic Methods" (Specialist Periodical Report, Royal Society of Chemistry) and "Annual Reports in Organic Synthesis" (Academic Press). Other relevant publications include "Compendium of Organic Synthetic Methods" (Wiley - Interscience) and "Chemistry of the Functional Groups", Supplement E : The Chemistry of Ethers, Crown Ethers, hydroxyl group and their sulphur analogues, ed. S. Patai, Parts 1 and 2, John Wiley, 1980.

The physical and thermodynamic properties of the lower alkanols are well established and are summarised in Table 1 (R.C. Wilhoit and B.J. Zwolinski, *J. Phys. Chem. Ref. Data Suppl.*, 1973, **2(1)**, 1; A.S. Kertes and C.J. King, *Chem. Rev.*, 1987, **87**, 687.

The acidities and basicities of alcohols have been discussed by Wilkinson (*loc. cit.*). Using a statistical method to analyse the data for many reactions, the cation and anion solvating tendencies of alcohols have been quantified (C.G. Swain *et al.*, *J. Amer. Chem. Soc.*, 1983, **105**, 502). The dipolarity, polarizability, hydrogen-bonding acidity and basicity of alcohols have been estimated using mainly electronic spectral measurements (R.W. Taft, *Prog. Phys. Org. Chem.* 1981, **13**, 485 and references therein). Several other properties of alcohols have been tabulated and discussed by Y. Marcus ("Ion Solvation", John Wiley, 1985).

TABLE 1

Property	MeOH	EtOH	n-PrOH	i-PrOH	n-BuOH	t-BuOH
Freezing point, ^{o}C	-97.7	-114.1	-126.2	-88.5	-89.3	25.8
Boiling point, ^{o}C	64.70	78.29	97.20	82.26	117.66	82.42
Vapor pressure, mmHg, $25^{o}C$	125.4	59.8	20.9	45.2	6.18	42.0
Density, g cm^{-3}, $25^{o}C$	0.7866	0.7851	0.7998	0.7813	0.8060	0.7812
Refractive index, $25^{o}C$	1.3265	1.3594	1.3837	1.3752	1.3971	1.3852
Dipole moment, debye	1.66	1.69	1.68	1.66	1.66	
Dielectric constant, $20^{o}C$	32.8	24.6	19.5	18.6	18.0	17.5
ΔH^{o}_{f}(liq), kJ mol^{-1}, $25^{o}C$	-239.0	-277.0	-304.0	-317.9	-327.1	-359.2
ΔG^{o}_{f}(liq), kJ mol^{-1}, $25^{o}C$	-166.8	-174.2	-170.6	-180.3	-162.5	-184.7
ΔH^{o}_{f}(liq), kJ mol^{-1}, $25^{o}C$	37.4	42.3	47.5	45.4	52.4	46.6
C_{p}^{o} (liq), J deg^{-1} mol^{-1}, $25^{o}C$	81.2	112.0	141.0	150.9	177.0	220.1
S^{o} (liq), J deg^{-1} mol^{-1}, $25^{o}C$	127.2	161.0	194.6	180.6	226.4	192.9

The ^{1}H.m.r. chemical shift of the hydroxylic proton signal in alkanols is usually in the range 0.5-4.5 δ and is dependent on the solvent, temperature and concentration, as a result of intermolecular hydrogen bonding and proton exchange. The signal is removed on addition of $^{2}H_{2}O$ and is often broad. Interpretation of ^{1}H.m.r. spectra is simplified by adding lanthanide paramagnetic shift reagents because downfield shifts are magnified when the hydroxyl group is associated with the reagent (A.F. Cockerill *et al.*, *Chem. Rev.*, 1973, **73**, 553). H.L. Goering *et al.* (*J. Amer. Chem. Soc.*, 1974, **96**, 1493) have described the use of an optically active lanthanide shift reagent to distinguish between the enantiomers of 2-butanol.

Empirical shift correlation relationships are useful signal assignment aids. Systematic studies of the ^{13}C.m.r. spectra of simple alkanols have been reported and a linear correlation noted between the substituent shifts for alkanols and those for the corresponding alkanes (J.D. Roberts *et al.*, *J. Amer. Chem. Soc.*, 1970, **92**, 1338). The procedure of D.M. Grant and E.G. Paul (*J. Amer. Chem. Soc.*, 1964, **86**, 2984) allowed prediction of shifts with satisfactory accuracy. A regression analysis was performed on the ^{13}C.m.r. shifts of 18 isomeric C_{6} alkanols and a good linear correlation was obtained (K.L. Williamson *et al.*, *J. Amer. Chem. Soc.*, 1974, **96**, 1471). These workers also studied the effects of *tris*(dipivaloylmethanato)europium(III) on the ^{1}H.m.r. spectra, and *tris*-(dipivaloylmethanato)ytterbium(III) on the ^{13}C.m.r. spectra of alkanols and applied the McConnell-Robertson equation to the measured shifts (H.M. McConnell and R.E. Robertson, *J. Chem. Phys.*, 1958, **29**, 1361), enabling determination of the average solution conformations. Other workers have extended the ranges of alkanol substituent shift correlations (A. Ejchart, *Org. Magn. Reson.*, 1977, **9**, 351) and examined solvent effects (C. Konno and H. Hikino, *Tetrahedron*, 1976, **32**, 325). Allyl, alkadiynyl and propargyl alcohols have also been studied (W. Wenkert *et al.*, *Org. Magn. Reson.*, 1975, **7**, 51; M.T.W. Hearn, *ibid.*, 1977, **9**, 141; *idem.*, *Tetrahedron*, 1976 **32**, 115). Primary, secondary and tertiary alkanols are readily distinguished because the carbon bearing the hydroxyl group is deshielded by oxygen and the ^{13}C.m.r. signal occurs at 50-70 ppm downfield from tetramethylsilane. Using the off-resonance partial decoupling procedure, characteristic triplets, doublets and singlets are obtained (Table 2).

TABLE 2

Compound	Chemical Shift in ppm			
	C-1	C-2	C-3	C-4
$CH_3CH_2CH_2CH_2OH$	62.9 (t)	36.0	20.3	15.2
$CH_3CH_2CH(OH)CH_3$	22.6	68.7 (d)	32.0	9.9
$(CH_3)_3COH$	69.9 (s)	32.7		

The position and appearance of the OH stretch band in the solution infrared spectra of alkanols is sensitive to hydrogen bonding. Intermolecular association produces a broad band (3200-3400 cm^{-1}) and a sharp free OH stretch occurs in the range 3590-3650 cm^{-1}, the relative intensities depending on the extent of hydrogen bonding. The frequencies and half-bandwidths of the OH stretch bands can be used to distinguish primary, secondary and tertiary alkanols (J.H. van der Maas and E.T.G. Lutz, *Spectrochim. Acta*, **30A**, 2005).

The electron impact mass spectra of alkanols are characterized by weak molecular ion regions due to ready elimination of water by rearrangement, induced partly by electron impact, and partly by thermal decomposition. The spectra are therefore usually similar to those of the corresponding alkenes (H. Hoshino *et al.*, *Bull. Chem. Soc. Japan*, 1973, **46**, 3043). Since the molecular ion is important for molecular weight determination and the quantitative GLC-MS analysis of mixtures, softer ionization techniques may be employed such as chemical ionization (F.H. Field, *J. Amer. Chem. Soc.*, 1970, **92**, 2672; R. Orlando *et al.*, *Org. Mass Spectrom.*, 1987, **22**, 579; B. Munson *et al.*, *ibid.*, 1987, **22**, 606) or volatile derivatives may be prepared (M.E. Rose and R.A.W. Johnstone, "Mass Spectrometry for Chemists and Biochemists", Cambridge University Press, 1982). Alternatively, mixture analysis is possible by using an on-line vapour phase dehydrogenation micro-reactor (Cu, 300°C) located between the mass spectrometer and the gas chromatograph (A.I. Mikaya *et al.*, *ibid.*, 1983, **18**, 99).

2. *Preparation of saturated monohydric alcohols (alkanols)*

(a) Reduction methods

(i) Carboxylic acids. Borane-THF rapidly and quantitatively reduces aliphatic carboxylic acids at 25°C or below. The reaction is selective under these conditions and amide, amino, hydroxyl, ester, halogeno, ketone, lactone, nitrile and nitro groups are not reduced (N.M. Yoon *et al.*, *J. Org. Chem.*, 1973, **38**, 2786; H.C. Brown and T.P. Stocky, *J. Amer. Chem. Soc.*, 1977, **99**, 8218; B.T. Cho and N.M. Yoon, Bull. *Korean Chem. Soc.*, 1982, **3**, 149).

$$3RCO_2H \longrightarrow (RCH_2O)_3B \xrightarrow{H_2O} 3RCH_2OH$$

Sodium salts of carboxylic acids are rapidly reduced to primary alcohols with two equivalents of borane-THF (N.M. Yoon and B.T. Cho, *Tetrahedron Lett.*, 1982 **23**, 2475). Borane-dimethylsulphide readily reduces aliphatic carboxylic acids at 25°C, but warming and the addition of trimethoxyborane are required for the reduction of benzoic acid (C.L. Lane *et al.*, *J. Org. Chem.*, 1974, **39**, 3052). Catecholborane reduces carboxylic acids and sodium salts to alcohols on heating (G.W. Kabalka *et al.*, *J. Org. Chem.*, 1977, **42**, 512).

(ii) Esters. Esters are reduced to primary alcohols in THF or diethylether using metal hydroborates : $NaBH_4$ is less effective than $LiBH_4$ or $Ca(BH_4)_2$ (H.C. Brown *et al.*, *J. Org. Chem.*, 1982, **47**, 4702).

$$RCO_2R' \xrightarrow[\text{(3) } 100^\circ C,\ 0.5-2\ h \text{ (4) hydrolysis}]{\text{(1) } LiBH_4,\ Et_2O \text{ (2) toluene}} RCH_2OH$$

When lithium tetrahydroborate is used trimethylborate and, especially, B-methoxy-9-borabicyclo[3.3.1]nonane have pronounced catalytic activity in this reaction (H.C. Brown and S. Narasimhan, *J. Org. Chem.*, 1984, **49**, 3891). Methyl and ethyl esters are reduced to primary alcohols using Me_2NBH_3Na or $t\text{-}BuNHBH_3Na$ (R.O. Hutchins *et al.*, *J. Org. Chem.*, 1984, **40**, 2438). Esters are also reduced to primary alcohols using sodium tetrahydroborate and $(HSCH_2)_2$ (N.C. Guida *et al.*, *Bull. Chem. Soc. Japan*, 1984, **57**, 1948). Benzoyl esters are converted to alcohols using NaH_2PO_2 and palladium-charcoal (R. Sala *et al.*, *Tetrahedron Lett.*, 1984, **25**, 4565).

(iii) Acyl halides. Acyl chlorides are reduced to primary alcohols using tetrabutylammonium octahydrotriborate and manganese(II) chloride (W.H. Tamblyn *et al.*, *Tetrahedron Lett.*, 1983, **24**, 4953).

$$CH_3(CH_2)_4COCl \xrightarrow[MnCl_2\,.\,4H_2O,\ THF]{n\text{-}Bu_4N^+\ B_3H_8^-} CH_3(CH_2)_4CH_2OH \quad (61\%)$$

(iv) Aldehydes and ketones. Borohydride exchange resin reduces aldehydes and ketones to the corresponding alcohols. Reduction is more rapid in methanol than ethanol and aldehydes are reduced more rapidly than ketones (H.C. Gibson and F.C. Baily, *J. Chem. Soc. Chem. Commun.*, 1977, 815; N.M. Yoon *et al.*, *Tetrahedron Lett.*, 1983, **24**, 5367). Aldehydes and ketones are also reduced using hydrous zirconium oxide and 2-propanol. Hydrous zirconium oxide is an alternative to aluminium isopropoxide in the Meerwein-Ponndorf-Verley reduction. This reagent is easily removed by filtration and is prepared by reacting zirconium oxychloride with aqueous alkali. Ketones are less reactive than aldehydes (H. Matsushita *et al.*, *Chem. Lett.*, 1985, 731).

$$CH_3(CH_2)_4CHO \xrightarrow[\text{OH}^-\text{, i-PrOH, reflux, 1h}]{ZrOCl_2 . 8H_2O} CH_3(CH_2)_4CH_2OH \quad (83\%)$$

Aldehydes and ketones are reduced to the corresponding alcohols using tetrabutylammonium octahydrotriborate (W.H. Tamblyn *et al.*, *loc. cit.*). Aldehydes are reduced to primary alcohols using pyridine-borane and alumina at room temperature. Higher temperatures are required in the absence of alumina (J.H. Babler and S.J. Sarussi, *J. Org. Chem.*, 1983, **48**, 4416). Aldehydes are selectively reduced to primary alcohols at room temperature using tributylamine, formic acid and $RuCl_2(PPh_3)_3$ in THF, followed by neutralization.

$$CH_3(CH_2)_8CHO \xrightarrow[Bu_3N\text{, }HCO_2H]{RuCl_2(PPh_3)_3} CH_3(CH_2)_9OH \quad (86\%)$$

Under these conditions such groups as nitro, keto, alkene, ester, tertiary amide and acetal are not affected (B.T. Khai and A. Arcelli, *Tetrahedron Lett.*, 1985, **26**, 3365). Aldehydes are reduced to primary alcohols using sodium tetrahydroborate, tetra-n-butylammonium bromide, benzene, 1-3 hours at room temperature (C.S. Rao *et al.*, *Ind. J. Chem.*, Sect. B, 1986, **25**, 626; M.R. Euerby and R.D. Waigh, *Synth. Commun.*, 1986, **16**, 779). Aldehydes are also reduced using a silylated immobilised organotin catalyst (S.A. Matlin and P.S. Gandham, *J. Chem. Soc. Chem. Commun.*, 1984, 798).

$$RCHO + Me_3SiO\text{–}Si(OMe)_2\text{–}(CH_2)_6\text{–}Sn(Bu)(OMe)_2 + Me_3SiO\text{–}(Si(Me)(H)\text{–}O)_x\text{–}SiMe_3 \xrightarrow[\text{reflux}]{MeOH} RCH_2OH$$

Tin catalysts such as $(Ph_3Sn)_2O$ and Ph_2SnO with butanol have been used to reduce aldehydes (J.D. Wuest and B. Zacharie, *J. Org. Chem.*, 1984, **49**, 166). Various reagents are reported to reduce ketones to secondary alcohols including tri-n-butyltin hydride (M. Degueil-Castaing *et al.*, *J. Org. Chem.*, 1986, **51**, 1672), diphenylamine-borane (C. Camacho *et al.*, *Synthesis*, 1982, 1027), isopropanol and nickel (G.P. Boldrini *et al.*, *J. Org. Chem.*, 1985, **50**, 3082) and diphenylantimony hydride with aluminium chloride (Y. Huang *et al.*, *Tetrahedron Lett.*, 1985 **26**, 5171). Reductive methylation of aldehydes and ketones has been reported, ketones reacting more slowly than aldehydes (M.T. Reetz *et al.*, *Chem. Ber.*, 1985, **118**, 1421).

$$CH_3(CH_2)_4CHO \xrightarrow[0^oC,\ Et_2O,\ 0.5\ h]{CH_3Ti(O\ i\text{-}Pr)_3} CH_3(CH_2)_4CH(OH)CH_3$$

$$CH_3(CH_2)_5COCH_3 \xrightarrow[22^oC,\ Et_2O,\ 48h]{CH_3Ti(O\ i\text{-}Pr)_3} CH_3(CH_2)_5C(CH_3)_2OH$$

(v) **Asymmetric reduction of aldehydes and ketones.** Accounts of asymmetric reduction methods are found in A. Pelter, K. Smith and H.C. Brown, "Borane Reagents", Academic Press, London, 1988; S.G. Wilkinson in "Comprehensive Organic Chemistry", D.H.R. Barton and W.D. Ollis, eds, vol.I, Chap. 4.1; M.M. Midland, *Chem. Rev.*, 1989, **89**, 1553; M.M. Midland in "Asymmetric Synthesis", J.D. Morrison, ed., vol.2, 45, Academic Press, New York, 1983.

Ketones are reduced asymmetrically by treatment with (1S, 2R)-6,6-dimethylbicyclo[3.1.1]heptan-2-ylmethylaluminium-dichloride (G. Giacomelli *et al.*, *J.Org.Chem.*, 1984, **49**, 310).

$$RCOR' \xrightarrow[(2)\ hydrolysis]{(1)\ (pinanyl)CH_2AlCl_2,\ Et_2O,\ room\ temperature} RCH(OH)R'$$

Chiral boranes have been used extensively for the asymmetric reduction of aldehydes and ketones. Isopinocampheylborane ($IpcBH_2$) and diisopinocampheylborane (Ipc_2BH) were used with moderate success to reduce prochiral ketones (H.C. Brown and A.K. Mandal, *J. Org. Chem.*, 1977, **42**, 2996; *ibid.*, 1984, **49**, 2558).

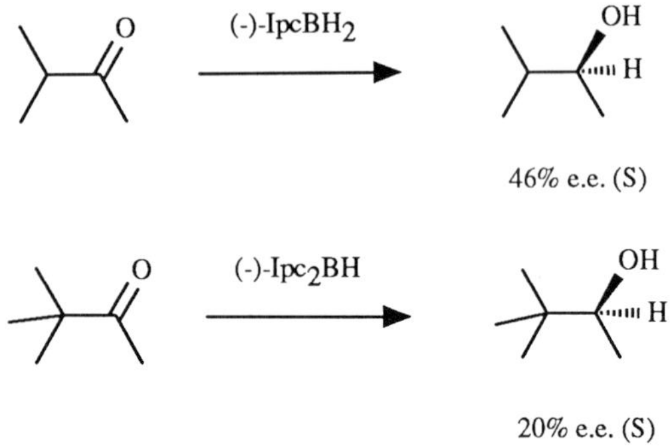

(-)-IpcBH$_2$

OH

H

46% e.e. (S)

(-)-Ipc$_2$BH

OH

H

20% e.e. (S)

B-isopinocampheyl-9-borabicyclo[3.3.1]nonane (R-Alpine-Borane) readily reduces aldehydes and acetylenic ketones with high enantiomeric excesses (M.M. Midland *et al.*, *J. Amer. Chem. Soc.*, 1977, **99**, 5211; *ibid.*, 1979, **101**, 2352; *ibid.*, 1980, **102**, 867).

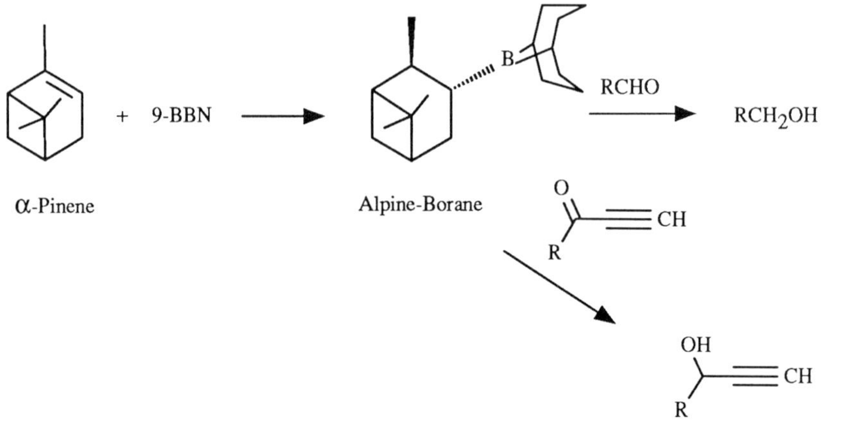

+ 9-BBN

RCHO

RCH_2OH

α-Pinene

Alpine-Borane

Unfortunately, Alpine-Borane appeared to be unsatisfactory for reducing dialkylketones and this was found to be due to dissociation of the reagent. The use of neat conditions suppressed the dehydroboration side reaction and increased reaction rates and optical purities (H.C. Brown and G.G. Pai, *J. Org. Chem.*, 1985, **50**, 1384). *cis*-Myrtanyl-9-borabicyclo-[3.3.1]nonane formed from β-pinene is effective at reducing hindered acetylenic ketones (M.M. Midland and J.I. McLoughlin, *J. Org. Chem.*, 1984, **49**, 4101).

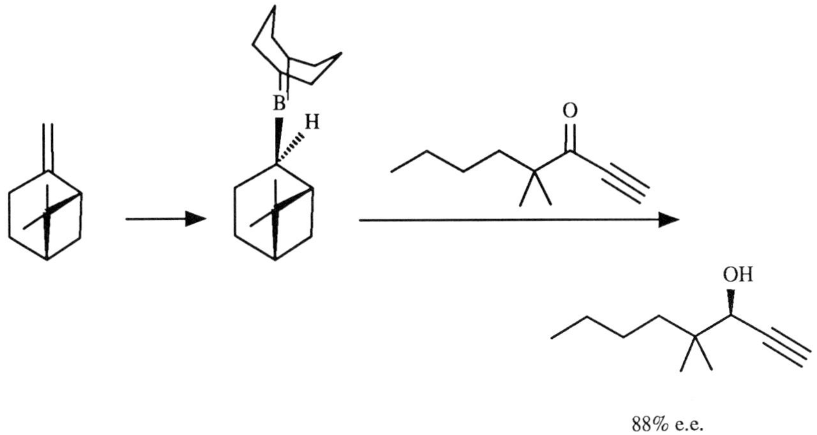

B

H

O

OH

88% e.e.

A modified borohydride reagent, potassium 9-*O*-(1,2:5,6-di-*O*-isopropylidene-α-D-glucofuranosyl)-9-boratabicyclo[3.3.1]-nonane (K-9-*O*-DIPGF-9-BBNH) derived from α-D-glucofuranose, as a chiral auxiliary, has been synthesised. This reagent reduces aliphatic ketones with good to excellent optical inductions at -78°C in THF (H.C. Brown *et al.*, *J. Org. Chem.*, 1986, **51**, 1934).

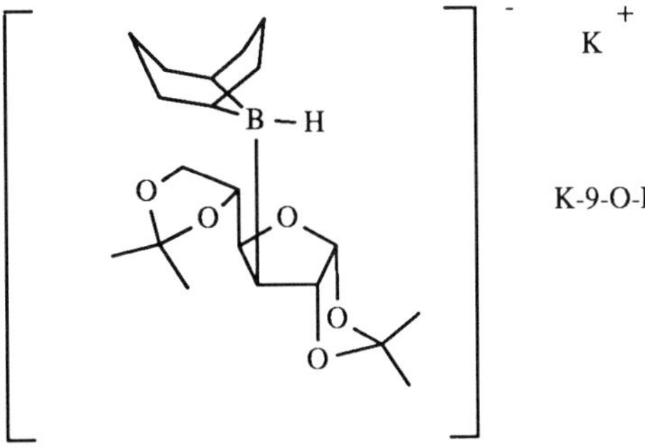

K^+

K-9-O-DIPGF-9-BBNH

-78°C

THF

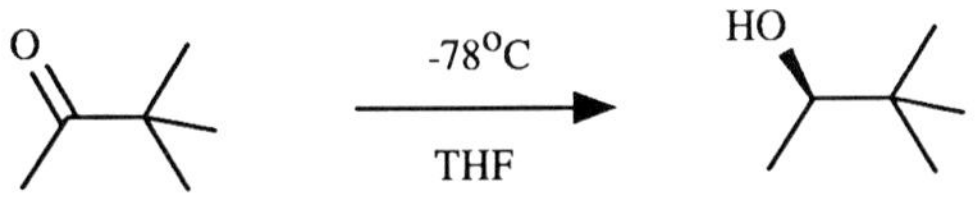

70% e.e.

The effect of placing a chlorine atom on boron in the reducing agent has been investigated. Ipc_2BCl was prepared by reacting Ipc_2BH with hydrogen chloride in ether. Ipc_2BCl has improved characteristics, reducing α-tertiaryalkyl ketones relatively rapidly under mild conditions with high optical inductions.

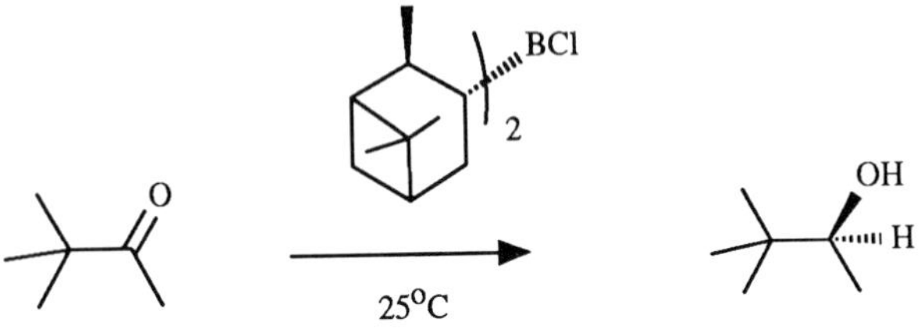

BCl

2

O

OH

H

25^oC

95% e.e.

2-pentanone has been asymmetrically reduced to 2-pentanol using an alcohol dehydrogenase enzyme in buffered aqueous isopropanol solution containing mercaptoethanol (E. Keinan *et al.*, *J. Amer. Chem. Soc.*, 1986, **108**, 162).

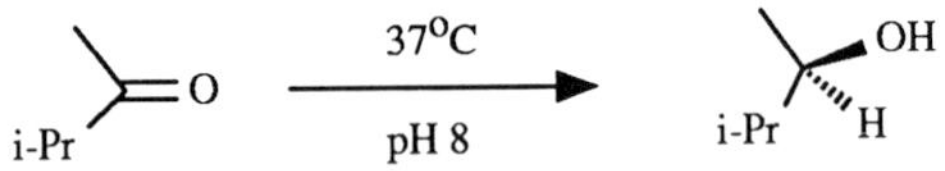

37^oC

i-Pr O

pH 8

OH

i-Pr H

61%

86% e.e.

Ketones are asymmetrically reduced to secondary alcohols using yeast and aqueous glucose, or yeast and water (M. Bucciarelli *et al.*, *Synthesis*, 1983, 897). Ketones are reduced by chiral aminoboranes (S.Itsuno *et al.*, *J. Org. Chem.*, 1984, **49**, 555).

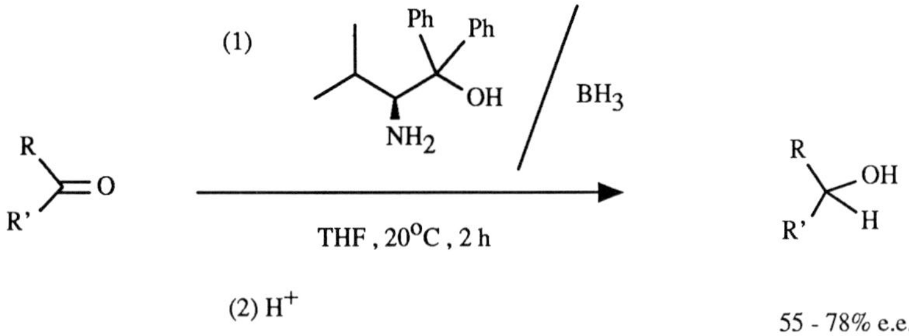

(1) Ph Ph

OH

NH_2

BH_3

R

O

R'

THF, 20^oC, 2 h

R

OH

R' H

(2) H^+

55 - 78% e.e.

R, R' = alkyl

(vi) Epoxides. Lithium selectride ($LiBH(CHMeEt)_3$) rapidly reduces epoxides to give the more hindered alcohols (S. Krishnamurthy *et al.*, *J. Amer. Chem. Soc.*, 1973, **95**, 8486). Ring-opening of epoxides to yield alcohols is also effected by lithium tetrahydroaluminate-aluminium chloride (K.N. Gurudutt and B. Ravindranath, *Synthesis*, 1983, 888). Sodium hydrogen telluride ring-opens epoxides by an S_N2 process to give telluro-alcohols which, by reduction with nickel boride, afford alcohols (D.H.R. Barton *et al.*, *Tetrahedron Lett.*, 1985, **26**, 6197).

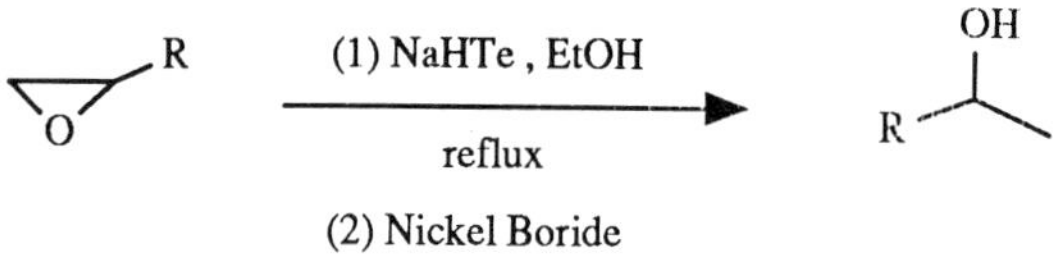

(1) NaHTe, EtOH, reflux
(2) Nickel Boride

Epoxides are reduced to alcohols using calcium metal in ethylenediamine (EDA) (R.A. Benkeser *et al.*, *J. Org. Chem.*, 1986, **51**, 3391).

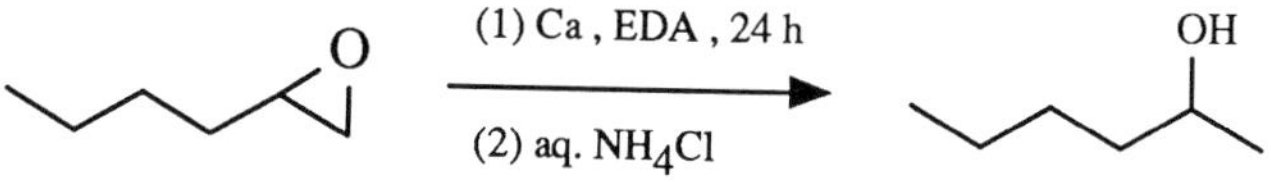

(1) Ca, EDA, 24 h
(2) aq. NH_4Cl

(91%)

(b) **Substitution Methods**

Bromoalkanes are converted to alcohols by reaction with the anion of methyldimesitylborane, followed by oxidation (A. Pelter *et al.*, *Tetrahedron Lett.*, 1983, **24**, 627).

$$Mes_2BCH_3 \xrightarrow{BuLi} Mes_2BCH_2^-$$

$$CH_3(CH_2)_5Br \xrightarrow[(2)\ Oxidation\ (MeOH,\ NaOH,\ H_2O_2)]{(1)\ Mes_2BCH_2^-} CH_3(CH_2)_5CH_2OH \quad (96\%)$$

Chloroalkanes are converted to the corresponding alcohols using sodium formate, tetrabutylammonium bromide, followed by alkali (H.A. Zahalka and Y. Sasson, *Synthesis*, 1986, 763).

$$CH_3(CH_2)_7Cl \xrightarrow[(2)\ aq.\ NaOH,\ room\ temperature]{(1)\ NaOCHO,\ N\text{-}Bu_4NBr} CH_3(CH_2)_7OH$$

Bromoalkanes react with water in the presence of mercury(II) perchlorate (formed from mercury(II) oxide and aqueous perchloric acid) and 1,2-dimethoxyethane, to give the corresponding alcohols (A. McKillop and M.E. Ford, *Tetrahedron*, 1974, 30, 2467).

$$R\text{-}Br + Hg(ClO_4)_2 + H_2O \longrightarrow R\text{-}OH$$

e.g. $R = n\text{-}C_8H_{17}$ (98%)

Primary iodoalkanes are oxidised by m-chloroperbenzoic acid (m-CPBA) in dichloromethane to give primary alcohols.

$$RCH_2CH_2I \xrightarrow[CH_2Cl_2,\ 0^oC]{m\text{-}CPBA} RCH_2CH_2OH$$

e.g. $R = CH_3(CH_2)_{13}$ (62%)

Primary iodides give exclusively displacement products but secondary iodides give a mixture of products resulting from displacement, elimination and α-carbon oxidation. Tertiary iodides give displacement and elimination products. (H.J. Reich and S.L. Reake, *J. Amer. Chem. Soc.*, 1978, **100**, 4888; T.L. Macdonald *et al.*, *ibid.*, 1980, **102**, 7760; R.C. Cambie *et al.*, *J. Chem. Soc. Perkin Trans. I, 1980, 822)*.

(c) **Organoborane reactions**

The use of chiral organoboranes for reducing aldehydes and ketones was outlined in 2 (a). Organoboranes are oxidised by amine-N-oxides such as trimethylamine-N-oxide and pyridine-N-oxide to give trialkyl borates which are hydrolysed to alcohols. The oxidation occurs with retention of configuration and may be carried out in anhydrous conditions (G.W. Kabalka and H.C. Hedgecock, *J. Org. Chem.*, 1975, **40**, 1776; G.W. Kabalka and S.W. Slayden, *J. Organomet. Chem.*, 1977, **125**, 273).

$$3Me_3NO + R_3B \longrightarrow 3Me_3N + (RO)_3B$$

$$(RO)_3B \xrightarrow{H_2O} 3ROH$$

Secondary and tertiary alcohols are obtained by bromination of organoboranes followed by oxidation with alkaline hydrogen peroxide (H.C. Brown and Y. Yamamoto, *J. Amer. Chem. Soc.*, 1971, **93**, 2796; C.F. Lane, *Intra-Sci. Chem. Reports*, 1973, **7**, 133; C.F. Lane and H.C. Brown, *J. Amer. Chem. Soc.*, 1971, **93**, 1025; *Synthesis*, 1972, 303; Y. Yamamoto and H.C. Brown, *J. Org. Chem.*, 1974, **39**, 861).

$$(RCH_2)_3B \xrightarrow[h\nu]{Br_2,\,H_2O} (RCH_2)B(OH){-}C(H)(CH_2R)(R) \xrightarrow[H_2O_2]{NaOH} RCH_2CH(OH)R$$

Cyanoborate salts, in the presence of acylating agents, undergo three migrations of organyl groups to yield intermediates which can be oxidised to tertiary alcohols (A. Pelter *et al.*, *J. Chem. Soc., Perkin Trans. I*, 1975, 129).

$$R_3B + NaCN \longrightarrow R_3\overset{-}{B}\text{-}CN\ Na^+ \xrightarrow{(CF_3CO)_2O}$$

$$R_3\overset{-}{B}{-}C{\equiv}\overset{+}{N}{-}COCF_3 \longrightarrow CF_3COO{-}B(CR_3){-}O{-}C(CF_3){=}N{-}COCF_3 \xrightarrow[H_2O]{NaOH} R_3COH$$

Alkali metal cyanotriorganylborates are stable salts which may be recrystallized from water. However, on treatment with electrophiles they become susceptible to rearrangement. The reaction with trifluoroacetic anhydride occurs readily below room temperature.

Organoboranes react with suitable anion bearing leaving groups to give intermediates which can be oxidised to give primary, secondary or tertiary alcohols depending on the number of leaving groups (E. Negishi *et al.*, *J. Org. Chem.*, 1975, **40**, 814; T. Mukaiyama *et al.*, *Bull. Chem. Soc. Japan*, 1972, **45**, 2244).

$$\text{9-BBN-R} \xrightarrow{LiCH_2SMe} \left[\text{9-BBN}\overset{-}{B}(R)(CH_2SMe) \right] Li^+$$

$$\xrightarrow{MeI} \text{9-BBN}\overset{-}{B}(R)(CH_2\overset{+}{S}Me_2) + LiI$$

$$\xrightarrow{\text{single migration}} \text{9-BBN-}CH_2R + Me_2S$$

$$\xrightarrow{\text{oxidation}} RCH_2OH$$

In the presence of oxygen, trialkylboranes react with formaldehyde to yield dialkylalkoxyboranes which give primary alcohols with water (N. Miyaura *et al.*, *J. Amer. Chem. Soc.*, 1972, **94**, 6549). An improved procedure for the synthesis of aldehydes and primary alcohols via carbonylation of organoboranes using potassium triisopropoxyhydroborate has been reported (H.C. Brown *et al.*, *Synthesis*, 1979, 701).

$$\text{9-BBN-R} + CO + K^+(\text{i-PrO})_3{}^-BH \longrightarrow \left[\text{9-BBN-B-C(R)(H)(OB(i-PrO)_3)} \right]^- K^+$$

$$\xrightarrow{H_2O_2,\ pH\ 7} RCHO$$

$$\xrightarrow{NaOH,\ H_2O,\ EtOH,\ reflux} RCH_2OH$$

Secondary alcohols are prepared by treating organoboranes with lithium trimethoxyhydroaluminate and carbon monoxide (J.L. Hubbard and H.C. Brown, *Synthesis*, 1978, 676). The availability of diisopinocampheylborane in optically pure form and using a reaction temperature of -25°C has improved the asymmetric inductions obtained in the hydroboration of *cis*-alkenes (H.C. Brown *et al.*, *J. Org. Chem.*, 1982, **47**, 5065; H.C. Brown and B. Singaram, *ibid.*, 1984, **49**, 945).

BH 2 100% e.e. -25°C B 2 H oxidation HO H 98.4% e.e.

Ipc_2BH is not a suitable reagent for asymmetric hydroboration of *trans*-alkenes and trisubstituted alkenes as optical inductions are poor but monoisopinocampheylborane ($IpcBH_2$) gives better results.

(1) $IpcBH_2$, -25°C (2) oxidation HO H 73% e.e.

(1) $IpcBH_2$, -25°C (2) oxidation HO H 53% e.e.

$IpcBH_2$, itself, is not easily prepared by hydroboration of α-pinene but is obtained from Ipc_2BH (H.C. Brown *et al.*, *J. Org. Chem.*, 1978, **43**, 4395; H.C. Brown and P.K. Jadhav, *J. Org. Chem.*, 1981, **46**, 5047; H.C. Brown and N.M. Yoon, *J. Amer. Chem. Soc.*, 1979, **99**, 5514; H.C. Brown *et al.*, *J. Org. Chem.*, 1982, **47**, 5074).

BH 2 (1) TMEDA (2) $BF_3 . Et_2O$ BH_2

TMEDA = N,N,N',N'-tetramethylethylenediamine

Chiral boronic esters have emerged as important intermediates in asymmetric synthesis. Hydroboration of alkenes with $IpcBH_2$, followed by reaction with acetaldehyde, hydrolysis and cyclisation yields chiral 1,3,2-dioxaborinanes.

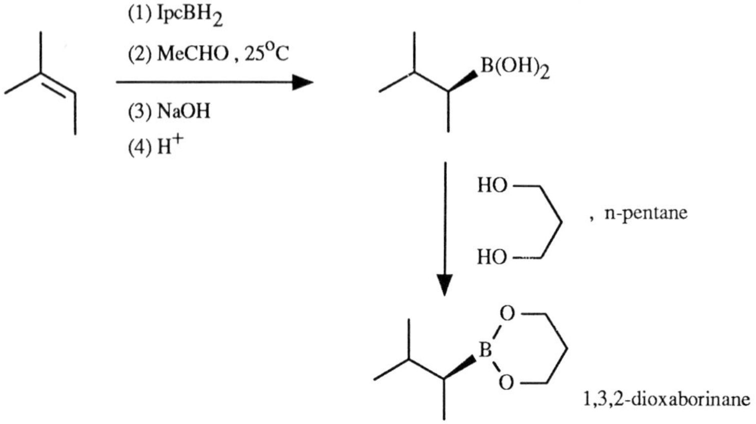

(1) $IpcBH_2$
(2) MeCHO, 25°C
(3) NaOH
(4) H^+

$B(OH)_2$

HO, HO, n-pentane

1,3,2-dioxaborinane

These chiral boronic esters, when treated with methoxy(phenylthio)methyl lithium (MPML) followed by mercury (II) chloride, furnish the homologated α-methoxyalkyl derivatives which on oxidation give α-chiral aldehydes which, in turn, are readily reduced to the corresponding alcohols with high optical induction (H.C. Brown *et al.*, *J. Amer. Chem. Soc.*, 1985, **107**, 4980; M. Srebnik and P. Veeraraghavan, *Aldrichimica Acta.*, 1987, **20**, 9).

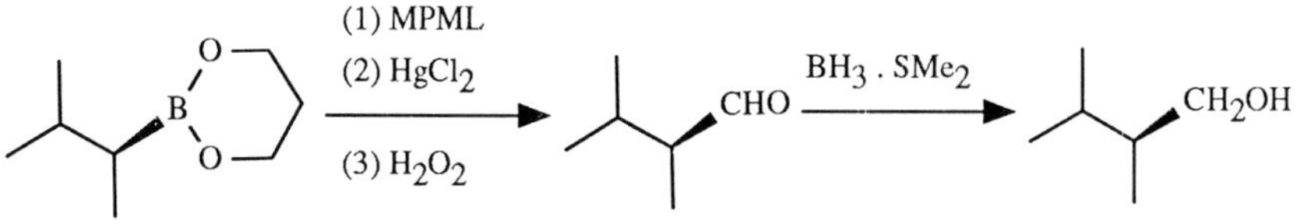

(1) MPML
(2) $HgCl_2$
(3) H_2O_2

CHO $\xrightarrow{BH_3 . SMe_2}$ CH_2OH

3. *Reactions of saturated monohydric alcohols (alkanols)*

(a) **Oxidation**

(i) **Metal-based oxidants.** A review of alcohol oxidation by chromium based reagents has appeared : G. Cainelli and G. Cardillo in "Chromium Oxidations in Organic Chemistry", Chap. 4, 118, Springer-Verlag, Berlin, 1984.

Amberlyst A-26 anion exchange resin (Rohm and Haas) which contains quaternary ammonium chloride groups, can be converted to the $HCrO_4$ form by reaction with aqueous chromium trioxide. This modified resin is used to oxidise alkanols to aldehydes and ketones in inert solvents. There is little further oxidation of aldehydes to carboxylic acids and work-up is greatly simplified - the resin is filtered and the filtrate evaporated to remove solvent (G. Cainelli *et al.*, *J. Amer. Chem. Soc.*, 1976, **98**, 6737).

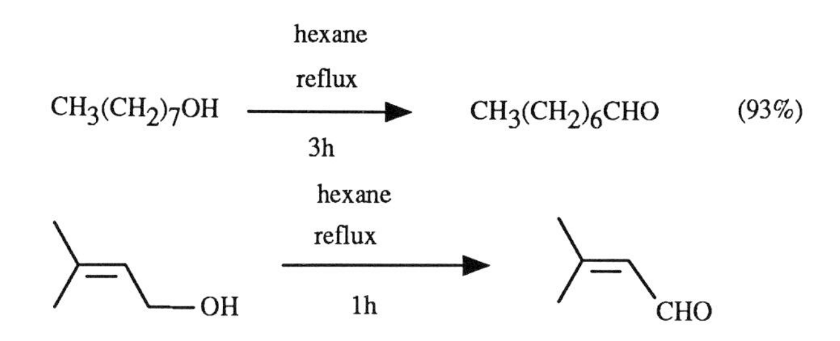

In contrast, zinc dichromate trihydrate oxidises primary and secondary alkanols, but allyl alcohols much less readily (H. Firouzabadi *et al.*, *Synthesis*, 1986, 285).

Bis-phosphonium dichromate is a mild oxidant for primary and secondary alcohols (H.J. Christau *et al.*, *Tetrahedron Lett.*, 1986, **27**, 1775). Chromic acid in dimethylsulphoxide readily oxidises primary alcohols to aldehydes (Y.S. Rao and R. Filler, *J. Org. Chem.*, 1974, **39**, 3305).

$$CH_3(CH_2)_7OH \xrightarrow{<70^{\circ}C} CH_3(CH_2)_6CHO$$

(80%)

Chromium trioxide in hexamethylphosphortriamide (HMPT) readily oxidises primary and secondary alkanols to aldehydes and ketones (G. Cardillo *et al.*, *Synthesis*, 1976, 394).

$$CH_3(CH_2)_{11}OH \xrightarrow{CrO_3,\ HMPT} CH_3(CH_2)_{10}CHO$$

(80%)

Chromium trioxide, in the presence of 18-crown-6 or 12-crown-4 in dichloromethane at room temperature, oxidises primary and secondary alkanols to aldehydes and ketones (I. Ganboa *et al.*, *J. Chem. Res. Synop.*, 1984, **3**, 92) and chromium trioxide - graphite reagent has also been used for this purpose (J.M. LaLancette *et al.*, *Canad. J. Chem.*, 1972, **50**, 3058; H.B. Kagan, *Chem. Tech.*, 1976, **6**, 510). Chromium trioxide- di-pyridine complex in dichloromethane prepared *in situ* is a good oxidant for acid-labile alcohols giving high yields of aldehydes and ketones. A large (6 to 10-fold) excess of this reagent is required (R.W. Ratcliffe, *Org. Synth.*, 1976, **55**, 84; E. Piers and P.M. Worster, *Canad. J. Chem.*, 1977, **55**, 733). Chromylchloride in dichloromethane, pyridine, t-butanol mixture has been used to oxidise primary and secondary alcohols to aldehydes and ketones. Much less oxidant is required than in the case of CrO_3/pyridine (K.B. Sharpless and K. Akashi, *J. Amer. Chem. Soc.*, 1976, **97**, 5927).

$$2\text{t-BuOH} + 3\ \text{pyridine} \xrightarrow[(2)\ CrO_2Cl_2]{(1)\ CH_2Cl_2,\ -78^oC} O{=}Cr(=O)(\text{O-t-Bu})_2$$

ditertiarybutylchromate

$$\xrightarrow{\text{alcohol}} \text{aldehyde or ketone}$$

Chromium trioxide-3,5-dimethylpyrazole complex oxidises primary and secondary alkanols in dichloromethane rapidly at room temperature to aldehydes and ketones (E.J. Corey and G.W.J. Fleet, *Tetrahedron Lett.*, 1973, 4499). Halogeno-chromate-tertiary amine complexes are generally efficient reagents for this conversion and are relatively stable, crystalline solids e.g., pyridinium chlorochromate (PCC) (E.J. Corey and J.W. Suggs, *Tetrahedron Lett.*, 1975, 2647).

$$CH_3(CH_2)_9OH \xrightarrow[1.5h,\ RT]{PCC,\ CH_2Cl_2} CH_3(CH_2)_8CHO$$

(92%)

Pyridinium fluorochromate is more suitable for acid sensitive alkanols being less acidic than the chlorochromate (M.N. Bhattacharjee *et al.*, *Synthesis*, 1982, 588). Quinolinium chlorochromate reacts much more rapidly with primary alkanols than secondary (J. Singh *et al.*, *Chem. Ind. (London)*, 1986, 751; C.S. Rao *et al.*, *Ind. J. Chem.*, *Sect. B*, 1986, **25**, 324). Secondary alcohols are oxidised to ketones using peroxyacetic acid with 2,4-dimethylpentane-2,4-diol chromate as catalyst. Yields are high and reaction times short (E.J. Corey *et al.*, *Tetrahedron Lett.*, 1985, 5855).

OH ; O ; O ; Cr ; O ; O ; , $MeCO_3H$; CH_2Cl_2, CCl_4 ; 0^oC, 0.35h ; O ; (99%)

Bis(benzyltrimethylammonium)dichromate in hexamethylphosphortriamide oxidises primary alcohols to aldehydes (X. Huang and C.C. Chen, *Synthesis*, 1982, 1091). Secondary alcohols are oxidised to ketones using pyridinium dichromate (PDC) and *bis*trimethylsilylperoxide (K. Oshima *et al.*, *Tetrahedron Lett.*, 1983, 2185).

$$CH_3(CH_2)_5CH(OH)(CH_2)_2CH_3 \xrightarrow[CH_2Cl_2\text{, RT, 1.5h}]{\text{PDC}, (CH_3)_3SiOOSi(CH_3)_3} CH_3(CH_2)_5CO(CH_2)_2CH_3 \quad (83\%)$$

PDC in dichloromethane at room temperature oxidises primary alkanols to aldehydes (E.J. Corey and G. Schmidt, *Tetrahedron Lett.*, 1979, **19**, 399).

Active manganese dioxide is an efficient mild oxidant and its use has been reviewed (A.J. Fatiadi, *Synthesis*, 1976, 65, 133).

$$CH_3(CH_2)_3OH \xrightarrow{MnO_2} CH_3(CH_2)_2CHO \quad (70\%)$$

Nickel peroxide (the Nakagawa reagent) is a powerful oxidant and oxidises primary alkanols to carboxylic acids. It can also be used catalytically for this purpose using sodium hypochlorite as oxidant (M.V. George and K.S. Balachandran, *Chem. Rev.*, 1975, **75**, 491; T.L. Ho and T.W. Hall, *Synth. Commun.*, 1975, **5**, 309). Secondary alcohols are oxidised to ketones using bromobenzene as oxidant and a palladium catalyst (Y.Tamaru *et al.*, *Tetrahedron Lett.*, 1979, **19**, 1401).

$$CH_3(CH_2)_5CH(OH)CH_3 + C_6H_5Br \xrightarrow[\text{dimethyl formamide, } 110^{o}C,\ 4.5h]{Pd(PPh_3)_4,\ K_2CO_3} CH_3(CH_2)_5COCH_3\ (84\%) + C_6H_6 + HBr$$

Secondary alcohols are oxidised to ketones using palladium (II) chloride and potassium carbonate in carbon tetrachloride (H. Nagashima *et al.*, *Tetrahedron*, 1985, **41**, 5645). Primary and secondary alkanols are oxidised to aldehydes and ketones in high yields using palladium(II)acetate and sodium bicarbonate under solid - liquid phase transfer conditions (B.M. Choudary *et al.*, *Tetrahedron Lett.*, 1985, **26**, 6257).

$$CH_3(CH_2)_2OH \xrightarrow[\text{PhI, dimethyl formamide, } Bu_4N^+Cl^-]{Pd(OAc)_2,\ NaHCO_3} CH_3CH_2CHO\ (97\%)$$

Secondary alkanols are oxidised to ketones using samarium(II) iodide in tetrahydrofuran at room temperature (J. Collin *et al.*, *Nouv. J. Chem.*, 1986, **10**, 229) and by silver(I) ferrate in benzene (H. Firouzabadi *et al.*, *Synth. Commun.*, 1986, **16**, 211). Secondary alkanols are oxidised to ketones using tertiarybutyl peroxide with molybdenum hexacarbonyl catalyst. Addition of cetylpyridinium chloride (CPC) increases the rate (K. Yamawaki *et al.*, *Synthesis*, 1986, 59).

$$CH_3(CH_2)_5CH(OH)CH_3 \xrightarrow[MgSO_4,\ \text{benzene, CPC, reflux, 24h}]{t\text{-}BuOOH,\ Mo(CO)_6} CH_3(CH_2)_5COCH_3\ (70\%)$$

Lead(IV) acetate and manganese(II) acetate in refluxing benzene oxidise primary alkanols to aldehydes (M.L. Mihailovic *et al.*, *Tetrahedron Lett.*, 1986, **27**, 2287).

$$CH_3(CH_2)_6OH \xrightarrow{Pb(OAc)_4,\ Mn(OAc)_2} CH_3(CH_2)_5CHO$$

Secondary alcohols are oxidised to ketones using sodium bromate and ruthenium trichloride catalyst under phase transfer conditions (Y. Yamamoto *et al.*, *Tetrahedron Lett.*, 1985, **26**, 2107).

$$CH_3CH(OH)CH_3 \xrightarrow[H_2O,\ \text{Aliquat } 336]{RuCl_3.(H_2O)_n \quad NaBrO_3,\ CHCl_3,} (CH_3)_2CO$$

Primary alkanols are photooxidised to ketones using platinised titanium dioxide as catalyst, but secondary alkanols react very slowly (F. Hussein *et al.*, *Tetrahedron Lett.*, 1984, **25**, 3363). Tetra-n-propylammonium perruthenate, $(Pr_4N)(RuO_4)$ is a selective, mild oxidant for primary and secondary alkanols (W.P. Griffith *et al.*, *J. Chem. Soc. Chem. Commun.*, 1987, 1625; W.P. Griffith and S.V. Ley, *Aldrichim. Acta.*, 1990, **23**, 13). Secondary alkanols can be oxidised selectively in the presence of primary alkanols by sodium bromate with a cerium catalyst (H. Tomioka *et al.*, *Tetrahedron Lett.*, 1982, **23**, 539; S. Kanemoto *et al.*, *Bull. Chem. Soc. Japan*, 1986, **59**, 105).

$$CH_3(CH_2)_7CH(OH)(CH_2)_2CH_3 \xrightarrow[NaBrO_3,\ aq.\ CH_3CN \quad 80^oC,\ 0.8h]{(NH_4)_2Ce(NO_3)_6} CH_3(CH_2)_7CO(CH_2)_2CH_3 \quad (94\%)$$

Sodium bromite in acetic acid oxidises secondary alkanols to ketones. Oxidative esterification of primary alcohols occurs (Y. Ueno *et al.*, *Synthesis*, 1983, 815).

$$CH_3CH(OH)(CH_2)_4CH_3 \xrightarrow[H_2O,\ RT]{NaBrO_2,\ CH_3COOH} CH_3CO(CH_2)_4CH_3 \quad (82\%)$$

Primary alkanols are converted to aldehydes in high yield by oxidation with 3-iodosobenzoic acid and a ruthenium catalyst (P. Muller and J. Godoy, *Helv. Chim. Acta*, 1983, **66**, 1790).
Secondary alkanols are dehydrogenated using Raney nickel in benzene but primary alkanols are unreactive under these conditions (M.E. Kraft and B. Zorc, *J. Org. Chem.*, 1986, **51**, 5482).

$$CH_3CH(OH)(CH_2)_5CH_3 \xrightarrow[\text{benzene, reflux, 3h}]{\text{Raney Nickel}} CH_3CO(CH_2)_5CH_3 \quad (93\%)$$

Secondary alkanols are oxidised to ketones using peracetic acid as oxidant and cobalt (III) acetate catalyst in acetic acid. Sodium bromide accelerates the reaction. Primary alkanols do not react (T. Morimoto *et al.*, *J. Chem. Soc., Perkin II*, 1983, 1949).

$$CH_3(CH_2)_4CH(OH)CH_3 \xrightarrow[\text{NaBr, }60^oC\text{, 1h}]{CH_3COOH,\ Co(OAc)_2} CH_3(CH_2)_4COCH_3 \quad (67\%)$$

Primary alkanols are oxidised to aldehydes in high yields using tertiary butylperoxide and *bis*(2,4,6-trimethylphenyl)-diselenide in refluxing benzene (I. Kuwajima *et al.*, *J. Org. Chem.*, 1982 **47**, 837).

$$CH_3(CH_2)_9OH \xrightarrow[(2,4,6\text{-}Me_3C_6H_2Se)_2]{\text{t-BuOOH, benzene}} CH_3(CH_2)_8CHO \quad (92\%)$$

(ii) **Non-metallic oxidants.** 4-methoxy-1-oxo-2,2,6,6-tetramethylpiperidinium chloride in dichloromethane oxidises primary and secondary alkanols rapidly (T. Miyazawa *et al.*, *J. Org. Chem.*, 1985, **50**, 1332.

$$\text{4-methoxy-2,2,6,6-tetramethylpiperidine-1-oxoammonium } Cl^- \xrightarrow[\text{5 min}]{CH_2Cl_2,\ 25^oC} \text{4-methoxy-2,2,6,6-tetramethyl-1-hydroxypiperidine} + HCl$$

$$+ \quad CH_3(CH_2)_2CH(OH)CH_2CH_3 \qquad\qquad + \quad CH_3(CH_2)_2COCH_2CH_3 \quad (92\%)$$

Secondary alcohols have been oxidised by a similar reaction using metachloroperbenzoic acid (MCPBA) and tetramethyl-piperidine, the latter being oxidised to the oxammonium salt (B. Ganem, *J. Org. Chem.*, 1975, **40**, 1988).

$$\text{2,2,6,6-tetramethylpiperidine (N–H)} \xrightarrow[\text{m-CPBA}]{\text{oxidation}} \text{N–OH} \xrightarrow[\text{m-CPBA}]{\text{oxidation}} \text{N}^+\text{=O } X^-$$

$$\text{N}^+\text{=O } X^- \xrightarrow{RR'CHOH} RR'CO$$

An electrooxidative variant of this reaction has been reported (M.F. Semmelhack *et al.*, *J. Amer. Chem. Soc.*, 1983, **105**, 4492). Alcohols are resistant to oxidation by peracids but, in the presence of mineral acid catalyst, oxidation of secondary alkanols by MCPBA occurs (J.A. Cella *et al.*, *Tetrahedron Lett.*, 1975, **15**, 4115).

$$CH_3(CH_2)_5CH(OH)CH_3 \xrightarrow[\text{THF, RT, 1h}]{\text{m-CPBA, HCl}} CH_3(CH_2)_5COCH_3 \quad (75\%)$$

Insoluble polymeric thioanisole has been used with chlorine for alcohol oxidation. Cross-linked polystyrene is modified to incorporate thiomethyl groups at the 4-positions (G.A. Crosby *et al.*, *J. Amer. Chem. Soc.*, 1975, **97**, 2232).

$$\text{(P)-C}_6\text{H}_5 \xrightarrow[\text{Br}_2,\ \text{CCl}_4]{\text{Tl(OAc)}_3} \text{(P)-C}_6\text{H}_4\text{-Br} \xrightarrow[(2)\ \text{CH}_3\text{SSCH}_3]{(1)\ \text{n-BuLi}} \text{(P)-C}_6\text{H}_4\text{-SCH}_3$$

$$\text{CH}_3(\text{CH}_2)_7\text{OH} \xrightarrow[\text{(P)-C}_6\text{H}_4\text{-SCH}_3,\ -25^{\circ}\text{C}]{\text{Cl}_2,\ \text{CH}_2\text{Cl}_2} \text{CH}_3(\text{CH}_2)_6\text{CHO}\ (95\%)$$

Chlorine-pyridine complex has been used to oxidise primary and secondary alkanols (J. Wicha and T. Zarecki, *Tetrahedron Lett.*, 1974, **14**, 3059).

$$\text{CH}_3(\text{CH}_2)_5\text{OH} \xrightarrow[\text{CHCl}_3,\ <30^{\circ}\text{C},\ 40\ \text{min}]{\text{Cl}_2\text{-pyridine}} \text{CH}_3(\text{CH}_2)_4\text{CHO}\ (85\%)$$

Alkanols are oxidised by conversion to their trimethylsilyl derivatives which are treated with trityltetrafluoroborate (M.E. Jung, *J. Org. Chem.*, 1976, **41**, 1479).

$$\text{RR'CHOH} \xrightarrow[(\text{Me}_3\text{Si})_2\text{NH},\ \text{pyridine},\ 25^{\circ}\text{C}]{\text{Me}_3\text{SiCl}} \text{RR'CH-O-SiMe}_3$$

$$\downarrow\ \text{hydride transfer};\ \text{Ph}_3\text{C}^+\text{BF}_4^-,\ \text{CH}_2\text{Cl}_2,\ 25^{\circ}\text{C}$$

$$\text{RR'CO} \xleftarrow{\text{H}_2\text{O}} \text{R(R')C}^+\text{-OSiMe}_3\ \ \text{BF}_4^- \ + \text{Ph}_3\text{CH}$$

Secondary alkanols are oxidised to ketones using calcium hypochlorite in aqueous acetic acid - acetonitrile mixture. In contrast, primary alkanols yield esters in which the acid and alcohol portions are derived from the alcohol (S.O. Nwaukwa and P.M. Keehn, *Tetrahedron Lett.*, 1982, 23, 35).

$$CH_3CH_2CH(OH)CH_2CH_3 \xrightarrow[CH_3CN,\ 0^oC]{aq.\ Ca(OCl)_2,\ CH_3COOH} CH_3CH_2COCH_2CH_3$$

$$CH_3(CH_2)_5OH \xrightarrow[CH_3CN,\ 0^oC]{aq.\ Ca(OCl)_2,\ CH_3COOH} CH_3(CH_2)_4COO(CH_2)_5CH_3$$

Periodinane (12-I-5) is a mild, selective oxidant for primary and secondary alkanols (D.B. Bess and J.C. Martin, *J. Org. Chem.*, 1983, **48**, 4155).

12-I-5 (I(OAc)$_3$ benziodoxolone) + RR'CHOH ⟶ (I-OAc benziodoxolone) + RR'CO + 2AcOH

Primary alkanols are oxidised to aldehydes using N-chlorosuccinimide in diisopropyl thioether at 0°C. Secondary alkanols afford ketones using the same reagents but at -78°C (K.W. Kim *et al.*, *J. Chem. Soc. Chem. Commun.*, 1984, 762). Primary and secondary alkanols are oxidised using dimethylsulphoxide, which has been activated with trifluoroacetic anhydride or oxalylchloride, followed by treatment with base (A.J. Mancuso and D. Swern, *Synthesis*, 1981, 165).

$$CH_3(CH_2)_5CH(OH)CH_3 \xrightarrow[(2)\ Et_3N,\ -60^oC;\ (3)\ H_2O,\ RT]{(1)\ DMSO,\ (CF_3CO)_2O,\ CH_2Cl_2,\ -60^oC} CH_3(CH_2)_5COCH_3$$

(75%)

Intermediate dimethylsulphonium salts are produced which react with alkanol and base.

$$(CH_3)_2S{=}O + (CF_3CO)_2O \xrightarrow{CH_2Cl_2} (CH_3)_2\overset{+}{S}{-}OCOCF_3 \;\; {}^{-}OCOCF_3$$

$$\xrightarrow{RR'CHOH} (CH_3)_2\overset{+}{S}{-}O{-}CHRR'$$

$$(CH_3)_2\overset{+}{S}{-}O{-}CHRR' \xrightarrow[H_2O]{base} Me_2S + RR'C{=}O$$

(b) **Substitution**

(i) **Formation of haloalkanes.** Alkanols are converted in high yields into bromo- or iodoalkanes by a one-step reaction with N,N$'$-carbonyldiimidazole and an excess of a reactive halocompound such as allyl bromide or iodomethane (T. Kamijo *et al.*, *Chem. Pharm. Bull.*, 1983, **31**, 4189).

$$CH_3(CH_2)_{11}OH \xrightarrow[\text{0.5h at 25°C then 2h reflux}]{\text{N,N'-carbonyldiimidazole; } CH_2{=}CHCH_2Br,\ CH_3CN} CH_3(CH_2)_{11}Br \quad (97\%)$$

Primary alkanols are converted into chloroalkanes by decarboxylation of their chloroformate esters. The esters are formed by reacting the alkanol with phosgene (R. Richter and B. Tucker, *J. Org. Chem.*, 1983, **48**, 2625).

$$R{-}OH \xrightarrow{COCl_2} R{-}O{-}COCl \xrightarrow{HCON(CH_3)_2} H{-}\overset{+}{C}(OCOOR){-}N(CH_3)_2 \;\; Cl^-$$

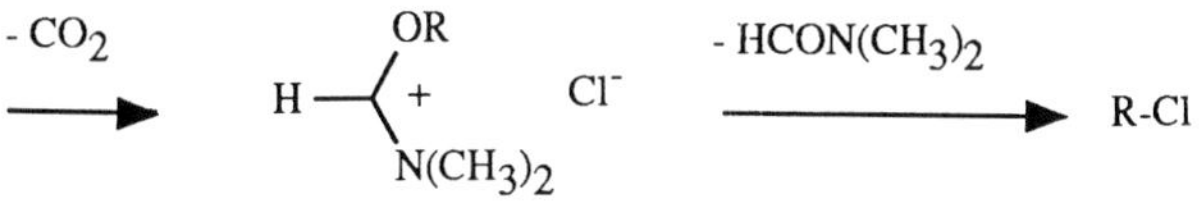

$$\xrightarrow{-CO_2} H-C^{+}(OR)(N(CH_3)_2)\ Cl^- \xrightarrow{-HCON(CH_3)_2} R\text{-}Cl$$

Iodoalkanes are formed from alkanols using sodium iodide and trimethylsilyl polyphosphate(PPSE) (T. Imamoto *et al.*, *Synthesis*, 1983, 460).

$$CH_3(CH_2)_7OH \xrightarrow[10h,\ 25^oC]{NaI,\ PPSE} CH_3(CH_2)_7I$$

(83%)

Bromoalkanes are obtained by reacting alcohols with polymer-supported triphenylphosphine dibromide or with polymer-supported triphenylphosphine and carbon tetrabromide in chloroform (P. Hodge and E. Khoshdel, *J. Chem. Soc. Perkin I*, 1984, 195).

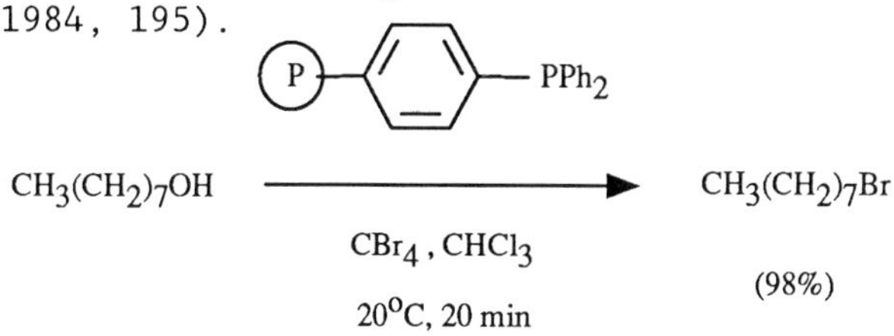

$$CH_3(CH_2)_7OH \xrightarrow[CBr_4,\ CHCl_3;\ 20^oC,\ 20\ min]{(P)-C_6H_4-PPh_2} CH_3(CH_2)_7Br$$

(98%)

Diphenylphosphinated ethylene oligomers (DPE) and carbon tetrachloride convert alkanols to chloroalkanes (D.E. Bergbreiter and J.R. Blanton, *J. Chem. Soc. Chem. Commun.*, 1985, 337).

$$CH_2{=}CH_2 \xrightarrow[hexane]{RLi} \xrightarrow{ClPPh_2} \text{-[}CH_2CH_2\text{]}CH_2CH_2PPh_2$$

DPE

$$R\text{-}OH \xrightarrow[toluene,\ 90^oC]{DPE,\ CCl_4} R\text{-}Cl$$

Secondary alkanols are converted into the corresponding bromides, with overall retention of configuration, in a two-step double inversion process via a phenylselenide intermediate (M. Sevrin and A. Krief, *J. Chem. Soc. Chem. Commun.*, 1980, 656).

(1) PhSeCN , n-Bu_3P / 20°C , THF / or / (2) $MeSO_2Cl$ / Et_2O , 0°C / PhSeNa , THF

R'R(H)C–OH → PhSe–C(H)R'R → (Br_2 , Et_3N , CH_2Cl_2 , 20°C) → R'R(H)C–Br

Tributyldiiodophosphorane and triphenyldiiodophosphorane, prepared *in situ* from the corresponding phosphines and iodine, are generally able to convert primary and secondary alkanols into iodoalkanes at room temperature in diethyl ether or benzene containing hexamethylphosphortriamide (R.K. Haynes and M. Holden, *Aust. J. Chem.*, 1982, **35**, 517).

$$Ph_3PI_2 + R\text{-}OH \longrightarrow Ph_3\overset{+}{P}\text{-}OR + I^- + HI$$

$$Ph_3\overset{+}{P}\text{-}OR + I^- \longrightarrow R\text{-}I + Ph_3PO$$

Highly pure t-butyl and t-pentyl bromides and iodides are prepared by reacting the alcohol with aqueous halogen acid and the corresponding lithium or calcium halide (H. Masada and Y. Murotani, *Bull. Chem. Soc., Japan*, 1980, **53**, 1181).

$$\text{t-BuOH} \xrightarrow[0^\circ C,\ 2h]{HBr,\ LiBr} \text{t-BuBr} \quad (87\%) \ 99.8\% \text{ pure}$$

Iminium salts formed from phosphoryl chloride and dimethylformamide (DMF) readily convert alcohols to chloroalkanes (M. Yoshihara *et al.*, *Synthesis*, 1980, 746).

$$\text{R-OH} \xrightarrow[DMF,\ RT,\ 24h]{dry\ CHCl_3,\ POCl_3} \text{R-Cl}$$

Alcohols are readily converted to iodoalkanes *via* the trimethylsilylethers (T. Morita *et al.*, *Synthesis*, 1979, 379).

$$\text{R-OH} + \text{Me}_3\text{SiCl} \xrightarrow[\text{benzene, reflux, 2h}]{\text{pyridine}} \text{R-OSiMe}_3$$

$$\text{R-OSiMe}_3 + \text{Me}_3\text{SiCl} + \text{NaI} \xrightarrow[\text{CH}_3\text{CN, 1h}]{70-75^\circ\text{C}} \text{R-I}$$

In a similar manner, alcohols are converted to the corresponding bromoalkane by reaction with trimethylsilylchloride and lithium bromide in acetonitrile (G.A. Olah *et al.*, *J. Org. Chem.*, 1980, **45**, 1638).

(ii) **Formation of esters and ethers.** A one-step procedure for the tosylation of secondary alcohols with inversion, using zinc tosylate, diethylazocarboxylate and triphenylphosphine, has been described (I. Galynker and W.C. Still, *Tetrahedron Lett.*, 1982, **23**, 4461).

OH (secondary alcohol) $\xrightarrow[\text{DEAD, benzene, RT}]{(\text{TsO})_2\text{Zn, Ph}_3\text{P}}$ OTs (inverted) (84%)

Alcohols are formylated using benzoyl chloride and dimethyl formamide, followed by aqueous acid (J. Barluenga *et al.*, *Synthesis*, 1985, 426).

$$\text{CH}_3\text{CH}_2\text{CH(CH}_3\text{)CH}_2\text{OH} \xrightarrow[\text{(2) dil. aq. H}_2\text{SO}_4]{\text{(1) PhCOCl, Me}_2\text{NCHO}} \text{CH}_3\text{CH}_2\text{CH(CH}_3\text{)CH}_2\text{OCHO} \quad (65\%)$$

Alcohols are inverted *via* the ester intermediate, using methane sulphonyl chloride, triethylamine and caesium acetate followed by hydrolysis (J.W. Huffman and R.C. Desai, *Synth. Commun.*, 1983, **13**, 553). Methoxymethyl (MOM) ethers of primary and secondary alcohols are obtained in high yield by reaction with dimethoxymethane and iodotrimethylsilane. The methoxymethyl group is a useful protecting group for alcohols being stable towards strong alkali, Grignard reagents and lithium tetrahydroaluminate but readily removed by mild acid treatment (G. Olah *et al.*, *Synthesis*, 1983, 896).

$$CH_3(CH_2)_8OH \xrightarrow[CH_2(OMe)_2]{Me_3SiI} CH_3(CH_2)_8OCH_2OCH_3 \quad (94\%)$$

β-methoxyethoxymethyl (MEM) ethers are obtained by reaction of an alcohol with MEM salt (E.J. Corey *et al.*, *Tetrahedron Lett.*, 1976, **16**, 809).

$$CH_3OCH_2CH_2OCH_2Cl \xrightarrow{Et_3N} CH_3OCH_2CH_2OCH_2\overset{+}{N}Et_3 \quad Cl^-$$

$$\downarrow \text{R-OH, NaH, THF}$$

$$CH_3OCH_2CH_2OCH_2OR$$

Methoxythiomethyl (MTM) ethers are formed by reaction of an alcohol with sodium hydride, sodium iodide and methylthiochloromethane in dimethoxyethane (E.J. Corey and M.G. Bock, *Tetrahedron Lett.*, 1975, **15**, 3269.
Triisopropylsilyl (TIPS) ethers are produced by reacting an alcohol with triisopropyl silylchloride and imidazole in DMF. These derivatives are relatively stable under hydrolytic conditions (R.F. Cunico and L. Bedell, *J. Org. Chem.*, 1980, **45**, 4797).

$$\text{R-OH} \xrightarrow[\text{imidazole, RT; 12 - 20h}]{\text{TIPS-Cl, DMF}} \text{R-OSiPr}^i_3$$

Trimethylsilyl ethers are prepared in high yield by reacting alcohols with allyltrimethyl silane in acetonitrile using 4-toluenesulphonic acid as catalyst. This method is advantageous in that the propene by-product is volatile and only a small amount of catalyst is required (T. Morita *et al.*, *Tetrahedron Lett.*, 1980, **21**, 835).

$$CH_3(CH_2)_6OH + CH_2{=}CHCH_2SiMe_3 \xrightarrow[70-80^oC,\ 2h]{TsOH,\ CH_3CN} CH_3(CH_2)_6OSiMe_3 + CH_2{=}CHCH_3 \quad (95\%)$$

4. Preparation of unsaturated monohydric alcohols

(a) **Reduction of aldehydes and ketones**

α,β-unsaturated ketones are reduced using sodium hydride, zinc chloride and magnesium bromide to give α,β-unsaturated alcohols. Use of sodium hydride, sodium alkoxide and nickel (II) acetate yields the saturated ketone (L. Mordenti *et al.*, *J. Org. Chem.*, 1979, **44**, 2203).

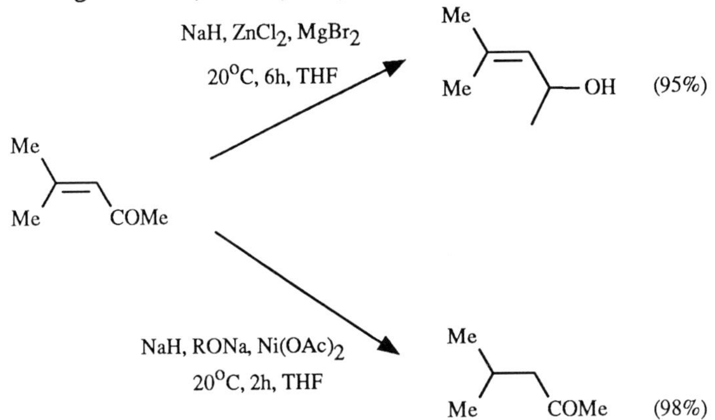

α,β-unsaturated ketones are reduced to α,β-unsaturated alcohols using triisobutylalane in pentane (A.M. Caporusso *et al.*, *J. Org. Chem.*, 1982, **47**, 4640).

$$CH_2{=}CHCOCH_3 \xrightarrow[\text{(2) aq.}NH_4Cl]{\text{(1) }{}^{i}Bu_3Al\text{, 10min, pentane, RT}} CH_2{=}CHCH(OH)CH_3\ (86\%) + CH_2{=}CHCH(OH)\text{-}{}^{i}Bu\ (14\%)$$

α,β-unsaturated aldehydes are reduced to α,β-unsaturated alcohols with formic acid, triethylamine and a ruthenium catalyst in tetrahydrofuran (B.T. Khai and A. Arcelli, *Tetrahedron Lett.*, 1985, **26**, 3365).

CHO $\xrightarrow[Et_3N\text{, HCOOH, RT}]{RuCl_2(PPh_3)_3\text{, THF}}$ OH (99%)

Triphenyltin formate catalyses the reduction of α,β-unsaturated aldehydes. The reaction involves hydride transfer in a mechanism analogous to that of the Meerwein-Ponndorf-Verley reduction (J.D. Wuest and B. Zacharie, *J. Org. Chem.*, 1984, **49**, 166).

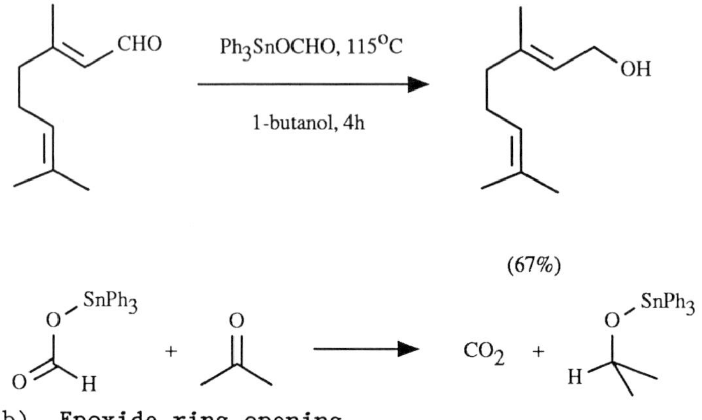

CHO $\xrightarrow[\text{1-butanol, 4h}]{Ph_3SnOCHO,\ 115^oC}$ OH

(67%)

$SnPh_3$ O, O, H + O $\longrightarrow$ CO_2 + $SnPh_3$ O, H

(b) **Epoxide ring opening**

Epoxides are converted to allyl alcohols in high yield by reaction with diethylaluminium-2,2,6,6-tetramethylpiperidine (A. Yasuda *et al.*, *Bull. Chem. Soc., Japan*, 1979, **52**, 1705).

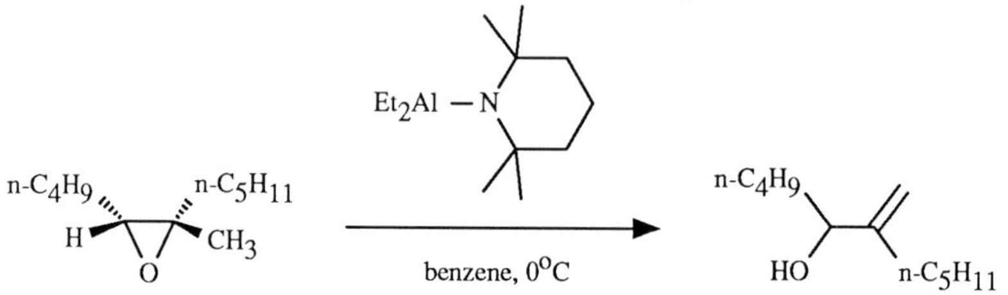

Et_2Al-N

n-C_4H_9, n-C_5H_{11}, H, CH_3, O $\xrightarrow{\text{benzene, }0^oC}$ n-C_4H_9, HO, n-C_5H_{11}

Vinyl oxiranes react with organoboranes and a catalytic amount of oxygen followed by treatment with water to give allyl alcohols. (Allenyl methanols are produced in a similar manner from ethynyloxiranes) (A. Suzuki *et al.*, *Synthesis*, 1973, 305.

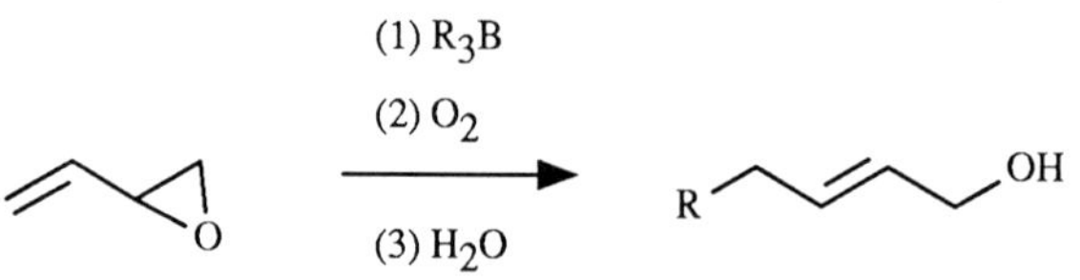

O $\xrightarrow[\text{(3) }H_2O]{\text{(1) }R_3B\text{ (2) }O_2}$ R, OH

Epoxides react with trimethylsilyltriflate and 2,5-lutidine followed by 1,5-diazabicyclo[5.4.0]undecene-5 (DBU) to give α,β-unsaturated alcohols (S. Murata *et al.*, *J. Amer. Chem. Soc.*, 1979, **101**, 2738).

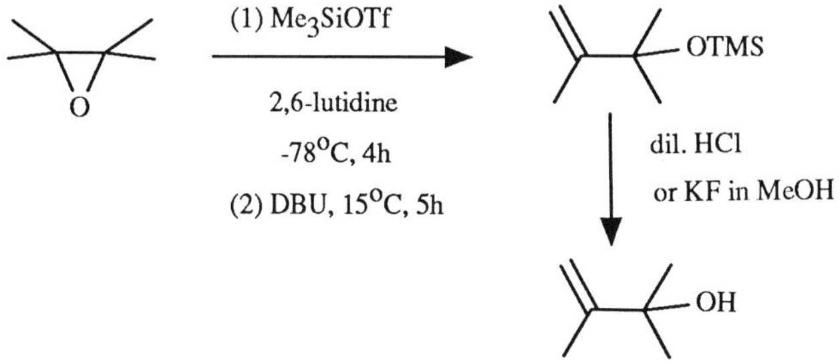

(1) Me_3SiOTf
2,6-lutidine
-78°C, 4h
(2) DBU, 15°C, 5h
OTMS
dil. HCl
or KF in MeOH
OH

Vinyl oxirane is ring-opened with methylation to give homoallylic alcohols by treatment with an organolithium reagent plus a lanthanoid amide or alkoxide in ether at -78°C, followed by quenching with aqueous ammonium chloride (I. Mukerji *et al.*, *Angew. Chem. Int. Ed. Engl.*, 1986, **25**, 760).

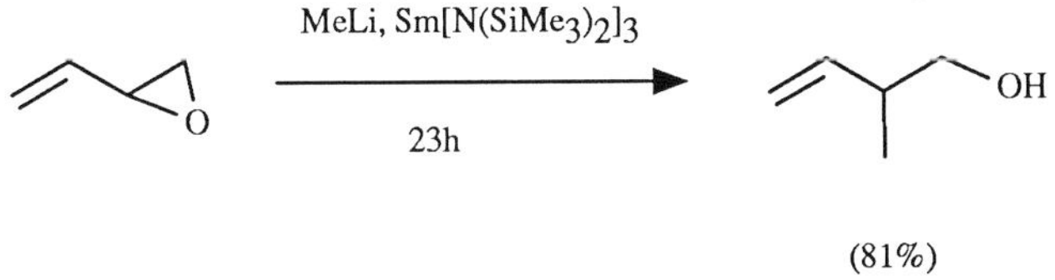

MeLi, $Sm[N(SiMe_3)_2]_3$
23h
OH
(81%)

(c) **From alkenes, alkynes and their derivatives**

In boiling tetrahydrofuran, B-alkenyl-9-BBN derivatives react by addition with aldehydes to give products which on oxidation yield allyl alcohols with retention of configuration of the allyl group (P. Jacob, III, and H.C. Brown, *J. Org. Chem.*, 1977, **42**, 579).

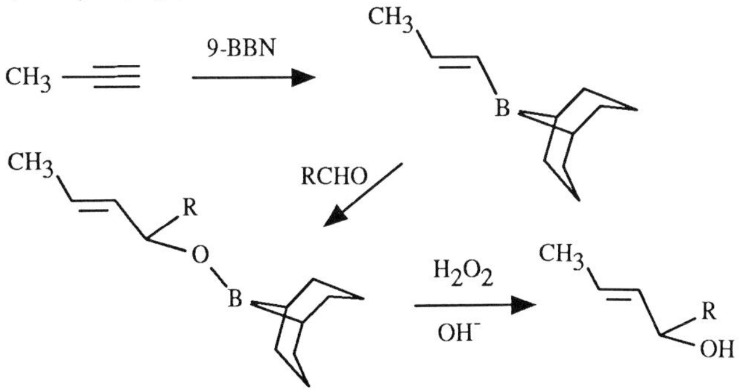

CH_3
9-BBN
CH_3
B
RCHO
CH_3
R
O
B
H_2O_2
OH^-
CH_3
R
OH

B-allyldiisopinocampheylborane reacts with aldehydes to produce secondary homoallyl alcohols with high optical purities (H.C. Brown and P.K. Jadhav., *J. Amer. Chem. Soc.*, 1983, **105**, 2092).

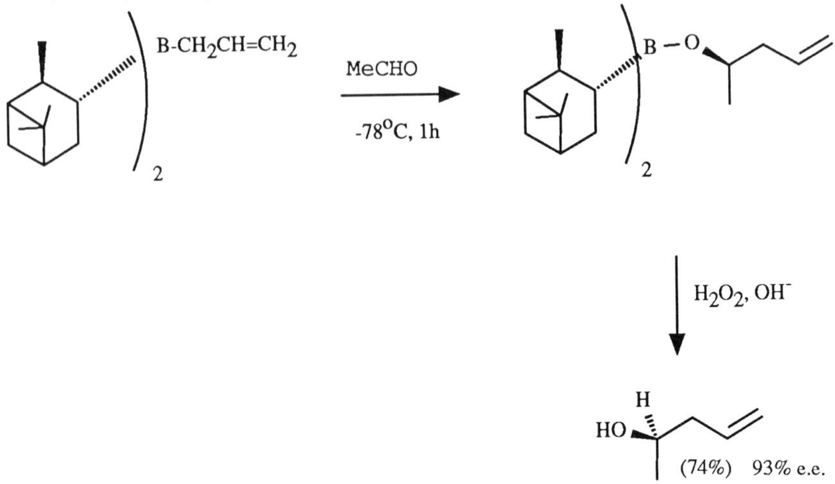

$B-CH_2CH=CH_2$ MeCHO -78°C, 1h B—O 2 2

H_2O_2, OH^-

H HO (74%) 93% e.e.

Alkenes are oxidised to allyl alcohols using tertiary butyl hydroperoxide, selenium dioxide and silica gel (H. Shirahama *et al.*, *Chem. Lett.*, 1981, 1703). Aldehydes react with allyl bromides in the presence of zinc, saturated aqueous ammonium chloride and tetrahydrofuran or, with tin, aqueous tetrahydrofuran with sonication, to give homoallylic alcohols (C. Petrier *et al.*, *Tetrahedron Lett.*, 1985, 1449.

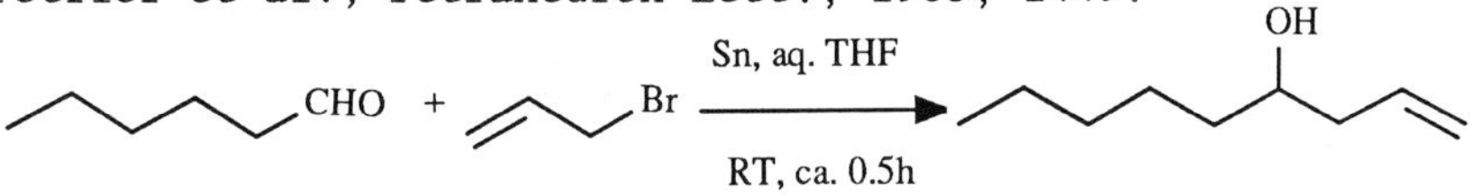

CHO + Br Sn, aq. THF RT, ca. 0.5h OH

(100%)

Allyl halides react with aldehydes or ketones in the presence of zinc, aqueous tetrahydrofuran with sonication to give homoallylic alcohols (C. Petrier and J.L. Luche, *J. Org. Chem.*, 1985, **50**, 910). Vinyl halides readily react with aldehydes and ketones in the presence of lithium and diethyl ether with sonication (J.L. Luche and J.C. Damiano, *J. Amer. Chem. Soc.*, 1980, **102**, 7926).

$$CH_3CH{=}CHBr + n\text{-}C_5H_{11}CHO \xrightarrow[\text{sonication 40 mins}]{Li,\ Et_2O} n\text{-}C_5H_{11}CH(OH)CH{=}CHCH_3$$

(95%)

$$CH_2=C(Br)CH_3 + n\text{-}C_5H_{11}CHO \xrightarrow[\text{sonication 40 mins}]{\text{Li, } Et_2O} n\text{-}C_5H_{11}CH(OH)C(CH_3)=CH_2 \quad (94\%)$$

Organolithium and organozinc reactions promoted by sonication offer synthetic advantages over classical methods with regard to selectivity, yields and rates (J.L. Luche *et al.*, *Ultrasonics*, 1987, **25**, 40). But-2-enyl lithium reacts with aldehydes *via* allylic boronate complexes to produce homoallylic alcohols (Y. Yamamoto *et al.*, *J. Chem. Soc. Chem. Commun.*, 1980, 1072).

BEt_3, -78°C; $\bar{B}Et_3$ Li^+; Li^+

MeCHO

OH, Me, Me (78%) 85% threo 15% erythro

Allylselenides are readily oxidised to allyl alcohols by reaction with aqueous hydrogen peroxide, dichloromethane and pyridine. The allylselenides are obtained from α-selenoaldehydes by reaction with methylenetriphenyl phosphorane (T. DiGiamberardino *et al.*, *Tetrahedron Lett.*, 1983, 3413).

$$RCH(SeR^1)_2 \xrightarrow[\text{DMF}, \; CH_2=PPh_3]{\text{BuLi}} RCH(SeR^1)CH=CH_2 \xrightarrow{H_2O_2, \text{ pyridine}, \; CH_2Cl_2} RCH=CHCH_2OH$$

Lithium propargylide reacts with organoboranes followed by an aldehyde and oxidation to give propargyl allyl alcohols or allenyl allyl alcohols depending on the conditions (T. Leung and G. Zweifel, *J. Amer. Chem. Soc.*, 1974, **96**, 5620; G. Zweifel *et al.*, *J. Amer. Chem. Soc.*, 1978, **100**, 5561).

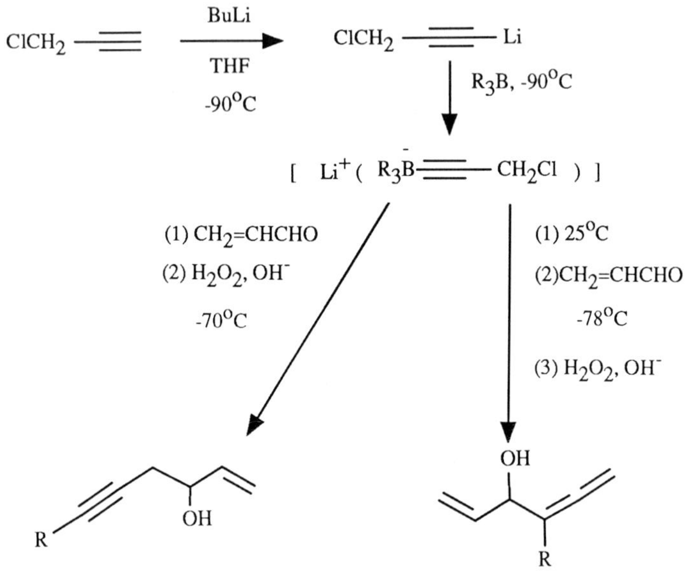

$ClCH_2C{\equiv}CH$ → (BuLi, THF, -90°C) → $ClCH_2C{\equiv}C{-}Li$ → (R_3B, -90°C) → $[Li^+(R_3\bar{B}{-}C{\equiv}C{-}CH_2Cl)]$

(1) $CH_2{=}CHCHO$ (2) H_2O_2, OH^- -70°C

(1) 25°C (2) $CH_2{=}CHCHO$ -78°C (3) H_2O_2, OH^-

OH R OH R

α,β-unsaturated ketones react with a titanium alkoxy ate complex in tetrahydrofuran. The complex is prepared by reacting allylmagnesium bromide with titanium isopropoxide (M.T. Reetz *et al.*, *Chem. Ber.*. 1985, **118**, 1421).

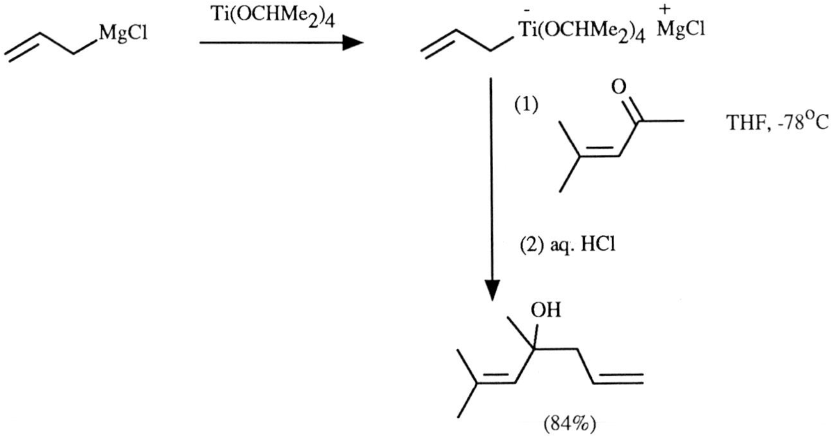

MgCl → ($Ti(OCHMe_2)_4$) → $\bar{T}i(OCHMe_2)_4\ \overset{+}{M}gCl$

(1) O THF, -78°C

(2) aq. HCl

OH

(84%)

5. *Reactions of alkenols*

(a) **Oxidation**

Some of the reactions described earlier for oxidising alkanols can also be used for alkenols. These include the methods of M.L. Mihailovic *et al.*, G. Cardillo *et al.*, and A.J. Fatiadi. Allylic alcohols are oxidised to α,β-unsaturated ketones using silver(II)dichromate-pyridine complex (H. Firouzabadi *et al.*, Synth. Commun., 1984, 14, 89). Allylic alcohols are converted to α,β-unsaturated aldehydes using potassium ferrate, benzyltriethylammonium chloride, benzene and 10% aqueous sodium hydroxide under phase-transfer conditions (K.S. Kim *et al.*, *Synthesis*, 1984, 866).

$$CH_2{=}CHCH_2OH \xrightarrow[\substack{PhCH_2\overset{+}{N}Et_3\ Cl^- \\ 10\%\ aq.\ NaOH \\ 25^oC,\ 0.1h}]{K_2FeO_4,\ benzene} CH_2{=}CHCHO\ (96\%)$$

Benzyltrimethylammonium tetrabromooxomolybdate catalyses the oxidation of secondary allylic alcohols to α,β-unsaturated ketones by tertiarybutylperoxide (Y. Mauyama *et al.*, *Tetrahedron Lett.*, 1984, 4417).

$$\text{(alkenol, OH)} \xrightarrow[60^oC,\ benzene,\ 20h]{PhCH_2\overset{+}{N}Me_3\ {}^-OMoBr_4} \text{(enone, O)}\ (60\%)$$

Allylic alcohols are oxidised by oxygen and ruthenium dioxide at 70°C in 1,2-dichloroethane (M. Matsumoto and N. Watanabe, *J. Org. Chem.*, 1984, **49**, 3435).

$$\text{(allylic alcohol, OH)} \xrightarrow[70^oC,\ ClCH_2CH_2Cl]{O_2,\ RuO_2} \text{(CHO)}\ (76\%)$$

The ruthenium-catalysed reaction of allylmethylcarbonate with allylic alcohols yields α,β-unsaturated aldehydes and ketones. The product carbonyl compounds are easily isolated because the other products, methanol, propene and carbon dioxide are volatile (I. Minami *et al.*, *Tetrahedron Lett.*, 1986, 1805).

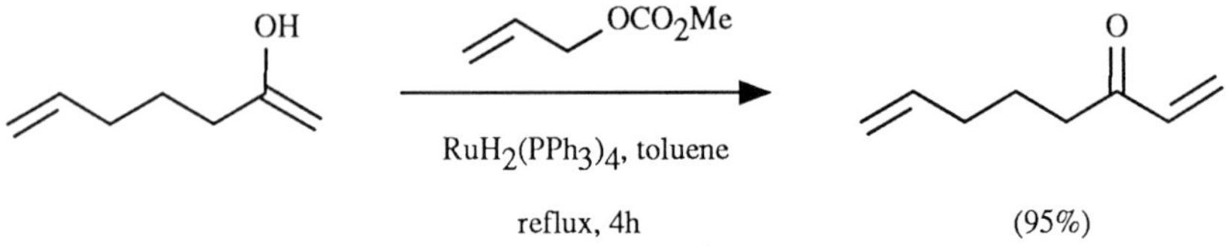

OH → OCO_2Me, $RuH_2(PPh_3)_4$, toluene, reflux, 4h → O (95%)

Mild oxidation of allylic alcohols has been described using Jones reagent (chromic acid in aqueous acetone) at 0°C (K. Harding *et al.*, *J. Org. Chem.*, 1975, **40**, 1664).

(b) **Epoxidation**

The Sharpless epoxidation of allylic alcohols mentioned in *chapter 4b* has been reviewed (A. Pfenninger, *Synthesis*, 1986, 89). The reaction time of the Sharpless epoxidation is greatly reduced when catalytic amounts of metal hydride and silica gel are added (W. Zhi-min *et al.*, *Tetrahedron Lett.*, 1985 **26**, 6221). Addition of molecular sieve also accelerates the epoxidation and permits catalytic amounts of tartrate ester and titanium reagent to be used (J.M. Klunder *et al.*, *J. Org. Chem.*, 1986, **51**, 3710; S. Y. Ko and K.B. Sharpless, *ibid*, 1986, **51**, 5413).

(c) **Substitution**

Allylic alcohols are converted to allylic chlorides using trimethylsilyl chloride and calcium carbonate (M. Lissel and K. Drechsler, *Synthesis*, 1983, 314).

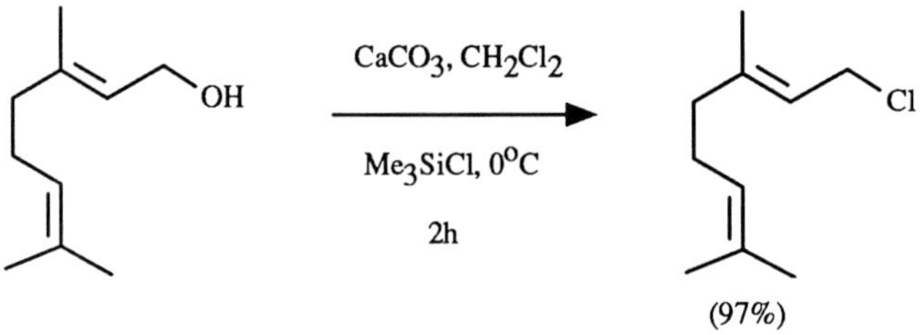

OH → $CaCO_3$, CH_2Cl_2, Me_3SiCl, 0°C, 2h → Cl (97%)

(d) **Rearrangements**

Allylic alcohols rearrange when treated with *bis*(trimethylsilyl)peroxide and a vanadium catalyst (S. Matsubara *et al.*, *Tetrahedron Lett.*, 1983, **24**, 3741).

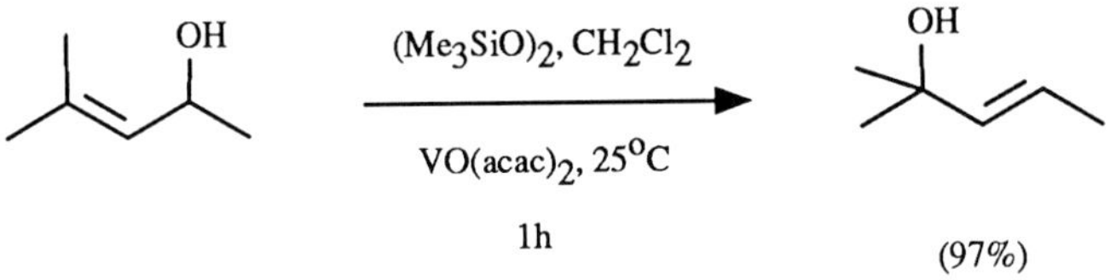

$(Me_3SiO)_2$, CH_2Cl_2

$VO(acac)_2$, 25°C

1h

(97%)

Allylic alcohols rearrange when treated with amylselenonitriles and tributylphosphine followed by oxidation with hydrogen peroxide (D.J. Clive *et al.*, *J. Chem. Soc. Chem. Commun.*, 1978, 770).

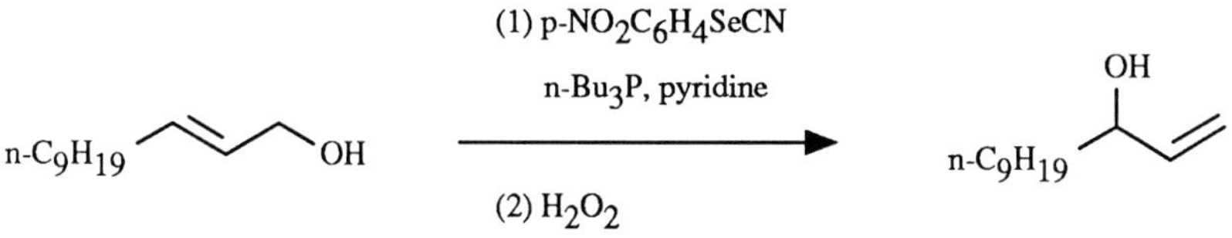

n-C_9H_{19} OH → (1) p-$NO_2C_6H_4SeCN$, n-Bu_3P, pyridine; (2) H_2O_2 → n-C_9H_{19} OH

(85%)

α-allenic alcohols rearrange to α,β-unsaturated ketones when treated with aqueous acid (R. Gelin *et al.*, *Bull. Soc. Chim. France*, 1972, 720).

$$MeCH(OH)CH{=}C{=}CH_2 \xrightarrow{aq.\ H_3PO_4} MeCH{=}CHCOMe$$

(70%)

6. *Preparation and reactions of alkynols*

Propargyl alcohols are produced when α,β-alkynic ketones are reduced with the adduct of 9-borabicyclo[3.3.1]nonane with nopol benzyl ether (NB-Enantrane), (M.M. Midland and A. Kazubski, *J. Org. Chem.*, 1982, **47**, 2814).

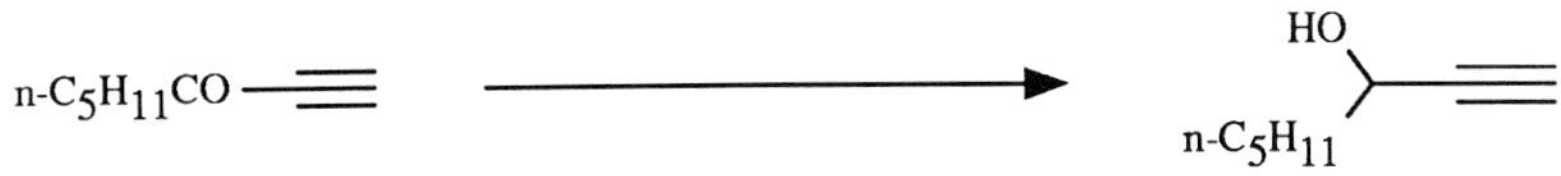

n-$C_5H_{11}CO$–C≡CH → HO, n-C_5H_{11}

(74%) 95%e.e.

A chiral alkynic alcohol is obtained when allenylboronic acid is treated with a dialkyltartrate and an aldehyde (R. Haruta *et al.*, *J. Amer. Chem. Soc.*, 1982, **104**, 7667).

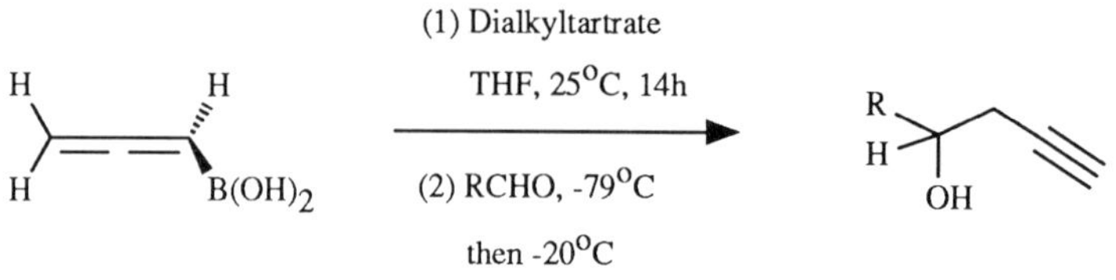

(1) Dialkyltartrate
THF, 25°C, 14h
(2) RCHO, -79°C
then -20°C

Conjugated enynes are reduced to δ-ynols using 9-BBN, followed by oxidation with alkaline hydrogen peroxide. Double bonds are reduced more rapidly than triple bonds with 9-BBN at 25°C in tetrahydrofuran or diethyl ether. This makes the formation of B-alkynic-9-BBN derivatives possible and these are oxidised to alkynols (C.A. Brown and R.A. Coleman, *J. Org. Chem.*, 1979, **44**, 2328).

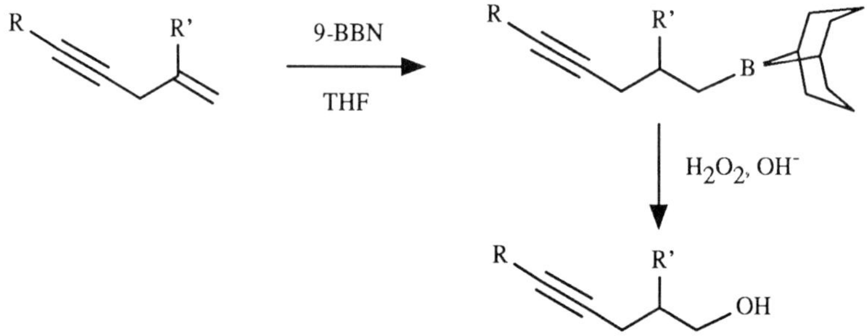

9-BBN
THF
H_2O_2, OH^-

α,β-alkynic ketones are reduced to α,β-alkynols using a chiral lithium tetrahydroaluminate-N-methylephedrine-3,5-dimethylphenol complex (J.P. Vigneron and V. Bloy, *Tetrahedron Lett.*, 1979, **19**, 2683). B-isopinocampheyl-9-borabicyclo[3.3.1]nonane (Alpine Borane) as mentioned elsewhere in this report, reduces α,β-alkynic ketones to α,β-alkynols of high optical purity (M.M. Midland *et al.*, *J. Amer. Chem. Soc.*, 1980, **102**, 867).

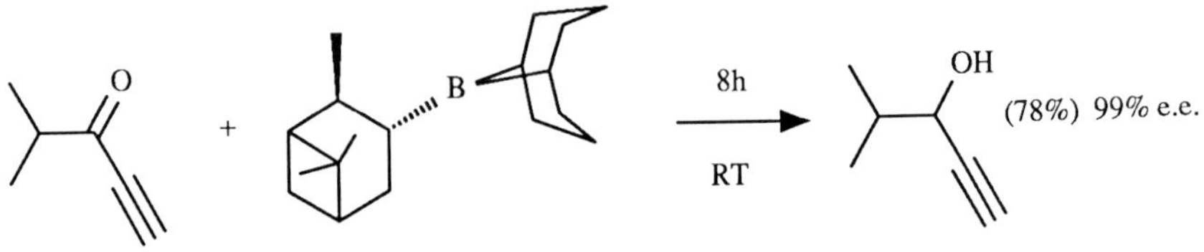

8h
RT
(78%) 99% e.e.

Alkynols are isomerised by treatment with potassium 3-amino-propylamide in 1,3-diaminopropane, followed by hydrolysis. The triple bond in the product is terminal and at the end remote from the hydroxyl group (C.A. Brown and A. Yamashita, *J. Chem. Soc. Chem. Commun.*, 1976, 959; J.C. Lindhoudt *et al.*, *Tetrahedron Lett.*, 1976, **16**, 2565).

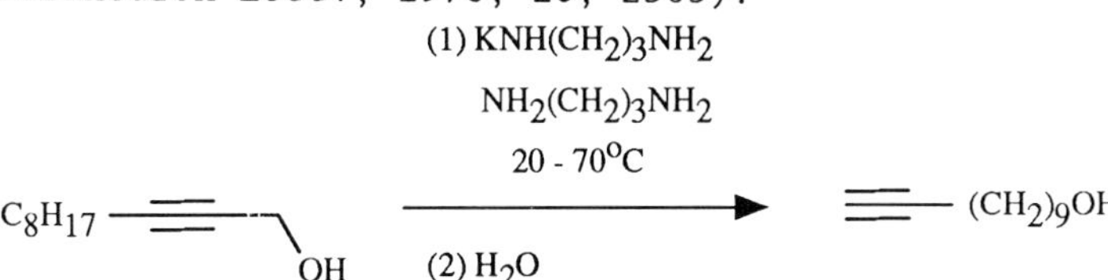

Second Supplements to the 2nd Edition of Rodd's Chemistry of Carbon Compounds, Vol. IA/B, edited by M. Sainsbury
© 1991 Elsevier Science Publishers B.V., Amsterdam

Chapter 4b

MONOHYDRIC ALCOHOLS AND RELATED COMPOUNDS: ACYCLIC ETHERS, HYDROPEROXIDES, PEROXIDES AND ESTERS OF INORGANIC ACIDS

C.W.THOMAS

1. *Preparation of dialkyl ethers*

(a) **Williamson-type substitutions**

Several variants of the Williamson synthesis have been reported.

(i) Ethers with one tertiary alkyl group can be prepared by treating a haloalkane or dialkyl sulphate with an alcohol and methylsulphinyl carbanion (B. Sjöberg and K. Sjöberg, *Acta. Chem. Scand.*, 1972, **26**, 275).

$$\text{t-BuOH} \xrightarrow[(CH_3O)_2SO_2]{^-CH_2SOCH_3} \text{t-BuOMe} \quad (78\%)$$

(ii) Alcohols react with haloalkanes or dialkylsulphates in dimethylsulphoxide in the presence of potassium hydroxide (D.R. Benedict, *et al.*, *Synthesis*, 1979, 428; R.A.W. Johnstone and M.E. Rose, *Tetrahedron*, 1979, **35**, 2169).

$$\text{i-PrOH} \xrightarrow[\text{KOH, DMSO}]{\text{PrI}} \text{i-PrOPr} \quad (31\%)$$

$$CH_2{=}CHCH_2OH \xrightarrow[\text{KOH, DMSO; }50^oC,\ 5h]{(CH_3O)_2SO_2} CH_2{=}CHCH_2OCH_3 \quad (70\%)$$

(iii) Methyl ethers of alcohols are formed by reacting the alcohol with methyl iodide and potassium fluoride - alumina under mild conditions (T. Ando *et al.*, *Bull. Chem. Soc. Japan*, 1982, **55**, 2504).

$$CH_3(CH_2)_7OH \xrightarrow[\text{KF-alumina, RT, 40h}]{CH_3I,\ CH_3CN} CH_3(CH_2)_7OCH_3 \quad (90\%)$$

(iv) Ethers may be synthesised by reacting alcohols with haloalkanes using potassium hydroxide, polyethyleneglycol, a phase transfer catalyst such as tetrabutylammonium chloride with ultrasonication (R.S. Davidson *et al.*, *Ultrasonics*, 1987, **25**, 35).

(v) The *O*-methylation of alcohols with dimethylsulphate in dioxan or triglyme with solid potassium hydroxide and small amounts of water represents a useful method for synthesising methyl ethers. A stoichiometric amount of dimethylsulphate is used to avoid product isolation problems (D. Achet *et al.*, *Synthesis*, 1986, 642).

$$CH_3(CH_2)_3OH \xrightarrow[65^oC,\ 1.5h]{\text{dioxan}} CH_3(CH_2)_3OCH_3 \quad (94\%)$$

(vi) Alcohols in cyclohexane give their methyl ethers in high yields using dimethylsulphate and neutral alumina. The alumina is easily removed after the reaction by filtration (H. Ogawa *et al.*, *Bull. Chem. Soc. Japan*, 1986, **59**, 2481).

(vii) Di-t-alkyl ethers are difficult to prepare but can be obtained in satisfactory yields by treating a tertiary halo-alkane with silver(I) carbonate or oxide (H. Masada and T. Sakajiri, *Bull. Chem. Soc. Japan*, 1978, **51**, 866).

$$\text{t-BuCl} \xrightarrow[\text{pentane, } 20^oC,\ 20h]{Ag_2CO_3} (\text{t-Bu})_2O \quad (73\%)$$

(viii) Haloalkanes react with alcohols in dichloromethane using mercury(II) oxide and tetrafluoroboric acid (J. Barluenga *et al.*, *Synthesis*, 1983, 53).

$$R^1OH + R^2X \xrightarrow[HBF_4,\ RT,\ 1h;\ (2)\ KOH]{(1)\ CH_2Cl_2,\ HgO} R^1OR^2 \quad (55\%)$$

$R^1 = CH_3CH_2(CH_3)CH$, $X = Br$

$R^2 = CH_3(CH_2)_7$

(ix) Copper (I) alkoxides, formed by the reaction between methylcopper (I) and alcohols, react with haloalkanes to give ethers (G.M. Whitesides *et al.*, *J. Amer. Chem. Soc.*, 1974, **96**, 2829).

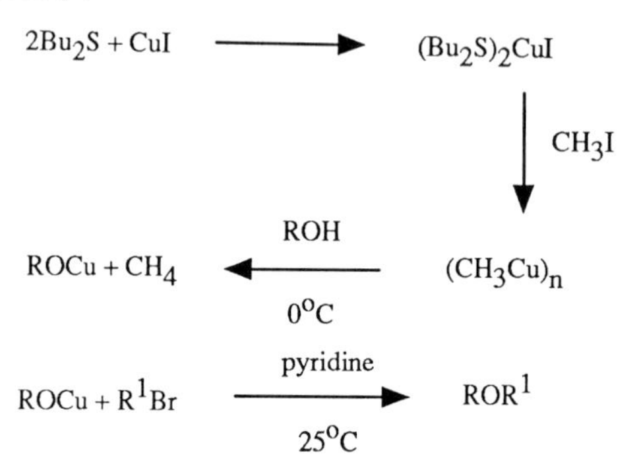

(x) Thallium(I) alkoxides may also be used for ether synthesis (H. Kalinowski *et al.*, *Angew. Chem. Int. Ed. Eng.*, 1975, **14**, 762).

$$R^1OH \xrightarrow[C_6H_6]{TlOEt} R^1OTl \xrightarrow[CH_3CN]{RX} R^1OR$$

Reaction occurs readily at low temperature between primary haloalkanes and primary or secondary alcohols. The method is suitable for alkylating alcohols without racemization.

(xi) Mercury(II) salts catalyse the reaction of methanol (or ethanol) with bromoalkanes (A. McKillop and M.E. Ford, *Tetrahedron*, 1974, **30**, 2467).

$$RBr + R^1OH \xrightarrow{Hg(ClO_4)_2} ROR^1$$

(b) **Miscellaneous ether syntheses**

Other ether syntheses include the following

(i) Alcohols react with diazoalkanes in the presence of rhodium(II) acetate to give the corresponding ether (R. Paulissen *et al.*, *Tetrahedron Lett.*, 1973, **14**, 2233).

$$EtOH + N_2CHCOOEt \xrightarrow[25^oC]{Rh_2(OAc)_4} EtOCH_2COOEt\ (88\%)$$

or by using silica gel (K. Ohno *et al.*, *Tetrahedron Lett.*, 1979, **20**, 4405).

$$\text{t-AmOH} \xrightarrow[\text{silica gel}]{CH_2N_2} \text{t-AmOMe} \quad (54\%)$$

(ii) Di-t-butylether has been prepared by direct attack of t-butanol on the t-butyl cation (at -80°C in SO_2ClF), (G.A. Olah *et al.*, *Synthesis*, 1975, 315).

$$\text{t-BuCl} \xrightarrow[SO_2ClF,\ -70^oC]{SbF_5} \text{t-Bu}^+ \xrightarrow[-80^oC]{\text{t-BuOH}} \text{t-Bu}_2O$$

(iii) Oxidation of alkylphenylselenides with m-chloroperbenzoic acid in methanol at room temperature affords the corresponding alkyl methyl ethers in high yield (S. Uemura *et al.*, *J. Chem. Soc. Chem. Commun.*, 1983, 1501).

$$CH_3(CH_2)_{10}CH_2SePh \xrightarrow[CH_3OH,\ 1h]{\text{m-CPBA}} CH_3(CH_2)_{10}CH_2OMe \quad (ca.100\%)$$

(iv) Aldehydes and ketones are converted into ethers by treating them with trityl perchlorate and triethylsilane (T. Kato *et al.*, *Chem. Lett.*, 1985, 743).

$$RCHO \xrightarrow[CH_2Cl_2,\ 0^oC,\ 10min]{TrClO_4,\ Et_3SiH} RCH_2OCH_2R \quad (88\%)$$

Conditions are mild, the reaction time is short and only a catalytic quantity of trityl perchlorate is required.

(v) Aldehydes and ketones are converted to ethers by treatment with alcohol and triethyl silane in the presence of strong acid (M.P. Doyle *et al.*, *J. Amer. Chem. Soc.*, 1972, **94**, 3659.

$$RCOR^1 + R^2OH \xrightarrow[CF_3COOH]{Et_3SiH} RCH(OR^2)R^1$$

(vi) Esters are reduced to ethers using trichlorosilane and di-t-butylperoxide as reagents (S.W. Baldwin and S.A. Haut, *J. Org. Chem.*, 1975, **40**, 3885; Y. Nagata *et al.*, *J. Org. Chem.*, 1973, **38**, 795).

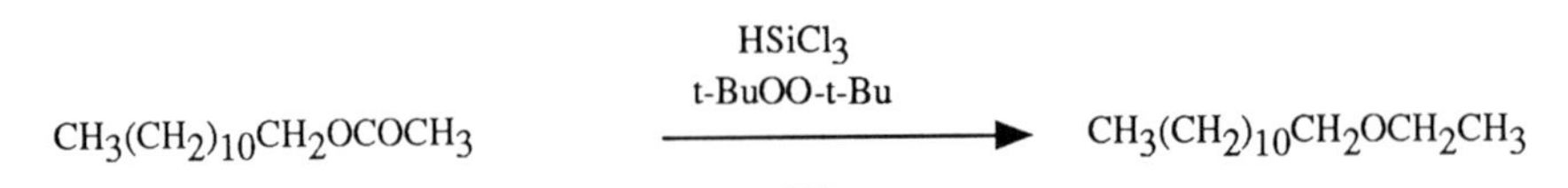

(vii) Thionoesters, prepared from carboxylic esters and 2,4-*bis*(4-methoxyphenyl)-1,3,2,4-dithiadiphosphetane-2,4-disulphide, are readily converted into ethers using Raney nickel (S.L. Baxter and J.S. Bradshaw, *J. Org. Chem.*, 1981, **46**, 831).

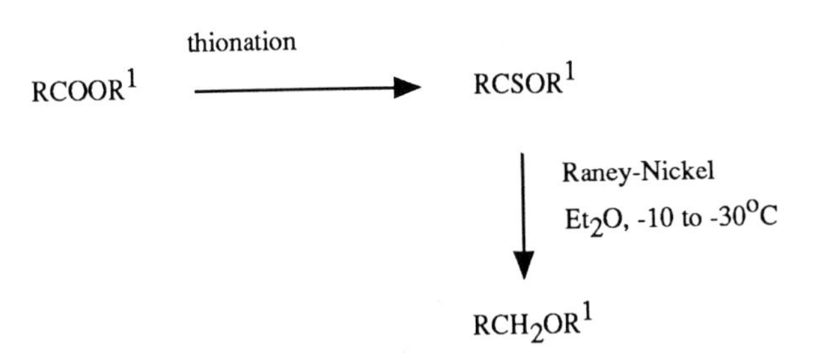

2. *Reactions of dialkylethers*

(a) **Rearrangement**

The Wittig rearrangement has been reviewed (G. Wittig, *Bull. Soc. Chim. France*, 1971, 1921). Silyl and germanyl ethers undergo the Wittig rearrangement and in the case of the silyl ether the reaction is reversible. The reaction proceeds with inversion of configuration in both cases (A. Wright and R.W. West, *J. Amer. Chem. Soc.*, 1974, **96**, 3222).

(b) **Ether fission**

Various methods for ether fission have been reported.

(i) $RCH_2OR^1 + R^2COCl + NaI \longrightarrow RCH_2I + R^2COOR^1$

(A Oku *et al.*, *Tetrahedron Lett.*, 1982, **23**, 681).

(ii) $ROR^1 \longrightarrow ROH + R^1Br$

using Me_2B-Br, Et_3N and $ClCH_2CH_2Cl$ (Y. Guirdon *et al.*, *Tetrahedron Lett.*, 1983, **24**, 2969).

(iii) $ROR^1 \longrightarrow ROAc$

using Me_3SiCl and Ac_2O where R′ = Me and methoxyethoxymethyl (N.C. Barua *et al.*, *Tetrahedron Lett.*, 1983, **24**, 1189).

(iv) $$ROR^1 + Me_2BBr \xrightarrow[-78^oC]{} ROH$$

where R′ = methoxymethyl, methoxyethoxymethyl and methoxythiomethyl (Y. Quindon *et al.*, *Tetrahedron Lett.*, 1983, **24**, 3969).

(v) $$ROR \xrightarrow[Ac_2O]{Tl(NO_3)_2} ROCOCH_3$$

(E. Mincione and F. Lanciano, *Tetrahedron Lett.*, 1980, **21**, 1149).

(vi) $$ROR^1 \xrightarrow{PhNH_3^+ \ ^-OTs} ROH$$

where R′ = methoxyethoxy methyl or methoxymethyl (H. Monti *et al.*, *Synth. Commun.*, 1983, **13**, 1021).

(vii) $ROR^1 \longrightarrow ROH$

using trityl tetrafluoroborate, R′ = methoxythiomethyl (P.K. Chowdhury *et al.*, *Tetrahedron Lett.*, 1983, **24**, 4485).

(viii) The reductive cleavage of acetals and ketals is readily accomplished using borane-THF (B. Fleming and H.I. Bolker, *Can. J. Chem.*, 1974, 888).

(c) **Oxidation**

(i) Ethers are converted to aldehydes and ketones by reaction with sodium bromate and ceric ammonium nitrate catalyst (G. A. Olah *et al.*, *Synthesis*, 1980, 897).

$$RR^1CHOR^2 \longrightarrow RR^1CO$$

(ii) Ethers are converted to aldehydes using 1-chlorobenzotriazole, followed by hydrolysis (P.M. Pojer, *Aust. J. Chem.*, 1979, **32**, 2787).

$$RCH_2OR^1 \longrightarrow RCHO$$

3. *Preparation and reactions of alkenic ethers*

Enol ethers (vinyl ethers) are produced by the base-promoted addition of alcohols to allenes or alkynes (V.E. Wiersum and T. Niewenhuis, *Tetrahedron Lett.*, 1973 **14**, 2581). The Horner-Wittig reaction using α-haloethers provides a useful route to vinyl ethers (C. Earnshaw *et al.*, *J. Chem. Soc. Chem. Commun.*, 1977, 314).

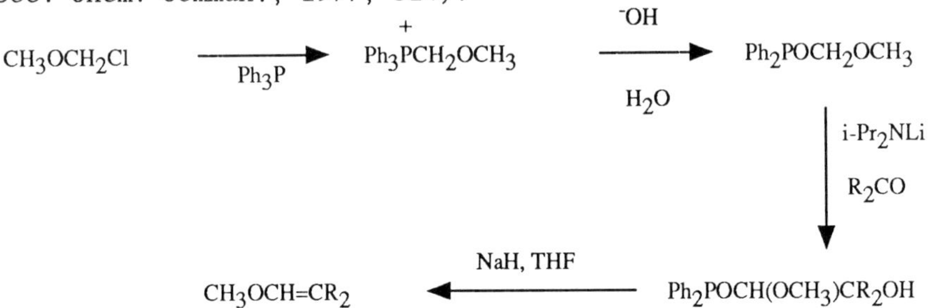

Allyl ethers are cleaved by rhodium complexes (eg. RhCl $(PPh_3)_3$) to give alcohols under neutral aprotic conditions (E.J. Corey and J.W. Suggs, *J. Org. Chem.*, 1973, **38**, 3224). Vinyl ethers undergo fission when treated with a Grignard reagent and a nickel complex catalyst (E. Wenkert *et al.*, *J. Amer. Chem. Soc.*, 1979, **101**, 2246).

$$CH_3O{-}CH{=}CH{-}(CH_2)_4CH_3 \xrightarrow[Ni(PPh_3)_2Cl_2]{CH_3MgBr} CH_3{-}CH{=}CH{-}(CH_2)_4CH_3$$

At low temperature, alkyl allyl or vinyl ethers may be rapidly metallated with s-butyllithium to give the equivalent homo-enolate anions. These anions react with electrophiles to give α- or γ-substituted products, the regioselectivity depending on the nature of the alkyl group, counter ion and the electrophile. Alkylation of t-butylallyl ether with 1-iodohexane gives mainly the γ - product (D.A. Evans *et al.*, *J. Amer. Chem. Soc.*, 1974, **96**, 556).

$$t\text{-}BuO{-}CH_2CH{=}CH_2 \xrightarrow{s\text{-}BuLI} [t\text{-}BuO{-}CH{=}CH{-}CH_2]^-\,Li^+ \xrightarrow{C_6H_{13}I} t\text{-}BuO{-}CH{=}CH{-}CH_2{-}C_6H_{13}$$

The Claisen rearrangement has been reviewed (S.J. Rhoads and N.B. Raulins, *Org. Reactions*, 1975 **22**, 1). In the Claisen rearrangement of allylethynyl ethers the intermediate ketenes undergo further reaction (J.A. Katzenellenborg and T. Utawanit, *Tetrahedron Lett.*, 1975, **15**, 3275).

$$C_3H_7CH{=}CHCH_2OC{\equiv}CH \longrightarrow O{=}C{=}CH\text{-}CH(C_3H_7)CH{=}CH_2 \searrow NH_2COCH_2CH(C_3H_7)CH{=}CH_2$$

The rearrangement of transient vinyl ethers produced by transvinylation or transacetylation, followed by elimination, extends the scope of the reaction (D.J. Faulkner and M.R. Peterson, *J. Amer. Chem. Soc.*, 1973, **95**, 553).

4. *Preparation of dialkyl peroxides and alkyl hydroperoxides*

A three volume review "Organic Peroxides" (Ed. D. Swern), Wiley-Interscience, New York, Vol I (1971), Vol II (1971) and Vol III (1972), is available. Further reviews to be found are those by R.A. Sheldon in "The Chemistry of Peroxides" (Chemistry of Functional Groups), (ed. S. Patai), Wiley, New York, 1983, chap. 6, p. 161, and D. Swern in "Comprehensive Organic Chemistry" (D.H.R. Barton and W.D. Ollis, eds, Pergamon, Oxford, 1979, chap. 4).

(a) The reaction of primary and secondary trifluoromethane sulphonates with the potassium salt of t-butylhydroperoxide and sodium carbonate provides a facile method for preparing unsymmetrical dialkyl peroxides (M.F. Salomon *et al.*, *J. Org. Chem.*, 1976, **41**, 3983).

(b) Dialkyl peroxides are synthesised by the peroxide transfer reaction between alkyl trifluoromethane sulphonates and germanium or tin peroxide (M.F. Salomon and R.G. Salomon, *J. Amer. Chem. Soc.*, 1979, **101**, 4290).

$$\text{(CH}_3)_2\text{CH-OTf} + Bu_3SnOOSnBu_3 \xrightarrow[20^{o}C]{0.1h} (CH_3)_2CH\text{-O-O-}CH(CH_3)_2$$

(c) Dialkyl peroxides may be prepared by treating primary or secondary bromoalkanes or alkyl tosylates with potassium superoxide (KO_2) in the presence of a crown ether (R.A. Johnson *et al.*, *J. Amer. Chem. Soc.*, 1978, **100**, 7960).

$$CH_3(CH_2)_5Br \xrightarrow[\text{dicyclohexyl 18-crown-6}]{\text{benzene, } KO_2} CH_3(CH_2)_5\text{-O-O-}(CH_2)_5CH_3 \quad (71\%)$$

(d) Primary, secondary and tertiary alkyl hydroperoxides and dialkylperoxides can be obtained by treating a bromo- or iodoalkane with hydrogen peroxide or a hydroperoxide in the presence of silver trifluoroacetate (P.G. Cookson *et al.*, *J. Chem. Soc. Chem. Commun.*, 1976, 1022).

$$R^2OOH + R^1X + AgOCOCF_3 \longrightarrow R^2OOR^1 + AgX + CF_3COOH$$

R^2 = H or alkyl X = Br or I

$$Pr_3CBr + H_2O_2 \xrightarrow[\text{pentane, } 0^oC]{AgOCOCF_3} \underset{(60\%)}{Pr_3COOH}$$

(e) The absorption of oxygen by trialkylboranes can be controlled to yield the diperoxy derivatives which on oxidation give alkyl hydroperoxides.

$$R_3B + 2O_2 \xrightarrow[-78^oC]{THF} (RO_2)_2BR \xrightarrow[0^oC]{H_2O_2} 2RO_2H + ROH$$

Alkyl dichloroboranes also react with oxygen to yield alkyl hydroperoxides (H.C. Brown and M.M. Midland, *J. Amer. Chem. Soc.*, 1971, **93**, 4078; M.M. Midland and H.C. Brown, *ibid.*, 1973, **95**, 4069).

$$RBCl_2 + O_2 \xrightarrow{-18^oC} RO_2BCl_2 \xrightarrow{H_2O} RO_2H$$

(f) In the presence of mercury(II) acetate, t-butylhydroperoxide reacts with monosubstituted alkenes in dichloromethane to yield 2-t-butylperoxyalkyl mercury(II) acetates, which when treated with sodium tetrahydroborate give secondary alkyl t-butylperoxides in high yields (D.H. Ballard and A.J. Bloodworth, *J. Chem. Soc. C*, 1971, 945).

$$RCH{=}CH_2 + Hg(OAc)_2 \xrightarrow{t\text{-}BuOOH} t\text{-}BuOOCHRCH_2HgOAc \xrightarrow{NaBH_4} t\text{-}BuOOCHRCH_3$$

(g) Brominolysis of the products of peroxymercuration of alkenes with t-butylhydroperoxide, mercury(II) acetate and aqueous perchloric acid gives β-bromoalkylperoxides (A.J. Bloodworth and J.L. Courtneidge, *J. Chem. Soc. Perkin Trans. I*, 1981, 3258).

$$CH_3CH{=}CHCH_3 \longrightarrow CH_3CHBrCH(OOBu\text{-}t)CH_3 \quad (93\%)$$

(h) Peroxymercuration/demercuration of phenylcyclopropane with mercury(II) trifluoroacetate and t-butylhydroperoxide followed by sodium tetrahydroborate gives the peroxide, PhCH(OOBu-t)Et (A.J. Bloodworth and J.L. Courtneidge, *J. Chem. Soc. Perkin Trans. 1*, 1982, 1807).

(i) Di-t-butyltrioxide is obtained in 20% yield when solid di-t-butyldiperoxymono carbonate is photolysed at -78°C for 20 hours (P.D. Bartlett and M. Lahav, *Israel J. Chem.*, 1972, **10**, 101).

$$\text{t-BuOO(CO)OOBu-t} \longrightarrow CO_2 + \text{t-BuOOOBu-t}$$

The product is unstable above -30°C.

5. *Reactions of dialkyl peroxides and alkyl hydroperoxides*

(a) **Various oxidations**

(i) t-Butylhydroperoxide has found increasing use, especially in combination with a metal catalyst, as a selective oxidant in organic reactions. This subject has been reviewed (K.B. Sharpless and T.R. Verhoeven, *Aldrichim. Acta.*, 1979, **12**, 63; R.A. Sheldon, *loc. cit*).

(ii) Hydrocarbons have been oxidised to ketones e.g., indane to 1-indanone (J. Muzart, *Tetrahedron Lett.*, 1987, **28**, 2131).

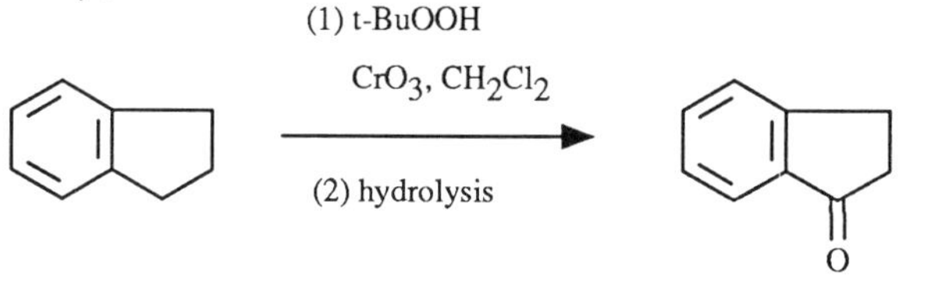

(1) t-BuOOH

CrO_3, CH_2Cl_2

(2) hydrolysis

(60%)

(iii) Alcohols are oxidised to ketones, e.g., 9-hydroxyfluorene to fluorenone (J. Muzart, *Tetrahedron Lett.*, 1987, **28**, 2133).

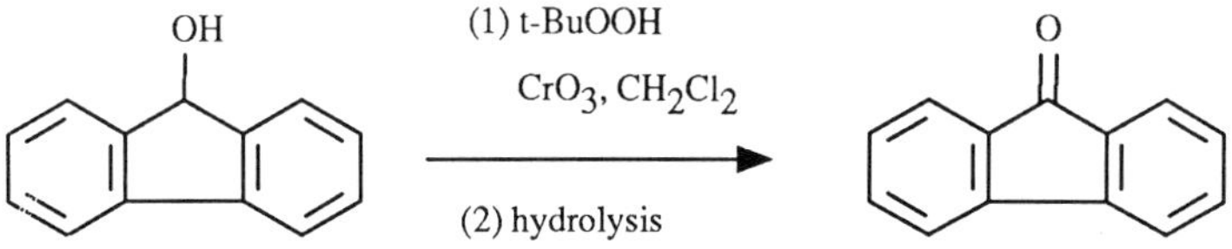

OH

(1) t-BuOOH

CrO_3, CH_2Cl_2

(2) hydrolysis

O

(iv) α,β-unsaturated ketones and esters have been selectivity oxidised to the corresponding 1,3-diketones and β-ketoesters (J. Tsuji *et al.*, *Chem. Lett.*, 1980, 257).

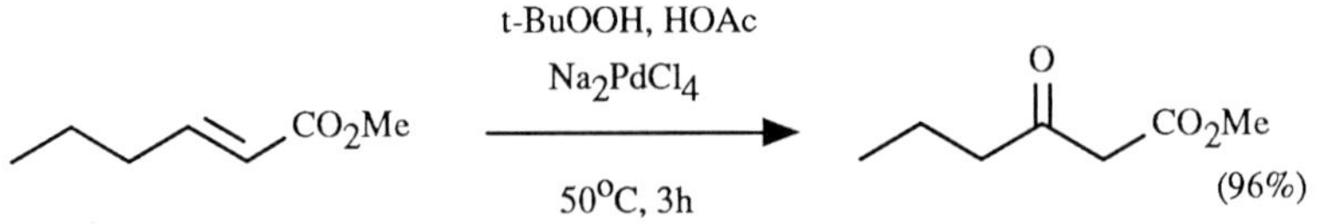

t-BuOOH, HOAc, Na_2PdCl_4, 50°C, 3h: CO_2Me → CO_2Me (96%)

(v) Terminal alkenes are selectively oxidised to methyl ketones (H. Mimoun, *J. Mol. Catal.*, 1980, **7**, 1 ; H. Mimoun *et al.*, *J. Amer. Chem. Soc.*, 1980, **102**, 1047).

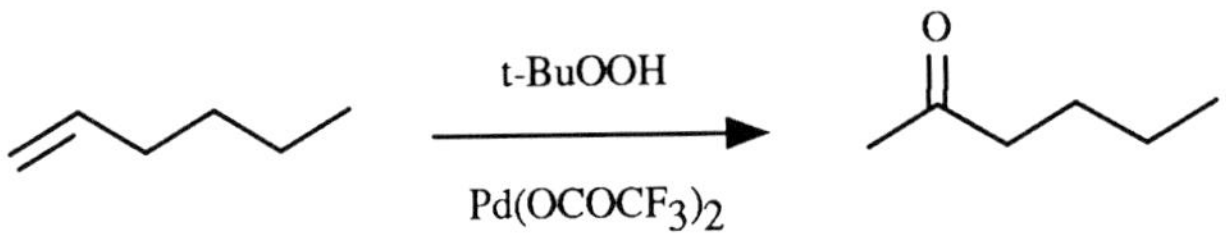

t-BuOOH, $Pd(OCOCF_3)_2$

(vi) Allylic oxidation of suitable alkenes can be accomplished (M.A. Umbreit and K.B. Sharpless, *J. Amer. Chem. Soc.*, 1977, **99**, 5526).

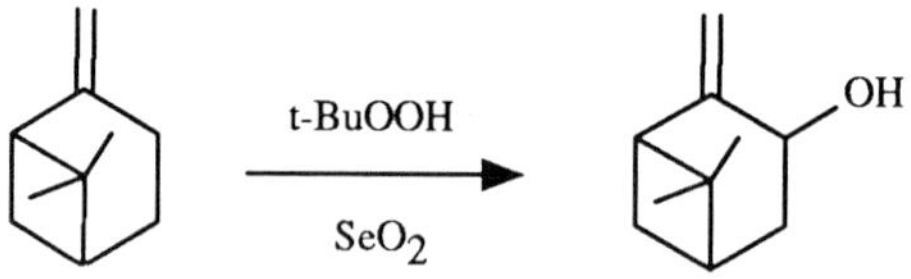

t-BuOOH, SeO_2 → OH

(vii) Alkynes have been oxidised to alkynols (B. Chabaud and K.B. Sharpless, *J. Org. Chem.*, 1979, **44**, 4202).

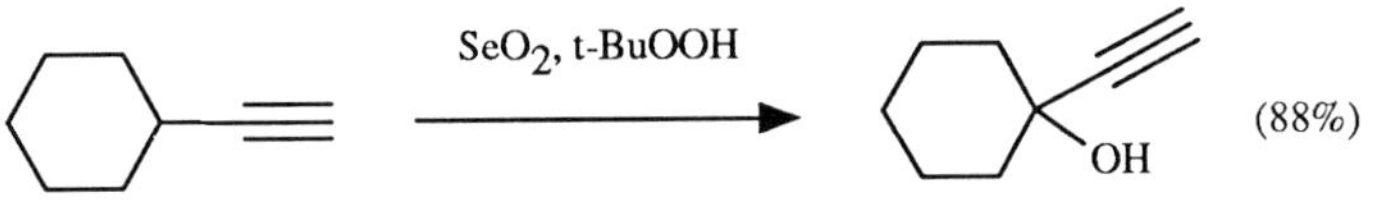

SeO_2, t-BuOOH → OH (88%)

(viii) Alkynes can also be oxidised to α-diketones (M. Schroder, *Chem. Rev.*, 1980, **80**, 187).

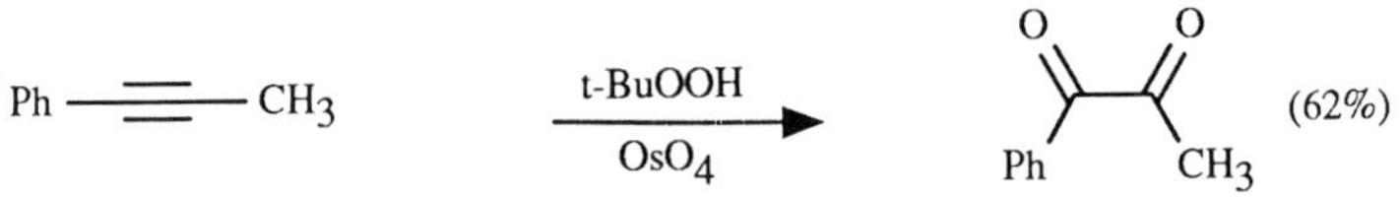

Ph—≡—CH_3 → t-BuOOH, OsO_4 → Ph–CO–CO–CH_3 (62%)

(ix) Cyclic ketones are oxidised to lactones (G.S.P. Field, *British Patent.*, 1975, 1, 413, 417); *Chem. Abstr.*, 1976, **84**, 73672 q).

(x) The oxidation of cyclohexanone in the presence of formic acid to produce 2-hydroxycyclohexanone has been reported (R. Rosenthal and G.A. Bonetti, *U.S. Patent*, 1973, 3, 755, 453; *Chem. Abstr.*, 1973, **79**, 125957e).

(xi) Trisubstituted alkenes are oxidised to 2-hydroxyketones with excess alkylhydroperoxide and a molybdenum catalyst (G.A. Tolstikov *et al.*, *J. Org. Chem.*, *USSR*, 1971, **8**, 1204; *J. Gen. Chem.*, *USSR*, 1973, **43**, 2058; U.M. Dzhemilev *et al.*, *Proc. Acad. Sci.*, *USSR*, *Chem. Sect.*, 1971, **196**, 79).

OH
O

(xii) Vicinal dihydroxylation of alkenes has been accomplished using t-BuOOH/OsO_4 in the presence of acetone (K.B. Sharpless and K. Akashi, *J. Amer. Chem. Soc.*, 1978, **98**, 1986; K. Akashi *et al.*, *J. Org. Chem.*, 1978, **43**, 2063).

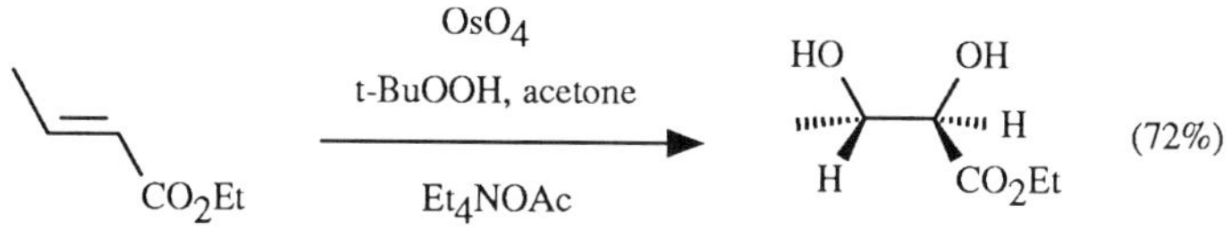

OsO_4
t-BuOOH, acetone
Et_4NOAc
CO_2Et
HO OH
H
H CO_2Et
(72%)

This has the advantage of avoiding the overoxidation tendency of the OsO_4/H_2O_2 method.

(xiii) Alcohols with remote double bonds are formed when alkyl hydroperoxides are treated with iron(II) sulphate and copper(II) acetate in acetic acid (Z. Cekovic and M.M. Green, *J. Amer. Chem. Soc.*, 1974, **96**, 3000; Z. Cekovic *et al.*, *Tetrahedron*, 1979, **35**, 2021).

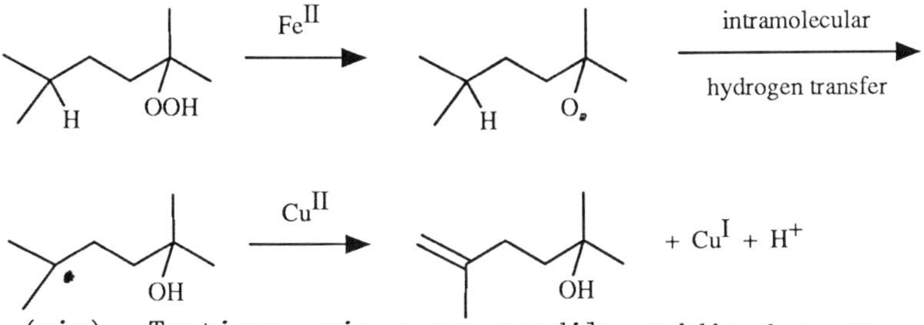

Fe^{II}
H OOH
H O•
intramolecular
hydrogen transfer

Cu^{II}
OH
OH
$+ Cu^{I} + H^{+}$

(xiv) Tertiary amines are readily oxidised to N-oxides (G.A. Tolstikov *et al.*, *J. Gen. Chem. USSR*, 1973, **43**, 1350).

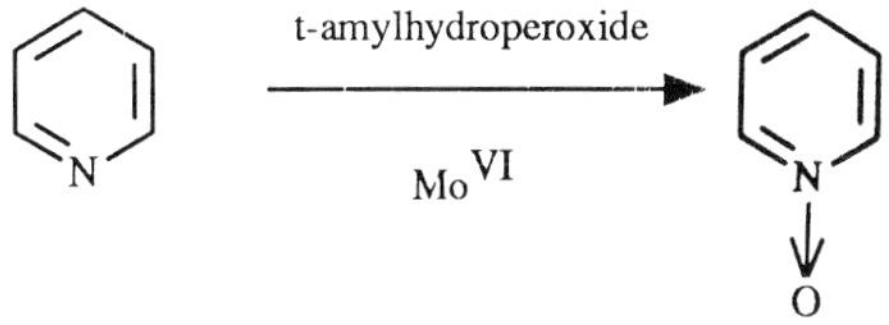

N
t-amylhydroperoxide
Mo^{VI}
N
O

(xv) Imines are oxidised to oxaziridines (G.A. Tolstikov *et al.*, *Tetrahedron Lett.*, 1971, **12**, 2807; G.A. Tolstikov *et al.*, *J. Org. Chem. USSR*, 1971, **8**, 1200).

$$R_2C{=}N{-}R \xrightarrow[Mo^{VI}]{\text{t-amylhydroperoxide}} R_2C(O)N{-}R \text{ (oxaziridine)}$$

(xvi) Sulphides are oxidised to the corresponding sulphoxides using t-butylhydroperoxide in the presence of a molybdenum catalyst. Excess hydroperoxide gives the sulphone, and tungsten, titanium and vanadium catalysts may also be used (F. DiFuria *et al.*, *Tetrahedron Lett.*, 1976, **17**, 4637; G.A. Tolstikov *et al.*, *J. Gen. Chem., USSR,* 1971, **41**, 1896; *idem.*, *Bull. Acad. Sci., USSR, Div. Chim. Sci.*, 1972, **21**, 2675; R. Curci *et al.*, *J. Chem. Soc., Perkin Trans. 2*, 1974, 752; R. Curci *et al.*, *ibid.*, 1977, 576, R. Curci *et al.*, *ibid.*, 1978, 979).

$$R_2S \longrightarrow R_2SO \longrightarrow R_2SO_2$$

(xvii) Nitroalkanes are converted to carbonyl compounds (P.A. Bartlett *et al.*, *Tetrahedron Lett.*, 1977, **18**, 331).

(xviii) t-Butylhydroperoxide in anhydrous conditions has been used to oxidise nucleoside phosphites to phosphates (Y. Hayakawa *et al.*, *Tetrahedron Lett.*, 1986, **27**, 4191).

(xix) t-Butylhydroperoxide, iron(II) sulphate, propanal and sulphuric acid have been used to acylate quinoxaline (G.P. Gardini and F. Minisci, *J. Chem. Soc. C*, 1970, 929; 1971, 1747).

$$\text{quinoxaline} \xrightarrow[\substack{C_2H_5CHO,\ 5-15^{o}C \\ 4M\ H_2SO_4,\ 10\ min}]{\text{t-BuOOH, }FeSO_4} \text{2-}COC_2H_5\text{-quinoxaline (73\%)}$$

(xx) Vicinal glycols undergo oxidative cleavage with alkylhydroperoxide and ruthenium(III) chloride (Mitsui Petrochemical Company, *Japanese Patent*, 1980, 80, 102, 528; *Chem. Abstr.*, 1981, **94**, 46774z).

HO OH → 2 (acetone)

(xxi) Benzyltrimethylammonium tetrabromooxomolybdate ($PhCH_2NMe_3O$-$MoBr_4$) catalyses various reactions of t-butyl-hydroperoxide including the following (Y. Masuyama *et al.*, *Tetrahedron Lett.*, 1984, **25**, 4417).

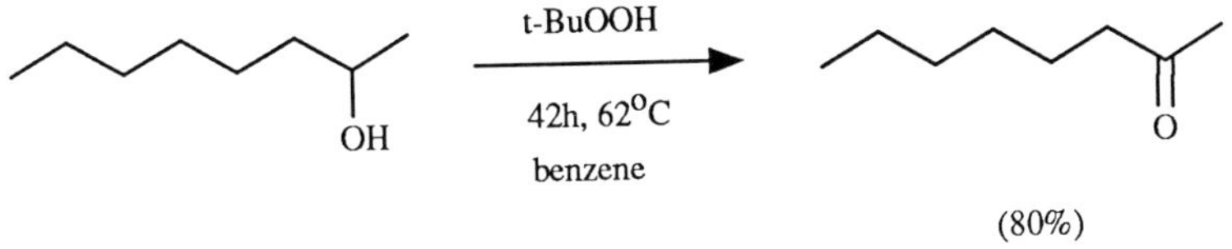

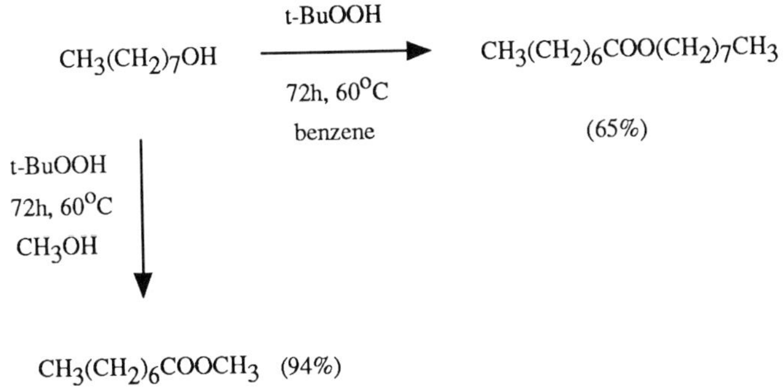

OH → (t-BuOOH, 42h, 62°C, benzene) → O (80%)

$CH_3(CH_2)_7OH$ → (t-BuOOH, 72h, 60°C, benzene) → $CH_3(CH_2)_6COO(CH_2)_7CH_3$ (65%)

$CH_3(CH_2)_7OH$ → (t-BuOOH, 72h, 60°C, CH_3OH) → $CH_3(CH_2)_6COOCH_3$ (94%)

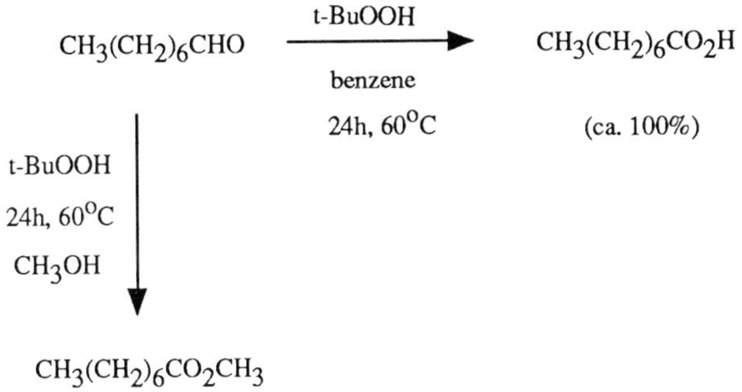

$CH_3(CH_2)_6CHO$ → (t-BuOOH, benzene, 24h, 60°C) → $CH_3(CH_2)_6CO_2H$ (ca. 100%)

$CH_3(CH_2)_6CHO$ → (t-BuOOH, 24h, 60°C, CH_3OH) → $CH_3(CH_2)_6CO_2CH_3$

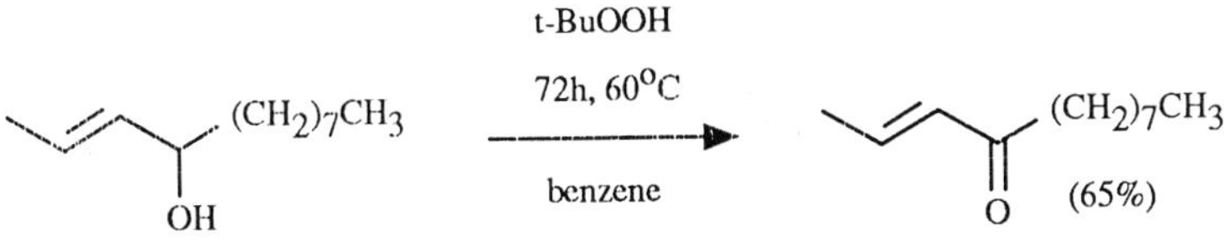

$(CH_2)_7CH_3$, OH → (t-BuOOH, 72h, 60°C, benzene) → $(CH_2)_7CH_3$, O (65%)

(xxii) Allylic methyl compounds are oxidised to allylic alcohols using t-butylhydroperoxide and selenium dioxide supported on silica gel (B.R. Chabra *et al.*, *Chem. Lett.*, 1981, 1703).

(b) **Alkene epoxidation**

(i) t-Butylhydroperoxide in the presence of a vanadium catalyst can be used for selective epoxidation of allyl alcohols (K.B. Sharpless and R.C. Michaelson, *J. Amer. Chem. Soc.*, 1973, **95**, 6136).

t-BuOOH; $VO(acac)_2$; OH; O; OH

(ii) Nickel(II) phthalocyanine catalyses the epoxidation of cyclohexane with t-butylperoxide (A.L. Stautzenberger, *U.S. Patent*, 1976, 3, 931, 249; *Chem. Abstr.*, 1976, **84**, 105379t).

(iii) Propylene oxide is manufactured by catalysed epoxidation of propene with t-butylhydroperoxide (R. Landau *et al.*, *Chem. Tech.*, 1979, 602).

$$CH_3CH{=}CH_2 \xrightarrow{\text{t-BuOOH}} CH_3\text{-(epoxide)} + \text{t-BuOH}$$

(iv) Unsaturated aldehydes undergo catalysed epoxidation using t-butylhydroperoxide, e.g., citral (R.A. Sheldon, *J. Mol. Catal.*, 1980, **7**, 107).

CHO; t-BuOOH; Ti^{IV}/SiO_2; O; CHO

(v) The epoxidation of allylic alcohols with t-butylhydroperoxide and $VO(acac)_2$ exhibits remarkable *syn*-stereoselectivity due to the participation of the hydroxyl group (T. Itoh *et al.*, *J. Amer. Chem. Soc.*, 1979, **101**, 159, R.B. Daniel and G.H. Whitham, *J. Chem. Soc., Perkin Trans. I*, 1979, 953).

OH; $VO(acac)_2$; ^{t}butylOOH; OH; O; 98%; +; OH; O; 2%

(vi) Epoxidations of allyl alcohols using t-butylhydroperoxide and titanium(IV) isopropoxide in the presence of (+) or (-)diethyltartrate as chiral ligand afford epoxides with high enantiomeric excesses (T. Katsuki and K.B. Sharpless, *J. Amer. Chem. Soc.*, 1980, **102**, 5974; B.E. Rossiter *et al.*, *J. Amer. Chem., Soc.*, 1981, **103**, 464).

$(CH_2)_9CH_3$ … OH $\xrightarrow[\text{(-) diethyltartrate, } -40^{o}C]{\text{t-BuOOH, Ti(OPr-i)}_4}$ O, $(CH_2)_9CH_3$ … OH

(80%) 91%e.e.

(vii) Selective epoxidation of allylic alcohols has been achieved using dibutyltinoxyperoxide which itself is obtained by reacting dibutyltin oxide with t-butylhydroperoxide, e.g., geraniol (S. Kanemoto *et al.*, *Tetrahedron Lett.*, 1986, **27**, 3387). The reaction exhibits high stereo- and regioselectivity.

OH $\xrightarrow[\text{(2) 3h at } 60^{o}C \text{; (3) aq. } NH_4Cl]{\text{(1) t-BuOOH, benzene; } Bu_2SnO}$ O, OH (83%)

(viii) The asymmetric epoxidation of allylic alcohols (the Sharpless epoxidation) has been reviewed (A. Pfenninger, *Synthesis*, 1986, 89).

6. *Preparation of alkyl nitrates*

Nitrations using alkyl nitrates have been reviewed: H. Feuer in "The chemistry of amino, nitroso and nitrocompounds and their derivatives", Supplement F, S. Patai Ed., John Wiley, Chichester, 1982.

(a) Primary and secondary alkyl nitrates are formed in good yields by the reactions of bromoalkanes with mercury(I) nitrate in 1,2-dimethoxyethane (A. McKillop and M.E. Ford, *Tetrahedron*, 1974, **30**, 2467).

$$CH_3(CH_2)_8Br \longrightarrow CH_3(CH_2)_8ONO_2 \quad (99\%)$$

(b) Alkyl sulphonates are converted into alkyl nitrates by reaction with tetrabutyl ammonium nitrate or with a nitrate ion exchange resin.

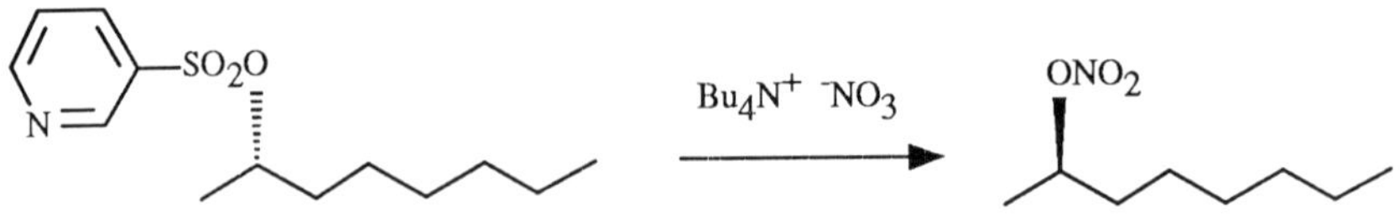

SO_2O Bu_4N^+ $^-NO_3$ ONO_2

(90%)

OSO_2Me Amberlyst A-26 NO_3^- form toluene ONO_2

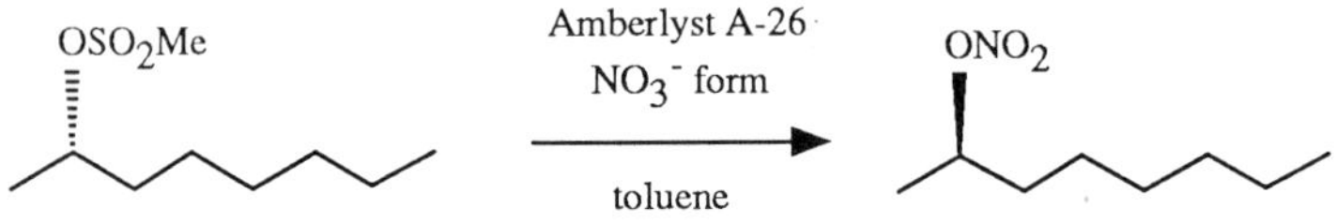

(70%)

Yields are high and competing elimination is negligible. The reaction involves inversion so this can be used as a method for inverting alcohols, the nitrates being readily reduced to alcohols with Zn/acetic acid or H_2/Pd/C/10% aqueous methanol (G. Cainelli *et al.*, *Tetrahedron Lett.*, 1985 **26**, 3369).

(c) Alcohols are converted to alkyl nitrates by transfer nitration using N-nitrocollidinium tetrafluoroborate under neutral conditions (G.A. Olah *et al.*, *Synthesis*, 1978, 452).

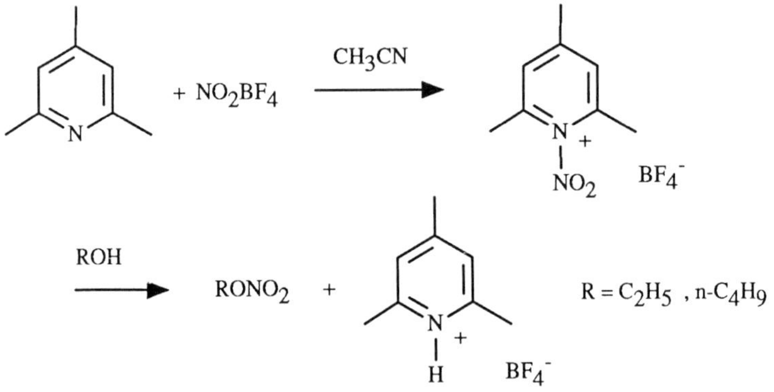

\+ NO_2BF_4 CH_3CN N^+ NO_2 BF_4^-

ROH $RONO_2$ + N^+ H BF_4^- $R = C_2H_5$, $n\text{-}C_4H_9$

(d) Primary amines are converted into alkyl nitrates by reacting with triphenylpyrylium nitrate in dry ether to give the N-alkylpyrylium nitrate which on heating forms the alkyl nitrate (A.R. Katritsky and L. Marzorati, *J. Org. Chem.*, 1980, **45**, 2515).

(e) Primary amines give alkyl nitrates on treatment with dinitrogentetraoxide at -78°C in the presence of an amidine base such as 1,5-diazabicyclo[5.4.0]undec-5-ene, e.g., octylamine is converted to octylnitrate (81%). Nitrate esters are useful intermediates in synthesis, since the hydroxy group is protected but can be regenerated quantitatively on reduction under mild conditions (D.H.R. Barton and S.C. Narang, *J. Chem. Soc., Perkin I*, 1977, 1114).

(f) Tertiarybutyl nitrite in $CFCl_3$, when irradiated at *ca.* 400 nm in the presence of oxygen yields tertiarybutyl nitrate (90%) (A. Mackor and T.J. de Boer, *Rec. Trav. Chim.*, 1970, **89**, 164). Steroidal nitrites also react in this way (D.H.R. Barton *et al.*, *J. Chem. Soc., Perkin I*, 1975, 2252).

$$\text{t-BuONO} \longrightarrow \text{t-BuONO}_2$$

7. *Reactions of alkyl nitrates*

(a) Alkyl nitrates react with lithium azide to give alkyl azides (P. Margaretha *et al.*, *Angew. Chem.*, 1971, **83**, 410).

$$(CH_3)_3C\text{-}ONO_2 \xrightarrow[\text{DMF, RT, 8h}]{LiN_3} (CH_3)_3C\text{-}N_3 \quad (75\%)$$

(b) Alkyl nitrates in the presence of base can be used to nitrate ketones, amides, carboxylic esters, sulphones, sulphonic esters and toluenes to give the salts of the nitrocompounds (H. Feuer in "Industrial Laboratory Nitrations", ed. L.F. Albright and C. Hanson, A.C.S. Symposium Series No. 22, A.C.S., Washington, p.160, 1976).

(c) 2-Oxazolines, bearing alkyl side chains at C-2, have been nitrated at the α-carbon to the nucleus, using base plus an alkyl nitrate (H. Feuer *et al.*, *Heterocyclic Chem.*, 1986, **23**, 825).

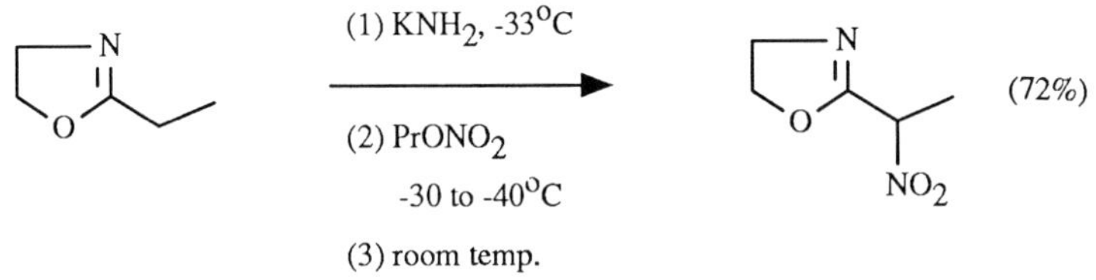

(1) KNH_2, -33°C
(2) $PrONO_2$ -30 to -40°C
(3) room temp.

2-ethyl-2-oxazoline → 2-(1-nitroethyl)-2-oxazoline (72%)

(d) Nitric acid esters react with nitriles in the presence of acid to give N-acylamines. This is a variant of the Ritter reaction (I.K. Moiseev *et al.*, *Khim - Farm. Zh.*, 1976, **10**, 32; *Chem. Abstr.*, 1976, **85**, 46053c).

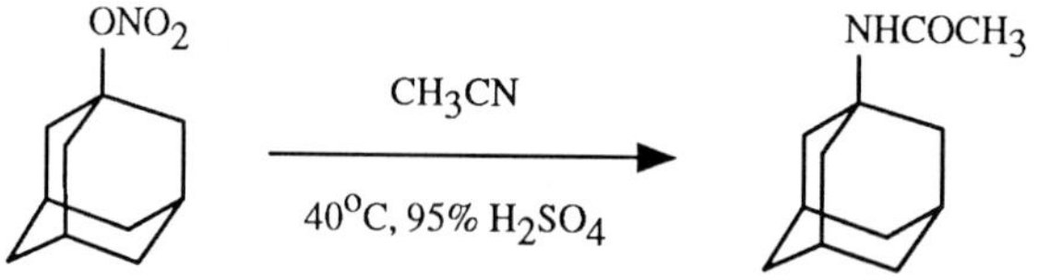

ONO_2 → $NHCOCH_3$

CH_3CN, 40°C, 95% H_2SO_4

1-adamantyl nitrate → 1-acetamido adamantane (80%)

(e) Alkyl nitrates, when refluxed with lithium bromide in acetone, afford bromoalkanes (M. Breugelmans and M.J.O. Anteunis, *Bull. Soc. Chim. Belg.*, 1975, **84**, 1197).

$$p\text{-}NO_2C_6H_4CH(ONO_2)CH_2CH_2ONO_2 \xrightarrow[\text{reflux, 40h}]{\text{LiBr, acetone}} p\text{-}NO_2C_6H_4CH(Br)CH_2CH_2Br$$

8. *Preparation of alkyl phosphites*

(a) An improved preparation of tri-t-butylphosphite from phosphorus trichloride has been reported. By using an excess of tertiary butanol and triethylamine the product can be purified by distillation. It does, however, decompose at 115°C to give phosphorous acid. Previously, due to acid and phosphochloridite impurities, the product had been undistillable owing to decomposition at 50°C to di-t-butylphosphite. Diethyl-t-butylphosphite has been prepared and shown to be stable up to 170°C, at which temperature elimination of isobutene occurs (T.K. Gazizov *et al.*, *J. Gen. Chem., USSR*, 1987, **57**, 414).

(b) β-chloroethyl esters of phosphorous acid undergo transesterification with alcohols (N.K. Ouchinnikova *et al.*, *Zh. Org. Khim.*, 1975, **11(9)**, 1839; *Chem. Abstr.*, 1976, **84**, 4401p).

$$P(OCH_2CH_2Cl)_3 \xrightarrow{\text{ROH}} P(OR)_3$$

(c) Diethyl 2-phenylethynyl phosphonite reacts with alcohols in the presence of sodium alkoxides to give trialkyl phosphites (A.F. Peshkov *et al.*, *Zh. Obshch. Khim.*, 1983, **53(2)**, 479; *Chem. Abstr.*, 1983, **98**, 198356k).

(d) Trialkyl phosphites are prepared by reacting alcohols with phosphine or white phosphorus, and base, in a mixed solvent (M. Kant *et al.*, *Ger. (East) Patent*, 1985, DD231, 074; *Chem. Abstr.*, **105**, 209205b).

$$MeOH \xrightarrow[\text{CCl}_4\text{, Et}_3\text{N, 55}^\circ\text{C}]{\text{P, DMF}} P(OMe)_3$$

(80%)

9. *Reactions of alkyl phosphites*

(a) The Arbuzov reaction has been reviewed (A.K. Bhattacharya and G. Thyagarajan, *Chem. Rev.*, 1981, **81**, 415). Attack by phosphorus on carbonyl carbon appears to be the first step in the reaction of trimethylphosphite with *trans*-but-2-enoyl chloride. However, the products isolated depend on the relative proportions of the reactants : *trans*-but-2-enoyl phosphonate is obtained when there is an excess of acid chloride and the diphosphonate ester is obtained when equimolar quantities are used. N.m.r. evidence is presented to support the intermediacy of the oxaphospholen (A. Szpala *et al.*, *J. Chem. Soc. Perkin I*, 1981, 1363).

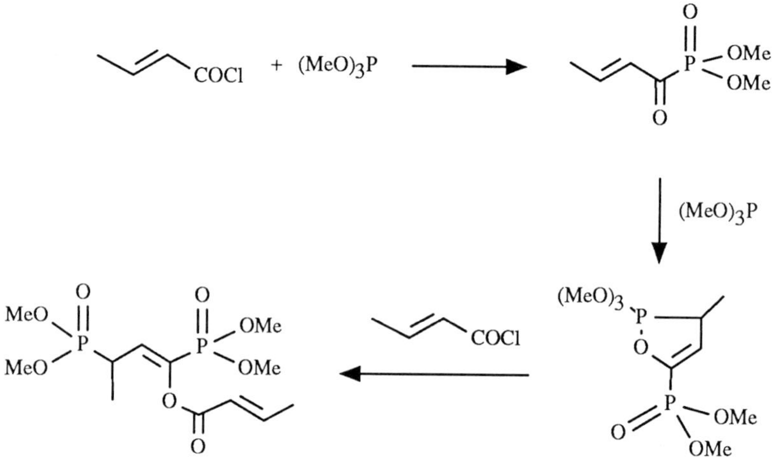

A detailed mechanistic study of the thermal reaction between triethylphosphite and carbon tetrachloride has been reported. It is concluded that nucleophilic attack at chlorine is involved (S_NCl) (S. Bakkas *et al.*, *Tetrahedron*, 1987, **43**, 501).

$$(EtO)_3P + CCl_4 \longrightarrow (EtO)_2P(O)CCl_3 + EtCl$$

Copper(I) halide complexes with trialkyl phosphites can be used to prepare vinyl phosphonates (G. Axelrad *et al.*, *J. Org. Chem.*, 1981, **46**, 5200; S. Banerjee *et al.*, *Phosphorus Sulfur*, 1983, **15**, 15).

$$(^iPrO)_3P + PhCH{=}CHCl \xrightarrow[200^oC]{CuCl} PhCH{=}CH{-}P(O)(O^iPr)_2 + {}^iPrCl$$

In a mechanistic study of the Arbuzov reactions of mixed trialkylphosphites, it was found that the relative yields of the products depended on the nature of the alkyl groups and on the solvent polarity, suggesting that the second step in the reaction has S_N1 character. If R = t-Bu, II is formed exclusively. In other cases, I and II are formed with the amount of II increasing with increasing solvent polarity (D. Cooper *et al.*, *J. Chem. Research (S)*, 1983, 234).

$$(EtO)_2P\text{-}OR + MeI \longrightarrow \underset{I}{(EtO)_2P(O)Me} + \underset{II}{EtO(Me)P(O)OR}$$

Examples of concomitant Arbuzov and substitution reactions are found in the reactions of some 4-chlorobutyramides with triethylphosphite (P.A. Gurevich *et al.*, *J. Gen. Chem.*, *USSR*, 1986, **56**, 1100).

$$R_2NCO(CH_2)_3Cl \xrightarrow[160-170^oC]{(EtO)_3P} R_2NCO(CH_2)_3O\text{-}P(OEt)_2 + R_2NCO(CH_2)_3P(O)(OEt)_2$$

α-hydroxyphosphonates are formed, in addition to vinyl phosphates and dehalogenated ketones, in the reactions of trimethylphosphite (in methanol) or triethylphosphite (in ethanol) with 1-chloro-, 1-bromo-, and 1,1-dichloroacetophenones. Triisopropyl phosphite (in propan-2-ol) gives only the vinyl phosphate. Ketophosphonates are not produced under the conditions used. Thus trimethylphosphite and 1-chloroacetophenone in methanol at 0^oC give

$$(MeO)_2P(O)\text{-}O\text{-}C(=CH_2)C_6H_5 \quad (MeO)_2P(O)\text{-}O\text{-}C(OH)(CH_2Cl)C_6H_5 \quad C_6H_5COCH_3$$

55 mole % 35% 10%

The results are consistent with the initial formation of a betaine intermediate:

$$(MeO)_3\overset{+}{P}-O-C(C_6H_5)(CH_2Cl)-O^-$$

(G. Keglevich *et al.*, *Phosphorus Sulfur*, 1987, **29**, 341).

Intermediates have been isolated in the reaction between neopentylphosphites and 1-halogenoacetophenones (I. Petnehazy *et al.*, *Tetrahedron*, 1983, **39**, 4229).

$$(RO)_3\overset{+}{P}CH_2COPh\ {}^-Br \quad \text{and} \quad RO(Ph)_2\overset{+}{P}\text{-O-}(Ph)C{=}CH_2\ {}^-Cl$$

A new route to alkoxyphosphonates has been described using trimethylphosphite and a vinyl ether (W.W. Epstein and M. Garrossian, *Phosphorus Sulfur*, 1988, **35**, 349).

$$EtOCH{=}CH_2 + (MeO)_3P \xrightarrow[\text{RT}]{\text{HCl, Et}_2\text{O}} EtO(Me)CH\text{-}P(O)(OMe)_2 \quad (94\%)$$

Vinyl halides react with trialkylphosphites to give vinylphosphonates (R. Dittrich and G. Hägele, *Phosphorus Sulfur*, 1981, **10**, 127).

$$F_2C{=}CFX + (RO)_3P \longrightarrow (RO)_2P(O)CF{=}CFX + (RO)_2P(O)CF{=}CFP(O)(OR)_2$$

2-ketophosphonic esters are prepared by reacting chloroepoxides with trialkyl phosphites (J. Gasteiger and C. Herzig, *Tetrahedron Lett.*, 1980, **21**, 2687).

$$\text{i-Pr(Cl)C}\overset{O}{—}\text{CH}_2 \xrightarrow{P(OMe)_3} {}^iPrCOCH_2P(O)(OMe)_2 + MeCl \quad (92\%)$$

Triethylphosphite and dicyanomethane react unusually to produce a phosphonate and ethane (B.M. Gladshtein *et al.*, *J. Gen. Chem., USSR*, 1987, **57**, 2051).

$$(EtO)_3P + CH_2(CN)_2 \xrightarrow[\text{reflux}]{CH_3CN} (EtO)_2P(O)CH(CN)_2 + C_2H_6$$

The product of the reaction of triethylphosphite with iodine is not diiodotriethoxyphosphorane, $(EtO)_3PI_2)$, as previously reported, but is in fact diethylphosphoroiodate formed by an Arbuzov reaction (D. Cooper and S. Trippett, *Tetrahedron Lett.*, 1979, **19**, 1725; A. Skowronska *et al.*, *Tetrahedron Lett.*, 1980, **21**, **321**).

(b) Carbonyl fluoride is a mild fluorinating agent for trialkylphosphites (O.D. Gupta and J.M. Shreeve, *J. Chem. Soc. Chem. Commun.*, 1984, 416).

(c) Pinacol thiocarbonate reacts with triethylphosphite at 150°C to give the alkene (W.T. Borden *et al.*, *Tetrahedron Lett.*, 1973, **13**, 3161).

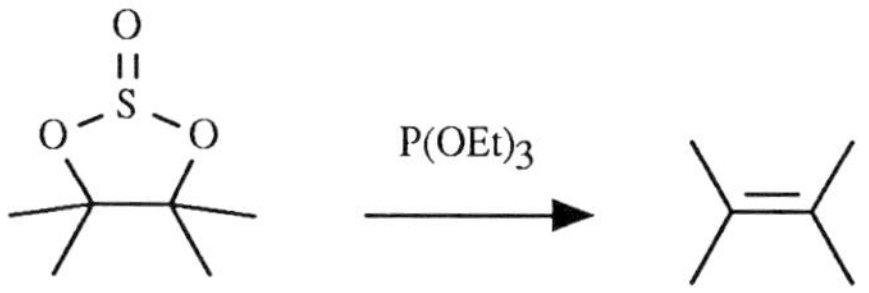

O
S
O O
$P(OEt)_3$

(d) Desulphurization reactions using phosphite esters have been reviewed (J.I.G. Cadogan and R.K. Mackie, *Chem. Soc. Rev.*, 1974, **3**, 87). [2.2]paracyclophane has been synthesised (M. Brink, *Synthesis*, 1975, 807).

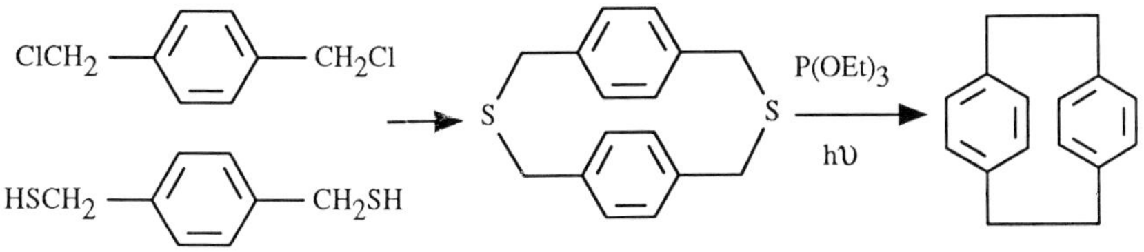

$ClCH_2$ CH_2Cl
$HSCH_2$ CH_2SH
S S
$P(OEt)_3$
hυ

(e) Copper(I) iodide complex with trimethylphosphite catalyses the decomposition of diazocarbonyl compounds (R. Pelliccari and P. Cogolli, *Synthesis*, 1975, 269; S. Julia *et al.*, *Compt. Rend.*, 1971, **272(c)**, 1898; B.W. Peace *et al.*, *Synthesis*, 1971, 658).

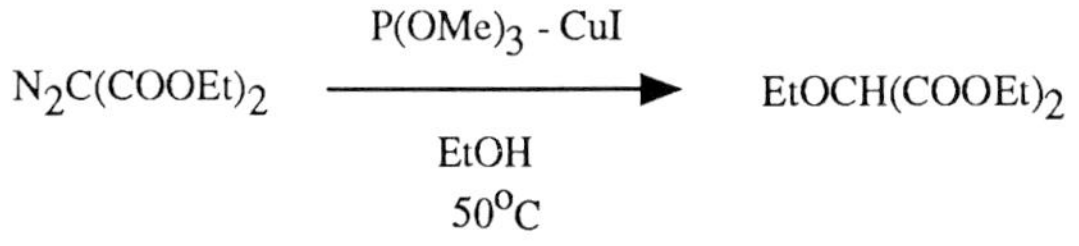

$$N_2C(COOEt)_2 \xrightarrow[\text{EtOH, } 50^oC]{P(OMe)_3 - CuI} EtOCH(COOEt)_2$$

$$\text{cyclohexene} + N_2C(COOMe)_2 \xrightarrow{P(OMe)_3\text{ - CuI}} \text{7,7-bis(COOMe)bicyclo[4.1.0]heptane}$$

The chemistry of phosphite - metal complexes has been reviewed (S.D. Robinson, MTP Review of Science, Series One, Inorg. Chem., vol. 6, M.J. Mays Ed., p. 121, Butterworth, London, 1972 and Series Two, vol. 6, p.71, 1975).

(f) Diethylphosphorocyanidate, prepared from cyanogen bromide and triethylphosphite, is used for synthesising amides. The method has been applied to peptide synthesis with little racemization (S. Yamada *et al.*, *Tetrahedron Lett.*, 1973, 1595).

$$RCOOH + R^1NH_2 \xrightarrow[Et_3N]{NCP(O)(OEt)_2} RCONHR^1$$

(g) The use of phosphite esters in the reductive cyclisation of nitro compounds has been reviewed (J.I.G. Cadogan, *Acc. Chem. Res.*, 1972, **5**, 303). 1H-Azepines are produced from benzenes by reaction with triethylphosphite and nitrosobenzene in trifluoroethanol (R.J. Sunberg and R.H. Smith, *Tetrahedron Lett.*, 1971, **11**, 267).

$$C_6H_5NO + \text{benzene} \xrightarrow{P(OEt)_3} C_6H_5\text{—N(azepine)}$$

(h) Primary amines are synthesised by reacting triethylphosphite with an alkyl azide (A. Koziara *et al.*, *Synthesis*, 1985, 202).

$$R\text{-}Br \xrightarrow[\text{phase transfer}]{NaN_3} R\text{-}N_3 \xrightarrow[RT]{P(OEt)_3} R\text{-}N{=}P(OEt)_3 \xrightarrow{HCl} RNH_3^+\ Cl^-$$

Triethylphosphite reacts with *trans*-10-azidodihydrophenanthrene-9-ol to give the corresponding imine (M. Weitzberg *et al.*, *J. Org. Chem.*, 1980, **45**, 4252).

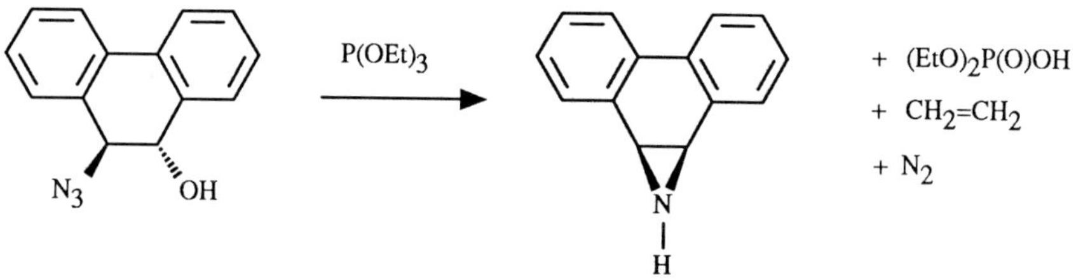

$$\text{N}_3\text{, OH-dihydrophenanthrene} \xrightarrow{P(OEt)_3} \text{aziridine (N–H)} + (EtO)_2P(O)OH + CH_2{=}CH_2 + N_2$$

(i) Trialkyl phosphites react with oximes to give α-ethoxyaminophosphonates (M.P. Osipova *et al.*, *J. Gen. Chem., USSR*, 1982, **52**, 392).

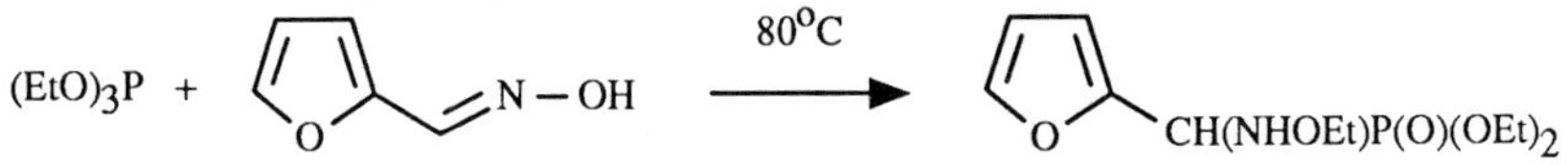

$$(EtO)_3P + \text{2-furyl-CH=N-OH} \xrightarrow{80^oC} \text{2-furyl-}CH(NHOEt)P(O)(OEt)_2$$

(39%)

(j) Schiff bases, prepared from 2-alkenals and t-butylamine, react exothermically with triethylphosphite in the presence of formic acid followed by hydrochloric acid to give 2-formyl- alkylphosphonates (M.P. Teulade and P. Savignac, *Synthesis*, 1987, 1037).

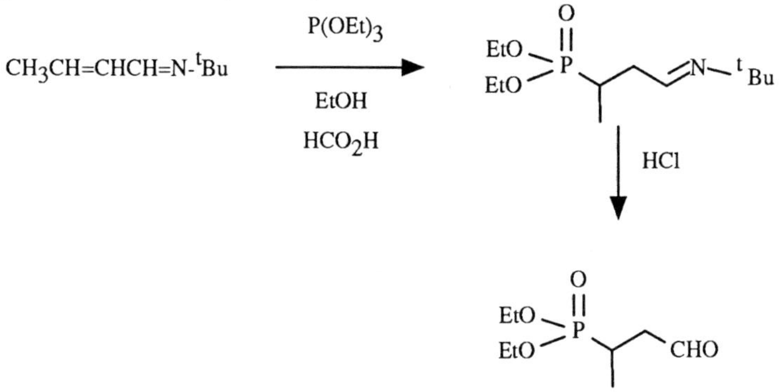

$$CH_3CH{=}CHCH{=}N\text{-}^tBu \xrightarrow[EtOH,\ HCO_2H]{P(OEt)_3} (EtO)_2P(O)CH(CH_3)CH_2CH{=}N\text{-}^tBu \xrightarrow{HCl} (EtO)_2P(O)CH(CH_3)CH_2CHO$$

(60%)

(k) Dimethylacetylene dicarboxylate and an excess of trimethylphosphite give a *bis*-ylid. Other products may be obtained depending on the conditions (M.T. Boisdon and J. Barrans, *J. Chem. Soc. Chem. Commun.*, 1988, 615; J.C. Tebby *et al.*, *J. Chem. Soc. Chem. Commun.*, 1981, 420; R. Burgada *et al.*, *Tetrahedron Lett.*, 1981, **22**, 3393).

2 P(OMe)$_3$ + MeOOC—≡—COOMe ⟶ (MeO)$_3$P=C(COOMe)–C(COOMe)=P(OMe)$_3$

(l) The reaction of triethylphosphite with maleic anhydride gives an ylid (D.B. Denney and D.Z. Denney, *Phosphorus Sulfur*, 1982, **13**, 315).

(EtO)$_3$P$^+$–[furanone ring]–O$^-$

(m) The addition reactions of esters of phosphorus(III) acids to unsaturated systems have been reviewed (A.N. Pudovik and I.V. Konovalova, *Synthesis*, 1979, 81). 2,5-diphenylfuran is formed when *trans*-dibenzoylethylene reacts with trimethylphosphite (O.P. Madan and C.P. Smith, *J. Org. Chem.*, 1965, **30**, 2284) and attempts to use this type of reaction as a general furan synthesis have been reported (M.J. Haddadin *et al.*, *J. Org. Chem.*, 1979, **44**, 494).

Ph, H, Ph, Ph, O, O —P(OEt)$_3$, triglyme, reflux, 2-3h⟶ Ph, Ph, Ph, O (66%)

(n) Trialkyl phosphites react with dialkylaroyl phosphonates to give carbenes which react with further phosphite to give quasi-ylids (D.V. Griffiths and J.C. Tebby, *J. Chem. Soc., Chem. Commun.*, 1986, 871).

ArC(O)–P(O)(OR)$_2$ + (R^{1}O)$_3$P ⇌ (85°C) Ar–C$^-$(O–P$^+$(OR^1)$_3$)–P(O)(OR)$_2$ ⟶

$$\text{Ar-CH}_2\text{-P(O)(OR)}_2 + O{=}P(OR^1)_3 \xrightarrow{(R^1O)_3P} \text{Ar(}(R^1O)_3P{=})C\text{-P(O)(OR)}_2$$

(o) Trialkylphosphites are dealkylated using 4-toluenesulphonic acid (Y. Nitta *et al.*, *Chem. Pharm. Bull.*, 1986, **34**, 2710).

$$3\,TsOH + P(OMe)_3 \xrightarrow[\text{reflux, 24h}]{\text{toluene}} 3\,TsOMe + HP(O)(OH)_2 \quad (81\%)$$

(p) Trialkylphosphites react with diphenyliodinium iodide in the presence of copper salts to give arylphosphonates (A.G. Varvoglis, *Tetrahedron Lett.*, 1972, **12**, 31).

$$Ph_2I^+\,I^- + P(OEt)_3 \longrightarrow (EtO)_2P(O)Ph + 4\text{-Ph-I-}C_6H_4I \;+$$

10. *Preparation of alkyl phosphates*

Texts and reviews in this area include : B.J. Walker, "Organophosphorus Chemistry", Penguin, London, 1972; J. Emsley and D. Hall, "The Chemistry of Phosphorus", Harper and Row, London, 1976; Organophosphorus Chemistry, Specialist Periodical Report, Royal Society of Chemistry, vol. 1, 1970 to vol. 20, 1989.

(a) Trialkylphosphates are readily prepared by alkylating dialkylphosphates with haloalkanes in acetonitrile in the presence of caesium fluoride (H. Takaku *et al.*, *Chem. Pharm. Bull.*, 1983, **31**, 2157).

$$(CH_3O)_2P(O)OH + CH_3I \xrightarrow[CH_3CN,\ \text{RT, 6h}]{CsF} (CH_3O)_3PO \quad (88\%)$$

(b) Mixed trialkylphosphates are obtained by transesterification (K.K. Ogilvie *et al.*, *J. Amer. Chem. Soc.*, 1977, **99**, 1277).

$$n\text{-PrO}(\text{Cl}_3\text{CCH}_2\text{O})\text{P(=O)OEt} + \text{CH}_3(\text{CH}_2)_7\text{OH} \xrightarrow[\substack{80^\circ\text{c} \\ 6\ \text{days}}]{\text{CsF}} n\text{-PrO}(\text{CH}_3(\text{CH}_2)_7\text{O})\text{P(=O)OEt}$$

(c) Di-t-butyl-N,N-diethylphosphoramidite has been prepared and is a stable, distillable reagent which phosphorylates alcohols when activated with 1-tetrazole (J.W. Perich and R.B. Johns, *Synthesis*, 1988, 142).

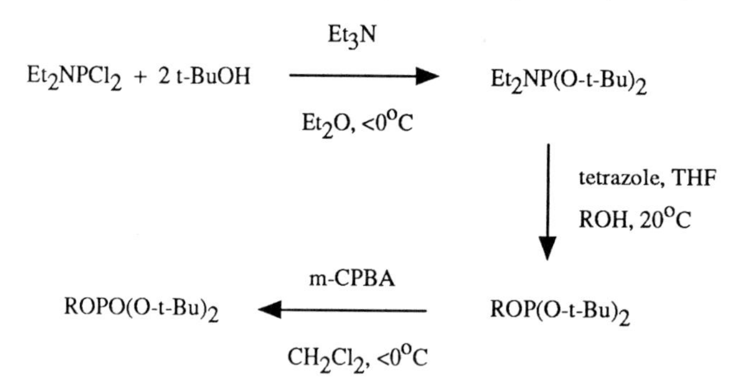

(d) The use of the Wadsworth-Emmons reaction for preparing enantiomeric phosphates has been reviewed (W.J. Stec, *Acc. Chem. Res.*, 1983, **16**, 411).

(e) Mixed trialkylphosphites are synthesised by oxidative phosphorylation of alcohols with dialkylphosphonates and copper(II) chloride (Y. Okamoto *et al.*, *Bull. Chem. Soc., Japan*, 1988, **61**, 3359).

$$(\text{EtO})_2\text{P(O)H} \xrightarrow[\text{O}_2,\ 50^\circ\text{C},\ 5\text{h}]{\text{PrOH, CuCl}_2} (\text{EtO})_2\text{P(O)O Pr} \quad (78\%)$$

(f) Phosphoric triesters are obtained from phosphorous triesters by oxidation with *bis*(trimethylsilyl)peroxide (BSPO) and aluminium chloride (L. Wozniak *et al.*, *Tetrahedron Lett.*, 1985, **26**, 4965).

$$(\text{RO})_3\text{P} \xrightarrow[\text{RT, 1h}]{\text{BSPO, AlCl}_3} (\text{RO})_3\text{P=O}$$

(g) Tetramethylammonium di-t-butylphosphate reacts with primary and secondary alkyl iodides or bromides in aprotic solvents to give the corresponding trialkylphosphates (A. Zwierak and K. Kluba, *Tetrahedron*, 1971, **21**, 3163).

(h) Nucleoside phosphites are oxidised to phosphates by treatment with BSPO, trimethylsilyltriflate and triethylamine at -20°C (Y. Hayakawa *et al.*, *Tetrahedron Lett.*, 1986, **27**, 4191).

(i) Mixed dialkylphosphates are obtained *via* pyridinium phosphate betaines (T. Hata *et al.*, *Tetrahedron Lett.*, 1970, **10**, 3505; *Bull. Chem. Soc., Japan*, 1971, **44**, 232; *Chem. Pharm. Bull.*, 1971, **19**, 687, 696).

CH_2Cl ; NO_2 ; O ; O ; P ; OEt ; OH ; N ; N+ ; O- ; $nC_5H_{11}OH$; RT, 2 days ; 90°C, 6h ; + ; $n\text{-}C_5H_{11}O$; (81%)

(j) Pyrophosphoric acid (from P_2O_5 and phosphoric acid) converts primary and secondary alcohols at 90°C to dihydrogen phosphates which are isolated as the barium or anilinium salts (H. Yamaguchi *et al.*, *Bull. Chem. Soc., Japan*, 1981, **54**, 1891).

(k) Phosphate monoesters are obtained by photochemical dearylation (R.A. Finnegan and J.A. Matson, *J. Chem. Soc. Chem. Commun.*, 1975, 928).

$$(4\text{-}MeOC_6H_4O)_2P(O)OPr^i \xrightarrow[EtOH]{h\nu} (HO)_2P(O)OPr^i \quad (99\%)$$

(l) Phosphate monoesters are produced by phosphorylation of alcohols with diethylphosphorochloridodithionite in diethyl ether in the presence of dimethylaniline. The products are isolated as their anilinium salts (H. Takaku and Y. Shimada, *Tetrahedron Lett.*, 1972, **12**, 411).

$$CH_3(CH_2)_2OH + Cl\text{-}P(SEt)_2 \xrightarrow[3h,\ 0^oC,\ Et_2O]{} CH_3(CH_2)_2OP(SEt)_2$$

$$CH_3(CH_2)_2OP(SEt)_2 \xrightarrow{I_2,\ acetone,\ H_2O} CH_3(CH_2)_2OP(O)(OH)_2$$

(m) Phosphate monoesters are produced by reacting alcohols with trimethylsilylpolyphosphate which is itself obtained by reacting diphosphorus pentoxide with hexamethylsiloxane (Y. Okamoto *et al.*, *Bull. Chem. Soc., Japan*, 1985, **58**, 3393).

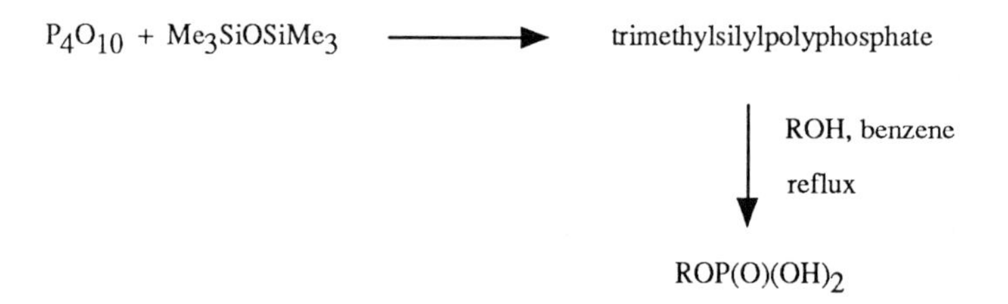

(n) P'-*O*-ethyl-P'-thyrophosphate*tris*(trialkylammonium) salt is used for phosphorylating alcohols (P.M. Cullis and A.J. Rous, *J. Amer. Chem. Soc.*, 1985, **107**, 6721).

$$EtO\text{-}P(=O)(S^-)\text{-}O\text{-}P(=O)(O^-)\text{-}O^- \xrightarrow[EtOH]{MeI} EtO\text{-}P(=O)(SMe)\text{-}O\text{-}P(=O)(O^-)\text{-}O^-$$

$$\xrightarrow{\text{spontaneous decomposition}} EtO-P(=O)(SMe)-O^- + PO_3^- \xrightarrow{ROH} RO-P(=O)(O^-)-O^-$$

(o) The use of phosphoric acid mono-, di- and triesters for phosphorylation has been discussed in a general review of phosphorylation (L. Soltin, *Synthesis*, 1977, 737).

(p) Trialkylphosphates are obtained from dialkyl hydrogen phosphites, lithium chloride and an alcohol by constant current electrolysis at a carbon anode (H. Ohmori *et al.*, *Chem. Pharm. Bull.*, 1979, **27**, 1700).

$$(EtO)_2P(O)H + PrOH \xrightarrow[\text{electrolysis}]{LiCl} (EtO)_2P(O)OPr \quad (64\%)$$

11. *Reactions of alkyl phosphates*

(a) Trialkylphosphates are dealkylated using p-toluenesulphonic acid (Y. Nitta *et al.*, *Chem. Pharm. Bull.*, 1986, **34**, 2710).

$$3\ TsOH + (EtO)_3P{=}O \xrightarrow[\text{reflux, 48h}]{\text{toluene}} 3\ TsOEt + PO(OH)_3 \quad (89\%)$$

(b) Amberlite 200C cation (sulphonic acid) exchange resin catalyses the dealkylation of trialkylphosphates (Y. Nitta and Y. Arakawa, *ibid.*, 1986, **34**, 3121).

$$(EtO)_2P(O)O^tBu \xrightarrow[\text{benzene, RT, 15h}]{\text{Amberlite 200C}} (EtO)_2P(O)OH \quad (77\%)$$

(c) When polyphosphoric acid and trimethylphosphate are heated at 190°C for 4 hours a 36-39% yield of a mixture of hydrocarbons is obtained. The mixture contains about 200 compounds of which 140 have been identified by GLC-MS. 2-methylhexane and 3-methylhexane are major components (D.E. Pearson, *J. Chem. Soc. Chem. Commun.*, 1974, 397).

(d) Trialkyl phosphates may be used to alkylate thymine and uracil (K. Yamauchi and M. Kinoshita, *J. Chem. Soc. Perkin I*, 1973, 391).

(e) Trimethylphosphate is effective at esterifying carboxylic acids and is less toxic than dimethylsulphate (M.M. Harris and P.K. Patel, *Chem. Ind. (London)*, 1973, 1002).

COOH
COOH
$(MeO)_3PO$
aq. NaOH
COOMe
COOMe
(90%)

(f) Trialkylphosphates are good solvents for the nitration and halogenation of aromatic compounds. When halogenation is conducted in trimethylphosphate, hydrogen halide is rapidly converted to the halomethane and this is important for acid-sensitive substrates. 1,3,5-*tris*(t-butyl)benzene (which dealkylates when treated with bromine in carbon tetrachloride and is unreactive towards bromine in acetic acid) is brominated with bromine in trimethylphosphate (D.E. Pearson *et al.*, *Synthesis*, 1976, 621).

t-Bu t-Bu
t-Bu
Br_2, $(MeO)_3PO$
64°C, 24h
Br
t-Bu t-Bu
t-Bu
(59%)

(g) The mechanism of hydrolysis of phosphate esters has been reviewed: J. Emsley and D. Hall, "The Chemistry of Phosphorus", Harper and Row, London, 1976, chap. 8, p. 312; G.R.J. Thatcher and R Kluger, *Adv. Phys. Org. Chem.*, 1989, **25**, 99; F.H. Westheimer *et al.*, *J. Amer. Chem. Soc.*, 1988, **110**, 181).

(h) Phosphoric esters are prepared from phosphate triesters by reaction with butyllithium to give the α-lithiophosphonate which is treated with iodomethane (M.P. Teulade and P. Savignac, *Tetrahedron Lett.*, 1987, **28**, 405).

$$(EtO)_3PO + PrCH_2Li \longrightarrow [(EtO)_2P(O)CH(Pr)Li] \xrightarrow{MeI} (EtO)_2P(O)CH(Pr)Me$$

(67%)

(i) Enol phosphates are cleaved by trialkyl alanes in the presence of *tetrakis*(triphenylphosphine)palladium (K. Takai *et al.*, *Tetrahedron Lett.*, 1980, **21**, 2531).

(j) Aldehydes and ketones react with allylic phosphates and samarium(II) iodide to give unsaturated alcohols (S. Araki *et al.*, *J. Organometallic Chem.*, 1987, **333**, 329).

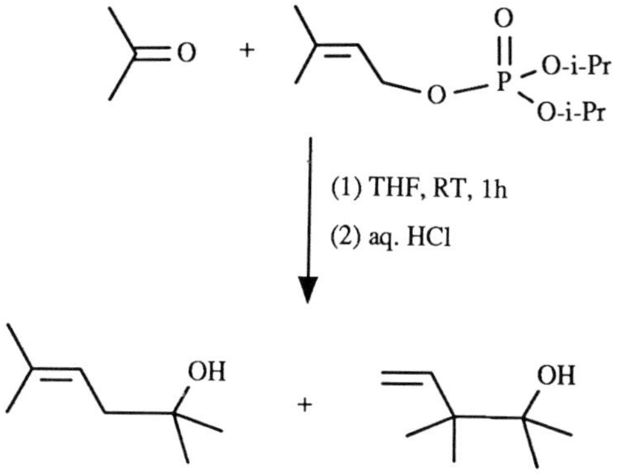

O-i-Pr, O-i-Pr

(1) THF, RT, 1h

(2) aq. HCl

OH + OH

64% 36%

(93% yield)

(k) The reaction between diethylenolphosphate esters and bromotrimethylsilane yields *bis* (trimethylsilyl) esters which, on ethanolysis, give the free enol phosphoric acids. The acids are isolated as their anilinium salts. (T. Hata *et al.*, *Synthesis*, 1979, 189).

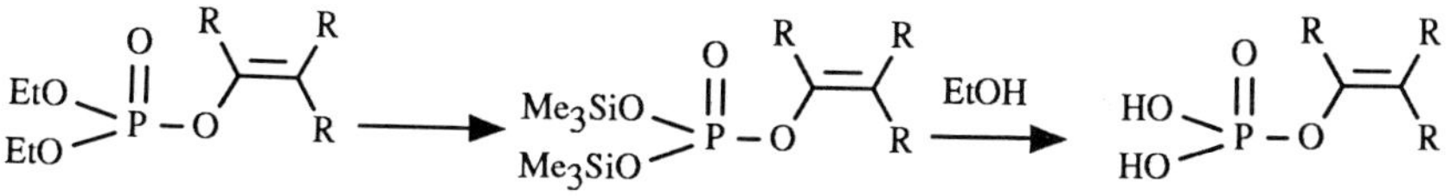

$$(EtO)_2P(O)-O-C(R)=C(R)R \longrightarrow (Me_3SiO)_2P(O)-O-C(R)=C(R)R \xrightarrow{EtOH} (HO)_2P(O)-O-C(R)=C(R)R$$

Second Supplements to the 2nd Edition of Rodd's Chemistry of Carbon Compounds, Vol. IA/B, edited by M. Sainsbury
© 1991 Elsevier Science Publishers B.V., Amsterdam

Chapter 5

SULPHUR ANALOGUES OF THE ALCOHOLS AND THEIR DERIVATIVES

D.W.BROWN

The sections and subsections in this chapter follow the pattern adopted in Chapter 5, Vol. 1B and the Supplement to the Second Edition.

Particularly relevant periodical publications, books and reviews include (i) Chemical Society Specialist Periodic Reports, The Royal Society of Chemistry, London, in six volumes between 1969 and 1980 in which the Organic compounds of Sulphur, Selenium and Tellurium are reviewed, (ii) The Chemistry of Functional Groups, series ed. S. Patai, John Wiley & Sons, Chichester, and (iii) Comprehensive Organic Chemistry, ed. D.H.R.Barton and W.D.Ollis, Vol. 3, ed. N.Jones 1979, Pergammon Press, Oxford. The relevant, individual books and chapters of the above will be referred to under the appropriate sections.

1. Thiols, mercaptans or hydrosulphides.

"The Thiol Group" parts 1 and 2 appear in the series edited by S. Patai (see above). The chemistry of this functional group is covered in "Comprehensive Organic Chemistry", G.C.Barret Vol.3, and in Specialist Periodical Reports, G.C.Barret, Ch. 2. A review of thiol chemistry (J.Norell and R.P.Loughan, Encycl. Chem. Technol., 3rd edn., Vol. 22, ed. M.Grayson and D.Eckroth, Wiley, New York, 1983) has appeared describing the occurrence, physical and thermodynamic properties, reactions, preparations, toxicities and industrial uses of thiols. Alkynyl thiols are featured in a review (R.Mayer and H.Kroeber, *Z. Chem.*, 1975, **15**, 91) containing 91 references.

Preparation

(1) The pyridine thione below is used under mild conditions to introduce the thiol group in halide displacement reactions (P.Molina *et al.*, *Tetrahedron Letters*, 1985, **26**, 469).

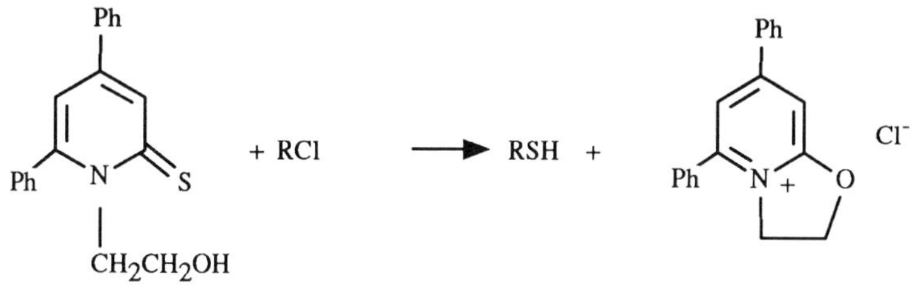

$$\text{1-(2-hydroxyethyl)-4,6-diphenylpyridine-2-thione } (N\text{-}CH_2CH_2OH) + RCl \longrightarrow RSH + \text{oxazolo[3,2-a]pyridinium (Ph, Ph) } Cl^-$$

(2) Halide ion is also displaced by thiosulphonate ion to give thiosulphonates which can be reduced to the corresponding thiols by LAH (G.K.Cooper, D.P.Bloxham and M.Webster, *J. Chem. Res.*, Synop. 1982, **4**, 104).

(3) 4-(Dimethylamino) pyridine *N*-oxide has been shown to be an excellent catalyst for the thione to thiol rearrangement of xanthates (K.Harano *et al.*, *Heterocycles*, 1983, **27**, 2327).

$$ROCS_2R \xrightarrow{\text{cat.}} RSCOSR \xrightarrow{\text{hydrol.}} 2RSH + CO_2$$

(4) Aminothiols have been synthesized by reduction of thiazetidines with Vitride (E.Meyle, H.H.Otto, *Arch. Pharm.*, 1987, **320**, 571).

$$\text{R-substituted } N\text{-}R^1\text{-thiazetidine } S,S\text{-dioxide} \longrightarrow R^1NHCHRCH_2SH$$

Also a series of α,ω-aminothiols have been prepared as bidentate ligands(J.L.Corbin *et al.*, *Inorg. Chem.*, 1984, **23**, 3404).

(5) Arene and alkanesulphonic acids are converted to polysulphides, RS_nR, by P_2S_5. Reduction of these polysulphides with LAH or $NaBH_4$ gives the corresponding thiols (S.Oae and H.Togo, *Tetrahedron Lett.*, 1982, **23**, 4701).

(6) Tungstate and other catalysts have been evaluated for the conversion of methanol to methane thiol with hydrogen sulphide (A.V.Mashkina *et al.*,

React. Kinet. Catal. Lett., 1988, **36**, 159 and references therein).

(7) The enantioselective reduction of thiiranes by chiral boranes give, in the presence of LiCl, high enantiomeric excesses of chiral thiols, $R^1R^2CHCH_2SH$ (J.S.Cha *et al.*, *Bull. Korean Chem. Soc.*, 1988, **9**, 23). Chiral syntheses of α-branched α-mercaptocarboxylic acids are described the latter *via* 1,3-oxathiolanones derived from naturally occurring chiral hydroxy-acids (B.Strijtveen and R.M.Kellogg, *Tetrahedron*, 1987, **43**, 5039; D.Seebach, R.Naef and G.Calderari, *Tetrahedron*, 1984, **40**, 1313).

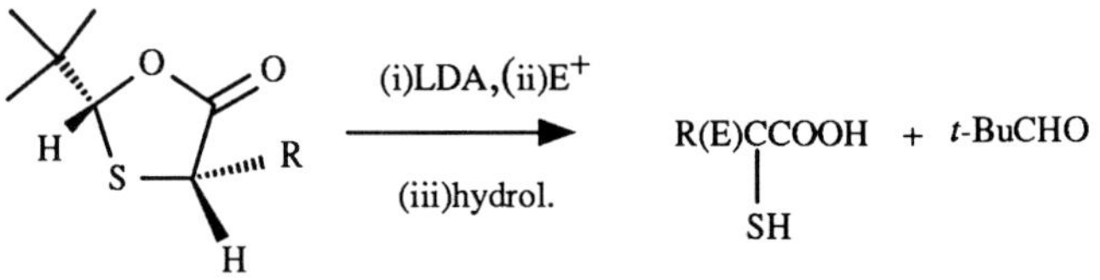

$$\text{(oxathiolanone)} \xrightarrow[\text{(iii)hydrol.}]{\text{(i)LDA,(ii)E}^+} \underset{\text{SH}}{\text{R(E)CCOOH}} + t\text{-BuCHO}$$

(8) *t*-Thiols are prepared by reaction of diaza compounds with thiols (P.J.Giddings *et al.*, *J. Chem. Soc., Perkin Trans.*, 1982, 2757).

$$R^1R^2C{:}N^+{:}N^- + CH_2{:}CHCH_2SH \longrightarrow R^1R^2C(SH)CH_2CH{:}CH_2$$

Reactions

(1) Variations on the alkylation of metal mercaptides include the Zn(II) catalysed reaction (B.Rajanikanth and B.Ravindranath, *Indian J. Chem. Sect.B*, 1984, **23**, 1043) shown below.

$$R^1SH + RBr \xrightarrow{Zn^{2+}} R^1SR + HBr$$

Similarly high yields of α-alkylthio carbonyl compounds have been obtained in light petrol or toluene containing Aliquat 336 and potassium carbonate (M.Lissel, *J. Chem. Res. Synop.*, 1982, 286).

$$RCOCH_2Cl + EtSH \longrightarrow RCOCH_2SEt + HCl$$

Evidence is put forward for a radical mechanism for the formation of perfluoroalkyl sulphides (A.E.Feiring, *J.Fluorine Chem.*, 1984, **24**, 191).

$$CF_3(CF_2)_5I + CF_3(CF_2)_3SNa \longrightarrow C_6F_{13}SC_4F_9 + NaI$$

(2) Thiols are esterified by thiocarboxylic acids to give alkyl thioesters using as catalyst sodium salts of oxo acids of sulphur (T.Yamanchi *et al.*, *Chem. Express*, 1986, **1**, 591). Polyphosphoric ester has also been used as a catalyst for this reaction (T.Imamoto *et al.*, *Bull. Chem. Soc. Jpn.*, 1982, **55**, 2303). Malonic half-thiol esters are produced from the corresponding half-acid using $Mg(OEt)_2$ as base (S.Ohta and M. Okamoto, *Tetrahedron Lett.*, 1981, **22**, 3245).

$$EtSH + CH_2{:}CHCH_2\underset{\displaystyle COOEt}{CH}COOH \longrightarrow CH_2{:}CHCH_2\underset{\displaystyle COSEt}{CH}COOEt$$

Anion exchange methods have been employed whereby mono or dicarboxylic acid chlorides are passed over polymer-bound butane thiolate to give *S*-butyl thioesters (G.Cainelli *et al.*, *Gazz. Chim. Ital.*, 1983, **113**, 523).

(3) Mercaptals or thioacetals may be prepared in high-yielding, Lewis-acid mediated reactions (V.Kumar and S.Dev, *Tetrahedron Letters*, 1983, **24**, 1289).

$$Me(CH_2)_5CHO + 2EtSH \xrightarrow{TiCl_4, -10^\circ C} Me(CH_2)_5CH(SEt)_2$$

A mild condition for such preparations utilizes a Pt(II) complex as catalyst and 1,1-dihalides as carbonyl equivalents (P.Page *et al.*, *Tetrahedron Lett.*, 1988, **29**, 4477).

The hemithioacetal shown below can be generated by the reaction between *t*-butyl lithium and *O*-ethylthioformate (E.Vedejis *et al.*, *J. Amer. Chem. Soc.*, 1986, **108**, 2985.). This product undergoes thermal decomposition to the first aliphatic thioaldehyde observed under normal laboratory conditions.

$$Me_3CCH(SH)OEt \xrightarrow{xylene, reflux} Me_3CCHS$$

(4) Reactions of thioacetals. Carbanions may be generated α- to thioketals as well as to thioacetals using tributylstannyllithium as base in aldol reactions or alkylations (T.Takeda *et al.*, *Chem. Lett.*, 1985, 1149).

$$RCR^1(SPh)_2 + R^2CHO \xrightarrow{Bu_3SnLi} R^2CH(OH)CRR^1(SPh)$$

The use of thioacetals as carbanion sources in synthesis is sometimes complicated by difficulties in deprotection of the aldehyde function. This problem is eased by the use of dimethyl dithio-*S,S*-dioxide acetals as protective groups which are more readily removed by hydrolysis (K.Ogura *et al., Chem. Lett.*, 1982, 813).

(5) In the presence of $Me_2(MeS)S^+BF_4^-$ thioacetals act as anion acceptors in aldol-type processes (T.Sato and B,Trost, *J. Amer. Chem. Soc.*, 1985, **107**, 719).

$$RCH(SR^1)_2 + CH_2{:}CHCH_2SnR_3 \longrightarrow RCH(SR^1)CH_2CH{:}CH_2$$

2. Sulphides or thioethers

Dimethyl sulphide is used extensively in conjunction with boranes (H.C.Browm and S.U.Kulkarni, *J. Org. Chem.,* 1979, **44**, 2417; 2422) for the reduction of many functional groups including esters and nitriles (to aldehydes), carboxylic acids (to alcohols), alkenes in hydroboration reactions and aldehydes in reductive formylation of amines. The reaction of carbenes having a neighbouring thioether group have been reviewed (G.K.Taylor, *Tetrahedron*, 1982, **38**, 2751). A review with 121 references has appeared covering the reactions of acrolein with alcohols, mercaptans and phenols (R.C.Morris, Acrolein, ed. C.W.Smith, Heuthig, Heidelberg, 1975). Sulphur-containing crown ethers have been prepared using standard routes (see R.Okazaki *et al., Tetrahedron Lett.*, 1982, **23**, 4973).

(a) Dialkyl sulphides

Preparation

(1) Bis(tributyltin) sulphide is an effective and general sulphur transfer agent (D.N.Harpp *et al., Synthesis*, 1987, 1122).

$$2RX + (Bu_3Sn)_2S \longrightarrow RSR + 2Bu_3SnX$$

R= alkyl, allyl, acyl, formyl, benzyl, etc.,

(2) Trimethylchloro or bromosilane-dimethyl sulphoxide reagent systems react with alkenes (F.Bellesia *et al., J. Chem. Res. Synop.*, 1987, 238) giving 2-haloalkyl methyl sulphides.

$$RCH{:}CH_2 \xrightarrow{\text{TMSCl , DMSO}} RCHClCH_2SMe$$

(3) Thiosilanes may be utilized for the preparation of symmetrical or unsymmetrical dialkyl sulphides (W.Ando *et al.*, *Synth.Commun.*, 1982, **12**, 627).

$$R^1SSiMe_3 + R^2X \xrightarrow{NaOMe} R^1SR^2 + Me_3SiOMe + NaX$$

(4) Unsymmetrical sulphides may be conveniently prepared by treating 2-mercapto-1,3-benzoxazole sequentially with equivalents of different alkyl halides in the presence of sodium hydroxide (J.C.Sih and D.R.Graber, *J. Org. Chem.*, 1983, **48**, 3842).

$$\text{2-mercapto-1,3-benzoxazole} \xrightarrow[\text{(ii)}R^1X,NaOH]{\text{(i)}RX,NaOH} RSR^1 + \text{benzoxazol-2(3H)-one}$$

(5) Michael addition reactions lead to many γ-functionalized thioethers, e.g. (J.H.Liu *et.al.*, *Technol. Rep. Osaka. Univ.*, 1983, **33**, 1703; 445.)

$$RSH + CH_2{:}C(Me)COOR^1 \longrightarrow RSCH_2CH(Me)COOR^1$$

(6) Allylsilanes react with monothioacetals in the presence of Lewis-acid catalysts to give homoallyl sulphides (H.Nishiyama *et al.*, *J. Chem. Soc. Chem. Commun.*, 1982, 459).

$$\text{Allyl-SiR}_3 + CH_2(OR^1)SPh \longrightarrow \text{Allyl-CH}_2SPh + R^1OSiR_3$$

(7) Vinyl sulphides are efficiently reduced to alkyl sulphides by diimide (R.M.Moriarty *et al.*, *Synth. Commun.*, 1987, **17**, 703) and by combinations of silanes and Lewis-acids (T.Takeda *et al.*, *Chem. Lett.*, 1984, 1219)

(8) Much impoved yields (93-100%) over those previously published result for sulphide preparations from *S,S*-dialkyl dithiocarbonates conducted under phase-transfer catalysis conditions (I.Degani *et al.*, *Synthesis*, 1983, 630). R^1 and X represent a wide range of substrate and leaving group respectively.

$$(RS)_2CO + R^1X \xrightarrow{Bu_4NBr, aq\ KOH} RSR^1 + RSCOX$$

(9) A similar phase-transfer system is used to obtain unsymmetrical sulphides from thioiminium salts (P.Singh *et al., Indian J. Chem.*, 1983, Sect.B, **22**, 484, see also T.Takido and K.Itabashi, *Synthesis*, 1987, 817).

$$MeC(SR){:}NH.HX + R^1X \xrightarrow[\text{transfer}]{\text{phase}} RSR^1 + MeC(X){:}NH.HX$$

(10) *O*-Allyl *S*-alkyl dithiocarbonates are converted to alkyl allyl sulphides by treatment with catalytic amounts of $Pd(PPh_3)_4$ (P.R.Auburn, *J. Chem. Soc. Chem. Commun.,* 1986, 146).

$$CH_2{:}CHCH_2OCS_2Me \xrightarrow{CHCl_3, 25^\circ C} CH_2{:}CHCH_2SMe + COS$$

Chirality is maintained at the allylic carbon in appropriate cases.

Reactions

(1) Alkyl phenyl sulphides are transformed into methyl carboxylates by the action of sulphuryl chloride followed by aqueous methanol (C.C.Fortes *et al., J. Chem. Soc. Chem. Commun.*, 1982, 857).

$$R(CH_2)_7SPh \longrightarrow R(CH_2)_5CH{:}C(Cl)SPh \longrightarrow R(CH_2)_6COOMe$$

(2) With chloramine T, under phase-transfer catalysis conditions, alkyl phenyl sulphides are converted into *N*-tosylsulphilimines (T.Yamamoto and D.Yoshida, *Org. Prep. Proced. Int.*, 1988, **20**, 271).

$$PhSCH_2CH_2R \xrightarrow[Bu_4NCl]{\text{chloramine T}} PhS(CH_2CH_2R){:}NTos$$

(3) Sulphides possessing α-H atoms react with phenyl azide possibly *via* the corresponding *N*-phenylsulphimides to give 2-substituted anilines (L.Benati *et al. J. Chem. Soc. Chem. Commun.*, 1982, 763.).

$$PhN_3 + Me_2S \xrightarrow{155^\circ C} o\text{-}C_6H_4(SMe)NH_2$$

(b) Alkenyl Sulphides

A review containing 305 references concerning the synthesis and properties of divinyl sulphide has appeared (A. Trofimov and S. V. Amosova, *Sulfur Rep.,* 1984, **39**, 323). Crown ethers containing (2)-dithioethene units have been synthesised (W. Schroth *et al., Z. Chem.,* 1985, **25**, 138).

Preparation

Vinyl sulphides are prepared (1) by the elimination of RSH from dithioacetals where R^1= Ph, Pr, Me, $CH{:}CH_2$, $C(Me){:}CH_2$ and R^2= H, Me (T.Oida *et al., Bull. Chem. Soc. Jpn.*, 1983, **56**, 959).

$$(RS)_2CHCHR^1R \xrightarrow{Cu(II),R_3N} RSCH{:}CR^1R$$

P_2I_4 Also catalyses the reaction and also that of orthothioesters which gives ketene thioacetals $MeCH{:}C(SMe)_2$ (J.N.Denis and A.Krief, *Tetrahedron Lett.*, 1982, **23**, 3407), and (2), by alkylation of trimethylsiloxy thiols (D.Harpp *et al., Tetrahedron Lett.*, 1985, **26**, 1795).

$$RR^1CH{:}C(SH)OSiMe_3 + R^2X \longrightarrow RR^1C{:}CHSR^2$$

(3) Unsymmetrical divinyl sulphides are derived from the readily-available (see above) divinyl sulphide (V.A.Potapov *et al., Sulfur Lett.*, 1985, **3**, 151).

$$(CH_2{:}CH)_2S \xrightarrow[\text{(ii) PhC}\equiv\text{CH}]{\text{(i) Na,NH}_3} (Z)\text{-}CH_2{:}CHSCH{:}CHPh$$

(4) *(Z,Z)*-Bis 3-acrylate sulphides are produced (B.A.Trofimov and A.N.Valivola, *Sulfur Lett.*, 1985, **3**, 189) as follows:

$$HC{\equiv}CCOOH + H_2S + NH_3 \longrightarrow S(CH{:}CHCOO^-)_2.2NH_4^+$$

(5) Sequential treatment of $(Me_2CHCHMe)_2BCI{:}CHR$ with R^1SMgBr, BuLi, and aq.NaOH in HMPT gives good to very good yields of (E)-vinyl sulphides, $R^1SCH{:}CHR$ (M.Hoshi *et al., J. Chem. Soc. Chem. Commun.*, 1985, 1068).

(6) *S,S*-Dialkyl-dithiocarbonates may be used as a source of thiolate ion in Michael addition reactions (S.Cadamuro *et al., Synthesis,* 1986, 1070). Michael adducts, e.g. β-sulphenylacrylic acids, are also described (K.Miyamoto and Y.Inoue, *Chem. Abs.*, 1977, **88**, 169966x).

$$(R_2S)_2CO + PhC{\equiv}COOH \longrightarrow Ph(RS)C{:}CHCOOH$$

(7) Benzene thiol shows a different reactivity from that of its anion in additions to conjugated allenic ketones (T.Sugita *et al., J. Org. Chem.*, 1987, **52**, 3789).

$$MeCH{:}C{:}CHCOEt \xrightarrow{PhS^-} MeCH{:}C(SPh)CH_2COEt$$

$$MeCH{:}C{:}CHCOEt \xrightarrow{PhSH} MeCH_2C(SPh){:}CHCOEt$$

(8) Dicyanoacetic ester alkali-metal salts give acrylates on reaction with thiols (R.Neidlein and D.Kikelj, *Chem. Ber.*, 1988, **121**, 1817).

$$MC(CN)_2COOR + R^1SH \longrightarrow (E)\text{-}H_2NC(SR^1){:}C(CN)COSR^1$$

(9) (Z)- or (E)-1-Bromo-2-phenylthioethenes cross-couple stereospecifically in the presence of transition metal catalysts with Grignard reagents (V.Fiandanese *et al., J. Organomet. Chem.*, 1986, **312**, 343).

$$(Z)\text{ or }(E)\text{-}PhSCH{:}CHBr + RMgBr \longrightarrow (Z)\text{ or }(E)\text{-}PhSCH{:}CHR$$

Reactions

(1) Alkenyl sulphides undergo $TiCl_4$ catalysed aldol-type reactions with silyl enol ethers (T.Takeda, *Tetrahedron Lett.*, 1986, **27**, 3029).

$$R^1R^2{:}CRSPh + R^4CH{:}CR^3OTMS \longrightarrow R^3COCHR^4CR(SPh)CHR^1R^2$$

(c) Alkynyl Sulphides

Preparation

(1) Mono-and bis (alkylthio) butenynes are formed from the reaction of hexachlorobutadiene with thiols (C.Ibis, *Liebigs Ann. Chem.*, 1987, 1009).

$$(CCl{:}CCl_2)_2 + RSH \longrightarrow CCl_2{:}CClC{\equiv}CSR + ClC(SR){:}CClC{\equiv}CSR$$

(2) Successive dehydrochlorination and dehydrochorination of 2,2-dichlorothioacetals with rearrangement gives good yields of alkynyl disulphides (R.Nagashima *et al., Chem. Pharm. Bull.*, 1983, **31**, 3306).

$$RSCH(SR^1)CCl_3 \longrightarrow RSC(SR^1){:}CCl_2 \xrightarrow{BuLi} RSC{\equiv}CSR^1$$

3. Derivatives of alkane sulphenic acids

Sulphenic acids and their derivatives are the subject of a review by D.R.Hogg in Comprehensive Organic Chemistry, Vol. 3, 1979, 261, containing 179 references. The reduction methods for sulphenic and sulphinic acids have been surveyed (S.Oae and H.Togo, *Kagaku (Kyoto).*, 1983, **38**, 201). 39 References are cited.

Preparation

(1) Sulphenic acids $CH_2{:}CH(CH_2)_nSOH$, where n=2-5, are generated by thermolysis of their corresponding *t*-butyl sulphoxides (D.N.Jones *et al., J. Chem. Soc., Perkin Trans. 1,* 1977, 1574). These rearrange under the reaction conditions to give cyclic sulphoxides. Me_3CSOH is produced similarly and is trapped by reaction with terminal alkenes as a sulphoxide.

$$^tBu_2SO \xrightarrow{140^\circ C} {}^tBuSOH \xrightarrow{RCH:CH_2} RCH_2CH_2S(O)^tBu$$

A series of sulphoxides, $RS(O)CMe_3$, on vacuum pyrolysis onto a cold

(-196°C) finger are converted to their sulphenic acids. On thawing these decompose as shown or may be intercepted as their addition products (F.A.Davis *et al., Tetrahedron Lett.*, 1978, 97).

$$RSOH \longrightarrow RS(O)SR + RSSR + RSO_2R$$

$$RSOH + CH{\equiv}CCOOMe \longrightarrow RS(O)CH{:}CHCOOMe$$

(2) Sulphenic esters, prepared by the reaction of sulphenyl chlorides with alcohols, are reported to undergo hydrolysis to the corresponding sulphenic acids by the action of water in THF at room temperature (F.Maurer *et al., Chem. Abs.*, 1985, **104**, 148309u).

(3) Esters may also be made *via* sulphenyl iodides (see K.Desai *et al., J. Inst. Chem.*, 1983, **55**, 85).

$$RSH + I_2 \longrightarrow RSI \xrightarrow{NaOR^1} RSOR^1$$

(4) Ketenes act as substrates for sulphenyl halides giving amides or carboxylic acids (N.Kondrashov *et al., Izv.Akad.Nauk SSSR,Ser.Khim.*, 1987, 2358).

$$FCCl_2CF_2SCl + CH_2{:}C{:}O \xrightarrow{H_2O} FCCl_2CF_2SCH_2COOH$$

$$FCCl_2CF_2SCl + CH_2{:}C{:}O \xrightarrow{NH_3} FCCl_2CF_2SCH_2CONH_2$$

(5) Trifluoromethanesulphenyl chloride reacts with orthoesters possessing α-hydrogen atoms (W.L.Mendelson *et al., J. Org. Chem.*, 1983, **48**, 298).

$$2\ CF_3SCl + RCH_2C(OR^1)_3 \longrightarrow (CF_3S)_2CRC(OR^1)_3$$

$$\longrightarrow (CF_3S)_2CHRCOOH$$

(6) Sulphenyl chlorides are produced by reaction of thiolacetates with sulphuryl chloride (S.Thea and G.Cevasco, *Tetrahedron Lett.*, 1988, **29**, 2865).

$$RSAc + SO_2Cl_2 \longrightarrow RSCl + AcCl + SO_2$$

(7) Sulphenyl chlorides can also be prepared from compounds possessing

acidic hydrogens. (H.Fah, *Chem. Abs.*, 1986, **105**, 152554h, D. Martinetz, Z. *Chem*, 1980, **20**,139).

$$R^1RNH + SCl_2 \longrightarrow R^1RNSCl + HCl$$

$$Me_2CHCN + SCl_2 \longrightarrow Me_2(CN)CSCl + HCl$$

R^1=alkyl, R=alkyl or alkoxycarbonyl

A review collecting bibliographic data for thiosulphenyl chlorides (chlorodisulphides,) RSSCl has appeared (E.K.Moltzen and A.Senning, *Sulfur Letts.*, 1986, **4**, 169).

4. Dialkyl disulphides or alkyldithioalkanes

Preparation

(1) Unsymmetrical disulphides are prepared by the action of phosphonium salts on symmetrical disulphides (M.Masui *et al., J. Chem. Soc. Chem. Commun.*, 1984, **66**, 843),
(2) by the displacement of the nitroso group from thionitrites (S.Oae *et al., J. Chem. Soc. Perkin Trans.1,* 1978, 913) and
(3) by displacement of sulphite from alkylsulphites (M.E.Alonso and H.Aragona, *Org. Synth.*, 1978, **58**, 147).

$$Ph_3P + (RS)_2 \longrightarrow Ph_3P^+SRX^- \xrightarrow{R^1SH} RSSR^1$$

$$RSH + N_2O_4 \longrightarrow RSNO \xrightarrow{R^1SH} RSSR^1$$

$$RSSO_3Na + R^1SNa \longrightarrow RSSR^1$$

(4) Treatment of thiols RSH with CCl_3SO_2Cl gives disulphides, $(RS)_2$ in very high yields (T.L.Ho, *Synth. Commun.*, 1977, **7**, 363).
(5) The oxidation of ethane thiol to diethyl sulphide is catalysed by Co(II) phthalocyanines (U.Huendorf *et al., Heterog. Catal.*, 1987, 6th, Pt.2, 73). More traditional routes which appear to have been improved include:
(6) the oxidation of thiols by halogens (M.Goodrow *et al., Tetrahedron Lett.,* 1982, **32**, 3231; J.Nakayama *et al., Sulphur Lett.*, 1982, **1**, 25 and J.Drabowicz and M.Mikolajczyk, *Synthesis*, 1980, 32),

(7) polysulphide mediated reactions of RX compounds (S.Chorbadzhiev, *Chem. Abs.*, 1983, **101**, 130174z; H.Yamaguchi *et al., Nippon Kagaku Kaishi*, 1982, 801; *Chem. Abs.*, **97**, 143983m; V.Ceausescu *et al., Chem. Abs.*, 1986, **106**, 175779v) and,

(8) a variation on the polysulphide reaction used above which employs Li_2S_2 (T.A.Hase and H Perakyla, *Synth. Commun.*, 1982, **12**, 947 and A.J.Gladysz *et al., J. Chem. Soc. Chem. Commun.*, 1978, 838).

Reactions

(1) The reduction of disulphides has been reviewed with 32 references [S.Oae and H.Togo, *Kagaku (Kyoto)*, 1983, **38**, 48; *Chem. Abs.*, **98**, 197543g]

(2) Disulphides add *anti* to alkene double bonds in reactions catalysed by BF_3 (M.Caserio *et al., J. Org. Chem.*, 1985, **50**, 4390), by Mn(III) or by Pb(IV) salts (A.Bewick *et al., J. Chem. Soc., Perkin Trans.1*, 1039), and by anodic oxidation in ethanonitrile to give acetamido sulphenylated alkanes.

$$RCH{=}CHR^1 + (R^2S)_2 \xrightarrow{BF_3} R^2SCHRCH(SR^2)R^1$$

$$n\text{-}C_6H_{13}CH{=}CH_2 + RSSR \xrightarrow[CF_3COOH]{Pb^{4+}, CH_2Cl_2} n\text{-}C_6H_{13}CH(OCOCF_3)CH_2SR$$

$$n\text{-}C_6H_{13}CH{=}CH_2 + RSSR \xrightarrow{[O], CH_3CN} n\text{-}C_6H_{13}CH(NHAc)CH_2SR$$

(3) Thiolate displacement reactions from disulphides are induced (a) by KCN and iodine (S.Kawamura, *Chem. Abs.*, 1983, **102**, 5680f),

$$RSSR + KCN \longrightarrow 2RSCN$$

$$RSSR + Me_2CNO_2^-\ Na^+ \longrightarrow Me_2CNO_2SR$$

$$RSSR + LiNR^1_2 \longrightarrow RSNR^1_2$$

(b) by carbanion species (W.Bowman *et al., Tetrahedron Lett.*, 1977, 4519), (c) in the formation of sulphenamides (H.Ikehira and S.Tanimoto, *Synthesis*, 1983, 716) and (d) in their reduction to their corresponding thiols by K(2-propoxy)$_3$BH (H.C.Brown *et al.*, *Synthesis,* 1984, 498).

5 *Polysulphides*

Preparation

(1) A general synthesis of trisulphides, RS_3R, where R= alkyl or aryl, utilizes the thiols, RH, and diimidazol-l-yl sulphide as reagent (A.Banerji and P.G.Kalena, *Tetrahedron Lett.*, 1980, **21**, 3003).

(2) Unsymmetrical alkyl allyl trisulphides are prepared in the following way (J.Kominato, *Chem. Abs.*, 1985, **105**, 114610p).

$$(CH_2{:}CHCH_2S)_2 \xrightarrow{MeCO_3H} CH_2{:}CHCH_2S(O)SCH_2CH{=}CH_2$$

$$\xrightarrow{EtSSH} CH_2{:}CHCH_2SSSEt$$

(3) The simultaneous passage of H_2S and ethyne through DMSO containing LiOH gives rise to to a mixture of Et_2S_2, Et_2S_4 and Et_2S_6 (B.A.Trofimov *et al., Zh. Org. Chim.*, 1985, **21**, 2209).

(4) Trisulphides of the type shown are made from the appropriate cyclic thiosulphonate. When the trisulphides are heated in water they are converted to the corresponding disulphides (J.D.Macke and L.Field, *J. Org. Chem.*, 1988, **53**, 6).

$$\underbrace{(CH_2)_nSO_2}_{S} + Na_2S \xrightarrow[n=3\text{-}5]{} S[S(CH_2)_nSO_2Na]_2$$

(5) A novel route to dialkylpolysulphides employs lithium alkyls (J.F.Boscato *et al., Tetrahedron Lett.*, 1980, **21**, 1519. If the carbanion is in large excess only the monothio ether is produced.

$$EtCHMeLi + S_8 \longrightarrow Me_2CHCH_2S_nCH_2CHMe_2 + Li_2S_x$$

Equimolecular quantities of reagents give a mixture where n=1-4.

(6) Regardless of reactant ratios, a range of alkyl thiols reduce SO_2 giving a 7:3 ratio of the corresponding dialkyl di-and trisulphides (F.Akiyama, *J. Chem. Soc. Perkin Trans.1*, 1978, 1046).

(7) A straightforward route to symmetrical tetrasulphides involves methoxycarbonylhydrazones of ketones or aldehydes (R.Ballini, *Synthesis,*

1982, 834).

$$RR^1{:}NNHCOOMe \xrightarrow{H_2S,MeOH,AcOH} (RR^1CH)_2S_4$$

The wide use of these polysulphides as lubricant additives is described together with a synthesis of trisulphides (Y.Koyama, H.Nakazawa, T.Yamashita and H.Moriyama, *Chem. Abs*., 1983, **100**, 67836d).

$$RSH \xrightarrow{MgO,S,30\text{-}100^{\circ}C} RS_3R$$

Mixtures of tri and tetrasulphides are produced from the reactions of thiols, nitrososulphides or disulphides with the oxidants $FeCl_3$ or $CuCl_2$ (Wako Pure Chemical Ind. Ltd., *Chem. Abs*., 1980, **94**, 15195g).

6. Alkylthiosulphuric acids

Preparation

(1) *S*-(3-Aminopropyl)thiosulphuric acid is formed, rather than the corresponding cyclic amine, by treatment of $H_2N(CH_2)_3Br.HBr$ with an aqueous solution of thiosulphate (E.Sanjust and A. Rinaldi, *Biol-Soc. Ital. Biol. Sper.*,1984, **60**, 245; *Chem. Abs* **101**, 23912s).

(2) Bunte salts (those of thiosulphuric acid) containing the amino-group are also prepared in good yield by reaction of amino-thiols with chlorosulphonic acid (T.Tanaka, H.Nakamur and Z.Tamura, *Chem. Pharm. Bull*., 1974, **22**, 2725).

$$HSCH_2N^+H_3.X^- + ClSO_3H \longrightarrow H_3N^+CH_2SSO_3H + HCl$$

Reaction

(1) The thiosulphate salt $CH_2{:}CH(CH_2)_3SSO_3^-$ affords 2-(ethoxymethyl) tetrahydrothiophene when treated with iodine in ethanol (R.Rieke, S.E.Bales and L.C.Roberts, *J. Chem. Soc. Chem. Commun.,* 1972, 974).

7. Sulphonium compounds

A number of reviews of sulphonium salts have appeared since the first supplement; these include: (i) G.C.Barrett, Comprehensive Organic Chemistry, 1979, Vol. 3, 105, Pergamon, Oxford, England, (ii) C.J.Stirling

(Ed.), The Chemistry of the Sulphonium Group Pt. 1, C.J.Stirling (Ed.), 1981, 385, John Wiley & Sons, Chichester, England. Increasing use of these salts has been made as phase transfer agents/reagents (see B.Badet, M.Julia and M.Ramirez-Munoz, *Synthesis*, 1980, 926; S.Kondo, Y.Takeda and K.Tsuda, *Synthesis*, 1988, 5403).

Preparation and reactions

(1) Trimethylsulphonium chloride is conveniently prepared as shown below (B.Byrne, L.Lafleur and M.Louise, *Tetrahedron Lett.*, 1986, **27**, 1233).

$$Me_2S + ClCOOMe \longrightarrow Me_3S^+Cl + CO_2$$

The preparation of the mono and dibromo analogues $Me_2S^+CH_2Br.Br^-$ and $Me_2S^+CHBr_2Br^-$ are also described (T.Laird and H Williams, *J. Chem. Soc. C,* 1971, 3471).

(2) α,ω-Hydroxysulphonium salts are prepared from their corresponding cyclic ethers (K.Okuma, S.Nakamura and H.Ohta, *Heterocycles,* 1987, **6**, 2343).

$$(CH_2)_nO + Me_2S \xrightarrow{CF_3SO_3H} Me_2S^+(CH_2)_nOH$$

(3) Allyl and benzyl sulphonium salts may be prepared from allyl silyl ethers as follows (E.Vedejs and J.Eustache, *J. Org. Chem.*, 1981, **6**, 3353).

$$R^1SR + CH_2{:}CHCH_2OTMS \xrightarrow{TMSSiOSO_2CF_3} CH_2{:}CHCH_2S^+R^1R$$

(4) A general method for the preparation of sulphonium salts is available starting from alkenes (H.Brosshard, *Helv. Chim. Acta.*, 1972, **55**, 37).

$$R_2S + CH_3CH{:}CHCH_3 \xrightarrow{H_2SO_4} R_2S^+CH(CH_3)CH_2CH_3$$

(5) The reaction between DMSO and epoxides results in the formation of vicinal hydroxysulphonium salts (M.A.Khuddus and D.Swern, *J. Amer. Chem. Soc.*, 1973, **95**, 8393).

$$RR^1COCR^2R^3 + Me_2SO \xrightarrow{H^+} RR^1C(OH)CR^2R^3(OS^+Me_2)$$

(6) Reactions of dimethylsulphonium methide with enol ethers of β-diketones give 2,4-disubstituted furans (C.M.Harris *et al., J.Org.Chem.,* 1974, **39**, 72).

$$PhCOC(R)=C(OMe)Ph + Me_2S^+\bar{C}H_2 \longrightarrow \longrightarrow$$ 2-Ph-3-R-4-Ph-furan (R, Ph, Ph, O)

(7) Dimethylbromosulphonium bromide is used as a mild reagent for the conversion of alcohols to bromoalkanes (N.Furukawa, *J. Chem. Soc. Chem. Commun.*, 1973, 212).

$$Me_2S^+Br.Br^- + ROH \longrightarrow Me_2SO + RBr + HBr$$

(8) The preparation of allenic sulphonium salts and their conversion to furans by a general reaction with β-keto esters is described (J.W.Batty, P.D.Howes and C.J.M.Stirling, *J. Chem. Soc. Perkin Trans.1,* 1973, 65).

$$Me_2S^+CH{=}C{=}CHR^1 + RCOCH_2COOEt \longrightarrow$$ furan (^{1}R, COOEt, O, R)

(9) Sulphonium salts are generated by Pummerer rearrangement of sulphoxides; they react with thiols to give thioacetals and with sulphides to give β-thiosulphonium salts (R.Tanikaga *et al. J. Chem. Soc. Chem. Commun.*, 1980, 41).

$$RS(O)CH_2R^1 \xrightarrow{TFA} RSC^+HR^1 \xrightarrow{R^2SH} RSCHR^1SR^2$$

$$RSC^+HR^1 \xrightarrow{R^2SR^3} RSCHR^1S^+R^2R^3$$

(10) $Et_3O^+BF_4^-$ may be used as alkylating agent to give sulphonium salts. These are converted by strong base into cyano-sulphonium methylides

(D.Jeckel and J.Gosselk, *Tetrahedron Lett.*, 1972, 2101).

$$RSCH_2CN \longrightarrow RS^+(Et)CHCN \xrightarrow[CHCl_3]{NaOH} R(Et)S{:}CHCN$$

(11) Vinyl sulphonium salts are used as precursors to cyclopropanes (K.Negoro, T.Agawa and T.Agawa, *J. Chem. Soc. Perkin Trans. 1*, 1979, 1490).

$$PhCH{:}CHS^+Me_2 + LiCHRR^1 \longrightarrow$$

Ph, R[1], R (cyclopropane)

(12) Stable sulphonium ylides are generated by reacting sulphoxides with acetylene dicarboxylates (T.J.Chow, U.K.Tan and S.M.Peng, *Synth. Commun.*, 1988, **18**, 519).

$$RO_2CC{\equiv}CCO_2R + R^1SO_2 \longrightarrow RO_2CC^-(S^+R^1)COCO_2R$$

(13) Silyl ylides are employed in the synthesis of aldehydes *via* the Pummerer reaction (P.J.Kocienski, *J. Chem. Soc. Chem. Commun.*, 1980, 1096).

$$R^1R^2C{:}CRCH_2S^+C^-HSiMe_3 \longrightarrow CH_2{:}CHRC(R^1R^2)(SMe)SiMe_3$$

$$\longrightarrow CH_2{:}CHRC(R^1R^2)CHO$$

(14) *S*-Allyl sulphonium compounds undergo proton abstraction to give cyclopropanes (Y.Yano *et al.*, *Phosphorus Sulphur*, 1976, **1**, 25) and butadienyl sulphonium salts undergo annelation reactions with enolates to give dihydroarenes (M.E.Garst, *J. Org. Chem.*, 1979, **44**, 1578).

$$R_2S^+CH_2CH{:}CHR^1 \xrightarrow{NaH,THF} R_2S + \underbrace{CH_2CH_2CHR^1}$$

$$CH_2{:}CHCH{:}CHS^{+}Me_2 + R^1CH{:}C(R)O^- \longrightarrow Me_2S + \text{(epoxide of cyclohexadiene bearing } R, R^1\text{)}$$

(15) Acetylenic or allenic sulphonium ylides are prepared by a reaction between phenyl propargyl or phenyl allenyl thioethers and carbenes. They undergo [2,3]-sigmatropic rearrangements to form respectively allenes and conjugated dienes (P.A.Grieco *et al., J. Org. Chem.* 1974, **39**, 199).

$$N_2C(COOMe)_2 \longrightarrow {:}C(COOMe)_2 + RC{\equiv}CCH_2SPh \longrightarrow$$

$$RC{\equiv}CCH_2S^{+}(Ph)C^{-}(COOMe)_2 \longrightarrow CH_2{:}C{:}C(R)C(SPh)(COOMe)_2$$

8. Dialkyl sulphoxides or alkylsulphinylalkanes

A review with 112 references on sulphoxides, but mainly concerned with dimethyl sulphoxide, has appeared in the Kirk-Othmer "Encycl. Chem. Technol"., 3rd edn., Vol. 22, pp. 64-67, eds. M.Grayson and D.Eckroth, Wiley, New York, 1983. In recent years much attention has been directed towards the enantiospecific preparation of sulphoxides from sulphides. The oxidising systems utilised include Ru(II) complexes dispersed on montmorillonite, various fungi, $NaIO_4$/(-)(*R*)-menthol, microbial and enzymatic systems, *Pseudomonas oleovorens,* chloroperoxidase, bovine serum albumin/$NaIO_4$, and *Mortierella isabellina* NRRL1757.

A review of the use of optically pure vinyl sulphoxides is devoted to their use in asymmetric synthesis (G.Posner *et al., Pure Appl. Chem.* 1981, **53**, 2307). Cyclic polysulphoxides have been prepared and used as phase transfer reagents for alkylation reactions (N.Furnkawa *et al., Chem. Lett.*, 1982, 1421).

A long series of papers by C.R.Johnson deals with the chemistry of sulphoxides (see C.R.Johnson and E.R.Janiga, *J. Amer. Chem. Soc.,* 1973, **95**, 7692).

Violent explosions have been experienced during the oxidation of tetrahydrothiophene to tetramethylene sulphoxide by 37% H_2O_2, a reaction previously having been conducted without incident over many years (H.C.Koppel, *Chem. Eng. News,* 1974, **52**, 3).

Preparations

(1) Methyltrifluoromethyl sulphoxide is readily prepared (S-L.Yu, D.T.Sauer and M.Shreeve, *Inorg. Chem.*, 1974, **13**, 484).

$$Hg(SCF_3)_2 + MeI \longrightarrow MeSCF_3 + Hg(SCF_3)I$$

(2) α-Phosphonylsulphoxides are prepared in good yield by by phosphite displacement of halide followed by oxidation of the phosphonylsulphide intermediate.

$$RSCH_2Cl + (R^1O)_3P \longrightarrow \longrightarrow RS(O)CH_2P(O)OR^1$$

(3) β-Enamino sulphoxides result from the reaction between sulphoxides and DMF in the presence of strong base,followed by treatment of the resulting metallate with amines (R.Kawecki and L.Kozerski, *Tetrahedron*, 1986, **42**, 1469).

$$Me_2SO + Me_2NCHO \xrightarrow{BuLi} MeS(O)CH{:}CHO^-Li^+$$

$$\xrightarrow[(2)\ HCl]{(1)\ Me_2NH} MeS(O)CH{:}CHNMe_2$$

(4) Dialkyl sulphoxides are prepared from alkyl aryl sulphoxides with retention of configuration (J.P.Lockard, C.W.Schroeck and C.R.Johnson, *Synthesis*, 1973, 485).

$$ArS(O)R + R^*Li \longrightarrow R^*S(O)R + ArLi$$

(5) Regiochemical control can be exercised in the formation of α-chloro dialkyl sulphoxides from sulphoxides (K.Ogura *et al.*, *Chem. Lett.*, 1980, 1587).

$$RR^1CClS(O)Me \xleftarrow{NCS} RR^1CHS(O)Me \xrightarrow[pyr]{NCS} RR^1CHS(O)CH_2Cl$$

Reactions

(1) The oxidation of sulphides to sulphoxides is often followed in synthetic sequences by their pyrolysis to the corresponding alkene (H.J.Reich and J.M.Renga, *J. Chem. Soc., Chem. Commun.*, 1974, 135).

$$PhCH_2SCH_2Br + RCOCH(Li)R^1 \longrightarrow RCOCHR^1CH_2SCH_2Ph$$

$$\xrightarrow{HIO_4} RCOCHR^1CH_2S(O)CH_2Ph \xrightarrow{\Delta} RCOCR^1{:}CH_2 + PhCH_2SOH$$

(2) β-Hydroxysulphoxides undergo pyrolysis to afford methyl ketones in high yields (J.Nokami, N.Kunieda and M.Kinoshita, *Tetrahedron Letts*, 1975, 2841).

$$PhS(O)CH_2CH(OH)R \xrightarrow{155^\circ C} CH_3COR + PhSOH$$

(3) β-Ketosulphoxides are widely used in synthesis as anion sources in alkylation and acylation reactions (see P.A.Bartlett, *J. Amer. Chem. Soc.*, 1976, **98**, 3305).

$$RCOCH_2S(O)Me \xrightarrow{base} RCOC^-HS(O)Me \xrightarrow{R^1X} RCOCHR^1S(O)Me$$

(4) Dialkyl sulphoxides react with diethylaminosulphur trifluoride to give α-fluoro thioethers (J.R.McCarthy *et al.*, *J. Amer. Chem. Soc.*, 1985, **107**, 735).

$$RCH_2S(O)R^1 \xrightarrow{Et_2NSF_3, ZnI_2} RCHFS(O)R^1$$

(5) Cyclic sulphinate esters are prepared from *tert*-butylhydroxy alkyl sulphoxides for the following cases where n = O-3 (N.K.Sharma, F.de Reinach-Hirtzbach and T.Durst, *Can. J. Chem.*, 1976, **54**, 3012).

$$Me_3CS(O)CH_2(CH_2)_nOH \xrightarrow{SO_2Cl_2} \underline{S(O)CH_2(CH_2)_nO}$$

(6) The deoxygenation of sulphoxides to their corresponding sulphides can be effected by a large number of reagents including $(EtO)_2P(O)SeH$, CF_3COOH/NaI, $(Me_3Si)_2S$, $NaIO_4$, molybdenum complexes, 2-phenoxy- or 2-chloro-1,3,2-benzodioxaphosphole, sodium bis(2-methoxyethoxy) aluminium hydride, I_2/pyridine/SO_2, Bu_3SnH, $NaCNBH_3$, $TiCl_3$, $HBCl_2$, B_2S_3, SiS_2, P_4S_{10}, $PhSeSiMe_3$, Me_3SiCl/Zn, thiophosphoryl bromide, PI_3, and thexylchloroborane/methyl sulphide complex.

(7) The reverse reaction has likewise been extensively studied and can be accomplished using, amongst other reagents, $Tl(NO_3)_3$, acyl nitrates, $SO_2Cl_2/H_2O/SiO_2$, Ce(IV) salts, PhIO and Bu_4NIO_4.

(8) The sulphoxide group can be employed in a direct synthesis of alkenes *via* lithium *N*-isopropylcyclohexylamide mediated alkylation-elimination reactions. This conversion may be achieved more directly as follows (B.Trost and A.J.Bridges, *J. Org. Chem.*, 1975, **40**, 2014):

$$PhS(O)CH_2R + PhCH_2Br \longrightarrow PhS(O)CH(R)CH_2Ph$$

$$\longrightarrow RCH{:}CHPh$$

Alternatively the dianion shown below may be generated (P.A.Grieco, D.Boxler and C.S.Pognonowski, *J. Chem. Soc. Chem. Commun.*, 1974, 497).

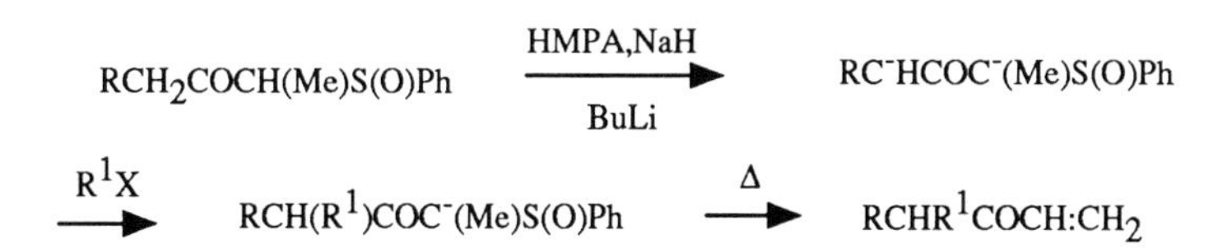

(9) Carbon-sulphur bond cleavage is observed in the reaction between sulphoxide and *N*-chlorosuccinimide in those cases where R is able to form a carbonium ion (F.Jung and T.Durst. *J.Chem. Soc. Chem. Commun.*, 1973, 4).

$$RS(O)R^1 \xrightarrow{NCS,EtOH} R^1S(O)OEt + RCl$$

9 Alkanesulphinic Acids

The formation and reactions of alkane sulphinic acids, their halides, anhydrides, esters, amides and thioesters have been reviewed (C.J.M.Stirling, *Int. J. Sulfur Chem.*, 1971, Part B, **6**, 277).

Sulphinic acids are reported as being unstable (but see below) and are usually prepared as their sodium salts.

Preparation

(1) A general method of preparation of sulphinic acids of the structure RCH_2SO_2H starts from the corresponding sulphone (P.Messinger and H Greve, *Liebigs Ann. Chem.*, 1977, 1457).

$$(RCH_2)_2SO_2 + NaCN \text{ or } NaSPh \longrightarrow RCH_2SO_2Na + RCH_2CN \text{ or } RCH_2SPh$$

(2) A second, similar, general method uses an ethoxide or an alkyl sulphide as nucleophile in a regiospecific cleavage of a precursor phthalimidomethyl alkyl sulphoxide (M.Uchino, K.Suzuki and M.Sekiya, *Synthesis*, 1977, 794).

$$\text{Phthm-}CH_2SO_2R \xrightarrow{EtO^-} \text{Phthm-}CH_2OEt + RSO_2Na$$

$$\text{Phthm-}CH_2SO_2R \xrightarrow{EtS^-} \text{Phthm-}CH_2SEt + RSO_2Na$$

(3) Stable sulphinic acids arise from an ene reaction between substituted allenes and SO_2 (G.Capozzi *et al.*, *Tetrahedron Lett.*, 1980, **34**, 3289).

$$Me_2C{:}C{:}CR_2 + SO_2 \longrightarrow H_2C{:}CMeC(SO_2H){:}CR_2$$

(4) Reaction of fluoro sulphonyl halides with hydrazine gives sulphinic salts which, on strong acid treatment, liberate the free acids (H.W.Roesk, *Angew Chem. Ind. Ed. Engl.*, 1971, **10,** 810).

$$CF_3SO_2F \xrightarrow{(i)NH_2NH_2,\ (ii)H_2SO_4} CF_3SO_2H$$

The corresponding chlorosulphonyl chloride is converted to the acid by the action of H_2S (U.Schöllkopf and P.Hilbert, *Liebigs Ann. Chem.*, 1973, 1061).

(5) High yields of ethyl, propyl and butyl sulphinic acids result from the oxidation of the corresponding thiols with 2-equivalents of *m*-chloroperoxybenzoic acid (W.G.Filby, K.Geunther and R.D.Penzhorn, *J.*

Org. Chem., 1973, **38**, 4020).

Reactions

(1) Sulphinic acids, as their sodium salts, may be esterified in very good yields using $R_3O^+BF_4^-$ (R=Me,Et) (P.K.Srivastava and L.Field, *Phosphorus Sulfur*, 1985, **25**, 161). Esterification of the acids by alcohols can be catalysed by 2-chloro-1-methyl- pyridinium iodide / Et_3N in CH_2Cl_2 or by $EtO_2CN{:}NCO_2Et$-Ph_3P in dry benzene. Sequential treatment of sulphinic acids by benzylamine then H^+, R^1OH gives the esters *via* the sulphinamides (M.Forukawa *et al.*, *Synthesis*, 1978, 441).

$$RSO_2H + PhCH_2NH_2 \longrightarrow RSONHCH_2Ph \xrightarrow{R^1OH} RSO_2R^1$$

(2) Dicyclohexylcarbodiimide is an effective dehydrating agent in the formation of sulphinamides using 2-chloro-1-methylpyridinium iodide as a coupling agent (M.Furukawa and T.Okawara, *Synthesis*, 1976, 339).

$$RSO_2H + R^1R^2NH \xrightarrow{\text{DCC,dioxane}} RSONR^1R^2$$

Sulphinamides are also prepared by the action of hydroxylamines, R^1R^2NOH, upon sulphinyl chlorides (K.Hovius and J.B.Engberts, *Tetrahedron Letts.*, 1972, 181).

(3) Benzyl-, phenyl- and adamantyl- sulphinyl cyanides are produced by the oxidation of the appropriate thiocyanates, and in the adamantyl case also by the action of NaCN on the corresponding sulphinyl chloride (A.Boerma-Markerink *et al.*, *Synth. Commun.*, 1975, **5**, 147).

$$RSCN \xrightarrow{\text{MCPBA}} RS(O)CN$$

(4) Sulphinic acids are transfomed into sulphonyl nitrates in their reaction with N_2O_4. These are brown, crystalline, unstable compounds (S.Oae, K.Shinhama and H.Yong, *Tetrahedron Lett.*, 1979, 3307).

$$RSO_2H + N_2O_4 \longrightarrow RS(O_2)ONO_2$$

(5) Thiosulphinyl chlorides are prepared from thiols by the action of

sulphuryl chloride (J-H.Youn and R.Herrmann, *Synthesis,* 1987, 72), or from their corresponding thiol acetates by the same reagent (S.Thea and G.Cevasco, *Tetrahedron Lett.,* 1987, **28**, 5193). They are converted into thiosulphinate esters with organotin thiolates (D.N.Harpp, T.Aida and T.H.Chan, *Tetrahedron Lett.*, 1983, **24**, 5173).

$$RS(O)Cl + R_3SnSR^1 \longrightarrow RS(O)SR^1$$

10 Dialkyl sulphones

Sulphones have assumed considerable importance in synthesis during the period covered by this supplement. This is reflected in the series of papers entitled "Organic Synthesis with Sulphones", the 48th of which appeared in 1988 (M.Julia et al., *Tetrahedron,* 1988, **44**, 111). The synthetic utility of sulphones depends, amongst other properties, upon their ready introduction into molecules (by sulphinate substitution, or by sulphide substitution, or through addition reactions, followed by oxidation of the sulphur atom), their ability to stabilize carbanions, to undergo Micheal addition, and rearrangement reactions. Finally, when necessary, they are easily removed.

Relevant reviews include the following: "Recent Developments in Sulphone Chemistry" P.Magnus, *Tetrahedron,* 1977, **33**, 2019, a review with 198 references; and "Sulphones", T.Durst, in Comprehensive Organic Chemistry, 1979, Vol.3, 171, ed. D.N.Jones, Pergamon, Oxford, with 85 references. These provide valuable surveys of the chemistry of the functional group. "Polyfunctional Sulphones" are also described (T.Durst, *ibid.*, Vol.3, 197).

Sulphonyl diazomethanes are recommended as stable and safe substitutes for diazomethane (Y.C.Kuo, T.Aoyama and T.Shioiri, *Chem. Pharm. Bull.* 1982, **30**, 526).

Preparation

(1) The major route to sulphones is through oxidation of sulphides and sulphoxides. Oxidants used for this purpose include: Ir, Ru and Rh complexes in the presence of oxygen , MCPBA, bromine water in the presence of base, $KMnO_4$, $NaBO_4$ and $HOC(CF_3)_2OOH$, and peroxytrifluoro acetic acid.

(2) Sulphinates readily rearrange into sulphones (J.B.Hendrickson and P.L.Skipper, *Tetrahedron,* 1976, **32**, 1627). This is the basis of many sulphone syntheses, the sulphinates often being prepared by substitution of sulphonyl halides by carbanions.

$$CF_3SOCl + ROH \longrightarrow ROS(O)CF_3 \longrightarrow RSO_2CF_3$$

(3) Allyl or vinyl sulphones are produced via iodosulphonylation reactions (T.Kobayashi *et al., Chem. Lett.*, 1987, 1209).

$$RSO_2I + CH_2{:}CHCH_2R^1 \longrightarrow RSO_2CH_2CHICH_2R^1$$

$$RSO_2CH_2CHICH_2R^1 \xrightarrow{DBU} RSO_2CH_2CH{:}CHR^1$$

$$RSO_2CH_2CHICH_2R^1 \longrightarrow RSO_2CH{:}CHCH_2R^1$$

(4) Such sulphones may also be prepared from allyl silanes, or vinyl silanes (J.Pillot, J.Dunogues and R.Calas, *Synthesis*, 1977, 469).

$$TMSCH_2CH{:}CH_2 + RSO_2Cl \longrightarrow CH_2{:}CHCH_2SO_2R$$

$$TMSCH{:}CH_2 + RSO_2Cl \longrightarrow TMSCHClCH_2SO_2R$$

$$\xrightarrow{-HX} TMSCH{:}CHSO_2R$$

(5) A general procedure for the synthesis of vinyl sulphones involves a Peterson olefination reaction (S.V.Ley and N.S. Simkins, *J. Chem. Soc. Chem. Commun.*, 1983, 1281).

$$TMSCH_2SO_2Ph \xrightarrow{(i)LDA,(ii)RCHO} RCH{:}CHSO_2Ph$$

(6) Sulphonomercuration of substituted alkynes gives access to vinyl and alkynyl sulphones (P.Rajakumar and A.Kannan, *J. Chem. Soc. Chem. Commun.*, 1989, 154).

$$R^1C\;CCOOMe \xrightarrow{HgCl_2,RSO_2Na} HgCR^1{:}C(SO_2R)COOMe$$

$$HgCR^1{:}C(SO_2R)COOMe \xrightarrow{NaOH} R^1CH{:}C(SO_2R)COOMe$$

$$HgCR^1{:}C(SO_2R)COOMe \xrightarrow{Br_2} R^1C{\equiv}CSO_2R$$

Both vinyl and ethynyl sulphones undergo Michael reactions to give alkyl and vinyl sulphones respectively.

(7) Primary, secondary and tertiary nitro compounds undergo Pd(0)-catalysed allylic substitution by RSO_2Na (R.Tamura *et al.*, *J. Org. Chem.*, 1986, **51**, 4375).

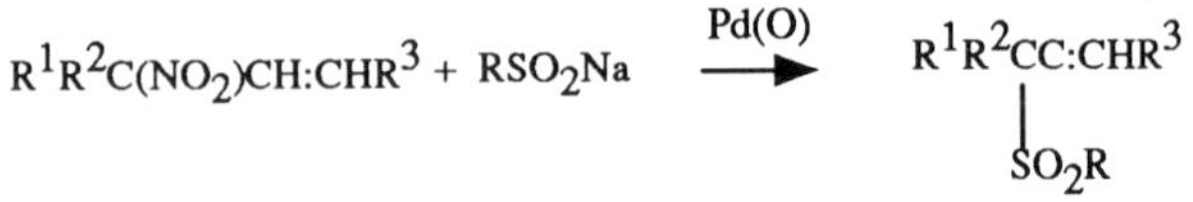

$$R^1R^2C(NO_2)CH{:}CHR^3 + RSO_2Na \xrightarrow{Pd(O)} R^1R^2C(SO_2R)C{:}CHR^3$$

(8) The potential of allyl sulphones as nucleophiles can be reversed by their ionization with Pd(0). Combining both possibilities, they become the functional equivalent of a 1,3-dipole e.g. $PhSO_2C^-HCH_2C^+HR$ (B.M.Trost, N.R.Schmuff and M.J.Miller, *J. Amer. Chem. Soc.*, 1980, **102,** 5979).

(9) γ,δ-Unsaturated sulphones are the product of Pd(II) catalysed addition of sulphinate to dienes (Y.Tamaru, *et al.*, *J. Org. Chem.*, 1983, **48**, 4669).

$$EtCH{:}CHCH{:}CH_2 \xrightarrow{(i)PdCl_2,(ii)RSO_2Na} EtCH{:}CH_2CH_2SO_2R$$

(10) The thermal sulphinate-sulphone rearrangement of allyl sulphinates may proceed with a 1,3-shift (K.Hiroi, R.Kitayama and S.Sato, *Chem. Pharm. Bull.*, 1984, **32**, 2628).

$$CH_3C_6H_4S(O)OCH_2CH{:}CRR^1 \longrightarrow CH_3C_6H_4SO_2CRR^1CH{:}CH_2$$

(11) Aryl-alkyl sulphones are the products of the Friedel-Crafts reaction between p-xylene and sulphonyl fluorides using $AlCl_3$ as catalyst (J.A.Hyatt and A.W.White, *Synthesis,* 1984, 214).

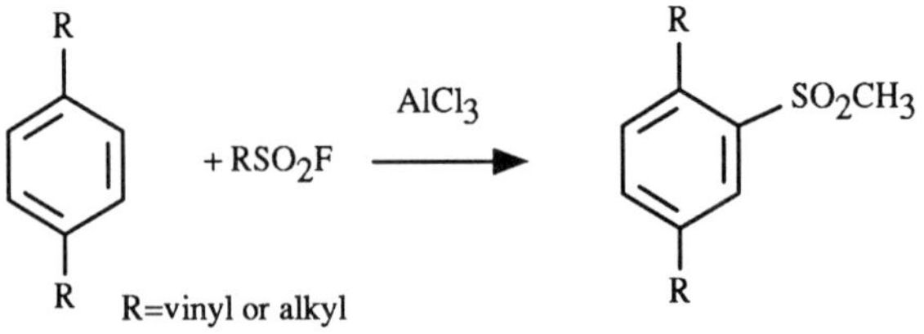

R
+ RSO_2F $\xrightarrow{AlCl_3}$ SO_2CH_3
R R=vinyl or alkyl R

(12) A variation on the use of substitution reactions for the preparation of alkynic sulphones is the addition-elimination process shown below (T.Miora and M.Kobayashi, *J. Chem. Soc. Chem. Commun.*, 1982, 438).

$$CH_3C_6H_4SO_2SePh + PhC{\equiv}CH \longrightarrow CH_3C_6H_4SO_2CH{:}C(Ph)SePh$$

$$\xrightarrow{H_2O_2} CH_3C_6H_4SO_2C{\equiv}CPh$$

(13) Sulphur dioxide gives sulpholenes as 1,4-cycloadducts in its reaction with butadienes (F.Borg-Visse, F.Dawans and E.Mareschal *Synthesis,* 1979, 817), and with 1,6-trienes gives 2,7-dihydrothiepin-1,1-dioxides (W.L.Mock and J.H.McCausland, *J. Org. Chem.,* 1976, **41**, 242).

$$RCH{:}CR^1CR^2{:}CR^3CR^4{:}CH_2 + SO_2 \longrightarrow RCHCR^1{:}CR^2CR^3{:}CR^4CH_2 \text{ (C-1 and C-6 bridged by } SO_2)$$

Reactions

(1) The sulphone functionality can be removed (i) oxidatively, to produce ketones in very good yields (D.R.Little and S.O.Myong, *Tetrahedron Lett.,* 1980, **21**, 3339; also see N.Ono, 1981 below):

$$RCH(SO_2Ph)R^1 \xrightarrow{(i)LDA,(ii)Mo(CO)_5} RCOR^1$$

(ii) by reductive desulphonylation (G.de Chirico *et al., J. Chem. Soc. Chem. Commun.,* 1981, 523); (iii) by hydrogenolysis of the sulphonyl group by treatment of the sulphone with a Grignard reagent in the presence of Ni or Pd catalysts (J.L.Fabre and M.Julia, *Tetrahedron Lett.,* 1983, **24**, 4311); (iv) by treatment with C_8K (D.Savoia, C.Trombini and A.Umani-Ronchi, *J. Chem. Soc. Perkin Trans* 1, 1977, 123); (v) by $Na_2S_2O_4$ and $NaHCO_3$ in water (M.Julia and J.P.Stacino, *Bull.Soc.Chim. Fr.,* 1985, 831); (vi) by nucleophilic displacement using *N*-benzyl-1,4-dihydronicotinamide as hydride source (N.Ono *et al., J. Chem. Soc. Chem.Commun.,* 1981, 71), with new alkene bond formation as in the reductive elimination reaction of β-nitrosulphones (N.Ono *et al., Tetrahedron Lett.,* 1978, 763); (vii) by the Ramberg-Bäcklund extrusion process (see later).

$$PhSO_2\,CH_2CRR^1CH{:}CHR \xrightarrow{NaHg} MeCRR^1CH{:}CHR$$

$$Me_2C(NO_2)CEt(CN)SO_2C_6H_4Me \xrightarrow[DMF]{Na_2S} Me_2C{:}C(Et)CN$$

(2) Vinyl and alkynyl sulphones undergo Michael addition or nucleophilic displacement reactions depending upon the nature of the reactants and the catalyst used (R.L.Smorada and W.E.Truce, *J. Org. Chem.*, 1979, **44**, 344; V.Fiadanese, G.Marchese and F.Naso, *Tetrahedron Lett.*, 1978, 5131; M.Julia *et al. Int. Congr. Sev. - Excerpta Med.*, 457 (Stereosel.Synth.Nat.Prod.,163); J.L.Fabre, M.Julia and J.N.Verpeaux, *Bull. Soc. Chim. Fr.*, 1985, 772).

$$R^1C{\equiv}CR \xleftarrow{R^1MgX} PhSO_2C{\equiv}CR \xrightarrow{R_2^1CuLi} PhSO_2\,CH{:}CRR^1$$

(3) Sulphone carbanions are generated by proton abstraction with *inter alia* BuLi or LDA. In the case of allyl sulphones, aqueous NaOH is a strong enough base (T.Radwan-Pytlewski and A.Jonczyk, *J. Org. Chem.*, 1983, **48**, 910) and the anion so produced acts as a carbon nucleophile in alkylation, Michael and aldol-type reactions. Vinyl sulphones appear to act as α-carbanions in reactions with aldehydes, the reaction proceeding by an addition-elimination process (P.Auvray, D.Knochel and J.F.Normant, *Tetrahedron,* 1988, **44**, 6095).

$$PhSO_2\,CH{:}CH_2 + RCHO \xrightarrow{DABCO} H_2C{:}C(SO_2Ph)CH(OH)R$$

(4) The selective reduction of alkenic bonds conjugated with the sulphonyl group is possible (M.Sekiya and K.Nanjo, *Chem. Pharm. Bull.*, 1979, **27**, 198) using the Et_3N-HCOOH azeotrope in DMF.

(5) Oxidative dimerization of allylic sulphones occurs mainly by 3,3′-coupling (G.Büchi and R.M.Freidinger, *Tetrahedron Lett.*, 1985, **26**, 5923).

$$R_2CH{:}CHCH_2\,SO_2Ph \xrightarrow[(ii)I_2]{(i)LDA} \begin{matrix} R_2CH{:}CH\ CH\ SO_2Ph \\ | \\ R_2CH{:}CH\ CH\ SO_2Ph \end{matrix}$$

$$R_2CH{:}CHCH_2\,SO_2Ph \xrightarrow[(ii)FeCl_3]{(i)LDA} \begin{matrix} R_2CHCH{=}CHSO_2Ph \\ | \\ R_2CHCH{=}CHSO_2Ph \end{matrix} \xrightarrow{Cope} \begin{matrix} R_2CH{:}CH\ CH\ SO_2Ph \\ | \\ R_2CH{:}CH\ CH\ SO_2Ph \end{matrix}$$

The product of the 3,3′-coupling undergoes the Cope rearrangement to give *threo*-vicinal disulphones which are also prepared by I_2 catalysed 1,1′-coupling of the starting allyl sulphone.

(6) The α-position of alkyl sulphones is deactivated towards chlorination by sulphuryl chloride. The reaction is said to proceed *via* hydride abstraction (I.Tabushi, Y.Tamaru and Z.Yoshida, *Tetrahedron,* 1974, **30**, 1457).

$$Et_2SO_2 \xrightarrow{SO_2Cl_2} CH_2ClCH_2SO_2Et \; (70\%) + (CH_2ClCH_2)_2SO_2 \; (25\%) + CH_3CHClSO_2Et \; (5\%)$$

(7) The Ramberg-Bäcklund SO_2 extrusion reaction was referred to earlier and constitutes one means of removing the sulphone functionality after the latter's use as an activating group. The basic systems used in the Ramberg-Backlund reaction include KOH in CCl_4 and NaOEt in ethanol.The synthesis of dienoic acids illustrates this (see P.A.Grieco and D.Boxler *Synth. Commun.,* 1975, **5**, 315).

$$t\text{-ButylSOCH}_2\text{COOMe} + RR^1C(OH)CH{:}CH_2 \xrightarrow{NCS}$$

$$CH_2{:}CHCRR^1SO_2CH_2COOMe \xrightarrow{\Delta} RR^1C{:}CHCH_2SO_2CH_2COOMe$$

$$\xrightarrow{KOH} RR^1C{:}CHCH{:}CHCOOH$$

Cyclic dienes are formed similarly from cyclic 2-carbethoxy sulphones, in this case by reaction with NaH in hexachloroethane (E.Vedejs and S.P.Singer, *J. Org. Chem.*, 1978, **43**, 4831).

$$\underbrace{CH_2SO_2\overset{COOEt}{\overset{|}{C}}HCH_2CH{:}CH(CH_2)_4}_{\text{ring}} \longrightarrow \underbrace{CH_2\overset{COOEt}{\overset{|}{C}}{:}CHCH{:}CH(CH_2)_4}_{\text{ring}}$$

(8) β-Iminosulphones are formed from the reaction between sulphonyl carbanions and nitriles (M.Muraoka *et al., J. Chem. Soc., Perkin Trans.*1, 1978, 1017).

$$LiCH_2SO_2Me + RCN \longrightarrow RC({:}NH)CH_2SO_2Me$$

(9) Allyl sulphones undergo stannolysis (Y.Ueno, S.Aoki and M.Okawara, *J. Chem. Soc. Chem. Commun.*, 1980, 683).

$$CH_2{:}CHCH_2SO_2C_6H_4Me \xrightarrow[BF_3OEt_2]{Bu_3SnH} CH_2{:}CHCH_2SnBu_3 + MeC_6H_4SO_2H$$

(10) Both allenic and alkynic sulphones react with 1,2-bis nucleophiles to give 1,3-heterocycles (M.Cinquini, F.Cozzi and M.P.Franco, *J. Chem. Soc., Perkin Trans. 1*, 1979, 1430).

$$MeC_6H_4SO_2CH{:}C{:}CH_2 \quad \text{or} \quad MeC_6H_4SO_2C{\equiv}CCH_3 + \text{(-)-ephedrine}$$

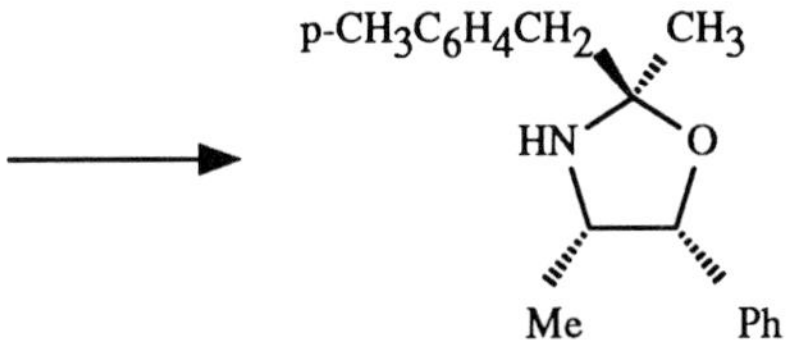

11 Alkanesulphonic Acids and Derivatives

Perfluorosulphonic esters (P.J.Stang, M.Hanack and L.R.Subramanian, *Synthesis,* 1982, 85) and the corresponding acids (R.A.Geunther, Kirk-Othmer Encycl. Chem. Technol., 1980, 3rd edn., Vol.10, 952) have been reviewed.

Preparations

(1) Sulphonic acids are prepared in high yields and purities by oxidation of thiols with HNO_3, and by the oxidation of thiols or disulphides with H_2O_2 using methane sulphonic acid as catalyst (G.Schreyer, F.Geiger and J.Hensel, *Chem.Abs.*,1977, **87**, 133888P).

(2) β-Chloro sulphonic esters and hence acids result from the photo-initiated reaction between methyl chlorosulphonate and 1-alkenes (M.S.Heller, D.P.Lorah and C.P.Cox, *J. Chem. Eng. Data.*, 1983, **28**, 134).

$$Me(CH_2)_3CH{:}CH_2 + MeOSO_2Cl \longrightarrow Me(CH_2)_3CHClCH_2SO_3Me$$

(3) γ-Aminosulphonic acids are the products of the ring-opening reaction of sultones by amines (I.Zeid and I.Ismail, *Justus Liebigs Ann. Chem.*,

1974, (41), 667), and α-aminosulphonic acids from the reactions between α-aminoacids and sulphur trioxide (S.Jinachitra and A.J.MacLeod, *Tetrahedron*, 1979, **35**, 1315).

$$RNH_2 + \underline{OCH_2CH_2CH_2SO_2} \longrightarrow RNHCH_2CH_2CH_2SO_3H$$

$$RCH(NH_2)COOH + SO_3 \longrightarrow RCH(NH_2)SO_3H$$

(4) α,β-Unsaturated sulphonic acids are prepared by the Wittig-Horner reaction (J.C.Carretero, M.Demillequand and L.Ghosez, *Tetrahedron,* 1987, **43**, 5125), by dehydrochlorination of the β-chloro acids above or, in a non-general reaction between alkenes and SO_3 where γ-sultones are an alternative product (M.D.Robbins and C.D.Broaddus, *J.Org.Chem.,* 1974, **39**, 2459).

$$(EtO)_2P(O)CH_2SO_3R + R^1COR^2 \longrightarrow R^1R^2C{:}CHSO_3R$$

$$CH_3CH_2C(CH_3){:}CH_2 + SO_3 \longrightarrow CH_3CH{:}C(CH_3)CH_2SO_3H$$

$$(CH_3)_2CHC(CH_3){=}CH_2 + SO_3 \longrightarrow (CH_3)_2\underline{CCH_2CH_2S(O)_2O}$$

Low temperature N.M.R. studies indicate that β-sultones are initially formed in all reactions of simple and perfluoro olefins with SO_3 (D.W.Roberts *et al., Tetrahedron Lett.,* 1987, **28**, 3383).

(5) Treatment of enolizable ketones with $ClSO_3SiMe_3$, followed by acid gives free β-keto sulphonic acids (K.Hoffmann and G.Simchen, *Synthesis,* 1979, 699).

$$RR^1CHCOR^2 + ClSO_3SiMe_3 \longrightarrow R^2COCRR^1SO_3SiMe_3$$

$$\longrightarrow R^2CORR^1SO_3H$$

Reactions

(1) Sulphonic acids may be esterified using methyl or ethyl orthoformates (A.A.Padmapriya, G.Just and N.G.Lewis, *Synth. Commun.,* 1985, **15**, 1057), and protected as a 2,2,2-trifluroethyl ester by 2,2,2-trifluoro diazoethane (C.O.Meese, *Synthesis,* 1984, 1041).

(2) The reaction between alkyl benzyl sulphides and chlorine in H_2O-HAc gives alkylsulphonyl chlorides (R.F.Langer, *Can. J. Chem.*, 1976, **54**, 498) and that between perfluoro iodides, SO_2 and Cl_2in contact with a Zn-Cu couple (H.Blancoll, P.Moreau and A.Commeyras, *J. Chem. Soc. Chem. Commun.*, 1976, 855) affords perfluorosulphonyl chlorides.

(3) A direct conversion of alkyl thiols to the corresponding sulphonyl fluorides is possible (C.Comninellis, P.Javet and E.Plattner, *Synthesis,* 1974, 887).

$$RSH + NO_2 + HF \longrightarrow RSO_2F$$

(4) Difluoromethane disulphonyl fluoride is the product of the electrochemical fluorination of methane disulphonyl fluoride (E.J.O'Sullivan *et al., J. Electrochem. Soc.*, 1985, **132**, 2424). Hydrolysis of this product by $Ba(OH)_2$ or LiOH yields the disulphonic acid.

$$CH_3SO_2F \longrightarrow CF_2(SO_2F)_2 \longrightarrow CF_2(SO_3H)_2$$

(5) 2-Hydroxyethane sulphonyl chloride is a relatively stable compound, the chemical properties of which have been surveyed (J.F.King and J.H.Hillhouse, *J. Chem. Soc., Chem. Commun.,* 1981, 295).

Reactions

(1) Aliphatic sulphonic acids, together with sulphinic acids, thiols, sulphonates and thiolsulphonates, are converted to their corresponding iodoalkanes by Ph_3P-I_2 (S.Oae and H.Togo, *Synthesis,* 1981, 371), whereas iodide-BF_3 reduces sulphonic acids and sulphonyl derivatives to disulphides (G.A.Olah *et al., J. Org. Chem.,* 1981, **46**, 2408). Sulphonate esters undergo reduction to alkanes when $NaBH_4$ is used as reductant under phase-transfer conditions in the presence of lipophilic quaternary ammonium or phosphonium salts (F.Rolla, *J. Org. Chem.,* 1981, **46**, 3909), or when sodium cyanoborohydride in hexamethylphosphoroamide is the reducing system employed (R.O.Hutchins *et al., J. Org. Chem.,* 1977, **42**, 82).

12 Alkane thiosulphonic acids

Preparation

(1) Alkane sulphonic esters are prepared by reaction of the appropriate sulphonyl bromides with thiols (M.G.Ranasinge and P.L.Fuchs, *Synth. Commun.,* 1988, **18,** 227), and

$$CH_3C_6H_4SO_2Br + RSH \longrightarrow CH_3C_6H_4SO_2SR$$

(2) by oxidation of thiosulphinates with sodium periodate (Y.H.Kim, T.Takata and S.Oae, *Tetrahedron Lett.*, 1978, 2305).

$$RS(O)SR^1 \xrightarrow{NaIO_4} RSO_2SR^1$$

Second Supplements to the 2nd Edition of Rodd's Chemistry of Carbon Compounds, Vol. IA/B, edited by M. Sainsbury
© 1991 Elsevier Science Publishers B.V., Amsterdam

*Chapter 7*a

ALIPHATIC ORGANOMETALLIC AND ORGANOMETALLOIDAL COMPOUNDS (DERIVED FROM METALS OF GROUPS IA, IIA/B, IIIA, IVA, VA, AND VIA

A.J.FLOYD

1. Introduction

The huge volume of work on organometallic chemistry published in the past 17 years, since the last supplement was written, bears witness to the continuing rapid growth in this area. In fact, two new specialist journals (*Organometallics and Appl. Organomet. Chem.*) were started in 1982 and 1987 respectively. Any treatment of such a vast output, must of necessity be *highly selective.* A major impetus, to the study of the organo-derivatives of main group elements and transition metals, has been the discovery of valuable reagents for use in organic synthesis. Indeed, during this period the synthetic organic chemist has been provided with an extensive array of organometallic reagents and catalysts, the reactions of which are characterised by excellent chemico-, regio-, and stereo-selectivity under generally mild conditions. It is this aspect of the chemistry of aliphatic organometallic and organometalloidal compounds, which will be stressed in this review.

a) *Bibliography*

References to the literature are given, as far as possible, in the section appropriate to the individual elements. However, many books, reviews, articles etc. are more general in scope, so are surveyed below.

i) General articles. Numerous books have appeared covering various aspects of organometallic chemistry, but pride of place must go to *Comprehensive Organometallic Chemistry*, G.Wilkinson, F.G.A.Stone, and E.W.Abel, Eds., Pergamon, Oxford, 1982. This 9 volume work does indeed constitute a comprehensive survey of the field upto the early 1980's. In the same series *Comprehensive Coordination Chemistry*, (G.Wilkinson, R.D.Gillard, and J.E.McCleverty, Eds. Pergamon, Oxford, 1987) and *Comprehensive Organic Chemistry* (Vol. 3, D.H.R.Barton and W.D.Ollis, Eds., Pergamon, Oxford, 1979) also contain much that is relevant. Further multi-volume works, which should be mentioned are Gmelin, *Handbook of Inorganic Chemistry*, 8th Edn., Springer-Verlag, Berlin, 1986; *Dictionary of Organometallic Chemistry,* 3 Volumes, J.Buckingham, Ed., Chapman and Hall, London, 1984; and *The Chemistry of the Metal-Carbon Bond,* 4 Volumes, F.R.Hartley, and S.Patai, Eds., Wiley, Chichester, 1982-1987.

This latter work deals with various topics in considerable detail, thus Vol.1: *The Structure, Preparation, Thermochemistry, and Characterization of Organometallic Compounds*; Vol. 2: *The Nature and Cleavage of Metal-Carbon Bonds; Vol.3: Carbon-Carbon Bond Formation using Organometallic Compounds*; Vol.4: *The Use of Organometallic Compounds in Organic Synthesis.*

Additionally many useful introductory texts have been written, including *Organometallics*, Ch.Elschenbroich and A.Salzer, VCH, Weinheim, 1989; *Principles of Organometallic Chemistry,* P.P.Powell, 2nd End., Chapman and Hall, London, 1988; *An Introduction to Organometallic Chemistry*, A.W.Parkins and R.C.Poller, Macmillan, London, 1986; *Basic Organometallic Chemistry*, I.Haiduc and J.J.Zuckerman, Walter de Gruyter, Berlin, 1985. The classic inorganic chemistry text (F.A.Cotton and G.Wilkinson, *Advanced Inorganic Chemistry*, 5th Edn., Wiley, New York, 1988) has recently appeared in a new edition, which of course gives wide ranging coverage of the chemistry of the organometals and metalloids.

The topic has been the subject of innumerable reviews during the past two decades. Annual surveys covering important developments are published regularly (e.g. *Specialist Periodical Reports*, The Royal Society of Chemistry, London, since 1971; *J. Organomet. Chem.*). There are a number continuing series which review recent developments of particular relevance to this chapter (see *J. Organomet. Chem. Library*, *Adv. Organomet. Chem.*, *Organomet. React.*, *Org. React.*, and *Top. Phosphorus Chem.*). The uses of organometallic reagents in organic synthesis have received regular attention (in, for example, *Aldrichimica Acta*).

ii) Methods of Preparation. Much useful information is contained in the general books and reviews already listed. Detailed practical and experimental information is provided in a number of texts (Houben-Weyl, *Methoden der Organischen Chemie,* G.Thieme, Stuttgart; G.Brauer, *Handbuch der Präparativen Anorganischen Chemie,* 3 Volumes, F.Enke, Stuttgart, 1975, 1978, 1981; *Experimental Organometallic Chemistry - A Practicum in Synthesis and Characterization,* A.L.Wayda and M.Y.Darensbourg, Eds., A.C.S. Sym. Series No. 357, Washington D.C., 1987; *Organometallic Synthesis*, 4 Volumes, J.J.Eisch and R.B.King, Eds., Academic Press, New York, 1965, 1981, Elsevier, Amsterdam, 1986, 1988; L.Brandsma and H.Verkruijsse, *Preparative Polar Organometallic Chemistry*, Spring, Berlin, 1988). There are a few continuing series (*Preparative Inorganic Reactions*, Wiley, New York; *Inorganic Syntheses*, Wiley, New York; *Inorganic Reactions and Methods,* J.J.Zuckerman, Ed., VCH, Weinheim), which contain material applicable to the preparation of organometallic derivatives. Special techniques have been employed to improve the preparation of organometallic compounds. Thus the use of metal atom and metal vapour in the synthesis of organometallic compounds has been discussed (P.L.Timms and T.W.Turney, *Adv. Organomet. Chem.*,

1977, **15**, 53; J.R.Blackborrow and D.Young, *Metal Vapour Synthesis in Organometallic Chemistry*, Springer-Verlag, Berlin, 1979). The preparation of highly reactive metal powders for use in organometallic synthesis has been reviewed (R.D.Rieke *et al.*, *ACS Symp. Ser.*, 1987, **333**, 223).

iii) Analysis. Spectroscopic techniques have made significant advances in recent years (R.S.Drago, *Physical Methods in Chemistry*, W.B.Saunders, Philadelphia, 1977; E.A.V.Ebsworth, D.W.H.Rankin, and S.Cradock, *Structural Methods in Inorganic Chemistry*, Blackwell, Oxford, 1987), indeed, since 1967, the progress has been reviewed annually (*S.P.R., Spectroscopic Properties of Inorganic and Organometallic Compounds,* Royal Society of Chemistry, London).

Infrared and Raman Spectra continue to be of great value in structure elucidation (K.Nakamoto, *Infrared and Raman Spectra of Inorganic and Coordination Compounds.*, 3rd Edn., Wiley, New York, 1978; M.Y.Darensbourg and D.J.Darensbourg, Infrared Determination of Stereochemistry in Metal Complexes, *J. Chem. Ed.*, 1970, **47**, 33 and 1974, **51**, 787; S.F.Kettle, The Vibrational Spectra of Metal Carbonyls, *Top. Curr. Chem.*, 1977, **71**, 111). *Mössbauer spectroscopy*, which is not so widely employed, can often provide useful structural information (see *Mössbauer Spectroscopy Applied to Inorganic Chemistry*, G.J.Long, Ed., Plenum, New York, 1984). The use of *electron spin resonance* has been well described (R.Kirmse and J.Stach, *ESR-Spektroskopie, Anwendungen in der Chemie,* Akademic Verlag, Berlin, 1985). *Photoelectron spectroscopy* has also found wide application (C.Cauletti and C.Furlani, *Comments Inorg. Chem.*, 1985, **5**, 29; J.C.Green, *Struct. Bonding*, 1981,**43** 37). *Nuclear magnetic resonance spectroscopy* has seen number of significant advances in recent years, which have had a profound influence on the study of organometallic compounds (see B.E.Mann, *Adv. Organomet. Chem.*, 1988, **28**, 397; W.von Philipsborn, *Pure Appl. Chem.*, 1986, **58**, 513; R.Benn and A.Rufinska, *Angew. Chem. Int. Ed. Engl.*, 1986, **25**, 861). High-field Fourier-transform multinuclear N.M.R. (R.Benn and H.Gunther, *ibid.*, 1983, **22**, 350; J.B.Lambert and F.G.Riddell, *The Multinuclear Approach to N.M.R. Spectroscopy*, Reidel, Boston, 1983), combined with new pulse sequences (C.Brevard and R.J.Schimpf, *J. Magn. Reson.*, 1982, **47**, 528), proton polarisation (C.Brevard, G.C.van Stein, and G.van Koten, *J. Amer. Chem. Soc.*, 1981, **103**, 6746), and two-dimensional techniques (C.Brevard, *et al., ibid.*, 1983, **105**, 7059; P.J.Domaille and W.H.Knoth, *Inorg. Chem.*, 1983, **22**, 818) have permitted the study of nuclei [e.g. ^{31}P, ^{13}C, ^{2}H, ^{103}Rh, ^{183}W, ^{29}Si, ^{119}Sn, ^{27}Al, ^{109}Ag, ^{51}V, ^{57}Fe] of importance to organometallic chemistry.

The development of *soft ionisation* methods in *mass spectrometry* (H.D.Beckey and H.R.Schulten, *Angew. Chem. Int. Ed. Eng.*, 1975, **14**, 403), especially F.A.B. (fast atom bombardment), have allowed the observation of large, fragile, non-volatile compounds (M.I.Bruce and M.J.Liddell, *Appl. Organomet. Chem.*, 1987, **1**, 191; J.M.Miller, *Adv. Inorg. Chem. Radiochem.*, 1984, **28**, 1); just the requirements for many organometallic compounds. The use of *X-rays* has been the method of choice for structure determination. Improved methods of recent years have resulted in >85% of all organometallic structures being determined in this way (M.I.Bruce, *Chem. in Britain,* 1984, 908). *Neutron diffraction* has increased in importance, especially for metal hydrides (R.G.Teller and R.Bau, *Struct. Bonding*, 1981, **44**, 1).

Chromatography, as a method of analysis and also for separation and purification, has contributed greatly to the advancement of knowledge in the organometallic field (J.C.McDonald, *Inorganic Chromatographic Analysis*, Wiley, New York, 1985). The techniques of *gas chromatography* (T.R.Crompton, *Gas Chromatography of Organometallic Compounds*, Plenum, New York, 1982) and *high performance liquid chromatography* (A.Casoli *et al.*, *Chem. Rev.*, 1989, **89**, 407) have been successfully applied to organometallic compounds.

iv) General properties and reactions. Structure and bonding relationships between the organo-derivatives of the main-group elements have been throughly reviewed (M.E.O'Neill and K.Wade, in *Comprehensive Organometallic Chemistry*, Vol. 1). The reaction mechanisms of organometallic compounds have been the subject of much study (e.g. D.S.Matteson, *Organometallic Reaction Mechanisms of the Nontransition Elements*, Academic, New York, 1974; J.K.Kochi, *Organometallic Mechanisms and Catalysis,* Academic, New York, 1978; J.D.Atwood, *Mechanisms of Inorganic and Organometallic Reactions,* Brooks/Cole, Monterey, 1985). The subject of π-bonding to main-group elements has been reviewed (P.Jutzi, *Adv. Organomet. Chem.*, 1986, **26**, 217). The topics of photochemistry (G.L.Geoffroy and M.S.Wrighton, *Organometallic Photochemistry*, Academic, New York, 1979 and *J. Chem. Ed.*, 1983, **60**, 861), electron-transfer reactions (N.G.Connelly and W.E.Geiger, *Adv. Organomet. Chem.*, 1984, **23**, 1; 1985, **24**, 87), high energy processes (K.S.Suslick, Ed., *A.C.S. Symp. Ser.*, 1987, **333**), and free radicals (M.F.Lappert and W.E.Lednor, *Adv. Organomet. Chem.*, 1976, **14**, 345; P.R.Jones, *ibid.*, 1977, **15,** 273; K.J.Klabunde, *Chemistry of Free Atoms and Particles,* Academic, New York 1980) in organometallic chemistry have received attention. Substantial benefits have been claimed for the effects of ultrasound in promoting reactions, including those of organic derivatives of mercury, copper, lithium, and zinc (R.F.Abdulla, *Aldrichimica Acta*, 1988, **21**, 31). The effects of organometallic compounds on the environment have been documented (J.S.Thayer, *Organometallic Compounds and Living Organisms*, Academic, New York, 1984; *Organometallic Compounds in the Environment, Principles and Reactions,* P.J.Craig, Ed., Longman, Harlow, 1986).

2. Group IA: Lithium, sodium, potassium, rubidium, caesium

The organic chemistries of the alkali metals have been subject to several reviews (e.g. J.L.Wardell, in *Comprehensive Organometallic Chemistry*, Vol.1; D.Seebach and K.-H.Geiss, in *New Applications of Organometallic Reagents in Organic Synthesis,* D.Seyferth, Ed., Elsevier, Amsterdam, 1976; M.Schlosser, *Polare Organometalle*, Springer Verlag, Berlin, 1973; Houben-Weyl, *Methoden der Organischen Chemie,* Vol. 13/1, 4th End., Thieme, Stuttgart, 1970). The use of lithium, sodium, and potassium derivatives in metalation reactions has been covered (J.Hartmann and M.Schlosser, *Helv. Chim. Acta*, 1976, **59**, 453; N.S.Narasimhan and R.S.Mali, *Synthesis,* 1983, 957). Ultrasound aids dispersion of sodium or potassium metal during metalation procedures (G.Gau and S.Marques, *J. Amer. Chem. Soc.*, 1976, **98**, 1538). Indeed, ultrasonics have found use in many reactions of alkali metal derivatives (J.L.Luche, *Ultrasonics*, 1987, **25**, 40; C.Einhorn, C.Allavena, and J.L.Luche, *J. Chem. Soc., Chem. Commun.*, 1988, 333; J.C.de Souza-Barboza, C.Petrier, and J.L.Luche, *J. Org. Chem.*, 1988, **53**, 1212). X-ray structural analysis data of Group IA organometallic derivatives has been published (W.N.Setzer and P.v.R.Schleyer, *Adv. Organomet. Chem.*, 1985, **24**, 353; C.Schade and P.v.R.Schleyer, *ibid.*, 1987, **27**, 169).

The applications of organo-derivatives of the alkali metals, to organic synthesis have continued apace (see J.L.Wardell, in *The Chemistry of the Metal-Carbon Bond*, Vol. 4, F.R.Hartley, Ed., Wiley, Chichester, 1987; B.J.Wakefield, in *Comprehensive Organometallic Chemistry,* Vol.7). The main uses have concerned organolithiums, which are discussed below.

a) Lithium

The chemistry of organolithium compounds has been the subject of a number of books (B.J.Wakefield, *Organolithium Methods*, Academic, London, 1988 and *The Chemistry of Organolithium Compounds*, Pergamon, Oxford, 1974 provide good examples). The preparation of lithium derivatives has been reviewed quite recently (W.F.Bailey and J.J.Patricia, *J. Organomet. Chem.*, 1988, **352**, 1; L.Loughmann and J.Trekovac, *Collect. Czech. Chem. Commun.*, 1988, **53**, 76). The use of arene radical ions, especially those based on aminonaphthalenes to aid their removal after reaction (K.S.Y.Lau and M.Schlosser, *J. Org. Chem.*, 1978, **43**, 1595), has proved useful for the reductive metalation of phenyl thioethers (Eqn. 1).

$$2\,ArH^{\cdot-}\,Li^{+} + PhSR \longrightarrow RLi + PhSLi + 2\,ArH \qquad (1)$$

This procedure allows the preparation of some organolithiums, otherwise difficult to obtain (T.Cohen and M.Bhupathy, *Acc. Chem. Res.*, 1989, **22**, 152). The application of modern N.M.R. spectroscopy to organolithium compounds has also been described (H.Günther *et al.*, *Angew. Chem. Int. Ed. Engl.*, 1987, **26**, 1212).

Organolithiums have found very wide application in organic synthesis (A.Krief, *Tetrahedron*, 1980, **36**, 2531; E.-i.Negishi, *Organometallics in Organic Synthesis*, Vol.1, Wiley, New York, 1980).

i) <u>Carbon-hydrogen bond formation</u>

Reaction with proton sources leads to the formation of C-H bonds, and such reactions can, of course, be used for the introduction of deuterium or tritium isotopes (B.J.Wakefield, in *Comprehensive Organometallic Chemistry*). Alkyl, benzyl, and aryl lithiums are all replaced with good regiospecficity (J.Barluenga, J.Florez, and M.Yus, *J. Chem. Soc., Perkin Trans. 1,* 1983, 3019; E.M.Kaiser, *Tetrahedron*, 1983, **39**, 2055). Alkenyllithiums lead to retention of configuration (L.Duhamel and J.M.Poirier, *J. Amer. Chem. Soc.*, 1977, **99**, 8356), often as the sole product (M.Duraisamy and H.M.Walborsky, *ibid.,* 1984, **106**, 5305(see Eqn. 2).

H Br 1) 2eq.^{t}BuLi -90°C/THF 2) MeOD H D (2)

ii) <u>Carbon-carbon bond formation</u>

a) *From organic halides.* This cross-coupling reaction (see Eqn. 3) has been reviewed (E.-i.Negishi, *Organometallics in Organic Synthesis*, Vol.1, Wiley, New York, 1980). The procedure has been developed for use in chiral syntheses and good enantimeric excesses have been achieved (A.I.Meyers, *Pure Appl. Chem.*, 1979, **51**, 1255; A.I.Meyers *et al., J. Amer. Chem. Soc.*, 1981, **103**, 3081; 3088)(see Eqn. 4).

$$RM + {}^{1}RX \rightarrow R^{1}R + MX \qquad (3)$$

H

H

Ph

O

+ H_2N Ph → N

OMe

OMe

1) LDA/THF/-20°C

O

R

2) RX

3) H_3O^+ 65% yield; 85%ee (4)

b) *Alkenes from carbonyl compounds*. Alkenes can be readily generated (see Eqn. 5); and the group X, which is eliminated can be: R_3Si [the *Peterson reaction* (D.J.Peterson, *J. Org. Chem.*, 1968, **33**, 780)]; RS, (O), $RS(O)_2$, or Se groups (A.Krief, *Tetrahedron*, 1980, **36**, 2531); NR_2 (D.J.Peterson, *Organomet. Chem. Rev.*, 1972, **7**, 295); or PR_2, $(R_2N)P(O)$, $(RO)_2P(O)$, or $(RO)_2P(S)$ [the *Wadsworth-Emmons reaction* (W.S.Wadsworth, *Org. React.* 1977, **25**, 73)]. There is a (*E*)-alkene selectivity, especially with aldehydes. Amides can replace aldehydes and ketones in a similar process (T.Agawa *et al.*, *Chem. Letters*. 1980, 335; *Bull. Chem. Soc. Jpn.*, 1982, **55**, 1205)(see Eqn. 6).

$$Y(X)CHLi + {}^1R,RCO \rightarrow Y(X)CHC^1R,ROLi \rightarrow YCH{=}C^1R,R \qquad (5)$$

$$Me_3SiCHLiSPh + HCONMe_2 \xrightarrow[2)\ H_2O]{1)\ THF/0^\circ C} PhSCH{=}CHNMe_2$$

65% yield (E)

(6)

c) *From alkenes and alkynes*. Lithium derivatives can be used to generate carbenes , which add to alkenes with retention of stereochemistry, yielding cyclopropanes (D.J.Nelson, *J. Org. Chem.*, 1984, **49**, 2059; R.A.Moss and R.C.Munjal, *Synthesis*, 1979, 425)(see Eqn. 7). Alkynes can give rise to cyclopropenes (R.A.Olofson, K.D.Lotts, and G.N.Barber, *Tetrahedron Lett.*, 1976, 3779). Krief (*Tetrahedron*, 1980, **36**, 2531) lists suitable carbene sources.

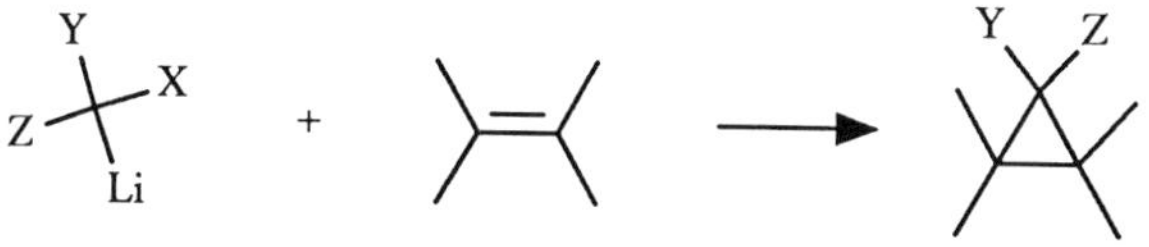

(7)

d) *Epoxides from carbonyl compounds.* Carbenes generated this way (A.Krief, *Tetrahedron*, 1980, **36**, 2531; E.-i.Negishi, *Organometallics in Organic Synthesis*, Vol.1, Wiley, New York, 1980) will also add to aldehydes and ketones yielding epoxides (see Eqn. 8). The group X eliminated, although frequently halogen, can be RS, RS(O), RSe, etc. Furthermore, aziridines (C.R.Johnson and G.F.Katekar, *J. Amer. Chem. Soc.*, 1970, **92**, 5733) and episulphides (A.I.Meyers and M.E.Ford, *J. Org. Chem.*, 1976, **41**, 1735) have been similarly obtained, by addition to imine and thiocarbonyl groups respectively (Eqn. 9).

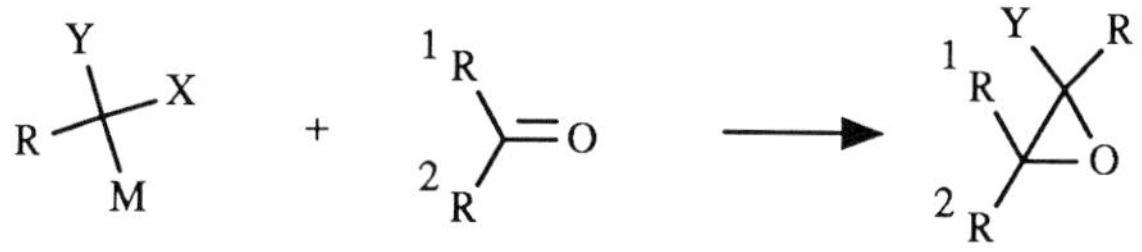

(8)

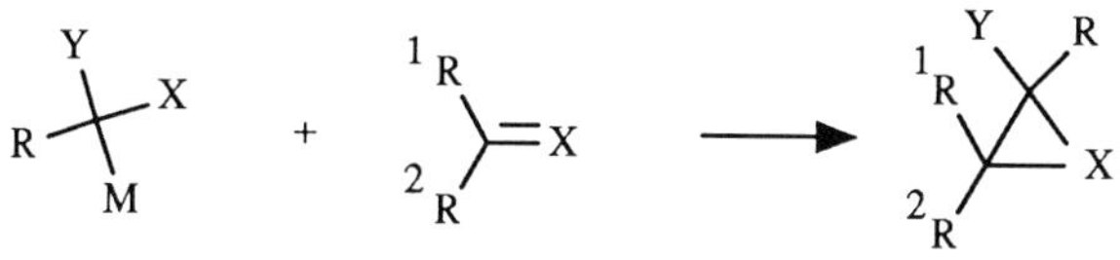

X=NH or S

(9)

e) *Aldehydes and ketones from carboxylic acid derivatives.* Carbene insertion in RCOX has been fully reviewed (A.Krief, *Tetrahedron*, 1980, **36**, 2531; E.-i.Negishi, *Organometallics in Organic Synthesis*, Vol.1, Wiley, New York, 1980; B.J.Wakefield, in *Comprehensive Organometallic Chemistry,* Vol.7).

f) *Aldehydes and ketones from acyl anion equivalents.* This has proved to be a powerful synthetic procedure, based on the use of acyl anion (1) equivalents, such as (2) (D.Seebach and E.J.Corey, *J. Org. Chem.*, 1975, **40**, 231) or (3) (B.T.Gröbel and D.Seebach, *Chem. Ber.*, 1977, **110**, 867), in which the usual electrophilic character of acyl carbon atoms has been reversed (umpolung). The reagent (3) is a masked acyl anion equivalent in the form of an enol. These types of reagent (see Eqn. 10) have become important for the synthesis of aldehydes and ketones (D.Seebach, *Angew. Chem. Int. Ed. Engl.*, 1979, **18**, 239). Enol equivalents can be generated *via* the Wadsworth-Emmons procedure (see Eqn. 11), resulting in a useful extension of the method (C.Earnshaw, C.J.Wallis, and S.Warren, *J. Chem. Soc., Chem. Commun.*, 1977, 314). Similar results can be obtained more directly (see Eqns. 12 and 13), using metal enolates (E.-i.Negishi, *Organometallics in Organic Synthesis*, Vol.1, Wiley, New York, 1980) or homoenolates (W.C.Still and T.L.MacDonald, *J. Amer. Chem. Soc.*, 1974, **96**, 5561).

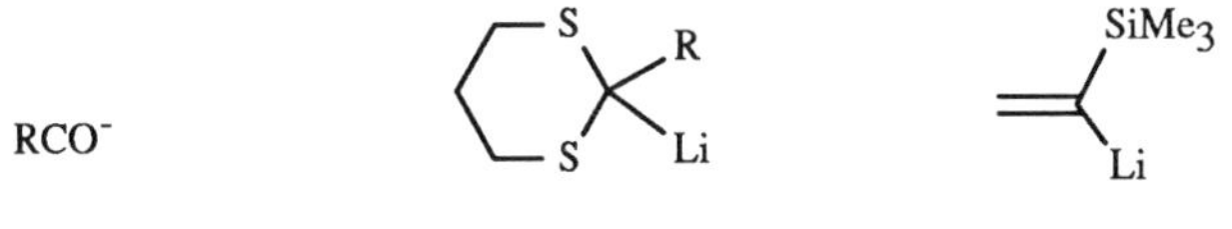

RCO^-

(1)

S S R Li

(2)

$SiMe_3$ Li

(3)

S S BuLi THF/-75°C S S H Li $PhCH_2CH_2Br$

S S H CH_2CH_2Ph NBS $MeCN/H_2O$ $PhCH_2CH_2CHO$

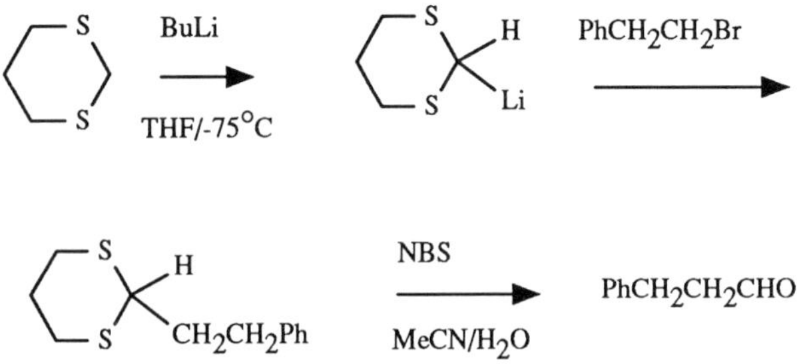

(10)

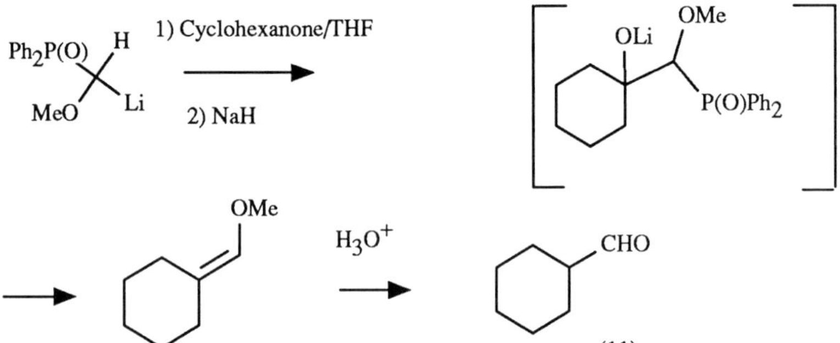

$Ph_2P(O)$ H MeO Li 1) Cyclohexanone/THF 2) NaH [OLi OMe $P(O)Ph_2$]

OMe H_3O^+ CHO

(11)

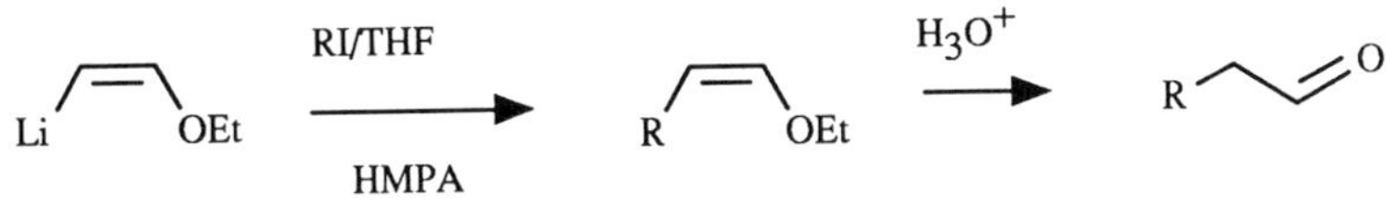

RI/THF, HMPA; H_3O^+

(12)

^{s}BuLi; MeI

H_3O+

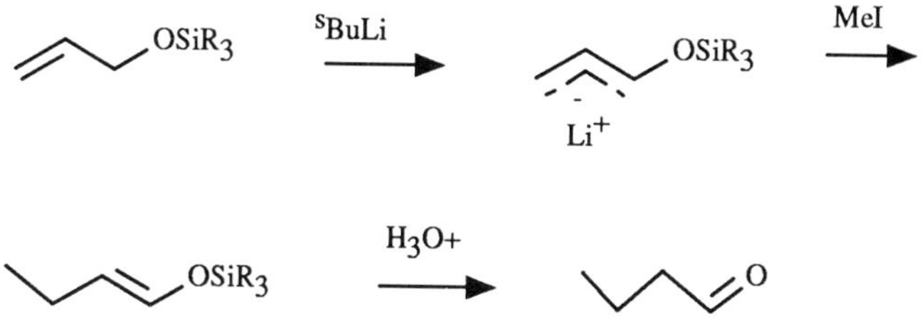

(13)

g) *Carboxylic acids via CO_2 insertion.* The insertion of carbon dioxide into carbon-lithium bonds is well covered in earlier reviews (B.J.Wakefield, in *Comprehensive Organometallic Chemistry,* Vol.7).

h) *Carboxylic acids from enolate anion equivalents.* The malonic ester synthesis can involve lithio-derivatives (E.-i.Negishi, *Organometallics in Organic Synthesis*, Vol.1, Wiley, New York, 1980). Oxazolines are amongst the most important substrates for carboxylic acid synthesis (see Eqn. 14), *via* lithiation (A.I.Meyers and E.D.Mihelich, *Angew. Chem. Int. Ed. Engl.*, 1976, **15**, 270).

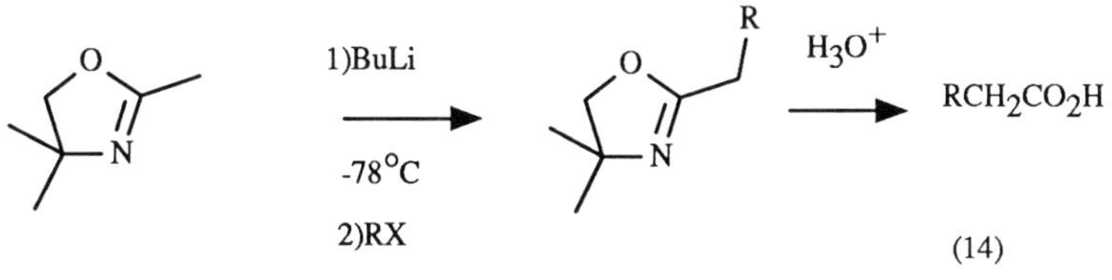

1)BuLi, -78°C, 2)RX; H_3O^+ → RCH_2CO_2H

(14)

i) Nitriles from cyanates, cyanogen, halides, etc. Alkyl lithiums will react with cyanates (R.E.Murray and G.Zweifel, *Synthesis*, 1980, 150,) or cyanogen halides (R.A. van der Welle and L.Brandsma, *Rec. Trav. Chim. Pays-Bas*, 1973, **92**, 667) to yield nitriles.

iii) Carbon-nitrogen bond formation

Acid azides (e.g. tosyl azide, 4-Me-C_6H_4-SO_2N_3) react with organolithium derivatives to yield organoazides, which can be conveniently reduced to primary amines (J.N.Reed and V.Snieckus, *Tetrahedron Letters*, 1983, **24**, 3795; P.Spagnolo, P.Zanirato, and S.Gronowitz, *J. Org. Chem.*, 1982, **47**, 3177).

iv) Carbon-sulphur bond formation

The cleavage of disulphides with organolithium derivatives leads to a convenient synthesis (see Eqn. 15) of mixed sulphides (*Chemistry of the Carbon-Metal bond,* Vol. 2, S.Patai and F.R.Hartley, Eds., Wiley, Chichester, 1985). The method has proved valuable for the preparation of vinyl sulphides (H.Neumann and D.Seebach, *Chem. Ber.*, 1978, **111**, 2785)(see Eqn. 16).

$$\mathrm{RSSR} \xrightarrow{{}^{t}\mathrm{BuLi}} \mathrm{RS^-} \xrightarrow{\mathrm{R^1X}} \mathrm{RSR^1} \qquad (15)$$

Br-alkene $\xrightarrow{1)\ ^{t}\mathrm{BuLi},\ -80^{\circ}\mathrm{C};\ 2)\ \mathrm{PhSSPh}}$ SPh-alkene, 93% yield (16)

3. Group IIA: beryllium, magnesium, calcium, strontium, barium

Group IIB: zinc, cadmium, mercury

The organic chemistry of the group II elements has been comprehensively surveyed in *Comprehensive Organometallic Chemistry* (G.Wilkinson, F.G.A.Stone, and E.W.Abel, Eds., Pergamon, Oxford, 1982). The relevant volumes are Vol.1 (N.A.Bell: Beryllium; W.E.Lindsell: Magnesium, Calcium, Strontium, and Barium) and Vol. 7 (B.J.Wakefield: Alkali and Alkaline Earth Metals; W.Carruthers: Zinc, Cadmium, Mercury, Copper, Silver and Gold). The application to organic synthesis of Group IIA derivatives have received detailed discussion (C.L.Raston and G Salem, *The Chemistry of the Metal-Carbon Bond*, Vol.4, F.R.Hartley, Ed., Wiley, Chichester, 1987). General coverage of the organo-derivatives of Group II have been provided in a number of reference books and review articles (G.E.Coates and K.Wade, *Organometallic Compounds*, Vol. 1 Part 1, 4th Edn., Groups I-III, Chapman and Hall, London, 1981; B.G.Gowenlock and W.E.Lindsell, Alkaline Earths, *J. Organomet. Chem. Libr.*, 1977, **3**, 1; M.A.Coles: Be, Ca, Sr, Ba, *MTP Int. Rev. Sci., Inorg. Chem. Sci., Inorg. Chem., Ser. Two,* 1975, **4**, 359; K.C.Bass: Zn, Cd, Hg, *ibid., Ser. One,* 1972, **4**, 41; L.G.Kuzmina, N.G.Bokii, and Yu.T.Struchkov: Zn, Cd, Hg, *Russ. Chem. Rev. Engl. Transl.*, 1975, **44**, 73; Houen-Weyl, *Methoden der Organischen Chemie, Vol.* 13/2a and 13/2b, *Metallorganische Verbindungen*: a) Be, Mg, Ca, Sr, Ba and b) Hg, Georg. Thieme, Stuttgart, 1973/1974). Gas chromatographic details have been reviewed in the Russian literature (V.A.Chernoplekova, V.M.Sakharov, and K.I.Sakodyneskii, *Russ. Chem. Rev. Engl. Transl.*, 1973, **42**, 1063).

a) Beryllium

The expense and toxicity of beryllium compounds has limited their use. The diorganoberyllium species tend to be air sensitive and spontaneously ignite to yield toxic beryllium oxide. They are usually prepared by metal exchange reactions from RMgX, RLi, R_2Hg or R_3B in ether solution (N.A.Bell, in *Comprehensive Organometallic Chemistry*, see above); and being good Lewis acids they form etherates. The lithium method (see Eqn. 17) has been used successfully for dialkynyl derivatives (G.E.Coates and B.R.Francis, J. *Chem. Soc., A*, 1971, 474). The borane method (see Eqn. 18) although slow, can be carried out at room temperature (*idem., ibid.*, 1971, 1308; G.Wiegand and K.H.Thiele, Z. *Anorg. Allg. Chem.*, 1974, **405**, 101). Acylberyllium bromides can be prepared from carboxylic acid bromides (B.G.Gowenlock and W.E.Lindsell, see above), which undergo some unusual reactions (e.g. I.I.Lapkin and T.N.Povarnitsyna, *Zh. Obshch. Khim.*, 1968, **38**, 99; I.I.Lapkin and G. Ya.Zinnatullina, *ibid.*, 1969, **39**, 1132; I.I.Lapkin, N.F.Tenenboim, and N.E.Eustafeeva, *ibid.*, 1971, **41**, 1554), including the conversion of arylaldehydes and arylnitro compounds to alkenes and amides respectively.

$$2\,RLi + BeX_2 \xrightarrow{Et_2O} R_2Be + 2\,LiX \qquad (17)$$

(R=alkyl or aryl)

$$R_3B + Et_2Be \longrightarrow R_2Be + Et_2RB \qquad (18)$$

b) Magnesium

The chemistry of organo-derivatives of magnesium is well covered in the reviews of Group IIA derivatives already quoted (see above), in addition mention should be made of an article in the MTP series (N.A.Bell, *MTP Int. Rev. Sci., Inorg. Chem. Ser. 2*, 1975, **4**, 381).

i) Organomagnesium halides

Grignard reagents still find extensive use in synthesis (W.E.Lindsell, *Comprehensive Organometallic Chemistry*, see above). Work continues to be published concerning their structures and mechanisms of reaction (T.Holm, *Acta Chem. Scand.*, 1983, **B37**, 567; M.Dagonneau, *Bull. Soc. Chim. Fr.*, 1982, 269; L.M.Lawrence and G.M.Whitesides, *J. Amer. Chem. Soc.*, 1980, **102,** 2493; E.C.Ashby and J.T.Laemmle, *Chem. Rev.*, 1975, **75**, 521; F.Bickelhaupt, *Angew. Chem. Int. Ed. Engl.*, 1974, **13**, 419; H.Normant, *Pure. Appl. Chem.*, 1972, **30**, 463).

Preparation. There have been some useful developments (W.E.Lindsell, see above):
a) The preparation of highly reactive magnesium metal powder (R.D.Rieke *et al., Org. Synth.*, 1980, **59**, 85; T.P.Burns and R.D.Rieke, *J. Org. Chem.*, 1983, **48**, 4141; R.D.Rieke *et al., A.C.S., Symp. Ser.*, 1987, **333,** 223; Y.H.Lai, *Synthesis,* 1981, 585) permits the use of low temperatures (D.Steinbom, *J. Organomet. Chem.*, 1979, **182**, 313) and thus the synthesis of thermally less stable Grignard reagents .

A reactive slurry can also be obtained, by condensing magnesium vapour in THF. This has been used to avoid rearrangement (E.P.Kundig and C.Perret, *Helv. Chim. Acta*, 1981, **64**, 2606) during the preparation of cyclopropyl -methylmagnesium bromide (see Eqn. 19). However, deliberate use of an intramolecular rearrangement in synthesis has been achieved *via* a *magnesium-ene* reaction (W.Oppololzer, R.Pitteloud, and H.F.Strauss, *J. Amer. Chem. Soc.*, 1982, **104**, 6476)(see Eqn. 20).

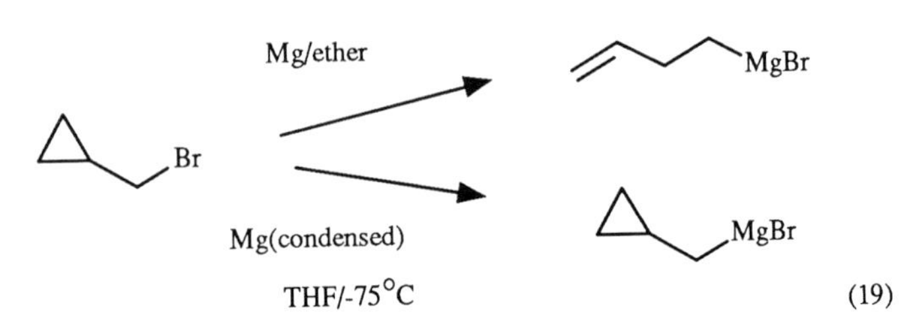

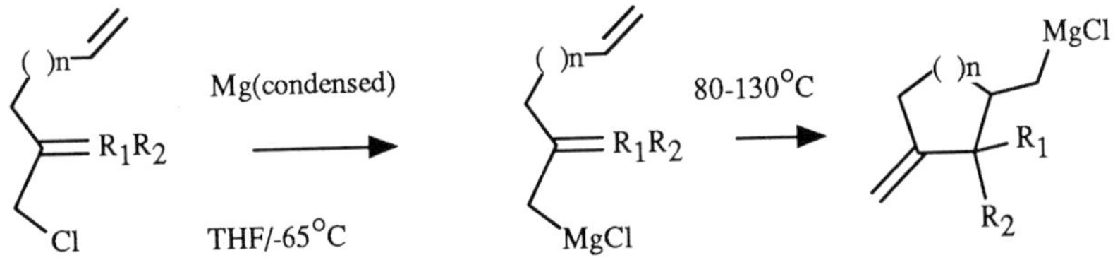

(20)

The use of magnesium-anthracene systems (B.Bogdanovic, *Acc. Chem. Res.*, 1988, **21**, 261; C.L.Raston and G.Salem, *J. Chem. Soc., Chem. Commun.*, 1984, 1702) are useful in reducing coupling, especially with benzylic Grignard reagents.

b) Another approach to the synthesis of unstable Grignard reagents involves their *in situ* generation (Barbier reaction) in the presence of a reactant (C.Blomberg and F.A.Hartog, *Synthesis,* 1977, 18); some related calcium systems are also discussed. Although not always successful, it has been used (A.Miodownik *et al., Synth. Commun.*, 1981, **11**, 241) to good effect in certain cases (see Eqn. 21).

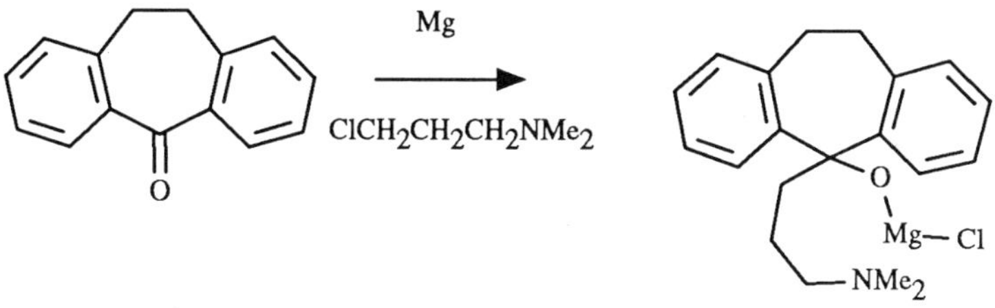

(21)

c) Additions to alkenes and alkynes can lead to useful new Grignard reagents (J.V.N.Vara Prasad and C.N.Pillai, *J. Organomet. Chem.*, 1983, **259**, 1). The processes (see Eqn. 22) can be regio- and stereo-selective (H.Lehmkuhl *et al., Chem. Ber.*, 1983, **116**, 2447; A.M.Moiseenkov, B.A.Czeskis, and A.V.Semenovsky, *J. Chem. Soc., Chem. Commun.*, 1982, 109).

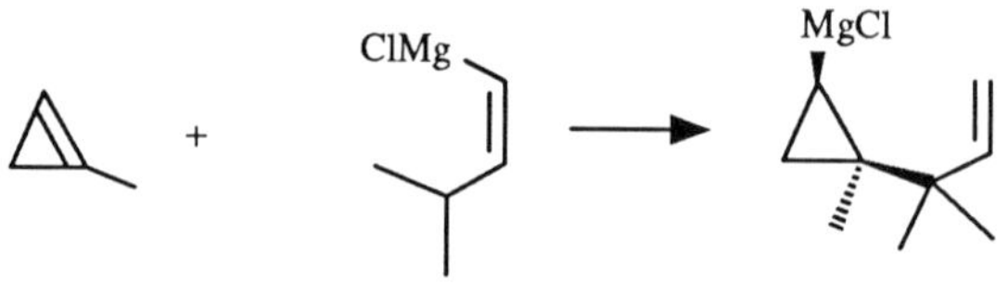

ClMg MgCl

(22)

Catalysis with $TiCl_2Cp_2$ allows the addition of Mg-H bonds to both alkenes and alkynes (F.Sato, H.Ishikawa, and M.Sato, *Tetrahedron Letters*, 1982, **22**, 85)(see Eqn. 23). This reaction proceeds *via* a titanocene hydride.

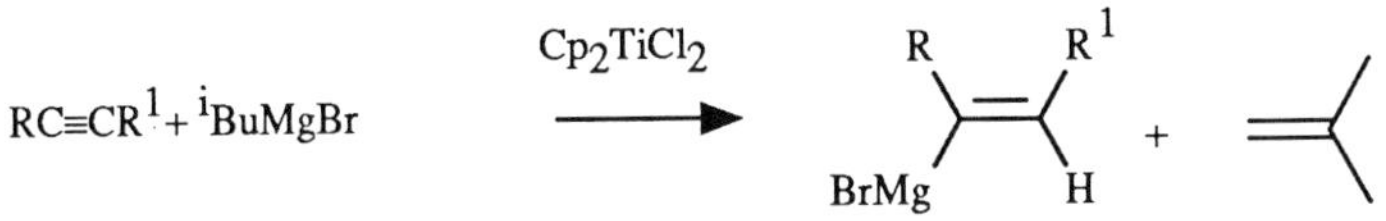

$$RC{\equiv}CR^1 + {}^iBuMgBr \xrightarrow{Cp_2TiCl_2} (BrMg)RC{=}CR^1(H) + (CH_3)_2C{=}CH_2$$

(23)

d) Salt elimination from organometallics, this method is usually applied to lithium derivatives, when magnesium exchange can improve yields or change products in further reactions (H.Hommes, H.D.Verkruijsse, and L.Brandsma, *Tetrahedron Letters*, 1981, **22**, 2495)(see Eqn. 24) .

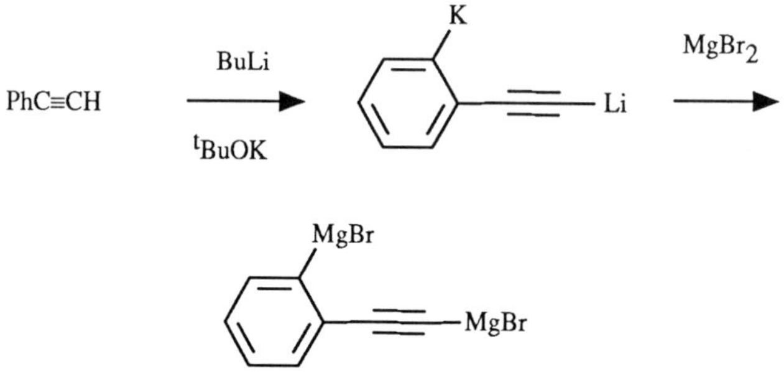

$$PhC{\equiv}CH \xrightarrow[{}^tBuOK]{BuLi} o\text{-}K\text{-}C_6H_4C{\equiv}CLi \xrightarrow{MgBr_2} o\text{-}BrMg\text{-}C_6H_4C{\equiv}CMgBr$$

(24)

e) Metal-halogen exchange can be useful (see Eqn. 25) when electronegative substituents are present (J.Villieras, *Organomet. Chem. Rev. A*, 1971, **7**, 81).

$$CHCl_3 + {}^iPrMgCl \xrightarrow[-80^{\circ}C]{THF/ether} Cl_2CHMgCl + {}^iPrCl$$

(25)

f) Metalation of acidic hydrogens (see Eqn. 26) is promoted by use of tetrahydrofuran (THF) and hexamethylphosphoramide (HMPA) as solvents. The reaction is commonly employed with terminal acetylenes (R.Rossi, A.Carpita, and M.G.Quirici, *Gazz. Chim. Ital.*, 1981, **111**, 173; P.Babin, J.Dunogues, and F.Duboudin, *J. Heterocycl. Chem.*, 1981, **18**, 519; H.J.Bestmann *et al.*, *Tetrahedron Letters*, 1982, **23**, 4007) or substituted toluenes (P.Canonne, G.B.Foscolos, and G.Lemay, *Tetrahedron Letters*, 1980, **21**, 155).

$$XH + RMgY \rightarrow XMgY + RH \qquad (26)$$

R is frequently ^{t}Bu

Applications to Organic Synthesis. For comprehensive surveys see reviews listed above (W.E.Lindsell; B.J.Wakefield,; W.Carruthers, in *Comprehensive Organometallic Chemistry;* C.L.Raston and G.Salem, in *Chemistry of the Metal-carbon Bond, Vol. 4;* K.Nutzel, in Houben-Weyl, *Methoden der Organischen Chemie,* Vol. 13/2a; J.V.N.Vara Parsad and C.N.Pilla, *J.Organomet. Chem.*, 1983, **259**, 1). The number and variety of uses are enormous, and only some important and useful applications will be highlighted here.

Grignard reagents

a) *Aldehydes and ketones.* The steric control of attack by Grignard reagents on prochiral carbonyl groups has received attention (E.C.Ashby and J.T.Laemmle, *Chem. Rev.*, 1975, **75**, 521; E.C.Ashby, J.Bowers, and R.Depriest, *Tetrahedron Letters*, 1980, **21**, 3541). An example (see Eqn. 27) involves the use of a chromium complex (J.M.LaMarche and B.Laude, *Bull. Soc. Chim. Fr., Part. II*, 1982, 97).

Ph MeMgBr Ph

$(OC)_3Cr$ O $(OC)_3Cr$ HO

(27)

b) *Conjugated enones.* 1,2-Addition is kinetically preferred, but since these reactions are reversible 1,4-addition is often the final result (B.J.Wakefield, see above), especially with hindered ketones (R.Sjoholm and V.Stahlstrom, *Finn. Chem. Letters.*, 1982, 46). Some cases of 1,3-addition known, although generally in low yield (L.Jalander, *Acta. Chem. Scand., Ser. B*, 1981, **35**, 419). Copper halides have been found to increase 1,4-additon (see Eqn. 28), so that it becomes exclusive in the case of ketones (K.E.Stevens and L.A.Paquette, *Tetrahedron Letters*, 1981, **22**, 4393). Steric control may be high, when a chiral centre is already present.

allylMgBr, CuI/THF/-78°C

(28)

c) *Carboxylic esters.* Controlled addition will yield ketones, however, excess reagent leads to tertiary alcohols. As a result, α-silylated esters (see Eqn. 29) afford alkenes (G.L.Larson and D.Hernandez, *Tetrahedron Letters*, 1982, **23**, 1035). While trihalogeno-acetates are converted into magnesium enolates (B.Rague, Y.Chapleur, and B.Castro, *J. Chem. Soc., Perkin Trans. 1*,, 1982, 2063) (see Eqn. 30).

$$Me_3SiCH_2CO_2Et + 2\ PhCH_2MgCl \longrightarrow CH_2{=}C(CH_2Ph)_2$$

(29)

$$CX_3CO_2Me \xrightarrow[\text{THF/-78°}]{^{i}PrMgCl} X_2C{=}C(OMgCl)(OMe)$$

(30)

d) *Carboxylate salts*. Treatment of a carboxylic acid with 1 equivalent of a Grignard reagent gives a salt, which will react with a second equivalent of the reagent to yield a ketone, after hydrolysis. In the case of formic acid (see Eqn. 31) this can provide a route to aldehydes (F.Sato *et al.*, *Tetrahedron Letters*., 1980, **21**, 2869).

$$HCO_2H + RMgX \xrightarrow{THF} HCO_2MgX \xrightarrow{RMgX} RCH(OMgX)_2 \xrightarrow{H^+} RCHO \quad (31)$$

e) *Imines*. Although imines generally react slowly, *N*-silylimines (see Eqn. 32) are useful precursors of primary amines (D.J.Hart and K.Kanai, *J. Org. Chem.*, 1982, **47**, 1555; D.J.Hart *et al.*, *J. Org. Chem.*, 1983, **48**, 289).

$$RCHO \xrightarrow{LiN(SiMe_3)_2} RCH{=}NSiMe_3 \xrightarrow[2)\ H_2O]{1)\ R^1MgX} RR^1CHNH_2 \quad (32)$$

Conjugated enamines undergo 1,4-addition to give the thermodynamic product; in appropriate cases this can be achieved with stereochemical control (see Eqn. 33) (H.Kagen *et al.*, *Tetrahedron*, 1981, **37**, 3951).

$$\text{[cyclopentenyl-CH=N-C(H)(Bu-t)(CO_2Bu-t)]} \xrightarrow[2)\ H^+]{1)\ RMgX} \text{[2-R-cyclopentane-1-CHO (H, H)]} \quad (33)$$

f) *Azides*. The addition of Grignard reagents to nitrogen-nitrogen multiple bonds is often complicated, however, addition to azides (see Eqn. 34) has been used as a convenient route to diazo-compounds (S.Mori *et al.*, *Chem. Pharm. Bull.*, 1982, **30**, 3380).

$$(PhO)_2P(O)N_3 \xrightarrow{Me_3SiCH_2MgCl} Me_3SiCHN_2 \quad (34)$$

g) *Displacement reactions*. The cleavage of carbon-halogen bonds with coupling (see Eqn. 35) is favoured by strongly coordinating solvents (e.g. HMPA) and adjacent electron donating groups (H.J.Bestmann *et al., Liebigs Ann. Chem.*, 1982, 536; J.P.Quintard, B.Elissondo, and M.Pereyre, *J. Org. Chem.*, 1983, **48**, 1559; R.W.Hoffmann and B.Landmann, *Tetrahedron Letters*, 1983, **24**, 3209).

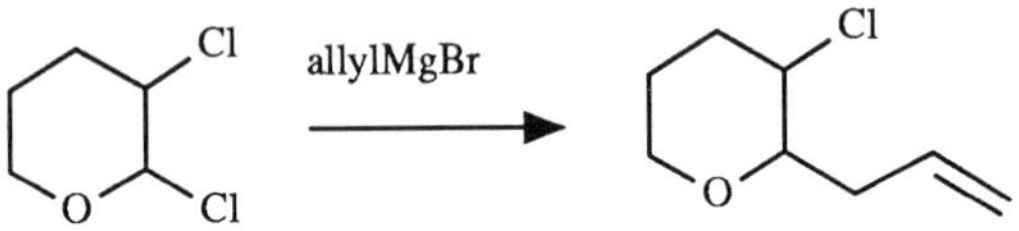

(35)

Cleavage of carbon-oxygen bonds is effective with acetals and orthoesters. Hemiacetals require excess reagent to react with the free hydroxyl function (G.Rousseau and N.Slougui, *Tetrahedron Letters*, 1983, **24**, 1251) (see Eqn. 36). Paraformaldehyde gives homologation with Grignard reagents (M.L.Mancini and J.F.Honek, *Tetrahedron Letters*, 1982, **23**, 3249). Cleavage of oxiranes has been thoroughly discussed (B.J.Wakefield see above). It is interesting that attack by allyl magnesium halides (see Eqn. 37) can occur through either C-1 or C-3 of the allyl unit, but regioselectivity can be induced when a cuprous halide catalyst is used (G.Linstrumelle, R.Lorne, and H.P.Dong, *Tetrahedron Letters*, 1978, 4069). Cleavage of carbon-sulphur bonds in thioethers is more difficult often requiring nickel or copper catalysis. A useful example (see Eqn. 38) is provided by the cleavage of a vinyl thioether, (H.Okamura, M.Miura, and H.Takei, *Tetrahedron Letters,* 1979, 43; Y.Ikeda *et al., J. Amer. Chem. Soc.*, 1982, **104**, 7663; H.Takei *et al., Chem. Letters*. 1980, 1209).

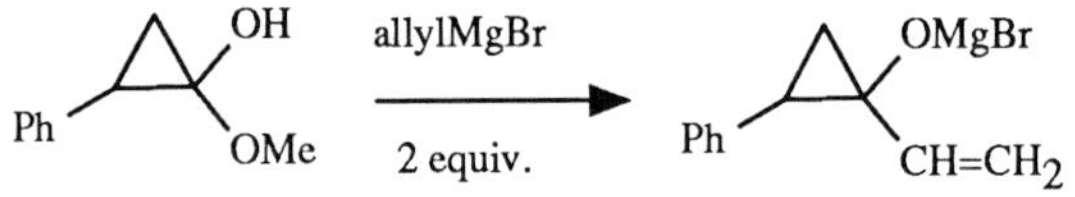

(36)

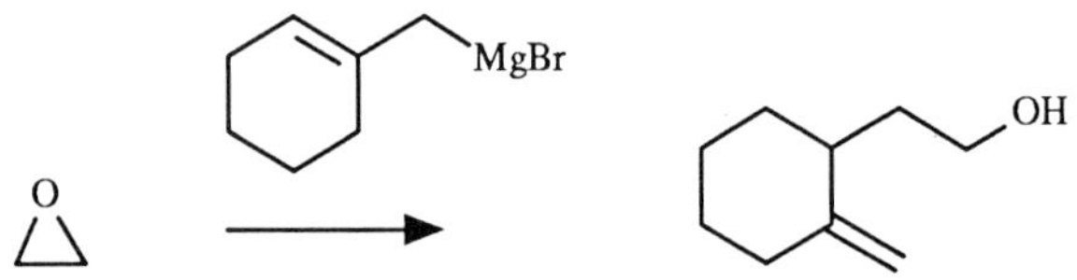

CuI → (37)

MeMgI, $NiCl_2(PPh_3)_2$ cat. benzene, Δ (38)

The stereospecific cleavage of vinyl sulphoxides to yield vinyl magnesium halides is a potentially valuable process (K.Ogura, K.Arai, and G.Tsuchihashi, *Bull. Chem. Soc. Jpn.*, 1982, **55**, 3669)(see Eqn. 39).

Ph, SMe, SOMe —EtMgCl→ Ph, SMe, MgCl (39)

Phosphorus-sulphur ester bonds are readily cleaved with retention of configuration (M.Moriyama and W.G.Bentrude, *Tetrahedron Letters*, 1982, **23**, 4547)(see Eqn. 40), while phosphorus-oxygen bonds are less prone to fission and do so with considerable racemisation.

i-PrO(Me)P(O)SMe —$PhCH_2MgCl$→ i-PrO(Me)P(O)CH_2Ph (99% ee) —n-PrMgBr→ n-Pr(Me)P(O)CH_2Ph (10% ee) (40)

h) *Elemental sulphur, selenium, and tellurium* all react with Grignard reagents (see Eqn. 41), allowing the formation of further products *via* hydrolysis (R.Pratap, R.Castle, and M.J.Lee, *J. Heterocycl. Chem.*, 1982, **19**, 439), oxidation (M.R.Detty *et al., J. Amer. Chem. Soc.*, 1983, **105**, 875), or reaction with electrophiles (F.Wudl and D.Nalewajek, *J. Organomet. Chem.*, 1981, **217**, 329; M.Evers and L.Christiaens, *Chem. Scr.*, 1981, **18**, 143).

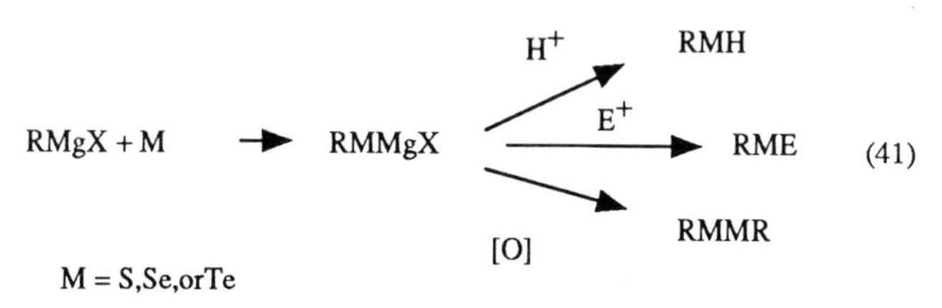

ii) Organomagnesium halides and transition metals The interaction of transition metal salts and complexes has increased the synthetic potential of Grignard reagents. The past 10-20 years have seen significant developments in this area.

Normant Copper Reagents. These have proved very valuable in synthesis (J.F.Normant, *J. Organomet. Chem. Libr.*, 1976, **1**, 219 and *Pure Appl. Chem.*, 1978, **50**, 709). The reagents ($RCuMgX_2$; R_2CuMgX; $RR'CuMgX$) are formed from CuX and RMgX in a stoichiometric ratio (J.F.Normant and A.Alexakis, *Synthesis,* 1981, 841). Improved purity and solubility have been attained by use of Me_2S. CuX (H.O.House *et al., J. Org. Chem.*, 1975, **40**, 1460), or Me_2S as a co-solvent (A.Marfat *et al., J. Amer. Chem. Soc.*, 1977, **99**, 253). The use of LiX or MgX_2 may enhance reactivity, while ZnX_2 and CdX_2 probably displace magnesium completely. The related argento-complexes [R_2AgMgX] are generally used in the presence of LiBr (H.Westmijze, H.Kleijn, and P.Vermeer, *J. Organomet. Chem.*, 1979, **172**, 377).

a) *Addition reactions*. In contrast to Grignard reagents Normant reagents readily add to carbon-carbon multiple bonds (J.F.Normant and A.Alexakis - see above); thus terminal alkynes react with primary alkyl Normant reagents with high stereo- and regiospecificity (see Eqn. 42). The use of THF solvent allows secondary and tertiary groups to be added, but with less specficity. When $R'=SiMe_3$, then the addition (see Eqn. 43) occurs with the opposite regioselectivity (M.Obayashi, K.Utimoto, and H.Nozaki, *Tetrahedron Letters*, 1977, 1805).

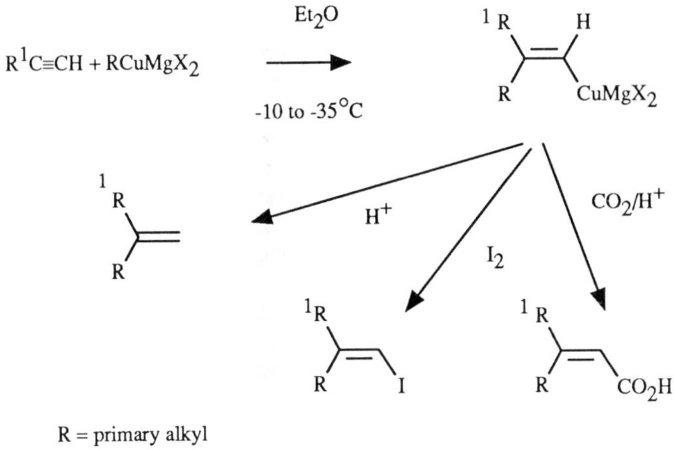

$$R^1C\equiv CH + RCuMgX_2 \xrightarrow[-10\ to\ -35^\circ C]{Et_2O} (R^1)(R)C{=}CH(CuMgX_2)$$

H^+ → $(R^1)(R)C{=}CH_2$; I_2 → $(R^1)(R)C{=}CHI$; CO_2/H^+ → $(R^1)(R)C{=}CHCO_2H$

R = primary alkyl

(42)

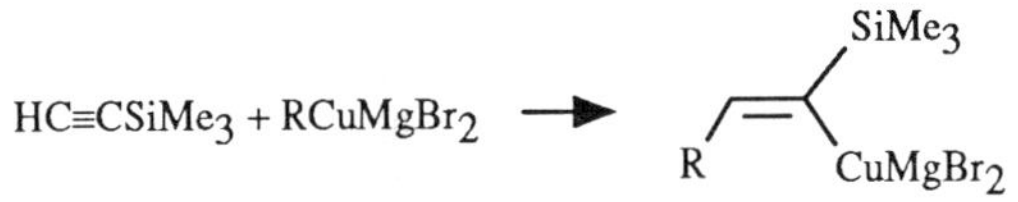

$$HC\equiv CSiMe_3 + RCuMgBr_2 \rightarrow RCH{=}C(SiMe_3)(CuMgBr_2) \qquad (43)$$

Differences in the stereochemistry of addition (see Eqn. 44) have also been noted between Normant reagents and the related argento-complexes (H.Westmijze, H.Kleijn, and P.Vermeer, *Synthesis*, 1978, 454 and *Tetrahedron Letters,* 1979, 3327).

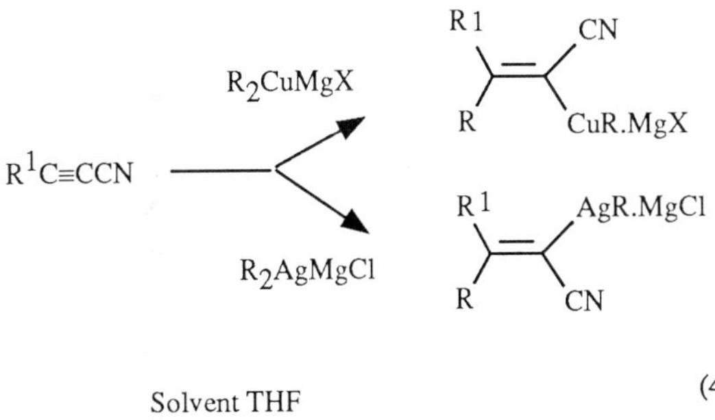

$R^1C\equiv CCN$ + R_2CuMgX → $(R^1)(R)C{=}C(CN)(CuR.MgX)$

$R^1C\equiv CCN$ + $R_2AgMgCl$ → $(R^1)(R)C{=}C(AgR.MgCl)(CN)$

Solvent THF

(44)

The presence of good leaving groups permits the generation of allenes, in which chiral induction is observed (G.Tadema *et al., Tetrahedron Letters*, 1978, 3935). With a dienyne (see Eqn. 45), attack takes place at the alkyne carbon (G.Balme, M.Malacria, and J.Gore, *J. Chem. Res. (M)*, 1981, 2869). Conjugated enones undergo smooth 1,4-additions (see Eqn. 46) in high yield (K.J.Shea and P.Q.Pham, *Tetrahedron Letters*, 1983, **24**, 1003). Furthermore, good stereochemical control (see Eqn. 47) can be obtained in suitable conditions (G.Stork and E.W.Logusch, *Tetrahedron Letters*, 1979, 3361), while by using a chiral ligand (see Eqn. 48), chiral induction becomes a significant feature of the reaction (F.Leyendecker, F.Jesser, and D.Laucher, *Tetrahedron Letters*, 1983, **24**, 3513).

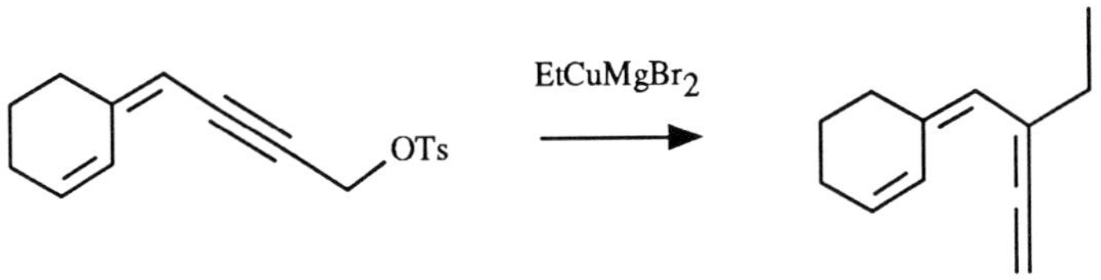

OTs $\xrightarrow{EtCuMgBr_2}$

(45)

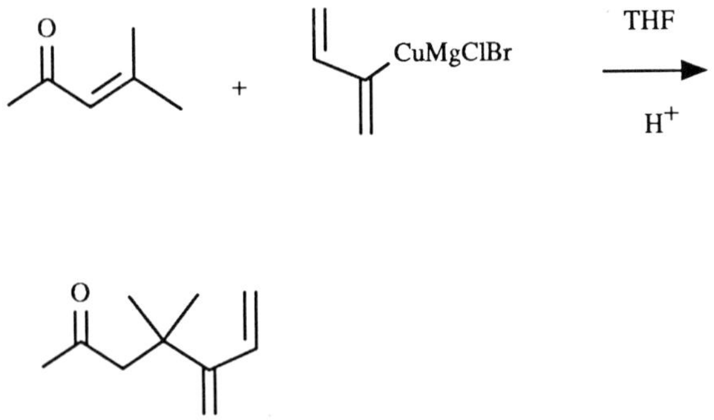

O + CuMgClBr $\xrightarrow[H^+]{THF}$

O

(46)

O O $\xrightarrow[THF]{RCuMgBr_2}$ O O O R

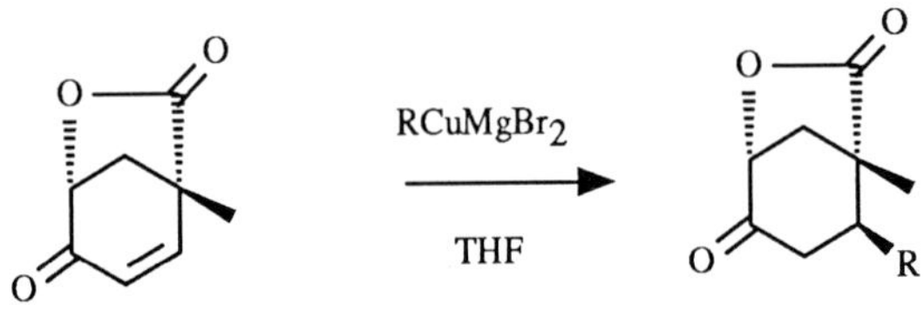

(47)

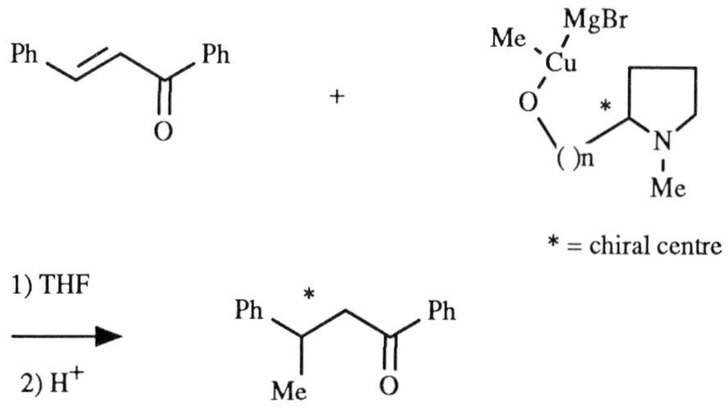

+ Me–Cu(MgBr)–O–( )n– (N-methylpyrrolidin-2-yl)*

* = chiral centre

1) THF

2) H^+

(48)

Even inactivated dienes will react with these reagents giving intermediates which further combine with electrophiles such as carbonyl compounds and alkyl halides (J.F.Normant, G.Cahiez, and J.Villieras, *J. Organomet. Chem.*, 1975, **92**, C28) (see Eqn. 49).

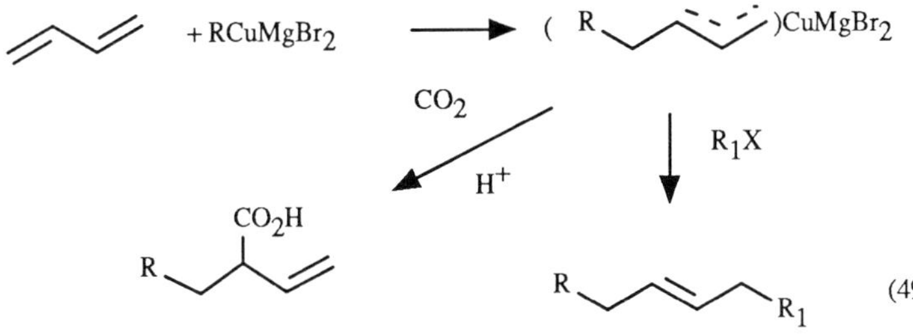

$+ RCuMgBr_2$ → $(R\ldots)CuMgBr_2$

CO_2, H^+ → CO_2H

R_1X

(49)

b) *Ring openings*. The application to opening oxirane rings (see Eqn. 50) can be achieved in excellent yield with good control of both regio- and stereochemistry (R.Baker, D.C.Billington, and N.Ekanayake, *J. Chem. Soc., Perkin Trans. 1*, 1983, 1387).

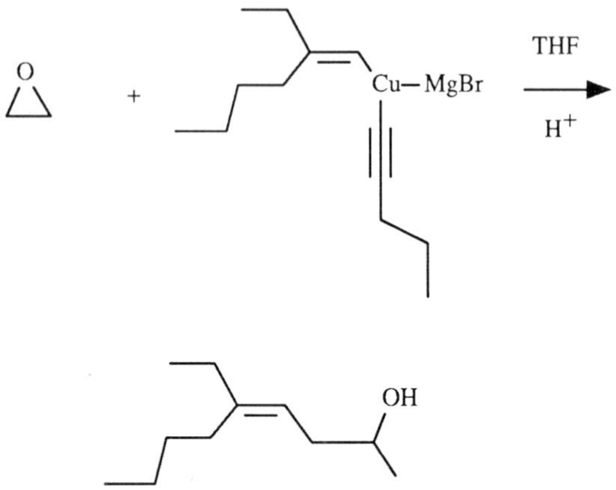

THF
Cu—MgBr
H^+

OH

(50)

c) *Acid chlorides*. The use of Normant reagents (see Eqn. 51) allows the ready formation of even highly hindered ketones, which are difficult to prepare from Grignard reagents (J.E.Dubois, M.Boussu, and C.Lion, *Tetrahedron Letters*, 1971, 829). Similar conversions can be achieved using $L_2Rh(CO)Cd$ catalysis of Grignard reagents (L.S.Hegedus *et al., J. Am. Chem. Soc.*, 1975, 97, 5448).

$$R^1COCl + RCuMgX_2 \rightarrow R^1COR \tag{51}$$

d) *Coupling reactions*. A range of useful coupling reactions have been developed based on Normant reactions. Thus the formation of enynes (D.W.Knight and A.P.Knott, *J. Chem. Soc., Perkin Trans. 1*, 1982, 623), allylic coupling (V.Calo, L.Lopez, and G.Pesce, *J. Organomet. Chem.*, 1982, **231**, 179) and homocoupling *via* air oxidation (J.F.Normant *et al., ibid.*, 1974, **77**, 269) are all illustrative of the procedures (see Eqns. 52, 53, and 54).

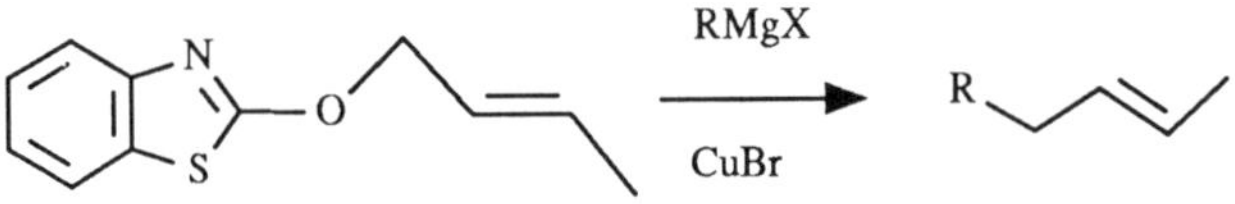

N S O
RMgX
CuBr
R

(52)

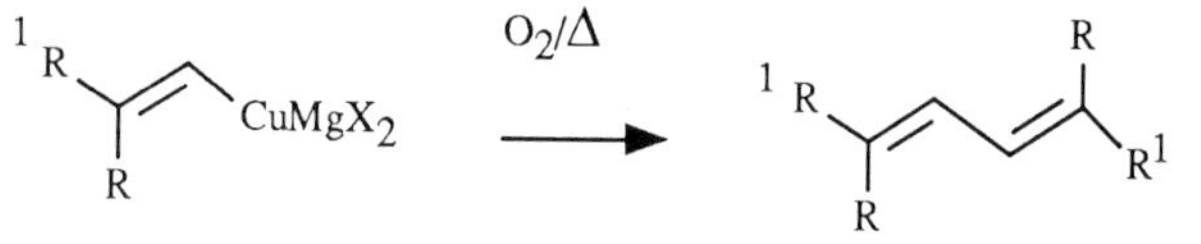

(53)

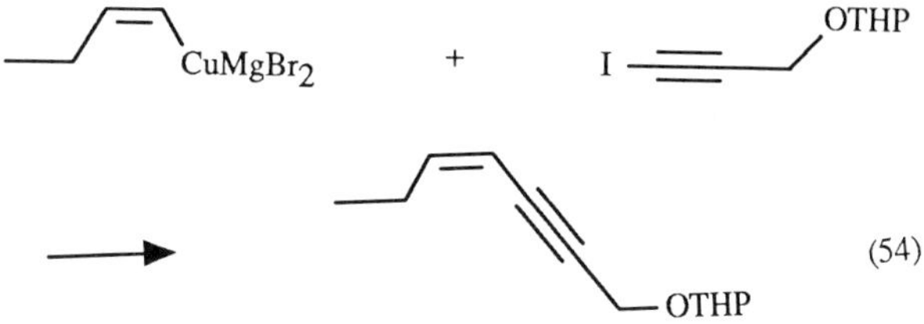

THP = tetrahydropyranyl

(54)

e) *Carboxylation and Carbonylation.* Normant reagents and their manganese (II) equivalents (G.Friour *et al., Bull. Soc. Chim. Fr., Pt. 2,* 1979, 515) react with carbon dioxide to give carboxylate salts, in a similar manner to Grignard reagents. Grignard reagents carbonylate to yield aldehydes on treatment with pentacarbonyl iron (M.Yamashita *et al., Bull. Chem. Soc. Jpn.*, 1982, **55**, 1663).

Catalysis using Transition Metals. The use of copper(I) and cobalt(II) salts to catalyse the reactions of Grignard reagents is well known (G.H.Posner, *Org. React.* 1972, **19**, 1 and 1975, 22, 253; L.F.Elsom, J.D.Hunt, and A.McKillop, *Organomet. Chem. Rev (A)* 1972, **8**, 135). The past 15 years has seen many useful developments.

a) *Addition reactions.* Alkynes undergo *anti*-addition in excellent yield, when the reactions are catalysed by cuprous iodide (B.Jousseaume and J.G.Duboudin, *J. Organomet. Chem.*, 1975, **91**, C1) (see Eqn. 55). While dichlorobis(cylcopentadienyl)titanium (IV) catalysis leads to *syn*-addition, although an alkyl group of the Grignard reagent is not transferred (F.Sato *et al., J. Chem. Soc., Chem. Commun.*, 1982, 1126)(see Eqn. 56). With Bu^iMgCl reduction results in high *Z*-selectivity, however, deuteriolysis reveals low regiospecificity (F.Sato, H.Ishikawa, and M.Sato, *Tetrahedron Letters,* 1981, **22**, 85). Similarly a 1:1 mixture of bis(acetylacetoneto)nickel(II) plus diisobutylhydroaluminium(III), catalyses additions to alkynes (see Eqn. 57), again without transfer of the Grignard alkyl group. The regio- and stereo-selectivity of the processes are excellent (B.B.Snider, M.Karass, and R.S.E.Conn, *J. Amer. Chem. Soc.*, 1978, **100**, 4624). In a related reaction using MeMgBr, the methyl group is transferred (see Eqn. 58) and stereospecificity is reduced (B.B.Snider and M.Karras, *J. Org. Chem.*, 1982, **47**, 4588).

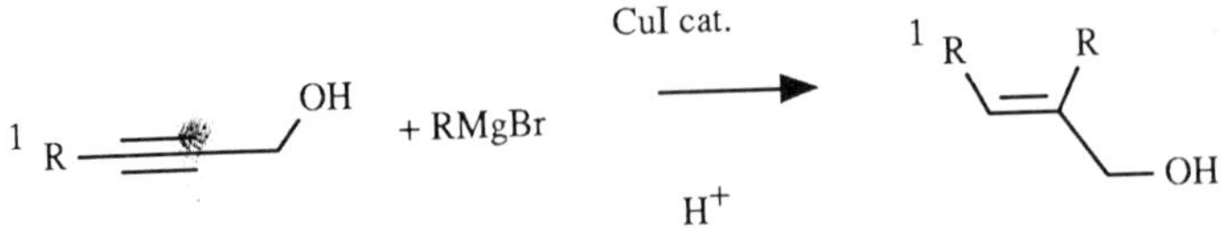

CuI cat.

1 R, OH, + RMgBr, H^+, 1 R, R, OH

(55)

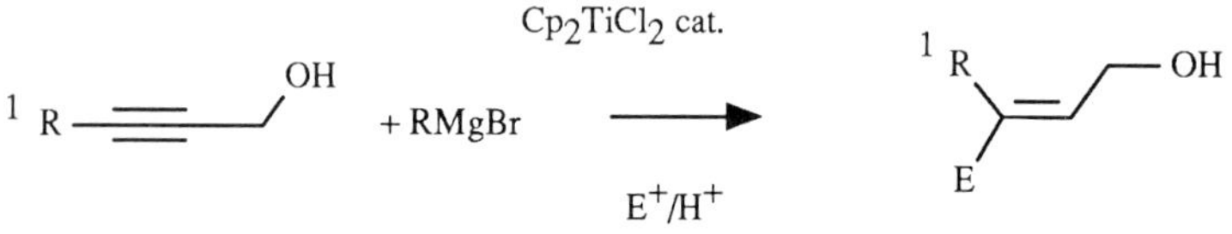

Cp_2TiCl_2 cat.

1 R, OH, + RMgBr, E^+/H^+, 1 R, OH, E

(56)

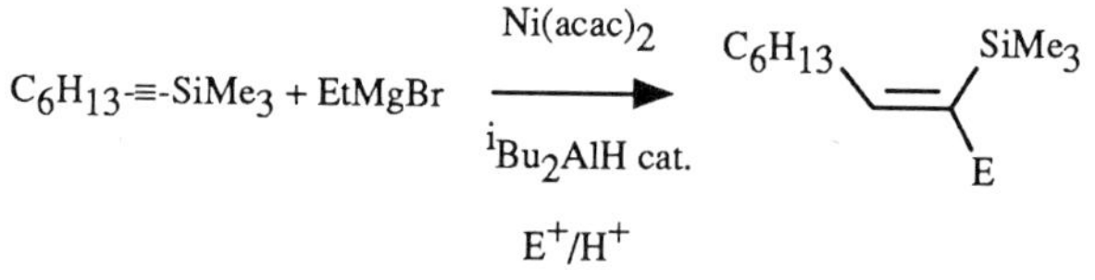

C_6H_{13}-≡-$SiMe_3$ + EtMgBr

$Ni(acac)_2$

$^{i}Bu_2AlH$ cat.

E^+/H^+

C_6H_{13}, $SiMe_3$, E

(57)

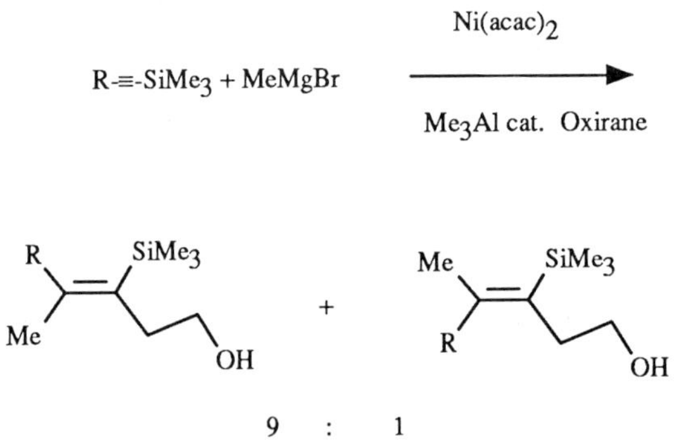

R-≡-$SiMe_3$ + MeMgBr

$Ni(acac)_2$

Me_3Al cat. Oxirane

R, $SiMe_3$, Me, OH + Me, $SiMe_3$, R, OH

9 : 1

(58)

Catalysed addition of Grignard reagents to conjugated enones has been reviewed (G.N.Posner, *Org. React.*, 1975, **22**, 253). An interesting application to synthesis (see Eqn. 59) is provided by the use of a vinyl magnesium reagent (B.M.Trost and B.P.Coppola, *J. Amer. Chem. Soc.*, 1982, **104**, 6879). Related additions to enonates may occur with steric control (W.H.Pirkle and P.E.Adams, *J. Org. Chem.*, 1980, **45**, 4117)(see Eqn.60).

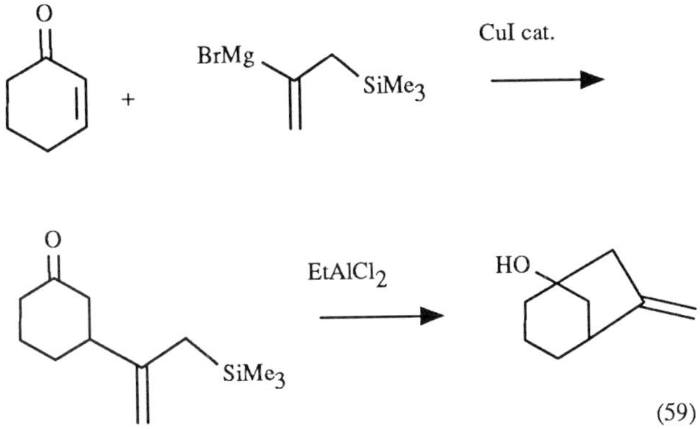

O + BrMg SiMe$_3$ —CuI cat.→

O SiMe$_3$ —$EtAlCl_2$→ HO

(59)

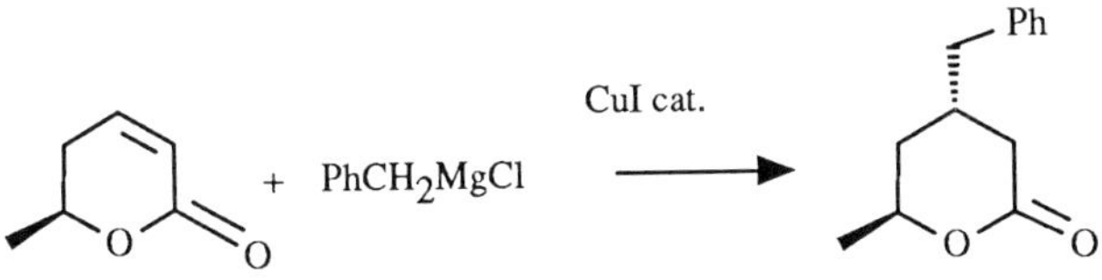

O O + $PhCH_2MgCl$ —CuI cat.→ Ph O O

(60)

Titanium tetrachloride or nickel dichloride can be employed to catalyse additions to 1-enes (H.Felkin and G.Swierczewski, *Tetrahedron*, 1975, **31**, 2735)(see Eqn. 61) .

$$RCH{=}CH_2 + PrMgBr \overset{TiCl_4}{\rightleftarrows} RCH_2CH_2MgBr + CH_3CH{=}CH_2 \quad (61)$$

b) *Ring Opening*. The use of cuprous halide catalysts permits oxirane ring opening (see Eqn. 62), under straightforward reaction conditions (C.Huynh, F.Derguini-Boumechal, and G.Linstrumelle, *Tetrahedron Letters*, 1979, 1503).

O + RMgX —(CuI cat.; H^+)→ OH, R (62)

c) *Coupling reactions*. The use of nickel salts to catalyse the cross coupling of organohalides and Grignard reagents has been extensively reviewed (P.W.Jolly, in *Comprehensive Organometallic Chemistry, Vol. 8*, see above). $Ni(acac)_2$, or $NiCl_2L$ (L = bidentate phosphine) are suitable catalysts, and alkenyl and aryl halides are readily coupled. An illustrative example can be found in a reaction of 1,2-dichloroethene under these conditions (V.Ratovelomanana and G.Linstrumelle, *Tetrahedron Letters*, 1981, **22**, 315)(see Eqn. 63). Chiral induction has been achieved, by use of a chiral ligand bound to the Nickel catalyst (T.Hayashi *et al.*, *Tetrahedron Letters*, 1980, **21**, 79). Palladium (0), palladium (II), and copper (I) compounds will catalyse similar processes (S.Nunomoto, Y.Kawakami, and Y.Yamashita, *Bull. Chem. Soc. Jpn.*, 1981, **54**, 2831 and *J. Org. Chem.*, 1983, **48**, 1912; E.Erdik, T*etrahedron*, 1984, **40**, 641) (see Eqn. 64) .

Cl, Cl + BuMgCl —($Ni(PPh_3)_4$)→ Bu, Cl (63)

MgCl + ArI —($Pd(PPh_3)_4$ cat.)→ Ar

(64)

The Ni, Pd, and Cu complexes will similarly catalyse the coupling of alcohols, ethers, thioethers, selenides, silyl ethers, tosylates, etc. with Grignard reagents (P.W.Jolly; W.Carruthers, in *Comprehensive Organometallic Chemistry*; E.Erdik, *Tetrahedron*, see above). Some examples illustrate (see Eqns. 65-68) how useful these processes can be in synthesis (T.Hayashi, M.Konishi, and M.Kumada, *J. Organomet. Chem.*, 1980, **186**, C1; G.D.Crouse and L.A.Paquette, *J. Org. Chem.*, 1981, **46**, 4272; Y.Ikeda *et al.*, *J. Amer. Chem. Soc.*, 1982, **104**, 7663; T.Hayashi *et al.*, *Synthesis*, 1981, 1001).

$$CH_2{=}CH_2OH + R(Me)CHMgBr \xrightarrow[\text{cat.}]{PdCl_2(dppf)} CH_2{=}CHCH(Me)R \qquad (65)$$

dppf = 1,1′-bis(diphenylphosphino)ferrocene

OSiMe$_3$ + 2 EtMgBr $\xrightarrow[\text{cat.}]{NiCl_2(PPh_3)_2}$ (66)

SPh + MeMgBr $\xrightarrow[\text{cat.}]{NiCl_2(PPh_3)_2}$ (67)

OP(O)(OEt)$_2$ + Me$_3$SiCH$_2$MgCl $\xrightarrow[\text{cat.}]{NiCl_2(PPh_3)_2}$ SiMe$_3$ (68)

c) Calcium, strontium, barium

Reviews of the chemistry and preparation of Ca, Sr, and Ba -organo-derivatives have already been listed above (W.E.Lindsell, in *Comprehensive Organometallic Chemistry*; B.G.Gowenlock and W.E.Lindsell, *J. Organomet. Chem. Libr.*; G.E.Coates and K.Wade, *Organometallic Compounds*, Vol. 1, Pt. 1; Houben-Weyl, *Methoden der Organischen Chemie, Vol. 13/2a*; M.A.Coles, *MTP Int. Rev. Sci.*) Mono- and di-organo derivatives are known for all three elements. They tend to be air and moisture sensitive, but mainly thermally stable, except for alkyl barium and strontium iodides. In the preparation of calcium derivatives, the purity of the calcium used is important (N.Kawabata, A.Matsumara, and S.Yamashita, *Tetrahedron*, 1973, **29**, 1069).

Reactivity. Their usefulness in synthesis is limited, because the more readily accessible lithium and magnesium compounds perform similar reactions.

Conjugated enynes. Reactions with conjugated enynes have been reported (L.N.Cherkasov, S.I.Radchenko, and K.V.Bal'yan, *Zh. Org. Khim.*, 1971, **7**, 2623) to give both 1,2- and 1,4-additions (see Eqn. 69). Regiospecific 1,4-addition is usually the case when R′=alkyl, $Me_2C(OH)$, RS, allyl, vinyl, Me_2NCH_2 (L.N.Cherkasov and R.S.Razina, *ibid.*, 1973, **9**, 17), but alkoxy groups may be displaced during 1,4-addition (L.N.Cherkasov, *Zh. Vses. Khim. Ova.*, 1977, **22**, 468; *Chem. Abs.*, 1977, **87**, 200702)(see Eqn. 70). Although organo- calcium, barium, and strontium halides react with carbonyl compounds by 1,2-addition, the products are often complicated. Since reduction reactions also occur (M.Chastrette and R.Gauthier, *J. Organomet. Chem.*, 1974, **66**, 219). Carboxylation of aryl-calcium or strontium halides leads to carboxylic acids in high yield, but the corresponding alkyl derivatives afford mixtures of ketones and carboxylic acids (D.Bryce-Smith and A.C.Skinner, *J. Chem. Soc.*, 1963, 577).

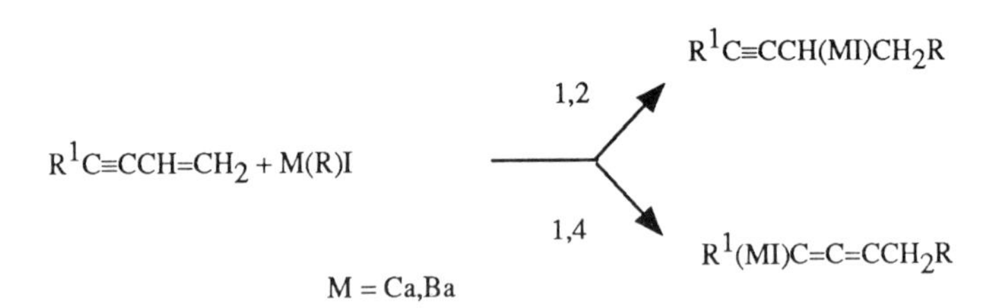

(69)

$$ROCH_2C{\equiv}CCH{=}CH_2 + EtCaI \longrightarrow PrCH{=}C{=}CHPr \qquad (70)$$

d) Zinc

Zinc alkyls were amongst the earliest organometallic compounds prepared, however, they have not been used in synthesis as much as the more reactive and more easily handled Grignard reagents. The main sources of information on the preparation and chemistry or organozinc compounds have been listed above (K.C.Bass, MTP Int. Rev. Sci.; W.Carruthers, in *Comprehensive Organometallic Chemistry*; L.G.Kuzmina, N.G.Bokii, and Yu.T.Struchkov, *Russ. Chem. Rev.*). Additionally, more specific articles regarding organozinc compounds have been published (G.J.M. Van der Kerk, *Russ. Chem. Rev. Engl. Transl.*, 1972, **30**, 389; J.Fukawa and N.Kawabata, *Adv. Organomet. Chem.*, 1974, **12**, 83).

Uses in Synthesis. The main uses of organozinc reagents often involve processes, in which the reagents undoubtedly intervene but are not usually isolated.

i) Reformatsky reaction (see Eqn. 71). All α-halo-esters react, however, their reactivity follows the sequence Cl<Br<I. Iodides are not readily available, so α-bromoesters are the most commonly used substrates (M.W.Rathke, *Org. React.*, 1975, **22**, 423). α-Bromoamides, such as *N,N*-dimethylbromoacetamide, afford the corresponding β-hydroxyamides (R.C.Poller and D.Silver, *J. Organomet. Chem.*, 1978, **157**, 247), while esters as well as aldehydes or ketones also react. Organozinc bromide intermediates, such as (4) and (5), are involved (W.R.Vaughan and H.P.Knoess, *J. Org. Chem.*, 1970, **35**, 2394). The organozinc intermediates can be preformed and dimethoxyethane is a good solvent for use in this procedure. The two-step sequence is particularly valuable if the carbonyl compound has a readily reducible function (e.g. NO_2) or one (e.g. NR_2) which would react with the bromoester. A continuous flow method has been used, in which the reagents are mixed and then passed through a column of granulated zinc (J.F.Ruppert and J.D.White, *J. Org. Chem.*, 1974, **39**, 269). Use of activated zinc ($ZnCl_2K$, THF) allows the reaction to proceed at low temperatures (R.D.Rieke and S.J.Uhm, *Synthesis*, 1975, 452). It is worth noting that a very convenient alternative to the Reformatsky reaction exists, thus the α-lithio derivative of *t*-butyl acetate (stable enough to be weighed in air) readily yields similar products with carbonyl compounds.

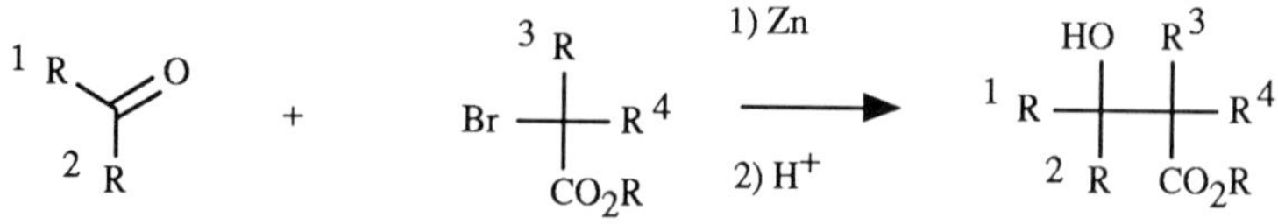

(71)

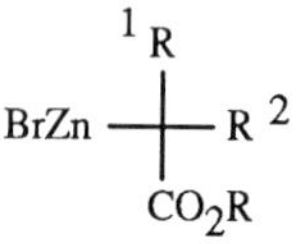

(4) 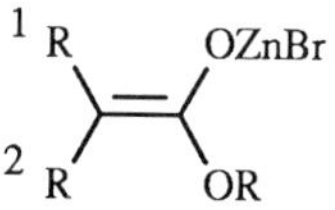 (5)

Allyl, benzyl, and propargyl bromide react with carbonyl compounds in the presence of zinc to yield alcohols. This Reformatsky-like process has been applied to the synthesis of artemisia alcohol, where an allylic rearrangement reaction is involved (J.F.Ruppert and J.D.White, *J. Org. Chem.*, 1976, **41**, 550)(see Eqn. 72).

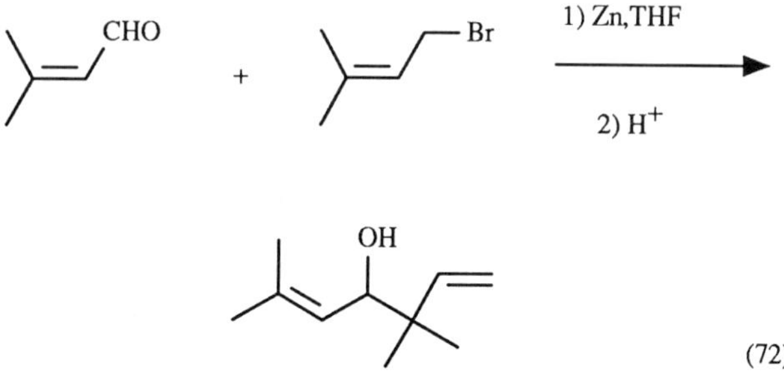

(72)

ii) The Simmons-Smith reaction. Zinc activated as a Zn-Cu couple reacts with di-iodomethane in the presence of alkenes (see Eqn. 73) to form cyclopropanes (H.E.Simmons *et al.*, *Org. React.*, 1973, **20**, 1). This versatile cyclopropane synthesis proceeds *via* organozinc intermediates [e.g. ICH_2ZnI, $(ICH_2)_2$ Zn. ZnI_2], to give a *syn*-addition product. A most interesting example of its use comes from the synthesis of (-) valeranone (E.Wenkert and D.A.Berges, *J. Amer. Chem. Soc.*, 1967, **89**, 2507), in which addition occurs on the least hindered face of the double bond (see Eqn. 74). The use of Zn-Ag couples generally improves yields (J.M.Davis and J.M.Conia, *Tetrahedron Letters*., 1972, 4593).

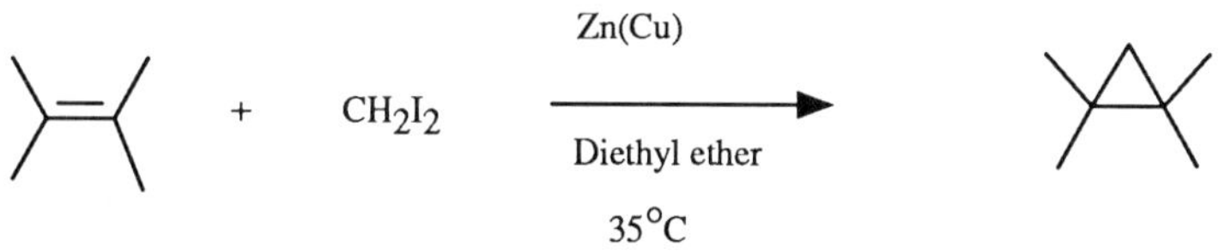

+ CH_2I_2 $\xrightarrow[\text{Diethyl ether, } 35^{\circ}C]{Zn(Cu)}$

(73)

HO OMe iPr $\xrightarrow[Zn(Cu)]{CH_2I_2}$ HO OMe iPr

1) CrO_3
2) Wolff-Kishner
3) HCl

HO O iPr

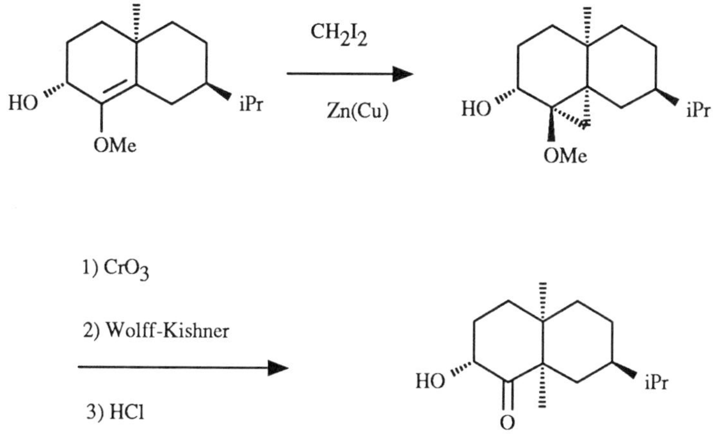

(-) valeranone

(74)

A further improvement results from the use of diethylzinc (J.Furukawa, N.Kawabata, and J.Nishimura, *Tetrahedron*, 1968, **24**, 53). The soluble intermediate (ICH_2ZnEt) allows the reaction to proceed in homogenous solution, which is faster and cleaner than the heterogeneous Zn-Cu couple method. This procedure can be applied to vinyl ethers, which polymerise in the presence of a Zn-Cu couple. Diethylzinc will react with 1,1-di-iodoalkanes, bromoform, and iodoform (see Eqns. 75 and 76), as well as with diiodomethane (J.Nishimura, N.Kawabata, and J.Furukawa, *Tetrahedron,* 1969, **25**, 2647; S.Miyano, Y.Matsumoto, and H.Hashimoto, *J. Chem. Soc., Chem. Commun.*, 1975, 364). Reactions with some aromatic compounds are possible (J.Nishimura *et al., Tetrahedron*, 1970, **26**, 2229)(see Eqn. 77) .

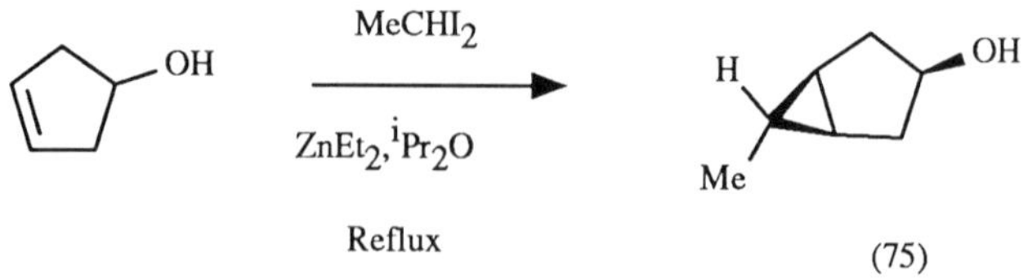

$$\xrightarrow[\text{ZnEt}_2, {}^{i}\text{Pr}_2\text{O} \quad \text{Reflux}]{\text{MeCHI}_2}$$

OH, H, Me, OH

(75)

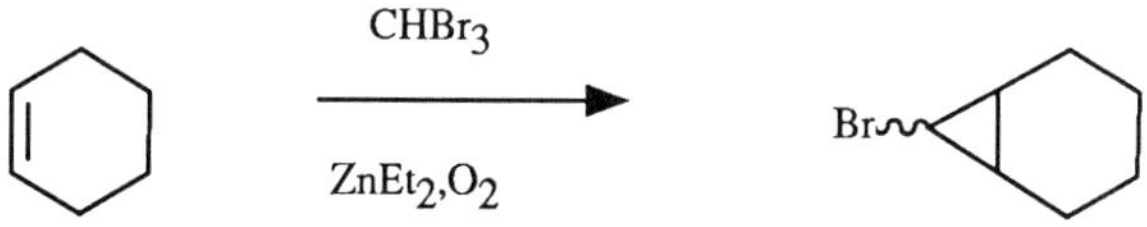

$$\xrightarrow[\text{ZnEt}_2, \text{O}_2]{\text{CHBr}_3}$$

Br

(76)

$$\xrightarrow[\text{ZnEt}_2]{\text{MeCHI}_2}$$

R, Me, H, R

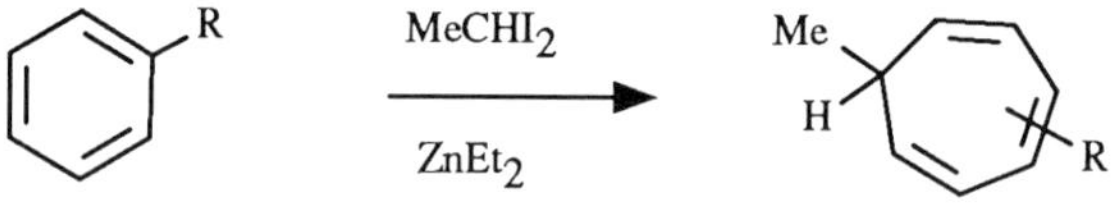

R=alkyl

(77)

Some imines with electron-withdrawing substituents also react yielding aziridines (P.Baret, H.Buffet, and J.L.Pierre, *Bull. Soc. Chim. Fr.*, 1972, 825).

iii) Miscellaneous reactions. Diazomethane and zinc iodide in diethyl ether or $(PhCO_2CH_2)_2Zn$ can be used for the cyclopropanation of alkenes (D.Wendisch, in Houben-Weyl, *Methoden der Organischen Chemie,* Vol. 4/3, see above). $Zn(CHCl_2)_2$ prepared from dichloromethane has been used to form chloro-substituted cyclopropanes (G.Köbrich and H.R.Merkle, *Chem. Ber.*, 1966, **99**, 1782)(see Eqn. 78).

$$CH_2Cl_2 \xrightarrow[\text{2) } ZnCl_2/\text{-}74^{\circ}C]{\text{1) BuLi/THF/-}65^{\circ}C} Zn(CHCl_2)_2$$

Cl

(78)

Dialkylzinc or alkylzinc halides can be coupled with acid chlorides, a process which can be more suitable than those using diakylcadmium reagents (F.Wingler, in *Houben-Weyl, Methoden der Organischen Chemie*, Vol. 7/2a, see above) (see Eqn. 79).

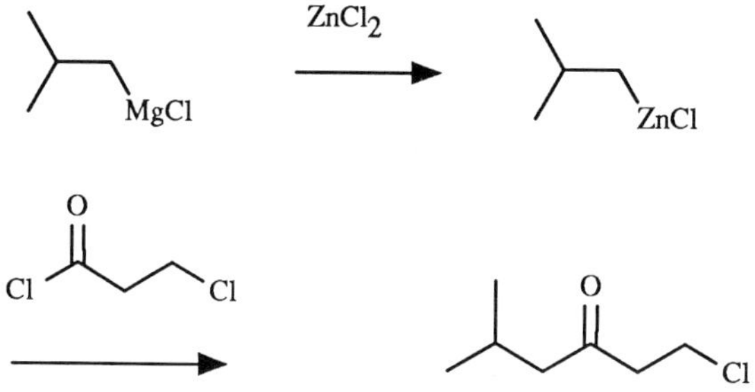

MgCl $\xrightarrow{ZnCl_2}$ ZnCl

O

Cl Cl

O

Cl

(79)

Similar couplings with aryl halides are possible (see Eqn. 80), but require catalysis with nickel or palladium complexes (E.-i.Negishi, A.O.King and N.Okukado, *J. Org. Chem.*, 1977, **42**, 1821).

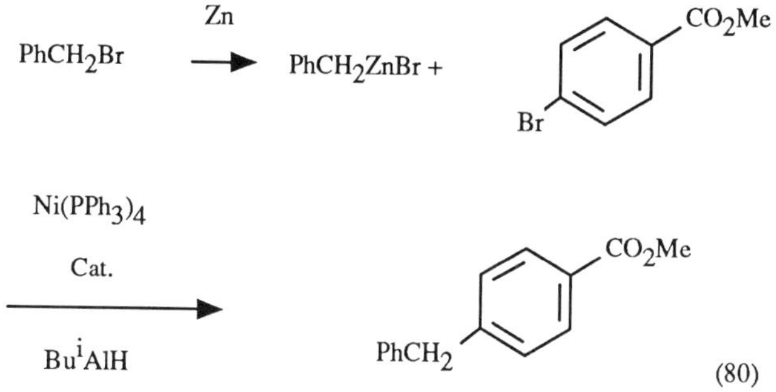

$PhCH_2Br \xrightarrow{Zn} PhCH_2ZnBr$ + CO_2Me Br

$Ni(PPh_3)_4$

Cat.

Bu^iAlH

CO_2Me

$PhCH_2$

(80)

e) Cadmium

Organocadmium compounds resemble organozinc compounds, but are generally less reactive (D.A.Shirley, *Org. React.* 1954, **8**, 28; N.I.Sheverdina and K.A.Kocheshkov, in *The Organic Compounds of Zn and Cd,* from the series *Methods of Elemento-Organic Chemistry,* Vol. 3, A.N.Nesmeyanov and K.A.Kocheshkov, Eds., North Holland, Amsterdam, 1963; D.Wendisch, in Houben-Weyl, *Methoden der Organischen Chemie*, Vol. 4/3 see above). They have advantages over zinc alkyls, namely their lack of spontaneous ignition and their appreciably faster rate of reaction with acid chlorides than with ketones. This latter behaviour means they can be used for the synthesis of ketones, with little amounts of *tertiary*-alcohols being formed as by-products. However, *secondary*- and *tertiary*-alkylcadmium compounds are difficult to prepare in good yield. As with organo-zinc compounds, the use of organo-cadmium compounds has been limited by the success of lithium and magnesium derivatives in organic synthesis.

Preparations. These are well covered in the reviews mentioned above. The most common procedure (see Eqn. 81) utilises Grignard reagents. Salt-free dialkylcadiums are not as reactive as those prepared *in situ.* Magnesium salts catalyse reactions with both acid halides and carbonyl compounds, while lithium bromide is more selective favouring the former (J.Kollonitsch, *J. Chem. Soc.*, (A), 1966, 453 and 456).

$$2RMgX + CdCl_2 \xrightarrow{\text{diethyl ether}} R_2Cd + 2MgXCl \qquad (81)$$

Allylic cadmium reagents cannot be prepared from Grignard reagents, but can be formed from boranes (see Eqn. 82); they are very reactive towards carbonyl compounds (K.H.Thiele and J.Köhler, *J. Organomet. Chem.*, 1967, **7**, 365; D.Abenhaim, E.Henry-Basch, and P.Freon, *C.R. Seances Acad. Sci., Ser. C.*, 1967, **264**, 1313).

$$2\,(allyl)_3B + 3\,Me_2Cd \longrightarrow 3(allyl)_2Cd + 2Me_3B$$

$$(allyl)_2Cd + R^1RCO \longrightarrow R^1RC(OH)allyl$$

>80% yield

$$R{=}H, Me, Et;\ {}^1R{=}Et, Pr, Ph \qquad (82)$$

Synthetic Uses. These are reviewed (A.R.Jones and P.J.Desio, *Chem. Rev.*, 1978, **78**, 491). It is interesting to note that organocadmium compounds tend to give more 1,4-addition in reactions with conjugated enones than the corresponding lithium or magnesium reagents (O.Eisenstein *et al.*, *C.R. Seances Acad. Sci., Ser. C*, 1972, **274**, 1310).

f) Mercury

This review will cover either well defined aliphatic organomercury compounds or reactions in which they are believed to be intermediates. The synthetic use of organomercurials has seen a great expansion of interest in the 1970's and 1980's (R.C.Larock a) *The Organomercury Solvomercuration-Demercuration Reactions in Organic Synthesis*, Springer-Verlag, Berlin, 1986; b) *Organomercury Compounds in Organic Synthesis,* Springer-Verlag, Berlin, 1985; c) *Tetrahedron*, 1982, **38**, 1713; and d) *Angew. Chem. Int. Ed. Engl.*, 1978, **17**, 27; W.Carruthers, in *Comprehensive Organometallic Chemistry* - see above). Review articles also include:) A.J.Bloodworth, in *The Chemistry of Mercury*, C.A. McAuliffe, Ed., Macmillan, London, 1977; K.-P.Zeller, H.Straub, and H.Laditsdke, in Houben-Weyl, *Methoden der Organischen Chemie, Vol. 13/2b*, E.Muller, Ed., Thieme, Stuttgart, 1974). This interest has been advanced by the ease with which mercury can be introduced, followed by its ready replacement by a wide variety of reagents. The catalysis of the replacement of organomercury, by transition metals, has been another important development (I.P.Beletskaya, *J. Organomet. Chem.*, 1983, **250**, 551; V.I.Sokolov and O.A.Reutov, *Coord. Chem. Rev.*, 1978, **27**, 89). The fact that many organomercurials are toxic should not be overlooked (D.L.Rabenstein, *Acc. Chem. Res.*, 1978, **11**, 101 ; *J. Chem. Ed.*, 1978, **55**, 292; K.H.Falchuk, L.J.Goldwater, and B.L.Vallee, in *The Chemistry of Mercury*, C.A. McAuliffe, Ed., Macmillan, London, 1977). The N.M.R. spectra and the structural analysis of organomercury compounds has been reviewed (V.S.Petrosyan and O.A. Reutov, *J. Organomet. Chem.*, 1976, **76**, 123).

Preparations. These are well covered in the reviews mentioned above. Mercury is most frequently introduced *via* the process (see Eqn. 83) known as *solvomercuration* (when HY is the solvent) or *oxymercuration* (when HY = H_2O, ROH, or RCO_2H). Thus the reaction of a mercury (II) salt with an alkene, in the presence of a nucleophile (HY or Y^-) leads to the formation of a β-substituted organomercury compound. Similar results can be obtained with alkynes (E.Winterfeldt, in *Chemistry of Acetylenes*, H.G.Viehe, Ed., Dekker, New York, 1969). Many salts have been used, although acetates and trifluoroacetates are amongst the most effective. The procedures show a high degree of regio- (Markownikov type) and stereo-selectivity (*anti*).

$$R\text{-}CH=CH_2 + HgX_2 + HY \longrightarrow R\text{-}CH(Y)\text{-}CH_2HgX + HX \quad (83)$$

Uses in Organic Synthesis. These can be divided into:

1) Displacement reactions

i) Carbon-hydrogen bond formation. Protonolysis can be useful in isotopic labelling. Deuteriolysis or protonolysis of dialkyl mercurials proceeds with significant racemisation, but with alkenyl mercurials replacement occurs with retention of configuration (M.Gifford and J.Cousseau, *J. Organomet. Chem.*, 1980, **201**, C1)(see Eqn. 84).

$$R(NCS)C=C(HgSCN)R^1 \xrightarrow{H_2SO_4} R(NCS)C=CHR^1 \quad (84)$$

Sodium borohydride can also be used to displace mercury, however, without control of stereochemistry (S.Uemura, H.Miyoshi, and M.Okano, *J. Chem. Soc., Perkin Trans. 1*, 1980, 1098)(see Eqn. 85).

$$Ph(AcO)C=C(HgCl)Me \xrightarrow{NaBH_4} Ph(AcO)C=CHMe \quad (85)$$

Tritiation has been carried out *via* mercury derivatives (R.M.K.Dale *et al.*, *Nucleic Acid Res.*, 1975, **2**, 915)(see Eqn. 86).

$$\text{5-HgX-uracil (N-1 R)} \xrightarrow{NaBT_4} \text{5-T-uracil (N-1 R)} \quad (86)$$

In conjunction with *oxymercuration,* reductive displacement of mercury provides a convenient two-step sequence for the preparation of specially functionalised alkanes from alkenes. *Alcohols* can be prepared (see Eqn. 87) with useful selectivity (H.C.Brown *et al.*, *J. Org. Chem.*, 1972, **37**,

1941). Replacing water by alcohols, as nucleophiles, gives ethers (H.C.Brown and M.-H.Rei, *J. Amer. Chem. Soc.*, 1969, **91**, 5646), while hydrogen peroxide (see Eqn. 88) allows the formation of peroxides (A.J.Bloodworth and M.E.Loveitt, *J. Chem. Soc., Chem. Commun.*, 1976, 94). *Esters* and *lactones* can also be prepared in this fashion. The process is not confined to oxygen nucleophiles, thus azides, nitrites, and nitriles yield *azido*, *nitro*, and *amido-alkanes* respectively (W.Carruthers, in *Comprehensive Organometallic Chemistry*, see above). β-Nitromecurial compounds undergo a novel base induced elimination, to yield synthetically useful *nitroalkenes* (E.J.Corey and H.Estreicher, *J. Amer. Chem. Soc.*, 1978, **100**, 6294) (see Eqn. 89).

$$\xrightarrow[\text{2) } NaBH_4/OH^-]{\text{1) } Hg(O_2CCF_3)_2} \quad \text{OH} \tag{87}$$

$$\xrightarrow[H_2O_2]{Hg(NO_3)_2} \quad HgNO_3 \ldots O\text{–}O \ldots HgNO_3 \quad \xrightarrow[OH^-]{NaBH_4} \tag{88}$$

$$RCH{=}CH_2 \xrightarrow[H_2O]{HgCl_2/NaNO_2} RCH(NO_2)CH_2HgCl \xrightarrow{Et_3N} R(NO_2)CH{=}CH_2 \tag{89}$$

Aminomercuration can be achieved using secondary amines or anilines, although the strong Hg-N interaction greatly slows the rate of reaction (J.J.Perie *et al.*, *Tetrahedron Letters*, 1971, 4399, *Tetrahedron*, 1972, **28**, 675 and 701)(see Eqn. 90).

$$RCH{=}CH_2 \xrightarrow[\text{piperidine}]{HgCl_2} RCH(\text{N-piperidinyl}))CH_2HgCl \xrightarrow[OH^-]{NaBH_4} R(\text{N-piperidinyl})CHCH_3 \qquad (90)$$

The formation of carbocyclic rings under mild conditions has been exploited in the synthesis of racemic aphidicolin, which involves a stereospecific double ring closure reaction, followed by demercuration (E.J.Corey, M.A.Tius, and J.Das, *J. Amer. Chem. Soc.*, 1980, **102**, 1742(see Eqn. 91); see also M.Kurbanov *et al.*, *Tetrahedron Letters*, 1972, 2175; M.Julia and J.D.Fourneron, *Tetrahedron Letters*, 1973, 3429).

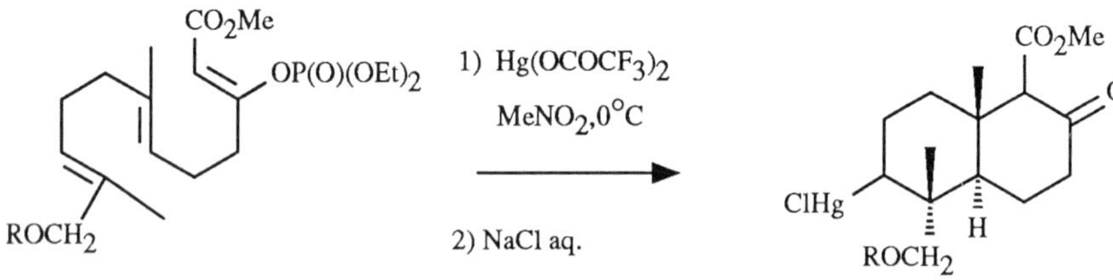

R=dimethyl^{t}butylsilyl (91)

Alkynes and cyclopropanes also undergo *oxymercuration-demercuration* procedures (A.J.Bloodworth, in *The Chemistry of Mercury*, see above). Acetoxymercuration of alkynes (see Eqn. 92) is a good method of preparing vinyl acetates (S.Uemura *et al., J. Chem. Soc., Chem. Commun.*, 1975, 548; R.J.Spear and W.A.Jensen, *Tetrahedron Letters*, 1977, 4535).

$$PhC{\equiv}CMe \xrightarrow[\text{2) KCl}]{\text{1) Hg(OAc)}_2\text{, HOAc}} \text{Ph(AcO)C=C(HgCl)Me} + \text{Ph(ClHg)C=C(OAc)Me}$$

$$\text{Ph(AcO)C=C(HgCl)Me} \xrightarrow{I_2/MeOH} \text{Ph(AcO)C=C(I)Me}$$

$$\text{Ph(ClHg)C=C(OAc)Me} \xrightarrow{HOAc} \text{PhCH=C(OAc)Me} \quad (92)$$

ii) Carbon-halogen bond formation. Can be useful for the formation of organo-halides, not readily formed by direct halogenation (see Eqn. 93).

$$R_2Hg + X_2 \rightarrow RHgX + RX$$
$$RHgX + X_2 \rightarrow HgX_2 + RX \quad (93)$$

Free radical processes can result in racemised or rearranged products. Bromine, which shows less tendency towards free radical reactions than iodine, will halogenate simple alkylmercurials at -65°C in pyridine, with good retention of configuration (F.R.Jensen *et al., J. Amer. Chem. Soc.*, 1960, **82**, 2466). An interesting example is provided by solvomercuration of a cyclopropane, followed by bromination (Yu.S.Shaborov *et al.*, *J.Organomet. Chem.*, 1978, **150**, 7)(see Eqn. 94) .

$$\text{cyclopropyl-Ar} \xrightarrow[\text{2) Br}_2]{\text{1) Hg(OAc)}_2\text{, MeOH}} \text{ArCH(OMe)CH}_2\text{CH}_2\text{Br} \quad (94)$$

Solvomercuration of terminal-alkenes, followed by bromination (see Eqn. 95) is a useful way to obtain anti-Markownikov hydrobromination, although internal alkenes give much reduced yields (J.J.Tufariello and M.M.Hovey, *J. Amer. Chem. Soc.*, 1970, **92**, 3221). Iodides can be similarly prepared using dicyclohexylborane (R.C.Larock, *J. Org. Chem.*, 1974, **39**, 834).

$$RCH{=}CH_2 \xrightarrow[3)\ Br_2]{1)\ BH_3\quad 2)\ Hg(OAc)_2} RCH_2CH_2Br \qquad (95)$$

Vinylmercurials are useful substrates for the formation of vinyl halides (J.F.Normant *et al., Synthesis*, 1974, 803)(see Eqn. 96). Control of stereochemistry in these displacements can be achieved by variation in reaction conditions. Thus bromine in the presence of air leads to almost complete retention when pyridine is the solvent, while in carbon disulphide inversion predominates (C.P.Casey, G.M.Whitesides, and J.Urth, *J. Org. Chem.*, 1973, **38**, 3406).

$$RC{\equiv}CH \xrightarrow[2)\ HgBr_2]{1)\ R^1CuMgBr_2} (R)(R^1)C{=}CH\text{–})_2Hg \xrightarrow{Br_2} (R)(R^1)C{=}CHBr \qquad (96)$$

Halogenation of aromatic and heterocyclic organomercurials has been widely employed, but falls outside the scope of this review.

iii) Carbon-oxygen bond formation. Ozonolysis yields *tertiary*-alcohols, ketones, or carboxyclic acids from *tertiary*-, *secondary*-, and *primary*-alkylmercurials respectively (P.E.Pike *et al., Tetrahedron Letters*, 1970, 2679); while peracids can be used to oxidise *secondary*-alkylmercurials (J.H.Robson and G.F.Wright, *Can. J. Chem.*, 1960, **38**, 1). Vinylmercuric salts decompose thermally leading to vinyl -esters and -ethers (D.J.Foster and E.Tobler, *J. Amer. Chem. Soc.*, 1961, **83**, 851)(see Eqn. 97).

$$(allyl)_2Hg \xrightarrow[\text{heat}]{\text{HOAr}} \text{allyl-O-Ar} \qquad (97)$$

The related solvolytic process (W.Treibs *et al., Liebigs Ann. Chem.*, 1953, **581**, 59) is convenient, in that inorganic products readily separate, but unfortunately yields are variable and isomeric mixtures may result. A good example of the process, which involves *addition demercuration*, is provided by the oxidation of methyl dec-8-enoate (S.Ranganathan, D.Ranganathan, and M.M.Mehrota, *Tetrahedron Letters*, 1978, 1851)(see Eqn. 98).

$$MeCH{=}CH(CH_2)_6CO_2Me \xrightarrow[\text{HOAc, heat}]{Hg(OAc)_2} AcOCH_2CH{=}CH(CH_2)_6CO_2Me \qquad (98)$$

Aryl and other groups may be added during the course of these oxidations (K.Ichikawa *et al., J. Amer. Chem. Soc.*, 1958, **80**, 6005; 1959, **81**, 3401, *J. Org. Chem.*, 1959, **24**, 1129; M.Julia, E.Colomer Gasquez, and R.Labia, *Bull. Soc. Chim. Fr.*, 1972, 4145)(see Eqn. 99).

$$\underset{(E)}{RCH{=}CH^1R} + {}^2RCO_2H + ArH \xrightarrow[Hg^{2+}]{BF_3} RCH(Ar)CH(OCOR^2)\,{}^1R \qquad (99)$$

However, especially valuable are processes based on *solvomercuration - palladation*, by which alkenes, or cyclopropanes, can be readily oxidised to ketones or ketals (G.T.Rodeheaver and D.F.Hunt, *J. Chem. Soc., Chem. Commun.*, 1971, 818; *Tetrahedron Letters*, 1972, 3595; R.J.Ouellette and C.Levin, *J. Amer. Chem. Soc.*, 1971, **93**, 471)(see Eqn. 100).

$$\underset{(E)}{RCH{=}CH^1R} \xrightarrow[\text{2) } LiPdCl_4 \text{ cat., DCM(3 eq.)}]{\text{1) } Hg(OAc)_2/MeOH} RCH_2CO^1R \qquad (100)$$

Good, stereospecific routes to enol acetates have also been developed which involve palladium catalysis (R.C.Larock, K.Oertle, and K.M.Beatty, *J. Amer. Chem. Soc.*, 1980, **102**, 1966)(see Eqn. 101).

$$\underset{\text{(R}^2\text{)(R)C=C(R}^1\text{)(HgCl)}}{} \xrightarrow[\text{PdOAc)}_2\text{ cat.}]{\text{Hg(OAc)}_2} \text{(R}^2\text{)(R)C=C(R}^1\text{)(OAc)} \qquad (101)$$

iv) <u>Carbon-sulphur bond formation</u>. Many useful processes also exist, which can often be applied to selenium or tellurium compounds (R.C.Larock, *Tetrahedron*, 1982, **38**, 1713). An interesting example involves photolysis of vinylmercurials in the presence of thiol anions (J.Hershberger and G.A.Russell, Synthesis, 1980, 475; *J. Amer. Chem. Soc.*, 1980, **102**, 7603)(see Eqn. 102).

$$\underset{(E)}{\text{RCH=CHHgX}} \xrightarrow[\text{R}^1\text{S}^-]{h\nu} \underset{(E)}{\text{RCH=CHSR}^1} \qquad (102)$$

v) <u>Carbon-nitrogen and carbon-phosphorus bond formation</u>. A large miscellany of methods exist for the formation of such linkages (R.C.Larock, see above).

vi) <u>Carbon-carbon bond formation</u>. Organomercurials generate new carbon-carbon bonds *via* alkylation, dimerisation, addition to alkenes or alkynes, carbonylation, and acylation (see R.C.Larock, above), however, in most cases catalysis by transition metals is desirable (see discussion below).

viii) <u>Miscellaneous displacement reactions</u>. A valuable method for the formation of enol-esters (see Eqn. 103), involves the indirect displacement of mercury from α-mercurated carbonyl compounds (H.O.House *et al., J. Org. Chem.*, 1973, **38**, 514; R.A.Olofsen, B.A.Bauman, and D.J.Wancowicz, *ibid.*, 1978, **43**, 752; Y.Kita *et al., ibid.*, 1980, 45, 4519; Y.Tamura *et al., Tetrahedron Letters*, 1978, 3737).

$R(^1RCO_2)C{=}CH_2$ ← ^1RCOCl — $RCOCH_2HgCl$ — $^1RCO_2Cl$ → $R(R^1OCO_2)C{=}CH_2$

$COCl_2$ → $R(CO_2Cl)C{=}CH_2$; $COCl_2$ → $(CH_2{=}C(R)O)_2CO$

(103)

2. Carbene-transfer reactions

The use of α-halomethylmercury compounds as a source of carbenes has been reviewed (D.Seyferth, *Acc. Chem. Res.*, 1972, **5**, 65; R.C.Larock, *Angew. Chem. Int. Ed. Engl.*, 1978, **17**, 27; *Houben-Weyl, Methoden der Organischen Chemie*, Vol. 4/3 and Vol. 13/2b, see above). The loss of halogen (see Eqn. 104) follows the sequence X=F<Cl<Br<I resulting in *carbenes*, which form cyclopropanes with alkenes (D.Seyferth and R.L.Lambert, Jr., *J. Organomet. Chem.*, 1969, **16**, 21; D.Seyferth and C.K.Haas, *J. Org. Chem.*, 1975, **40**, 1620). Thus iodine compounds are most reactive, requiring 1-4 days at room temperature or a few minutes at benzene reflux. With mixed halogen compounds, loss of the most reactive halogen leads to product(see Eqn. 105).

$$PhHgCXYZ + R_2C{=}CR_2 \xrightarrow[\text{or THF, } -PhHgX]{\text{benzene}} \text{cyclopropane (CZY)} \qquad (104)$$

$$CH_2{=}CHCN + PhHgCBrCl_2 \xrightarrow[\text{reflux}]{\text{benzene}} \text{1,1-dichloro-2-cyanocyclopropane} \qquad (105)$$

The reactions occur stereospecifically with *syn*-additon (D.Seyferth *et al., J. Amer. Chem. Soc.*, 1965, **87**, 4259) (see Eqn. 106).

Iodide ions catalyse these reactions, so that even $PhHgCF_3$ will react, releasing $:CF_2$ (D.Seyferth *et al., J Amer. Chem. Soc.*, 1967, **89**, 959; D.Seyferth and S.P.Hopper, *J. Org. Chem.*, 1972, **37**, 4070). The carbenes generated can also add to C=N, C=O and C=S bonds, although they generally require electron withdrawing substituents (D.Seyferth *et al., J. Org. Chem.*, 1974, **39**, 158; *J. Organomet. Chem.*, 1974, **67**, 341; *J. Org. Chem.*, 1972, **37**, 1537)(see Eqn. 107).

1,4-Additions have been recorded with diethyl diazocarboxylate and related compounds (D.Seyferth and Houng-min Shih, *J. Org. Chem.*, 1974, **39**, 2329; 2336) (see Eqn. 108).

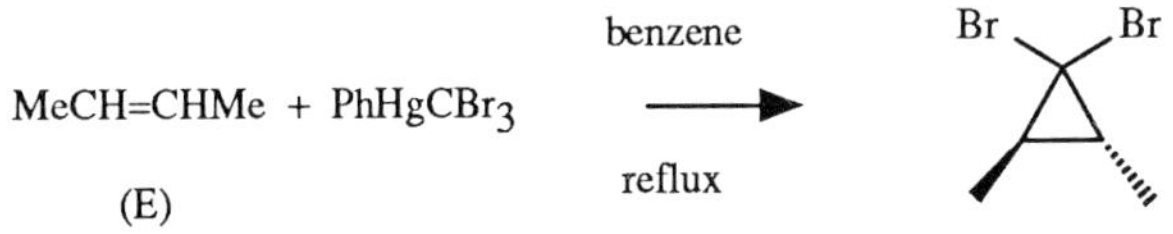

$$MeCH{=}CHMe + PhHgCBr_3 \xrightarrow[reflux]{benzene}$$ (E)

Br Br

(106)

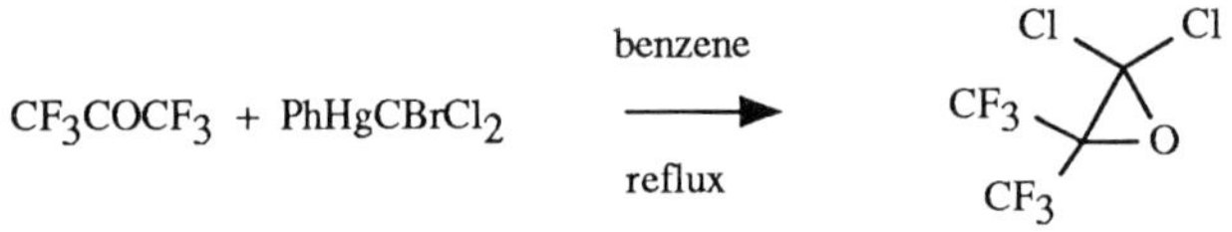

$$CF_3COCF_3 + PhHgCBrCl_2 \xrightarrow[reflux]{benzene}$$

Cl Cl CF3 O CF3

(107)

$$EtO_2CN{=}NCO_2Et + PhHgCBrCl_2 \xrightarrow[reflux]{benzene}$$

EtO2C Cl N N Cl O OEt

$$\longrightarrow (EtO_2C)_2N\text{-}N{=}CCl_2$$

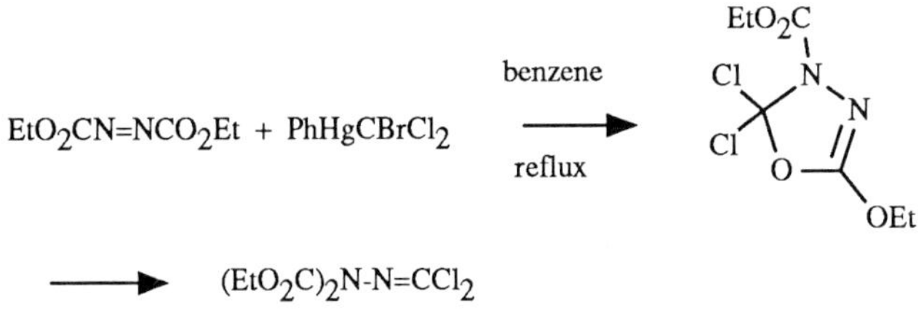

(108)

Acetylenes lead to cyclopropenes and cyclopropanones (D.Seyferth and R.Damrauer, *J. Org. Chem.*, 1966, **31**, 1660; E.V.Dehmlow, *J. Organomet. Chem.*, 1966, **6**, 296), although not with dialkyl substitution (see Eqn. 109).

The use of triphenylphosphine allows the synthesis of halo-alkenes, possibly *via* ylides (D.Seyferth *et al., J. Organomet. Chem.*, 1966, **5**, 267)(see Eqn. 110).

$$ArC{\equiv}CAr + PhHgCBrCl_2 \xrightarrow[\text{reflux}]{\text{benzene}} \text{1,1-dichloro-2,3-diarylcyclopropene (Cl, Cl; Ar, Ar)} \longrightarrow \text{2,3-diarylcyclopropenone (O; Ar, Ar)} \qquad (109)$$

$$RCHO + PhHgCHBrCl \xrightarrow[\text{benzene reflux}]{PPh_3} RCH{=}CHCl \ (E) \qquad (110)$$

3. Transition-metal Catalysed Reactions

i) Alkylations. α-Mercurated carbonyl compounds can transfer the organo-group, as a nucleophile in a directed aldol process (G.A.Artamkina, I.P.Beletskaya, and O.A.Reutov, *J. Organomet. Chem.*, 1972, **42**, C17)(see Eqn. 111). These reactions can be effectively catalysed by nickel tetracarbonyl (I.Rhee *et al., Chem. Letters*, 1979, 1435). Organo-groups can also be transferred as electrophiles (J.Barluenga and A.M.Mastral, *Anal. Quim.*, 1977, **73**, 1032, *Chem. Abs.*, 1978, 88, 105502K; J.Barluenga *et al., J. Chem. Soc., Perkin Trans. 1*, 1980, 1420)(see Eqn. 112). These processes can also be subject to catalysis (see Eqn. 113), although stoichiometric rhodium is required in this case (R.C.Larock and J.Hershberger, *J. Organomet. Chem.*, 1982, **225**, 31).

Copper reagents are more useful (see Eqn. 114), thus organocuprates work smoothly and can be used together with oxymercuration (R.C.Larock and R.D.Leach, *Organometallics*, 1982, **1**, 74).

$$RCOCH_2HgI + {}^1RCHO \xrightarrow[DMF]{Ni(CO)_4} \underset{(E)}{RCOCH{=}CH^1R} \quad (111)$$

$$RHgCl + C_6H_6 \xrightarrow{AlCl_3} PhR \quad (112)$$

$$RHgCl + MeRhI_2(PPh_3)_2 \xrightarrow[Heat]{HMPA} MeR \quad (113)$$

$$Me(CH_2)_3CHOHCH_2HgCl \xrightarrow[O_2/MeI]{Li_2CuMe_2} Me(CH_2)_3CHOHCH_2Me \quad (114)$$

ii) Coupling of alkenyl groups. Palladium and some rhodium complexes bring about rapid, stereospecific coupling of alkenylmercury halides to yield 1,3-dienes (R.C.Larock, *J. Org. Chem.*, 1976, **41**, 2241)(see Eqn. 115). The reaction can be made catalytic, with some loss of stereospecificity, using Pd(0) catalyst (R.C.Larock and J.C.Bernhardt, *ibid.*, 1977, **42**, 1680)(see Eqn. 116); however, rhodium reagents overcome this difficulty (see Eqn. 117).

MeO_2C HgCl $\xrightarrow[HPMA/0^\circ C]{PdCl_2/LiCl}$ MeO_2C CO_2Me

(115)

$$\underset{(E)}{(RCH{=}CH)_2Hg} \xrightarrow[MeCN]{Pd(PPh_3)_4\ cat.} \underset{(E,E)}{RCH{=}CH\text{-}CH{=}CHR} \quad (116)$$

$$\underset{(E)}{RCH{=}CHHgCl} \xrightarrow[\text{LiCl,THF}]{[ClRh(CO)_2]_2\ \text{cat.}} \underset{(E,E)}{RCH{=}CH{-}CH{=}CHR} \qquad (117)$$

The above processes all result in *head-to-head* dimers. By using non-polar reaction conditions *head-to-tail* dimerisation results, and is catalytic in the presence of $CuCl_2$ as reoxidant (R.C.Larock and B.Riefling, *ibid.*, 1978, **43**, 1468)(see Eqn. 118). 1,4-Dienes can be prepared (see Eqn. 119) by coupling with allyl halides (R.C.Larock, J.C.Bernhardt, and R.J.Driggs, *J. Organomet. Chem.*, 1978, **156**, 45).

$$\underset{(E)}{RCH{=}CHHgCl} \xrightarrow[\text{Benzene}/Et_3N/25^\circ C]{PdCl_2} \underset{(E,E)}{RCH{=}CH{-}CR{=}CH_2} \qquad (118)$$

$$\underset{(E)}{{}^tBuCH{=}CHHgCl} \xrightarrow[PdCl_2/LiCl/THF]{\text{allylchloride}} \underset{(E)}{{}^tBuCH{=}CH{-}CH_2CH{=}CH_2} \qquad (119)$$

Synthetically useful π-allylpalladium complexes can be prepared in high yield from alkenylmercury halides and alkenes with palladium chloride and lithium chloride in THF (R.C.Larock and M.A.Mitchell, *J. Amer. Chem. Soc.*, 1976, **98**, 6718). Aryl-, alkoxycarbonyl-, and alkyl- mercury halides, without β-hydrogen atoms, react differently with simple alkenes yielding substituted alkenes by elimination of PdH (R.F.Heck, *ibid.*, 1968, **90**, 5518, 1971, **93**, 6896). The intermediate organo-palladium derivative from the reaction of alkoxycarbonylmercury chlorides (see Eqn. 120) can be trapped using carbon monoxide, yielding substituted succinic esters (R.F.Heck, *ibid.*, 1972, **94**, 2712).

$$PhCH{=}CH_2 + ClHgCO_2Me \xrightarrow[CO/MeOH]{PdCl_2} PhCH(CO_2Me)CH_2CO_2Me \qquad (120)$$

iii) Acylation. 1-Alkenylmercuric halides can be acylated in the presence of tetrakis (triphenylphosphine) palladium/HMPA (K.Takagi *et al., Chem. Letters*, 1975, 951)(see Eqn. 121), however, aluminium chloride in CH_2Cl_2 also gives excellent yields (R.C.Larock and J.C.Bernhardt, *J. Org. Chem.*, 1978, **43**, 710). Dialkyl- and diaryl-mercurials will also react in this fashion, although only one organic group is utilised.

$$\underset{(E)}{PrCH{=}CHHgCl} + \underset{(E)}{MeCH{=}CHCOCl} \xrightarrow[DCM/25^{\circ}C]{AlCl_3} \underset{(E,E)}{Me\,CH{=}CHCOCH{=}CHPr} \qquad (121)$$

iv) Carbonylation. Direct carbonylation of organomercury compounds is difficult, leading to poor yields of carboxylic acids under forcing conditions (J.M.Davidson, *J. Chem. Soc., Chem. Commun.*, 1966, 126; *J. Chem. Soc. A*, 1969, 193). Palladium catalysts are effective with alkenylmercuric halides (see Eqn. 122), which form esters with carbon monoxide at atmospheric pressure in alcoholic solution (R.C.Larock, *J. Org. Chem.*, 1975, **40**, 3237). The addition of $CuCl_2$ or $FeCl_3$ as reoxidant, allows these reactions to become catalytic in $PdCl_2$ (S.M.Brailovski *et al., Zh. Org. Khim.*, 1977, **13**, 1158). The replacement of mercury occurs stereospecifically, with retention of configuration (see Eqn. 123), this has allowed the synthesis of the butenolide ring system (W.C.Baird, R.L.Hartgerink, and J.H.Surridge, *U.S. Patent*, 1975. 3,917,670).

$$\underset{(E)}{RCH{=}C(^{1}R)HgCl} \xrightarrow[LiPdCl_4,\ ^{2}ROH,\ -78\ to\ -25^{\circ}C]{CO(1atm.)} \underset{(E)}{RCH{=}C(^{1}R)CO_2{}^{2}R} \qquad (122)$$

HOCH$_2$C≡CH —($HgCl_2$ / NaClaq.)→ HO–CH$_2$–C(Cl)=CH–HgCl —($PdCl_2$(cat.) / CO/$CuCl_2$, ether)→ 4-chloro-butenolide (O, Cl, O)

(123)

Other catalysts have been used. Thus Wilkinson's catalyst effects carbonylation of phenyl-or alkyl-mercuric acetates (W.C.Baird, Jr., and J.H.Surridge, *J. Org. Chem.*, 1975, **40**, 1364)(see Eqn. 124). Rhodium compounds also catalyse the conversion of alkenylmercuric chlorides to the corresponding ketones (R.C.Larock and J.Hershberger, *ibid.*, 1980, **45**, 3840)(see Eqn. 125).

$$\mathrm{ROAc} \xrightarrow[\mathrm{CO/MeOH}]{\mathrm{ClRh(PPh_3)_3\ cat.}} \mathrm{RCO_2Me + RCO_2H} \qquad (124)$$

$$\underset{(E)}{\mathrm{RCH{=}C(^1R)HgCl}} \xrightarrow[\mathrm{CO/LiCl(2eqiv.)}]{\mathrm{[ClRh(CO)_2]_2}} \underset{(E,E)}{\mathrm{RCH{=}C(^1R)COC(^1R)C{=}CHR}} \qquad (125)$$

Nickel tetracarbonyl (Y.Hirota, M.Ryang, and S.Tsutsumi, *Tetrahedron Letters*, 1971, 1531) or dicobalt octacarbonyl are also recommended, efficiently bring about similar carbonylations of *primary*-alkyl or arylmercuric halides to ketones (D.Seyferth and R.J.Spohn, *J. Amer. Chem. Soc.*, 1968, **90**, 540; *ibid.*, 1969, **91**, 3037; D.Seyferth, J.S.Merola, and C.S.Eschback, *ibid.*, 1978, **100**, 4124) (see Eqn. 126).

$$\mathrm{2RHgBr} \xrightarrow[\mathrm{DMF/60\text{-}70^{\circ}C}]{\mathrm{Ni(CO)_4}} \mathrm{RCOR} \qquad (126)$$

4. Group IIIA: boron, aluminium, gallium, indium, thallium

General reviews of the organo-derivatives of group IIIA elements have appeared (G.Bahr *et al.*, in Houben-Weyl, *Methoden der Organischen Chemie,* Vol. 13/4: Al, Ga, In, Tl, E.Muller *et al.*, Eds., Thieme, Stuttgart, 1970; G.Zweifel, in *Comprehensive Organic Chemistry, Vol. 3,* D.H.R.Barton and W.D.Ollis, Eds., Pergamon, Oxford, 1979). The rearrangement reactions of unsaturated boron and aluminium compounds (J.J.Eisch, *Adv. Organomet. Chem.*, 1977, **16**, 67) and the gas chromatography of group IIIA organometallics (V.A.Chernoplekova, V.M.Sakharov, and K.I.Sakodynskii, *Russ. Chem. Rev. Engl. Trans.*, 1973, **42**, 1063) have also been covered.

a) Boron

The literature covering organoboron chemistry is enormous (I.Haiduc and J.J.Zuckerman, *Basic Organometallic Chemistry*, W.de Gruyer, Berlin, 1985). An attempt will be made to select some general reviews to guide the reader (*Comprehensive Organometallic Chemistry*, Vol. 3, G.Wilkinson, F.G.A.Stone, and E.W.Abel, Eds., Pergamon, Oxford, 1982; Houben-Weyl, *Methoden der Organischen Chemie*, Vol. 13/3, I and II: B, E.Muller *et al*., Eds., Thieme, Stuttgart, 1982-3; *Mellor's Modern Treatise on Inorganic and Theoretical Chemistry, Pt. 5, Boron*, P.Mellor Ed., Longmans, London, 1980, R.Thompson, Ed., 1982; *IMEBORON V,* A.Pelter, Ed., in *Pure Appl. Chem*., 1983, 55; T.P.Onak, *Organoborane Chemistry*, Academic, New York, 1975; *Boron Hydride Chemistry*, E.L.Muetterties, Ed., Academic, New York, 1975). The interest in organoboron chemistry has been stimulated by the prodigious efforts of H.C.Brown and his collaborators (H.C.Brown, *Chem. Eng. News*, 1981, **59**, 24), who have shown the synthetic potential of these reagents (A.Pelter, *Chem. Soc. Rev*. 1980, **11**, 191; H.C.Brown and B.Singaram, *Pure Appl. Chem*., 1987, **59**, 879; A.Pelter, K.Smith, and H.C.Brown, *Borane Reagents*, Academic, London, 1988). A special feature of organoboron chemistry has been their use in chiral synthesis (D.S.Matteson, *Synthesis,* 1986, 973; *Acc. Chem. Res*.,1988, **21**, 294; H.C.Brown, P.K.Jadhav, and B.Singaram, *Mod. Synth. Methods*, 1986, 4, 307; R.W.Hoffmann, *Pure. Appl. Chem*., 1988, 60, 123). The mass spectra (R.H.Cragg and A.F.Weston, *J. Organomet. Chem*., 1974, **67**, 161), N.M.R. spectra (N.Noth and B.Wrackmeyer, *N.M.R. of Boron Compounds*, Springer, Berlin, 1978), and photochemistry (R.F.Porter and L.J.Turbini, *Top. Curr. Chem*., 1981, 96) have been reviewed.

Preparations. A recent review of the synthesis of boranes (D.S.Matteson, in *Chemistry of the Metal-Carbon Bond*, Vol. 4, F.R.Hartley, Ed., Wiley, Chichester, 1987) is available. Only a brief survey of methods will be undertaken here.

i) Boronic acids and esters. These are commonly prepared from Grignard or lithium reagents (e.g. D.S.Matteson and K.H.Arne, *Organometallics*, 1982, **1**, 280)(see Eqn. 127). Trimethylboroxine-pyridine complexes can be useful for isolating and storing some boronic acids (D.S.Matteson, *J. Org. Chem*., 1962, **27**, 3712; 1964, **29**, 3399). Isopropyl esters are readily made from alkyl lithiums (H.C.Brown and T.E.Cole, *Organometallics*, 1983, **2**, 1316)(see Eqn. 128).

Tetra-alkyltin reagents have been used to yield $RBCl_2$ or R_2BCl from BCl_3 (K.Niedenzu, J.W.Dawson, and P.Fritz, *Inorg. Syn*. 1967, **10**, 126; N.Nöth and H.Vahrenkamp, *J. Organomet. Chem*., 1968, **11**, 399).

$$B(OMe)_3 + RMgX \text{ or } RLi \longrightarrow RB(OMe)_3^- MgX^+ \text{ or } Li^+ \xrightarrow{H_2O} RB(OH)_2 \quad (127)$$

$$B(O^iPr)_3 + RLi \rightarrow RB(O^iPr)_3^- Li^+ \xrightarrow{HCl} RB(O^iPr)_2 \quad (128)$$

ii) Alkoxydialkylboranes and trialkylboranes. The use of Grignard, lithium, or aluminium reagents (see Eqn. 129) generally fails to compete with hydroboration as a laboratory method (T.P.Onak, *Organoborane Chemistry*, see above). Application of zirconium catalysed organo-aluminium additions to acetylenes, as precursors to vinylboranes, is illustrated for the synthesis of a 9-borabicyclo[3,3,1]nonane (9-bbn) (E.-i.Negishi and L.D.Boardman, *Tetrahedron Letters*, 1982, **23**, 3327)(see Eqn. 130).

$$RB(OR^1)_2 + R^2MgX \longrightarrow R^2RB(OR^1)$$

$$BX_3 + RM \longrightarrow R_3B$$

$$(M{=}Li, MgX, AlR_2) \quad (129)$$

$$Me(CH_2)_5C{\equiv}CH \xrightarrow[Cp_2ZrCl_2]{Al_2Me_6} Me(CH_2)_5C(Me){=}CHAlMe_2 \text{ (E)}$$

$$\xrightarrow{\text{MeO-9-bbn}} Me(CH_2)_5C(Me){=}CH\text{-}9\text{-}bbn \text{ (E)} \quad (130)$$

iii) Hydroboration. By far the most common method of preparing boranes this was pioneered by H.C.Brown (Nobel Prize 1979). The scope of the process is considerable (H.C.Brown *et al., Organic Synthesis via Boranes*, Wiley, New York 1975).

Ethereal solvents increase reaction rate by assisting dissociation of diborane (B_2H_6+ Et_2O → $2BH_3$ OEt_2). The use of either BH_3. Me_2S complex (R.A.Braun, D.C.Braun, and R.M.Adams, *J. Amer. Chem. Soc.*, 1971, **93**, 2833; C.F.Lane, *J. Org. Chem.*, 1974, **39**, 1438; H.C.Brown, A.K.Mandal, and S.U.Kulkarni, *ibid.*, 1977, **42**, 1392), or the less smelly 1,4-oxathiane-borane (H.C.Brown and A.K.Mandal, *Synthesis*, 1980, 153) also provides a convenient source of BH_3. The sulphur compounds are readily removed by selective oxidation with sodium hypochlorite (H.C.Brown and A.K.Mandal, *J. Org. Chem.*,1980, **45**, 916).

a) *Regioselectivity*. This factor is mainly steric in origin, thus 1-alkenes yield predominately 1-substitution (see Eqn. 131). The use of a hindered RBH_2 or R_2BH (see Scheme 1) increases regioselectivity; these stable boranes are readily prepared (H.C.Brown *et al.*, see above). The borane 9-bbn has been particularly favoured, since useful further reaction tends to occur at the alkyl-boron bond leaving 9-bbn intact (E.F.Knights and H.C.Brown, *J. Amer. Chem. Soc.*, 1968, **90**, 5280; 5281; 5283).

$$RCH{=}CH_2 \xrightarrow{BH_3} (RCH_2CH_2)_3B + B(CHRMe)_3 \quad 94:6 \tag{131}$$

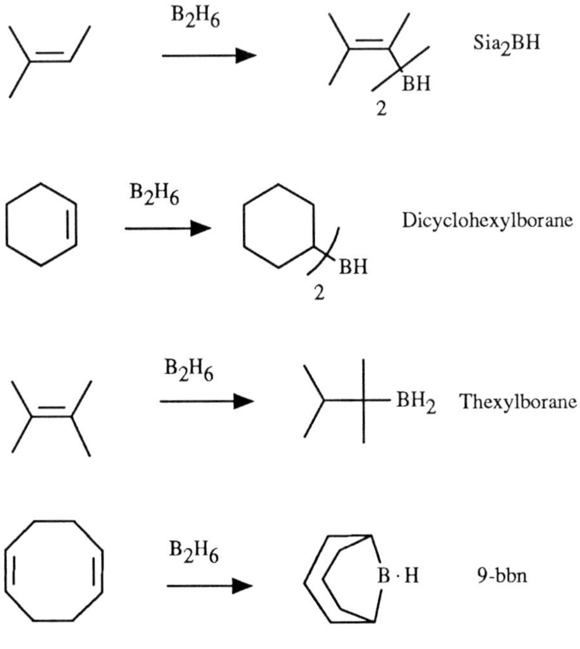

B_2H_6 → Sia_2BH

B_2H_6 → Dicyclohexylborane

B_2H_6 → BH_2 Thexylborane

B_2H_6 → B·H 9-bbn

Scheme 1

Electronic effects can operate. Thus vinyl chloride or vinyl trimethylsilane yield mixtures of 1- and 2-boryl addition products; however, the use of 9-bbn results in the sole formation of a 2-boryl product (H.C.Brown and S.P.Rhodes, *ibid.*, 1969, **91**, 2149; J.A.Soderquist and A.Hassner, *J. Organomet. Chem.*, 1978, **156**, C12).

b) *Stereoselectivity.* Hydroboration occurs *via syn*-addition (cyclic intermediates), which combined with the susceptibility of the reaction to steric effects has given it a central role for diastereoselective additions to alkenes and alkynes. An interesting example is provided by an addition reaction of thexylborane, which shows good diastereoselectivity at non-adjacent chiral centres and regioselective addition to two double bonds (W.C.Still and K.P.Darst, *J. Amer. Chem. Soc.*, 1980, **102**, 7385)(see Eqn. 132).

It should be noted, however, that such systems are complex and the reactions are sometimes difficult to control. Dimesitylborane (dmb) has been shown to be more effective than 9-bbn in additions to alkynes (see Eqn. 133) .

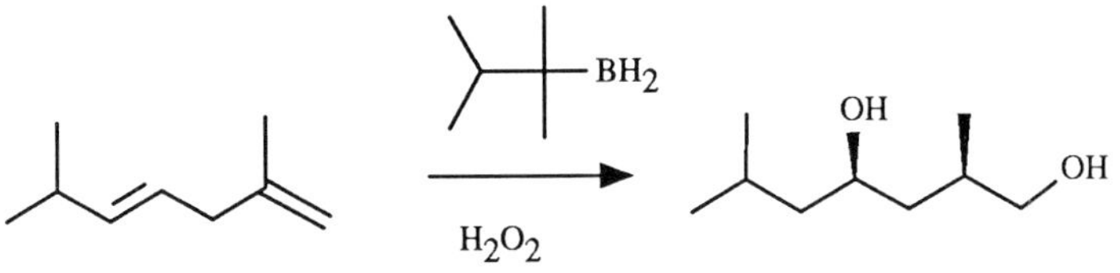

BH_2

OH

OH

H_2O_2

(132)

$$MeCH_2CH_2C{\equiv}CMe \xrightarrow{R_2BH} MeCH_2CH_2CH{=}C(Me)BR_2 + MeCH_2CH_2C(BR_2){=}CHMe$$

when R_2BH = dmb mol ratio 98:2;

when 9-bbn mol ratio 35:65

(133)

c) *Boronic esters and haloboranes.* The substituent groups in these types of compounds has an effect on their reactivity. Thus halogen deactivates, while oxygen greatly deactivates the B-H bond towards hydroboration. Nevertheless, these compounds provide useful alternatives to hindered alkyl groups as blocking functions, as the latter are sometimes not inert during further reactions of the borane.

d) *Boronic esters.* Simple dialkoxyboranes readily disproportionate (A.B.Bury and H.I.Schlesinger, *J. Amer. Chem. Soc.*, 1933, **55**, 4020), however, hindered analogues such as catecholborane hydroborate alkenes (D.S.Matteson and D.Majumdar, *Organometallics*, 1983, **2**, 1529), or alkynes (H.C.Brown, and S.K.Gupta, *J. Amer. Chem. Soc.*, 1971, **93**, 1816; *ibid.*, 1972, **94**, 4370; *ibid.*, 1975, **97**, 5249; H.C.Brown and J.Chandrasekharan, *J. Org. Chem.*, 1983, **48**, 5080) at temperatures of 65-100°C (see Eqn. 134). More reactive still is 1,3,2-dithiaborolane (S.Thaisrivongs and J.D.Wuest, *J. Org. Chem.*, 1977, **42,** 3243), which will react at around 50°C and can be hydrolysed to give a boronic acid (see Eqn. 135).

CO_2Me + B – H

→ B CO_2Me (134)

$MeCH_2CH_2CH_2C{\equiv}CH$ + H–B(S)(S) $\xrightarrow{50^{\circ}C}$

(135)

e) *Haloboranes*. $ClBH_2$ (from BCl_3 and $LiBH_4$) in diethyl ether react readily and completely with alkenes (H.C.Brown and N.Ravindran, *J. Amer. Chem. Soc.*, 1972, **94,** 2112; *ibid.*, 1976, **98**, 1785). The addition of THF (1 equivalent) allows the reaction to be stopped at ClBHR. Alkynes yield products with the *E*-configuration. Cl_2BH-Ether is very unreactive, but BCl_3 (1 equivalent) allows rapid reaction with either alkenes or alkynes (H.C.Brown and H.Ravindran, *ibid.*, 1976, **98**, 1798). Haloborane-Me_2S complexes are even more effective (H.C.Brown and H.Ravindran, *Inorg. Chem.*, 1977, **16**, 2938; *J. Org. Chem.*, 1977, **42**, 2533; *J. Amer. Chem. Soc.*, 1977, **99**, 7097). The bromoborane (see Eqn. 136) allows easy formation of *(Z)*-alkyl-1-enylboronic esters (H.C.Brown and T.Imai, *Organometallics*, 1984, **3**, 1392).

$$RC{\equiv}CBr \xrightarrow[Me_2S]{HBBr_2} R\text{–}CH{=}C(Br)BBr_2 \xrightarrow{^1ROH}$$

$$R\text{–}CH{=}C(Br)B(OR^1)_2 \xrightarrow{KHB(O^iPr)_3} (Z)\text{-}R\text{–}CH{=}CH\text{–}B(OR^1)_2$$

(136)

Uses in synthesis of boranes. The ability to introduce boron with regio- and stereo-selective control, allied to the ease of its replacement in a wide variety of processes, has led to boranes playing an important role in organic synthesis (*Comprehensive Organometallic Chemistry Vol. 7,* see above; D.S.Matteson, in *Chemistry of the Metal-Carbon, Vol. 4*, see above)

1) Oxidative replacement of boron

i) Hydrogen peroxide, this is the reagent of choice for oxidative cleavage of C-B bonds. The reaction is generally quantitative at 0°C, the C-B being replaced with retention of configuration, leading to effective *syn*-addition of H_20 across the double bond. The boron is commonly added to the least substituted site, thus overall addition (see Eqn. 137) tends to be anti-Markownikov (G.W.Kabalka and N.S.Bowman, *J. Org. Chem.*, 1973, **38**, 1607). Aldehydes or ketones result from alkenylboranes; but labile aldehydes may require buffered conditions, instead of the more usual alkaline media (D.S.Matteson and R.J.Moody, *J. Org. Chem.*, 1980, **45**, 1091).

Sodium perborate (D.S.Matteson and R.J.Moody, *loc. cit.*), 3-chloroperbenzoic acid (D.J.Pasto and J.Hickman, *J. Amer. Chem. Soc.*, 1968, **90**, 4445), and Mo-pyridine-HMPA (D.A.Evans, E.Vogel, and J.V.Nelson, *ibid.*, 1979, **101**, 6120) have all been tried as alternatives to hydrogen peroxide for the oxidative displacement process.

ii) Trimethylamine oxide is more selective (less reactive) than hydrogen peroxide. It generally favours alkyl-boron bonds, rather than aryl or alkenyl (R.P.Fisher *et al., Synthesis*, 1982, 127).

iii) Pyridinium chlorochromate is useful for the oxidation of *primary*-alkylboranes to aldehydes (C.G.Rao, S.U.Kulkarni, and H.C.Brown, *J. Organomet. Chem.*, 1979, **172**, C20; *Synthesis*, 1980, 151). and also borate esters, under strictly anhydrous conditions (*idem., Synthesis*, 1979, 702).

1) BD_3/THF; 2) H_2O_2/OH^- — D, OH, H, H (137)

2. Replacement with halogen

Bromine in DCM displaces boron by a free radical process with loss of stereochemistry (C.F.Lane and H.C.Brown, *J. Amer. Chem. Soc.*, 1970, **92,** 7212), while in MeONa/MeOH either bromine or iodine can be employed to cleave alkyl-boron bonds with predominant inversion (H.C.Brown and C.F.Lane, *ibid.*, 1970, **92**, 6660; H.C.Brown, M.W.Rathke, and M.M.Rogit, *ibid.*, 1968, **90**, 5038; H.C.Brown *et al., ibid.*, 1976, **98**, 1290). The most useful stereospecific replacement for alkenylboranes involves a halogenation-deborohalogenation sequence (D.S.Matteson and J.D.Liedtke, *ibid.*, 1965, **87**, 1526)(see Eqn. 138). The radio-labelled halogenation of alkenylboranes has been reviewed (G.W.Kabalka, *Acc. Chem. Res.*, 1984, **17**, 25).

$B(OR)_2$ — Br_2 → H, Br, $B(OR)_2$, Br — NaOH → Br (138)

3. Replacement with nitrogen

Introduction of the NH_2 group can be achieved using hydroxylamine sulphonic acid (M.W.Rathke *et al., J. Amer.Chem. Soc.*, 1966, **88**, 2870) or with chloramine formed *in situ* (G.W.Kabalka *et al., J. Org. Chem.*, 1981, **46**, 4296). Controlled syntheses of *secondary*-amines are possible using alkylazides and alkyldichloroboranes (H.C.Brown, M.M.Midland, and A.B.Levy, *J. Amer. Chem. Soc.*, 1973, **95**, 2394).

4. Replacement with hydrogen

Alkylboranes are usually smoothly protodeboronated using propanoic acid/diglyme at reflux (H.C.Brown and K.Murray, *J. Amer. Chem. Soc.*, 1959, **81**, 4108), it is possible to remove only one alkyl group using methanesulphonic acid (S.Trofimenko, *ibid.*, 1969, **91**, 2139). Alkenylboranes are more easily cleaved, thus acetic acid at 0°C is suitable (H.C.Brown and G.Zweifel, *ibid.*, 1959, **81**, 1512).

5. Replacement with mercury

Alkylboranes react with $Hg(OAc)_2$ at 0-25°C to yield alkylmercuric acetates (R.C.Larock and H.C.Brown, *ibid.*, 1970, **92**, 2467; J.D.Buhler and H.C.Brown, *J. Organomet. Chem.*, 1972, **40**, 265), which provides an effective anti-Markownikov addition of acetic acid to 1-alkenes (R.C.Larock, *J. Org. Chem.*, 1974, **39**, 834)(see Eqn. 139). Alkenylboranes react with retention of configuration (S.A.Kunda, R.S.Varma, and G.W.Kabalka, *Synth. Commun.*, 1984, **14**, 755).

$$RCH{=}CH_2 \xrightarrow[2)\ Hg(OAc)_2]{1)\ BH_3} RCH_2CH_2HgOAc \xrightarrow{I_2} RCH_2CH_2OAc \qquad (139)$$

6. Carbon-carbon bond formation

i) α-Haloalkylborates undergo rearrangement (see Eqn. 140), a process uniquely useful in directed chiral synthesis (D.S.Matteson and R.W.H.Mah, *J. Amer. Chem. Soc.*, 1963, **85,** 2599) (see below).

A powerful general synthetic procedure has been developed using $LiCHCl_2$ and boronic esters (D.S.Matteson and D.Majumdar, *Organometallics*, 1983, **2**, 1529; *J. Amer. Chem. Soc.*, 1980, **102**, 7588)(see Eqn. 141). The problem of diastereoisomer formation has been solved by use of chiral boronates (see below)

$$CH_2{=}CHB(OBu)_2 \xrightarrow[\text{radicals}]{Cl_3CBr} Cl_3CCH_2CH(Br)B(OBu)_2$$

$$\xrightarrow[-78\,^{\circ}C]{RMgX} Cl_3CCH_2CH(Br)B^{-}(R)(OBu)_2 \rightarrow$$

$$\xrightarrow[25C^{\circ}]{H^{+}} \begin{cases} Cl_3CCH_2CH(Br)BR(OBu) \\ Cl_3CCH_2CH(R)B(OBu)_2 \end{cases}$$

(140)

$$\xrightarrow[-100^{\circ}C]{LiCHCl_2} \quad \xrightarrow{0\text{-}25^{\circ}C} \quad \xrightarrow{LiCH_2CO_2{}^{t}Bu} \quad \xrightarrow{LiCHCl_2}$$

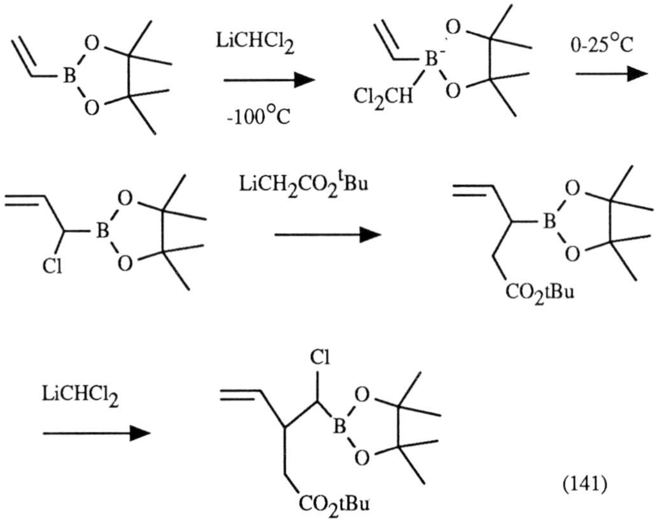

(141)

ii) α-Halotrialkylboranes are the initial products of free radical bromination of trialkylboranes (cleaved by HBr); in the presence of H_2O these intermediate rearrange (see Eqn. 142). *N*-Bromnosuccinimide is preferred as the reagent for this type of reaction (H.C.Brown and Y.Yamamoto, *J. Amer. Chem. Soc.*, 1971, **93**, 2796; *Synthesis*, 1972, 699).

A similar reaction (see Eqn. 143) is known with α-haloalkenylboranes (E.-i.Negishi *et al., J. Organomet. Chem.*, 1975, **92**, C4).

$$Et_3B \xrightarrow{Br_2} Et_2BCH(Me)Br \xrightarrow{H_2O} EtB(OH)CH(Me)Et \quad (142)$$

$$\underset{}{BuCHC{\equiv}CI} \xrightarrow{R_2BH} \underset{(Z)}{BuCH{=}C(I)BR_2} \xrightarrow{LiBHEt_3} \underset{(Z)}{BuCH{=}CHBR_2} \quad (143)$$

Anions derived from α-halo-esters (ketones, nitriles etc.) provide a neat way of generating α-halo-borate complexes (see Eqn. 144) which undergo rearrangement (H.C.Brown *et al., J. Amer. Chem. Soc.*, 1968, **90**, 818), the resultant alkylation takes place with retention of configuration.

$$R_3B + BrCH_2CO_2Et \xrightarrow{{}^tBuOK,{}^tBuOH} R_2B^-(R)CH(Br)CO_2Et$$

$$\longrightarrow R_2BCH(R)CO_2Et \longrightarrow R_2BO(OEt)C{=}CHR$$

$$\xrightarrow{Bu^tOH} R_2BO^tBu + RCH_2CO_2Et \quad (144)$$

iii) Carbonylation. The scope of this reaction has been reviewed (H.C.Brown, *Science*, 1980, **210**, 485). Various hydrides are effective in promoting the absorption of carbon monoxide (H.C.Brown, J.L.Hubbard, and K.Smith, *Synthesis*, 1979, 701; H.C.Brown and J.L.Hubbard, *J. Org. Chem.*, 1979, **44**, 467). Subsequent treatment leads to aldehydes or alcohols (see Eqn. 145).

iv) Cyanidation. Sodium cyanide and trifluoroacetic anhydride, followed by oxidation converts boranes to ketones or tertiary alcohols (A.Pelter, M.G.Hutchings, and K.Smith, *J. Chem. Soc., Perkin Trans. 1*, 1975, 142; A.Pelter, A.Arase, and M.G.Hutchings, *J. Chem. Soc., Chem. Commun.*, 1974, 346)(see Eqn. 146).

v) Thiol substituted carbanions. Reaction with boranes leads to the introduction of aldehyde equivalents (A.Mendoza and D.S.Matteson, *J. Organomet. Chem.*, 1978, **156**, 149), generally with retention of configuration (H.C.Brown and T.Imai, *J. Amer. Chem. Soc.*, 1983, **105**, 6285)(see Eqn. 147).

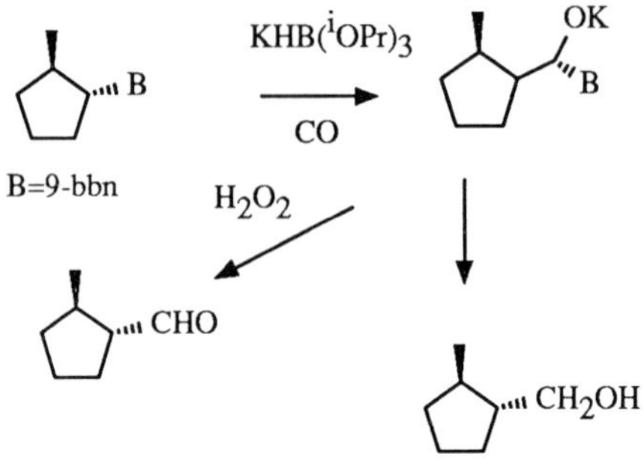

KHB(^{i}OPr)$_3$, CO; H_2O_2

B=9-bbn

(145)

$$R_2BCH{=}CHR^1 \xrightarrow[2)\ H_2O_2]{1)\ NaCN/\ (CF_3CO)_2O} RCOCH{=}CHR^1$$

(E) (E)

(146)

LiCH(SPh)OMe; H_2O_2

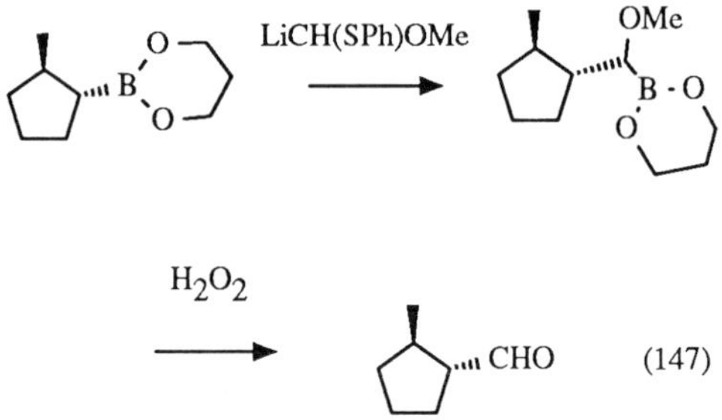

(147)

vi) Alkenyl and alkynyl boranes undergo a number of rearrangements leading to carbon-carbon bond formation (D.S.Matteson, in *The Chemistry of the Metal-Carbon Bond*, Vol. 4, Ch., 3, F.R.Hartley, Ed., Wiley, Chichester, 1987). The Zweifel alkene synthesis provides *Z* or *E*-alkenes (G.Zweifel, H.Arzoumanian, and C.C.Whitney, *J. Amer. Chem. Soc.*, 1967, **89**, 3652; G.Zweifel and H.Arzoumanian, *ibid.*, 1967, **89**, 5086)(see Eqns. 148 and 149). These procedures have found wide application in, for example, prostaglandin synthesis (D.A.Evans, *et al., J. Org. Chem.*, 1976, **41**, 3947). Many other electrophiles bring about alkenylborate rearrangement giving rise to a wide range of products (A.B.Levy and S.J.Schwartz, *Tetrahedron Letters,* 1976, 2201; A.B.Levy *et al., J. Organomet. Chem.*, 1978, **156**, 123) (see Eqn. 150).

$$R_2BH + R^1C{\equiv}CH \rightarrow R^1CH{=}CHBR_2 \xrightarrow{I_2/NaOH} R^1CH{=}CHR \ (Z) \quad (148)$$

$$R_2BH + R^1C{\equiv}CI \rightarrow R^1CH{=}CIBR_2 \xrightarrow[2)\ AcOH]{1)\ NaOMe} R^1CH{=}CHR \ (E) \quad (149)$$

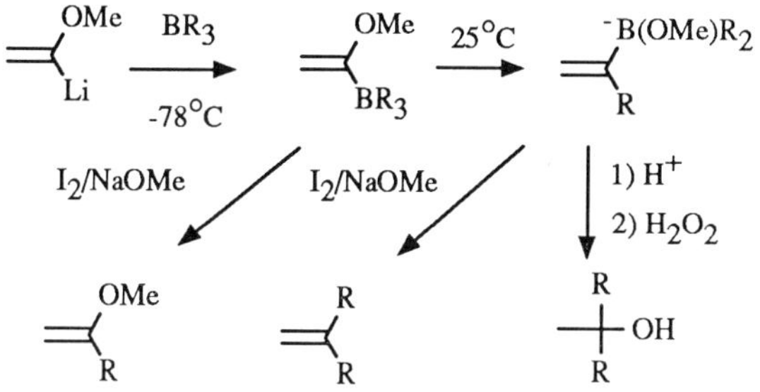

OMe, Li vinyl → (BR_3, -78°C) → vinyl(OMe)(BR_3) → (25°C) → vinyl($^-B(OMe)R_2$)(R)

I_2/NaOMe → vinyl(OMe)(R); I_2/NaOMe → vinyl(R)(R); 1) H^+ 2) H_2O_2 → $R_2C(OH)$

(150)

The use of aldehydes as electrophiles also yields interesting products (K.Utimoto, K.Uchida, and N.Nozaki, *Tetrahedron,* 1977, **33**, 1949)(see Eqn. 151). Alkynylboranes provide, amongst many other applications, a route to acetylenic ketones (H.C.Brown, U.S.Racherla, and S.M.Singh, *Tetrahedron Letters,* 1984, **25**, 2411)(see Eqn. 152).

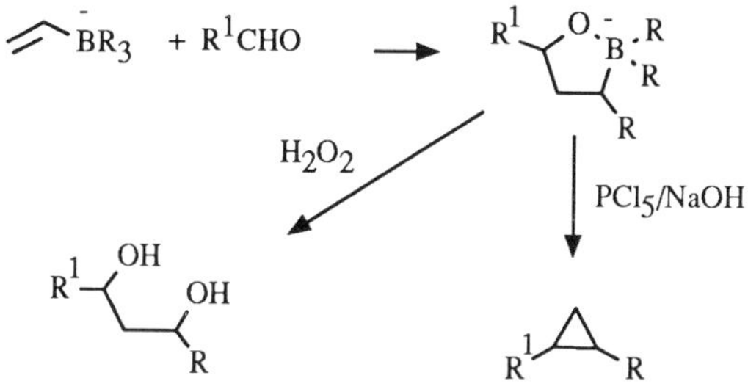

allyl-$\bar{B}R_3$ + R^1CHO → cyclic borate (R^1, O, B^-R_2, R)

H_2O_2 → $R^1CH(OH)CH_2CH(OH)R$; PCl_5/NaOH → cyclopropane (R^1, R)

(151)

$$RC{\equiv}CLi + BF_3.Et_2O \rightarrow RC{\equiv}CB^-F_3 \xrightarrow{(^1RCO)_2O} RC{\equiv}CCOR^1 \quad (152)$$

vii) Allyl borane chemistry has been reviewed (B.M.Mikhailov, *Pure Appl. Chem.*, 1983, **55**, 1439). The displacement of boron by carbonyl compounds occurs with allylic rearrangement, and in a stereoselective manner. This is an important procedure and may give rise to products exhibiting either *erythro-* (R.W.Hoffmann and H.J.Zeiss, *J. Org. Chem.*, 1981, **46**, 1309) or *threo*-geometries (Y.Yamamoto, H.Yatagai, and K.Maruyama, *J. Amer. Chem. Soc.*, 1981, **103**, 3229) (see Eqns. 153 and 154).

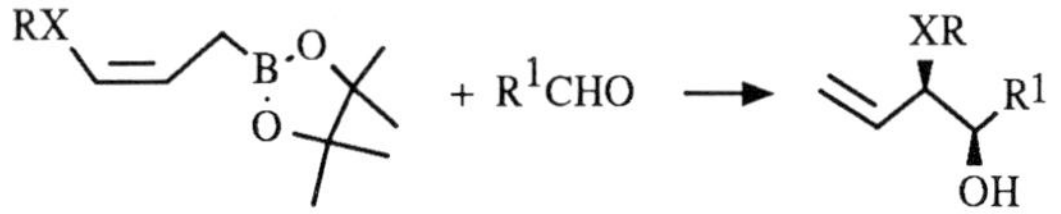

RX ... B(O,O) + R^1CHO → XR, R^1, OH (153)

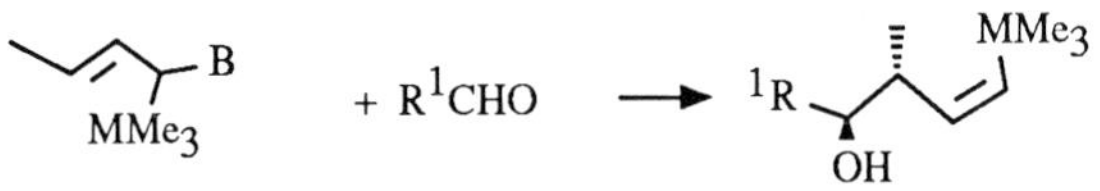

MMe$_3$, B + R^1CHO → ^{1}R, OH, MMe$_3$ (154)

M = Sn or Si

viii) Transition-metal catalysis promotes cross-coupling reactions (E.-i.Negishi, *Acc. Chem. Res.,* 1982, **15**, 340). Useful examples of reactions catalysed by palladium are illustrated below (N.Miyaura, K.Yamada, and A.Suzuki, *Tetrahedron Letters*, 1979, 3437) and copper (N.Miyaura, M. Itoh, and A.Suzuki, *Bull. Chem. Soc. Jpn.*, 1977, **50**, 2199) (see Eqns. 155 and 156).

$$\underset{(E)}{BuCH{=}CHBSia_2} + PhBr \xrightarrow[NaOEt]{Pd(PPh_3)_4} \underset{(E)}{BuCH{=}CHPh} \quad (155)$$

$$R_3B + \underset{(Z)}{BrCH{=}CHCO_2Et} \xrightarrow{MeCu} \underset{(Z)}{RCH{=}CHCO_2Et} \quad (156)$$

ix) β-Elimination of boron along with halogen has been used to effect stereospecific cyclopropane syntheses (H.L. Goering and S.L.Trenbeath, *J. Amer. Chem. Soc.*, 1976, **98**, 5016).

x) Radical alkylations using boranes lead to the formation of new carbon-carbon bonds (H.C.Brown, *Boranes in Organic Chemistry*, Cornell Press, Ithaca, 1972).

7. Chiral Synthesis

i) α-Chloroboronic esters derived from α-pinenediol have been used to achieve high diastereoselectivity (>95%) in synthesis (D.S.Matteson and R.Ray, *J. Amer. Chem. Soc.*, 1980, **102**, 7590; D.S.Matteson *et al.*, *Organometallics*, 1983, **2**, 1536; D.S.Matteson, K.M.Sadhu, and M.L.Peterson, *J. Amer. Chem. Soc.*, 1986, **108**, 810)(see Eqn. 157). However, better results have been obtained using (*R,R*)-butane-2,3-diol in place of the pinenediol, since this ancillary group is more readily removed (K.M.Sadhu *et al.*, *Organometallics*, 1984, **3**, 804) .

ii) Isopinocamphenylboranes have found use as chiral hydroborating agents (see A.K.Mandal, *Tetrahedron*, 1981, **37**, 3547; H.C.Brown, and P.K.Jadhav, in *Asymmetric Synthesis*, Vol. 2, H.C.Brown J.D.Morrison, Ed., Academic Press, New York, 1983). Diastereoselectivity can be essentially complete (A.K.Mandal, P.K.Jadhav, and H.C.Brown, *J. Org. Chem.*, 1980, **45**, 3543)(see Eqn. 158).

iii) Chiral sigmatropic rearrangements of allylboranes (R.W. Hoffmann, *Angew. Chem. Int. Ed. Eng.*, 1982, **21**, 555) occur with good stereoselectivity. An interesting example (see Eqn. 159) is accomplished with 96% ee (H.C.Brown and P.K.Jadhav, *Tetrahedron Letters*, 1984, **25**, 1215). Similar results have been observed with allenylboronic esters (R.Haruta, *et al.*, *J. Amer. Chem. Soc.*, 1982, **104,** 7667).

iv) Chiral reductions using boron reagents have been reviewed (M.M.Midland, in *Asymmetric Synthesis*, Vol.2, J.D.Morrison, Ed., Academic Press, New York, 1983).

Me_3NO / OsO_4 → HO, HO (pinanediol) ≡ HO–(R)–HO → R–B(O)(O)(R) → [1) $LiCHCl_2$/100°C; 2) $ZnCl_2$/25°C] → Cl, R, H–C–B(O)(O)(R)

(157)

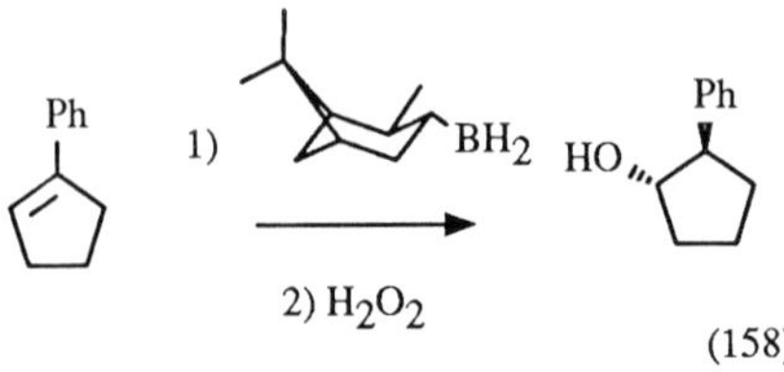

Ph 1) BH$_2$ HO Ph

2) H_2O_2

(158)

1) CHO -78°C

B

2

2) H_2O_2

OH

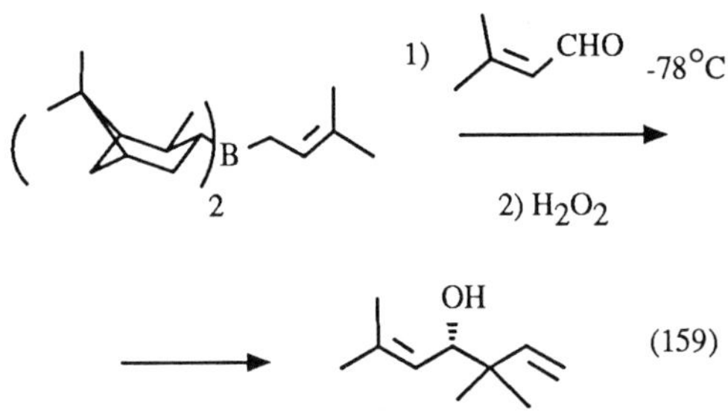

(159)

b) Aluminium

The chemistry of organoaluminium compounds is discussed in a number of reviews (B.B.Snider, *Acc. Chem. Res.*, 1980, **13**, 426; J.J.Eisch in Vol. 1 and J.R.Zietz, G.C.Robinson, and K.L.Lindsay in Vol. 7, *Comprehensive Organometallic Chemistry*, Pergamon, Oxford, 1980; G.Zwiefel, in *Comprehensive Organic Chemistry*, Vol. 3, D.Barton and W.D.Ollis, Eds., Pergamon, Oxford 1979; H.Yamamoto and H.Nozaki, *Angew. Chem. Int. Ed. Engl.*, 1978, **17**, 169).

Organoaluminium compounds are of great importance as catalysts in Ziegler-Natta polymerisation (H.Sinn and W.Kaminsky, *Adv. Organomet. Chem.*, 1980, **18**, 99; J.Boor, *Ziegler-Natta Catalysts and Polymerizations*, Academic, New York, 1978). They have also proved useful as reagents in organic synthesis. Although difficulties with preparation and handling have tended to limit their use, simple alkyl derivatives (R_3Al, R_2AlH, R_2AlCl, and $RAlCl_2$, where R = Me, Et or Bu^i) are commercially available (P.A.Chaloner, in *The Chemistry of the Metal-Carbon Bond*, Vol. 4, Ch. 4, F.R.Hartley, Ed., Wiley, Chichester, 1987; E.Winterfeldt, *Synthesis*, 1975, 617). Useful N.M.R. data for organoaluminium (and gallium) compounds has been published (J.W.Akitt, *Ann. Rev. N.M.R. Spectroscopy*, 1972, **5A**, 465).

Uses in Synthesis. As with boron, aluminium is fairly readily introduced and can be replaced by a wide range of electrophiles. This has made aluminium compounds useful in a number of synthetic roles.

1. Reactions with alkenes

i) Hydroalumination. Aluminium hydrides are well known to add to alkenes (K.Ziegler, *Angew. Chem.*, 1956, **68**, 721) with high selectivity for the primary carbon site (see Eqn. 160). These processes are facilitated by transition-metal (e.g. Ni, Ti, and Co) catalysis (E.C.Ashby and J.J.Lin, *J. Org. Chem.*, 1978, **43**, 2567). The initially formed alanes also react with a wide range of electrophiles (F.Sato *et al.*, *J. Organomet. Chem.*, 1977, **142**, 71). These processes are effectively catalysed by copper(II) salts (F.Sato *et al.*, *ibid.*, 1978, **157**, C30; *Chem. Letters*, 1978, 789;833; *ibid.*, 1979, 167; 623; K.Isagawa, *ibid.*, 1978, 1155). Selective reduction can be achieved in this way (J.Tsuji and T.Mandai, *ibid.*, 1977, 975)(see Eqn. 161). Zirconium complexes have been used to effect hydroalumination with $Bi^i{}_3AlCl$ (E.-i.Negishi and T.Yoshida, *Tetrahedron Letters.*, 1980, **21**, 1501).

$$RCH{=}CH_2 \xrightarrow[TiCl_4]{LiAlH_4} Li[(RCH_2CH_2)_4Al] \xrightarrow{X_2} RCH_2CH_2X \qquad (160)$$

$$CH_2{=}CH(CH_2)_3CH(OAl^iBu_2)CH{=}CH_2 \xrightarrow[2)\ H_2O]{1)\ LiAlH_4/Cp_2TiCl_2} CH_3(CH_2)_4CH(OH)CH{=}CH_2 \qquad (161)$$

ii) Carboalumination. Reactions of this type are more important with alkynes, but are also known to occur with alkenes (see Eqn. 162). Ti and Zr reaction catalysis is common (J.J.Barber, C.Wills, and G.M.Whitesides, *J. Org. Chem.*, 1979, **44**, 3603).

$$RCH{=}CH_2 + R^1{}_3Al \xrightarrow{Cp_2TiCl_2} R,R^1CHCH_2AlR^1{}_2 \longrightarrow R,R^1C{=}CH_2 + HAlR^1{}_2 \qquad (162)$$

2. Reactions with alkynes

i) Hydroalumination. Alkynes react more rapidly than alkenes (A.P.Kozikawa and Y.Kitigawa, *Tetrahedron Letters*, 1982, **23**, 2087) generally with good stereoselectivity (see Eqn. 163).

R_2AlH usually results in overall *syn*-addition, while with $RAlH_2$, AlH_3, $LiAlH_4$ etc. *anti*-addition results (W.Granitzer and A.Stutz, *Tetrahedron Letters*, 1979, 3145). Aluminium can be removed by treatment with water, leading to *syn*-reduction with Bu^i_2AlH (W.J.Gensler and J.J.Bruno, *J. Org. Chem.*, 1963, **28**, 21254).

$^iBu_2AlH$ → Al-i-Bu_2 (163)

ii) Carboalumination. These reactions are much slower, although Et_3Al reacts to yield *syn*-addition products (J.J.Eisch and J.M.Biedermann, *J. Organomet. Chem.*, 1971, **30**, 167) (see Eqn. 164). Transition-metal catalysis enhances the reactivity (D.E.VanHorn and E.-i.Negishi, *J. Amer. Chem. Soc.*, 1978, **100**, 2252; E.-i.Negishi, *Pure Appl. Chem.*, 1981, **53**, 2333), Ti and Zr complexes being especially useful. The resulting vinyl alanes readily undergo reaction with electrophiles leading to numerous useful products (see Eqn. 165). Initial reaction with *n*-butyl lithium generates "ate" complexes, which are even more reactive toward electrophiles (N.Okukado and E.-i.Negishi, *Tetrahedron Letters*, 1978, 2357).

Treatment with mercuric chloride (see Eqn. 166) provides a simple stereoselective route to organomercurials (E.-i.Negishi and L.D. Boardman, *Tetrahedron Letters.*, 1982, **23**, 3327).

Ph-C≡C-Ph → 1) Et_3Al; 2) H_2O → Ph, Ph (164)

3. Reaction with carbonyl compounds

i) Aldehydes and ketones.

a) *Organoalanes* are generally less effective than Grignard or lithium reagents for alkylating these compounds. However, the good selectivity shown by the readily available vinylalanes has made them popular reagents (H.Newman, *J. Amer. Chem. Soc.*, 1973, **95**, 4098)(see Eqn. 167). Allylalanes also react, but with allylic rearrangement (F.Barbot, *Bull. Soc. Chim. Fr.*, 1984, 83).

b) Bu^i_2AlH is an effective reducing agent for aldehydes and ketones (J.A.Oakleaf *et al.*, *Tetrahedron Letters*, 1978, 1645) (see Eqn. 168).

The high stereoselectivity of this large reagent has found considerable use. Indeed, increased selectivity has been achieved by addition of bulky coordinating hydroxy compounds (M.Suzuki, T.Kawagashi, and R.Noyori, *Tetrahedron Letters*., 1982, **23**, 5563), or conversion to the highly hindered "ate" complex, Li ($Bu^i_2Bu^tAlH$) by treatment with Bu^tLi (S.Kim, K.H.Ahn, and Y.W.Chung, *J. Org. Chem.*, 1982, **47**, 4581).

c) The formation of aluminium enolates provides access to directed aldol products (K.Maruoka *et al.*, *J. Amer. Chem. Soc.*, 1977, **99**, 7705; *Bull. Chem. Soc. Jpn.*, 1980, **53**, 3301; J.Tsuji *et al.*, *Tetrahedron Letters*, 1979, 2257; *Bull. Chem. Soc. Jpn.*, 1980, **53**, 1417).

$$MeO_2CCH_2CH(NR_2)CHO + {}^iBu_2AlCH{=}CHR^1 \longrightarrow$$
(E)

$$MeO_2CCH_2CH(NR_2)CH(OH)CH{=}CHR^1$$
(E) (167)

CHO $\xrightarrow{Bu^i_2AlH}$ CH_2OH

(168)

ii) Carboxylic acid derivatives.

a) *Carboxylic acids* can be reduced to ketones (Y.Yamamoto, H.Yatagi, and K.Maruyama, *J. Org. Chem.*, 1980, **45**, 195), or aldehydes (T.D.Hubert, D.P.Eyman, and D.F.Wiemer, *J. Org. Chem.*, 1984, **49**, 2279) using organoalanes.

b) *Acid chlorides* are effectively reduced to ketones (H.Reinheckel, K.Haage, and D.Jahnke, *Organomet. Chem. Rev.*, 1969, **4**, 47).

4. Reactions at carbon-oxygen σ-bonds

i) Epoxides are smoothly opened by organoaluminium compounds with inversion at the carbon atom attacked; "ate" complexes may give improved yields. Use of excess alane results in attack at the most substituted carbon atom (J.-L.Namy, G.Boireau, and D.Abenhaim, *Bull. Soc. Chim. Fr.*, 1971. 3191)(see Eqn. 169). Vinyl and alkynylalanes also react. The reactions of the latter are especially valuable in view of the lack of reactivity of the corresponding Grignard and lithium reagents. These processes often exhibit useful regio- and stereoselectivity (R.S.Matthews *et al., J. Org. Chem.*, 1983, **48**, 409)(see Eqn. 170). The use of Et_2AlCN to open oxiranes with introduction of a cyano group has found use in carbohydrate chemistry (A.Murbarak and B.Fraser-Reid, *J. Org. Chem.*, 1982, **47**, 4265). Dialkylaluminium amides yield aminoalcohols with oxiranes (L.E.Overman and L.A.Flippin, *Tetrahedron Letters*, 1981, **22**, 195), however, sterically crowded examples, such as diethylaluminium-2,2,6,6-tetramethylpiperidide (datmp) cause isomerisation to allylic alcohols (A.Yasuda, H.Yamamoto, and H.Nozaki, *Bull. Chem. Soc. Jpn.*, 1979, **52**, 1705; H.Yamamoto and H.Nozaki, *Angew. Chem. Int. Ed. Engl.*, 1978, **17**, 169)(see Eqn. 171). Similar isomerisations have been caused by Bu^i_2AlH and by Bu^i_3Al (E.Winterfeldt, *Synthesis*, 1975, 617).

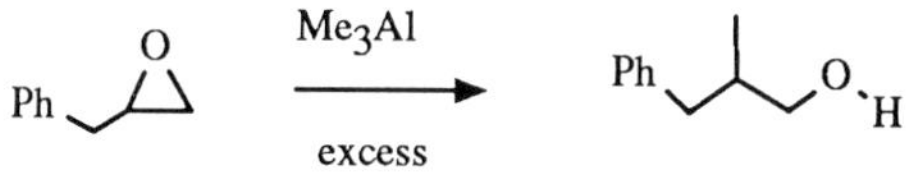

Me_3Al, excess

(169)

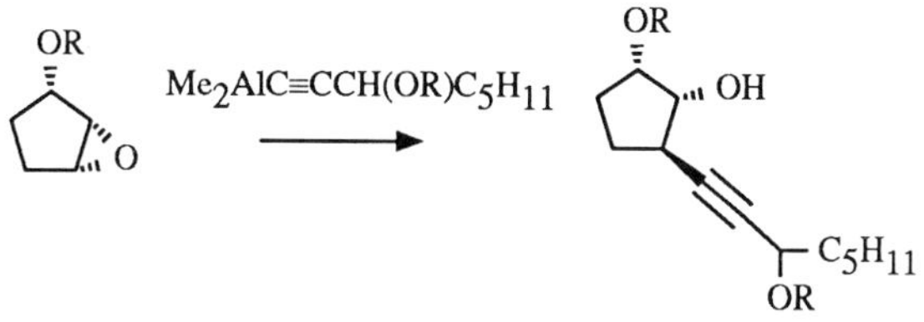

$Me_2AlC{\equiv}CCH(OR)C_5H_{11}$

(170)

datmp

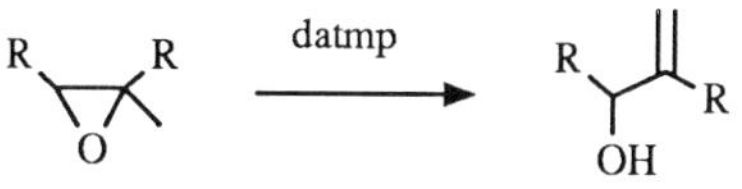

(171)

ii) Esters and lactones. Organoaluminiums displace carboxylate groups, usually with inversion at the carbon attacked (see Eqn. 172); Pd(0) catalysis promotes these reactions (H.Matsushita and E.-i.Negishi, *J. Chem. Soc., Chem. Commun.*, 1982, 160). Sulphonates in the presence of Pd(0) catalyst undergo similar displacements (C.G.M.Janssen and E.F.Godefroi, *J. Org. Chem.*, 1982, **47**, 3274) and vinyl phosphates (K.Takai *et al.*, *Bull. Chem. Soc. Jpn.*, 1984, **57**, 108).

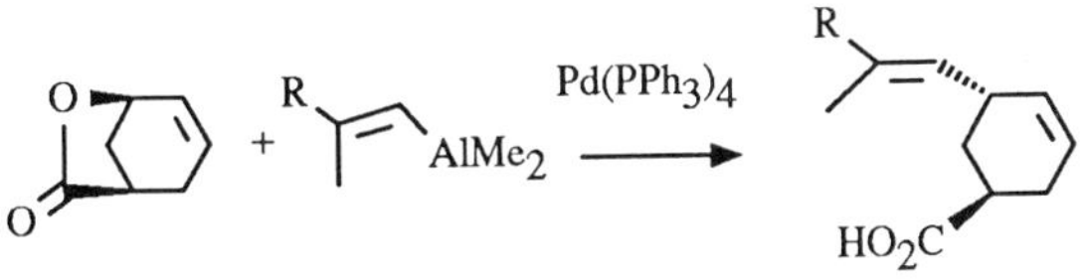

$Pd(PPh_3)_4$

(172)

5. Reactions at carbon-halogen bonds.

Alkyl halides react readily, while aryl halides (E.-i.Negishi, *Acc. Chem. Res.*, 1982, **15**, 340), allyl halides (H.Matsushita and E.-i.Negishi, *J. Amer. Chem. Soc.*, 1981, **103**, 2882), or vinyl halides (S.Baba and E.-i. Negishi, *J. Amer. Chem. Soc.*, 1976, **98**, 6720; N.Okykado *et al., Tetrahedron Letters,* 1978, 1027) couple in the presence of Ni(0) or Pd(0) catalysts (see Eqn. 173).

ML_4

R Al-i-Bu_2 + R I → R R

99% E,E; M=Pd

95% E,E; M=Ni

(173)

6. Aluminium-promoted rearrangement reactions

Organoaluminium compounds provide a range of Lewis acid acidities: $R_3Al<R_2AlCl<RAlCl_2$, with the added ability to scavenge protons (S.Miyajima and T.Inukai, *Bull. Chem. Soc. Jpn.*, 1972, **45**, 1553), this makes them useful catalysts for a range of Lewis acid mediated rearrangement reactions.

i) Cycloadditions. The role of aluminium catalysts in Diel-Alder reactions has been reviewed (H.Wollweber, *Diels-Alder Reactions*, Georg Theime, Stuttgart, 1972). The coordination of aluminium can lead to impressive control of stereochemistry (T.Cohen and Z.Kosarych, *J. Org. Chem.*, 1982, **47**, 4005)(see Eqn. 174)

ii) The ene and related reactions. Organo- aluminiums are effective catalysts for the ene reaction, in which formaldehyde is a commonly used "enophile" (B.B.Snider *et al., J. Org. Chem.*, 1983, **48**, 464; 3003)(see Eqn. 175).

iii) Pinacol-pinacolone rearrangements. The variations in Lewis acidity which are available allow conditions to be selected for stereospecific migration of alkyl groups (G.I.Tsuchihashi, K.Tomooka, and K.Suzuki, *Tetrahedron Letters*, 1984, **25**, 4253)(see Eqn. 176), and also of aryl or vinyl groups (K.Suzuki *et al., ibid.*, 1983, **24**, 4997; 1984, **25**, 3715).

$EtAlCl_2$

MeO SPh + O → MeO SPh O

(174)

1) CH_2O/Me_2AlCl

2) Ac_2O

AcO

1) $CH_2O/EtAlCl_2$

2) H_2O

HO HO

(175)

H MsO OH

Et_2AlCl

-78°C

H O

(176)

c) Gallium and Indium

The comparative rarity and high cost of these elements has limited the practical use of their organo-derivatives. Organogallium and organoindium chemistries are reminiscent of that of the organoaluminiums (D.G.Tuck, in *Comprehensive Organometallic Chemistry*, Vol. 1, G.Wilkinson, F.A.Stone, and E.W.Abel, Eds., Pergamon, Oxford, 1982; G.Bähr, *et al.*, in Houben-Weyl, *Methoden der Organischen Chemie*, 4th Ed., Vol. 13/4 : Al, Ga, In, Tl, E.Muller, Ed., Thieme, Stuttgart, 1970; I.A.Sheka, I.S.Chares, and T.T.Mityueva, *The Chemistry of Gallium*, Elsevier, Amsterdam, 1966)

d) Thallium

A large range of organothallium (III) and organothallium (I) compounds have been prepared and characterised (A.G.Lee, *The Chemistry of Thallium*, Elsevier, Amsterdam, 1971; H.Kurosawa, in *Comprehensive Organometallic Chemistry*, Vol. 1, G.Wilkinson, F.G.A.Stone, and E.W.Abel, Eds., Pergamon, Oxford, 1982). The mono-organothallium (III) compounds, easily obtained by thallation of aromatics, or by oxythallation of olefines etc., using thallium (III) salts, have proved valuable in organic synthesis (S.Uemura, in *The Chemistry of the Metal-Carbon Bond*, Vol.4, F.R.Hartley, Ed., Wiley, Chichester, 1987; A.McKillop and E.C.Taylor, in *Comprehensive Organometallic Chemistry*, see above). The nature of the thallium (III) salts used to generate the mono-organothallium (III) reagents effects their stability. However, thallium (III) trifluoroacetate (ttfa), or thallium (III) triacetate (tta) are commonly used, giving fairly stable reagents, which can be isolated. In many synthetic processes, organothallium compounds (with weak C-Tl bonds) are used directly, without isolation. In these cases thallium (III) nitrate (ttn) is often preferred.

Useful ^{1}H and ^{13}C N.M.R. data for organothallium compounds has been published (C.S.Hoad *et al., J. Organomet. Chem.*, 1977, **124**, C31; F.Brady *et al., ibid.*, 1983, **252**, 1; B.V.Cheesman and R.F.M.White, *Can. J. Chem.*, 1984, **62**, 521).

Uses of Thallium in Organic Synthesis

1. Oxythallation of alkenes, alkynes and allenes.

This procedure (see Eqn. 177) is a simple, direct method, yielding the corresponding mono-organothallium (III) compounds, which closely resembles oxymercuration (R.N.Butler, in *Synthetic Reagents*, Vol. 4, J.S.Pizey, Ed., Ellis Horwood, Chichester, 1977). The thallium compounds can be isolated, although they are less stable than the corresponding mercury compounds. It should be noted however, that the use of trifluoroacetate (ttfa), or nitrate (ttn), in these reactions leads to intermediates, which are not sufficiently stable to be isolated. Alkynes behave similarly (see Eqn. 178) and generally afford *anti*-addition products (S.Uemura *et al., Chem. Letters,* 1973, 545; R.K.Sharma and N.H.Fellers, *J. Organomet. Chem.*, 1973, **49**, C59). Non-symmetric alkynes, such as 1-phenylalkynes do not yield regiospecific addition products (S.Uemura *et al., Bull. Chem. Soc. Jpn.*, 1974, **47**, 2663)(see Eqn. 179).

Terminal alkynes tend to add thallium reagents regiospecifically and yield stable "dimeric" species (S.Uemura *et al., J. Chem. Soc., Perkin Trans. 1.*, 1981, 991)(see Eqn. 180). Allenes react regiospecfically, with thallium becoming attached to the central carbon atom (R.K.Sharma and E.D.Martinez, *J. Chem. Soc., Chem. Commun.*, 1972, 1129). In these isolatable thallium compounds, the thallium group [generally $Tl(OAc)_2$] may be subsequently replaced.

+ $Tl(OAc)_3$ $\xrightarrow{ROH}$ RO ... $Tl(OAc)_2$

(177)

R—≡—R $\xrightarrow{\text{tta/AcOH}}$ AcO(R)C=C(R)Tl(OAc)$_2$

(178)

Ph—≡—R $\xrightarrow{\text{tta/AcOH}}$ AcO(Ph)C=C(R)Tl(OAc)$_2$ (major)

+ (AcO)$_2$Tl(Ph)C=C(R)OAc (minor)

(179)

H—≡—R $\xrightarrow{\text{tta/AcOH}}$ AcO(R)C=C(Tl(OAc))$_2$C=C(R)OAc

(180)

a) *Alkoxy or acetoxy thalliums* undergo dethallation by heating in an appropriate solvent, although this is often accompanied by rearrangement (see Eqn. 181).

b) *Halides, cyanides, thiocyanates and selenocyanates* are obtained by refluxing the appropriate thalliums with the corresponding Cu(I), or K, salts in acetonitrile (S.Uemura *et al., Bull. Chem. Soc. Jpn.*, 1974, **47**, 920; *ibid.*, 1975, **48**, 1925)(see Eqn. 182). Reactions with KBr proceed smoothly in the presence of a catalytic amount of 2,6-dimethyl-18-crown-6- (A.J.Bloodworth and D.J.Lapham, *J. Chem. Soc., Perkin Trans. 1.*, 1981, 3265). Vinylthallium intermediates are similarly replaced, generally with retention of configuration (S.Uemura *et al., Bull. Chem. Soc. Jpn.*, 1974, **47**, 2663; *J. Chem. Soc., Perkin Trans. 1*, 1981, 991).

c) *Hydrogen* can be introduced using sodium-amalgam in water (K.C.Pande and S.Winstein, *Tetrahedron Letters*, 1964, 3393), or better with *N*-benzyl-1,4-dihydronicotinamide, as the reductant for the organothallium derivative (D.J.Pasto and J.A.Gontarz, *J. Amer. Chem. Soc.*, 1969, **91**, 719).

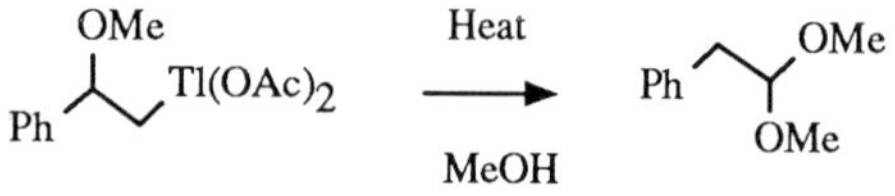

OMe Heat OMe
Ph Tl(OAc)$_2$ → Ph
MeOH OMe

(181)

OMe Cu_2X_2 or KX MeO
Ph Tl(OAc)$_2$ → Ph X
MeCN/reflux

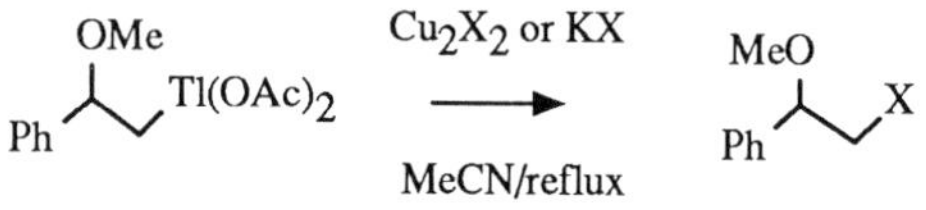

(182)

2. Oxidation of alkenes.

The use of tta is limited by the difficulty of predicting the outcome of the reactions: product mixtures are formed, and there is further variability depending upon solvent and the reaction conditions selected. Thallium intermediates are not usually isolated. A useful example (see Eqn. 183) involves hemiacetal formation (V.Simonidesz *et al., J. Chem. Soc., Perkin Trans 1*, 1980, 2572). Very rapid reactions result from the use of ttn, and there is usually good product selectivity (A.McKillop *et al., J. Amer. Chem. Soc.*, 1973, **95**, 3635). This reagent has been used in a prostaglandin synthesis (E.J.Corey and T.Ravindranthan, *Tetrahedron Letters,* 1971, 4753)(see Eqn. 184).

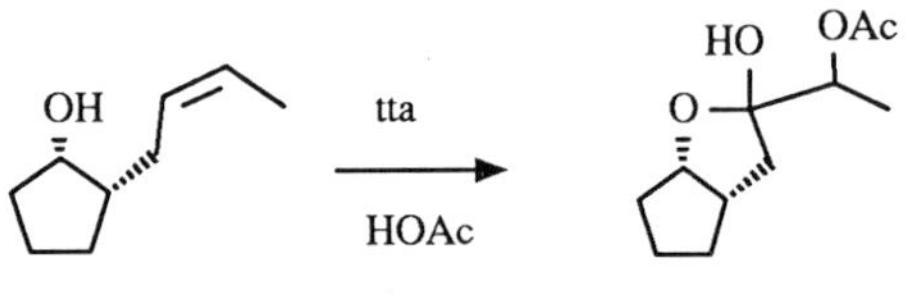

OH tta HO OAc
→ O
HOAc

(183)

O O ttn O O
→
$HClO_4$ aq. CHO

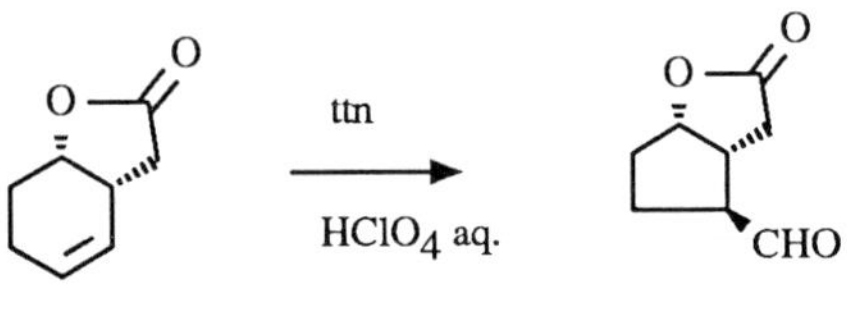

(184)

Cis-diols (see Eqn. 185) also result from reactions of this type, thus the consequence of the application of ttn as a reagent is not always predictable (M.J.Begley *et al., J. Chem. Soc., Perkin Trans.* 1, 1983, 883; S.Antus *et al., Chem. Ber.*, 1979, **112**, 3879).

The use of n-pentane as solvent is reported to result in the formation of nitrate esters as major reaction products (R.J.Ouellette and R.J.Bertsch, *J. Org. Chem.*, 1976, **41**, 2782)(see Eqn. 186).

Chalcones are oxidised slowly by tta (W.D.Ollis *et al., J. Chem. Soc.*, C, 1970, 125), but more rapidly, and more efficiently by ttn (A.McKillop *et al., J. Amer. Chem. Soc.*, 1973, **95**, 3641; E.C.Taylor *et al., J. Org. Chem.*, 1980, **45**, 3433)(see Eqn. 187).

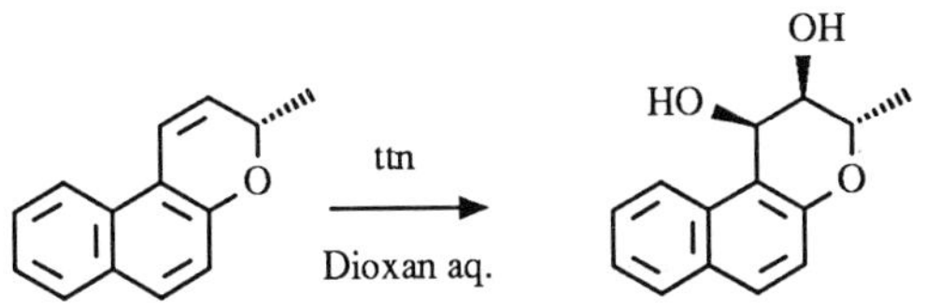

ttn, Dioxan aq.; HO, OH, O

(185)

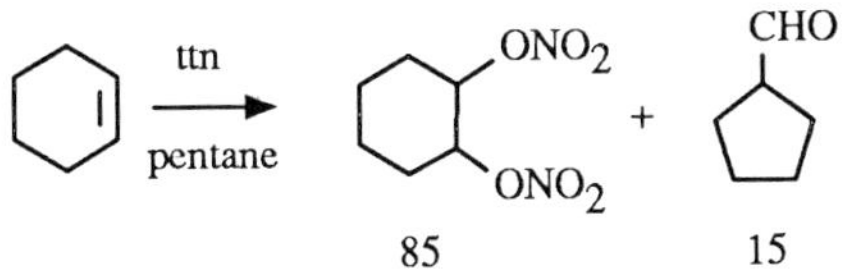

ttn, pentane; ONO_2, ONO_2, CHO; 85, 15

(186)

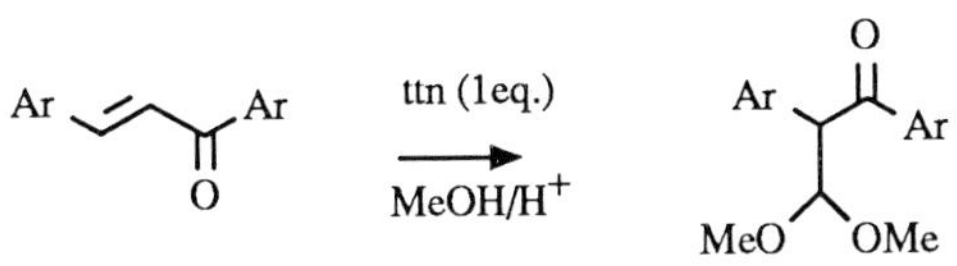

Ar, Ar, O; ttn (1eq.), MeOH/H^+; Ar, O, Ar, MeO, OMe

(187)

3. Oxidation of alkynes.

A wide range of products are possible using ttn, depending upon the solvent and the alkyne structure (A.McKillop *et al.*, *J. Amer. Chem. Soc.*, 1971, **93**, 7331; 1973, **95**, 1296). Terminal alkynes can give carboxylic acids or methyl ketones depending upon conditions employed (S.Uemura *et al.*, *Bull. Chem. Soc. Jpn.*, 1967, **40**, 1499).

4. Oxidation of ketones.

These reactions proceed *via* oxythallation of the enol tautomer leading to mixed products (M.Vollmerhaus and F.Huber, *Z.Naturforsch., Teil B*, 1981, **36**, 141)(see Eqn. 188). Control of reaction conditions can lead to useful products with, for example, cyclohexanones (see Eqn. 189). Aryl group migrations occur when acetophenones are oxidised (A.McKillop, B.P.Swann, and E.C.Taylor, *J. Amer. Chem. Soc.*, 1971, **93**, 4919; 1973, **95**, 3340)(see Eqn. 190) .

Enamine formation enhances yields of oxidation products, when compared to similar reactions with the parent ketones (M.E.Kuehne and T.J.Giacobbe, *J. Org. Chem.*, 1968, **33**, 3359)(see Eqn. 191).

$$RCOMe \xrightarrow[MeOH]{tta} RCOCH_2Tl(OAc)_2 + R^1CH[Tl(OAc)_2]COMe$$

$$RCOCH_2Tl(OAc)_2 \rightarrow RCOCH_2OAc \qquad R^1CH[Tl(OAc)_2]COMe \rightarrow R^1CH(OAc)COMe$$

(188)

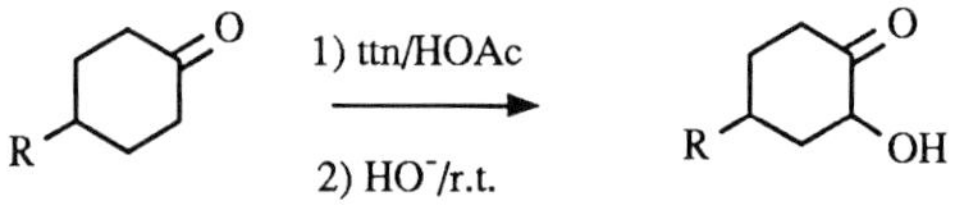

$$\text{4-R-cyclohexanone} \xrightarrow[2)\ HO^-/r.t.]{1)\ ttn/HOAc} \text{4-R-2-hydroxycyclohexanone}$$

(189)

$$ArCOMe \rightleftharpoons ArCH(OH){=}CH_2 \xrightarrow[H^+/MeOH]{ttn} Ar(HO)(MeO)C{-}CH_2Tl(NO_2)_2$$

$$\rightarrow ArCH_2CO_2Me$$

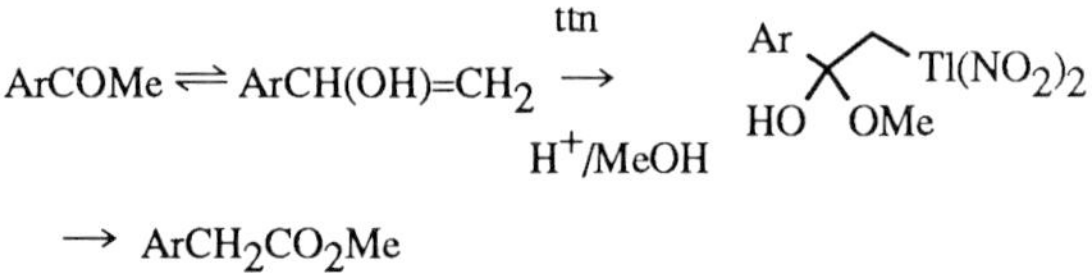

(190)

$$RCH_2COMe \longrightarrow \text{(morpholine enamine)} \xrightarrow[HOAc]{tta} RCH(OAc)COMe \qquad (191)$$

5. Group IVA: Silicon, Germanium, Tin, and Lead

The organo-derivatives of this group have found extensive application, leading to an extensive literature. General reviews covering these elements have been published (B.J.Aylett, *Organometallic Compounds*, 4th Ed., Vol.1, Chapman and Hall, London 1979, R.C.Poller, in *Comprehensive Organic Chemistry*, Vol. 3, D.H.R.Barton and W.D.Ollis, Eds., Pergamon, Oxford, 1979), plus numerous reports in their own journal (*Rev. Si, Ge, Sn, Pb Compounds,* 1972, **1**). Details of the mass spectra (V.Yu.Orlov, *Russ. Chem. Rev., Engl. Transl.*, 1973, **42**, 529) and I.R., Raman and N.M.R. (K.Licht and P.Reich, *Literature data for I.R., Raman, and N.M.R. Spectroscopy for Si, Ge, Sn and Pb-Organic Compounds*, VEBN Deutscher Verlag der Wissenschaften, Berlin, D.D.R., 1971) data for organo-group IV compounds are also available.

a) Silicon

The chemistry of organosilicon compounds is vast and there are a correspondingly large number of reviews (e.g. *The Chemistry of the Organic Silicon Compounds, Parts I and II*, S.Patai and Z.Rappoport, Eds., Wiley, Chichester, 1989; *Silicon Chemistry*, J.Y.Corey, E.J.Corey and P.P.Gaspar, Eds., Ellis Horwood, New York, 1988; S.Pawlenko, in Houben-Weyl, *Methoden der Organischen Chemie*, 4th Ed., Vol. 12/5, Thieme, Stuttgart, 1980; I.Fleming in *Comprehensive Organic Chemistry*, Vol. 3., D.H.R.Barton and W.D.Ollis, Eds., Pergamon, Oxford, 1979; *Carbon-Functional Organosilicon Compounds*, V.Cevalovsky and J.M.Bellama, Eds., Plenum, New York, 1984; T.A.Blinka, B.J.Helmer, and R.West, *Adv. Organomet. Chem.*, 1984, **23**, 193; R.J.P.Corriu, C.Gueria, and J.J.E.Moreau, *Top. Stereochem.*, 1984, **15**, 43; H.F.Schaefer III, *Acc. Chem. Res.*, 1982, **15**, 283). There has been extensive use of organosilicon derivatives in organic synthesis (e.g. E.W.Colvin, *Silicon Reagents in Organic Synthesis,* Academic, New York, 1988; in *The Chemistry of the Metal-Carbon Bond*, Vol.4, F.R.Hartley, Ed., Wiley, Chichester, 1987; E.Block and M.Aslam, *Tetrahedron*, 1988, **44**, 281; *J. Organomet. Chem.*, 1988, **341**, 315; A.Hosomi, *Acc. Chem. Res.*, 1988, **217**, 200; H.Reich, Ed., Tetrahedron, 1983, **39**, 839-1009; P.D.Magnus, T.Sarkar, and S.Djuric, in *Comprehensive Organometallic Chemistry*, G.Wilkinson, F.G.A.Stone, and E.W.Abel, Eds., Pergamon, Oxford, 1982).

Uses in Organic Synthesis

1. Rearrangement reactions

A general review of silicon rearrangement reaction has been written (A.G.Brook and A.R.Bassindale, in *Rearrangements in Ground and Excited States*, Vol. 2, P.de Mayo, Ed., Academic, New York, 1980).

i) 1,2-Rearrangement, known as the Brook rearrangement has been used to yield silyl enol ethers, in a stereo- and regio- specific manner (M.Kato *et al., J. Amer. Chem. Soc.*, 1984, **106**, 1773; I.Kuwajima, *J. Organomet. Chem.*, 1985, **285**, 137)(see Eqn. 192).

ii) 1,3-Rearrangement also gives rise to silyl enol ethers from β-ketosilanes (S.Sato, I.Matsuda, and Y.Izumi, *Tetrahedron Letters*, 1983, **24**, 2787 and 3855)(see Eqn. 193). The reverse process is known, and is useful for the formation of β-ketosilanes from sterically crowded silyl enol ethers (I.Kuwajima and R.Takeda, *Tetrahedron Letters*, 1981, **22**, 2381; E.J.Corey and C.Rücker, *ibid.*, 1984, **25**, 4345). The sila-Pummerer rearrangement is most useful synthetically (D.J.Ager, *ibid.*, 1981, **22**, 587; J.D.White, M.Kang, and B.G.Sheldon, *ibid.*, 1983, **24**, 4463)(see Eqn. 194).

iii) 1,4-Rearrangement can be base induced under mild conditions (I.Urabe and I.Kawajima, *ibid.*, 1983, **24**, 4241; F.Sato, M.Kusakabe, and Y.Kobayashi, *J. Chem. Soc., Chem. Commun.*, 1984, 1130; K.Suzuki, T.Ohkuma, and G.Tsuchihashi, *Tetrahedron Letters,* 1985, **26**, 861) to give the equivalent of an allyl anion (see Eqn. 195). The reverse migration is known (C.Rücker, *ibid.*, 1984, **25,** 4349).

$RCOSiMe_3$ —[1) CH_2=CHMgBr; 2) BuLi]→ R(OLi)(SiMe$_3$)C–CH=CH_2 →

Me_3SiO → Li (allyl anion, R) —[R^1CH_2X]→ $RC(OSiMe_3)$=$CHCH_2CH_2R^1$

(E)

(192)

RCHO $\xrightarrow{H_2C=C(SiMe_3)MgBr}$ R(OH)–C(=CH$_2$)SiMe$_3$ $\xrightarrow[\text{cat.}]{\text{Rh}}$ [R–CO–CH(CH$_3$)SiMe$_3$] $\longrightarrow$ R–C(OSiMe$_3$)=CH–CH$_3$ (193)

$PhXCH_2SiMe_3$ $\xrightarrow[2)E^+]{1)\ base}$ $PhXCHESiMe_3$ $\xrightarrow{[O]}$ $Ph(XO)CHESiMe_3$ $\longrightarrow$ $PhXCHEOSiMe_3$ $\xrightarrow{H^+}$ ECHO (X=S or Se) (194)

HO SiMe$_3$ $\xrightarrow[2)\ E^+]{1)\ BuLi}$ OSiMe$_3$ E (195)

2. Vinylsilanes

Methods of preparation are discussed in many of the reviews given above. Two commonly used routes involve a) vinyl iodides and *tris* (trimethyl silyl) aluminium (B.M.Trost and T.Yoshida, *Tetrahedron Letters*, 1983, **24**, 4895), or b) metalation, followed by carbo- or protio-demetalation of trimethysilylalkynes (F.Sato *et al., J. Chem. Soc., Chem. Commun.*, 1984, 1126; E.-i.Negishi and F.T.Luo, *J. Org. Chem.*, 1983, **48,** 1560). These vinylsilanes react with a wide variety of electrophiles, making them synthetically valuable (see Eqn. 196). The silicon β-effect generally controls regiochemistry, an example is provided (see Eqn. 197) involving an acetal electrophile (H.F.Chow and I.Fleming, *J. Chem. Soc., Perkin Trans 1*, 1984, 1815). The presence of an α-substituent capable of stabilising a carbonium ion (e.g. O or S, see Eqn. 198) can alter the normal regiospecificity (P.D.Magnus, D.A.Quagliato, and J.C.Huffman, *Organometallics*, 1982, **1**, 1240; P.D.Magnus and D.A.Quagliato, *J. Org. Chem.*, 1985, **50**, 1621). Protiodesilylation of vinylsilanes is readily achieved with aqueous HBF_4 in CH_3CN (R.E.Ireland and M.D.Varney, *J. Amer. Chem. Soc.*, 1984, **106,** 3668). The presence of electronegative substituents on silicon, allow fluoride ion mediated oxidation of vinyl silanes (K.Tamao, M.Kumada, and M.Maeda, *Tetrahedron Letters,* 1984, **25**, 321). The level of oxidation can be controlled, making this a useful oxidation procedure for terminal alkynes(see Eqn. 199).

$$CH_2{=}CHSi + E^+ \rightarrow {}^+CH_2CH(E)Si \rightarrow NuCH_2CH(E)Si$$

$$\searrow \quad \swarrow$$

$$CH_2{=}CHE \qquad (196)$$

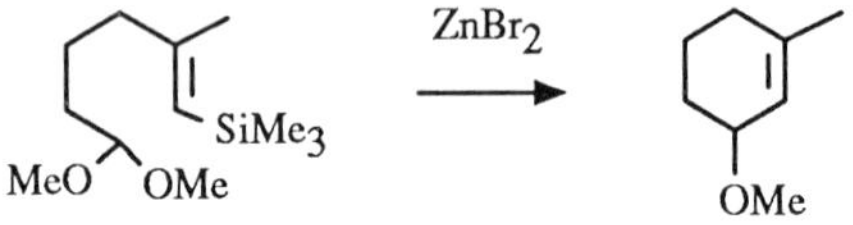

MeO OMe SiMe3 —ZnBr2→ OMe

(197)

O Cl + Me3Si SPh —AgBF4→ H O H SPh

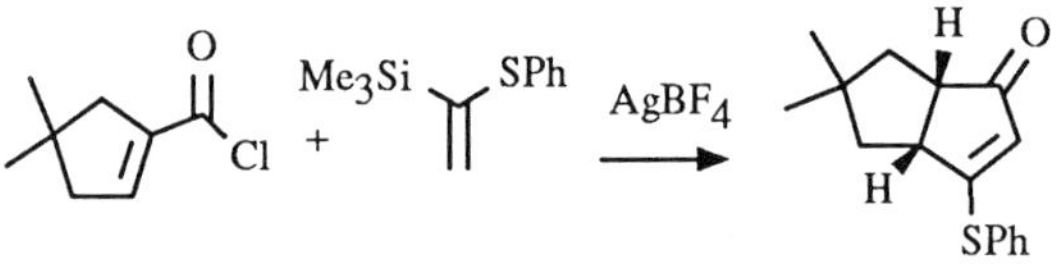

(198)

$$RC{\equiv}CH \xrightarrow[H_2PtCl_6]{(EtO)_2MeSiH} RCH{=}CHSiMe(OEt)_2 \ (E)$$

$$F^- \mid H_2O_2$$

$$RCH_2CHO \leftarrow \quad \rightarrow RCH_2CO_2H \qquad (199)$$

3. α,β-Epoxysilanes

Commonly prepared by epoxidation of vinylsilanes, or by reaction of aldehydes/ketones with $Me_3SiCHClLi$ (C.Burford *et al., Tetrahedron*, 1983, **39**, 867), or $Me_3SiC(CH_3)ClLi$ (F.Cooke, G.Roy, and P.Magnus, *Organometallics*, 1982, **1**, 893) etc., these compounds undergo α-ring-opening (no β-effect) with nucleophiles (see Eqn. 200). Subsequent Peterson elimination yields heteroatom substituted alkenes of specific configuration such as, for example, enamines (P.F.Hudrlik, A.M.Hudrlik, and A.K.Kulkarni, *Tetrahedron Letters*, 1985, **26**, 139).

$$\text{R-epoxide-SiMe}_3 \xrightarrow[\text{2) H}^+]{\text{1) Nuc}^-} \text{HO, SiMe}_3\text{; R, Nuc}$$

KH → H, H; R, Nuc

BF_3OEt_2 → H, Nuc; R, H

(200)

4. Allylsilanes

Allylsilanes can be prepared from alkynes using organometallic reagents (see E.-i.Negishi, F.T.Luo, and C.L.Rand, *ibid.*, 1982, **23**, 27; M.Bourgain-Commercon, J.P.Foulin, and J.E.Normant, *ibid.*, 1983, **24**, 5077), by nucleophilic displacement reactions of allyl halides with Me_3SiLi (J.G.Smith *et al., J. Org. Chem.*, 1984, **49**, 4112; B.M.Trost and D.M.T.Chan, *J. Amer. Chem. Soc.*, 1982, **104**, 3733; N.Shimuzu, F.Shibata, and Y.Tsuno, *Bull. Chem. Soc. Jpn.*, 1984, **57**, 3017), or displacement reactions of allyl alcohol derivatives with silylcuprate reagents (I.Fleming *et al.*, Synthesis, 1981, 560; *Tetrahedron Letters*, 1983, **24**, 4151; *J. Chem. Soc., Perkin Trans. 1*, 1984, 1805). Electrophilic attack on the double bond, controlled by a silicon β-effect, followed by the loss of the silyl group affords useful products with an allylic shift (see Eqn. 201). These reactions generally involve *anti*-stereoselectivitiy (T.Hayashi *et al., Tetrahedron Letters,* 1983, **24**, 5661; I.Fleming and N.K.Terrett, *ibid.*, 1983, **24**, 4253; H.Wetter and P.Scherer, *Helv. Chim. Acta.*, 1983, **66**, 118; G.Wickham and W.Kitching, *J. Org. Chem.*, 1983, **48**, 612), the reaction shown in Eqn. 202 illustrates this effect (H.F.Chow and I.Fleming, *Tetrahedron Letters,* 1985, **26**, 397; C.Santelli-Rouvier, *ibid.*, 1984, **25**, 4371).

$$CH_2{=}CHCH_2SiMe_3 \xrightarrow{E^+} [ECH_2{}^+CHCH_2SiMe_3] \longrightarrow ECH_2CH{=}CH_2 \qquad (201)$$

$$\text{R-CH=CH-CH(SiMe}_3\text{)R}^1 \xrightarrow{E^+} \text{R(E)CH-CH=CH-R}^1$$

(202)

An interesting use of allylsilanes (see Eqn. 203) permits the generation of both isomeric allyl alcohols from allylsilanes using PhSeCl or PhSCl (H.Nishiyama *et al., ibid.*, 1981, **22**, 5285; 5289; *ibid.*, 1982, **23**, 1267). The reactions of allylsilanes have been reviewed (H.Sakurai, *Pure Appl. Chem.*, 1982, **54**, 1). Chiral aldehydes show good diastereoselectivity in this addition to allylsilanes (M.T.Reetz and K.Kesseler, *J. Chem. Soc., Chem. Commun.*, 1984, 1079; M.T.Reetz and A.Jung, *J. Amer. Chem. Soc.*, 1983, **105**, 4833). Furthermore, metal complexation of allylsilane anions allows controlled diastereoselectivity with α-regioselectivity, when they are added to aldehydes (D.J.S.Tsai and D.S.Matteson, *Tetrahedron Letters*, 1982, **23**, 4597; F.Sato, Y.Suzuki, and M.Sato, *ibid.*, 1982, **23**, 4589; Y.Yamamoto, Y.Saito, and K.Maruyama, *J. Chem. Soc., Chem. Commun.*, 1982, 1326). Thus boron, aluminium, and titanium lead to *threo*-products, while tin favours *erythro*-selectivity. Elimination now provides terminal dienes of known configuration (see Eqn. 204).

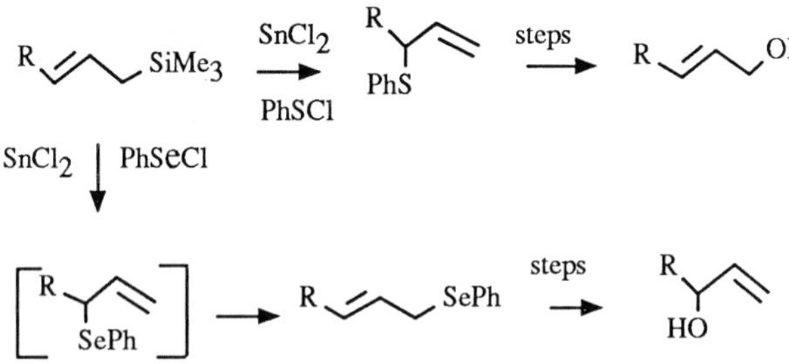

R SiMe$_3$ —[$SnCl_2$ / PhSCl]→ R(PhS)CH–CH=CH$_2$ —steps→ R–CH=CH–CH$_2$OH

$SnCl_2$ | PhSeCl ↓

[R(SePh)CH–CH=CH$_2$] → R–CH=CH–CH$_2$SePh —steps→ R(HO)CH–CH=CH$_2$

(203)

Me$_3$Si (allyl anion) —1) Bu_3SnCl/BF_3 2) RCHO→ R–CH(OH)–CH(SiMe$_3$)–CH=CH$_2$ —KH→ R–CH=CH–CH=CH$_2$

Me$_3$Si (allyl anion) —1) $EtAlCl_2$ 2) RCHO→ R–CH(OH)–CH(SiMe$_3$)–CH=CH$_2$ —KH→ R–CH=CH–CH=CH$_2$

$BF_3.Et_2O$ (crossed arrows)

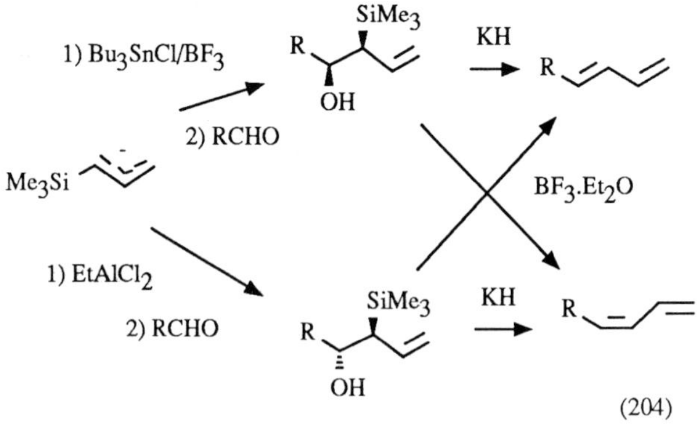

(204)

5. Organosilyl anions

Silyl cuprates react with conjugated enones to yield β-silyl enolates, which can be alkylated (see Eqn. 205), or protonated with good diastereoselectivity (W.Bernhard, I.Fleming, and D.Waterson, *ibid.*, 1984, 28; W.Bernhard and I.Fleming, *J. Organomet. Chem.*, 1984, **271**, 281; I.Fleming *et al., J. Chem. Soc., Chem. Commun.*, 1985, 318).

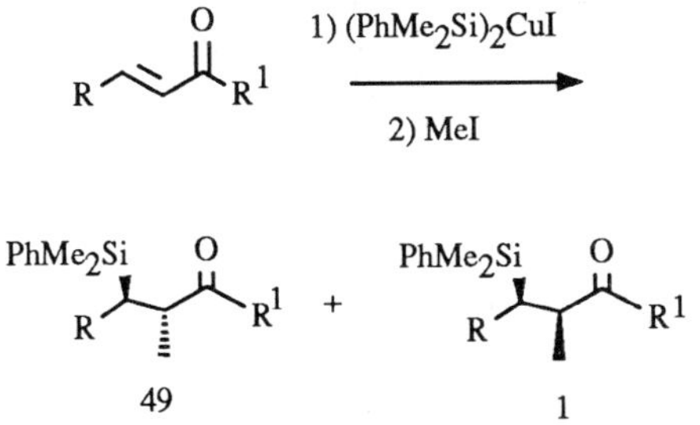

1) $(PhMe_2Si)_2CuI$

2) MeI

$PhMe_2Si$ … R^1 + $PhMe_2Si$ … R^1

49 1

(205)

6. β-Functional organosilanes - Peterson olefination

The presence of electronegative substituents β to silicon, permits a 1,2-elimination known as the Peterson olefination. This synthetically useful process has been reviewed (D.J.Ager, *Synthesis*, 1984, 384). With the commonly used β-hydroxy compounds, acidic catalysis leads to *anti*-elimination, while bases give largely *syn*-elimination. Strong base may result in protiodesilylation (P.F.Hudrlik, A.M.Hudrlik, and K.Kulkarni, *J. Amer. Chem. Soc.*, 1982, **104**, 6809). The β-hydroxysilanes are commonly generated by reacting α-metallosilanes with carbonyl compounds. Grignard reagents (e.g. P.A.Brown *et al., J.Chem. Soc., Chem. Commun.*, 1984, 253) or α-lithiosilanes (e.g. D.J.Ager, *Tetrahedron Letters*, 1981, **22**, 2923; T.Cohen *et al., J. Amer. Chem. Soc.*, 1984, **106**, 3245) being the metallic reagents most often encountered. Simple deprotonation of silanes is possible, when a carbanion-stabilising substituent is present (D.B.Tulshian and B.Fraser-Reid, *ibid.*, 1981, **103**, 474; G.A.Larson, J.Prieto, and A.Hernandez, *Tetrahedron Letters*, 1981, **22**, 1575)(see Eqn. 206).

β-Trimethylsilylethanol has been used as a protecting group (M.L.Mancini and J.F.Honek, *ibid.*, 1982, **23**, 3249) for "reactive hydrogens", since it can be easily removed *via* a fluoride ion or Lewis acid induced Peterson elimination (see Eqn. 207).

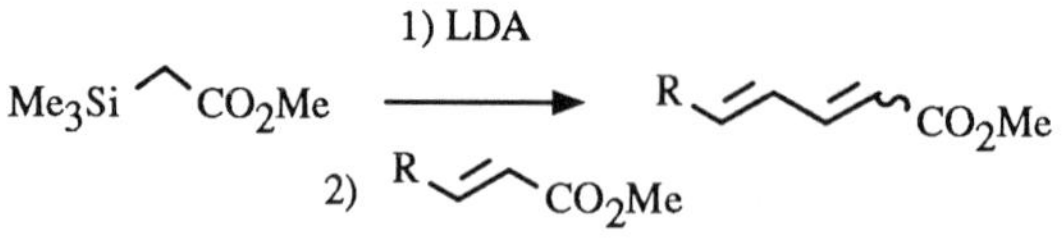

$$Me_3Si\frown CO_2Me \xrightarrow[2)\ R\diagup\!\!\!=\!\!\!\diagdown CO_2Me]{1)\ LDA} R\diagup\!\!=\!\!\diagdown\!\!\diagup\!\!=\!\!\sim CO_2Me$$

(206)

$$XCH_2CH_2SiMe_3 \xrightarrow{HY} XH + CH_2{=}CH_2 + Me_3SiY$$

(207)

7. Cyclopropyl- and cyclopropylcarbinyl- silanes

Some of the main contributions in this area have been summarised (L.A.Paquette, *Top. Curr. Chem.*, 1984, **119**, 1). An interesting example from this chemistry involves thermal rearrangement(see Eqn. 208).

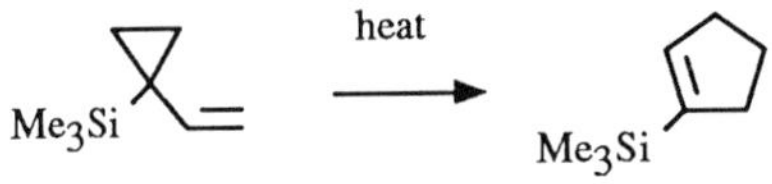

Me3Si [1-(trimethylsilyl)-1-vinylcyclopropane] $\xrightarrow{heat}$ Me3Si [1-(trimethylsilyl)cyclopentene]

(208)

8. Acylsilanes

The chemistry of acylsilanes has been reviewed (D.J.Ager, *Chem. Soc. Rev.*, 1982, **11**, 493); likewise the preparation of α,β-unsaturated acylsilanes (H.J.Reich *et al., Tetrahedron*, 1983, **39**, 949). Saturated acylsilanes can be prepared by application of the Brook rearrangement, followed by isomerisation (A.Hosomi, H.Hashimoto, and H.Sakurai, *J. Organomet. Chem.*, 1979, **175**, C1)(see Eqn. 209). Cleavage of the Si-C bond in acylsilanes leads to acyl anions, which can be trapped by various electrophiles (D.Schinzer and C.H.Heathcock, *Tetrahedron Letters*, 1981, **22**, 1881; A.Degl.Innocenti *et al., J. Chem. Soc., Chem. Commun.*, 1980, 1202; A.Ricci *et al., Tetrahedron Letters*, 1982, **23**, 577)(see Eqn. 210).

$$\text{R(}^1\text{R)C=CH–CH}_2\text{OSiR}^2_3 \xrightarrow[\text{2) Me}_3\text{SiCl}]{\text{1) Bu}^s\text{Li}} \text{R(}^1\text{R)C=CH–CH(SiR}^2_3\text{)OSiMe}_3$$

$$\xrightarrow[\text{MeOH}]{\text{10\%Pd/C}} \text{R(}^1\text{R)CH–CH}_2\text{–C(=O)SiR}^2_3 \qquad (209)$$

$$\text{RCOSiMe}_3 \xrightarrow{\text{F}^-/\text{H}_2\text{O}} \text{RCHO}$$
$$\text{RCOSiMe}_3 \xrightarrow{\text{F}^-/\text{E}^+} \text{RCOE}$$
$$\text{RCOSiMe}_3 \xrightarrow{\text{HO}^-/\text{H}_2\text{O}_2} \text{RCO}_2\text{H} \qquad (210)$$

9. <u>Miscellaneous</u>

There remains a colossal range of silyl reagents containing silicon-aliphatic carbon bonds, but in which reaction involves a silicon-heteroatom bond. These are excluded from this review, although a good introduction to their chemistry can be obtained from the various general and synthetic reviews already quoted.

b) Germanium

The chemistry of organogermanium compounds has received considerable attention in the 1970's and 1980's (e.g. M.Lesbré, P.Mazerolles, and J.Satgé, *The Organic Chemistry of Germanium*, Wiley, London, 1971; E.J.Bulten, in *MTP Int. Rev. Sci., Inorg. Chem., Ser. Two*, 1975, **4**, 246; R.C.Poller, in *Comprehensive Organic Chemistry*, Vol. 3, D.H.R.Barton and W.D.Ollis Eds., Pergamon, Oxford, 1979; P.Rivière, M.Rivière-Baudet, and J.Satgé, *Comprehensive Organometallic Chemistry*, Vol. 2, G.Wilkinson, F.G.A.Stone, and E.W.Abel, Eds., Pergamon, Oxford, 1982; W.P.Neumann, *Naturwissenschaften*, 1981, **68**, 354; K.C.Molloy and J.J.Zuckermann, *Adv. Inorg. Chem. Radiochem.*, 1983, **27**, 113; J.Satge, *Pure Appl. Chem.*, 1984, **56**, 137; M.Dräger, *Comments Inorg. Chem.*, 1986, **5**, 201). Germanium is a typical semi-metal from the centre of the periodic table, the chemistry of which resembles that of silicon. Organogermanium compounds have not yet found a great deal of practical use, nor application to organic synthesis.

c) Tin

The chemistry of organotin compounds has been the subject of extensive industrial and academic research, resulting from the wide range of technical and synthetic uses, to which they have been put (*Organotin Compounds: New Chemistry and Applications*, J.J.Zuckermann, Ed., *Advances in Chemistry Series*, No. 157, American Chemical Society, Washington, DC, 1976; Houben Weyl, *Methoden der Organischen Chemie*, Vol. 13/6, E.Muller, Ed., Theime, Stuttgart, 1978; B.J.Aylett, *Organometallic Compounds*, 4th Edn., Vol., Part 2, Chapman and Hall, London, 1979; J.W.Connolly and C.Hoff, *Adv. Organomet. Chem.*, 1981, **19**, 124; M.Veith and O.Recktenwald, *Top. Curr. Chem.*, 1982, **104**, 1; A.G.Davies and P.J.Smith, in *Comprehensive Organometallic Chemistry*, Vol. 2, G.Wilkinson, F.G.A.Stone, and E.W.Abel, Eds., Pergamon, Oxford, 1982; P.G.Harrison, *The Chemistry of Tin*, Elsevier, Amsterdam, 1984). A useful sourcebook of physical data covering the organometallic derivatives of germanium tin, and lead, has been published (P.G.Harrison, *Organometallic Chemistry of Germanium, Tin, and Lead*, Chapman and Hall, London, 1984). N.M.R. data for organotin compounds has also been summarised (P.J.Smith and A.P.Tupciauskas, *Annu. Rep. N.M.R. Spectroscopy*, 1978, **8**, 291).

Tin organyls have found large scale use as polyvinylchloride stabilizers and biocides (R.F.Bennett, *Ind. Chem. Bull.*, 1983, **2**, 171; P.A.Cuszack and P.J.Smith, *Rev. Si, Ge, Sn, Pb, Compounds.*, 1983, **7**, 1; C.J.Evans and R.Hill, *ibid.*, 1983, **7**, 57). The toxicity of tin compounds in the environment has been a cause for concern (J.J.Zuckermann *et al.*, in *Organometals and Organometal Occurrence and Fati in the Environment,* F.E.Brinckman and J.M.Bellama, Eds., *A.C.S. Sym. Ser.*, No. 82, American Chemical Society, Washington, DC, 1978; F.E.Brinckman, *J. Organomet. Chem. Libr.*, 1981, **12**, 319).

The use of organotin compounds in organic synthesis has boomed in recent years. The ready introduction of tin, combined with its replacement under mild conditions, to generate new carbon-carbon bonds regio- and stereo-specfically, has led to a multitude of applications (M.Pereyre and J.-P.Quintard, *Pure Appl. Chem.*, 1981, 53, 2401; P.J.Smith *et al., Chem. Ind. (London)*, 1984, 167, M.Pereyre, J.-P.Quintard, and A.Rahm, *Tin in Organic Synthesis*, Butterworths, London, 1987).

Preparation

There are three important routes for the formation of new Sn-C bonds.

i) Nucleophilic displacement of halide (or other leaving group) from tin. Organometallic reagents, commonly Grignard or lithium reagents, are used for this purpose (see Eqn. 211). Vinylstannanes can be generated in this way, the *in situ* generation of a vinyl anion leading to displacement of triflate anion with good regio- and stereo-control (E.J.Corey and T.M.Eckrich, *Tetrahedron Letters*, 1984, **25**, 2415)(see Eqn. 212).

$$R_3SnY + R^1M \rightarrow R_3SnR^1 + MY \qquad (211)$$

Y=halide, tosyl etc.

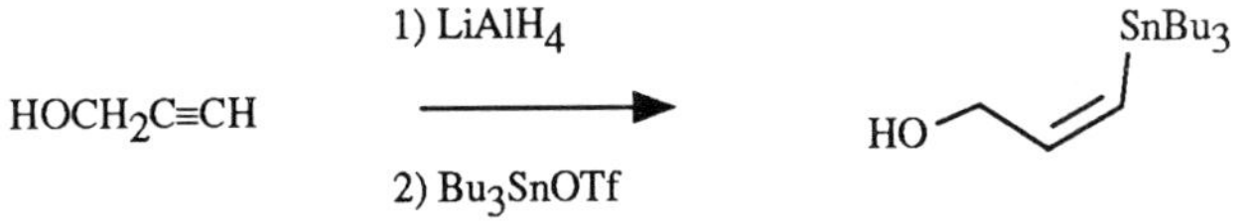

$$HOCH_2C{\equiv}CH \xrightarrow[2)\ Bu_3SnOTf]{1)\ LiAlH_4} HOCH_2CH{=}CHSnBu_3$$

(212)

Allyl acetates can be substituted in a highly chemo- and stereo-selective manner (B.M.Trost and J.W.Herndon, *J. Amer. Chem. Soc.*, 1984, **106**, 6835), when catalysed by palladium (see Eqn. 213).

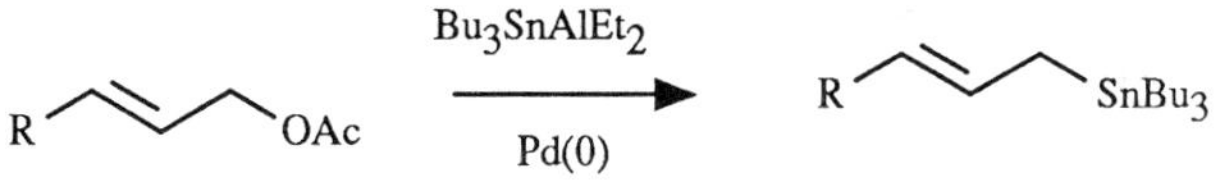

$$RCH{=}CHCH_2OAc \xrightarrow[Pd(0)]{Bu_3SnAlEt_2} RCH{=}CHCH_2SnBu_3$$

(213)

$$R_3SnM \xrightarrow{+\,{}^1RY} R_3SnR^1 + MY$$

$$R_3SnM \xrightarrow{+\,A{=}B,\ EY} R_3SnA\text{-}BE + MY$$

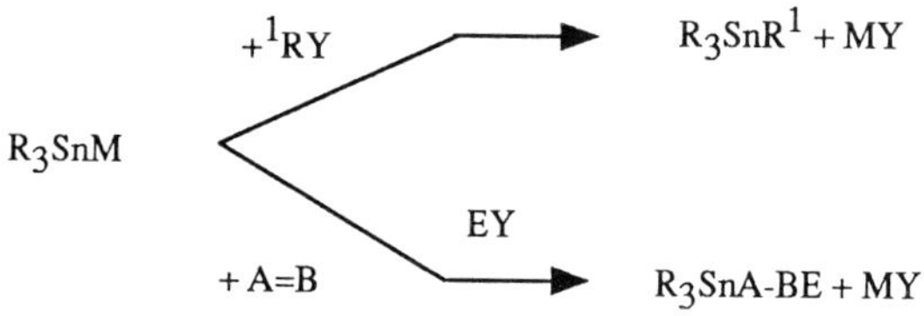

M = Li, Na, Mg, or Cu (214)

Alkynyl tin compounds are readily obtained by using alkynyl lithiums as nucleophiles (J.C.Bottaro, R.N.Hanson, and D.E.Seitz, *J. Org. Chem.*, 1981, **46**, 5221).

ii) Nucleophilic displacement or addition reactions of metallotin compounds (see Eqn. 214).

Displacement can be observed with a range of substrates, e.g. halides or tosylates (J.San Filippo and J.Silbermann, *J. Amer. Chem. Soc.*, 1981, **103**, 5588), orthoesters (see J.P.Quintard, B.Elissondo, and M.Pereyre, *J. Organomet. Chem.*, 1981, **212**, C31), etc. Oxiranes can be opened in a regiospecific manner(J.C.Lahourner and J.Valade, *ibid.*, 1971, **33**, C4)(see Eqn. 215). Vinyl halides undergo displacement, generally with retention of configuration (D.E.Seitz and S.-H.Lee, *Tetrahedron Letters*, 1981, 4909)(see Eqn. 216).

The nucleophilic addition process commonly involves carbonyl compounds (see Eqn. 217), e.g. heptanal (W.C.Still, *J. Amer. Chem. Soc.*, 1978, **100**, 1481).

$$\text{Et-oxirane} \xrightarrow{Bu_3SnMgCl} EtCH(OH)CH_2SnBu_3 \qquad (215)$$

$$Cl\text{-}CH{=}CH\text{-}CO_2Me \xrightarrow{Bu_3SnCu} Bu_3Sn\text{-}CH{=}CH\text{-}CO_2Me \qquad (216)$$

$$C_6H_{13}CHO \xrightarrow[2)\ EtOCH(Me)Cl]{1)\ Bu_3SnLi} C_6H_{13}CH(Bu_3Sn)OCH(Me)OEt \qquad (217)$$

However, addition to acetylenes has been achieved (E.Piers and J.M.Chong, *J. Chem. Soc., Chem. Commun.*, 1983, 934). Indeed, effective regio- and stereo-control can be obtained by varying the stannyl anion and the reaction conditions (E.Piers and H.E.Morton, *J. Org. Chem.*, 1980, **45**, 4263; E.Piers, J.M.Chong, and H.E.Morton, *Tetrahedron Letters*, 1981, **22**, 4905)(see Eqn. 218).

Use of the stannyl anion, in excess, can lead to bis(trialkylstannyl)ethenes (E.Piers and J.M.Chong, *J. Org. Chem.*, 1982, **47**, 1602).

$$RC{\equiv}CCO_2Et \xrightarrow[Me_2S]{Me_3SnCu.LiBr_2} (Me_3Sn)RC{=}CH(CO_2Et)$$

$$RC{\equiv}CCO_2Et \xrightarrow{Me_3SnCu(SPh)Li} (Me_3Sn)RC{=}CH(CO_2Et)$$

(218)

iii) Addition of tin hydrides to alkenes or alkynes. This is usually a free radical chain reaction promoted by AIBN (azobisisobutyronitrile), or light (H.G.Kuivila, *Synthesis*, 1970, 499; Y.Yamamoto, Kagaku, Zokan (Kyoto), 1982, **96**, 155; *Chem. Abs.*, 1982, **97**, 161834)(see Eqn. 219).

$$CH_2C{=}CHR^1 + R_3Sn^{\cdot} \rightleftharpoons R_3SnCH_2CH{\cdot}R^1$$

$$\xrightarrow{R_3SnH} R_3SnCH_2CH_2R^1 + R_3Sn^{\cdot}$$

(219)

The Uses of Organotin Compounds in Organic Synthesis

The cleavage of inactivated Sn-C bonds are of limited use synthetically (J.M.Fukuto and F.R.Jensen, *Acc. Chem. Res.*, 1983, **16**, 177). Halodemetalation in general (O.A.Reutov, *J. Organomet. Chem.* 1983, **250**, 145) and iododestannylation in particular (M.Julliard and M.Chanon, *Chem. Rev.*, 1983, **83**, 425) are useful reactions in synthesis. In fact, organotin compounds often provide a milder alternative to the corresponding organosilicons due to the Sn-C bond being weaker than the Si-C bond (E.W.Colvin, *Silicon in Organic Synthesis*, Butterworths, London, 1981).

1. Reducing agents

Tin hydrides ($R_{4-n}SnH_n$) are useful reagents for the reduction of organohalides, carbonyl compounds, conjugated alkenes, thio-, seleno- and telluro- compounds, nitro compounds, and for the demercuration of mercurated alkenes (M.Pereyre, J.-P.Quinford, and A.Rahm, *Tin in Organic Synthesis*, Butterworths, London, 1987).

2. Formation of C-H bonds (or C-D bonds)

Deuterolysis of vinyltins (see A.Alvanipour, C.Eaborn, and D.R.M.Walton, *J. Organomet. Chem.*, 1980, **201**, 233), or of allyltins (J.A.Mangravite, *J. Organomet. Chem. Libr.*, 1979, **7**, 45) provides a stereo-selective route to deuterated products (see Eqn. 220).

Straight forward protonolysis of allyltin compounds has found many uses, thus methylenecyclohexanes can be easily obtained (S.R.Wilson, L.R.Phillips, and K.J.Natalie, Jr., *J. Amer. Chem. Soc.*, 1979, **101**, 3340)(see Eqn. 221).

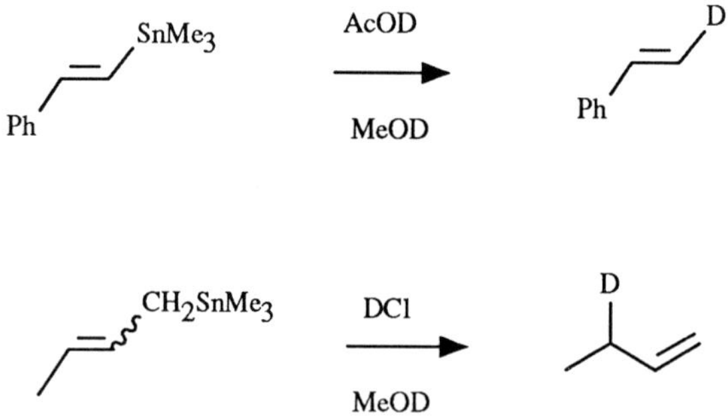

SnMe$_3$ Ph — AcOD / MeOD → D Ph

(220)

CH$_2$SnMe$_3$ — DCl / MeOD → D

SnMe$_3$ MeO$_2$C — HCl / MeOH → MeO$_2$C

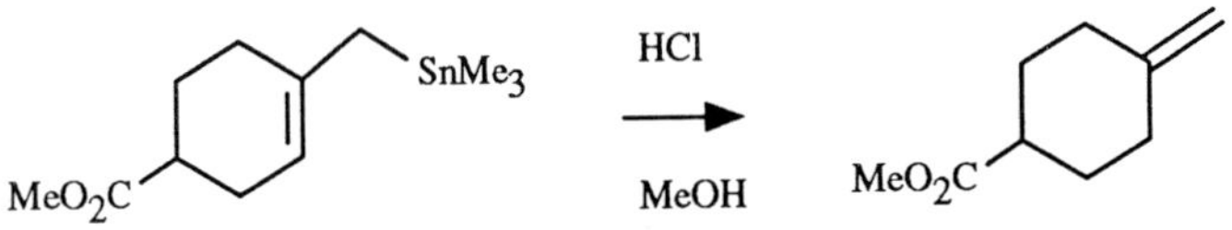

(221)

3. Formation of C-halogen bonds

Halodestannylation of vinyltin compounds has been known for sometime (D.Seyferth, *J. Amer. Chem. Soc.*, 1957, **79**, 2133; S.O.Rosenberg, A.I.Gibbons, *ibid.*, 1957, **79**, 2137). The vinylhalide products are formed with retention of configuration (P.Backelmans *et al., Bull. Soc. Chim. Belg.*, 1968, **77**, 85). The ability to generate vinyliodides, and thus vinyllithium and cuprate reagents with stereochemical control has been of great value in prostaglandin synthesis (R.Noyori and M.Suzuki, *Angew. Chem. Int. Ed. Engl.*, 1984, **23**, 847).

4. Formation of C-O bonds

The relatively weak Sn-C bonds can be oxidised under mild conditions (W.C.Still, *J. Amer. Chem. Soc.*, 1977, **99**, 4836)(see Eqn. 222).

Very mild oxidative cleavage of cyclohexanol rings has been achieved using iodosylbenzene (E.Ochiai *et al., J. Chem. Soc., Chem. Commun.*, 1984, 1007; 1985, 637), or lead tetraacetate (K.Nakataui and S.Isoe, *Tetrahedron Letters*, 1984, **25**, 5335; 1985, **26**, 2209) in reaction with their γ-trialkylstannyl derivatives(see Eqn. 223).

Allyl- and benzyltin derivatives have especially weak bonds, so very mild oxidation using MnO_2 or MCPBA (*meta*-chloroperbenzoic acid) is possible. This allows protection of conjugated enone systems (W.C.Still, *J. Amer. Chem. Soc.*, 1977, **99**, 4186; 1979, **101**, 2493)(see Eqn. 224).

Vinyltin compounds give rise to organotin epoxides (R.H.Fish and B.M.Broline, *J. Organomet. Chem.*, 1978, **159**, 255), which can be further

$$RCH(Br)Me \xrightarrow[2)\ CrO_3/py.(15\ eqv.)]{1)\ Me_3SnLi/THF} RCOMe \qquad (222)$$

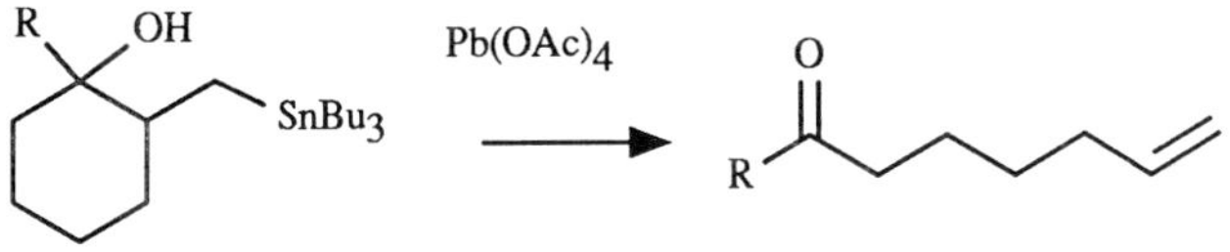

R, OH, SnBu₃ → Pb(OAc)₄ → R, O

(223)

1) Me_3SnLi 2) Me_3SiCl → Me_3Sn, $OSiMe_3$

[O] → O

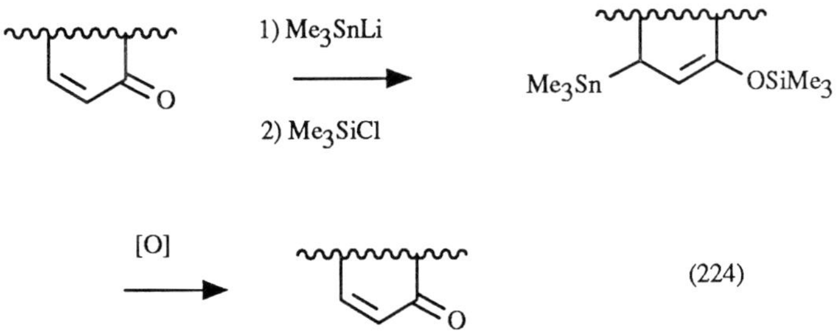

(224)

reacted to yield ketones (A.Nishida, M.Shibasaki, and S.Ikegami, *Tetrahedron Letters*, 1981, **22**, 4819).

5. Transmetalation

Lithiation of organotin compounds (see Eqn. 225) was first described in 1959 (D.Seyferth and M.A.Weiner, *J. Org. Chem.*, 1959, **24**, 1395). For best results the requirement is that R, R^2should be alkyl or aryl; while R^1 is allyl, benzyl, vinyl, α-heterosubstituted alkyl, etc. (D.J.Peterson, *Organomet. Chem. Rev. A*, 1972, **7**, 295). Transmetalation can also occur with other organometallic reagents (L.J.Krause and J.A.Morrison, *Inorg. Chem.*, 1980, **19**, 604; E.J.Bulten and H.A.Budding, *Revs. Si, Ge, Sn, Pb Compounds.*, 1978, Special issue, **103**).

$$R_3SnR^1 + R^2Li \longrightarrow R^1Li + R_3SnR^2 \qquad (225)$$

This transmetalation procedure often compares favourably with conventional methods of generating organolithium reagents; by avoiding lithium halide contamination and achieving regiospecificity (B.J.Wakefield, *The Chemistry of Organolithium Compounds*, Pergamon, Oxford, 1974).

a) *Vinyltin derivatives* have received much attention. An interesting use involves trapping vinylic anions, formed in association with by-products on lithiation, by addition of an organotin halide. After purification, the vinyltin compound can be used to regenerate a pure vinyl anion (M.P.Cooke, *J. Org. Chem.*, 1982, **47**, 4963). Retention of configuration on transmetalation has determined that this reaction has been extensively used for the regio- and specific introduction of vinyl groups (D.Seyferth and L.G.Vaughan, *J. Amer. Chem. Soc.*, 1964, **86**, 883; D.Seyferth, L.G.Vaughan, and R.Suzuki, J. Organomet. *Chem.*, 1964, **1**, 437)(see Eqn. 226), and in the synthesis of many classes of biologically important compounds, e.g. prostaglandins (E.J.Corey and R.H.Wollenberg, *J. Org. Chem.*, 1975, **40**, 2265).

Othp Bu3SnH Othp 1) BuLi 2) PrC≡CCu 3)

Bu3Sn O

Othp

O

(226)

(b) *Conjugated polyenic vinyltin compounds* react in similar fashion (E.Piers and H.E.Morton, 1980, **45**, 4263)(see Eqn. 227), the resultant lithium derivatives can be coupled with electrophiles, such as alkyl bromides and carbonyl compounds.

(c) *Non-conjugated polyenic vinyllithiums* can be used to assemble chains (E.J.Corey and T.M.Eckrich, *Tetrahedron Letters*, 1984, **25**, 2419) of nonconjugated double bonds, such as occur in arachidonic acid derivatives. The procedure involves transmetalation of the corresponding organotin compounds (see Eqn. 228). 1,1-Dibutyl-1-stannacyclohexa-2,5-diene is a useful reagent (E.J.Corey and J.Kang, *ibid.*, 1982, **23**, 1651)(for an example see Eqn. 229.

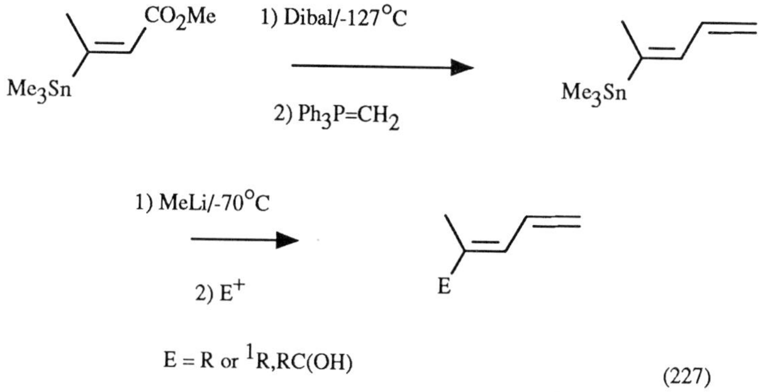

CO_2Me Me_3Sn — 1) Dibal/-127°C; 2) $Ph_3P=CH_2$ → Me_3Sn

1) MeLi/-70°C; 2) E^+ → E

E = R or [1]R,RC(OH)

(227)

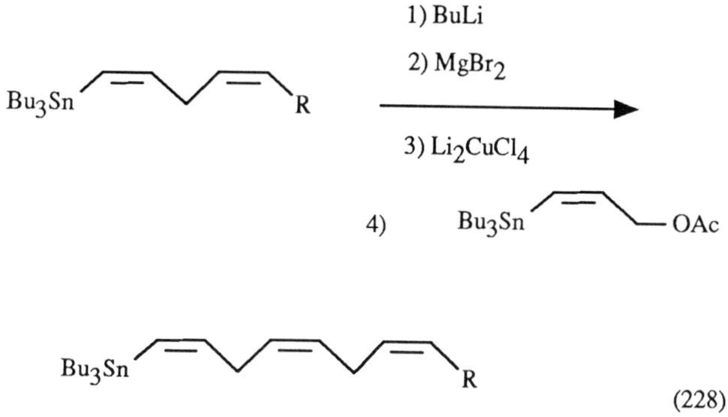

Bu_3Sn — R — 1) BuLi; 2) $MgBr_2$; 3) Li_2CuCl_4; 4) Bu_3Sn — OAc →

Bu_3Sn — R

(228)

(d) *α-Alkoxyvinyllithium derivatives* are important reagents in organic synthesis (J.A.Soderquist and G.J.H.Hsu, *Organometallics*, 1982, **1**, 830).

Although they can be obtained by direct lithiation of vinyl ethers, isolation and purification as an organotin derivative is advisable (see Eqn. 230). The α-alkoxylvinyllithium intermediate can be readily regenerated by transmetalation, without loss of the original configuration.

β-Alkoxyvinyltin compounds are readily obtained (K.S.Y.Lau and M.Schlosser, *J. Org. Chem.*, 1978, **43**, 1595; R.H.Wollenberg, K.F.Albizati, and R.Peries, *J. Amer. Chem. Soc.*, 1977, **99**, 7365; J.Ficini *et al., Tetrahedron Letters*, 1977, 3589), after transmetalation, they can be coupled with aldehydes to yield enals (see Eqn. 231). Coupling with alkyl bromides may also be accomplished, but requires prior conversion of the β-alkoxyvinyllithium to a cuprate. (*E*)-(β-Silylvinyl)lithium derivatives, derived from the corresponding stannanes (e.g. R.F.Cunico and F.J.Clayton, *J. Org. Chem.*, 1976, **41**, 1480), have found wide use in organic synthesis (E.W.Colvin, *Silicon in Organic Synthesis*, Butterworths, London, 1981).

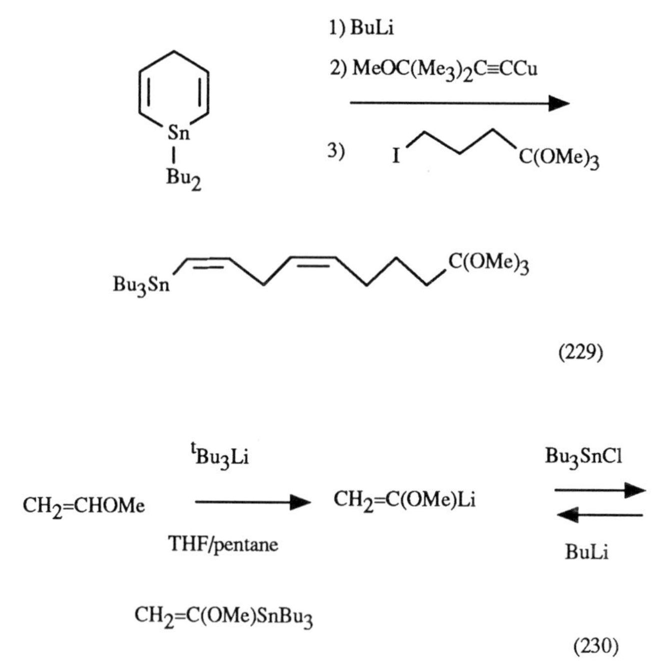

α-Heterosubstituted organotin compounds are precursors (U.Schöllkopf, *Angew. Chem. Int. Ed. Engl.*, 1970, **9**, 763) of a powerful group of synthetic reagents, the α-hetero-substituted carbanions (B.T.Gröbel and D.Seebach, *Synthesis*, 1977, 357; A.Krief, *Tetrahedron*, 1980, **36**, 2531). α-*Methoxymethyltin compounds* can be used to provide lithium or lithio-cuptrate analogues (A.Duchene, D.Mouko-Mpegna, and J.P.Quintard, *Bull. Soc. Chem. Fr.*, II, 1985, 787), thus accessing

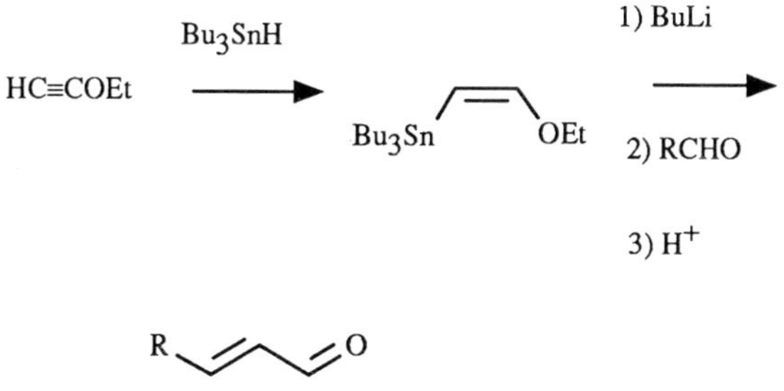

$$HC{\equiv}COEt \xrightarrow{Bu_3SnH} Bu_3Sn\text{-}CH{=}CH\text{-}OEt \xrightarrow[\text{2) RCHO; 3) } H^+]{\text{1) BuLi}} R\text{-}CH{=}CH\text{-}CH{=}O$$

(231)

Bu_3SnCH_2OMe — 1) BuLi; 2) cyclohexenone → 3-(methoxymethyl)cyclohexanone (O, OMe)

Bu_3SnCH_2OMe — 1) BuLi; 2) $CuI.Me_2S$; 3) cyclohexenone → 1-(methoxymethyl)cyclohex-2-en-1-ol (OH, OMe)

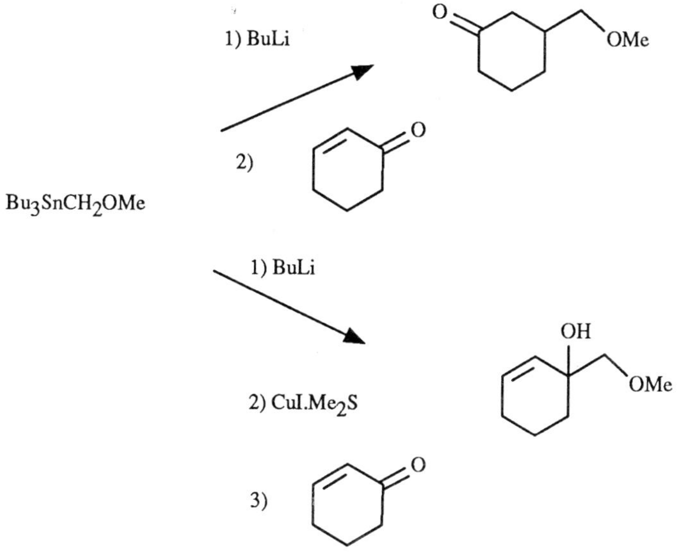

(232)

regio-controlled addition to conjugated enones (see Eqn. 232). Alcohols are conveniently converted into alkoxymethyl tins using iodobutyltributyltin (W.C.Still, *J. Amer. Chem. Soc.*, 1978, **100**, 1481; W.C.Still *et al.*, *Tetrahedron Letters*, 1979, 593). Allyl alcohols give tin derivatives, which undergo [2,3]-sigmatropic shifts on transmetalation (W.C.Still and A.Mitra, *J. Amer. Chem. Soc.*, 1978, **100**, 1927). Advantage of this rearrangement (see Eqn. 233) has been taken in natural product synthesis, e.g. for C18 *Cecropia* juvenile hormone (W.C.Still *et al.*, *Tetrahedron Letters*, 1979, 593). These α-oxygenerated carbanions show remarkable

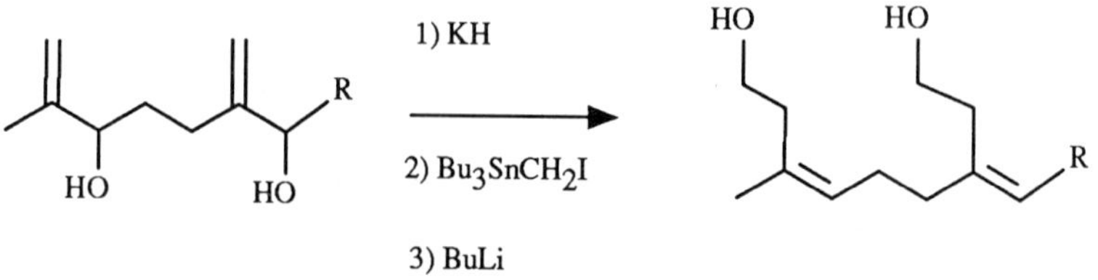

1) KH

2) Bu_3SnCH_2I

3) BuLi

(233)

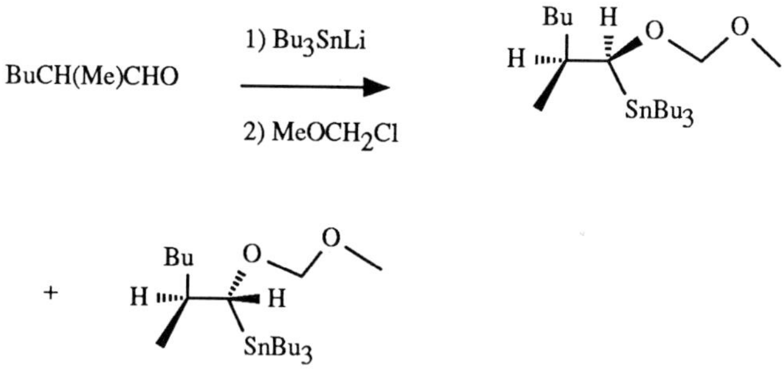

BuCH(Me)CHO

1) Bu_3SnLi

2) $MeOCH_2Cl$

(234)

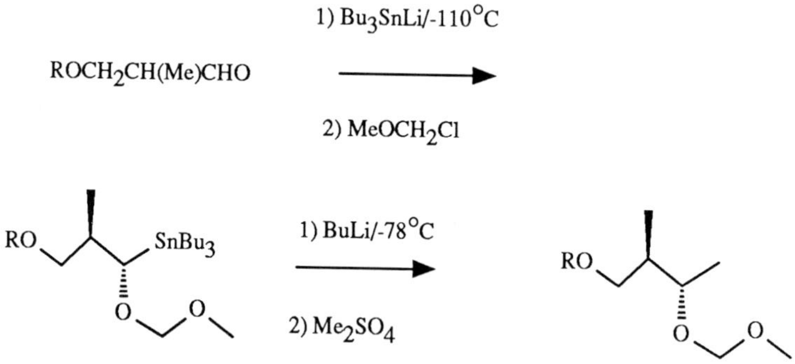

$ROCH_2CH(Me)CHO$

1) Bu_3SnLi/-110°C

2) $MeOCH_2Cl$

major isomer

1) BuLi/-78°C

2) Me_2SO_4

(235)

configurational stability , thus the diastereoisomers could be separated, transmetalated, and quenched with an electrophile (e.g. Me_3SiCl) at -78°C, with total retention of configuration (W.C.Still and C.Sreekumar, *J. Amer. Chem. Soc.*, 1980, **102**, 1201)(see Eqn. 234). Furthermore, if the stereochemistry of the stannylation reaction is subject to chelation control from a β-alkoxy group (see Eqn. 235), then a high degree of stereospecificity. can result.

α-Aminomethyltins have found use (see Eqn. 236) in the synthesis of β-aminoalcohols (D.J.Peterson, *J. Amer. Chem. Soc.*, 1971, **93**, 4027), although *N*-benzyl precursors need low temperatures (-65°C), if Wittig rearrangement is to be avoided (D.J.Peterson and J.F.Ward, *J. Organomet. Chem.*, 1974, **66**, 209). Peterson olefinations (*J. Org. Chem.*, 1968, **33**, 780) may be effectively performed (see Eqn. 237) using an organotin precursor (D.E.Seitz and A.Zapata, *Synthesis*, 1981, 557 ; *Tetrahedron Letters*, 1981, 3451). The readily available bis(trimethylstannyl)methane (E.J.Bulten, H.F.M.Grutter, and H.F.Martens, *J. Organomet. Chem.*, 1976, **117**, 329) is a methylenating reagent (E.Murayama *et al., Chem. Letters*, 1984, 1897).

$$Bu_3SnCH_2NMePh \xrightarrow[2)\ PhCHO]{1)\ BuLi/THF/0^{\circ}C} PhCH(OH)CH_2NMePh \qquad (236)$$

$$Me_3SiCH_2Cl \xrightarrow[hexane/0^{\circ}C]{Bu_3SnLi} Me_3SiCH_2SnBu_3 \xrightarrow[2)\ Ph_2CO]{1)\ BuLi}$$

$$(Ph)_2C(OH)CH_2SiMe_3 \xrightarrow{HOAc} (Ph)_2C{=}CH_2 \qquad (237)$$

$$Ph_3CH_2CH{=}CMe_2 \xrightarrow[3)\ \text{3-furylmethyl OTs}]{1)\ PhLi;\ 2)\ CuI} \text{3-(4-methylpent-3-enyl)furan} \qquad (238)$$

Allyltin compounds readily yield the corresponding lithium derivatives by transmetalation (D.Seyferth and M.A.Weiner, J. Org. Chem., 1959, 24, 1395; D.Seyferth and T.F.Jula, *J. Organomet. Chem.*, 1974, **66**, 195). These generally react to yield mixed products with and without allylic shift (D.Seyferth, K.R.Wursthorn, and R.E.Mammarella, *J. Org. Chem.*, 1977, **42**, 3104; D.Seyferth and R.E.Mammarella, *J. Organomet. Chem.*, 1979, **177**, 53), however, with *gem*-disubstituted allyl groups the reaction tends to occur at the last hindered position (R.A.Wiley, H.Y.Choo, and D.McClellan, *J. Org. Chem.*, 1983, **48**, 1106)(see Eqn. 238).

6. Metalation

Again this is normally lithiation, achieved either by halogenlithium exchange (e.g. T.Kauffman, R.Kriegesmann, and A.Woltermann, *Angew. Chem. Int. Ed. Engl.*, 1977, **16**, 862) or by deprotonation with lithium amides. The latter requires a α-heteroatom to be present, which can be sulphur (T.Kauffmann, R.Kriegesmann, and A.Hamsen, *Chem. Ber.*, 1982, **115**, 1818), or selenium (T.Kauffmann, *Angew. Chem. Int. Ed. Engl.*, 1982, **21**, 410). α-Halogeno-substitution can lead to organostannyl carbenes (R.A.Olofson, D.H.Hoskins, and K.D.Lotts, *Tetrahedron Letters.*, 1978, 1677), which can form cyclopropanes with double bonds (see Eqn. 239).

1) N Li / Et_2O

Me_3SnCH_2Cl → $SnMe_3$

2)

(239)

7. Formation of C-C bonds

The use of organotin compounds in the formation of new C-C bonds has increased dramatically since the mid-1970s, so that it is now completing with that of the organosilicons (M.Pereyre and J.C.Pommier, *J. Organomet. Chem. Libr.*, 1976, **1**, 161; Z.N.Parnes and G.I.Bolestova, *Synthesis*, 1984, 991). There have been two important developments contributing to this growth in use. First, the use of transition metal catalysis, especially palladium and rhodium, for increased efficiency in cross-coupling reactions of the main-group organometallic compounds (E.-i.Negishi, *Acc. Chem. Res.*, 1982, **15**, 340 and in *Current Trends in Organic Synthesis*, H.Nozaki, Ed., Pergamon, Oxford, 1983). Second, the use of Lewis acid catalysis for the control of regio- and stereo- specificity in addition reactions with carbonyl compounds (V.G.Kumar Das and C.K.Chu, in *The Chemistry of the Metal-Carbon Bond*, Vol. 3, F.R.Hartley and S.Patai, Eds., Wiley, Chichester, 1985).

i) Substitution reactions occur under very mild conditions, in view of the relatively weak Sn-C bond.

a) *Cross-coupling with allyltin compounds may* occur (see Eqn. 240) on heating, or more commonly *via* free radical initiation (J.Grignon, C.Servens, and M.Pereyre, *J. Organomet. Chem.*, 1975, **96**, 225; T.Migita, K.Nagai, and M.Kasngi, *Bull. Chem. Soc. Jpn.*, 1983, **56**, 2480; N.Ono, K.Zinmeister, and A.Kaji, *ibid.*, 1985, **58**, 1069; E.Block and M.Aslam, *J. Amer. Chem. Soc.*, 1983, **105**, 6165; J.H.Simpson and J.K.Stille, *J. Org. Chem.*, 1985, **50**, 1759). γ-Alkyl substitution of the allyl group, slows the reaction and may result in hydrogen abstraction, rather than coupling (G.E.Keck and J.B.Yates, *J. Organomet. Chem.*, 1983, **248**, C21). Lewis acids will also catalyse allylic coupling (see Eqn. 241 and 242), but the correct choice of catalyst is important. Thus $ZnCl_2$ effects coupling of allyl groups which have no γ-substitution (J.P.Godschalx and J.K.Stille, *Tetrahedron Letters*, 1983, **24**, 1905), while those with substituents require a catalyst such as bis(diethylaluminium)sulphate (A.Hosomi *et al.*, *J. Organomet. Chem.*,1985, **285**, 95).

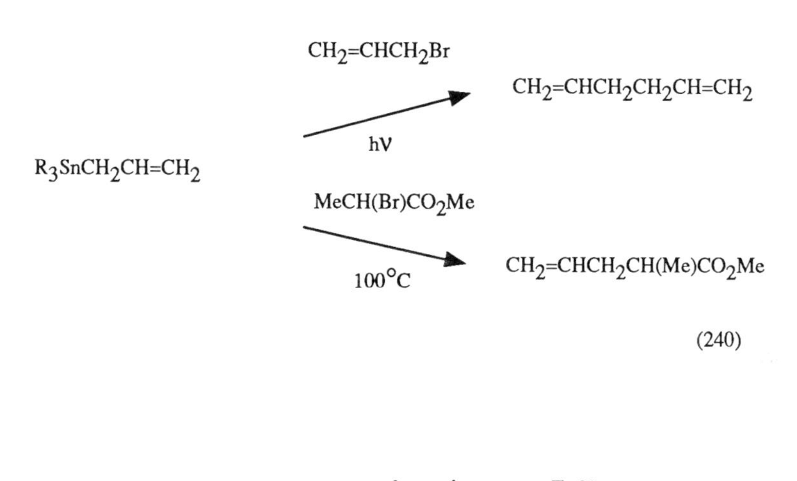

(240)

Br + Me3Sn

$ZnCl_2$

THF/65°C

(241)

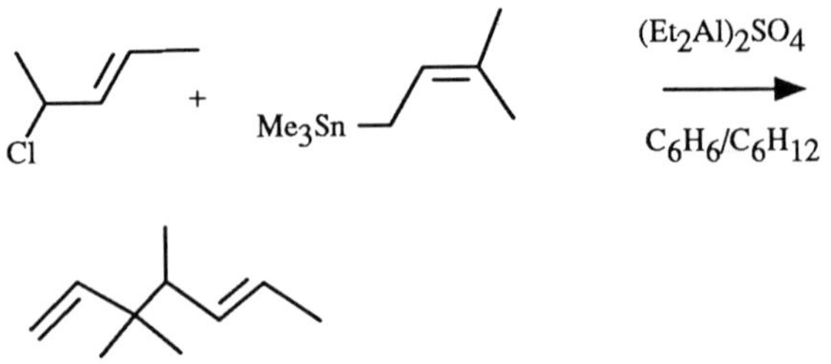

$(Et_2Al)_2SO_4$, C_6H_6/C_6H_{12}; Cl; Me$_3$Sn

(242)

The use of transition metal catalysts allows aryl bromides to be allylated (M.Kosugi *et al., Chem. Letters*, 1977, 301)(see Eqn. 243).

Allyl and benzyl halides can also be coupled using $BnPdCl(PPh_3)_2$ (J.P.Godschalx and J.K.Stille, *Tetrahedron Letters*, 1980, **21**, 2599)(see Eqn. 244), while vinyl bromides or triflates lead to nonconjugated dienes (W.J.Scott, G.T.Crisp, and J.K.Stille, *J. Amer. Chem. Soc.*, 1984, **106**, 4630).

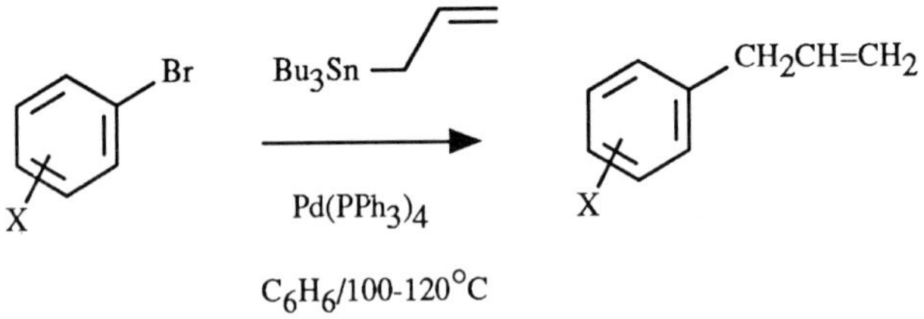

Br; X; Bu_3Sn; $Pd(PPh_3)_4$; C_6H_6/100-120°C; $CH_2CH{=}CH_2$; X

(243)

Br; + ; Me_3Sn; $BnPdCl(PPh_3)_2$; $CHCl_3$/65°C

major + minor

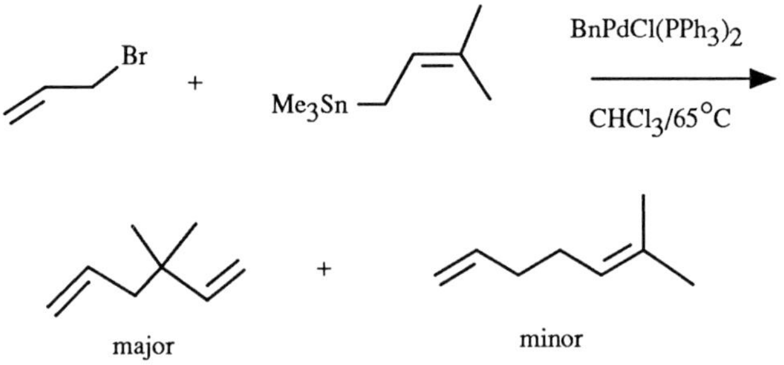

(244)

b) *Cross-coupling with alkyltin* compounds really requires transition metal catalysis (see Eqn. 245) to be effective. Methylation of aryl bromides (D.Milstein and J.K.Stille, *ibid.*, 1979, **101**, 4981 and 4992), or diazonium salts (K.Kikukawa *et al., J. Org. Chem.*, 1983, **48**, 1333) can be readily achieved.

Br — Me_4Sn, $BnPdCl(PPh_3)_2$, HMPA → X

(245)

c) *Cross-coupling with vinyl and alkynyltin compounds* can be achieved by free radical processes (G.A.Russell, H.Tashtousch, and P.Ngoviwatchai, *J. Amer. Chem. Soc.*, 1984, **106**, 4622)(see Eqn. 246). However, transition metal catalysis is more efficient leading to good yields under mild conditions (J.K.Stille, *Pure Appl. Chem.*, 1985, **57**, 1771). Replacement occurs with retention of configuration at the vinyl unit (J.Hibino *et al.*, *Tetrahedron Letters.*, 1984, **25**, 2151)(see Eqn. 247).

Ph, $SnBu_3$ — Pr^iHgCl, $h\nu/C_6H_6$ → Ph, Pr-i

E:Z = 92:8

(246)

RO, $SnBu_3$ — allylMgBr, $Pd(PPh_3)_4$, C_6H_6/heat → RO

(247)

d) *Cross-coupling with acyl halides* usually requires catalysis, and Lewis acids or ammonium salts have been used to yield ketones when reacted with organotin compounds (A.N.Kashin *et al., Zh. Org. Khim.*, 1980, **16**, 1569). There is, however, no real advantage over the use of organosilicon derivatives. Transition metals, commonly palladium or rhodium, are effective (M.Kosugi, Y.Shimizu, and T.Migita, *J. Organomet. Chem.*, 1977, **129**, C36; *Chem. Letters,* 177, 1423; D.Milstein and J.K.Stille, *J.*

Amer. Chem. Soc., 1978, 100, 3636; *J. Org. Chem.*, 1979, **44**, 1613), when inversion of configuration has been observed (J.W.Labadie and J.K.Stille, *J. Amer. Chem. Soc.*, 1983, **105**, 669)(see Eqn. 248).

Allyltin compound react readily with acyl halides (H.Sauo, M.Okawara, and Y.Ueno, *Synthesis*, 1984, 933), but Lewis acid catalysis is required (A.Gambaro, V.Peruzzo, and D.Marton, *J. Organomet. Chem.*, 1983, **258**, 291), this can be internal, with tin functioning as the Lewis acid (see Eqn. 249).

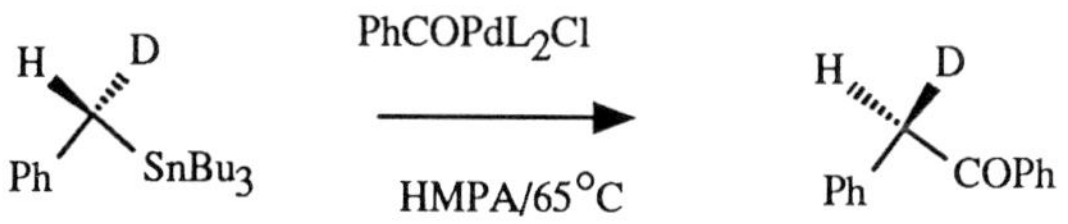

H, D, Ph, $SnBu_3$ —[$PhCOPdL_2Cl$; HMPA/65°C]→ H, D, Ph, COPh

(248)

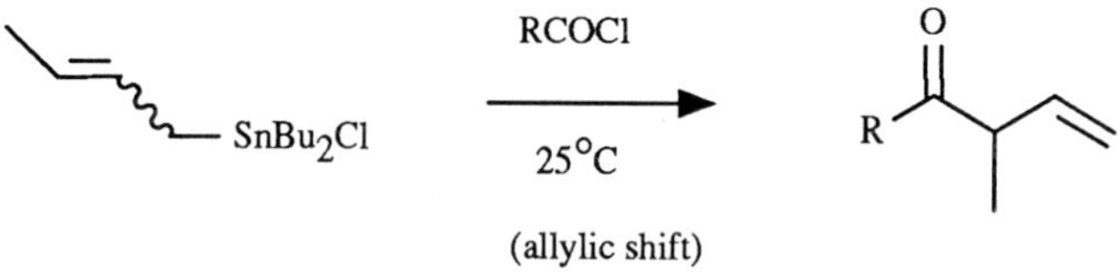

$SnBu_2Cl$ —[RCOCl; 25°C; (allylic shift)]→ O, R

(249)

Bu_3Sn, OEt —[1) PhCOCl, $BnPdCl(PPh_3)_2$, C_6H_6,90°C; 2) H^+]→ O, O, Ph

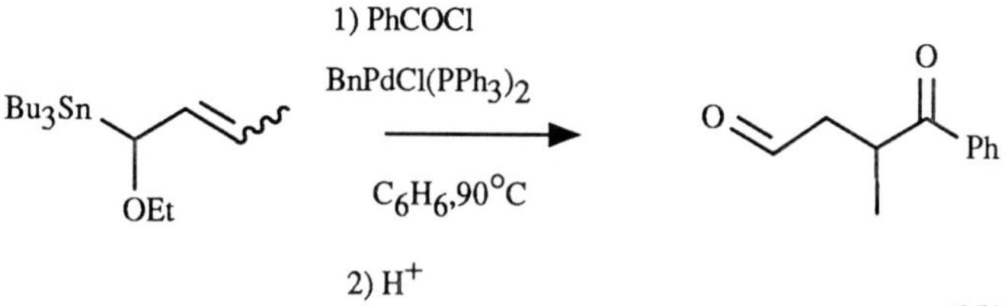

(250)

The use of transition metals promotes reaction under mild conditions, without unwanted isomerisation of β,γ-enones, which may be caused by Lewis acids (M.Pereyre, B.Elissondo, and J.P.Quintard, in *Selectivity-A Goal for Synthetic Efficiency*, Vol. 14, W.Bartmann and B.M.Trost, Eds., Verlag Chemie, Weinheim, 1984)(see Eqn. 250). For similar reasons vinyltin compounds are best coupled using transition metal catalysts, when considerable chemico-selectivity is possible (D.H.Rich, J.Singh, and J.H.Gardner, *J. Org. Chem.*, 1983, **48**, 432)(see Eqn. 251).

$$\text{AcNH-CH(CO}_2\text{Et)-(CH}_2)_5\text{-COCl} \xrightarrow[\text{BnPdCl(PPh}_3)_2,\ \text{HMPA / 56}^\circ\text{C}]{(\text{allyl})_4\text{Sn}} \text{AcNH-CH(CO}_2\text{Et)-(CH}_2)_5\text{-COCH=CH}_2 \qquad (251)$$

$$\text{PhI + Bu}_4\text{Sn} \xrightarrow[\text{HMPA/120}^\circ\text{C},\ \text{PhPdI(PPh}_3)_2\text{,cat.}]{\text{CO(30atm.)}} \text{PhCOBu} \qquad (252)$$

The possibility of decarbonylation during these processes may be avoided by the use of a carbon monoxide atmosphere (J.W.Labadie *et al., J. Org. Chem.*, 1983, **48**, 4634; *Tetrahedron Letters.*, 1983, **24**, 4283). Alkynyltin compounds react in a similar fashion, with or without Lewis acid catalysis (A.N.Kashin *et al., Zh. Org. Khim.*, 1980, **16**, 1569), although less reactive derivatives may benefit from palladium complex catalysis (J.Ackroyd, M.Karpf, and A.S.Dreiding, *Helv. Chim. Acta.*, 1985, **68**, 338).

e) *Cross-coupling with organic halides under carbon monoxide.* An atmosphere of carbon monoxide prevents decarbonylation during the reaction with acyl halides (see above), however, under pressure other organohalides may couple with carbonylation (M.Tanaka, *Tetrahedron Letters*, 1979, 2601)(see Eqn. 252). The use of triphenylarsine ligands avoids isomerisation (T.Kobayashi and M.Tanaaka, *J.Organomet. Chem.*, 1981, **205**, C27), while some nickel complexes are also effective (M.Tanaka, *Synthesis*, 1981, 47). An example, which illustrates the potential of these procedures is provided by the synthesis of (±)-$\Delta^{9(12)}$-capnellene (G.T.Crisp, W.J.Scott, and J.K.Stille, *J. Amer. Chem. Soc.*, 1984, **106**, 7500) (see Eqn. 253).

OTf Me_3Sn $SiMe_3$ $Ph(PPh_3)_4$ CO/LiCl O $SiMe_3$

$BF_3.Et_2O$ toluene 100°C H O 1) L-selectride 2) Tf_2NPh

H OTf Me_3Sn $SiMe_3$ $Ph(PPh_3)_4$ CO/LiCl O H $SiMe_3$

1) $BF_3.Et_2O$ 2) $H_2\backslash Pd$ 3) Wittig H H H

(253)

ii) Addition reactions

a) *Allylation of carbonyl compounds.* Since this process was first reported (B.Konig and W.P.Neumann, *Tetrahedron Letters,* 1967, 495)(see Eqn. 254), it has been extensively developed to become a powerful stereoselective synthetic method. The use of Lewis acid catalysts was required to increase yields and shorten reaction times (Y.Yamamoto, K.Maruyama, and K.Matsumoto, *J. Chem. Soc., Chem. Commun.*, 1983, 489). Conversion of the allyltin compound to a more Lewis acidic halogen derivative leads to internal catalysis (V.Peruzzo and G.Tagliavini, *J. Organomet. Chem.*, 1978, **162**, 37; A.Gambaro *et al., ibid.*, 1980, **197** 45; , 1981, **204**, 191). The allyltin halide may be generated *in situ*, leading regio- and stereo-specifically to a (*Z*)-allyl alcohol (A.Gambaro *et al., ibid.*, 1982, **231**, 307)(see Eqn. 255).

$$4\text{-R-}C_6H_4CHO \xrightarrow[\text{2) } CH_2(CO_2H)_2]{\text{1) allylSnEt}_3} 4\text{-R-}C_6H_4CH(OH)CH_2CH{=}CH_2 \qquad (254)$$

$$Bu_3Sn\text{-}CH_2CH{=}CHCH_3 + Bu_2SnCl_2 \longrightarrow [Bu_2SnCl\text{-}CH(CH_3)CH{=}CH_2] \xrightarrow{RCHO} RCH(OH)CH_2CH{=}CHCH_3 \qquad (255)$$

External Lewis acid, such as $BF_3.Et_2O$, catalyse additions under mild conditions (K.Maruyama and Y.Naruta, *J. Org. Chem.*, 1978, **43**, 3796; *Letters,* 1978, 431; *ibid.*, 1979, 881; 885; Y.Naruta, S.Ushida, and Y.Maruyama, *ibid.*, 1979, 919; Y.Naruta, *J. Amer. Chem. Soc.*, 1980, **102**, 3774).

Allylic rearrangement is normally observed, although the use of $AlCl_3/Pr^iOH$ is unpredictable and sometimes no rearrangement occurs (Y.Yamamoto, N.Maeda, and K.Maruyama, *J. Chem. Soc., Chem. Commun.*, 1983, 742; Y.Yamamoto *et al., Tetrahedron*, 1984, **40**, 2239). Allyl acetates can be converted directly into homoallylic alcohols without isolation of intermediates (B.M.Trost and J.W.Henderson, *J. Amer. Chem. Soc.*, 1984, **106**, 6835), using Pd(0) mediation (see Eqn. 256). Indeed, a palladium catalyst allows the formation of oxiranes from α-haloketones, or aldehydes (I.Pri-bar, P.S.Pearlman, and J.K.Stille, *J. Org. Chem.*, 1983, **48**, 4629)(see Eqn. 257).

$$4\text{-}BrC_6H_4CH_2CH_2C(CH_3)(OAc)CH{=}CH_2 \xrightarrow[\text{2) EtCHO, } BF_3.Et_2O]{\text{1) } Et_2AlSnBu_3\text{, } Pd(PPh_3)_4} 4\text{-}BrC_6H_4CH_2CH_2C(CH_3)(CH{=}CH_2)CH(OH)CH_2CH_3 \qquad (256)$$

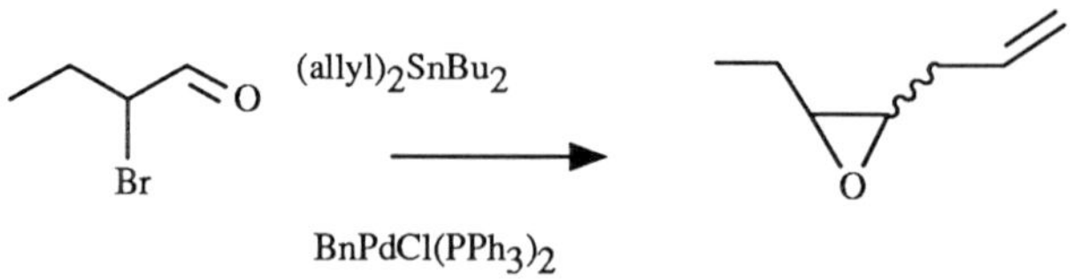

O
Br
(allyl)$_2$SnBu$_2$
BnPdCl(PPh$_3$)$_2$
O

(257)

The *stereochemistry* of these allylation has been the subject of several reviews (D.Hoppe, *Angew. Chem. Int. Ed. Engl.*, 1984, **23,** 932; R.W.Hoffman, *ibid.*, 1982, **21**, 555; Y.Yamamoto and K.Maruyama, *Heterocycles*, 1982, **18**, 357). Two different conventions, for naming products, has led to some confusion at times: *erythro-* and *threo-* isomers (C.H.Heathcock *et al., J. Org. Chem.*, 1980, **45,** 1066), or *syn-* and *anti-* isomers (S.Masamune *et al., Angew. Chem. Int. Ed. Engl.*, 1980, **19**, 557), the corresponding structures are given in Eqn. 258.

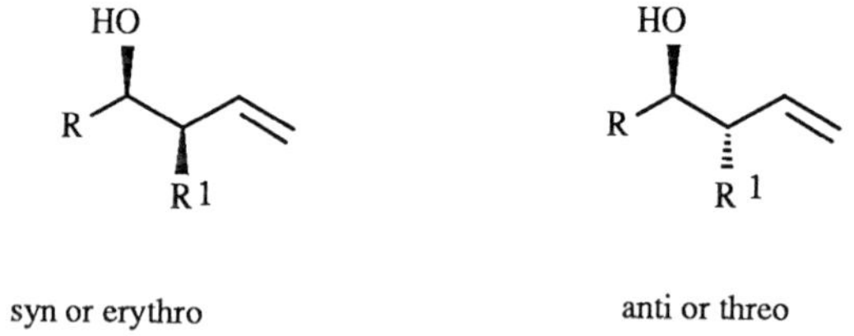

HO
R
R1
syn or erythro
HO
R
R 1
anti or threo

(258)

Diastereoselective addition to enantiotopic faces of γ-substituted allyltins is stereoselective in the absence of a catalyst (C.Servens and M.Pereyre, *J. Organomet. Chem.*, 1972, **35**, C20)(see Eqn. 259). The use of BF_3Et_2O as catalyst, however, promotes *erythro-* selectivity for the additions between crotyltin and aldehydes, from both (*E*)- and (*Z*)-configurations (Y.Yamamoto *et al., Tetrahedron*, 1984, **40**, 2239; *J. Amer. Chem. Soc.*, 1980, **102**, 7107)(see Eqn. 260).

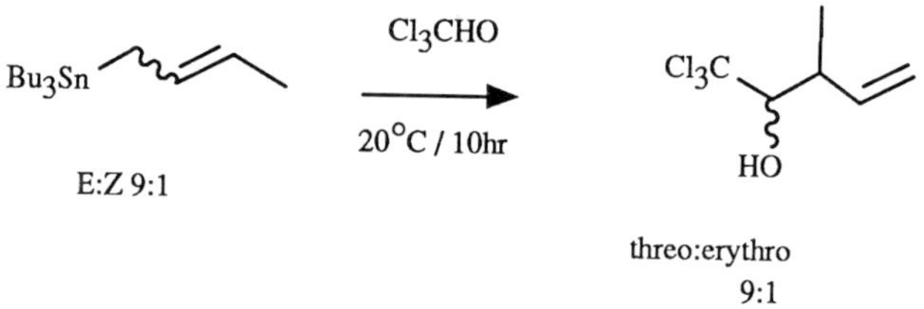

Bu$_3$Sn
E:Z 9:1
Cl$_3$CHO
20°C / 10hr
Cl$_3$C
HO
threo:erythro
9:1

(259)

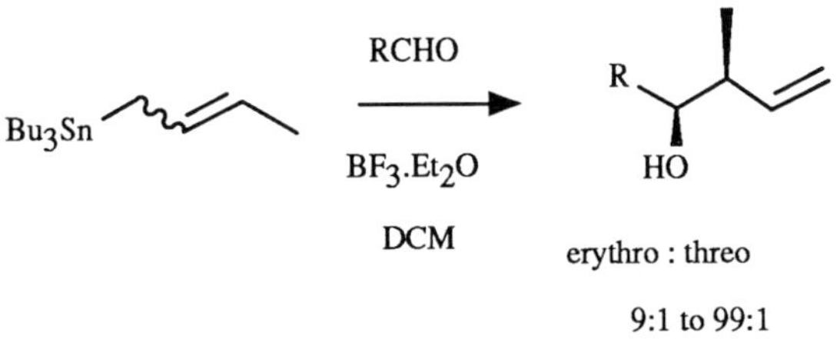

RCHO

Bu_3Sn → $BF_3.Et_2O$ DCM

R, HO

erythro : threo

9:1 to 99:1

(260)

Similar selectivity is found with many allyltins, although not for pyruvates (Y.Yamamoto, T.Komatsu, and K.Maruyama, *J. Chem. Soc., Chem. Commun.*, 1983, 191). While (*E*)-cinnamyltriphenyltin gives solely the *threo*-isomer (M.Koreeda and Y.Tanaka, *Chem. Letters.*, 1982, 1299). It is interesting that for analogous organo-silicon derivatives *threo*-selectivity is the general rule (T.Hayashi *et al., Tetrahedron Letters*, 1983, **24**, 2865). For cyclohexane carboxaldehyde and crotyltin additions catalysed by $TiCl_4$, the selectivity is controlled by the order of addition; thus crotyltributyltin added to aldehyde and $TiCl_4$ is a "normal" reaction while the addition of crotyltributyltin and $TiCl_4$ to the aldehyde is an "inverse" process (G.E.Keck *et al., Tetrahedron Letters*. 1984, **25**, 3927) (see Eqn. 261).

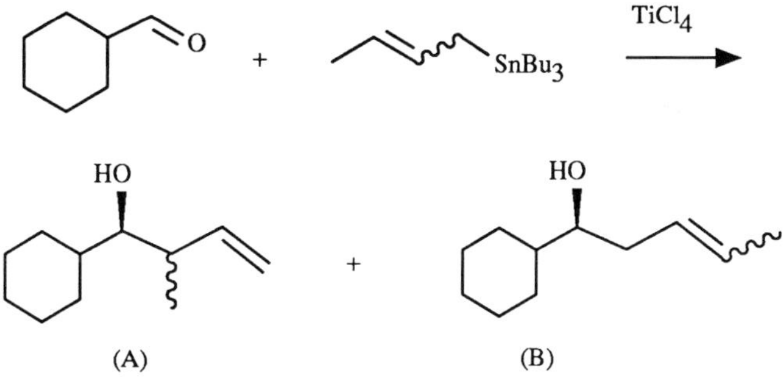

O + $SnBu_3$ → $TiCl_4$

HO + HO

(A) (B)

'normal addition' with 1.05equiv.of reagent and 1.05equiv.

of Lewis acid gives >97% (A)[erythro:threo 93:7] and <3% (B)

'inverse addition' with 2equiv.of reagent and 2equiv.

of Lewis acid gives >95% (A)[erythro:threo 5:95] and <5% (B)

(261)

Diastereoselective addition to diastereotopic faces of allyltin compounds, generally results (see Eqn. 262) in allylation from the less hindered face (Y.Naruta, S.Ushida, and Y.Maruyama, *Chem. Letters*., 1979, 919; E.C.Ashby and J.T.Laemmle, *Chem. Revs*., 1975, **75**, 521; M.Gaudemar, *Tetrahedron*, 1976, **32**, 1689). The presence of an α-alkoxy or α-siloxy unit in the aldehyde allows remarkable stereo-selectivity depending upon the reaction conditions (G.E.Keck and E.P.Boden, *Tetrahedron Letters,* 1984, **25**, 1879)(see Eqn. 263). These reactions involve the principle of "chelation" or "non-chelation" control which has been reviewed (M.T.Reetz, *Angew. Chem. Int. Ed.* Engl., 1984, **23**, 556).

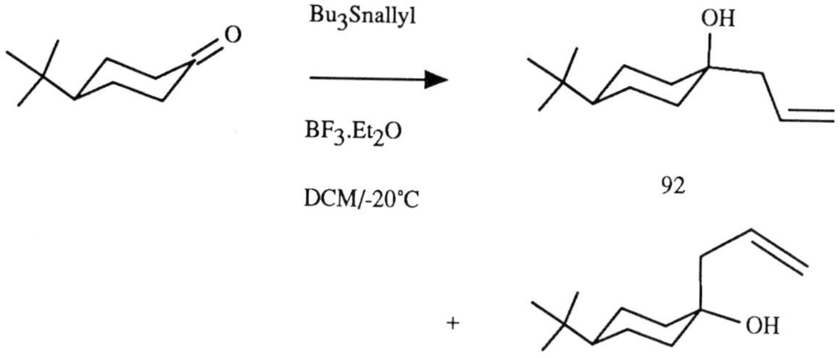

$Bu_3Snallyl$

$BF_3.Et_2O$

DCM/-20°C

OH

92

\+

OH

8 (262)

OR^1 + Bu_3Sn — Lewis acid, DCM →

^1RO ... OH + ^1RO ... OH

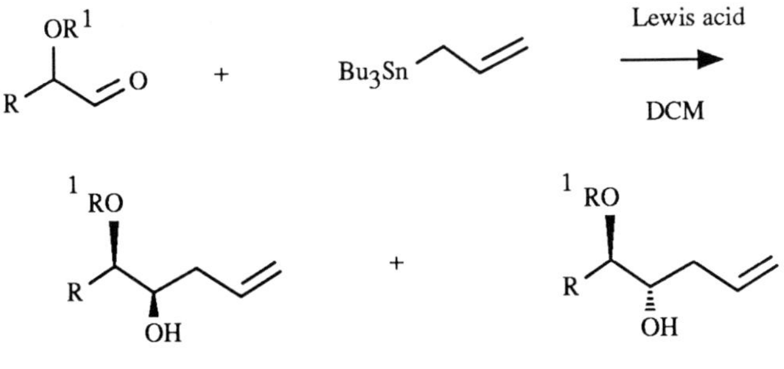

(A) (B)

Lewis acid = $TiCl_4$ >250:1 (A):(B)

Lewis acid = $BF_3.Et_2O$ 5:95 (A):(B)

(263)

β-Alkoxyaldehydes are subject to similar "chelation control" during $SnCl_4$ catalysed allyltin addition reactions (G.E.Keck and D.E.Abbott, *Tetrahedron Letters*, 1984, **25**, 1883). When crotyltin compounds are added to α- or β- alkoxyaldehydes "chelation" control can be exercised over both developing chiral centres, leading to *double diastereoselection* (G.E.Keck and E.P.Boden, *ibid.*, 1984, **25**, 1879; see also G.E.Keck and D.E.Abbott, *ibid.*, 1984, **25**, 1883)(see Eqn. 264).

OBn R O + Bu_3Sn — $MgBr_2$, DCM/-22°C →

BnO R OH + BnO R OH

syn-syn >90 syn-anti <10

(264)

Enantioselection has been accomplished by use of either a chiral aldehyde (e.g. Y.Yamamoto, N.Maeda, and K.Maruyama, *J. Chem. Soc., Chem. Commun.*, 1983, 774) or a chiral allyltin derivative (see V.J.Jephcote, A.J.Pratt and E.J.Thomas, *ibid.*, 1984, 800).

b) *Alkylation of carbon-carbon double bonds.* Lewis acid catalysis will bring about the addition of alkyl groups to alkenes (Z.N.Parnes *et al., ibid.*, 1980, 748)(see Eqn. 265).

Me_4Sn, CF_3CO_2H/AlX_3 →

(265)

Internal addition to conjugated enones has been used for carbocyclisation (T.L.Macdonald and S.Mahalingam, *J Amer. Chem. Soc.*, 1980, **102**, 2113)(see Eqn. 266), although the cyclisation is inhibited by steric hindrance. This type of cyclisation has been applied to other functionalities (T.L.McDonald, S.Mahalingam, and D.E.O'Dell, *ibid.*, 1981, **103**, 6767)(see Eqn. 267).

SnMe3 · TiCl4(1%) · DCM · cis:trans · 40°C/2min. 93:7 · -78°C/30min. 33:67

(266)

1) SnCl4/DCM · 2) SOCl2/LiBr/MeCN

(267)

iii) Elimination reactions

a) *α-Elimination reactions.* A number of organotin compounds behave as dihalocarbene precursors, for example, triorganostannyl trichloroacetates (H.C.Clark and C.J.Wills, *J. Amer. Chem. Soc.*, 1960, **82**, 1888; F.M.Arbrecht, W.Tronich, and D.Seyferth, *J.Amer.Chem.Soc.*, 1969, **91**, 3218) (see Eqn. 268). There is some question as to whether carbene species are intermediate in these reactions (P.M.Warner and R.D.Herold, *Tetrahedron Letters*, 1984, **25**, 4897). An interesting α-elimination leads to reductive etherification of aryl aldehydes: a simple route to benzylic ethers, even in the presence of phenolic groups (J.-P.Quintard, B.Elissondo, and D.Mouko-Mpegna, *J. Organomet. Chem.*, 1983, **251**, 175)(see Eqn. 269) .

$Cl_3CCO_2SnMe_3$ · 140°C · diglyme

(268)

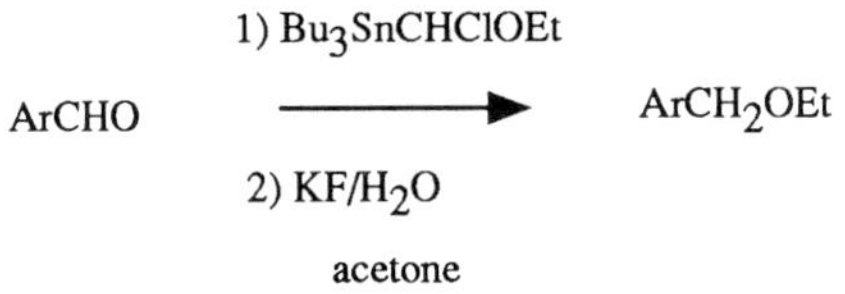

$$ArCHO \xrightarrow[\text{2) KF/H}_2\text{O, acetone}]{\text{1) Bu}_3\text{SnCHClOEt}} ArCH_2OEt$$

(269)

b) β-*Elimination reactions.* Acid catalysed deoxystannylation of β-stannylalcohols leads stereospecifically to alkenes (D.D.Davis and C.E.Gray, *J. Organomet. Chem.*, 1969, **18**, P1; *J. Org. Chem.*, 1970, **35**, 1303); following an *anti*-elimination pathway (W.Kitching *et al., ibid.*, 1978, **43**, 898; R.H.Fish and B.M.Broline, *J. Organomet. Chem.*, 1978, **159**, 255). By contrast (see Eqn. 270), thermal elimination takes a *syn*-pathway (T.Kauffmann, *Angew. Chem. Int. Ed. Engl.*, 1982, **21**, 410; T.Kauffmann, R.Kriegesmann, and R.Hamsen, *Chem. Ber.*, 1982, **115**, 1818). An interesting extension (see Eqn. 271) provides for the stereo-controlled conversion of vinylsulphones to vinylsilanes (M.Ochiai, T.Ukita, and E.Fujita, *J. Chem. Soc., Chem. Commun.*, 1983, 619).

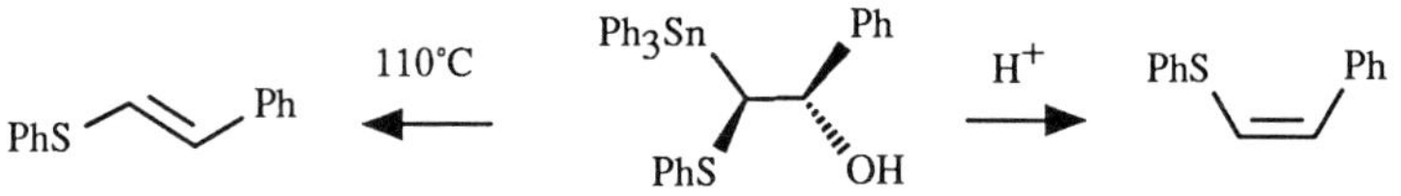

PhS–CH=CH–Ph (E) ← 110°C — Ph_3Sn, PhS, Ph, OH β-stannylalcohol — H^+ → PhS–CH=CH–Ph (Z)

(270)

R–CH=CH–SO_2Ph → 1) Bu_3SnLi; 2) Me_3SiCl → R, $SiMe_3$, Bu_3Sn, SO_2Ph adduct

→ SiO_2, $CHCl_3$, reflux → R–CH=CH–$SiMe_3$ (E) >90%

→ SiO_2, $CHCl_3$, 25°C → R–CH=CH–$SiMe_3$ (Z) >96%

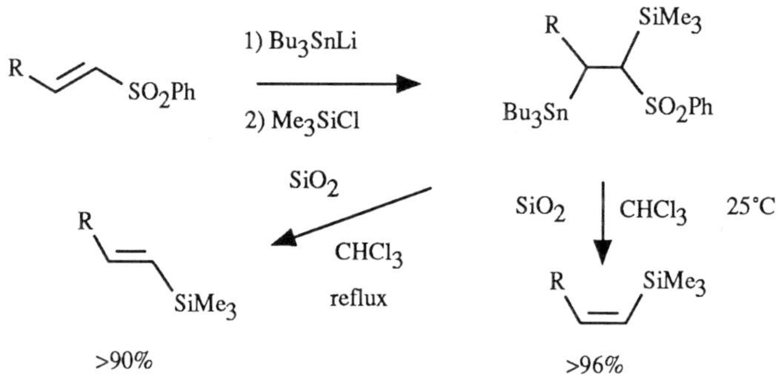

(271)

c) γ-*Elimination reactions*. 1,3-Deoxystannylation provides a convenient route to cyclopropanes (H.G.Kuivila and N.M.Scarpa, *J. Amer. Chem. Soc.*, 1970, **92**, 6990; D.D.Davis, R.L.Chambers, and H.T.Johnson, *J. Organomet. Chem.*, 1970, **25**, C13; I.Fleming and C.Urch, *ibid.*, 1985, **285**, 173). Cyclopropylcarbinyl halides can also be prepared in a related process (see Eqn. 272), although no elimination is involved (D.J.Peterson, M.D.Robbins, and J.R.Hansen, *ibid.*, 1974, **73**, 237). β-Stannylpropanal affords access to a number of possible cyclopropane syntheses, some of which do involve 1,3-elimination reactions (Y.Ueno, M.Ohta, and M.Okawara, *Tetrahedron Letters*, 1982, **23**, 2577)(see Eqn. 273).

Bu_3Sn (but-3-enyl) $\xrightarrow{X_2/DCM}$ cyclopropyl–CH_2X (272)

Bu_3Sn (propanal) $\xrightarrow{RLi}$ Bu_3Sn–CH₂CH₂CH(R)OH $\xrightarrow[THF]{SOCl_2/py.}$ cyclopropyl–R (273)

d) Lead

The chemistry of organolead compounds has been reviewed (Houben-Weyl, *Methoden der Organischen Chemie*, Vol. 13/7, E.Muller *et al.*, Eds., Thieme, Stuttgart, 1975; R.C.Poller, in *Comprehensive Organic Chemistry*, Vol. 3, D.H.R.Barton and W.D.Ollis, Eds., Pergamon, Oxford, 1979; P.G.Harrison, in *Comprehensive Organometallic Chemistry*, Vol. 2, G.Wilkinson, F.G.A.Stone, and E.W.Abel, Eds., Pergamon, Oxford, 1982). The technical use of lead alkyls as gasoline additives has led to much concern regarding lead in the environment (J.M.Ratcliffe, *Lead in Man and the Environment*, Halsted Press, New York, 1981). Organolead derivatives have not found extensive use in organic synthesis.

6. Group VA: phosphorus, arsenic, antimony, and bismuth

The organochemistry of phosphorus, a nonmetal, is often treated separately from those of arsenic and antimony, metalloids, and bismuth, a metal. For the purposes of this review, however, phosphorus is considered as a metalloid. Various aspects of the organochemistry of the main-group V elements have been reviewed (e.g. D.Hellwinkel, *Top. Curr. Chem.*, 1983, 109, 1; B.J.Aylett, *Organometallic Compounds, 4th Edn., Vol. 1, Part 2*, Chapman and Hall, London, 1979; Houben-Weyl *Methoden der Organischen Chemie*, E.Muller *et al.*, Eds., G.Thieme, Stuttgart, 1978; F.Bickelhaupt and H.Vermeer, *Method. Chim.*, 1978, **78B**, 549; H.Schmidbaur, *Adv. Organomet. Chem.*, 1976, **14**, 205; E.Maslowsky, Jr., *J. Organomet. Chem.*, 1974, **70**, 153; G.O.Doak and L.D.Freedman, *Synthesis,* 1974, 328; R.Luckenbach, *Dynamic Stereochemistry of Pentacoordinated Phosphorus and Related Elements*, Georg Thieme, Stuttgart, 1973), including their role as ligands with transition metals (e.g. O.J.Scherer, *Angew. Chem. Int. Ed. Engl.*, 1985, **24**, 924; C.A.McAuliffe and W.Levason, *Phosphine, Arsine, and Stibine Complexes of the Transition Elements,* Elsevier, Amsterdam, 1979, *Transition-Metal Complexes of Phosphorus, Arsenic, and Antimony Ligands*, MacMillan, New York, 1973; D.I.Hall, J.H.Ling, and R.S.Nyholm, *Struct. Bonding*, 1973, 15, 3; R.J.Cross, *M.T.P. Int. Rev. Sci., Inorg. Chem., Ser. Two*, 1974, **5**, 147).

a) Phosphorus

The organochemistry of phosphorus continues to flourish, excellent books have been written (B.J.Walker, *Organophosphorus Chemistry*, Penguin, London, 1972; J.Emsley and D.Hall, *The Chemistry of Phosphorus*, Harper and Row, London, 1976) and numerous aspects of the subject reviewed. These include the synthesis of phosphines (M.Baudler, *Angew. Chem. Int. Ed. Engl.*, 1987, **26**, 419); ylides by desilylation (E.Vedejs and F.G.West, *Chem. Rev.*, 1986, **86**, 941); reactive ylides (B.E.Maryanoff and A.B.Reitz, *Phosphorus Sulphur*, 1986, **27**, 167); reactions with quinones (A.A.Kutyrev and V.V.Moskva, *Russ. Chem. Revs.*, 1987, **56**, 1028); alkoxy and aryloxy phosphoranes (L.N.Markovskii, N.P.Kolesnik, and Yu.G.Shermolovich, *ibid.*, 1987, **56**, 894); preparation of chiral tervalent phosphorus compounds (A.Mortreux e*t al., Bull. Soc. Chim. Fr.*, 1987, 631); phosphorus π-bonding (E.E.Schweiger *et al., J. Org. Chem.*, 1987, **52**, 1810; **52**, 1810; H.Sun, D.A.Hovat, and W.T.Borden, *J. Amer. Chem. Soc.,* 1987, *109*, 5275), and nucleophilic substitution at phosphorus (H.Corriu, *Phosphorus Sulphur*, 1986, **27**, 1). An excellent overview of the period under review can be obtained from the *Specialist Periodical Reports* of the Chemical Society (e.g. *Organophosphorus Chemistry, S.P.R.*, B.J.Walker and J.B.Hobbs, Eds., R.S.C., London, 1989, **20**), or by consulting the *Proceedings of the*

International Conferences on Phosphorus (e.g. 10th Conference, Bonn, 1986, *Phosphorus Sulphur*, 1987, **30**, 3-850). The preparation of organophosphorus compounds is also well covered (see G.M.Kosolapoff and L.Maier, *Organic Phosphorus Compounds*, Vol. 1-7, Wiley, New York, 1973). The use of organophosphorus compounds in organic synthesis (*Organophosphorus Reagents in Organic Synthesis*, J.I.G.Cadogan, Ed., Academic Press, London, 1979; *Comprehensive Organic Chemistry,* Vol. 2, D.H.R.Barton and W.D.Ollis, Eds., Pergamon, Oxford, 1979) has been dominated by olefination reactions (I.Gosney and A.G.Rowley, *The Wittig Reaction*); D.J.H.Smith, *P=O Activated Olefinations*, in *Organophosphorus Reagents in Organic Synthesis*, see above). Indeed these reactions have been widely used in natural product synthesis (H.J.Bestmann and O.Vostrowsky, *Top. Curr. Chem.*, 1983, **109**, 85), and in industrial processes (H.Pommer and P.C.Thieme *ibid.*, 1983, **109**, 165).

Uses in Synthesis of Phosphoranes

1. Wittig Reaction.

First reported in 1953 (G.Wittig and G.Geissler, *Liebigs Ann. Chem.*, 1953, **580**, 44) the reaction was described by Wittig as a "Variationen zu einem Thema von Staudinger" (see Eqn. 274).

$$Ph_3P^+CH_3\,Br^- \xrightarrow[Et_2O]{PhLi} \left[Ph_3P^+CH_2^- \leftrightarrow Ph_3CH{=}CH_2 \right] \xrightarrow{Ph_2CO} Ph_2C{=}CH_2 + Ph_3PO \qquad (274)$$

The reaction has since become a powerful synthetic method for olefination, due to a) its *generality,* a wide range of carbonyl compounds and phosphoranes will react, b) its *regiospecificity*, the carbon-carbon double always occupies the site to the original carbonyl function, and c) control of *stereochemistry*, variation in reagents and reaction conditions allow control over the configuration of the alkene generated. The preparations of the necessary quaternary phosphonium salts are well covered (G.M.Kosolapoff and L.Maier, *Organic Phosphorus Compounds*, see above), while suitable bases for the generation of the ylides (or phosphoranes) appear in the references quoted above.

The nature of the ylide is important and affects the outcome of the process. The reactivity of ylides depends upon the stabilisation afforded the anionic carbon centre by any substituents present (see Figure 1).

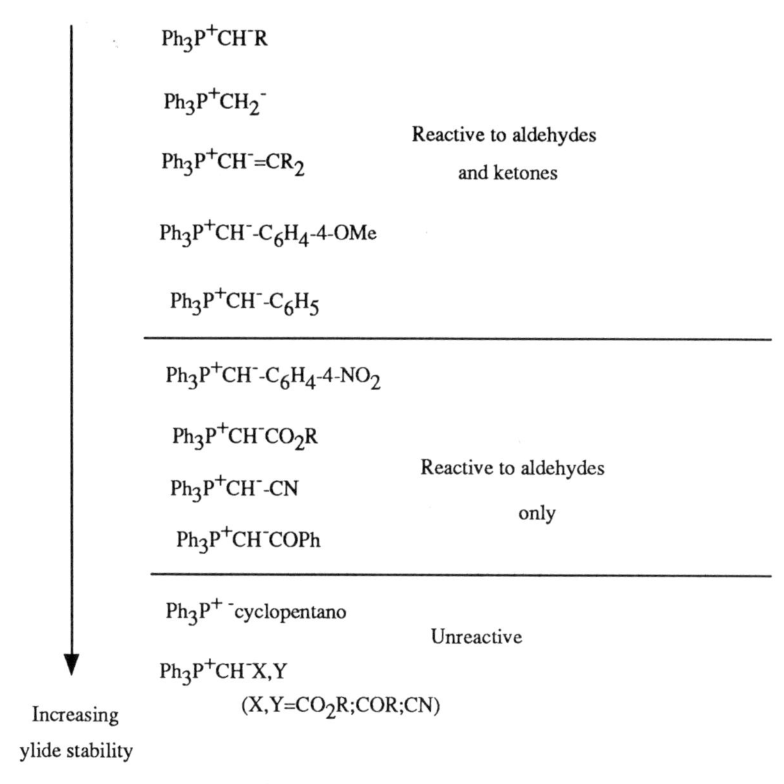

Figure 1

i) Stable ylides favour the formation of (E)-alkenes. Thus stabilised ylides yield predominantly (E)-alkenes (e.g. E.Vedejs, T.Fleck, and S.Hara, *J. Org. Chem.*, 1987, **52**, 4637)(see Eqn. 275). Indeed, alkyl groups

$MePh_2P^+CH^-CO_2Et$ + (cyclohexyl)CHO → (cyclohexyl)CH=CHCO_2Et

(E):(Z) =95:5

(275)

attached to phosphorus tend to increase (E)-selectivity (D.E.Bissing, *J. Org. Chem.*, 1965, **30**, 1296), to aryl groups (see Eqn. 276). However, substitution at the anionic carbon in the ylide tends to reduce selectivity (A.J.Speziale and K.W.Ratts, *J. Amer. Chem. Soc.*, 1963, **85**, 2790)(see Eqn. 277). In general, ketones only react under conditions of acid catalysis or relatively high temperature (D.L.Roberts *et al., J. Org. Chem.*, 1968, **33**, 3566). Then stereoselectivity is low (see Eqn. 278). Recently it has been recognised that the application of high pressures can lead to increased reaction rates in difficult Wittig processes, such as those between stabilised ylides and ketones (N.S.Isaacs and O.H.Abed, *Tetrahedron Letters.*, 1986, **27**, 995; B.J.Walker, in *Organophosphorus Chemistry, S.P.R.*, B.J.Walker Ed., R.S.C., London, 1987, **18**, 304; W.G.Dauben and J.J.Takasugi, *Tetrahedron Letters.*, 1987, **28**, 4377). Furthermore α-hydroxyketones exhibit greater reactivity towards ester-stabilised ylides (P.Garner and S.Ramakanth, *J. Org. Chem.*, 1987, **52**, 2629). Solvents are important, and (E)-selectivity is highest with aprotic solvents; while protic solvents or added soluble salts increase the percentage of (Z)-isomer formed (I.Gosney and A.G.Rowley *loc.cit.*)

$R_3P^+CH_2CO_2Et$ X^- (X = Br or BPh_4) —[1) NaOEt/EtOH/25°C; 2) PhCHO]→ Ph–CH=CH–CO_2Et

	(E) : (Z)
R=Ph	85 : 15
R=Bu	95 : 5
R=cyclohexyl	100 : 0

(276)

Ph_3P=C(X)–C(=O)Ph —[ArCHO; $CHCl_3$; 24hrs/25°C]→ ArCH=C(X)–C(=O)Ph (Z) + ArCH=C(X)–C(=O)Ph (E)

X = H 100 : 0

X = Br 35 : 65

(277)

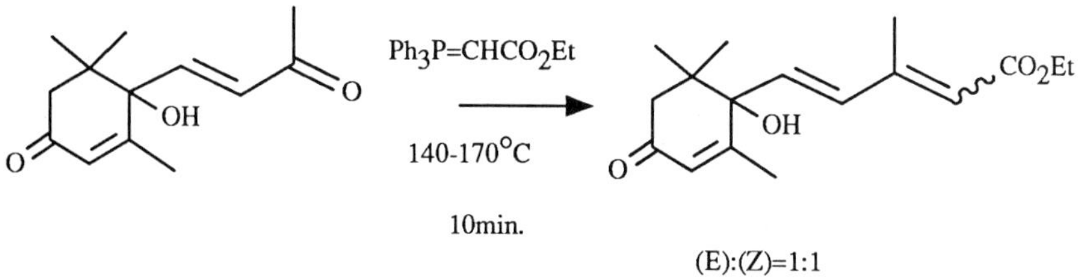

Ph$_3$P=CHCO$_2$Et

140-170°C

10min.

(E):(Z)=1:1

(278)

ii) Reactive ylides, in contrast, tend to favour (Z)-alkenes. The (Z)-selectivity is variable, although it can be strikingly high for reactions in polar aprotic solvents (e.g. M.I.Dawson and M.Vasser, *J. Org. Chem.*, 1977, **42**, 2783)(see Eqn. 279). Non-polar solvents under "salt-free" conditions can also led to high (Z)-preference (R.J.Anderson and C.A.Hendrick, *J. Amer. Chem. Soc.*, 1975, **97**, 4327; R.M.Boden, *Synthesis*, 1975, 784; H.J.Bestmann, W.Stransky, and O.Vostrowsky, *Chem. Ber.*, 1976, **109**, 1694). Unbranched aliphatic aldehydes tend to favour a high (Z)-selectivity, while the use of conjugated enals, aromatic aldehydes or ketones reduces selectivity (see B.G.James and G.Pattenden, *J. Chem. Soc., Perkin Trans. 1,* 1976, 1476) (see Eqn. 280).

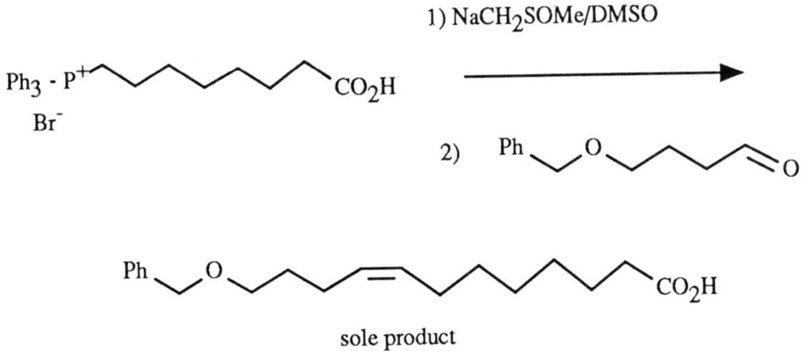

1) NaCH$_2$SOMe/DMSO

2) Ph–CH$_2$–O–(CH$_2$)$_3$–CHO

sole product

(279)

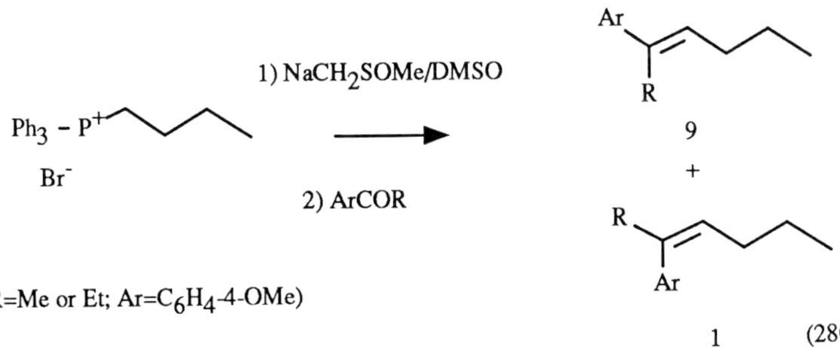

1) NaCH$_2$SOMe/DMSO

2) ArCOR

(R=Me or Et; Ar=C$_6$H$_4$-4-OMe)

9 + 1

(280)

Methods have been developed, which allow (E)-alkenes to be obtained from reactive ylides (see Eqn. 281). In the *Schlosser-Wittig procedures* a Wittig reaction is carried out at low temperature, in the presence of lithium salts, this affords an intermediate betaine which may be sufficiently stable to be α-lithiated. The resultant β-oxaylide, can then undergo rapid equilibration leading to (E)-selectivity in the product (M.Schlosser and K.F.Christmann, *Angew. Chem. Int. Ed. Engl.*, 1966, **5**, 126). In the related

Ph_3–P^+ - - O^- – LiX PhLi Ph_3–P^+ - - O^- – LiX

"betaine"

H^+

$-Ph_3PO$

(281)

Snoopy reaction the intermediate β-oxaylides are reacted with electrophiles (e.g. aldehydes). Subsequent elimination of Ph_3PO yields trisubstituted alkenes (M.Schlosser and K.F.Christmann, *Synthesis*, 1969, 38). In this way, allylic alcohols can be synthesised in a stereo-controlled fashion (E.J.Corey, P.Ulrich, and A.Venkateswarlu, *Tetrahedron Letters*, 1977, 3231)(see Eqn. 282).

$Ph_3P^+CH^-Me$ PhCHO BuLi

PhCHO

(282)

Acylsilanes, which tend to behave as sterically modified aldehydes, react with ylides in a highly (Z)-selective manner, although the carbon skeleton of the resultant alkene is of *trans*-geometry (J.A.Soderquist and C.L.Anderson, *ibid.*, 1988, **29** 2425; 2777)(see Eqn. 283). Unfortunately yields tend to be reduced by competitive acylsilane enolate formation.

$$RCOSiMe_3 + Ph_3P{=}CHR^2\,LiX \longrightarrow (R)(Me_3Si)C{=}CHR^2$$

(283)

Ylide anions, which are usually more reactive than the related ylides, also generate (E)-alkenes on reaction with aldehydes (H.J.Cristau *et al.*, *Phosphorus Sulphur*, 1987, **30**, 135; E.G.McKenna and B.J.Walker, *Tetrahedron Letters.*, 1988, **29** 485)(see Eqn. 284). The increased reactivity of these reagents makes them suitable for reaction with ester or amide carbonyl groups, which generally are more difficult substrates for the Wittig reaction (P.J.Murphy and J.Brennan, *Chem. Soc. Rev.*, 1988, **17**, 1).

$$Ph_3\text{-}P^+\text{-}C(\text{-}CH^-R)_2\ Li^+ \xrightarrow{PhCHO} PhCH{=}CHR$$

(284)

iii) Ylides of intermediate reactivity tend to give rise to mixtures of (E)- and (Z)-alkenes (e.g. G.Märkel and A.Merz, *Synthesis*, 1973, 295)(see Eqn. 285). There are three important groups of moderate reactivity ylides:

$$Ph_3\text{-}P^+CH_2Ph\ Cl^- \xrightarrow[\text{50\% NaOH/H}_2\text{O, DCM}]{PhCH{=}CHCHO} PhCH{=}CHCH{=}CHPh$$

(E):(Z)

64:36

(285)

a) *Allylic ylides* typically lead to mixed products (see Eqn. 286), although the problem can be avoided by careful choice of reagents (G.Goto *et al.*, *Chem. Letters*, 1975, 103)(see Eqn. 287). A further problem concerns loss of configuration about the allylic double bond (J.E.Johansen and S.Liaaen-Jensen, *Acta. Chem. Scand.*, 1974, **B28**, 301) and/or allylic rearrangement (E.J.Corey and B.W.Erickson, *J. Org. Chem.* 1974, **39**, 821). (see Eqn. 288). High (E)-selectivity has been achieved by use of allylphosphonium salts (see Eqn. 289), formed by replacement of an allylic

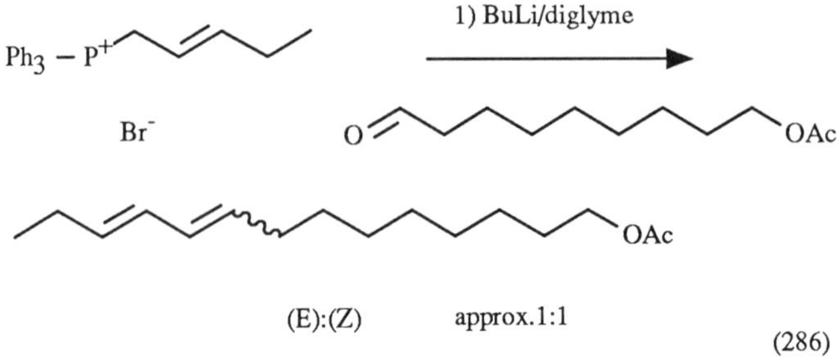

1) BuLi/diglyme

Ph_3-P^+ ... Br^- ... O ... OAc

OAc

(E):(Z) approx.1:1

(286)

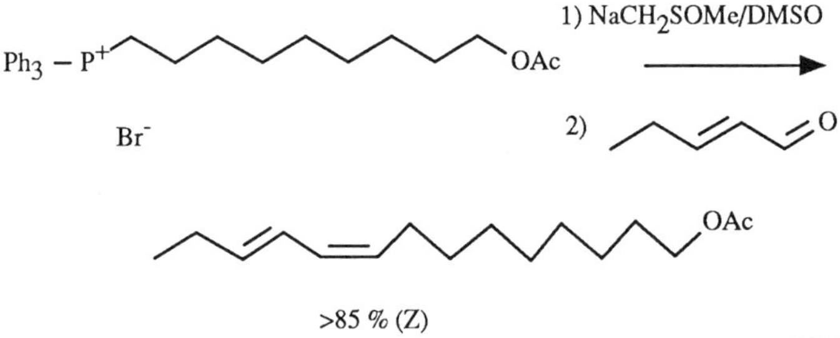

1) $NaCH_2SOMe/DMSO$

Ph_3-P^+ ... OAc ... Br^-

2) ... O

OAc

>85 % (Z)

(287)

Ph_3-P^+ ... CO_2Me RCHO DCM/25°C R ... CO_2Me

43% mixed isomers

+ R CO_2Me + R CO_2Me

10% 47%

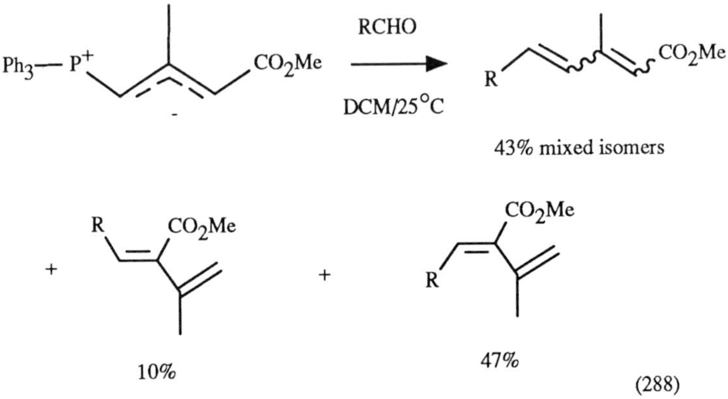

(288)

nitro-function by tributylphosphine under palladium (0) catalysis (R.Tamura *et al., J. Org. Chem.*, 1987, **52**, 4121). The use of triphenylphosphine in this process results in much lower (E)-selectivity.

b) *Benzylic ylides*, in which substitution of the aryl ring effects both ylide reactivity and the stereochemical outcome of the reaction (see Eqn. 290). Thus electron withdrawing substituents decrease reactivity, with resultant increased (E)-selectivity. Electron withdrawal in the phosphonium substituents, however, increases reactivity resulting in increased (Z)-selectivity (A.Johnson and V.L.Kyllingstad, *J. Org. Chem.*, 1966, **31**, 334). Alkyl substituents at phosphorus generally lead to enhanced (E)-preference. Solvents have a small effect, with protic or polar aprotic types favouring (Z)-isomers.

c) *Propargylic ylides* may isomerise yielding allenic products (E.J.Corey and R.A.Ruden, *Tetrahedron Letters.*, 1973, 1495)(see Eqn. 291), although this is not always the case (K.Eiter and H.Oediger, *Liebigs Ann. Chem.*, 1965, **682**, 62). The use of a terminal trimethylsilyl group to overcome this rearrangement has been applied in natural product synthesis (M.Ahmed *et al.*, *J. Chem. Soc., Perkin Trans. 1*, 1974, 1981).

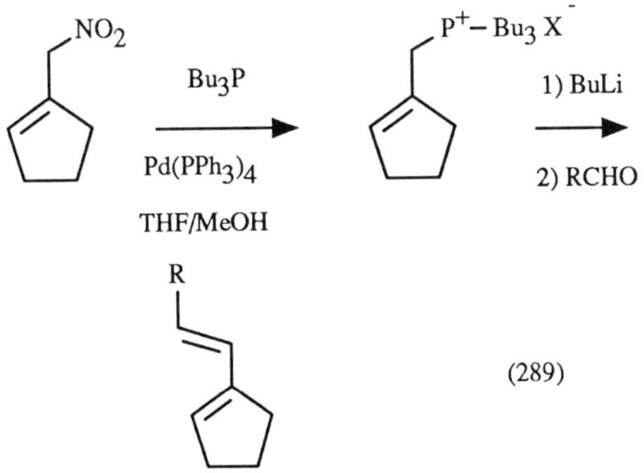

NO_2 Bu_3P, $Pd(PPh_3)_4$, THF/MeOH $P^+{-}Bu_3\,X^-$ 1) BuLi 2) RCHO   R

(289)

$$Ar_3P^+CH_2Ar^1 \xrightarrow[\text{2) Ar''CHO}]{\text{1) BuLi /benzene/25°C}} Ar^1CH{=}CHAr^2$$

Ar	Ar¹	Ar²	(E)	(Z)
Ph	4--NO_2Ph	4-MeO-Ph	74	26
Ph	4-MeO-Ph	4--NO_2Ph	56	44
4-Cl-Ph	4-MeO-Ph	4--NO_2Ph	20	80

(290)

$$Ph_3P^+CH_2C\equiv CH\ Br^- \xrightarrow[\text{2) cyclohexanecarbaldehyde}]{\text{1) BuLi/THF/0°C}} \text{cyclohexyl-CH=C=CH}_2 \quad (291)$$

2. P=O activated olefinations

The first processes of this type were reported in 1958 (L.Horner, H.Hoffmann, and H.G.Wippel, *Chem. Ber.*, 1958, **91**, 61)(see Eqn. 292). The phosphonate version (W.S.Wadsworth and W.D.Emmons, *J. Amer. Chem. Soc.*, 1961, **81**, 1733) has been developed into an important alkene synthesis (W.S.Wadsworth, *Jr., Org. Reactions*, 1978, **25**, 73; W.J.Stec, *Acc. Chem. Res.*, 1983, **16**, 411). Related processes involve phosphinates, $(R)R^1P(O)CH_2R^2$ (L.Horner *et al., Chem. Ber.*, 1962, **95**, 581);

$$R_2P(O)CH_2Ph \xrightarrow[\text{2) } Ph_2CO]{\text{1) } NaNH_2\text{/benzene}} PhCH=CPh_2 + R_2P(O)ONa \quad (292)$$

(R=OEt or Ph)

phosphonamides, $R_2N)_2P(O)CH_2R^1$ (E.J.Corey and G.T.Kwiatkowski, *J. Amer. Chem. Soc.*, 1964, **86**, 5652; 1966, **88**, 5653; 1968, **90**, 6816; E.J.Corey and D.E.Cane, *J. Org. Chem.*, 1969, **34**, 3053); and phosphonothionates, $(RO)_2P(S)CH_2R^1$ (J.Michalski and S.Musierowicz, *Tetrahedron Letters*, 1964, 1187). These reactions have a number of advantages, which have contributed to their success for olefination. They are a) regiospecific like Wittig reagents, b) more nucleophilic than Wittig ylides, and thus react with a wider range of carbonyl compounds under milder conditions, c) readily prepared from trialkyl phosphites *via* Michaelis-Arbusov procedures, and (d) side-reactions are less common than with the Wittig reaction. Furthermore, the phosphonic acid by-products are more readily removed than phosphine oxides,

Although (E)-selectivity is normally the case, considerable control over the stereochemical outcome of the reactions can be achieved by, for example, using cyclic phosphonates which favours (Z)-alkenes (E.Breur and D.M.Bannett, *ibid.*, 1977, 1141; B.Deschamps, J.P.Lampin, and

F.Mathey, *ibid.*, 1977, 1137). In fact, control over reaction conditions can have a remarkable effect on the configuration of the product (A.Redjel and J.Seyden-Penne, *Tetrahedron Letters*, 1974, 1733)(see Eqn. 293).

$$(EtO)_2P(O)CH(Me)CN \xrightarrow[2)\ PhCHO]{1)\ KO^tBu} PhCH{=}C(CN)Me$$

-75°C (Z):(E) 80:20

+65°C (Z):(E) 10:90

(293)

By the use of phosphonates containing chiral ester groups it has been possible to achieve the diastereoselective synthesis of 3-oxacarbacyclin precursors (H.J.Gais *et al., ibid.*, 1988, **29**, 1773; H.Rehwinkel, J.Skupsch, and H.Vorbruggen, *ibid.*, 1988, **29**, 1775)(see eqn 294).

R H

R^1O OR^1

(294)

Carbanions can be formed using Li bases, however, there is evidence that this may slow olefin formation, when compared with, for example, the use of $KOBu^t$ (P.Perriot, J.Villieras, and J.F.Normant, *Synthesis,* 1978, 33). Carbanions, such as $R_2P(O)CH^-R^1$ (R^1=alkyl or H) react rapidly with carbonyl compounds (see Eqn. 295), but only slowly eliminate $R_2P(O)O^-$ to give olefines in the absence of activating substituents, especially if R = alkoxy (C.Earnshaw, C.J.Wallis, and S.Warren, *J. Chem. Soc., Chem. Commun.*, 1977, 314; E.Schaumann and F.Grabley, *Liebigs Ann. Chem.*, 1977, 88). The Corey procedure ($R=NMe_2$), however, leads to ready elimination. Steric effects reduce the reactivity of the carbanion, thus $(RO)_2P(O)CR^1R^2$ fails to react with ketones (I.Shahak, J.Almog, and E.D.Bergmann, *Israel J. Chem.*, 1969, **7**, 585). Improved methods for carbanion formation, such as electrolytic methods (O.J.Scherer, P.Quintus, and W.S.Sheldrick, *Chem. Ber.*, 1987, **120**, 1183), or treatment with barium oxide under sonochemical conditions (O.J.Scherer *et al., ibid.*, 1987, **120**, 1463 and 1885) can lower reaction temperatures and improve yields.

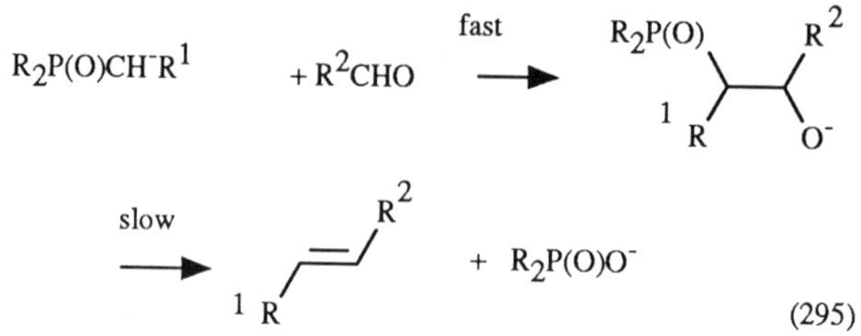

$R_2P(O)CH^-R^1$ $+ R^2CHO$ $\xrightarrow{\text{fast}}$ $R_2P(O)$ R^2 R^1 O^-

$\xrightarrow{\text{slow}}$ R^1 R^2 $+ R_2P(O)O^-$

(295)

The preparation of the intermediate β-hydroxylalkylphosphorus compounds (e.g. 296) has been used to synthesis a range of products, including pheromones (A.D.Buss *et al.*, *J. Chem. Soc., Perkin Trans. 1*, 1987, 2569), allylamines (D.Cavalla, W.B.Cruse, and S.Warren, *ibid.*, 1987, 1883) etc.

Stereochemical control can be obtained by preparation of *erythro-* or *threo-* 2-hydroxyalkylphosphorus derivatives (A.D.Buss and S.Warren, *ibid.*, 1985, 2307). The cyclic phosphine oxide (297) can be stereoselectively reduced. Thus $NaBH_4/CeCl_3$ gives the *threo*-2-hydroxy- and $NaBH_4$ the *erythro*-2-hydroxy derivatives, respectively, as the major products (J.Elliot and S.Warren, *Tetrahedron Letters.*, 1986, **27**, 645).

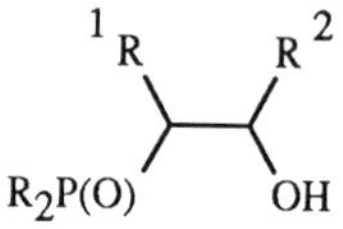

R^1 R^2 $R_2P(O)$ OH

(296)

R^1 R^2 P O O

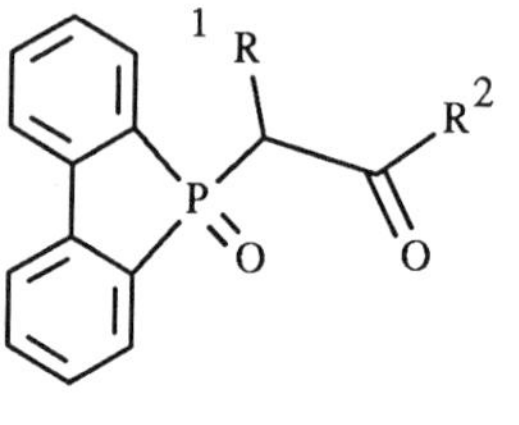

(297)

A recent modification, which allows increased diastereoselectivity in the synthesis of *erythro*-2-hydroxyalkyl derivatives (296, R = 2-methoxyphenyl), depends on the presence of *ortho*-substituted aryl groups on the phosphine oxide (T.Kaufmann and P.Schwartze, *Chem. Ber.* 1986, **119**, 2150). Corey's modification can also be effected *via* the corresponding 2-hydroxyalkylphosphonodiamide (see Eqn. 298). Once again, good stereochemical control is achieved (R.A.Hill, G.S.Macauley, and W.S.MacLachlan, *J. Chem. Soc., Perkin Trans. 1*, 1987, 2009).

Conjugated enones and enals undergo competitive Michael additions (P.D.Landor, S.R.Landor, and O.Odyek, *J. Chem. Soc., Perkin Trans. 1*, 1977, 93), which in the case of enones often predominate over attack at the

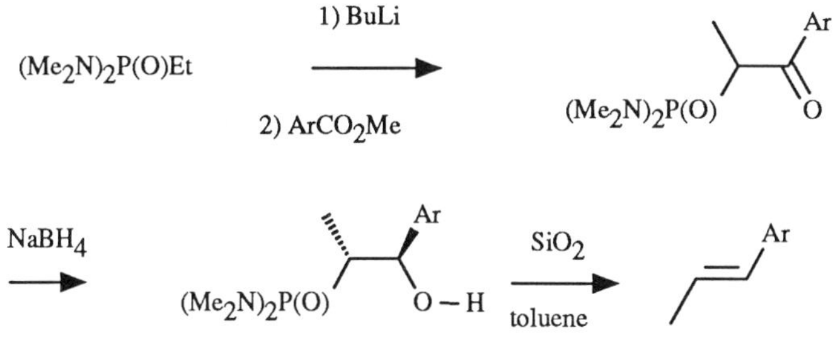

$(Me_2N)_2P(O)Et$ 1) BuLi 2) $ArCO_2Me$ → $(Me_2N)_2P(O)$ … Ar, O

$NaBH_4$ → $(Me_2N)_2P(O)$ … Ar, O – H → SiO_2, toluene → Ar

(298)

carbonyl centres (M.Cossentini *et al., Tetrahedron,* 1977, **33**, 409; B.Deschamps and J. Seyden-Penne, *ibid.*, 1977, **33**, 413). These Michael additions have been used for syntheses with chiral induction, and involve an allylic-shift (D.A.Hua, *J. Amer. Chem. Soc.*, 1987, **109**, 5026) (see Eqn. 299). These processes have found wide application in organic synthesis for the formation of conjugated enals, enones and enoic acid derivatives, stilbenes, allenes, dienes, enamines, vinyl halides, and vinyl sulphur compounds (see D.J.H.Smith, in *Organophosphorus Reagents in Organic Synthesis*, J.I.G.Cadogan, Ed., Academic Press, London, 1979). The mild conditions required for intramolecular olefination and cyclisation have led to their frequent use in natural product synthesis, for example, leukotrienes (e.g. Y.Guindon *et al., J. Org. Chem.*, 1988, **53**, 267), macrolides (e.g. J.A.Marshall and B.S.Dehoff, *Tetrahedron*, 1987, **43**, 4849), carotenoids (E.Kolling *et al., Angew. Chem. Int. Ed. Engl.*, 1987, **26**, 580), prostaglandins (e.g. R.C.Larock, M.H.Hsu, and K.Narayan, *Tetrahedron*, 1987, **43**, 2891), etc.

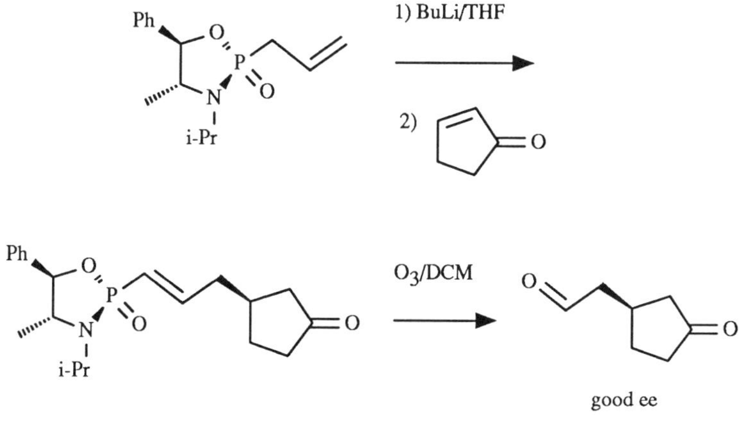

1) BuLi/THF 2) cyclopentenone → O_3/DCM → good ee

(299)

3. Reactions of phosphorus stabilised carbanions with electrophiles other than carbonyl compounds

i) Alkylation of ylides, or P=O stabilised carbanions can be achieved using alkyl halides. Reductive removal of the phosphorus group can lead to useful products, e.g. non-conjugated dienes (e.g. E.Axelrod, G.Milne, and E.E.van Tamelen, *J. Amer. Chem. Soc.*, 1970, **92**, 2139; K.Kondo, A.Negishi, and D.Tunemoto, *Angew. Chem. Int. Ed. Engl.*, 1974, **13**, 407)(see Eqn. 300).

α-Haloketones also react, however, transylidation and elimination follow, yielding conjugated enones (H.J Bestman and H.Schulz, *Angew. Chem.*, 1961, **73**, 620)(see Eqn. 301).

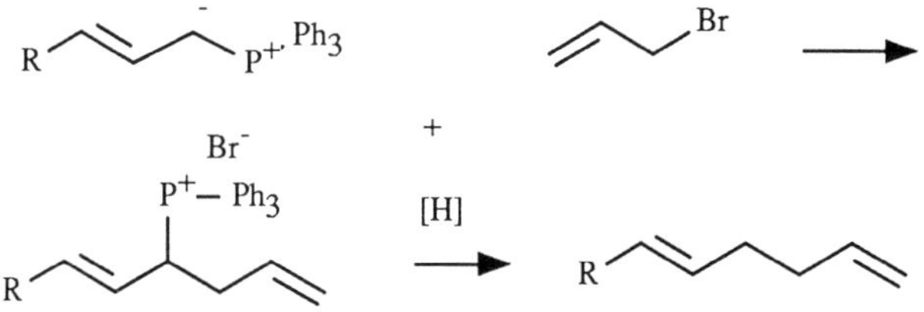

R–CH=CH–CH⁻–$P^{+}Ph_3$ + CH_2=CH–CH_2Br →

Br⁻ P^{+}–Ph_3 adduct: R–CH=CH–CH($P^{+}Ph_3$)–CH_2–CH=CH_2 —[H]→ R–CH=CH–CH_2–CH_2–CH=CH_2

(300)

$RCOCH_2X + Ph_3P^+CH^-CO_2Me$ →

$Ph_3P{=}CHCO_2Me$

$RCOCH_2CH(P^+Ph_3\ X^-)CO_2Me$ ⇌

$RCOCH_2CH{=}(PPh_3)CO_2Me$ —$-PPh_3$→ $RCOCH{=}CHCO_2Me$ (E)

+ $Ph_3P^+CH_2CO_2Me$ X^-

(301)

ii) Acylation of ylides is a useful route to 2-ketophosphonium salts. Once again, reductive cleavage can yield useful product ketones (H.J. Bestmann, *Angew. Chem. Int. Ed. Engl.*, 1965, **4**, 645).

Phosphonate anions behave similarly (N.Kreutzkamp, *Chem. Ber.*, 1955, **88**, 195); and it is worth noting that chloroformate esters provide substrates for carboxylic acids, although the process is wasteful of the phosphorane (A.Maercker and W.Theyson, *Liebigs Ann. Chem.*, 1972, **759**, 132) (see Eqn. 302).

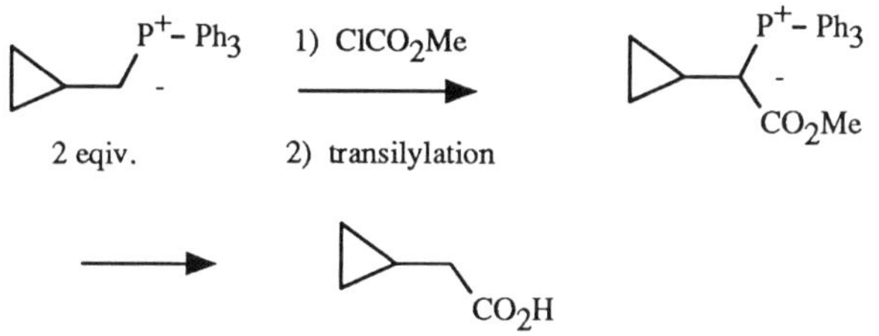

$$\text{cyclopropyl-CH}_2\text{-}\overset{-}{C}H\text{-}P^+Ph_3 \text{ (2 eqiv.)} \xrightarrow[\text{2) transilylation}]{\text{1) ClCO}_2\text{Me}} \text{cyclopropyl-CH}_2\text{-}\overset{-}{C}(P^+Ph_3)CO_2Me \longrightarrow \text{cyclopropyl-CH}_2\text{-}CH_2CO_2H$$

(302)

iii) Oxiranes react readily with phosphorus stabilised anions, leading to a variety of products (S.Trippett, *Quart. Rev.*, 1963, **17**, 406)(see Eqn. 303). The most useful process is cyclopropane formation which occurs when the group R′ is able to stabilise the carbanion (e.g. R.A. Izydore and R.G.Ghirardelli, *J. Org. Chem.*, 1973, **38**, 1790)(see Eqn. 304).

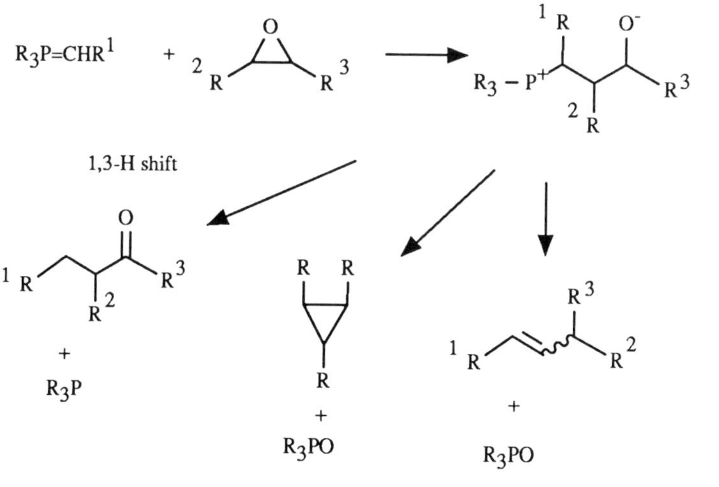

$$R_3P{=}CHR^1 + R^2\text{-oxirane-}R^3 \longrightarrow R_3-P^+\text{-}CH(R^1)\text{-}CH(R^2)\text{-}CH(O^-)R^3$$

1,3-H shift → $R^1CH_2CH(R^2)C(=O)R^3 + R_3P$

→ cyclopropane (R, R, R) $+ R_3PO$

→ $R^1CH{=}CH\text{-}CH(R^3)R^2 + R_3PO$

(303)

$$\text{2,3-dimethyloxirane} + (EtO)_2P(O)CH^-CO_2Et \longrightarrow \text{2,3-dimethylcyclopropane-1-}CO_2Et$$

(304)

4. Heterocyclic syntheses

This is a very large topic, which is covered here only in brief outline (for a fuller impression of its scope consult E.Zbiral, in *Organophosphorus Reagents in Organic Synthesis,* J.I.G.Cadogan, Ed., Academic Press, London, 1979).

i) Alkylidenephosphoranes

a) *Phosphacumulene ylides* undergo [4+2]-cycloadditions with isocyanates, isothiocyanates, vinylketones, etc. to form a range of heterocyclic types (see H.J.Bestmann, *Angew. Chem. Int. Ed. Engl.*, 1977, **89**, 361; L.Kniezo, P.Kristian, and J.Imrich, *Tetrahedron*, 1988, **44**, 543) (see Eqn. 305).

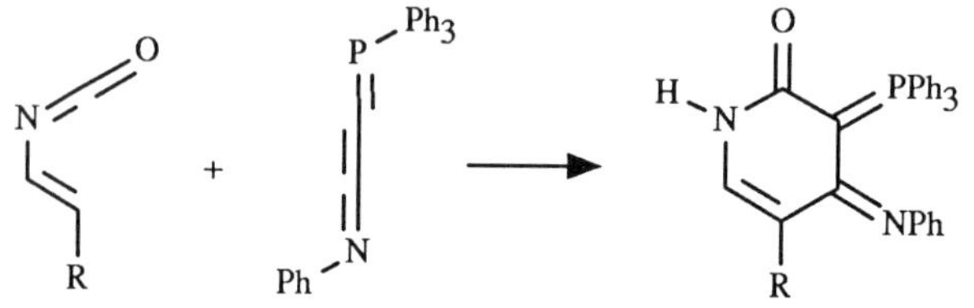

(305)

b) *1,3-Dipolar cycloadditions* can also be applied to heterocyclic ring formation, for example, 1,2,3-triazoles (P.Ykman, G.Labbé, and G.Smets, *Tetrahedron*, 1971, **27**, 5623). Similarly diazo compounds react to give pyrazoles (P.DallaCroce, *J. Chem. Soc., Perkin Trans. 1*, 1976, 619)(Eqn. 306).

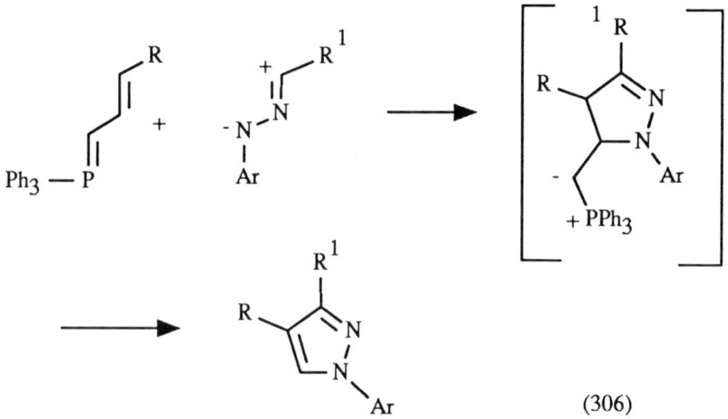

(306)

c) *Normal carbonyl olefination reactions.* The use of double Wittig reactions is one way of making heterocyclic rings (K.P.C.Vollhardt, *Synthesis*, 1977, 765). Difunctional ethers or thioethers are suitable substrates (H.J.Bestmann and E.Kranz, *Angew. Chem. Int. Ed. Engl.*, 1967, **6**, 81)(see Eqn. 307).

Internal Wittig reactions can be used for cyclisation (see Eqn. 308) resulting in heterocyclic products, e.g. 3-hydroxypyrroles (W.Flitsch, K.Hampel, and M.Hohenhorst, *Tetrahedron Letters,* 1987, **28**, 4395).

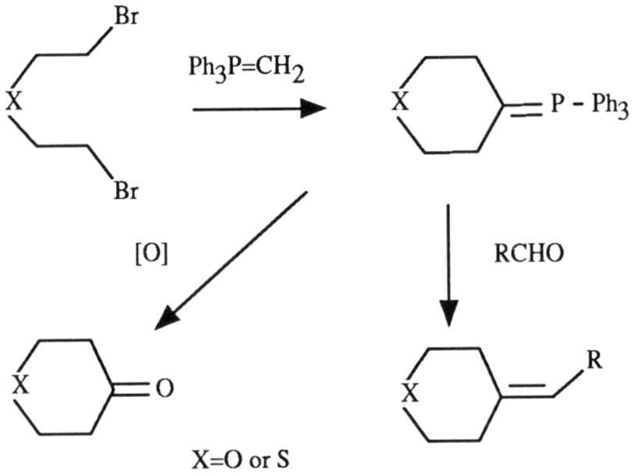

Br
X
Br
$Ph_3P=CH_2$
X
$=P-Ph_3$
[O]
RCHO
X
=O
X
R
X=O or S

(307)

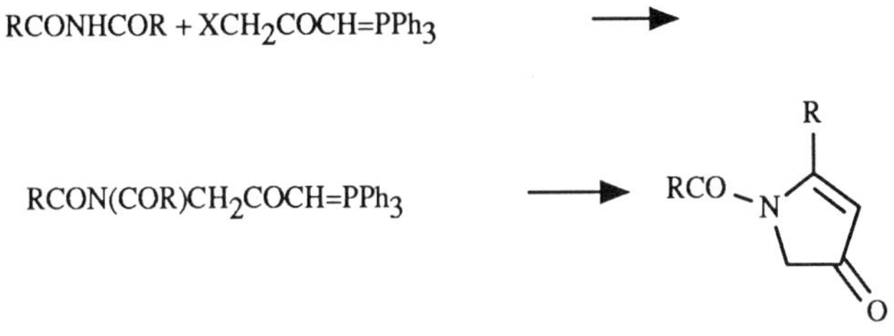

$RCONHCOR + XCH_2COCH=PPh_3 \longrightarrow$

$RCON(COR)CH_2COCH=PPh_3 \longrightarrow$ RCO–N (R, O)

(308)

ii) Alkenyl- and 3-oxoalkenylphosphonium salts

a) *Vinylphosphonium salts* react readily with carbonyl compounds containing an adjacent nucleophilic centre (see Eqn. 309). In this way pyrrolizines, furans, thiophenes, quinolines, pyrroles, and others have been synthesised. One example may serve to illustrate the procedure (M.E.Garst and Th.A.Spencer, *J. Org. Chem.*, 1974, **39**, 584)(see Eqn. 310).

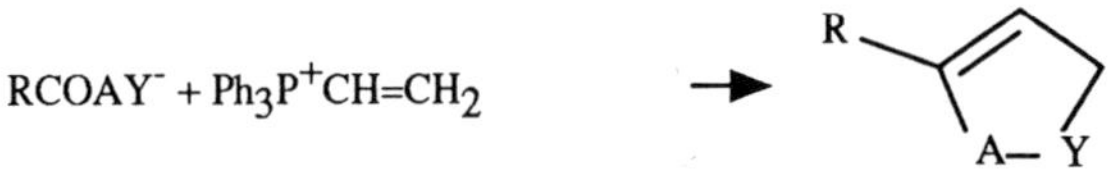

$$RCOAY^- + Ph_3P^+CH{=}CH_2 \longrightarrow$$

(309)

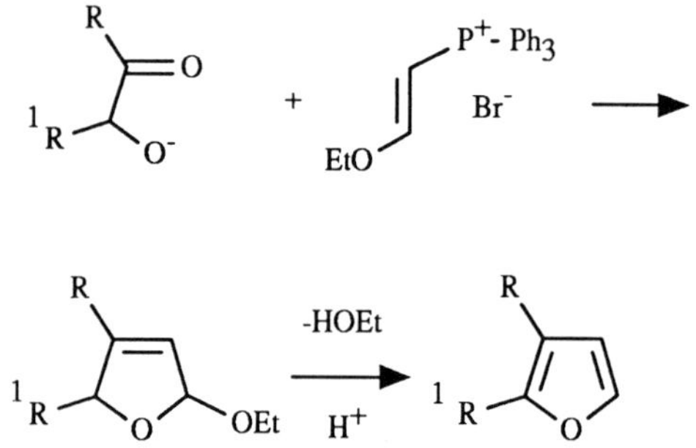

(310)

b) *3-Oxoalk-1-enylphosphonium salts*, which are available from 2-dichlorovinylketones and phosphines, show an "Umpolung" of normal enone reactivity. Thus the effect of the positive phosphorus atom leads to reverse Michael addition. This has been utilised in the synthesis of range of heterocycles, for example, the use of azide ions can lead to 1,2,3-triazoles (M.Rasberger and E.Zbiral, *Monatsh. Chem.*, 1969, **100**, 64)(see Eqn. 311).

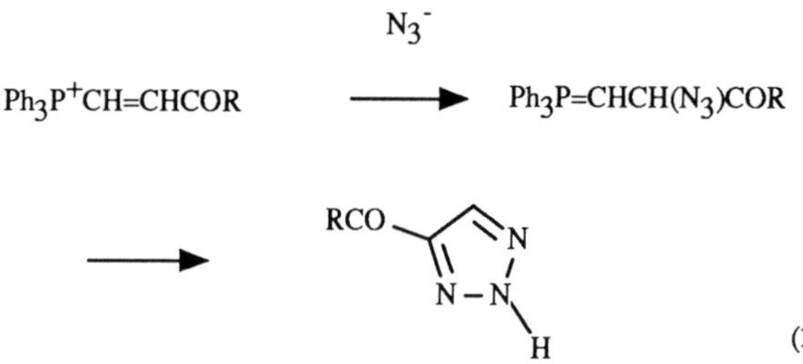

$$Ph_3P^+CH{=}CHCOR \xrightarrow{N_3^-} Ph_3P{=}CHCH(N_3)COR$$

(311)

5. Tervalent phosphorus-polyhalogenoalkane reagents

These reagents have found extensive use in organic synthesis (J.I.G.Cadogan and R.K.Mackie, *Chem. Soc. Rev.*, 1974, **3**, 87: R.Appel, *Angew. Chem. Int. Ed. Engl.*, 1975, **14**, 801). The most commonly used reagents have been tetrachloromethane or hexachloroethane in combination with triphenylphosphine. The R_3P/CCl_4 reagent involves the formation of intermediates which have a phosphorus-carbon bond, some of these compounds can be isolated (R.Appel *et al., Chem. Ber.*, 1976, **109**, 58; R.Appel and W.Morbach, *Synthesis*, 1977, 699)(see Eqn. 312). The R_3P/C_2Cl_6 reagent is a source of R_3PCl_2, which will effect many similar processes, however, this does not strictly fall within the scope of this discussion.

$$R_3P + CCl_4 \rightarrow [R_3P^+CCl_3]\,Cl^- \qquad (312)$$

i) Chlorination

a) *Alcohols and thiols* can be chlorinated, under very mild conditions (P.J.Kocienski, G.Cernigliaro, and G.Feldstein, *J. Org. Chem.*, 1977, **42**, 353)(see Eqn. 313), although secondary compounds may yield alkenes by elimination (R.Appel and H.-D.Wihler, *Chem. Ber.*, 1976, **109**, 3446). Addition of NaI (S.Miyano, H.Ushiyama, and H.Hashimoto, *Nippon Kagaku Kaishi,* 1977, 138), or NaCN (D.Brett, I.M.Downie, and J.B.Lee, *J. Org. Chem.*, 1967, **32**, 855) allows the formation of iodides or nitriles respectively.

$$[R_3P^+CCl_3]Cl^- + R^1XH \rightarrow [R_3P^+XR^1]Cl^- \rightarrow R_3PX + R^1Cl \qquad (313)$$

b) *Aldehydes and ketones* tend to yield mixed products with PPh_3/CCl_4, but $(Me_2N)_3P/CCl_4$ converts aldehydes into $RCH(OH)CCl_3$, or with excess reagent into the corresponding 1,1-dichloroalkenes, $RCH{=}CCl_2$ (J.C.Combert, J.Villieras, and G.Lavielle, *Tetrahedron Letters*, 1971, 1035).

c) *Carboxylic and phosphoric acids* are converted to acid chlorides under mild, essentially neutral conditions (J.B.Lee, *J. Amer. Chem. Soc.*, 1966, **88**, 3440; R.Appel and H.Einig, *Z. Anorg. Allg. Chem.*, 1975, **414**, 236).

d) *Ureas*, when *N,N,N*-trisubstitutred, these yield chloroformamidines under exceptionally mild conditions (R.Appel, K.D.Ziehn, and K.Warning, *Chem. Ber.*, 1973, **106**, 2093; 1974, **107**, 698)(see Eqn. 314).

$$R_3P + CCl_4 + RNHCONR^1R^2 \longrightarrow RN{=}CClNR^1R^2 + CHCl_3 + Ph_3PO \quad (314)$$

ii) Dehydrations

These processes have much in common with dehydrations effected by diethyl azodicarboxylate/triphenylphosphine (O.Mitsunobu, *Synthesis*, 1981, 1), with which they share the advantage of mild, essentially non-acidic, reaction conditions.

a) *Intramolecular* dehydrations can yield nitriles from amides or aldoximes (e.g. R.Appel, R.Kleinstück, and K.D.Ziehn, *Chem. Ber.*, 1971, **104**, 1030 and 2025), and isonitriles from *N*-substituted formamides (*idem.*, *Angew. Chem. Int. Ed. Engl.*, 1971, **10**, 132). The procedure also provides a mild high yielding, one-step route to aziridines (R.Appel and R.Kleinstuck, *Chem. Ber.*, 1974, **107**, 5)(see Eqn. 315) .

$$\text{R(HO)CH–CH(NHR}^2\text{)R}^1 \xrightarrow[Et_3N]{Ph_3P/CCl_4} \text{aziridine (N–R}^2\text{; C–R, C–R}^1\text{)} \qquad R^2 \neq H \quad (315)$$

b) *Intermolecular dehydration* can lead to esterification, or more usefully peptide syntheses (R.Appel, G.Baümer, and W.Strüver, *ibid.*, 1975, **108**, 2680; , 1976, **109**, 801; M.S.Manhas *et al., Synthesis*, 1976, 689). Little racemisation occurs during coupling and may be further suppressed by the presence of 1-hydroxybenzotriazole. It is worth noting that $(Me_2N)_3P/C_2Cl_6$ has been found to be very useful in peptide synthesis (B.Castro *et al., Synthesis*, 1976, 751; I.J.Galpin *et al., Tetrahedron*, 1976, **32**, 2417; R.Appel and L.Willms, *Chem. Ber.*, 1979, **112**, 1057 and 1064).

iii) Formation of P-N bonds

a) *Primary and secondary amines* yield stable aminophosphonium halides, which in the case of primary amines can be deprotonated to iminophosphoranes (R.Appel *et al., ibid.*, 1970, **103**, 3631)(see Eqn. 316).

$$R^1NH_2 \xrightarrow[CCl_4]{R_3P} [R_3P^+NHR^1]Cl^- \xrightarrow{base} R_3P{=}NR^1$$

(316)

The process has been applied to the phosphorylation of amino sugars (I.Pinker, J.Kovaćs, and A.Messmer, *Carbohydr. Res.*, 1977, **53**, 117). Sulphonamides and phosphoramidates behave similarly (R.Appel and H.Einig, *Z.Naturforsch.* 1975, **30b**, 134).

b) Arsenic, antimony, and bismuth

The organic chemistry of these elements has been reviewed (L.K.Peterson, *M.T.P. Int. Rev. Sci., Inorg. Chem., Ser. Two*, 1975, 4, 319; R.C.Poller, in *Comprehensive Organic Chemistry, Vol. 3,* D.H.R.Barton and W.D.Ollis, Eds., Pergamon, Oxford, 1979; J.L.Wardell, in *Comprehensive Organometallic Chemistry, Vol. 2,* G.Wilkinson, F.G.A.Stone and E.W.Abel, Eds., Pergamon, Oxford, 1982). *Arsenicals* continue to find use as pesticides (E.A.Woolson, *Arsenical Pesticides, A.C.S. Symp. Ser. No. 7,* A.C.S., Washington DC, 1975) and their impact on the environment has been considered (*Arsenic:Industrial, Biomedical, Environmental Perspectives,* W.H.Lederer and R.J.Fensterheim, Eds., Von Nostrand-Reinhold, New York, 1983). Aspects of the chemistry (I.N.Azerbae, Z.A.Abramova, and Yu. G.Bosyakov, *Russ. Chem. Rev.*, 1974, **43**, 657), stereochemistry (F.D.Yambushev and V.I.Savin, *ibid.*, 1979, **48**, 582), and preparation (L.C.Duncan, *Ann. Rep. Inorg. Gen. Synth.*, 1977, **5**, 111) of organoarsenic compounds have been discussed. Applications to the synthesis of organic compounds have mainly concerned arsonium ylides (Y.Huang and Y.Shen, *Adv. Organomet. Chem.*, 1982, **20**, 115; Y.-Z.Huang *et al., J. Org. Chem.*, 1987, **52**, 3558). Good results in natural product synthesis, have been obtained for ylides containing carboxylic acid, aldehyde, and ketone functions (Y.Wang *et al., Tetrahedron Letters.*, 1986, **27**, 4583; L.Shi *et al., ibid.*, 1987, **28**, 2155 and 2159)(see Eqn. 317). In these reactions there is considerable (E)-selectivity, however, treatment with I_2 is used to further reduce the amount of (Z)-alkene by equilibration.

The chemistry of *organoantimony* and *organobismuth* compounds has been reviewed (Sb: R.Okawara and Y.Matsumura, *Adv. Organomet. Chem.*, 1976, **14**, 187; V.K.Jain, R.Bohra, and R.C.Mehrotra, *Struct. Bonding*, 1982, 52, 147; L.D.Freedman and G.O.Doak, *J. Organomet. Chem.*, 1988, **351**, 25.

Bi: P.G.Harrison, *Organomet. Chem. Rev.*, 1970, **5**, 183; K.C.Moss and M.A.R.Smith, *M.T.P. Int. Rev. Sci., Inorg. Chem., Ser. Two*, 1975, **2**, 287; L.D.Freedman and G.O.Doak, *Chem. Rev.*, 1982, **82**, 15; *J. Organomet. Chem.*, 1988, **351**, 63). The aliphatic derivatives of neither antimony, nor bismuth have been much used for organic synthesis; although aromatic bismuth compounds have been studied by Barton and his co-workers (e.g. D.H.R.Barton *et al.*, *Tetrahedron*, 1988, **44**, 5661).

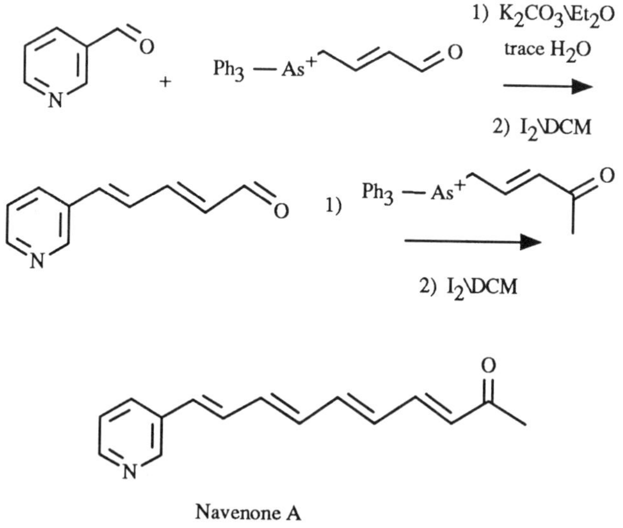

+ Ph_3 — As^+

1) K_2CO_3\Et_2O trace H_2O

2) I_2\DCM

1) Ph_3 — As^+

2) I_2\DCM

Navenone A

(317)

7. Group VIA : Selenium, Tellurium, Polonium

The organic compounds of selenium and tellurium resemble those of sulphur, in much of their chemistry. Selenium derivatives, like those of sulphur, have been increasingly used in organic synthesis (see P.D.Magnus, in *Comprehensive Organic Chemistry,* Vol. 3, D.H.R.Barton and W.D.Ollis, Eds., Pergamon, Oxford, 1979; periodic reviews, in *Organic Compounds of Sulphur, Selenium and Tellurium, S.P.R. Chem. Soc.*). So extensive is the use of organoselenium reagents, that it has been possible to devote two large volumes to their chemistry (*The Chemistry of Organic Selenium and Tellurium Compounds,* Vol. 1 and 2, S.Patai and Z.Rappoport, Eds., Wiley, Chichester, 1986 and 1987). These volumes cover a comprehensive range of topics, including detection and determination, stereochemistry, chiroptical properties, N.M.R., E.S.R., I.R., and mass spectrometry, plus an overview of their chemistry.

a) Selenium

The explosive growth of organoselenium chemistry has been documented in a number of excellent texts and review articles (e.g. H.J.Reich, in *Oxidation in Organic Chemistry*, Part C, W.S.Trahanowsky, Ed., Academic Press, New York, 1978; *Acc. Chem. Res.*, 1979, **12**, 22; D.Liotta, *ibid.*, 1984, **17**, 28; C.Paulmier, *Selenium Reagents and Intermediates in Organic Synthesis*, Pergamon, Oxford, 1986; *Organoselenium Chemistry*, D.Liotta, Ed., Wiley, New York, 1987). This growth is due to a number of valuable properties which selenium derivatives display, thus like sulphur, they exhibit useful chemo-, regio- and stereo- selectivities in their reactions. Furthermore, most reactions take place under very mild conditions, allowing kinetic and thermodynamic control to be exercised over the synthesis of regioisomers of functionally complex molecules, e.g. natural products. Although selenium reagents are more expensive than those of sulphur, they enjoy the advantage, that the weaker C-Se bond facilitates the easy removal of selenium, often leading to the creation of useful new functional groups. These factors have ensured the success of selenium intermediates in organic synthesis (see K.C.Nicolaou and N.A.Petasis, *Selenium in Natural Product Synthesis,* CIS, Philadelphia, 1984). The stench and toxicity of selenium compounds has caused some concern (R.J.Shamberger, in *The Biochemistry of Selenium*, E.Frieden, Ed., Plenum, New York, 1983). The use of polymer-bound selenium reagents (R.T.Taylor and L.A.Flood, *J. Org. Chem.*, 1983, **48**, 5160) could help to overcome these drawbacks. The application of pulsed Fourier transform N.M.R. spectroscopy for the observation of the ^{77}Se nucleus has been valuable for the understanding of organoselenium chemistry (for an introduction with many references, see C.Paulmier *loc. cit.* Ch.1).

The Uses of Selenium Reagents in Synthesis

1. Selenium stabilised carbanions.

Alkyllithium reagents tend to undergo Li/Se exchange rather than Li/H exchange with selenides (D.Seebach and N.Peleties, *Chem. Ber.*, 1972, **105**, 511; *Angew. Chem. Int. Ed. Engl.*, 1969, **8**, 450; D.Seebach and A.K.Beck, *ibid.*, 1974, **13**, 806). Lithium dialkylamides are less "selenophilic" and serve well for the deprotonation of selenium compounds (H.J.Reich, in *Organoselenium Chemistry*, D.Liotta, Ed., Wiley, New York, 1987). Selenoacetals and selenoketals are readily available, so their deselenation also gives access to α-lithioselenides (see Eqn. 318).

α-Lithioselenoxides are usually prepared *in situ* from selenides by oxidation with, for example, *meta*-chloroperbenzoic acid, followed by deprotonation, with a base such as lithium diisopropylamide (B.-T.Gröbel and D.Seebach, *Chem. Ber.*, 1977, **110**, 852) (see Eqn. 319). Both α-lithioselenides and α-lithioselenoxides are obtained by the addition of lithium reagents to vinylselenides, or selenoxides (see T.Kauffmann *et al.*,

$$R^1(R^2)C(SeR)_2 \xrightarrow[\text{THF/ -78°C}]{\text{BuLi}} R^1(R^2)C^-SeR$$

(R=Me or Ph)

(318)

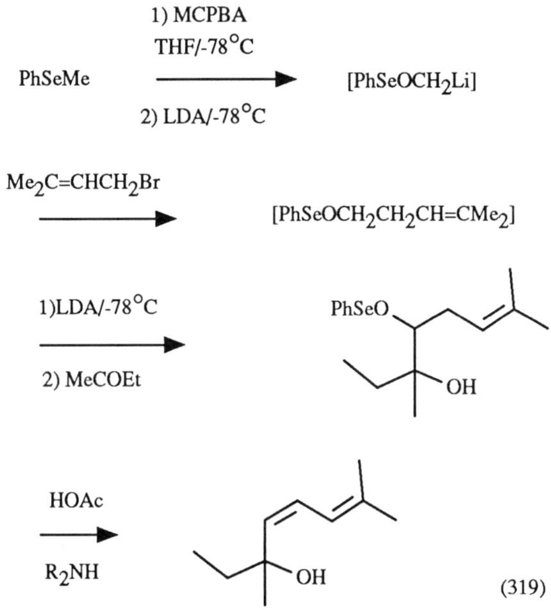

$$PhSeMe \xrightarrow[\text{2) LDA/-78°C}]{\text{1) MCPBA, THF/-78°C}} [PhSeOCH_2Li]$$

$$\xrightarrow{Me_2C{=}CHCH_2Br} [PhSeOCH_2CH_2CH{=}CMe_2]$$

$$\xrightarrow[\text{2) MeCOEt}]{\text{1) LDA/-78°C}}$$ PhSeO ... OH

$$\xrightarrow[R_2NH]{\text{HOAc}}$$ ... OH

(319)

Angew. Chem. Int. Ed. Engl.,1977, **16**, 710; S.Raucher and G.A.Koolpe, *J. Org. Chem.*, 1978, **43**, 4252; M.Shimizu and I.Kuwajima, *ibid.*, 1980, **45**, 4063; G.Zima, C.Barnum, and D.Liotta, *ibid.*, 1980, **45**, 2736).

The α-lithio-selenides and selenoxides are powerful nucleophiles, readily acylated, or alkylated (by S_N2 reactive halides and epoxides). They will add to conjugated enones to yield mainly 1,2-addition products in nonpolar solvents (e.g. Et_2O / THF), whereas more polar solvents (e.g. THF / HMPA, 1,2-dimethoxyethane) often favour 1,4-addition. These selenium stabilised carbanions, although good nucleophiles, are weak bases; thus little or no enolisation is observed, when they react with compounds containing enolisable carbonyl groups (W.Dumont and A.Krief, *Angew.*

Chem., 1975, **87**, 347; D.Van Ende, W.Dumont, and A.Krief, *ibid.*, 1975, **87**, 709). The selenium atom provides an opportunity for further functional group modifications, after generation of new C-C bonds at the carbanion. Indeed, this is a common feature of the selenium derivatives, generated by alternative procedures. These topics are discussed later. Methods for the replacement of selenium functions are now considered.

i) Oxidative *syn*-elimination and [2,3]sigmatropic rearrangement. Oxidative *syn*-elimination is the most frequently used procedure for the removal of selenium. Oxidation of selenides may be carried out using hydrogen peroxide (R.Walter and J.Roy, *J. Org. Chem.*, 1971, **36**, 2561; K.B.Sharpless and R.F.Lauer, *J. Amer. Chem. Soc.*, 1973, **95**, 2697), sodium periodate (K.B.Sharpless, R.F.Lauer, and A.Y.Teranishi, *ibid.*, 1973, **95**, 6137), peracids (*idem., ibid.*, H.J.Reich and co-workers, *J. Chem. Soc., Chem. Commun.*, 1975, 790; *J. Amer. Chem. Soc.*, 1975, **97**, 3250; *J. Org. Chem.* 1975, **40**, 3313), or ozone (D.N.Jones, D.Mundy, and R.D.Whitehouse, *J. Chem. Soc., Chem. Commun.*, 1970, 86). The resultant selenoxides readily undergo *syn*-elimination to give good yields of alkenes (D.Labar *et al., Tetrahedron Letters.*, 1978, 1141; H.J.Reich *et al., J. Org. Chem.*, 1978, **43**, 1697). An example (see Eqn. 320) involving an α-lithioselenide, shows good (E)-preference (K.B.Sharpless, R.F.Lauer, and A.Y.Teranishi, *J. Amer. Chem. Soc.*, 1973, **95**, 6137).

$$PhSeCH_2CO_2Et \xrightarrow[2)PhCH_2X/DMSO]{1)\ LDA} PhCH_2CH(SePh)CO_2Et$$

$$\xrightarrow{[O]} PhCH{=}CHCO_2Et \quad \text{(E)-isomer} \qquad (320)$$

Allylic, propargylic, or allenic selenoxides usually undergo [2,3] sigmatropic rearrangement, rather than *syn*-elimination (see Eqn. 321). The resultant selenenate esters are easily hydrolysed to the corresponding alcohols (see H.J.Reich, *J. Org. Chem.*, 1975, **40**, 2570). The presence of a silyl group can result in a sila-seleno Pummerer reaction (J.D.White, M.Kang, and B.G.Sheldon, *Tetrahedron Letters.*, 1983, **24**, 4539), instead of *syn*-elimination (see Eqn. 322). In these cases, phenylseleno -trimethylsilylmethane acts as a formyl anion equivalent (K.Sachdev and H.S.Sachdev, *Tetrahedron Letters.*, 1976, 4223; *ibid.*, 1977, 814; H.J.Reich and S.K.Shah, *J. Org. Chem.*, 1977, **42**, 1773).

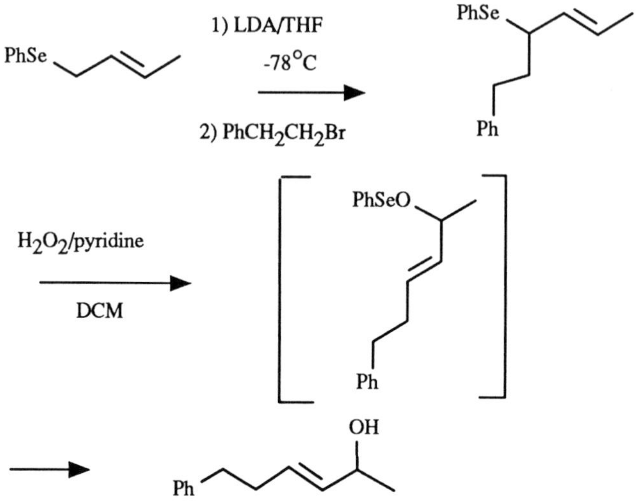

1) LDA/THF
-78°C
2) $PhCH_2CH_2Br$

H_2O_2/pyridine
DCM

(321)

1) LDA/THF
-78°C
2)

H_2O

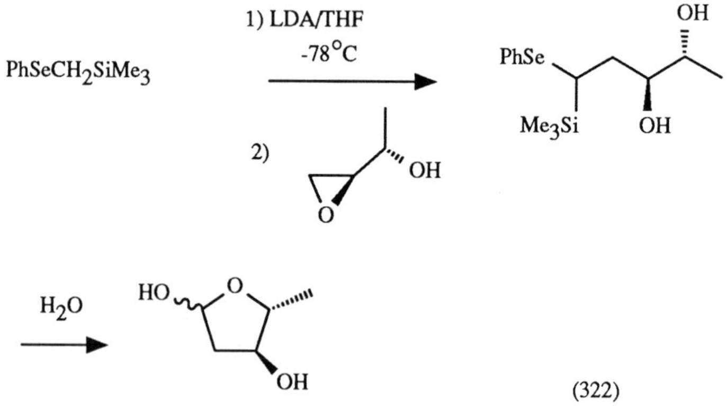

(322)

The conversion of a lithium reagent to another metal, such as tin, provides further control of regioselectivity (Y.Yamamoto *et al., J. Org. Chem.*, 1984, **49**, 1096; *ibid.*, 1983, **48**, 5408). An allylic example provides first regiocontrol, when the iminium salt attacks exclusively γ- to the tin grouping, followed by a [2,3]-sigmatropic rearrangement on oxidative removal of selenium (H.J.Reich, M.C.Schroeder, and I.L.Reich, *Israel J. Chem.*, 1984, **24**, 157) (see Eqn. 323).

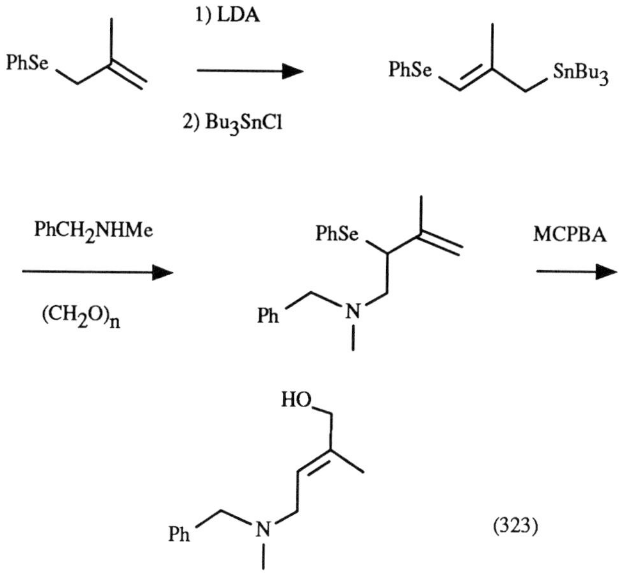
1) LDA
2) Bu_3SnCl
PhSe
PhSe
$SnBu_3$
$PhCH_2NHMe$
$(CH_2O)_n$
PhSe
Ph
N
MCPBA
HO
Ph
N

(323)

ii) Reductive elimination from β-hydroxyselenides. Treatment of the hydroxy function with activating reagents, e.g. TsOH / pentane; $HClO_4$ / ether; CF_3CO_2H / Et_3N; Me_3SiCl / NaI; Bu^tMe_2SiCl / Na-liq. NH_3; $SOCl_2$ / Et_3N etc., leads to alkenes *via* reductive elimination (S.Halazy and A.Krief, *J. Chem. Soc., Chem. Commun.*, 1979, 1136; H.J.Reich, F.Chow, and S.K.Shah, *J. Amer. Chem. Soc.*, 1979, **101**, 6638). These reactions are *anti*-eliminations (J.Rémion and A.Krief, *Tetrahedron Letters.*, 1976, 3743).

iii) Reconversion of lithium reagents. Selenoacetals or seleno-orthoesters can be used for the formation of one, two, or more carbanions consecutively at the same centre (H.J.Reich, S.K.Shah, and F.Chow, *J. Amer. Chem. Soc.*, 1979, **101**, 6648)(see Eqn. 324).

iv) Hydrolysis of selenoketals and vinylselenides. Hydrolysis is slower than for the corresponding sulphur compounds, but carbonyl compounds are still obtained (J.Lucchetti, and A.Krief, *Synth. Commun.*, 1983, **13**, 1153)(see Eqn. 325).

v) Base catalysed elimination of selenonium salts. Selenides are readily converted to selenonium salts on alkylation, which undergo elimination on treatment with base (S.Halazy and A.Krief, *Tetrahedron Letter*, 1979,

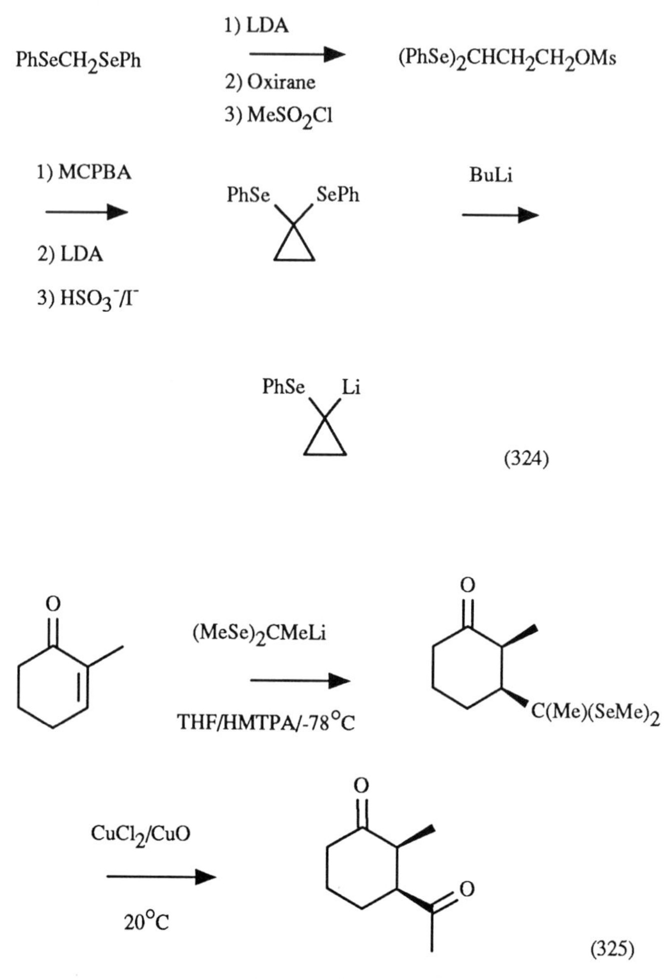

4233) (see Eqn. 326). If there is a β-hydroxy group present, then the alternative pathway to oxiranes predominates (D.Van Ende and A.Krief, *ibid.*, 1976, 457) (see Eqn. 327).

vi) Replacement of selenide groups by hydrogen. The reductive removal of selenium can be achieved using various reagents, e.g. Li reagents, Raney Ni, dissolving metals; Bu_3SnH, azobisisobutyronitrile; Pd, Et_3BH etc. An example (see Eqn. 328) involving the use of Bu_3SnH yields 3-ethylcyclohexanone (R.O.Hutchins and K.Learn, *J. Org. Chem.*, 1982, **47**, 4380). Alternatively, α-selenocarbonyl compounds are smoothly deselenated using soft nucleophiles, e.g. RS^-, RSe^-, in protic media (see T.Takahashi, H.Nagashima, and J.Tsuji, *Tetrahedron Letters.*, 1978, 799) (see Eqn. 329).

PhSe SePh → 1) BuLi/THF, 2) $C_9H_{19}I$/HMPA → PhSe C_9H_{19}

1)MeI/$AgBF_4$, 2) ^tBuOK/THF → C_8H_{17}

(326)

$MeSeCH_2SeMe$ → 1) BuLi, 2) Ph, O → Ph, OH, SeMe

1)MeI, 2) ^tBuOK/DMSO → Ph, O

(327)

$OSiMe_3$, SeMe, SeMe → 1) BuLi, 2) MeI → $OSiMe_3$, SeMe

1)TsOH, 2) Bu_3SnH/AIBN/80°C → O

(328)

$$RCOCH_2SePh \xrightarrow[CH_2=CHCH_2Br]{Bu^tOK/Bu^tOH} RCOCH(SePh)CH_2CH=CH_2$$

$$\xrightarrow[Et_3N]{PhSH} RCOCH_2CH_2CH=CH_2$$

(329)

vii) Replacement of selenide groups by alkyl groups. In the case of allyl selenides, the selenium may be displaced using dialkyl cuprates (H.J.Reich, *J. Org. Chem.*, 1975, **40**, 2570)(see Eqn. 330).

2. Electrophilic reagents.

Selenenyl halides and pseudohalides (RSeX), or diselenides (RSeSeR) are commonly employed to introduce selenium at a nucleophilic carbon centre. The use of aryl derivatives (R=Ar) helps with increased stability and reduced volatility (odour).

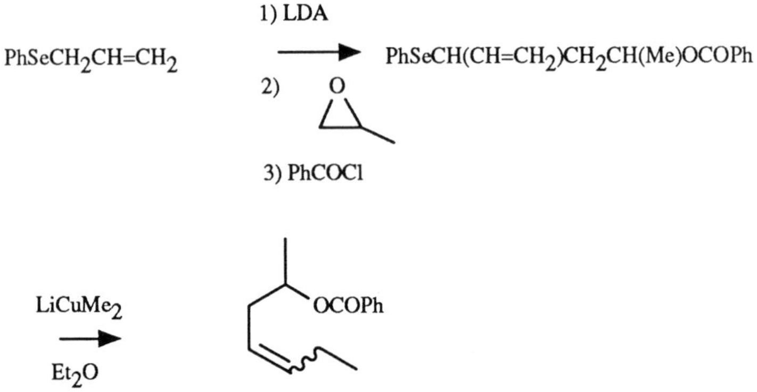

$$PhSeCH_2CH=CH_2 \xrightarrow[\text{2) propylene oxide; 3) PhCOCl}]{\text{1) LDA}} PhSeCH(CH=CH_2)CH_2CH(Me)OCOPh$$

$$\xrightarrow[Et_2O]{LiCuMe_2}$$ (structure: CH(Me)(OCOPh)CH_2CH=CHCH_2CH_3)

(330)

i) Reaction with aldehydes and ketones. *Acidic catalysis* leads to the introduction of selenium *via* the enol (see Eqn. 331), and oxidative removal of the selenium leads to enones with good (E)-preference (K.B.Sharpless, R.F.Lauer, and A.Y.Teranishi, *J. Amer. Chem. Soc.*, 1973, **95**, 6137; K.B.Sharpless and R.F.Lauer, *ibid.*, 1973, **95**, 2697). These selenenylations can occur with good chemo- and regio- selectivity (J.H.Zaidi and A.J.Waring, *J. Chem. Soc., Chem. Commun.*, 1980, 618) (see Eqn. 332).

$$RCOCH_2CH_2R^1 \xrightarrow[\text{EtAc, 25°C}]{\text{PhSeCl}} RCOCH(SePh)CH_2R^1 \xrightarrow{[O]} RCOCH{=}CHR^1 \text{ (E)} \qquad (331)$$

PhSeCl

PhSe

O_3

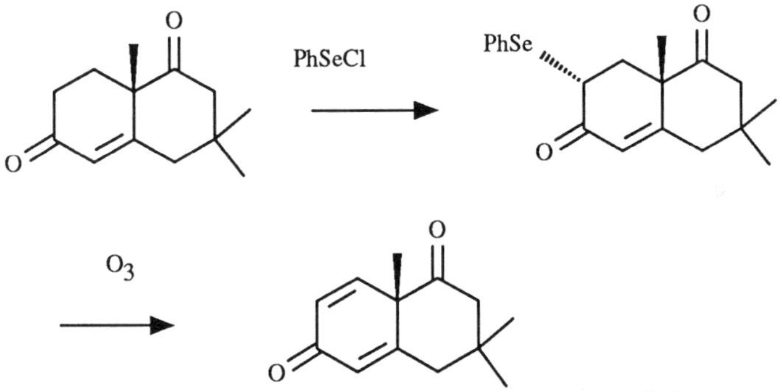

(332)

Basic catalysis. The use of lithium diisopropylamide (LDA) in tetrahydrofuran at -78°C, for example, allows the formation of the kinetic lithium enolates. These can be reacted with an electrophilic selenium reagent resulting in reaction at the least hindered position (H.J.Reich, J.M.Renga, and I.L.Reich, *J. Amer. Chem. Soc.*, 1975, **97**, 5434)(see Eqn. 333). The extra acidity of α-phenylselenoketones may then be exploited, and alkylation can be effected adjacent to the selenium atom (M.Nishizawa *et al., J. Chem. Soc., Chem. Commun.*, 1978, 76). The less expensive elemental selenium may be used leading to methylselenoketones on methylation (D.Liotta *et al., Tetrahedron Letters.*, 1980, **21**, 3643)(see Eqn. 334).

$$RCOCH_2CH_2R^1 \xrightarrow[\text{THF/-78°C}]{\text{LDA}} RC(OLi){=}CHCH_2R^1 \xrightarrow{\text{PhSeCl}}$$

$$RCOCH(SePh)CH_2R^1 \xrightarrow[\text{2) } R^2X]{\text{1) LDA}} RCOC(R^2)(SePh)CH_2R^1 \qquad (333)$$

$$RCOCH_2CH_2R^1 \xrightarrow[\text{2) Se}]{\text{1) LDA/THF/HMPA, -10 to -20}^\circ\text{C}} RCOCH(SeLi)CH_2R^1 \xrightarrow{\text{MeI}} RCOCH(SeMe)CH_2R^1 \quad (334)$$

Organocuprates (H.J.Reich, J.M.Renga, and I.L.Reich, *J. Amer. Chem. Soc.*, 1975, **97**, 5434), organozirconium and organoaluminium (J.Schwartz and Y.Hayasi, *Tetrahedron Letters.*, 1980, **21**, 1497), and Grignard reagents (H.E.Zimmerman and M.C.Hovey, *J. Org. Chem.*, 1979, **44**, 2331) can all be used to generate enolates from enones, with concomitant β-alkylation (see Eqn. 335). Selenenamides function as nucleophiles in reactions with enones before transferring a selenenyl group to the resultant enolate (H.J.Reich and J.M.Renga, *J. Org. Chem.*, 1975, **40**, 3313) (see Eqn. 336). In a related process, pyridine and PhSeCl bring about the same type of conversion (G.Zima and D.Liotta, *Synth. Commun.* 1979, **9**, 697). The greater kinetic acidity of aldehydes compared to ketones results in their selective reaction with selenamides (M.Jefson and J.Meinwald, *Tetrahedron Letters.*, 1981, **22**, 3561) (see Eqn. 337).

$$PhCOCH{=}CHR \xrightarrow[\text{2) PhSeBr/PhSeSePh}]{\text{1) LiMe}_2\text{Cu}} PhCOCH(SePh)CH(CH_3)R \xrightarrow{H_2O_2} PhCOCH{=}C(CH_3)R \quad (335)$$

ii) Reactions with esters and lactones. These processes require strong basic catalysis. The configuration of the resultant phenylselenenyl group is dependent upon the order of reaction in cyclic systems, and the requirement for a *syn*-elimination within the corresponding selenoxide ensures control over the structure of the product (see Eqn. 338). Unhindered conjugated (E)-enoate esters undergo α-selenenylation; while (Z)-enoate esters, or those which are too hindered, yield γ-selenenyl products (T.A.Hase and P.Kukkola, *Synth. Commun.*, 1980, **10**, 451).

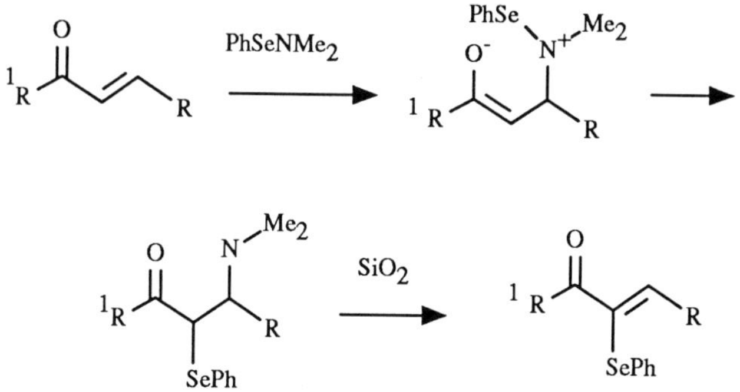

O
^1R R
PhSeNMe$_2$
PhSe N^+ Me$_2$
O^-
^1R R

O N Me$_2$
^1R R
SePh
SiO$_2$
O
^1R R
SePh

(336)

PhSeNEt$_2$

RCO(CH$_2$)$_4$CHO ⇄ RCO(CH$_2$)$_3$CH=CHO$^-$ +

PhSeNH$^+$Et

⟶ RCO(CH$_2$)$_3$CH(SePh)CHO

(337)

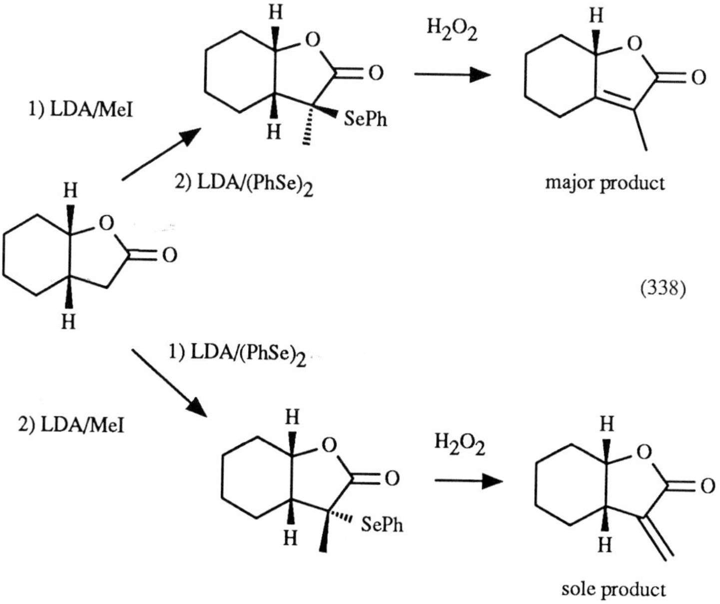

H
O
O
1) LDA/MeI
H SePh
2) LDA/(PhSe)$_2$
H_2O_2
H
O
O
major product

H
O
O
H

(338)

1) LDA/(PhSe)$_2$
2) LDA/MeI
H
O
O
H SePh
H_2O_2
H
O
O
H
sole product

iii) Reactions with enol acetates, enol silyl ethers, etc. Enol acetates are readily converted into α-phenylselenoketones using PhSeBr / CF_3CO_2Ag (D.L.J.Clive, *J. Chem. Soc. Chem. Commun.*, 1973, 695; H.J.Reich, *J. Org. Chem.*, 1974, **39**, 428), PhSeCl / aqueous CH_3CN (A.Toshimitsu *et al.*, *J. Chem. Soc., Chem. Commun.*, 1980, 412) or *via* a lithium enolate (H.J.Reich, J.M.Ranga, and I.L.Reich, *J. Org. Chem.*, 1974, **39**, 2133; J.P.Konopelski, C.Djerassi, and J.P.Raynaud, *J. Med. Chem.*, 1980, **23**, 722). The conversion of enol acetates nicely complements direct selenenylation (see Eqn. 339).

Enol silyl ethers are readily converted into α-selenoketones; for example, Danishefsky's diene has been prepared in a selenylated version (see Eqn. 340). Such a reagent is useful in Diels Alder reactions (S.Danishefsky, C.F.Yan, and P.M.McCurry, Jr., *J. Org. Chem.*, 1977, **42**, 1819, S.Danishefsky *et al.*, *J. Amer. Chem. Soc.*, 1979, **101**, 7001).

O Ph 1) LDA 2) PhSeCl 3) H_2O_2 O Ph

Ac_2O

OAc Ph 1) MeLi 2) PhSeBr 3) H_2O_2 O Ph

(339)

iv) Reactions with terminal alkynes. Acetylide carbanions readily yield 1-selenoalkynes (N.Petragnani, R.Rodriques, and J.V.Comasseto, *J. Organomet. Chem.*, 1976, **114**, 281; J.V.Comasseto, J.T.B.Ferreira, and N.Petragnani, *ibid.*, 1981, **216**, 287; S.Raucher, M.R.Hansen and M.A.Colter, *J. Org. Chem.*, 1978, **43**, 4885); the use of Cu_2CN_2 promotes these reactions (S.Tomoda, Y.Takeuchi, and Y.Nomura, *Chem. Letters.*, 1982, 253). 1-Selenoalkynes can be reduced with various reagents to give vinylselenides, with known configurations (see Eqn. 341).

v) Reactions involving phosphines. Alcohols react with ArSeCN, in the presence of phosphines, to yield selenides (P.A.Grieco, S.Gilman, and

OMe

PhSeCl

OMe

$Me_3SiCl/ZnCl_2$

Me_3SiO

O

Et_3N

SePh

OMe

Me_3SiO

SePh

(340)

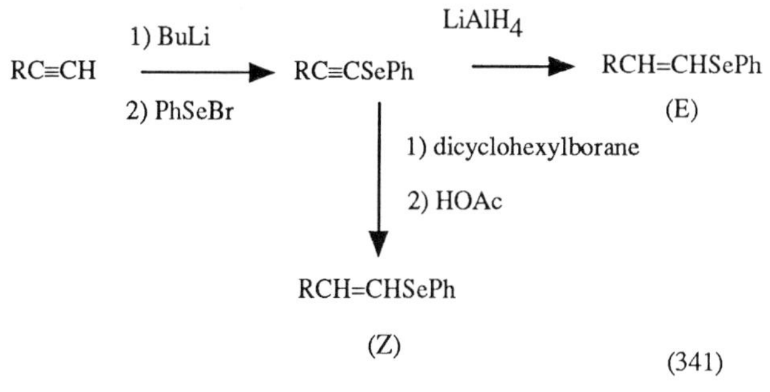

(341)

M.Nishizawa, *J. Org. Chem.*, 1976, **41**, 1485) (see Eqn. 342). Primary alcohols react more rapidly than secondary alcohols, which permits their selective selenenylation. This has found use in reactions with the 5-hydroxy function of ribose derivatives (H.Takaku, T.Nomoto, and K.Kimura, *Chem. Letters.*, 1981, 1221). Aldehydes react under similar conditions (P.A.Grieco and Y.Yokoyama, *J. Amer. Chem. Soc.*, 1977, **99**, 5231) (see Eqn. 343).

vi) Additions to alkenes. This process has been reviewed (G.H.Schmid and D.G.Garratt, in *The Chemistry of the Double-Bonded Functional Groups,* Supp. A, Part 2, S.Patai, Ed., Wiley, Chichester, 1977). The addition occurs with *anti*-stereochemistry, but the lack of good

$$\mathrm{ROH} \xrightarrow[\mathrm{ArSeCN}]{\mathrm{Bu_3P}} \mathrm{RSeAr} \qquad (342)$$

$$\mathrm{RCH_2CHO} \xrightarrow[\mathrm{ArSeCN}]{\mathrm{Bu_3P}} \mathrm{RCH_2CH(CN)SePh} \xrightarrow{\mathrm{H_2O_2}} \underset{(E)}{\mathrm{RCH{=}CHCN}} \qquad (343)$$

regiospecificity limits its use (J.M.Renga and H.J.Reich, *Org. Synth.*, 1980, **59**, 58; J.Bruhn, H.Heimgartner, and H.Schmid, *Helv. Chim. Acta.*, 1979, **62**, 2630). Addition in the presence of nucleophiles (e.g. H_2O, ROH, RCO_2H, azides, nitriles, etc.) can yield useful products (see T.Takahashi, H.Nagashima, and J.Tsuji, *Tetrahedron Letters*., 1978, 799) (see Eqn. 344). While nitriles in the presence of water and acid lead to amides (A.Tochimitsu *et al., J. Org. Chem.*, 1981, **46**, 4727)(see Eqn. 345).

$$\mathrm{CH_3(CH_2)_5CH{=}CH_2} \xrightarrow[\mathrm{EtOH}]{\mathrm{PhSeBr}} \mathrm{CH_3(CH_2)_5CH(OEt)CH_2SePh} \xrightarrow{\mathrm{NaIO_4}} \mathrm{CH_3(CH_2)_5COCH_2SePh} \qquad (344)$$

$$\mathrm{RCH{=}CHR^1} \xrightarrow[\mathrm{R''CN}]{\mathrm{PhSeCl}} \mathrm{RCH(SePh)CH(R^1)N{=}C(Cl)R^2} \xrightarrow{\mathrm{H^+}} \mathrm{RCH(SePh)CH(R_1)NHC(O)R^2} \qquad (345)$$

vii) Additions to enol ethers. Ready addition occurs in alcoholic solvents and leads to α-phenylselenoaldehydes, after hydrolysis of the initially formed acetals (R.Bandat and M.Petrzilka, *Helv. Chim. Acta.*, 1979, **62**, 1406; D.L.J.Clive, C.G.Russell, and S.C.Suri, *J. Org. Chem.*, 1982, **47**, 1632)(see Eqn. 346).

The use of allylic alcohols, provides a route to γ,δ-enolate esters *via* Claisen rearrangements (M.Petrzilka, *Helv. Chim. Acta.*, 1978, **61**, 2286; R.Pitteloud and M.Petrzilka, *ibid.*, 1979, **23**, 4067). A useful example of this procedure is found in a guaianolide synthesis (P.Metz and H.I.Schäfer, *Tetrahedron Letters.*, 1982, **23**, 4067)(see Eqn. 347).

$$CH_2{=}CHOEt \xrightarrow[EtOH]{PhSeCl} PhSeCH_2CH(OEt)_2 \xrightarrow{H^+} PhSeCH_2CHO \quad (346)$$

$$\text{(bicyclic allylic alcohol; OMe, H, HO)} \xrightarrow[PhSeBr]{CH_2{=}CHOEt} \text{(bicyclic ether; OMe, H, } PhSeCH_2CH(OEt)O)$$

$$\xrightarrow[\text{2) Claisen rearrangement}]{\text{1) } NaIO_4} \text{(bicyclic ester; OMe, H, } EtO_2C) \quad (347)$$

viii) Additions to alkynes. Addition can occur with either *syn-* or anti-stereochemistry, although, as for alkenes, poor regioselectivity has limited the use of the reactions (G.H.Schmid and D.G.Garratt, *Chem. Ser.*,1976, **10**, 76; G.H.Schmid, in *The Chemistry of the Carbon-Carbon Triple Bond,* S.Patai, Ed., Wiley, Chichester, 1978). The formations of α-phenylselenoketones are amongst the more important of applications of organoselenium chemistry (H.J.Reich, J. Org. Chem., 1974, 39, 428; H.J.Reich, J.M.Renga, and I.L.Reich, *J. Amer. Chem. Soc.*, 1975, **97**, 5434)(see Eqn. 348)

ix) Cycloaddition reactions. These important procedures involve the addition of electrophilic selenenyl reagents to double bonds in the presence of an internal nucleophile (e.g. OH, see Eqn. 349), which then facilitates cyclisation (K.C.Nicolaou *et al., J. Amer. Chem. Soc.*, 1980, **102**, 3784). The scope of this type of reaction has been reviewed (K.B.Sharpless *et al., Chem. Scr.*, 1975, **8A**, 9; H.J.Reich, *Acc. Chem. Res.*, 1979, **12**, 22; D.L.J.Clive, *Aldrichimica Acta*, 1978, **11**, 43; *Tetrahedron*, 1978, **34**, 1049).

PhC≡CMe $\xrightarrow{PhSeOCOCF_3}$ CF_3CO_2(Ph)C=C(Me)SePh $\xrightarrow{KOH}$ PhCOCH(SePh)Me

(348)

HO ... $\xrightarrow[Et_3N/DCM, -78°C]{PhSeCl}$ O ... SePh

$\xrightarrow{[O]}$ O ...

(349)

Phenylselenoetherification is a mild, efficient process complementary to *haloetherification.* In the latter base catalysed elimination occurs towards the oxygen atom, whilst the reverse is true of the oxidative elimination of selenium (K.C.Nicolaou, W.E.Barnette, and R.L.Magolda, *J. Chem. Res.*, (S), 1979, 202). Furthermore, reductive removal of selenium allows the formation of saturated cyclic ethers (see S.Ley and B.Lygo, *Tetrahedron Letters.*, 1982, **23**, 4625) (See Eqn. 350). The use of PhSeOH generated *in situ* allows 1,5-dienes to be cyclised (R.M.Scarborough, Jr. *et al., J. Org. Chem.*, 1979, **44**, 1742) (see Eqn. 351).

N-(PhSe)phthalimide, $ZnBr_2$; Raney Ni

(350)

$(PhSe)_2$, H_2O_2

(351)

Phenylselenolactonisation tends to be a milder process than the equivalent *halolactonisation* (M.D.Dowle and D.I.Davies, *Chem. Soc. Rev.*, 1979, **8**, 171). The procedure is more effective for 5-membered rings, rather than for 4- or 6-membered rings (K.C.Nicolaou *et al., J. Amer. Chem. Soc.*, 1979, **101**, 3884; D.Goldsmith *et al., Tetrahedron Letters.*, 1979, 4801) (see Eqn. 352).

Both phenylseleno- etherification and -lactonisation can be reversed using sodium in liquid ammonia (K.C.Nicolaou *et al., J. Chem. Soc., Chem. Commun.*, 1979, 83), or sodium iodide/chlorotrimethylsilane in acetonitrile (D.L.J.Clive and V.N.Kale, *J. Org. Chem.*, 1981, **46**, 231). This allows the reactions to be used for the simultaneous protection and further transformation of an alcohol, a carboxylic acid, or an alkene. An illustrative example employs geranyl acetate as the substrate (R.M.Scarborough, Jr. *et al., J. Org. Chem.*, 1979, **44**, 1742) (see Eqn. 353).

The cyclisation process can also be applied to thiols (see K.C.Nicholaou and Z.Lysenko, *Tetrahedron Letters.*, 1977, 1257), and to amines, as their urethane derivatives (D.L.J.Clive *et al., J. Org. Chem.*, 1980, **45**, 2120).

Phenyselenocarbocyclisation is also a known procedure (see Eqn. 354), an example involves acetonitrile as a nucleophile (A.Toshimitsu, S.Uemura, and M.Okano, *J. Chem. Soc., Chem. Commun.*, 1982, 82).

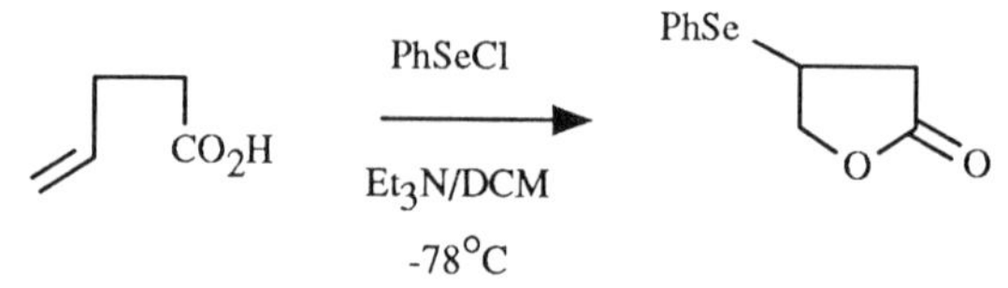

PhSeCl; Et_3N/DCM, -78°C

(352)

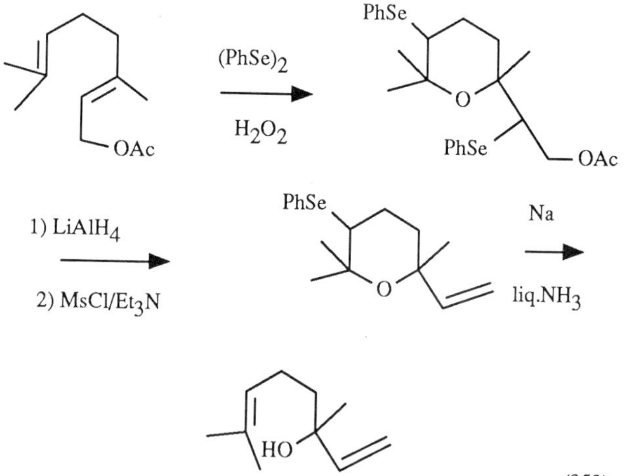

$(PhSe)_2$; H_2O_2

1) $LiAlH_4$; 2) $MsCl/Et_3N$

Na; liq.NH_3

(353)

$(PhSe)_2/I_2$; MeCN.aq./25°C

PhSe — NHCOMe

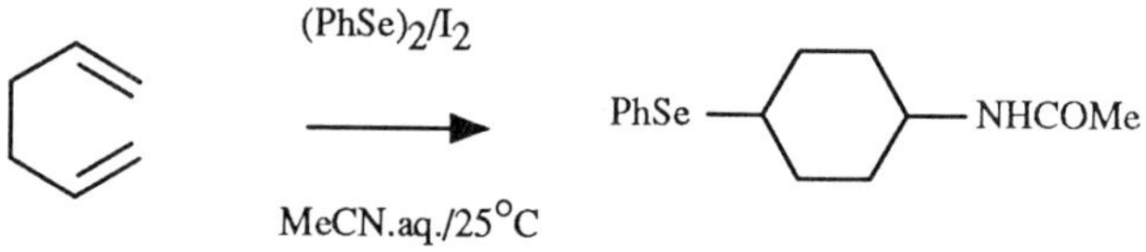

(354)

3. Nucleophilic Reagents

Metal phenylselenides are powerful nucleophiles, although their reactivity depends on solvent, counter ion, and any complexing agent present (D.Liotta, W.Markiewicz, and H.Santiesteban, *Tetrahedron Letter*, 1977, 4365; D.Liotta and H.Santiesteban, *ibid.*, 1977, 4369). The order of reactivity for typical systems is NaSePh / 18-crown-6 / THF > NaSePh / HMPA / THF > LiSePh / HMPA / THF > LiSePh /THF≅ LiSePh / Et_2O > PhSeSePh / $NaBH_4$ / EtOH / THF) (D.Liotta *et al.*, *J. Org. Chem.*, 1981,

13, 2605). For example, the borane complex $PhSeNa.BH_3$, from the $NaBH_4$ reduction of phenylselenide, opens oxiranes, but does not react with esters or lactones. Lithium alkylselenides are attractive reagents, and are readily prepared by reaction of an alkyllithium with selenols, they are quite soluble in diethyl ether or THF (A.B.Smith, III, and R.M.Scarborough, Jr., *Tetrahedron Letters*., 1978, 1649). Sodium and potassium phenylselenides are insoluble in THF, and HMPA is needed to solubilise them (D.Liotta, *Acc. Chem. Res.*, 1984, **17**, 28).

i) Reactions with oxiranes. Oxiranes are cleaved in ethanolic solution usually by a S_N2 mechanism (K.B.Sharpless and R.F.Lauer, *J. Amer. Chem. Soc.*, 1973, **95**, 2697) (see Eqn. 355). Conversion of the alcohol function to a good leaving group results in stereospecific deoxygenation *via* an *anti*-elimination process (J.M.Behan, R.A.W.Johnstone, and M.J.Wright, *J. Chem. Soc., Perkin Trans. 1*, 1975, 1216; A.M.Leonard-Coppens and A.Krief, *Tetrahedron Letters*, 1976, 1385)(see Eqn. 356).

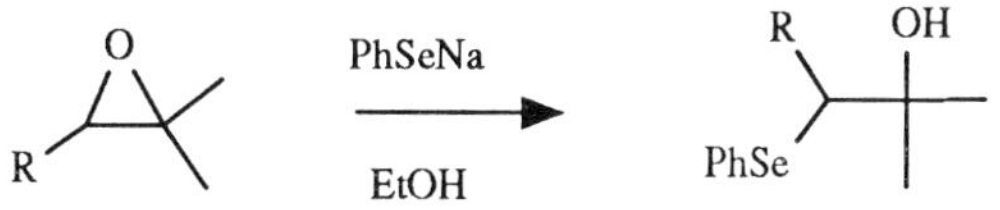

(355)

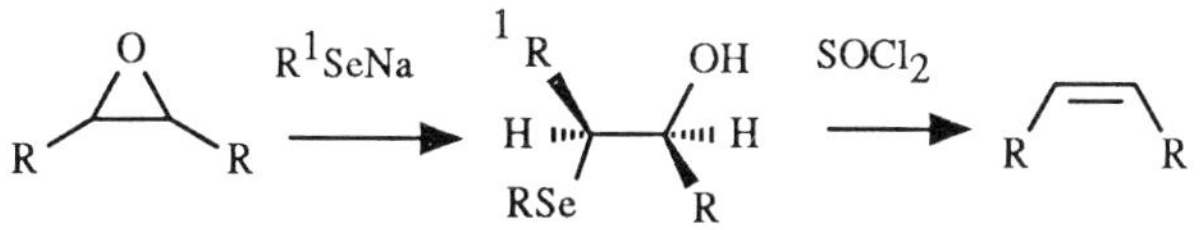

(356)

Direct deoxygenation of oxiranes using nucleophilic selenium reagents is known e.g. $Ph_3P{=}Se$ (D.L.J.Clive and C.V.Denyer, *J. Chem. Soc., Chem. Commun.*, 1973, 253), $Bu_3P{=}Se$ (T.H.Chan and J.R.Tinkelbine, *Tetrahedron Letters*., 1974, 2091), or KSeCN (J.M.Behan, R.A.W.Johnstone, and M.J.Wright, *J. Chem. Soc., Perkin Trans. 1*, 1975, 1216).

ii) Reactions with esters and lactones. Reactive preparations of sodium phenylselenide cleave many esters and lactones (see Eqn. 357), thus while PhSeNa / BH_3 fails to react in boiling ethanol, PhSeNa / THF / HMPA

$$\text{(γ-butyrolactone)} \xrightarrow{PhSeNa} PhSe(CH_2)_3CO_2H \qquad (357)$$

reacts at 65°C, or at room temperature when a catalytic amount of 18-crown-6 is present (D.Liotta, *Acc. Chem. Res.*, 1984, **17**, 28; D.Liotta, *et al., J. Org. Chem.*, 1981, **13**, 2605). This reaction normally involves attack on the carbinol carbon atom, and thus may prove an useful substitute to hydrolysis for hindered esters, where saponification is sluggish and requires attack at the carbonyl carbon.

iii) Displacement reactions with aliphatic halides, sulphonates, etc. Selenide anions react smoothly at both *primary* and *secondary* centres, in contrast to carbon nucleophiles, which often cause elimination, especially at the *secondary* sites. Allylic halides may react by either S_N2 or S_N2' processes; then oxidation yields selenoxides, which may undergo *syn*-eliminations to dienes (D.N.Jones, D.Mundy, and R.D.Whitehouse, *J. Chem. Soc., Chem. Commun.*, 1970, 86; K.B.Sharpless, W.M.Young, and R.F.Lauer, *Tetrahedron Letters.*, 1973, 1979), or [2,3]sigmatropic rearrangement to yield selenenate esters. These hydrolyse to allylic alcohols (K.B.Sharpless and R.F.Lauer, *J. Amer. Chem. Soc.*, 1972, **94**, 7154).

iv) Reactions with enals, enones, and enoic acid derivatives. Conjugate addition of arylselenols can usually be effected without catalysis (A.A.Anciaux et al., *Tetrahedron Letters.*, 1975, 1613;1617). Conjugated enones react with $PhSeSiMe_3/Me_3SiOTf$ yielding predominantly 1,4-addition products (see Eqn. 358). Such products are useful synthetically, and react, for example, with acetals or orthoesters (D.Liotta *et al., ibid.*, 1978, 5091; M.Suzuki, T.Kawagishi, and R.Noyori, *ibid.*, 1981, **22**, 1809).

v) Reactions with vinyl- and aryl halides. Phenylselenide anions replace vinyl halides with retention of configuration (M.Tiecco *et al., J. Org. Chem.*, 1983, **48**, 4289) (see Eqn. 359). Inactivated aryl halides react readily with excess MeSeLi to form arylmethylselenides, which are demethylated by excess reagent. Addition of alkyl halides leads to alkyl aryl selenides.

vi) Reaction with alcohols. Alcohols, which readily form carbonium ions (e.g. R_3COH, $PhCHROH$, $RCH{=}CH.CH_2OH$, etc.) yield selenides with phenylselenol under conditions of acid catalysis (M.Clarenbeau and A.Krief, *Tetrahedron Letters.*, 1984, **25**, 3625). This procedure complements the preparation of selenides from *primary*- and *secondary*-alkyl alcohols using $PhSeCN/R_3P$ (see above: Section 7.2.v).

$$CH_3COCH{=}CH_2 \xrightarrow[TMSOTf]{PhSeTMS} CH_3C(OTMS){=}CHCH_2SePh$$

$$\xrightarrow{PhCH(OMe)_2} CH_3COCH(CH_2SePh)CH(OCH_3)Ph \xrightarrow{[O]}$$

$$CH_3COC({=}CH_2)CH(Ph)OCH_3 \qquad (358)$$

$$PhSCH{=}CHCH_2Cl \xrightarrow{PhSeNa} PhSCH{=}CHCH_2SePh \qquad (359)$$

4. Oxidations

Selenium dioxide oxidation (see N.Rabjohn, in *Organic Reactions*, W.G.Dauben, Ed., Wiley, New York 1976) of alkenes, probably proceeds *via* allylic seleninic acids and involves a [2,3]-sigmatropic rearrangement (see Eqn. 360). Such processes can be improved and made catalytic in

$$RCH{=}C(CH_3)_2 \xrightarrow{SeO_2} RCH(SeO_2H)C(CH_3){=}CH_2 \longrightarrow$$

$$RCH{=}C(CH_3)CH_2OSeOH \xrightarrow{H_2O} RCH{=}C(CH_3)CH_2OH \qquad (360)$$

selenium dioxide, by the use of *tert*-butylhydroperoxide (M.A.Umbreit and K.B.Sharpless, *J. Amer. Chem. Soc.*, 1977, **99**, 5526). Indeed, imido analogues of selenium dioxide undergo allylic amination (K.B.Sharpless *et al., ibid.*, 1976, **98**, 269; K.B.Sharpless and S.P.Singer, *J. Org. Chem.*, 1976, **41**, 2540).

The use of *seleninic anhydrides* and *acids* (S.V.Ley, in *Organoselenium Chemistry,* D.Liotta, Ed., Wiley, New York, 1987) brings about oxidations with the concomitant generation of a carbon-selenium bond. For example, 1-alkenes may be converted into α-selenoketones (M.Shimizu, R.Takeda, and I.Kuwajima, *Tetrahedron Letters.*, 1979, 419)(see Eqn. 361).

$$RCH{=}CH_2 \xrightarrow[(PhSe)_2]{Ph(SeO)_2O} RCOCH_2SePh \qquad (361)$$

The regioselectivity of this process is reversed in some cases, leading to aldehydes (M.Shimizu, R.Takeda, and I.Kuwajima, *ibid.*, 1979, 3461; *Bull. Chem. Soc. Jpn.*, 1981, **54**, 3510)(see Eqn. 362). Vinyl selenides give similar products (A.Gavador and A.Krief, *J. Chem. Soc., Chem. Commun.*, 1980, 951). *Seleninic acid* generated *in situ* will add to alkenes (D.Labar, A.Krief, and L.Hevesi, *Tetrahedron Letters.* 1978, 3967)(see Eqn. 363).

$$RCH(OR^1)CH{=}CH_2 \longrightarrow RCH(OR^1)CH(SePh)CHO \qquad (362)$$

$$PhSeO_2H + H_3PO_2 \longrightarrow [RSeOH]$$

$$C_8H_{17}CH{=}CH_2 \longrightarrow C_8H_{17}(OH)CH{=}CHSePh \qquad (363)$$

b) Tellurium, and Polonium

The organic chemistry of tellurium has been reviewed (K.J.Irgolic, *The Organic Chemistry of Tellurium*, Gordon and Breach, New York, 1974; *J. Organomet. Chem.*, 1975, **103**, 91; *ibid.*, 1977, **130**, 411; *ibid.*, 1978, **158**, 235. The organo derivatives of these two elements have been little used in synthesis.

Second Supplements to the 2nd Edition of Rodd's Chemistry of Carbon Compounds, Vol. IA/B, edited by M. Sainsbury
© 1991 Elsevier Science Publishers B.V., Amsterdam

Chapter 7b

ALIPHATIC ORGANOMETALLIC AND ORGANOMETALLOID COMPOUNDS (COMPOUNDS DERIVED FROM THE TRANSITION METALS)

A.J.FLOYD

1. *Introduction and bibliography*

Organotransition metal chemistry has continued its explosive growth during the past 17 years covered by this supplement. This can be judged by consulting the annual surveys of the relevant literature, to be found in a number of journals (e.g. *Ann. Rep. Prog. Chem.*, Sect. B; *J. Organomet. Chem.*; *S.P.R., Organometallic Chemistry*). Clearly the space constraints of a chapter of this type require a highly selective approach. The transition metal complexes hold a central place in inorganic chemistry (see F.A.Cotton and G.Wilkinson, *Advanced Inorganic Chemistry*, 5th Edn., Wiley, New York, 1988) and have found wide use in industry (G.W.Parshall, *Organometallics*, 1987, **6**, 687; G.W.Parshall, W.A.Nugent, and T.V.Rajan-Babu, *Chem. Scr.*, 1987, **27**, 527). However, most important from an organic chemist's point of view is their extensive application to organic synthesis (e.g. see series of lectures arranged by M.L.H.Green and S.G.Davies, *The Influence of Organometallic Chemistry on Organic Synthesis: Present and Future, Philos. Trans. Royal Soc. London, A*, 1988, **326**, 501). This is the aspect of their chemistry, which will be stressed in this supplement, as a means of coming to terms with a representative selection from this vast amount of data.

The format used in the 2nd Edition will be retained in this supplement, thus subject will be treated according to the nature of the carbon-metal bonds *viz.* a) metal bonded to one carbon atom - σ-bonded complexes; b) metal bonded to two carbon atoms - alkene/alkyne complexes; c) metal bonded to three carbon atoms - π-allylic complexes; d) metal bonded to four or more carbon atoms - diene complexes. Initially, it will be necessary to discuss some sources of information and describe some reactions of transition-metal compounds, which do not fit within this format.

i) General articles

Several textbooks covering the topic of transition metal chemistry have been published. Good examples are *Organotransition Metal Chemistry: Applications to Organic Synthesis* (S.G. Davies, Pergamon, Oxford, 1982); *Metallo-organic Chemistry* (A.J.Pearson, Wiley, Chichester, 1985); *Principles and Applications of Organotransition Metal Chemistry* (J.P.Collman, L.S.Hegedus, J.R.Norton and R.G.Finke, University Science Books, California, 1987); and P.Powell *Principles of Organometallic Chemistry* (2nd Edn. Chapman and Hall, London, 1988). The bonding in transition metal complexes has been reviewed (R.Hoffman, *Science*, 1981, **21**, 995; *Angew. Chem. Int. Ed. Engl.*, 1982, **21**, 711) and discussed in the book *Orbital Interactions in Chemistry* (T.A.Albright, J.K.Burdett, and M.-H.Whangbo, Wiley, New York, 1985). Detailed coverage of all aspects of transition metal chemistry are to be found in three multi-volume works (*Gmelin's Handbook of Inorganic Chemistry*, Springer-Verlag, Berlin; *Comprehensive Organometallic Chemistry*, G.Wilkinson, F.G.A. Stone, and E.W.Abel, Eds., Pergamon, Oxford, 1982; *The Chemistry of the Metal-Carbon Bond*, Vol. 1. *The structure, Preparation, Thermochemistry, and Characterisation of Organometallic Compounds*"; Vol. 2: *The Nature and Cleavage of Metal-Carbon Bonds*; Vol. 3: *Carbon-Carbon Bond Formation Using Organometallic Compounds*; Vol. 4: *The Use of Organometallic Compounds in Organic Synthesis*, F.R.Hartley and S.Patai, Eds., Wiley, Chichester, 1982-7). These texts also cover main group metal chemistry in depth.

In recent years a number of reviews have appeared, which describe interesting and diverse aspects of transition metal chemistry, examples are gas-phase reactions of transition metal cations with alkanes (P.B.Armentrout and J.L.Beauchamp, *Acc. Chem. Res.*, 1989, **22**, 315), the role of electrophilic metal centres in the activation of carbon-carbon double and single bonds, and carbon-hydrogen bonds (A.Sen, *Acc. Chem. Res.*, 1988, **21**, 421), chemiluminescence of organometallic complexes (A.J.Lees, *Chem. Rev.*, 1987, **87**, 711), applications of ^{13}C N.M.R. spectroscopy to structure elucidation of organotransition metal complexes (P.W.Jolly, *Adv. Organomet. Chem.*, 1981, **19**, 257); the application of computers to organometallic chemistry problems (I.Theodosian, R.Barone, and M.Chanon, *Adv. Organomet. Chem.*, 1986, **26**, 165), and a survey of current techniques available for studying organometallic reaction intermediates, illustrated by the study of $Fe(CO)_4$ (M.Poliakoff and E.Wietz, *Acc. Chem. Res.*, 1987, **20**, 408).

ii) Organo-element chemistry

a) *Lanthanides* (H..Kagan and J.L.Namy, *Tetrahedron*, 1986, **42**, 6573; W.J.Evans, *Polyhedron*, 1987, **6,** 803; H.Schumann *et al., Polyhedron*,

1988, 7, 2307; Qi Shen *et al.*, *Pure Appl. Chem.*, 1988, **60**, 1251).
b) *Zirconium* (E.-i.Negishi, *Acc. Chem. Res.*, 1987, **20**, 65; E.-i.Negishi and T.Takahashi, *Synthesis*, 1988, 1).
c) *Zirconium, Hafnium* (D.J.Cardin, M.F.Lappert, and C.L.Raston, *Chemistry of Organo-zirconium and Organo-hafnium Compounds*, Ellis Horwood, Chichester, 1986.
d) *Zirconium, Molybdenum* (J.J.Schwartz *et al.*, *Pure. Appl. Chem.*; 1988, **60**, 65).
e) *Molybdenum, Tungsten* (R.R.Schrock, *Acc. Chem. Res.*, 1986, **19**, 342; W.E.Buhro and M.H.Chisholm, *Adv. Organomet. Chem.*, 1987, **27**, 311).
f) *Molybdenum, Tungsten, Rhenium* (E.R.Burkhardt *et al.*, *Pure Appl. Chem.*, 1988, **60**, 1).
g) *Rhenium* (W.A.Herrmann and J.Okuda, *J. Mol. Catal.*, 1987, **41**, 109).
h) *Ruthenium* (T.Mitsudo, Y.Hori, and Y.Watanabe, *J. Organomet. Chem.*, 1987, **334**, 157).
i) *Ruthenium, Rhodium* (B.Marciniec *et al.*, *J. Mol. Catal.*, 1987, **46**, 329).
j) *Rhodium, Iridium* (R.S.Dickson, *Organometallic Chemistry of Rhodium and Iridium*, Academic Press, London, 1983).
k) *Cobalt, Nickel, Palladium, Rhodium* (G.P.Chiusoli, *J. Mol. Catal.*, 1987, **41**, 75).
l) *Nickel* (J.M.Brown and N.A.Cooley, *Chem. Rev.*, 1988, 88, 1031).
m) *Palladium* (R.F.Heck, *Palladium Reagents in Organic Synthesis*, Academic Press, London, 1985).
n) *Iron* (*The Organic Chemistry of Iron, Vol. 2,* Von G.Koerner *et al.*, Eds., Academic Press, New York, 1981).

iii) Reactions of transition-metal compounds (including syntheses, catalysis, and hydrogenation)

a) *Homogeneous catalytic hydrogenation. The* subject has been reviewed on many occasions (e.g. A.J.Birch and D.H.Williamson, *Org. React.*, 1976, **24**, 1; B.R.James in *Comprehensive Organometallic Chemistry*, Vol. 8, p. 285, G.W.Wilkinson, F.G.A.Stone, and E.W.Abel, Eds., Pergamon, Oxford, 1982). The uses of Wilkinson's catalyst [J.F.Young *et al.*, *J. Chem. Soc., Chem. Commun.*, 1965, 131; R.S.Coffey, I.C.I. Ltd., *Brit. Pat.*, 1121642 (1965)] continue to expand. A significant development has been in connection with chiral hydrogenations (D.Valentine and J.W.Scott, *Synthesis*, 1978, 329; K.Harada in *Asymmetric Synthesis*, Vol. 5, Ch. 10, J.D.Morrison, Ed., Academic Press, New York, 1985). Enantiomeric excesses of *ca* 100% have been achieved. A good example is the reduction of α-acylamidocinnamate derivatives (see K.E.Koenig *ibid.*, Vol. 5, Ch. 3; W.S.Knowles, *Acc. Chem. Res.*, 1983, **16**, 106)(see Eqn.1). Chiral phosphorus ligands are used, which can contain either chiral carbon (e.g. CHIRAPHOS), or chiral phosphorus (e.g. DIPAMP) atoms.

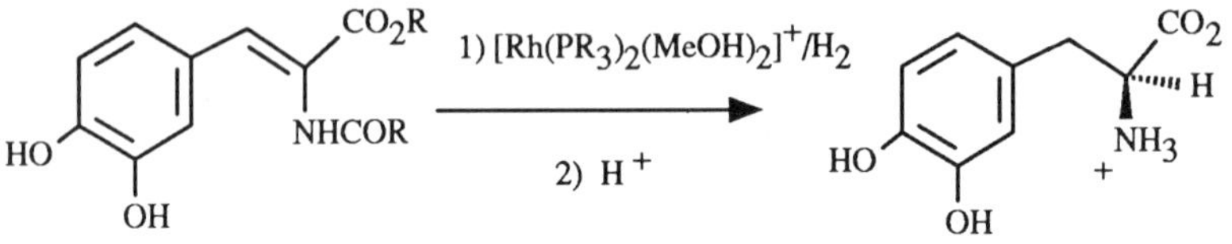

CO_2R 1) $[Rh(PR_3)_2(MeOH)_2]^+/H_2$ CO_2 H

HO NHCOR 2) H^+ HO $\overset{+}{N}H_3$

OH OH

L-Dopa

(1)

$(Ph_2P\text{-}C^*HCH_3)_2$
CHIRAPHOS

$(o\text{-}MeOPh(Ph)PCH_2)_2$
DIPAMP

Crabtree's iridium catalyst (R.H.Crabtree, *Acc. Chem. Res.*, 1979, **12**, 331) has also proved a highly effective and stereospecific catalyst for hindered alkenes. The catalyst is deactivated by polar solvents, with which it tends to complex strongly. Use has been made of this property in the reduction of a cyclohexene (see Eqn 2), where hydrogen is delivered from the same face as the hydroxyl group, albeit at a slow rate (R.H.Crabtree and M.W.Davis, *Organometallics*, 1983, **2**, 681).

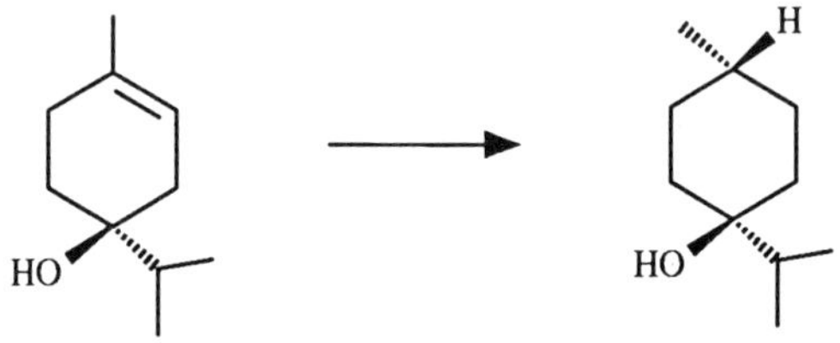

H

(2)

HO HO

99.9%

Arene chromiumtricarbonyl catalysts are useful for the selective hydrogenation of conjugated dienes, yielding Z-enes by 1,4-addition (J.R.Tucker and D.P.Riley, *J. Organomet. Chem.*, 1985, **279**, 49). An example (see Eqn. 3) is provided by the reduction of a hexa-2,4-dienoate ester (M.J.Mirbach, N.P.Tuyet, and A.Saus, *J. Organomet. Chem.*, 1982, **236**, 309).

$Cr(CO)_3$

H_2

(3)

b) *Hydrogen transfer reactions.* Birch and Williamson (see above) have discussed the use of reducing reagents other than hydrogen (e.g. NH_2NH_2, $NaBH_4$, ROH etc.). In some cases, a solvent such as DMSO, can be the hydrogen source (see Eqn. 4).

H_2IrCl_6

DMSO

(4)

c) *Hydrosilation.* The catalysed addition of Si-H bonds to carbonyl and carbon-carbon double bonds using soluble transition-metal compounds has been reviewed (J.L.Speier, *Adv. Organomet. Chem.*, 1978, **17**, 407). Substituted alkenes such as *Z*-2-pentene may react and isomerise to form n-alkyl silanes (see Eqn. 5) (A.J.Chalk, *J. Organomet. Chem.*, 1970, **21**, 207). It is interesting, that the less strongly coordinating *E*-pent-2-ene fails to react under these conditions.

$PhSiMe_2H$

$RhClL_3$

$SiMe_2Ph$

(5)

d) *Hydrocyanation.* This is an interesting process whereby hydrogen cyanide is added to alkenes or alkynes (B.R.James in *Comprehensive Organometallic Chemistry*, Vol. 8, G.Wilkinson, F.G.A.Stone, and E.W.Abel, Eds., Pergamon, Oxford, 1982). It is employed in Dupont's process for the conversion of 1,3-butadiene to adiponitrile, an important industrial example of homogeneous catalysis (C.A.Tolman *et al.*, *Adv.in Catal.*, 1985, **33**, 1). The addition is known to take place in the *syn* mode for both alkenes [W.R.Jackson *et al.*, *Organic Synthesis, an Interdisciplinary Challenge* (5th IUPAC Symp. on Organic Synthesis), Blackwell, London, 1986] and for alkynes (W.R.Jackson and C.G.Lovel, *Aust. J. Chem.*, 1983, **36**, 1975). The processes also tend to be regiospecific (see Eqn. 6).

HCN

$[(PhO)_3P]_4Ni$

CN

(6)

e) *Polymerisation of alkenes and alkynes.* These important industrial processes depend on transition-metal catalysts, however, the exact nature of the promoters are frequently closely guarded trade secrets (W.Keim, A.Behr, and M.Roper, in *Comprehensive Organometallic Chemistry*, G.Wilkinson, F.G.A.Stone, and E.W.Abel, Eds., Pergamon, Oxford, 1982).

f) *Carbon monoxide.* The utilisation of carbon monoxide in the synthesis of bulk chemicals, represents a vital role for transition-metal compounds in the chemical industry (G.W.Parshall, *Homogeneous Catalysis*, Wiley, New York 1980; J.Falbe, *New Syntheses with Carbon Monoxide*, Springer-Verlag, New York, 1980). The water-gas shift reaction, which produces hydrogen from the reaction of carbon monoxide and steam, involves transition metallacarboxylic acid intermediates (M.A.Bennett, *J. Mol. Catal.*, 1987, **41**, 1). Synthesis gas, a mixture of carbon monoxide and hydrogen is the basis of many industrial processes. Thus homogeneous transition-metal catalysis is used in the *oxo-reaction* to make aldehydes and alcohols and the *Monsanto process* to make acetic acid. While the *Fischer-Tropsch process* involves heterogeneous catalysis for the conversion of synthesis gas to hydrocarbons and alkenes (W.A.Herrmann, *Angew. Chem. Int. Ed. Engl.*, 1982, **21**, 117).

g) *Transition-metal hydrides.* These are useful reagents for the reduction of conjugated double bonds (e.g. those of unsaturated aldehydes, ketones, esters, nitriles) without reduction of the carbonyl function. An examples is $KHFe(CO)_4$ (R.Noyori, I.Umeda, and I.Ishigami, *J. Org. Chem.*, 1972, **37**, 1542)(see Eqn. 7).

$Fe(CO)_5 + KOH \rightarrow [(OC)_4FeCO_2H]^-K^+ \rightarrow [HFe(CO)_4]^-K^+$ (7)

The strongly basic conditions limit the use of these reactions, although they have been exploited in the α-alkylation of ketones (G.Cainelli, M.Panunzio, and A.Umani-Ronchi, *J. Chem. Soc., Perkin 1*, 1975, 1273) and β-ketoesters (M.Yamashita *et al., Bull. Chem. Soc. Jpn., 1978*, **51**, 835)(see Eqn. 8). Use of an aprotic solvent allows full reduction to the saturated alcohol (M.Yamashita *et al., Bull. Chem. Soc. Jpn.*, 1982, **55** 1329). The binuclear compound $NaHFe_2(CO)_8$ will accomplish similar reductions (J.P. Collman *et al., J. Amer. Chem. Soc.*, 1978, **100**, 1119). Further useful reactions of $KHFe(CO)_4$ include reduction of acid chlorides to aldehydes (T.E Cole and R.Pettit, *Tetrahedron Letters*, 1977, 781) and reductive amination of aldehydes and ketones (Y.Watanabe *et al., Bull. Chem. Soc. Jpn.,* 1976, 49, 1378; G.P.Boldrini, A.Umani-Ronchi, and A.Panunzio, *J. Organomet. Chem.*, 1979, **171**, 85). The latter process has been used to prepare piperidines from dialdehydes (Y.Watanabe *et al., Bull. Chem. Soc. Jpn.* 1976, **49**, 2302) (see Eqn. 9).

1 R R2 + R(3)CHO —base→ 2 R, 3 R, O, R1

—$KHFe(CO)_4$→ 2 R, 3 R, O, R1 (8)

CHO CHO + RNH_2 —$KHFe(CO)_4$, CO→ N–R (9)

The binuclear chromium cluster $NaHCr_2(CO)_{10}$ also cleanly reduces a similar range of conjugated double bond types (G.P.Boldrini, A.Umani-Ronchi, and A.Panunzio, *Synthesis*, 1976, 596). However, by contrast $HCr(CO)_5^-$ and HW $(CO_5)^-$ will reduce organic halides and acid chlorides (S.C.Kao and M.Y.Darensbourg, *Organometallics*, 1984, **3**, 646; 1601). Copper hydrides are effective reducing agents, and once again selective reduction of the conjugated double bonds of unsaturated ketones and esters is typical. Thus $[Pr\text{-}\equiv CuH]^-Li^+$ in fairly large excess will reduce enone double bonds, although not conjugated alkyne derivatives (R.K.Boeckman and R.Michalak, *J. Amer. Chem. Soc.*, 1974, **96**, 1623). The mixed cuprate $[BuCuH]^-Li^+$ is effective for the 1,4-reduction of enones at -40°C. At 25°C, however, this reagent also reduces aldehydes and ketones to alcohols, and primary, secondary and tertiary halides or tosylates, without elimination, to hydrocarbons (S.Masamune, G.S.Bates, and P.E.Georghiou, *J. Amer. Chem. Soc.*, 1974, **96**, 3686). $Cu_2I_2/2LiAlH(OMe)_3$ in tetrahydrofuran is a most effective reagent for the reduction of alkyl, vinyl, allyl and aryl halides to hydrocarbons (S.Masamune, P.A.Rossy, and G.S.Bates, *J. Amer. Chem. Soc.*, 1973, **95**, 6452), while Ph_3PCuBH_4 in acetone is good for the reduction of acid chlorides to aldehydes (G.W.J.Fleet, C.J.Fuller, and P.J.C.Harding, *Tetrahedron Letters.*, 1978, 1437; T.S.Sorrell and R.J.Spillane, *ibid.*, 2473; T.N. Sorrell and P.S.Pearlman, *J. Org. Chem.*, 1980, **45**, 3449). Palladium (0)-tinhydride reagents are also useful for reduction of acid chlorides (P.Four and F.Guibe, *J. Org. Chem.*, 1981, **46**, 4439; E.Keinan and P.A.Gleize, *Tetrahedron Letters.*, 1982, **23**, 477). Ashby has examined most first-row transition-metal halide/$LiALH_4$ mixtures as reducing systems. Nickel complexes were effective in the reduction of organic halides or tosylates, while cobalt complexes reduced alkenes (E.C.Ashby and J.J.Lin, *J. Org. Chem.*, 1978, **43**, 1263 and 2567). It has been shown that iron complexes with sodium hydride reduce aldehydes and ketones (T.Fujisawa, K.Sugimoto, and H.Ohta, *ibid.*, 1976, **41**, 1667). Palladium(0) complexes reduce allyl ethers, thioethers, trialkylsilyl ethers, and sulphones to the corresponding alkene with retention of stereo- and regio- chemistry (R.O.Hutchins and K.Learn, *ibid.*, 1982, **47**, 4380).

2. *Compounds with metal bonded to one carbon atom.*

These include σ-bonded alkyl and alkenyl, as well as, carbonyl and carbene complexes.

i) Alkyl and alkenyl copper complexes.

σ-Carbon to copper bonded complexes have found wide application in organic synthesis (J.F.Normant, *Pure Appl. Chem.*, 1978, **50**, 709; G.A. Posner, *An Introduction to Synthesis Using Organocopper Reagents*, Wiley, New York, 1980; E. Erdik, *Tetrahedron*, 1984, **40**, 641; B.H.Lipshultz, R.S.Wilhelm and J.A.Kozlowski, *Tetrahedron*, 1984, **40**, 5005).

a) *Symmetric organocuprates* [R_2CuM (M=Li or MgX]. These have been the most extensively used copper reagents, due no doubt to their ready formation from two equivalents of organolithium or Grignard reagent with one equivalent of anhydrous cuprous iodide. Two major uses have been found for these complexes: alkylation of organic halides or tosylates (G.H.Posner, *Org. React.*, 1975, **22**, 253), and the 1,4-alkylation of conjugated enones (*idem, ibid.*, 1972, **19**, 1). The organocuprate reagents are appreciably less nucleophilic than the lithium or magnesium compounds from which they are obtained, so they will react with halides in the presence of remote ester, cyano, and keto groups. Aldehydes react faster than halides, so cannot be present.

Acetylenic groups do not transfer in 1,4-additions, a property used to advantage in the mixed cuprate reagents (see below). Conjugated ketones react very rapidly and α,α'-, or β-substitution only slows the reactions moderately. Conjugated aldehydes suffer competitive 1,2-addition; a problem which can be overcome by the use of "higher order" cuprates, e.g. $Me_5Cu_3Li_2$, resulting in greater 1,4-addition (D.L.J.Clive, V.Farina, and P.L.Beaulieu, *J. Org. Chem.*, 1982, **47**, 2572). Conjugated carboxylic acids fail to react, while their esters are less reactive than conjugated ketones. The corresponding acetylenic esters are much more reactive. These reactions proceed *via* a vinyl copper intermediate, formed by a specific *syn*-addition. This can be reacted with a number of electrophiles (see Eqn. 10) resulting in the stereospecific synthesis of alkenes.

Chiral substrates often lead to significant asymmetric induction, thus addition has been observed with greater than 90% diastereoselectivity (W.R.Roush and B.M.Lesur, *Tetrahedron Letters.*, 1983, **24**, 231)(see Eqn. 11).

^1R—≡—CO_2R $\xrightarrow{R^2CuLi}$ (^1R)(^2R)C=C(CO_2R)(CuR^2Li)

$\xrightarrow{H^+}$ (^1R)(^2R)C=C(CO_2R)(H)

$\xrightarrow{I_2}$ (^1R)(^2R)C=C(CO_2R)(I)

(10)

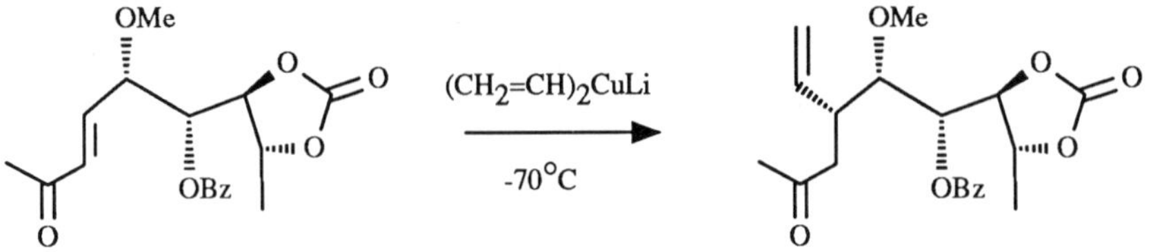

(11)

b) *Mixed cuprates*. The major drawback with symmetrical dialkyl cuprates lies in their inefficient transfer of alkyl groups. Only one of the two alkyl groups is transferred, while instability of the reagents often means that a considerable excess is required for complete reaction. This problem has been solved by using mixed cuprates of increased stability, with stabilising alkoxy, mercapto, cyano, or alkynyl groups, which are themselves not transferred (G.H.Posner, C.E.Whitten, and J.J.Stirling, *J. Amer. Chem. Soc.*, 1973, **95**, 7788). For example, the cyanocuprates have found application in opening allylic epoxides (J.P. Marino and M.G.Kelly, *J. Org. Chem.*, 1981, **46**, 4389)(see Eqn. 12). Higher order mixed organocuprates have also found use in organic synthesis (B.H.Lipshutz, *Synthesis*, 1987, 325).

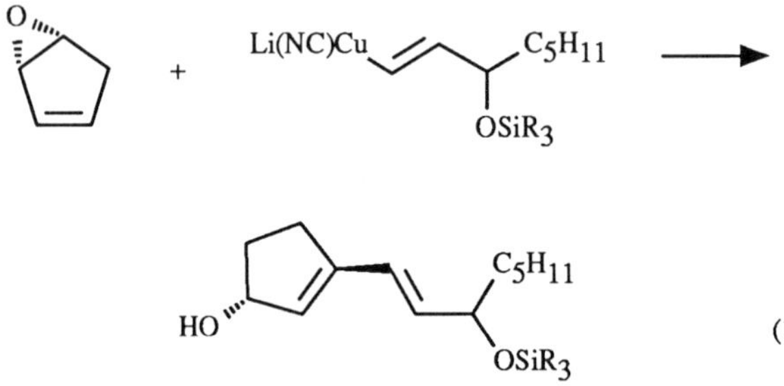

(12)

c) *Lewis acid complexes*. Another approach, to increased stability of the copper reagent, has been the use of Lewis acid complexes (e.g. $RCu.BF_3$). These are especially efficient in 1,4-conjugate additions to enones, even to those which are sterically crowded. Some conjugated carboxylic acids react as well (Y.Yamamoto *et al.*, *J. Org. Chem.*, 1982, **47**, 119). A $RCu.MgX_2$ complex is required for the alkylation of terminal alkynes, which occurs by *syn* addition to yield vinylcuprates in a regiospecific manner (see Eqn. 13).

These vinylcuprates are reactive, undergoing all the reactions common to organocuprates (A.Alexakis *et al.*, *Pure Appl. Chem.*, 1983, **55**, 1759). Internal alkynes, however, react readily with dialkyl cuprates, R_2CuLi again yielding vinyl cuprates, which, as expected, react with a range of electrophiles (C.Germon, A.Alexakis, and J.F.Normant, *Synthesis*, 1984, 40;43).

$$^{1}R{-}C{\equiv}C{-}H \; + \; RCuMgX_2 \longrightarrow (^{1}R)(R)C{=}C(H)(CuMgX_2) \qquad (13)$$

ii) Alkyl and alkenyl zirconium complexes.

The "carbocupration" of alkynes, mentioned above, is an example of a quite general method for the preparation of σ-alkyl or σ-alkenyl transition metal complexes (J.F.Normant and A.Alexakis, *Synthesis*, 1981, 841). Electron-poor transition metals in high formal oxidation state favour the formation of σ-alkyl complexes. Zirconium, amongst the early transition metals which fit this requirement, has found considerable use in organic synthesis (J.J.Schwartz and J.Labinger, *Angew. Chem. Int. Ed. Engl.*, 1976, **15**, 333).

The steric requirements of the most commonly employed biscyclopentadienylzirconium (IV) complex results in addition to the least crowded position on the alkene, irrespective of the initial position of the double bond (see Eqn. 14).

$$Cp_2Zr(H)Cl \; + \; \text{(hex-1-ene or hex-3-ene)} \longrightarrow Cp_2Zr(Cl)C_6H_{13} \qquad (14)$$

Alkynes undergo stereospecific *syn* addition, with the least hindered vinyl zirconium complex as the major product. Although the resultant alkyl/vinyl complexes can be isolated, they undergo a number of reactions under mild conditions, which makes them synthetically useful. Replacement of zirconium by hydrogen, bromine (D.W.Hart and J.J.Schwartz, *J. Amer. Chem. Soc.*, 1974, **96**, 8115), or hydroxyl (T.F.Blackburn, J.A.Labinger, and J.J.Schwartz, *Tetrahedron Letters*, 1975, 3041; T.Gibson, *ibid.*, 1982, **23**, 157) are all readily achieved. Insertion of carbon monoxide occurs at 25°C to yield acyl complexes (C.A.Bertelo and J.J.Schwartz, *J. Amer. Chem. Soc.*, 1975, **97**, 228 and 1976, **98**, 262), which can be converted to aldehydes (protonation), carboxylic acids (peroxides), ester (bromine in methanol), or acid bromides (*N*-bromosuccinimide).

The addition of trimethylaluminium to alkynes is catalysed by Cp_2ZrCl_2, to yield vinylalanes with high regio- and stereo- selectivity (D.E.Van Horne and E.-i.Negishi, *J. Amer. Chem. Soc.*, 1978, **103**, 4985). These vinylalanes are reactive and can lead to a variety of products (see Eqn. 15)

R—≡—H → (Cp_2ZrCl_2, Me_3Al) → R(Me)C=CH–$AlMe_2$ → (BuLi) → R(Me)C=CH–Al Me_2Bu

→ ($ClCO_2Et$) → R(Me)C=CH–CO_2Et

→ (CH_2O) → R(Me)C=CH–CH_2OH

→ (CO_2) → R(Me)C=CH–CO_2H

(15)

The synthetic utility of zirconium alkyls is increased by transmetalation, through which the zirconium is replaced by another metal of greater, or different, reactivity. Such transmetalation processes are generally reversible (see Eqn. 16); however, irreversible removal of the new metal alkyl can lead to efficient synthetic procedures.

$$RM + M^1X \rightleftarrows RM^1 + MX \qquad (16)$$

Thus although zirconium alkenyls fail to react satisfactorily with acyl halides to form ketones, preliminary treatment with $AlCl_3$ results in transfer of alkenyl to aluminium, and this is followed by a rapid reaction forming the ketone in high yield (D.B.Carr and J.J.Schwartz, *J. Amer. Chem. Soc.*, 1977, **99**, 638)(see Eqn. 17).

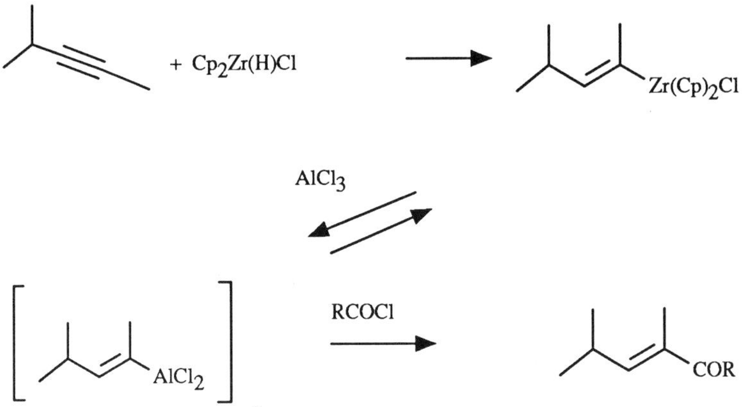

+ $Cp_2Zr(H)Cl$ → $Zr(Cp)_2Cl$

$AlCl_3$

$[\ AlCl_2\ ]_n$ —RCOCl→ COR

(17)

iii) <u>Alkyl and alkenyl palladium and nickel complexes</u>

a) *Transmetalation.* The most extensively used transmetalations involve electrophilic metals, such as mercury or thallium, as a route to late transition metal alkyls, especially palladium (R.C.Larock, *Tetrahedron*, 1982, **38**, 1713). It is worth noting a restriction to this type of chemistry, namely the absence of a hydrogen atom β- to the palladium site. If this is the case then Pd-H elimination is rapid. An example of a useful transmetalation procedure of this type is the synthesis of coumarins from arylacetylenes (R.C.Larock and L.W.Harrison, *J. Amer. Chem. Soc.*, 1984, **106**, 4218)(see Eqn.18).

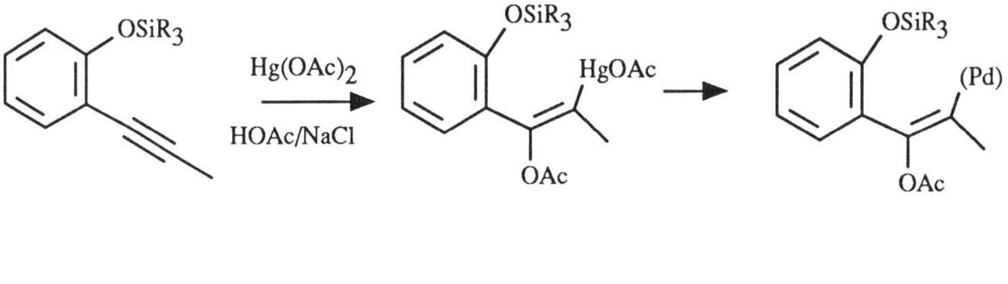

$OSiR_3$ —$Hg(OAc)_2$ / HOAc/NaCl→ $OSiR_3$, HgOAc, OAc → $OSiR_3$, (Pd), OAc

$$\xrightarrow{\text{CO/MeOH}} \text{[o-(R}_3\text{SiO)C}_6\text{H}_4\text{C(OAc)=C(CH}_3\text{)CO}_2\text{Me]} \xrightarrow{\text{HF/HOAc}} \text{[4-acetoxy-3-methylcoumarin]} \quad (18)$$

b) *Oxidative Addition-Transmetalation.* An extension (E.-i.Negishi, *Acc. Chem. Res.*, 1982, **15**, 340; J.K.Stille, *Angew. Chem. Int. Ed. Engl.* 1986, **25**, 508) to this transmetalation chemistry involves oxidative addition to a low-valent transition metal, often Ni(0) or Pd(0), followed by transmetalation by a main-group organometallic reagent to produce a dialkyl transition metal species. Reductive elimination can then lead to carbon-carbon bond formation (see Eqn.19), with obvious uses in organic synthesis.

$$\begin{array}{l} \text{M(0)} + \text{RX} \rightarrow \text{R-M(II)-X} \\ \text{R-M(II)-X} + \text{R}^1\text{M}^1 \rightarrow \text{R-M(II)-R}^1 + \text{M}^1\text{X} \\ \text{R-M(II)-R}^1 \rightarrow \text{R-R}^1 + \text{M(0)} \\ \text{M} = \text{Ni or Pd;} \quad \text{M}^1 = \text{Mg,Zn,Zr,Sn,B,Al,Li} \end{array} \quad (19)$$

An early example (K.Tamao *et al.*, *Bull. Chem. Soc. Jpn.*, 1976, **49**, 1958) of this type of process involves the catalysis of the selective cross-coupling of Grignard reagents and vinyl, or aryl halides by nickel phosphine complexes (see Eqn. 20).

$$\text{RMgX} + \text{CH}_2\text{=CHX} \xrightarrow{\text{NiL}_2\text{Cl}_2} \text{CH}_2\text{=CHR} + \text{MgX}_2 \quad (20)$$

c) *Oxidative Addition-Insertion.* The insertion chemistry of Pd(0) and, to a lesser extent, Ni(0) complexes has been much exploited in organic synthesis (R.F.Heck, *Pure App. Chem.*, 1978, **50** , 691). A wide range of reactions are possible (see Eqn. 21), all of which can be made catalytic.

$$L_2Pd(0) + RX \rightarrow RPdL_2X$$

(21)

An interesting example involves an intramolecular process leading to a $\Delta^{\alpha,\beta}$-butenolide (A.Cowell and J.K.Stille, *Tetrahedron Letters*, 1979, 133)(see Eqn. 22).

iv) Transition metal carbonyls

Important stoichiometric and catalytic reactions involving transition metal carbonyls are to be found in both industrial and laboratory organic synthesis (J.Falbe, *Carbon Monoxide in Organic Synthesis*, Springer Verlag, New York, 1970 and *New Synthesis with Carbon Monoxide*, Springer Verlag, New York, 1980; H. Alper, *J. Organomet. Chem. Library*, 1976, **1**, 305).

a) *Carbonylation reactions*. These have been reviewed (C.Narayagan and M.Periasamy, *Synthesis*, 1985, 253). The reaction of nickel carbonyl with allyl halides is very dependant upon the reaction conditions, the use of solvents of moderate polarity and excess carbon monoxide results in carbonylation (G.P.Chiusoli, *Acc. Chem. Res.*, 1973, **6**, 422)(see Eqn. 23).

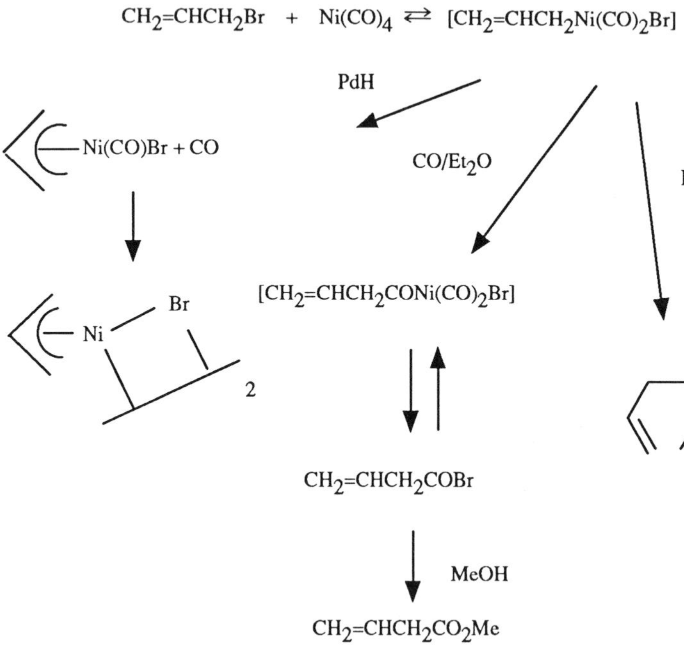

$CH_2{=}CHCH_2Br + Ni(CO)_4 \rightleftarrows [CH_2{=}CHCH_2Ni(CO)_2Br]$

PdH

$-Ni(CO)Br + CO$

CO/Et_2O

DMF

Ni Br

2

$[CH_2{=}CHCH_2CONi(CO)_2Br]$

$CH_2{=}CHCH_2COBr$

MeOH

$CH_2{=}CHCH_2CO_2Me$ (23)

Anionic metal carbonyls have found great use in carbonylation reactions, where their strong nucleophilic character can be exploited. Disodium tetracarbonylferrate, $Na_2Fe(CO)_4$, reacts with alkyl and acyl halides to yield anionic iron complexes, $RFe^-(CO)_4$ and $RCOFe^-(CO)_4$ respectively, which in turn can be converted to a wide range of products, e.g. RCHO, RCO_2H, RCO_2R^1, $RCONR^1R^{11}$, $RCOR^1$, etc. (J.P.Collman, *Acc. Chem. Res.*, 1975, **8**, 342). A neat example of this type of chemistry involves intramolecular olefin insertion (J.Y.Merour *et al.*, *J. Organomet. Chem.*, 1973, **51**, C24)(see Eqn. 24).

$NaCo(CO)_4$ also shows useful carbonylation chemistry (R.F.Heck in *Organic Synthesis via Metal Carbonyls*, Vol. 1, I.Wender and P.Pino, Eds., Wiley, New York, 1968).

Br $\xrightarrow{Na_2Fe(CO)_4}$ $Fe(CO)_4^-$ $\xrightarrow{PPh_3}$ $Ph_3PFe(CO)_3^-$ O → O

(24)

Metal acyl "ate" complexes, formed from the reaction of a neutral metal carbonyl and an organolithium reagent, are reactive species with a similar chemistry to the anionic acyliron complexes. For example, the acylnickel carbonylates, $RCONi(CO)^-_3Li^+$, undergo 1,4-addition with conjugated enones (E.J.Corey and L.S.Hegedus, *J. Amer. Chem. Soc.*, 1969, **91**, 4926). Toxicity and handling difficulties have restricted the use of these acylnickel carbonylates. In 1985 Hegedus developed an air-stable, crystalline acylcobalt complex, $Ph_3P(CO)_2(NO)Co$, which although somewhat less reactive was much easier to handle (L.S.Hegedus and R.J. Perry, *J. Org. Chem.*, 1985, **50**, 4955). This complex undergoes a range of reactions including 1,4-conjugate addition (see Eqn. 25).

$Ph_3P(CO)_2(NO)Co + RLi \longrightarrow [Ph_3(CO)(NO)CoCOR]^- \; Li^+$

O → R O O

(25)

b) *Decarbonylation.* The reverse process, whereby a carbonyl unit is removed from organic compounds, has found some use in synthesis. The Wilkinson hydrogenation catalyst, $(Ph_3P)_2RhCl$ is amongst the most efficient for this reaction converting aldehydes to the corresponding hydrocarbons (e.g. D.E.Iley and B.Fraser-Reid, *J. Amer. Chem. Soc.*, 1975, **97**, 2563) and aliphatic acid chlorides without β-hydrogen to alkyl chlorides (J.K.Stille and R.W.Fries, *J. Amer. Chem. Soc.*, 1974, **96**, 1514).

Aryl acid chlorides yield arylrhodium complexes (J.A.Kampmeier, R.M.Rodehorst, and J.B.Philip Jr., *J. Amer. Chem. Soc.*, 1981, **103**, 1847;

J.A.Kampmeier and S.Mahalingain, *Organometallics*, 1984, **3**, 489), α,β-unsaturated acid chlorides give vinyl phosphonium salts (J.A.Kampmeier, S.H.Harris, and R.M.Rodehorst, *J. Amer. Chem. Soc.*, 1981, **103**, 1478), while acid chlorides with β-hydrogen present give mainly alkenes (J.K.Stille, F.Huang, and T.M.Regan, *ibid.*, 1974, **96**, 1518).

c) *Metal Acyl Enolates.* Acylmetal complexes are a common feature of metal carbonyl chemistry, indeed, many of them are stable and readily isolatable. When hydrogen atoms are present α- to the carbonyl group, treatment with base will form the corresponding enolate anion, allowing entry to a range of *C*-alkylation products. The enolate chemistry of the cyclopentadienyl acyliron complex, $CpFe(CO)(PPh_3)COCH_3$ has been much studied (see L.S.Liebeskind and M.E.Welker, *Organometallics*, 1983, **2**, 194; S.G.Davies and G.J.Baird, *J. Organomet. Chem.*, 1983, **248**, C1). It has been shown that the chirality and bulk of the complex can be used to control the stereochemistry of these reactions (L.S.Liebeskind, M.E.Welker, and R.W.Fengl, *J. Amer. Chem. Soc.*, 1986, **108**, 6328; S.G.Davies *et al., Tetrahedron Letters*, 1984, **25**, 2709; *Bull. Soc. Chim. Fr.*, 1987, 608; *J. Chem. Soc., Chem. Commun.*, 1984, 956). While Bergman (K.H.Theopold, P.N.Becker, and R.G.Bergman, *J. Amer. Chem. Soc.*, 1982, **104**, 5250) has shown an efficient chiral induction in a related system (see Eqn. 26).

1) LDA; 2) ^{t}BuCHO; FeCl$_3$ (26)

v) Transition metal carbenes

The use of carbene complexes in organic synthesis has been reviewed (K.H.Dötz, *Angew. Chem. Int. Ed. Engl.*, 1984, **23**, 587; A.I.Sayatkovskii and B.D.Babitskii, *Russ. Chem. Rev.*, 1985, **53**, 672; M.A.Gallop and W.R.Roper, *Adv. Organomet.* Chem., 1986, **25**, 121; P.J.Brothers and W.R.Roper, *Chem. Rev.*, 1988, **88**, 1293). The book by K.H.Dötz *et al.* (*Transition Metal Carbene Complexes*, Verlag Chemie, Weinheim, 1983), also includes much valuable information on both their preparation and properties. Tungsten carbene complexes have been considered separately (J.Kress, A.Aguero, and J.A.Osborn, *J. Mol. Catal.*, 1986, **36**, 1).

Two classes of carbene ligands are recognised: first the "electrophilic" or "Fischer" type (E.O.Fischer, *Adv. Organomet. Chem.*, 1976, **14**, 1), which contain a heteroatom or other substituents capable of π-interaction with the carbene carbon, and generally suffer nucleophilic attack at this carbon. They may be regarded as singlet carbenes. The second type are "nucleophilic" or "Schrock" carbene ligands (R.R.Schrock, *Acc. Chem. Res.*, 1979, **12**, 98), These lack such substitution (they are generally alkylidenes) and require π-donation from the metal. "Schrock" carbene ligands are usually subject to electrophilic attack at the carbene carbon and may be regarded as triplet species. These distinctions are not always clear-cut, indeed the reactivity of any given carbene ligand depends upon the metal and what other ligands are present; methylene ligands in particular are capable of both "nucleophilic" and "electrophilic" behaviour (see chapter 2b).

a) *Electrophilic carbene complexes*

Reaction with alkynes. (Alkoxyaryl)chromium carbenes react with alkynes to yield 4-alkoxy-1-naphthols (K.H. Dötz, *Angew. Chem. Int. Ed. Engl.*, 1975, **14**, 644)(see Eqn. 27).

OMe
$(CO)_5Cr$
RC≡CR
50°C/-CO
OH
R
R
$(CO)_3Cr$
OMe

(27)

Tungsten complexes behave differently yielding indenes (H.C.Foley *et al.*, *J. Amer. Chem. Soc.*, 1983, **105**, 3064) (see Eqn. 28). While iron complexes have been shown to lead to pyrones (M.F.Semmelhack *et al.*, *J. Amer. Chem. Soc.*, 1984, **106**, 5363) (see Eqn. 29).

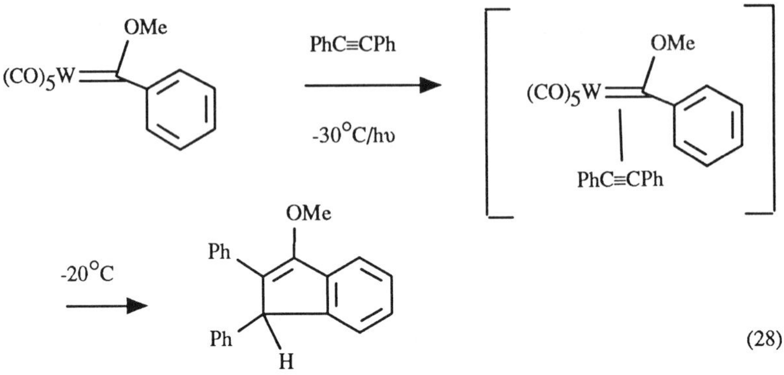

(28)

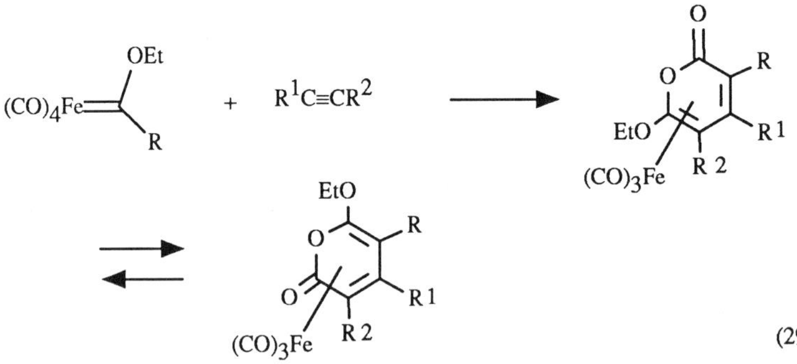

(29)

Reaction with alkenes. Olefin metathesis (see Eqn. 30) has proved important in industrial processes (J.C.Mol, *J. Mol. Catal.*, 1982, **15**, 35; *Chemtech.*, 1983, 250; R.J.Puddephatt, *Comments Inorg. Chem.*, 1982, **2**, 69) and has formed the subject of several books (K.J.Ivin, *Olefin Metathesis*, Academic Press, New York, 1983; V. Dragutan, A.T.Balaban, and M.Dimonie, *Olefine Metathesis and Ring Opening Polymerisation of Cycloolefins*, Wiley, New York, 1986). Catalysts have been found which are valuable for the synthesis of long-chain unsaturated esters (J.Tsuji and S.Hashiguchi, *J. Organomet. Chem.*, 1981, **218**, 69; D.Villemin, *Tetrahedron Letters.*, 1983, **24**, 2855). Preformed electrophilic carbene complexes more often yield cyclopropanes with alkenes. Especially good for this process are electrophilic, cationic carbene complexes of iron, which can be generated *in situ* from more stable precursors by heating, for example (S.Brandt and P.Helquist, *J. Amer. Chem. Soc.*, 1979 **101**, 6473; E.J.O'Connor and P.Helquist, *ibid.*, 1982, **104**, 1869)(see Eqn. 31). The introduction of a chiral phosphine ligand promotes significant chiral induction (M.Brookhart *et al., J. Amer. Chem. Soc.*, 1983, **105**, 6721)(see Eqn. 32).

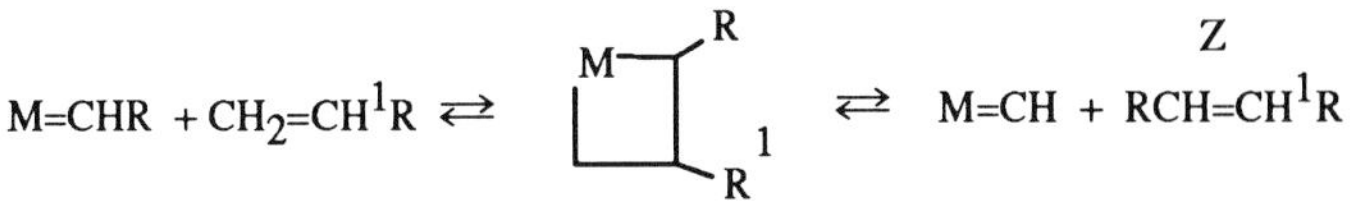

$$M{=}CHR + CH_2{=}CH^1R \rightleftarrows \text{[M-CHR-CHR}^1\text{-CH}_2\text{ metallacyclobutane]} \rightleftarrows M{=}CH + RCH{=}CH^1R \quad \text{(Z)} \qquad (30)$$

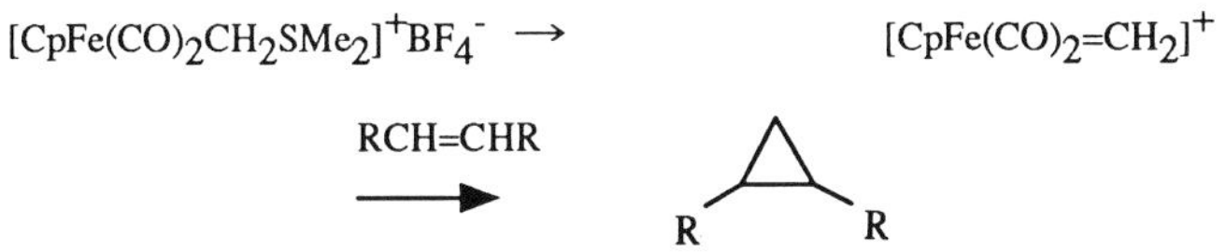

$$[CpFe(CO)_2CH_2SMe_2]^+BF_4^- \longrightarrow [CpFe(CO)_2{=}CH_2]^+ \xrightarrow{RCH=CHR} \text{1,2-R}_2\text{-cyclopropane} \qquad (31)$$

$$[CpFe(CO)(PPh_2R^*){=}CHMe]^+ \xrightarrow{PhCH=CH_2} \text{trans-1R,2R (3.5)} + \text{cis-1R,2S (1)}$$

75%yield, 83%ee

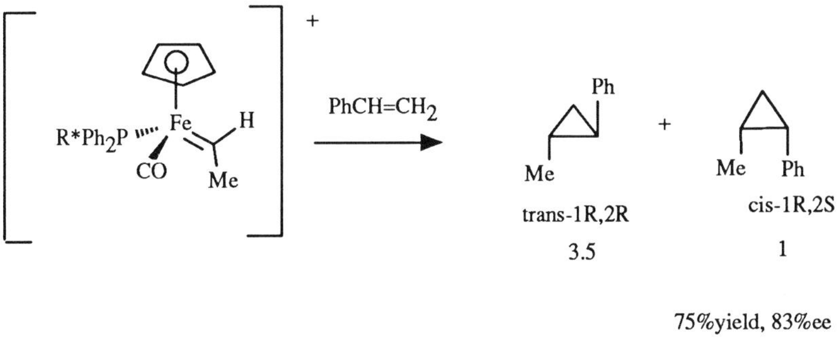

(32)

<u>Transition metal catalysed decomposition of diazo compounds</u>. Cyclopropanation, the most useful outcome of this procedure, has been reviewed (M.P.Doyle, in *Catalysis of Organic Reactions*, R.L. Augustine, Ed., Marcel Dekker, New York, 1985). These reactions probably involve metal-carbene intermediates.

<u>Reaction with imines</u>. *N*-Unsubstituted imines displace methoxy groups forming a new carbene complexes (E.O.Fischer *et al., J. Organomet. Chem.*, 1976, **113**, C31) (see Eqn. 33).

$$(CO)_5Cr{=}C(OMe)Me + Me_2C{=}NH \xrightarrow{-MeOH} (CO)_5Cr{=}C(N{=}CMe_2)Me \qquad (33)$$

However, *N*-substituted imines, for which this reaction is disfavoured, yield β-lactams stereospecifically. Thus the optically active thiazoline ester (see Eqn. 34) yields a single enantiomer of the corresponding penam derivative (L.S.Hegedus *et al., Tetrahedron,* 1985, **41**, 5833).

(34)

b) *Nucleophilic carbene complexes.* Tantalum and zirconium alkylidene complexes behave like organic ylides, such as Wittig reagents. Thus they convert a wide range of carbonyl compounds to alkenes, but unlike Wittig reagents, they react with esters and amides, to yield enol esters and enamines, respectively (R.R.Schrock, *Acc. Chem. Res.*, 1979, **12**, 98; F.W.Hartner, J.J.Schwartz, and S.M.Clift *J. Amer. Chem. Soc.*, 1983, **105**, 640). The most generally useful carbonyl-olefinating complex is Tebbe's reagent (F.N.Tebbe, G.W.Parshall, and G.S.Reddy, *J. Amer. Chem. Soc.*, 1978, **100**, 3611), which in the presence of pyridine is the equivalent of $Cp_2Ti = CH_2$ (see Eqn. 35). This complex efficiently methylenates aldehydes, ketones, lactones, esters, amides, and even carbonates (K.A.Brown-Wensley *et al., Pure Appl. Chem.*, 1983, **55**, 1733; L.F.Cannizzo and R.H.Grubbs, *J. Org. Chem.*, 1985, **50**, 2316).

$$Cp_2TiCl_2 + 2AlMe_3 \longrightarrow Cp_2-Ti(\mu\text{-}CH_2)(\mu\text{-}Cl)Al-Me_2 \xrightarrow{\text{pyridine}} py[Cp_2Ti{=}CH_2] \qquad (35)$$

3. *Compounds with metal bonded to two carbon atoms*

This section will be mainly concerned with π-bonded alkene and alkyne transition metal complexes, in which the normal reactivity of the double or triple bond is reversed. Attack by a nucleophile creates a new bond at an olefinic (or acetylenic) carbon atom, and generates a new metal-carbon bond capable of further reaction. It is not surprising therefore that these processes have been amongst the most valuable for the elaboration of complex organic molecules.

i) (η^2-Alkene) palladium (II) complexes

Numerous syntheses involving palladium(II) complexes have been devised and frequently reviewed (e.g. L.S.Hegedus, *Tetrahedron*, 1984, **40,** 2415). The processes can be made catalytic and a general scheme is shown below (see Eqn. 36).

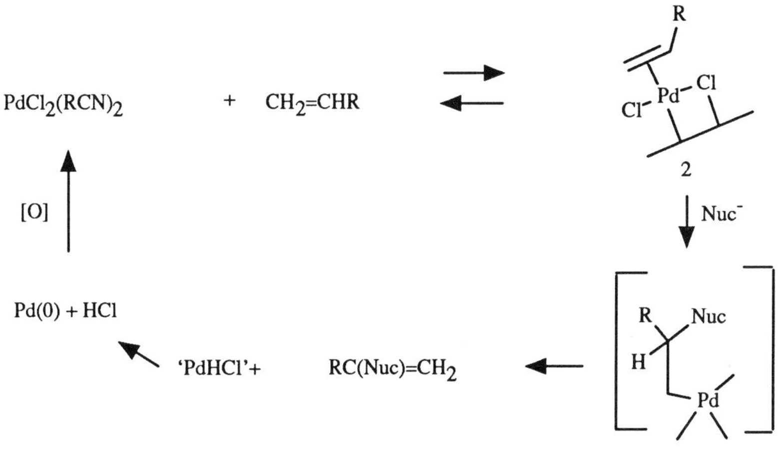

$PdCl_2(RCN)_2$ + CH_2=CHR ⇌ 2

2 → (Nuc^-) [R, Nuc, H, Pd] → RC(Nuc)=CH_2 + 'PdHCl' → Pd(0) + HCl → ([O]) $PdCl_2(RCN)_2$

(36)

Terminal alkenes, followed by *Z*-, and then *E*- disubstituted alkenes co-ordinate most strongly, in fact, *gem*-disubstituted and more highly substituted alkenes fail to react. When the nucleophile is water, ketones are formed and the process readily distinguishes between terminal and internal alkenes (see Eqn. 37).

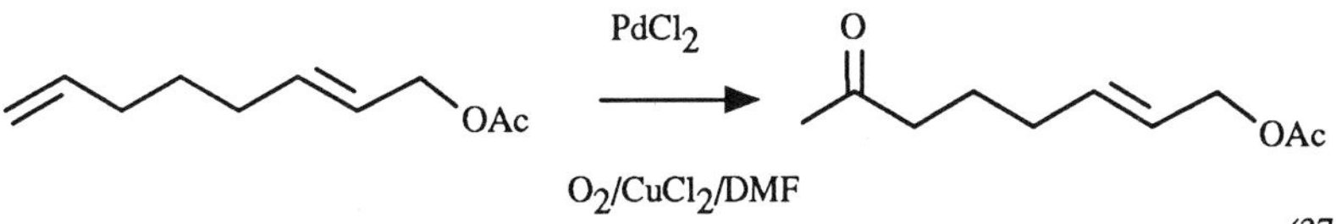

OAc —($PdCl_2$, O_2/$CuCl_2$/DMF)→ O, OAc

(37)

Internal hydroxyl nucleophiles allow intramolecular cyclisation

(T.Hosokawa, H.Ohkata, and I.Moritani, *Bull. Chem. Soc. Jpn.*, 1975, **48**, 1533)(see Eqn. 38).

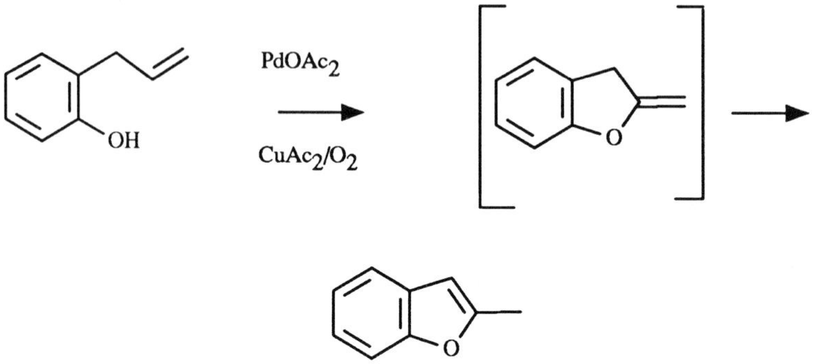

$PdOAc_2$, $CuAc_2/O_2$

(38)

Amines, as well as acting as nucleophiles, are potent ligands for palladium and form complexes. Further reactions may then be carried out effecting overall *syn*-additon (J.E.Bäckvall and E.E.Bjorkman, *J. Org. Chem.*, 1980, **45**, 2893)(see Eqn. 39).

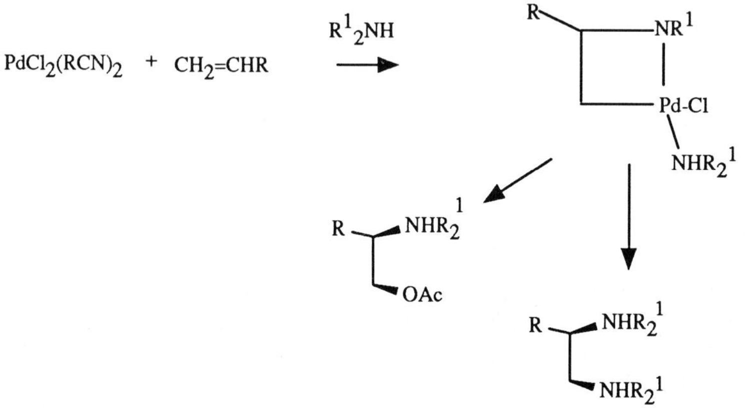

$PdCl_2(RCN)_2$ + CH_2=CHR $\xrightarrow{R^1{}_2NH}$

R, NR^1, Pd-Cl, $NHR_2{}^1$

R, $NHR_2{}^1$, OAc

R, $NHR_2{}^1$, $NHR_2{}^1$

(39)

Indeed, using a chiral tertiary amine as ligand and a chiral secondary amine as nucleophile, alkenes have been oxaminated with *ca* 60% ee (J.E.Bäckvall *et al.*, *Tetrahedron Letters.*, 1982, **23**, 943). Intramolecular aminations proceed without difficulty, leading to cyclisation *via* the most substituted terminus with alkylated side chains (see Eqn. 40). In the presence of a oxidising agent palladium (II) functions as a catalyst (L.E.Hegedus *et al.*, *J. Amer. Chem. Soc.*, 1978, **100**, 5800). The process can be extended to the more basic aliphatic amines, if they are derivatised as the less basic tosamides to avoid strong complexation of the palladium (II) reactant (L.S.Hegedus and J.M.Kearin, *J. Amer. Chem. Soc.*, 1982, **104**, 2444)(see Eqn. 41).

NH_2 — $PdCl_2(MeCN)_2$ (cat.), Benzoquinone → N–H (40)

NHTs — $PdCl_2(MeCN)_2$ (cat.), Benzoquinone → N–Ts (41)

Alkylation of alkenes can be achieved by choice of suitable ligands on palladium, which overcome the tendency of Pd(II) compounds to oxidise the alkylating carbanions (L.S.Hegedus *et al.*, *J. Amer. Chem. Soc.*, 1980, **102**, 4973). [3,3]-Sigmatropic rearrangements are catalysed by palladium(II) salts under mild conditions and with good stereospecificity. This chemistry has been reviewed (L.E.Overman, *Angew. Chem. Int. Ed. Engl.*, 1984, **23**, 579).

ii) (η-Alkene) iron (II) complexes

Unlike palladium(II) systems, the resultant σ-alkyliron complexes tend to be stable, requiring a further chemical reaction to remove the iron. Of course, it may be possible to make further use of the intermediate, before removing the iron, as in the following β-lactam synthesis which utilises the readily available η^5-cyclopentadienyl(isobutylene)iron dicarbonyl (S.R.Berryhill, T.Price, and M.Rosenblum, *J. Org. Chem.*, 1983, **48**, 158)(see Eqn. 42).

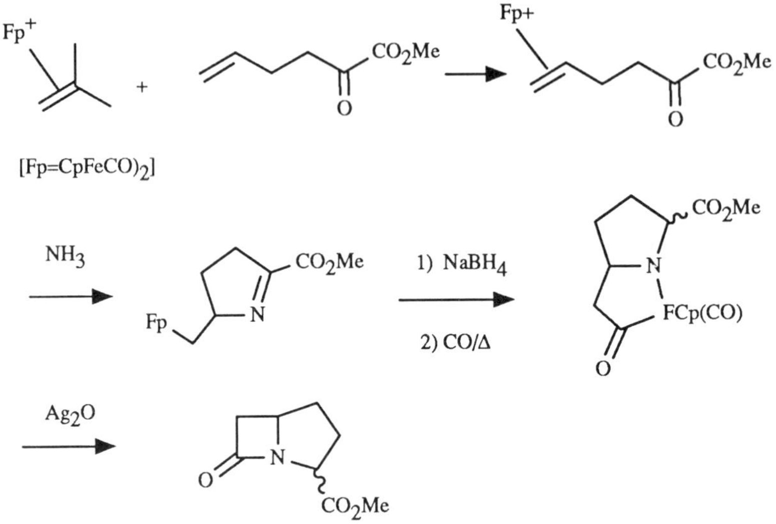

Fp$^+$ + → Fp+ CO_2Me

$[Fp=CpFe(CO)_2]$

NH_3 → Fp CO_2Me 1) $NaBH_4$ 2) CO/Δ → CO_2Me N FCp(CO)

Ag_2O → N CO_2Me (42)

η[1]-Allyl ligands attached to iron(II) react with alkenes, to form η[2]-alkene complexes, which are susceptible to nucleophilic attack (M.Rosenblum, *Acc., Chem. Res.*, 1974, **7**, 12). This chemistry has proved valuable in organic synthesis (A.Buchester, P.Klemarczyk, and M.Rosenblum, *Organometallics*, 1982, **1**, 1679)(see Eqn. 43).

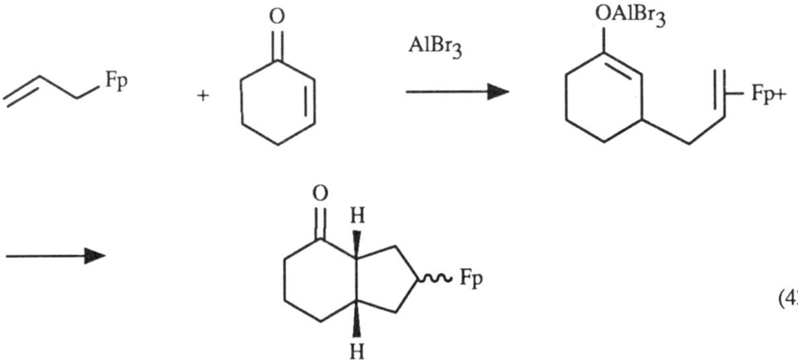

Fp + $AlBr_3$ → $OAlBr_3$ Fp+

→ H Fp H (43)

The iron complexes have also proved useful for the protection of double bonds towards electrophiles (P.F.Boyle and K.M.Nicholas, *J. Org. Chem.*, 1975, **40**, 2682). In dienes the iron will selectively coordinate with the less-substituted double bond (see Eqn. 44) and can be readily removed with sodium iodide.

Fp$^+$ + → Fp$^+$

1) H_2/Pd → 2) NaI (44)

iii) Transition-metal alkyne complexes

The chemistry of η^2-alkyne complexes is not so simple as for the corresponding alkene complexes. The alkyne-metal complexes are often reactive with a further alkyne molecule; while those that are stable lack sufficient reactivity to be used in synthesis. Despite this, systems have been devised that overcome these problems (K.Utimoto, *Pure. Appl. Chem.*, 1983, **55**, 1845)(see Eqn. 45).

R—≡—NH_2 $\xrightarrow[\text{reflux}]{PdCl_2/MeCN}$ R—N (45)

a) *$Co_2(CO)_8$ Complexes.* Dicobalt octacarbonyl forms very stable complexes with alkynes (R.S.Dickson and P.J.Fraser, *Adv. Organomet. Chem.*, 1974, **12**, 323), which have been used as protecting groups for alkynes (K.M.Nicholas and R.Pettit, *Tetrahedron Letters*, 1971, 3475). The protecting group is removed oxidatively with Fe^{3+} (see Eqn. 46).The cobalt-alkyne complexes are generally quite stable towards electrophiles, but undergo a useful reaction with alkenes at high temperatures (D.C.Billington and P.L.Pauson, *Organometallics*, 1982, **1**, 1560)(see Eqn. 47).

$Co_2(CO)_8$

Co– Co

$(OC)_3$ $(CO)_3$

1) BH_3

2) O_2

3i) Fe^{2+}

HO

(46)

Co– Co

$(OC)_3$ $(CO)_3$

+

OTHP

Δ

O

OTHP

(47)

b) *Metal-catalysed cyclooligomerisation.* The reactions of alkynes with carbon monoxide in the presence of transition metals has been reviewed (P.Pino and G.Braca, in *Organic Synthesis via Metal Carbonyls*, Vol. 2, p. 420, I.Wender and P.Pino, Eds., Wiley, New York, 1977). Among the most useful applications of this type of reaction, is the synthesis of 1,4-quinones from maleoyl- and phthaloyl-cobalt complexes (L.S.Liebeskind *et al.*, *Tetrahedron*, 1985, **41**, 5839).

Ni(0) salts form quite stable complexes with alkynes and carbon dioxide (see Eqn. 48). The resultant metallocycle will react with a variety of reagents to yield useful products (H.Hoberg, D.Schaefer, and G.Burkhardt, *J. Organomet. Chem.*, 1982, **228**, C21; H.Hoberg *et al.*, *ibid.*, 1984, **266**, 203; H.Hoberg and B.Aapotecher, *ibid.*, 1984, **270**, C15). Isocyanates have been shown to behave in a similar manner (H.Hoberg and B.W.Oster, *Synthesis*, 1982, 324; *J. Organomet. Chem.*, 1982, **234**, C35) (see Eqn. 49), and indeed when the reaction is mediated by an appropriate cobalt complex pyridones are formed (R.A.Earl and K.P.C.Vollhardt, *J. Amer. Chem. Soc.*, 1983, **105**, 6991; *J. Org. Chem.*, 1984, **49**, 4786) (see Eqn. 50). It is of interest that in the presence of Ni(0) the reaction leads to a different regioisomer.

Vollhardt has developed a use of this reagent for the cyclotrimerisation of alkynes, leading to a useful synthesis of polycyclic systems (K.P.C.Vollhardt, *Acc. Chem. Res.*, 1977, 10, 1; *Angew. Chem. Int. Ed. Engl.*, 1984, **23**, 539).

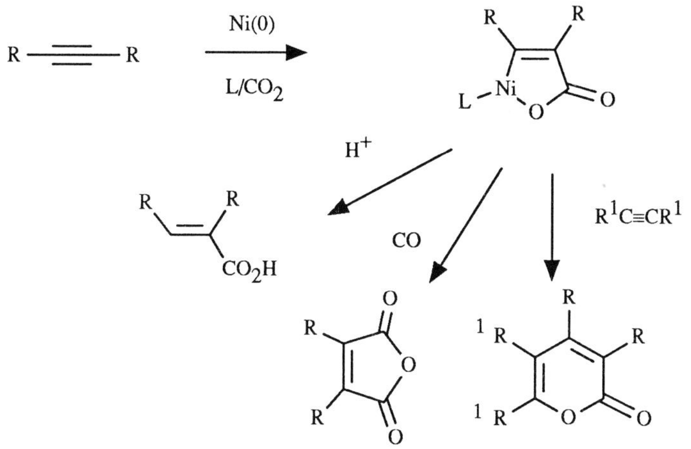

(48)

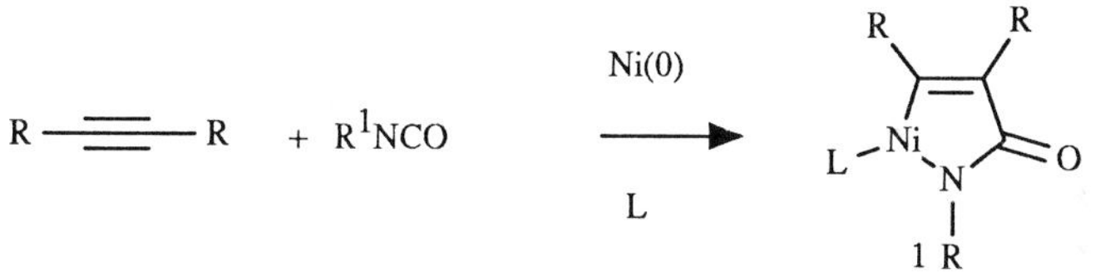

(49)

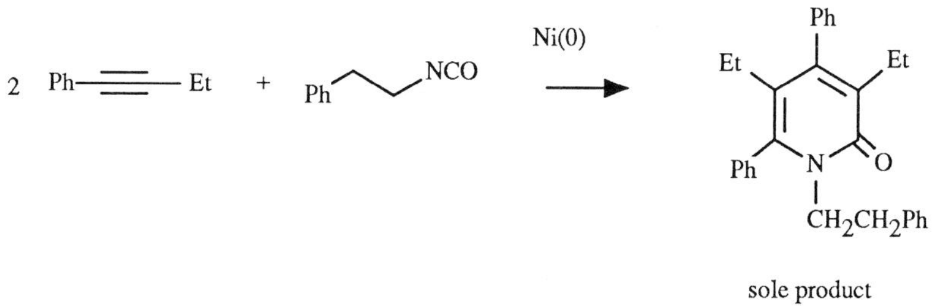

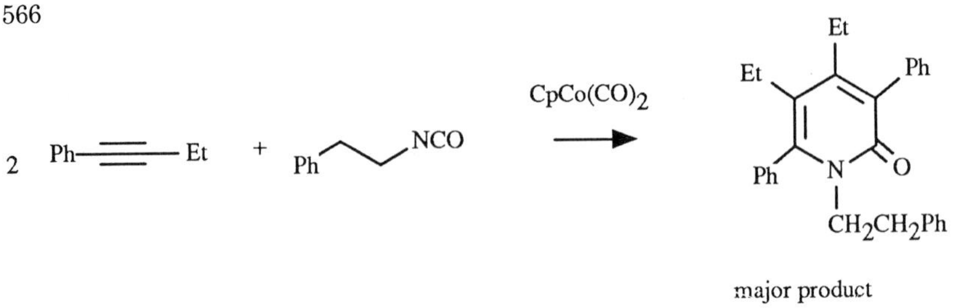

2 Ph—≡—Et + Ph⌒⌒NCO —CpCo(CO)$_2$→ (pyridone: Et, Et, Ph, Ph, N–CH$_2$CH$_2$Ph, =O)

major product

(50)

4. *Compounds with metal bonded to three carbon atoms*

This section will be concerned with η^3-allyl complexes, especially their application to organic synthesis, where palladium intermediates have received most attention.

i) (η^3-Allyl)palladium complexes

Dimerisation of conjugated dienes with the incorporation of a nucleophile can be achieved in the presence of palladium acetate/triphenyl phosphine catalysts (see Eqn. 51).

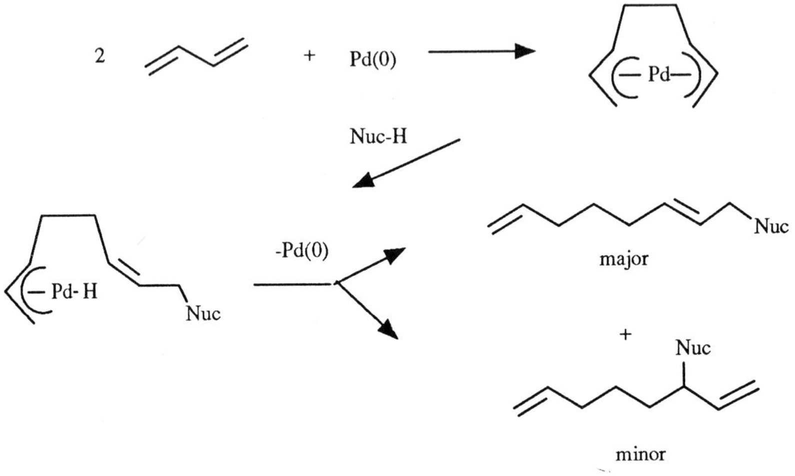

2 (butadiene) + Pd(0) ——→ (bis-allyl Pd complex)

Nuc-H

(Pd-H allyl complex with Nuc) —-Pd(0)→ (octadienyl–Nuc) major + (branched Nuc octadiene) minor

(51)

A wide range of nucleophiles (e.g. H_2O, ROH, RNH_2, RCO_2^-, etc.) can be used, all react smoothly (J.Tsuji, *Organic Synthesis with Palladium Complexes*, Springr-Verlag, Berlin, 1980 and *Pure Appl. Chem.*, 1982, **54**, 197; L.S.Hegedus, in *Comprehensive Carbanion Chemistry*, E.Buncel and T.Durst, Eds., Elsevier, Amsterdam, 1984). Carbonylation is also a useful option (N.T. Byron *et al. J. Chem. Soc., Perkin Trans 1*, 1984, 1643)(see Eqn. 52).

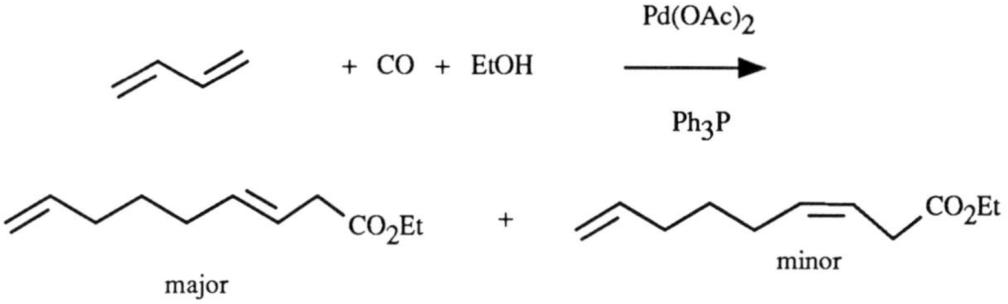

$Pd(OAc)_2$

+ CO + EtOH

Ph_3P

CO_2Et + CO_2Et

major minor

(52)

Alkenes react with palladium(II) salts to yield η^3-allylpalladium halides, which are air-stable, crystalline solids. Although themselves chemically unreactive, addition of strongly coordinating ligands (e.g. R_3P or DMSO) encourages attack of nucleophiles at the η^3-allyl ligand (J.Tsuji, *Bull. Chem. Soc. Jpn.*, 1973, **46**, 1896; B.M.Trost *et al., J. Amer. Chem. Soc.*, 1978, **100**, 3416). Since these complexes form more readily at non-conjugated double bonds, a useful synthetic selectivity is available (B.M. Trost *et al., ibid.*, 1978, **100**, 3426)(see Eqn. 53). Transmetalation with the more electropositive zirconium reagents, in the presence of good π-acceptor ligands, leads to coupling (see Eqn. 54)(J.E.McMurry and J.R.Matz, *Tetrahedron Letters.*, 1982, **23**, 2723; J.E.McMurry *et al., ibid.*, 1982, **23**, 1777). The requirement for stoichiometric amounts of palladium salts in η^3-allylpalladium halide reactions limits their use. However, the development of a catalytic process has lead to its wide spread application in organic synthesis (see Eqn. 55), (B.M.Trost, *Acc. Chem. Res.*, 1980, **13**, 385; *Pure Appl. Chem.*, 1981, 53, 2357; T.Yamamoto, O.Saito, and A.Yamamoto, *J. Amer. Chem. Soc.*, 1981, **103**, 5600; J.Tsuji, *Pure Appl. Chem.*, 1982, **54**, 197).

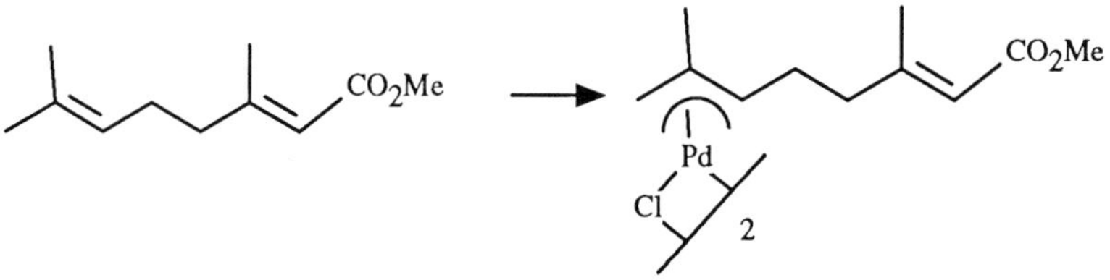

CO_2Me → CO_2Me

Pd

Cl

2

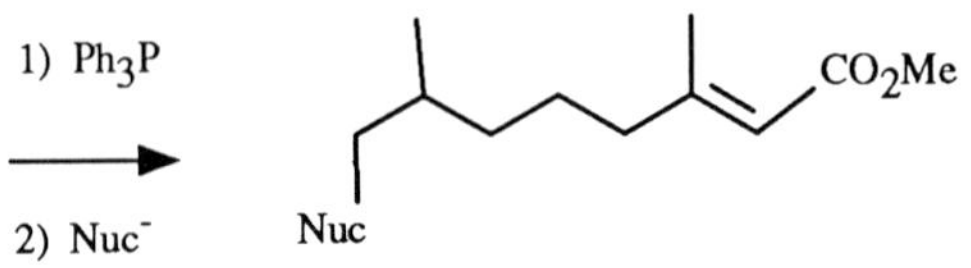

1) Ph_3P

2) Nuc^-

Nuc

CO_2Me

(53)

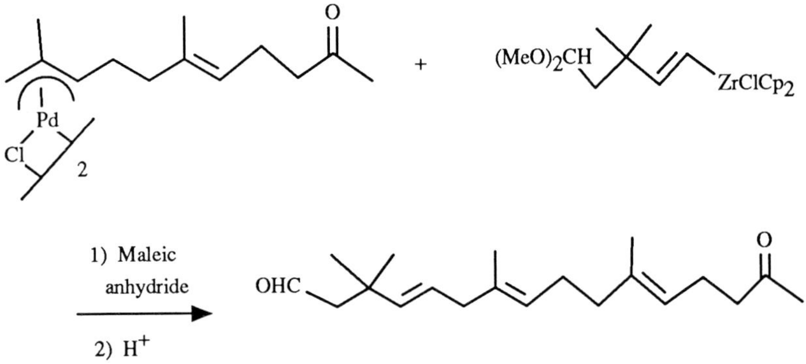

O

+

$(MeO)_2CH$

$ZrClCp_2$

Pd

Cl

2

1) Maleic anhydride

2) H^+

OHC

O

(54)

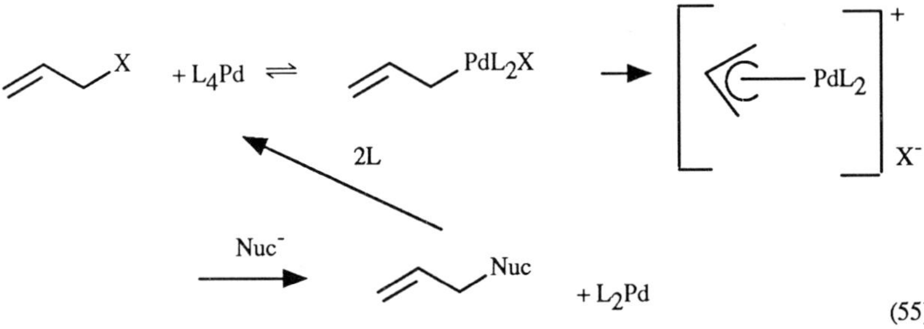

X

$+ L_4Pd \rightleftharpoons$

PdL_2X

PdL_2

+

X^-

2L

Nuc^-

Nuc

$+ L_2Pd$

(55)

A cyclopropane formation is given as a specific example of the procedure, in which selectivity between acetoxy and hydroxy groups as the site of reaction is indicated (J.P.Genet and F.Prau, *J. Org. Chem.*, 1981, **47**, 381)(see Eqn. 56).

HO OAc + ⁻ X Y $\xrightarrow{L_4Pd}$ HO X Y

$\xrightarrow[\text{2) NaOH/L}_4\text{Pd}]{\text{1) Ac}_2\text{O}}$ X Y (56)

A large variety of latent nucleophiles can be used in this procedure, including bifunctional allyl acetates useful for cyclopentene ring formations (B.M.Trost, *Chem. Soc. Rev.*, 1982, **11**, 141; *Angew. Chem. Int. Ed. Engl.*, 1986, **25**, 1)(see Eqn. 57) .

Me_3Si OAc $\xrightarrow{L_4Pd}$ Me_3Si $\overset{+}{Pd}L_2$ ⟶

$\overset{+}{Pd}L_2$ ⁻ $\xrightarrow[]{\text{Z}\ MeO_2CH{=}CHCO_2Me}$ MeO_2C CO_2Me (57)

The catalytic opening of allylic epoxides occurs regiospecifically at the alkene carbon atom *via* a stereospecific *syn* S_N2^1 process (B.M.Trost and G.A.Molander, *J. Amer. Chem. Soc.*, 1981, **103**, 5969)(see Eqn. 58).

O $\xrightarrow[CH_2XY]{L_4Pd}$ HO X Y (58)

Indeed, Trost has demonstrated (B.M.Trost and T.P.Klun, *J. Amer. Chem. Soc.*, 1981, **103**, 1864) the highly specific regio- and stereo-specific nature of these palladium catalysed additions of nucleophiles as the key step in a synthesis of the vitamin E side-chain (see Eqn. 59).

L_4Pd, $CH^-(CO_2Me)_2$; product: MeO_2C, CO_2Me, H, OR, H, CO_2H

95% ee (59)

ii) η^3-Allyl complexes of other metals

a) *Cobalt.* The cobalt carbonyl anion $[CO(CO)_4]^-$ reacts with organic halides, followed by carbonyl insertion to yield acylcobalt carbonyls $[RCOCo(CO)_4]$ (R.F.Heck, in *Organic Synthesis via Metal Carbonyls*, Vol. 1, I.Wender and P.Pino, Eds., Wiley , New York, 1968). These acylcobalt complexes react with conjugated dienes to yield η^3-allyl complexes, which can be alkylated by stabilised carbanions (L.S.Hegedus and Y.Inoue, *J. Amer. Chem. Soc.*, 1982, **104**, 4917; L.S.Hegedus and R.J.Perry, *J. Org. Chem.*, 1984, **49**, 2570)(see Eqn. 60).

$RCOCo(CO)_4$ +

$Co(CO)_4$; CH^-XY; R, X, Y

(60)

b) *Molybdenum and tungsten.* The reactions of allyl acetates are catalysed by molybdenum and tungsten complexes as for palladium(0) complexes, but with different regioselectivities (B.M.Trost and M.Lautens, *J. Amer. Chem. Soc.*, 1982, **104**, 5543, *ibid.*, 1983, **105**, 3343; *Organometallics*, 1983, **2**, 1687; B.M.Trost and M.H.Hung, *J. Amer. Chem. Soc.*, 1983, **105**, 7757). This is illustrated by the reactions of allyl acetates with stabilised carbanions. Molybdenum catalysis lies between that of palladium and tungsten in its regioselectivity, and it is sensitive to the steric requirements of the nucleophile (see Eqn. 61).

$$\text{R-CH=CH-CH}_2\text{OAc} \xrightarrow[\text{R}^1\text{CH}^-\text{XY}]{\text{Mo(CO)}_6} \text{R-CH=CH-CH}_2\text{-C(R}^1\text{)(X)(Y)}$$

$$\text{R-CH=CH-CH}_2\text{OAc} \xrightarrow[\text{CH}^-\text{XY}]{\text{L}_4\text{Pd}} \text{R-CH=CH-CH}_2\text{-CH(X)(Y)}$$

$$\text{R-CH=CH-CH}_2\text{OAc} \xrightarrow[\text{CH}^-\text{XY}]{\text{W(CO)}_6 \text{ or Mo(CO)}_6} \text{R-CH(CH=CH}_2\text{)-CH(X)(Y)} \qquad (61)$$

c) *Nickel.* η^3-Allynickel halide complexes are structurally similar to the corresponding palladium compounds, however, they behave differently. They are not subject to nucleophilic attack, but react as if they were nucleophiles *via* radical pathways. These reactions have proved synthetically valuable, for example (see Eqns. 62 and 63), in the alkylation of aromatic halides (K.Saito, S.Inoue, and M.Sato, *J. Chem. Soc., Chem. Commun.*, 1972, 953; *J. Chem. Soc., Perkin Trans 1*, 1973, 2289) and in addition reactions with aldehydes and ketones (L.S.Hegedus *et al.*, *J. Org. Chem.*, 1975, **40**, 593).

$$\text{1,4-di(OAc)-2-methyl-3-bromonaphthalene} + [(\eta^3\text{-C}_5\text{H}_9)\text{NiBr}]_2 \xrightarrow[20^\circ\text{C}]{\text{HMPA}} \text{1,4-di(OAc)-2-methyl-3-(3-methylbut-2-enyl)naphthalene} \qquad (62)$$

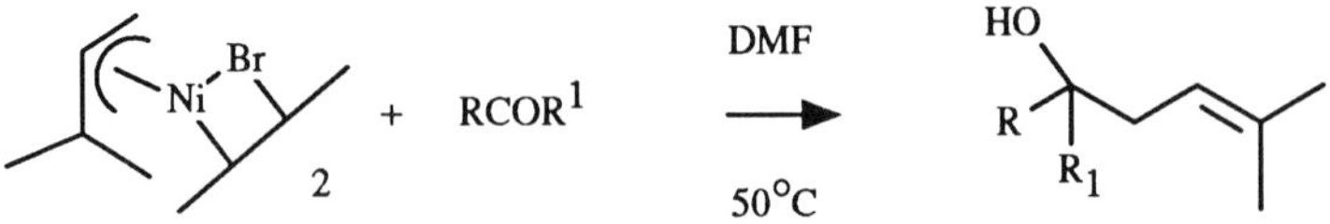

$$\left[\text{(2-methylallyl)NiBr}\right]_2 + RCOR^1 \xrightarrow[50^\circ C]{DMF} \text{HO, R, R}_1\text{-substituted homoallylic alcohol} \qquad (63)$$

5. *Compounds with metal bonded to four or more carbon atoms*

Arene complexes (see M.F.Semmelhack, *Pure Appl. Chem.*, 1981, **53**, 2379; D.Astruc, *Tetrahedron*, 1983, **39**, 4027) and cyclopentadienyl complexes (see W.E. Watts, in *Comprehensive Organometallic Chemistry*, Vol. 8, Ch. 59, G.Wilkinson, F.G.A.Stone, and E.W.Abel, eds., Pergamon, Oxford, 1982) fall outside the scope of this review but will be covered in a subsequent volume. Thus this section will be concerned mainly with acyclic diene and dienyl complexes and their applications to synthesis.

i) $(\eta^4$-Diene)metal complexes

a) *Palladium*. One of the main synthetic applications of palladium diene complexes has been in the catalysis of [3,3]-sigmatropic rearrangements (L.E.Overman, *Angew. Chem. Int. Ed. Engl.*, 1984, **23**, 579). These processes usually occur under very mild conditions and with good sterespecificity (see Eqn. 64).

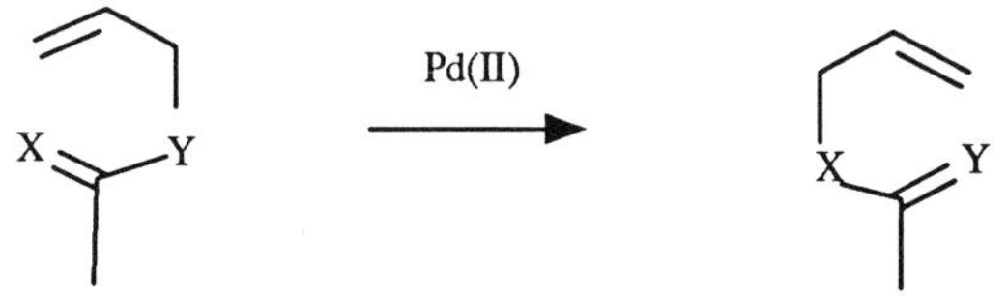

$$\xrightarrow{Pd(II)} \qquad (64)$$

X=CH_2, O, NR
Y=CH_2, O, S

The rearrangement of acetates (see Eqn. 65) has been shown to transfer chirality without racemisation (P.E.Grieco *et al., J.Amer.Chem.Soc.*, 1980, **102** 7587; P.E.Grieco *et al., J. Org. Chem.*, 1981, **46**, 5005). Cope rearrangements are also known (see Eqn. 66) (L.E.Overman, *loc. cit.*). Chelating alkene complexes also undergo nucleophilic substitution (see Eqn. 67), when the stabilising influence of the second double bond results in a fairly stable σ-alkyl complex compared with simple mono-alkenes. Use may be made of these complexes for further reactions, e.g. carbonylation (R.C.Larock and D.R.Leach, *J. Org. Chem.*, 1984, **49**, 2144).

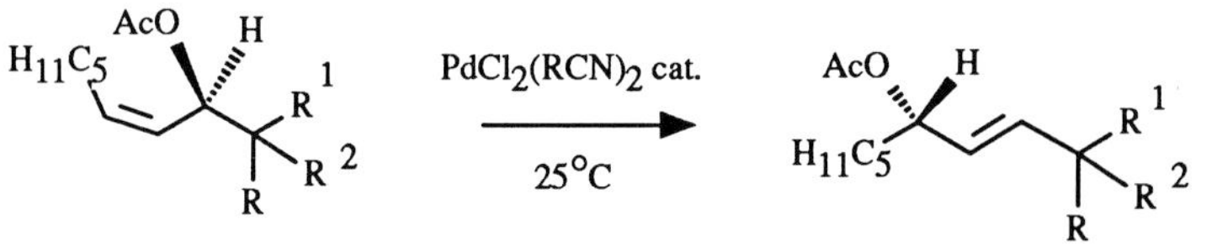

(65)

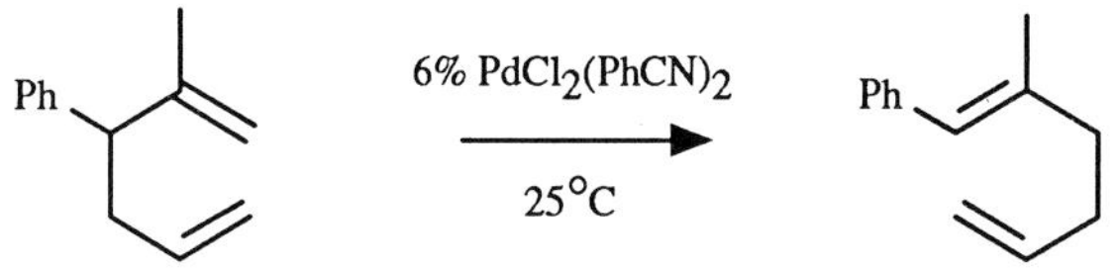

(66)

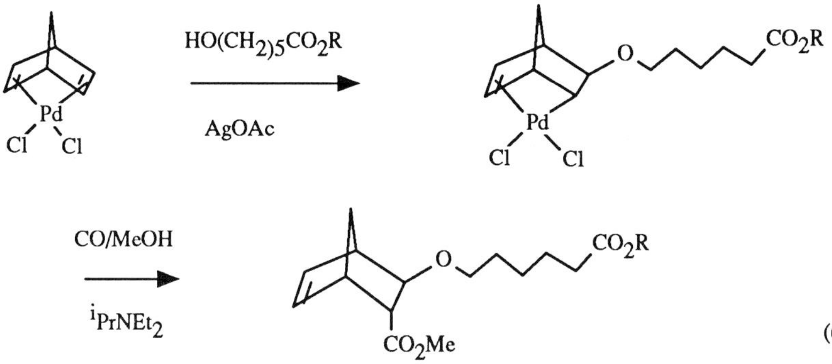

(67)

b) *Iron.* Chelating alkenes bind most strongly with palladium or rhodium, while conjugated dienes form stable complexes with iron(0). Commonly 1,3-dienes are reacted with iron carbonyls, such as $Fe(CO)_5$ or better $Fe_2(CO)_9$ (R.Pettit and G.F.Emerson, *Adv. Organomet. Chem.*, 1964, **1**, 1). This type of complexation has been used to protect the double bonds of a 2,4-dienal, during modifications to the aldehyde function (M.Frank-Neumann, D.Martina, and M.P.Heitz, *Tetrahedron Letters*, 1982, **23**, 3493) (see Eqn. 68).

Protection can be afforded to a conjugated diene grouping, whilst allowing normal conversions of an isolated double bond in the molecule. Thus complexation of the terpene myrcene can be followed by typical reactions at the free double bond (D.V.Banthorpe, H.Fitton, and J.Lewis, *J. Chem. Soc., Perkin Trans. 1*, 1973, 2051). Furthermore, electrophilic intramolecular reactions with the diene-$Fe(CO)_3$ group are still possible

(A.J.Birch and A.J.Pearson, *J. Chem. Soc., Chem. Commun.*, 1976, 601; A.J.Pearson, *Aust. J. Chem.* 1976, **29**, 1841) (see Eqn. 69).

CHO $\xrightarrow{Fe_2(CO)_9}$ CHO, $Fe(CO)_3$

$\xrightarrow[NaH]{(RO)_2P(O)CH_2CO_2Me}$ $Fe(CO)_3$, CO_2Me $\xrightarrow{Me_2CN_2}$

N=N, $Fe(CO)_3$, CO_2Me $\xrightarrow{\Delta\ 120^{\circ}C}$ $Fe(CO)_3$, CO_2Me

$\xrightarrow{FeCl_3}$ CO_2Me (68)

(η^4-1,3-Diene)iron tricarbonyls are generally unreactive to Diels-Alder reactions, catalytic hydrogenation, hydroxylation, etc., but react with electrophiles. Thus butadiene-$Fe(CO)_3$ can be acetylated to yield mixed stereoisomers, in proportions dependent upon the work-up procedure (see Eqn. 70). The regiochemistry of the acetylation of substituted acyclic dienes has been studied (R.E.Graf and C.P.Lillya, *J. Organomet. Chem.*, 1976, **122**, 377). It is of interest that (2-methoxybutadiene)iron tricarbonyl gives only one regioisomer as the product of the reaction (see Eqn. 71). This would not have been expected from a consideration of the electronic effects in the uncomplexed diene.

Nucleophilic reactions occur with active carbanions such as $LiCMe_2CN$ or $LiCHPh_2$, which add at an internal carbon atom (M.F.Semmelhack and H.T.M.Lee, *J. Amer. Chem. Soc.*, 1984, **106**, 2715; M.F.Semmelhack, J.W.Herndon, and J.P.Springer, *ibid.*, 1983, **105**, 2497). The chemistry of cyclohexadieneiron tricarbonyl complexes has plenty of synthetic potential, when nucleophilic attack is followed by carbonyl insertion and reaction with an electrophile (see Eqn. 72).

The use of acylic diene complexes complicates the picture, 1,3-butadiene giving a good yield of a cyclopentanone (see Eqn. 73). This results from the remaining double bond acting as an electrophile and entering into an insertion reaction at the carbonyl group. Such a reaction is inhibited when the double bond is substituted, leading instead to the corresponding aldehyde through protonation of the intermediate complex (M.F.Semmelhack, J.W.Herndon, and J.K.Liu, *Organometallics*, 1983, **2**, 1885; M.F.Semmelhack and M.T.H.Lee, *J. Amer. Chem. Soc.*, 1985, **107**, 1455).

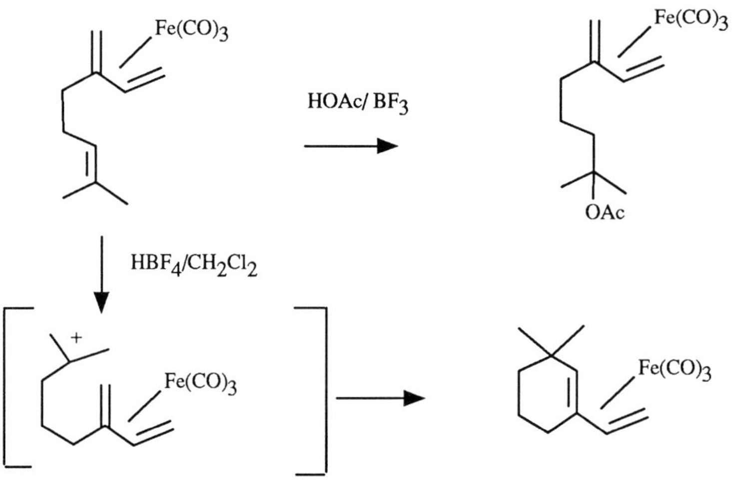

$Fe(CO)_3$ HOAc/ BF_3 $Fe(CO)_3$ OAc

HBF_4/CH_2Cl_2

+ $Fe(CO)_3$ $Fe(CO)_3$

(69)

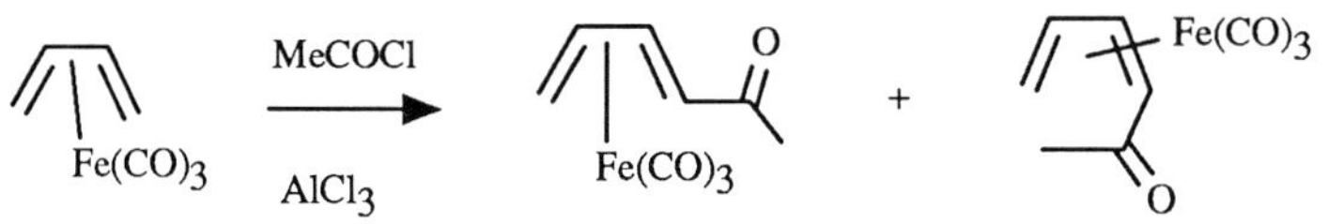

$Fe(CO)_3$ MeCOCl $AlCl_3$ O $Fe(CO)_3$ + $Fe(CO)_3$ O (70)

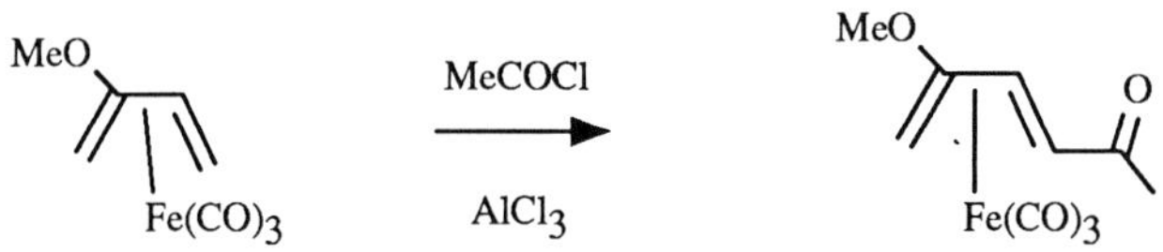

MeO $Fe(CO)_3$ MeCOCl $AlCl_3$ MeO O $Fe(CO)_3$ (71)

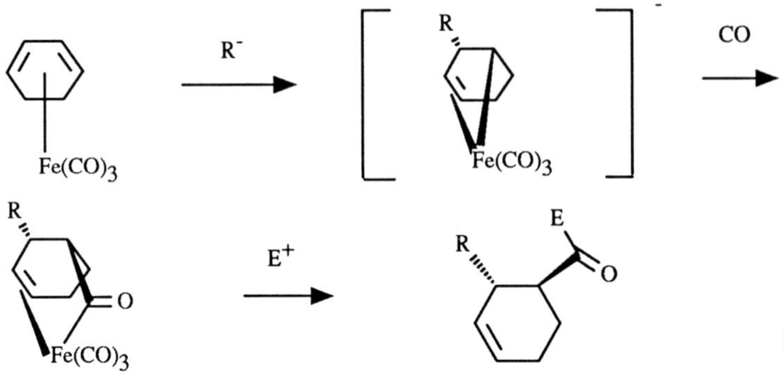

R⁻ CO E⁺ (72)

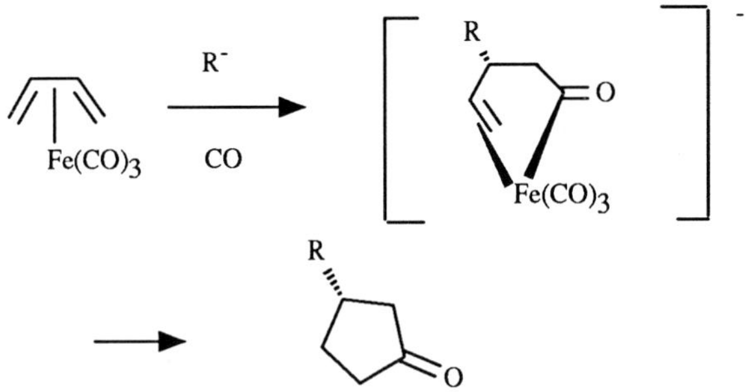

R⁻ CO (73)

ii) η^5-Dienylmetal complexes

a) *Iron.* (η^5-Dienyl)iron tricarbonyl complexes can be prepared by hydride ion abstraction from pentadienyl-$Fe(CO)_3$ compounds, but only when they have a *cisoid* structure (see Eqn. 74).This *cisoid* arrangement, naturally arising from cyclohexadienes, is not generally produced by reaction of 1,3-pentadienes with iron carbonyls, which prefer a *transoid* structure. The 2,4-pentadienol complex, although *transoid*, will yield cationic species on treatment with anhydrous HBF_4 (see Eqn. 75).

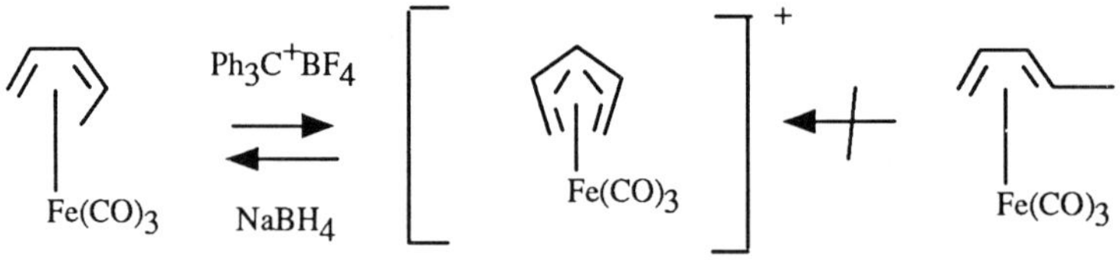

$Ph_3C^+BF_4$ / $NaBH_4$ (74)

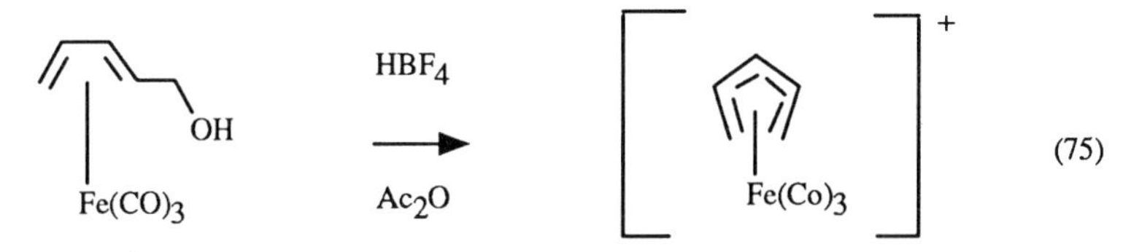

A good guide to the preparation and properties of these cationic iron complexes has been given by A.J.Pearson, in *Comprehensive Organometallic Chemistry*, Vol.8, Ch. 58, G.Wilkinson, F.G.A.Stone, and E.W.Abel, Pergamon, Oxford, 1982. Naturally these cationic diene complexes are much more reactive towards nucleophiles (see Eqn. 76) than the neutral 1,3-diene complexes.

$[Fe(Co)_3]^+$ —Nuc^-→ Nuc, $Fe(CO)_3$ (76)

Nuc =RNH,H,RO,$(RO_2C)_2CH$, etc.

The development of mild methods for the removal of iron from the resultant η^4-diene complexes has lead to them being much exploited for organic synthesis (A.J.Pearson, *Acc. Chem. Res.*, 1980, **13**, 463; *Pure Appl. Chem.*, 1983, **55**, 1767), especially those derived from cyclohexadienes by Birch reduction of aromatic compounds.

b) *Cobalt and Rhodium.* These metals will also form η^5- dienyl complexes, generally with a cyclopentadienyl ancillary ligand (see Eqn. 77). Nucleophiles generally attack the dienyl rather than the cyclopentadienyl ligand. Furthermore, the nucleophile may attack at the central carbon atom generating a non-conjugated 1,4-diene (P.Powell and L.J.Russell, *J. Chem. Res.*(S), 1978, 283; P.Powell, *J. Organomet. Chem.*, 1977, **165**., C43).

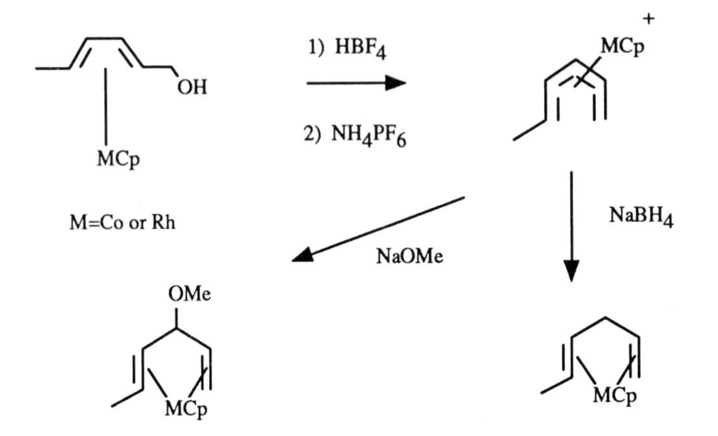

(77)

Guide to the Index

This index is constructed in a similar manner to the volume indexes of the first edition of the Chemistry of Carbon Compounds. However, to make the index easier to use, more descriptive entries have been made for the commonly occurring individual, and groups of chemicals.

The indexes cover primarily the chemical compounds mentioned in the text, and also include reactions and techniques, where named, and some sources of chemical compounds such as plant and animal species, oils, etc.

Chemical compounds have been indexed alphabetically under the names used by authors, editing being restricted to ensuring uniformity of entries under the same heading. In view of the alternative nomenclature that can often be used, a limited amount of cross-referencing has been done where it is considered to be helpful, but attention is particularly drawn to Convention 2 below.

For this and the succeeding volumes, the indexing conventions listed below have been adopted.

1. Alphabetisation

(a) A letter by letter alphabetical sequence is followed for entries, firstly for the main entry, followed by the descriptive entry.

(b) The following prefixes have not been counted for alphabetising:

n-	*o-*	*as-*	*meso-*	*C-*	*E-*
	m-	*sym-*	*cis-*	*O-*	*Z-*
	p-	*gem-*	*trans-*	*N-*	
	vic-			*S-*	
		lin-		*Bz-*	
				Py-	

Some prefixes and numbering have been omitted in the index, where they do not usefully contribute to the reference.

(c) The following prefixes have been alphabetised:

Allo	Epi	Neo
Anti	Hetero	Nor
Bis	Homo	Pseudo
Cyclo	Iso	

2. Cross references

In view of the many alternative trivial and systematic names for chemical compounds, the indexes should be searched under any alternative names which may be indicated in the main body of the text. Only a limited amount of cross-referencing has been carried out, where it is considered that it would be helpful to the user.

3. Derivatives

Simple derivatives are not normally indexed if they follow in the same short section of the text.

4. Collective and plural entries

In place of "– derivatives" the plural entry has normally been used. Plural entries have occasionally been used where compounds of the same name but differing numbering appear in the same section of the text.

5. Main entries

The main entry of the more common individual compounds is indicated by heavy type. Multiple entries, such as headings and sub-headings over several pages are shown by "–", e.g., 67–74, 137–139, etc.

Index

Acetaldehyde, manufacture from ethene, 69
Acetal electrophiles, 459
Acetals, reactions with Grignard reagents, 396
–, reductive cleavage, 315
Acetamidoadamantane, synthesis, 328
Acetamido sulphenylated alkanes, 356
Acetates, rearrangements, 572
Acetoxyallenes, 125
Acetoxy anhydrides, 73
Acetoxy group, migration, 453
β-Acetoxysulphones, 161
"Acetylene Zipper", 170
α,β-Acetylenic *N*,*N*-dimethylamides, 178
Acetylenic esters, 169, 207
Acetylenic ketenes, 205
Acetylenic ketones, 158, 201, 204
Acetylenic sulphonium ylides, 362
Acetylide carbanions, 171
Acid azides, 388
Acid chlorides, reactions with Normant copper reagents, 402
–, reduction, 448, 543, 544
–, reductive decarbonylation, 7
Acids, oxidative decarboxylation, 8
Acrolein, reactions with alcohols, 348
–, – with mercaptans, 348
–, – with phenols, 348
Acyclic diene complexes, 572, 574
Acyclic dienes, 575
Acyclic ethers, 310
α-Acylamidocinnamate derivatives, reduction, 539
N-Acylamines, synthesis, 328
–, – from nitriles, 328
Acylanion equivalents, 385
Acylanions, synthesis from acylsilanes, 464
α-Acyl-α-(arylseleno)phosphoranes, 161
Acylberyllium bromides, 389
Acyl chlorides, 267
Acylcobalt carbonyls, 570
Acylcobalt complexes, 553
Acyl fluorides, 267
Acyl halides, cross coupling, 481
–, reduction, 218
–, stannylation, 483
Acyl nickel carbonylates, 1,4-addition to conjugated enones, 553
Acylphosphoranes, 161
Acylsilanes, 169, 464, 498
Adamantyl nitrate, 328
Adamantylsulphinyl cyanides, 367
Addition-demercuration, 421
Adiponitrile, manufacture, 62
Alcohol dehydrogenase, 278
Alcohols, acidity, 269
–, basity, 269
–, chemical shift correlations, 270
–, chlorination, 296, 306, 511
–, ^{13}C NMR spectra, 271
–, general chemistry, 269
–, hydrogen bonding, 269
–, infrared spectra, 271
–, lanthanide shift reagents, 270
–, mass spectra (E.I.), 271
–, *O*-methylation, 310, 311
–, nitration, 326
–, NMR spectra, 270
–, oxidation, 319
–, oxidative phosphorylation, 338
–, phosphorylation, 340
–, polarisability, 269
–, reactions with *N*,*N*-carbonyldiimidazole, 294
–, – with hydrogen bromide, 296
–, – with phosgene, 294
–, – with trimethylsilylpolyphosphate, 340
–, see also alkanols
–, selenylation, 526
–, solvating power, 269
–, synthesis, 272
–, – from acylhalides, 273
–, – from aldehydes, 273–277

–, – from alkenes, 80, 283, 284, 301, 304, 416
–, – from boranes, 439
–, – from bromoalkanes, 279
–, – from carboxylic acids, 272
–, – from chloroalkanes, 279
–, – from epoxides, 279
–, – from esters, 273
–, – from ketones, 273–278
Aldehyde equivalents, 439
Aldehydes, α-alkylation, 79
–, asymmetric reduction, 275
–, catalysed addition reactions, 571
–, chlorination, 511
–, reactions with Grignard reagents, 393
–, – with tetraiodomethane, 317
–, reduction, 299, 300, 543, 544
–, reductive amination, 543
–, reductive methylation, 275
–, see also alkanals
–, selenylation, 522
–, synthesis, 297
–, – from acyl anion equivalents, 385
–, – from alkenes, 53, 54, 67
–, – from boranes, 439
–, – from carboxylic acid derivatives, 385
–, – from vinyl selenides, 536
–, – *via* the Pummerer reaction, 360, 361
Aldimines, 201
Aldoximes, dehydration, 512
Aliphatic halides, reactions with selenides, 534
Aliphatic sulphonic acids, 376
Aliphatic thioaldehydes, 347
Aliquat 336, 346
Alkadiynyl alcohols, NMR spectra, 271
Alkali metal cyanotriorganyl borates, 281
Alkanals, asymmetric reduction, 275, 276
–, decarbonylation by rhodium complexes, 7
–, reactions with *B*-alkenyl-9-BBN derivatives, 301
–, – with *B*-allyldiisopinocampheylborane, 292
–, – with allyl halides, 292
–, – with allyl phosphates, 343
–, – with butenyl lithiums, 303
–, – with silylated immobilised organotin catalyst, 274
–, – with trityl perchlorate/triethylsilane, 313
–, – with vinylhalides, 302
–, reduction with tetrabutylammonium octahydrotriborate, 274
–, – with zirconium oxide/propanol, 273
–, reductive methylation, 275
–, see also alcohols
–, synthesis from chiral boronic esters, 284
Alkanes, addition to alkenes, 22, 55
–, aromatisation, 15, 17
–, as alkylating agents, 30
–, autoxidation, 20
–, biochemical hydroxylation, 22
–, bromination, 32
–, bulk properties, 12
–, carbonylation, 29
–, chlorination, 31
–, ^{13}C NMR spectra, 12–14
–, conformation, 10, 11
–, conversion into ketones, 24
–, cracking, 15, 17
–, cyanation, 29, 30
–, cyclisation, 15
–, C–C bond fusion, 10
–, dehydrogenation, 15, 17
–, dimerization, 9
–, electrochemical reactions, 27
–, enthalpy, 10, 12
–, fluorination, 31
–, fragmentation, 15
–, gas chromatography, 14
–, halogenation, 31, 32
–, ^{1}H NMR spectra, 11, 12
–, hydrogenolysis, 18
–, hydroxylation with 4-nitroperoxybenzoic acid, 27
–, interconversions, 15
–, isomerization, 15, 16
–, oligomerization, 15
–, oxidation, 20
–, – by Co(III) ions, 26
–, – by metal oxo species, 22
–, – by ozone, 26, 27
–, – by Pd(II) ions, 26
–, – by vanadium complexes, 25
–, oxidative coupling, 18, 19
–, photochemical chlorination, 219

–, photochemical dimerization, 19
–, physical properties, 9
–, radical reactions, 17
–, reactions with metal ions, 29
–, – with nitronium ions, 33
–, – with transition metal complexes, 27–29
–, sonification, 17
–, sources, 1
–, synthesis from alkanols, 4
–, – from alkenes, 53, 416
–, – from alkyl halides, 4
–, – from carbonyl compounds, 6
–, – from carboxylic derivatives, 6
–, – from seleno compounds, 6
–, – from thio compounds, 6
–, – *via* organometallic coupling reagents, 8
–, thermochemical properties, 9
–, thermochemistry, 11
–, topology of liquid state, 12
–, viscosity, 12
–, water solubility, 12
–, X-ray analysis, 11
Alkanesulphinic acid amides, 365
Alkanesulphinic acid anhydrides, 365
Alkanesulphinic acid esters, 365
Alkanesulphinic acid halides, 365
Alkanesulphinic acids, 365–367, 374
–, conversion to polysulphides, 345
–, dehydration, 367
–, esterification, 367
–, reactions with benzylamine, 367
–, synthesis from alkyl sulphines, 366
–, – from chlorosulphonyl chlorides, 366
–, – from ene reactions, 366
–, – from fluoro sulphonyl halides, 366
–, – from sulphones, 366
Alkanesulphinic acid thioesters, 365
Alkanesulphinic chlorides, reactions with hydroxylamines, 367
Alkanethiosulphonic acids, 376
Alkanols, dehydrogenation by Raney nickel, 390
–, esterification, 398
–, etherisation, 297, 298
–, formylation, 297
–, halogenation by N,N^1-carbonyldiimidazole/halo compounds, 294
–, – by diphenylphosphinated ethylene oligomers/carbon tetrachloride, 295
–, – by halogen acids/calcium halides, 296
–, – by halogen acids/lithium halides, 296
–, – by methanesulphonyl chloride/bromine, 296
–, – by phenylselenyl cyanide/bromine, 296
–, – by phosphoryl chloride/dimethylformamide, 296
–, – by sodium iodide/trimethylsilyl polyphosphate, 295
–, – by sodium phenylselenate/bromine, 296
–, – by tributyldiiodo phosphorane, 296
–, – by trimethylsilyl chloride/lithium bromide, 297
–, – by trimethylsilyl chloride/sodium iodide, 297
–, – by triphenyldiiodophosphorane, 296
–, – by triphenylphosphine/carbon tetrabromide, 295
–, – by triphenylphosphine dibromide, 295
–, methoxymethyl ethers, 297
–, oxidation, 284–294
–, – by Amberlyst A-26, $HCrO_4$ form, 285
–, – by anodic methods, 291
–, – by bis(benzyltrimethylammonium) dichromate, 287
–, – by bisphosphonium dichromate, 285
–, – by bromobenzene, 288
–, – by calcium hypochlorite, 293
–, – by chlorine/pyridine, 292
–, – by *m*-chloroperbenzoic acid/tetramethylpiperidine, 291
–, – by *N*-chlorosuccinimide, 293
–, – by chromium reagents, 284
–, – by chromium trioxide/crown ethers, 286
–, – by chromium trioxide/3,5-dimethylpyrazole, 286
–, – by chromium trioxide–dipyridine complex, 286
–, – by chromium trioxide/hexamethylphosphortriamide, 285
–, – by chromylchloride, 286
–, – by dimethyl sulphoxide/oxalyl chloride, 293

–, – by dimethyl sulphoxide/trifluoroacetic anhydride, 293
–, – by halogenochromate/tertiary amines, 286
–, – by lead(IV) acetate, 289
–, – by manganese dioxide, 287
–, – by manganese(II) acetate, 289
–, – by metal based oxidants, 284
–, – by 4-methoxy-1-oxo-2,2,6,6-tetramethylpiperidinium chloride, 290
–, – by Nakagawa's reagent, 287
–, – by nickel peroxide, 287
–, – by non-metallic oxidants, 290–294
–, – by oxygen/platinised titanium dioxide, 289
–, – by palladium(II) chloride, 288
–, – by periodinane, 293
–, – by peroxyacetic acid/2,4-dimethylpentane-2,4-diol, 287
–, – by polymer bound reagents, 291, 292
–, – by pyridinium chlorochromate, 286
–, – by pyridinium dichromate/bistrimethylsilyl peroxide, 287
–, – by pyridium fluorochromate, 287
–, – by quinolinium chlorochromate, 297
–, – by samarium(II) iodide, 288
–, – by sodium bromate/cerium salts, 289
–, – by sodium bromate/ruthenium trichloride, 289
–, – by sodium bromite, 289
–, – by tertiarybutylperoxide/bis(2,4,6-trimethylphenyl) diselenide, 290
–, – by tertiarybutylperoxide/molybdenum hexacarbonyl, 288
–, – by tetrapropylammonium perruthenate, 289
–, – by thioanisole/chlorine, 291
–, – by tris(triphenylphosphine) 3-iodosobenzoic acid/ruthenium dichloride, 289
–, – by trityltetrafluoroborate, 292
–, – by zinc dichromate trihydrate, 285
–, phosphorylation, 338, 340
–, reactions with allyl phosphates, 343
–, – with pyrophosphoric acid, 339
–, see also alcohols
–, substitution, 294–298
–, synthesis from acyl halides, 273
–, – from aldehydes, 273
–, – from bromoalkanes, 279, 280
–, – from carboxylic acids, 272
–, – from chloroalkanes, 279
–, – from epoxides, 279
–, – from esters, 273
–, – from iodoalkanes, 280
–, – from ketones, 273–275, 278
–, – from organoboranes, 280–284
–, tosylation, 297
Alkanones, asymmetric reduction, 275, 276
–, nitration, 328
–, reactions with allyl halides, 302
–, – with trityl perchlorate/triethylsilane, 313
–, – with vinyl halides, 302
–, reduction with chiral aminoboranes, 278
–, – with 9-*O*-(1,2:5,6-di-*O*-isopropylidene-α-*D*-glucofuranosyl)-9-boratabicyclo[3.3.1]nonane, 277
–, – with diphenylamine/borane, 275
–, – with diphenylantimony hydride/aluminium trichloride, 275
–, – with isopropanol/nickel, 275
–, – with tetrabutylammonium octahydrotriborate, 274
–, – with tributyltin hydride, 274
–, – with yeast, 278
–, – with zirconium oxide/propanol, 273
–, reductive methylation, 275
Alkanoyl chlorides, reactions with hydroxypyridinethione/polyhalogenomethanes, 233, 288
2-Alkenals, reactions with tertiarybutylamine, 280, 335
Alkene, hydrogenation catalysts, 2
η^2-Alkene complexes, 562
(η-Alkene)iron(II) complexes, 561
(η^2-Alkene)palladium(II) complexes, 559
Alkenes, 34–83
–, addition of fluorine, 58
–, – of hydrogen halides, 56
–, – of phenylselenium azide, 65
–, – of phenylselenyl carbamate, 66
–, – of phenylselenyl chloride, 66
–, alkylation, 489
–, allylic hydrogen abstraction, 80

–, amination, 63
–, aminomercuration-demercuration, 65
–, aminomethylation, 63
–, aminopalladation, 66
–, aminotellurinylation, 66
–, autoxidation, 75
–, bromodemercuration, 79
–, chlorohydroxylation, 57
–, coupling reactions, 50
–, cyanamidoselenylation, 65
–, diesterification, 72
–, dihydroxylation, 72, 321
–, dimerisation, 50
–, ene reactions, 75, 76
–, epoxidation, 67, 324
–, halogenation, 58
–, – catalysis by transition metal salts, 62
–, homologation, 50
–, hydration, 34
–, hydroacylation, 78
–, hydroalumination, 53, 390, 445
–, hydroboration, 37, 229
–, hydroboration/coupling, 50
–, hydroboration/iodination, 37
–, hydrocyanation, 62, 63
–, hydroformylation, 21, 54
–, hydrogenation, 34, 49
–, hydroiodination with trimethylsilane and sodium iodide, 57
–, *cis*-hydroxylation, 73
–, hydroxy-palladation, 70
–, isomerisation, 48
–, mercuration, 79
–, metathesis, 50
–, –, catalysts, 51
–, methylation, 8
–, miscellaneous reactions, 76
–, nickel complexation, 62
–, nitration, 80
–, oligomerisation, 50
–, oxidation, 66, 320, 421
–, – by tertiarybutylhydroperoxide, 302
–, – by thallium salts, 454
–, – to alkynes, 162
–, – with potassium permanganate, 73
–, oxythallation, 452
–, ozonolysis, 66
–, peroxymercuration, 79, 80
–, phenylselenylation, 60
–, photo-assisted isomerisations, 48
–, photocatalytic dehydrogenation, 46
–, photooxygenation, 75
–, photosensitised oxidation, 75
–, reactions, 48
–, – with aldehydes, 78
–, – with alkyl hypohalites/boron trifluoride, 62
–, – with carbon monoxide, 72
–, – with 1,3-dipoles, 82
–, – with fluoroalkyl iodides, 58
–, – with hydrazoic acid, 64
–, – with hydrogen chloride, 72
–, – with maleic anhydride, 76
–, – with mercury(II) nitrate, 81
–, – with methylthiodimethylsulphonium salts, 81
–, – with nitronium tetrafluoroborate, 81
–, – with nitroso compounds, 81
–, – with oxygen, 72, 75
–, – with palladium nitro complexes, 81
–, – with palladium(II) salts, 567
–, – with phenylsulphenamides, 82
–, – with phenyltellurium tribromide, 61
–, – with sulphenamides, 81
–, – with tellurium tetrachloride, 60
–, – with tertiarybutylhydroperoxide/mercury(II) acetate, 263
–, – with thiols, 74
–, – with water, 72
–, reduction, 3, 469
–, selenosulphonation, 82
–, selenylation, 527
–, synthesis, 366, 497, 498
–, – by elimination reactions, 41
–, – from alcohols, 44
–, – from aldol compounds, 47
–, – from alkyl halides, 46
–, – from amines, 43, 44
–, – from boranes, 440
–, – from carbonyl compounds, 384
–, – from α-chlorocarboxylic acid chlorides, 40
–, – from Grignard reagents, 39
–, – from methanol, 45
–, – from *vic*-diols, 44
–, – *via* Wittig reaction, 3
–, trisubstituted, 498
–, vicinal dihydroxylation, 321

(*E*)-Alkenes, synthesis by Wittig reaction, 495
(*E*)-2-Alkenes, isomerisation, 49
–, synthesis from 1-alkenes, 48
Alkenols, oxidation by benzyltrimethylammonium tetrabromooxomolybdate/tertiarybutyl peroxide, 305
–, – by chromic acid/acetone/water, 306
–, – by Jones reagent, 305
–, – by oxygen/ruthenium dioxide, 305
–, – by potassium ferrate/benzyltriethylammonium chloride, 305
–, – by silver(II) dichromate/pyridine, 305
–, rearrangement by reaction with amylseleno nitriles/tributylphosphine, 307
–, – with aqueous acid, 307
–, – with bis(trimethylsilyl) peroxide/vanadium catalyst, 307
–, Sharpless epoxidation, 306
–, substitution reactions, 306
–, synthesis, 299–304
–, – from alkynes, 301–304
–, – from epoxides, 300, 301
Alkenones, oxidation, 319
2-Alkenylaryl sulphoxides, reactions with lithium tetramethylpiperidide (LTMP), 116
Alkenylboranes, 436, 437, 442
–, cleavage, 437
–, halogenation, 436
–, synthons for aldehydes and ketons, 435
Alkenyl chromiums, 264
Alkenyl copper complexes, 544
Alkenyl halides, dehydrohalogenation, 263
–, reduction, 264
Alkenyllithiums, 383
Alkenylmercurials, 416
Alkenylmercuric chlorides, 429
Alkenylmercuric halides, 426–428
–, acylation, 428
Alkenylnickel complexes, 549
Alkenylpalladium complexes, 549
Alkenylphosphonium salts, 509
Alkenyl sulphides, 352
–, reactions with silyl enol ethers, 352
Alkenylzirconium complexes, 492
(*E*)-1-Alkenylzirconiums, 262
α-Alkoxyaldehydes, 488, 489
B-Alkoxyalkylphenyltellurium dibromides, 61
Alkoxyallenes, 118, 142
–, reactions with *B*-alkylborabicyclononanes, 147
–, – with butyllithium and trimethylsilyl chloride, 146
(Alkoxyaryl)chromium carbenes, reactions with alkynes, 555
(Alkoxyaryl)chromium complexes, 555
Alkoxydialkylboranes, 431
Alkoxy group, migration, 453
Alkoxymethyltins, 474
Alkoxy-1-naphthols, 555
Alkoxyphosphonates, 332
Alkoxyphosphoranes, 493
α-Alkoxyvinyllithiums, 474
β-Alkoxyvinyllithiums, 474
Alkylallenes, 145
Alkylallyl sulphides reaction with tetrakistriphosphinepalladium(IV), 349
Alkylallyl trisulphides, synthesis from ethyne, 357
Alkylamines, synthesis, 63, 64
Alkylarylmercuric halides, carbonylation, 429
Alkylating agents, toxicity, 214
Alkylation, alkenes, 559
–, aromatic halides, 571
1,4-Alkylation, conjugated enones, 545
Alkylazides, 64
–, synthesis, 327
Alkylbenzyl sulphides, reactions with chlorine, 376
Alkylboranes, protodeboronation, 437
–, reactions with mercuric acetate, 437
Alkyl bromides, synthesis from alkenes, 53
Alkylcarbenes, 101
Alkylcopper complexes, 544
Alkyldichloroboranes, 437
–, reaction with oxygen, 318
Alkyl dihydrogenphosphates, synthesis, 339
Alkyldinitriles, 174

(*Z*)-Alkylenylboronic esters, 435
Alkyl halides, addition to alkenes, 55
–, analysis, 216–218
–, carcinogenicity, 215
–, elimination reactions, 46
–, gas chromatography, 218
–, hydrogenation, 6
–, in drinking water, 215
–, methylation, 8
–, photoaddition to alkenes, 55, 56
–, reactions with tetramethylammonium ditertiarybutyl phosphate, 338
–, toxicity, 214
Alkyl haloboranes, 38
Alkyl hydroperoxides, 23, 310, 321, 322
–, reactions with iron(II) sulphate, 321
–, synthesis, 317, 318
Alkylidenecarbenes, 101
–, reaction with chiral α-naphthylphenylmethylsilane, 102
2-Alkylidene-1,3-cyclopentenediones, 205
Alkylidenecyclopropanones, 103
3-Alkylidenefurans, 196
Alkylidenephosphoranes, 508
Alkyllithiums, 383
Alkylmercurials, halogenation, 419
Alkylmercuric acetates, 437
Alkylmercuric halides, 427
Alkylnickel complexes, 549
Alkyl nitrates, reactions with lithium azide, 327
–, synthesis from alcohols/*N*-nitrocollidinium tetrafluoroborate, 326
–, – from alkyl sulphonates/tetrabutylammonium nitrate, 326
–, – from amines/dinitrogen tetraoxide, 327
–, – from amines/triphenylpyrilium nitrate, 327
–, – from bromoalkanes/mercury(I) nitrate, 325
–, – from tertiarybutyl nitrite/oxygen, 327
Alkyl nitrites, synthesis from alkenes, 63
Alkylpalladium complexes, 549
Alkyl penta-3,4-dienoates, reactions with alumina, 153
Alkyl peroxides, 310
Alkylphenylselenides, oxidation with *m*-chloroperbenzoic acid, 313
Alkylphenyl sulphides, 350
Alkylphenyltellurides, oxidation with 3-chloroperbenzoic acid, 41
Alkylphenyltelluroxides, decomposition to alkenes, 41
Alkyl phosphates, reactions, 341–343
–, synthesis, 337–341
Alkyl phosphites, synthesis from alcohols, 329
Alkyl radical-amine ion radical recombinations, 45
Alkyl selenides, synthesis from alcohols, 535
Alkylsilanes, 541
Alkyl sulphides, oxidation, 364
Alkyl sulphinamides, 367
Alkyl sulphones, reactions with sulphuryl chloride, 373
Alkylsulphonyl chlorides, synthesis from alkyl sulphides and chlorine, 376
Alkylsulphonyl fluorides, synthesis from alkyl thiols, 376
Alkyl sulphoxides, pyrolysis, 364
Alkylthiobutenynes, 354
α-Alkylthio carbonyl compounds, 347
Alkyl thioesters, 347
Alkyltins, cross coupling reactions, 480
Alkyl titanium chlorides, dimethylation of ketones, 8
Alkyl tosylates, reduction, 5
Alkyltrimethylammonium salts, 43
Alkylzinc halides, reactions with acid chlorides, 413
–, – with aryl halides, 413
Alkylzirconium complexes, 547
Alkyne–allene isomerization, 199
Alkyne–carbene metal complexes, 207
Alkyne–iron complexes, 208
Alkynes, 156–213
–, acetoxymercuration, 418
–, acetoxysulphenylation, 177
–, acylation in presence of vanadium trichloride, 169
–, addition of hydrogen cyanide, 174
–, – of vinylzirconium complexes, 548
–, – reactions, 171, 403
–, alkylation, 166

–, carboalumination, 182, 445
–, carbometalation, 182
–, carbotitanation, 182
–, halogenation, 173
–, hydration, 174
–, hydroalumination, 182, 446
–, hydromagnesation, 182
–, isomerization, 170
–, mercuration, 184, 185
–, metal mediated reactions, 186
–, oxidation, 210
–, – by thallium salts, 75
–, – to alkynols, 320
–, oxidative rearrangement, 210
–, oxymercuration–demercuration, 418
–, oxythallation, 452
–, pericyclic reactions, 204
–, photochemical reactions, 202–204
–, reactions, 171
–, – with radicals, 186, 201
–, – with zirconium hydrides, 262
–, reduction, 3, 171
–, – by epitaxial palladium, 172
–, – by interlamellar montmorillonite–diphenylphosphine Pd(II) complex, 172
–, – by Ni-graphite, 171
–, – by niobium(V) chloride, 171
–, – by palladium dichloride/sodium borohydride, 172
–, – by Pd-graphite, 171
–, – by sodium aluminium hydride, 171
–, – by zinc–copper couple, 171
–, selenylation, 529
–, synthesis, 156
–, – from β-acetoxysulphones, 161
–, – from metal acetylides, 163
Alkynic acids, synthesis from 1,3-dicarbonyl compounds, 160
Alkynic ketones, synthesis, 441
Alkynoic acid derivatives, 169
Alkynols, 167
–, isomerisation by potassium 3-aminopropylamide, 309
–, synthesis from alkynones, 307, 308
–, – from allenylboronic acids, 308
–, – from enynes, 308
Alkynones, asymmetric reduction, 276
–, reduction by Alpine borane, 308
–, – by *B*-isopropinocamphenyl-9-borabicyclo[3.3.1]nonane, 308
–, – by lithium tetrahydroaluminate-*N*-methylephedrine/3,5-dimethylphenol, 308
–, – by *cis*-myrtanyl-9-borabicyclo[3.3.1]nonane, 276
–, – by NB-entrane, 307
–, – by nopol benzyl ether/9-borabicyclo[3.3.1]nonane, 307
–, synthesis from β-diketones, 160
Alkynylalanes, 448
Alkynylanilines, cyclisation, 198
Alkynylanisoles, 197
Alkynylboranes, 440
Alkynylborate complexes, 167
Alkynylborons, 167, 176
γ-Alkynylcarbonyl compounds, 160
Alkynylcuprate–dimethyl sulphide complexes, 165
2-Alkynylcyclobut-3-enones, thermal rearrangement, 205
Alkynyldialkylphosphates, 170
Alkynylene allenes, cyclisation, 206
Alkynyl epoxides, 196
Alkynyl ketones, 167
2-Alkynyl-5-methoxy-1,4-benzoquinones, 165
Alkynylnitriles, 182
Alkynyl(phenyl)iodinium phosphates, 170
Alkynylphosphoric diamides, 160
β-Alkynylselenoketones, synthesis, 161
Alkynylstannanes, 167–169, 204
Alkynyl sulphides, 353
Alkynyl sulphones, 369, 370
–, Michael additions, 372
–, nucleophilic displacement reactions, 372
–, reactions with 1,2-bisnucleophiles, 374
–, synthesis from phenylselenoalkenyl-sulphones, 162
Alkynylthiols, preparations, 344
Alkynyltin compounds, synthesis from alkynyl lithiums, 468
Alkynyltins, 481
Allenes, 115–155, 194, 363
–, addition of thiophenol, 133
–, ^{13}C NMR shifts, 115

–, cycloaddition and addition reactions, 131
–, geometries, 115
–, heats of formation, 115
–, metalation, 128
–, miscellaneous reactions, 139
–, oxythallation, 452
–, photoaddition reactions, 130
–, reactions, 128–140
–, – with acetylcobalt tetracarbonyl, 139
–, – with bis(phenyldimethyl)cuprate, 132
–, – with bis(phenylsulphonyl)methane, 138
–, – with 1,3-dipoles, 138
–, – with phenylselenyl bromide, 134
–, – with phenylselenyl chloride, 134
–, – with vinyl halides, 135
–, silylcupration with bis(dimethylphenylsilyl)copper lithiate, 130
–, structure and spectra, 115
–, synthesis, 116, 502, 505
–, thermal rearrangements, 130
–, vibrational circular dichroism spectra, 115
Allenic acids, 129
Allenic alanates, 129
Allenic alcohols, 184
–, conversion into furans, 147
α-Allenic alcohols, rearrangement, 307
Allenic amines, 151
Allenic boronates, 155
Allenic ketones, 135, 152
Allenic nitriles, 153
Allenic selenoxides, 517
Allenic sulphones, 153
–, reactions with 1,2-bisnucleophiles, 375
Allenic sulphonium salts, conversion into furans, 361
Allenic sulphonium ylides, 363
Allenylallyl alcohols, synthesis, 303, 304
Allenylboronic esters, 443
Allenylmethanols, synthesis, 306
Allyl acetates, 570
–, substitution, 467
Allyl alcohols, 102, 164, 498
–, epoxidation, 324, 325
–, NMR spectra, 271
–, reactions with silylcuprates, 461
–, – with trimethylsilyl chloride/calcium carbonate, 306
–, rearrangement, 307
–, synthesis, 303, 484
–, – from allylmethyl compounds, 324
–, – from allylsilanes, 462
O-Allyl *S*-alkyl ditriocarbonates, 350
Allyl anion equivalent, 458
Allylboranes, 442, 443
Allyl bromides, 184
Allyl cadmium reagents, 414
η^3-Allyl cobalt complexes, 570
η^3-Allyl complexes, 565
Allyl epoxides, ring opening, 569
Allyl ethers, cleavage by rhodium complexes, 316
–, metalation, 316
–, reduction, 544
Allyl ethynyl ethers, Claisen rearrangement, 316
Allyl halides, reactions with selenides, 534
–, stannylation, 579, 580
Allyl hydroperoxides, 75
Allyl hydroxyamines, 81
Allylic amination, 536
Allylic rearrangements, 264, 319, 410, 485, 499
Allylidene cyclopropanones, 103
η^3-Allylnickel halide complexes, 571
(η^3-Allyl)palladium complexes, 565
(π-Allyl)palladium complexes, synthesis from alkenylmercury halides, 427
(η^3-Allyl)palladium halides, 567
Allyl phosphates, reactions with aldehydes and ketones, 343
Allyl phosphonium salts, 499
Allylselenic acids, 535
Allylselenides, oxidation, 303
–, selenium displacement, 522
Allyl selenoxides, 517
Allylsilane anions, 462
Allylsilanes, 369
–, reactions, 462
–, – with monothioacetals, 349
–, synthesis, 461
Allyl sulphinates, sulphinate–sulphone rearrangement, 370
Allyl sulphones, 369

–, dimerisation, 372
–, ionisation with Pd(0), 370
–, stannolysis, 374
S-Allylsulphonium compounds, proton abstraction, 361
Allylsulphonium salts, synthesis from allylsilyl ethers, 359
Allyl tin compounds, 471
–, cross coupling reactions, 479
Allyl tin halides, synthesis, 484
Allyl tins, acylation, 482
–, addition reactions, 487, 488
–, lithiation, 477
Allylylides, 499
Aluminium compounds, 444
Aluminium phosphate, catalyst, 48
Amberlite 200C cation, 341
Amberlyst A26 (perbromide form), 173
Amides, dehydration, 512
–, nitration, 328
–, synthesis from nitriles, 528
Amidoalkanes, 417
Amines, reductive formylation, 348
–, synthesis from alkyl azides, 334
Aminoalcohols, synthesis, 448
β-Aminoalcohols, 477
Aminoalkynes, 200, 206
Aminocarbene complexes, 188
1-Aminocyclopropane-1-carboxylate, 35
Aminoisoxazoles, 153
Aminomalonate esters, 110
Aminomercuration, 417
α-Aminomethyltins, 477
Aminonaphthalenes, 382
Aminophosphonium halides, synthesis from amines, 512
S-(3-Aminopropyl) thiosulphuric acid, synthesis, 358
Aminosugars, phosphorylation, 523
α-Aminosulphonic acids, synthesis from α-amino acids and Sulphan, 375
γ-Aminosulphonic acids, synthesis from sultones, 374
Aminothiols, synthesis, 345
Anatoxin-a, 151
Anilines, 206, 207
–, synthesis, 350
Anionic metal carbonyls, 552
Anion solvation, 221
Anti-Markownikov, addition of acetic acid to 1-alkenes, 437
Aphidicolin, 418
Arbuzov reaction, 330, 331
Arene chromiumtricarbonyl catalysts, 540
Arene complexes, 572
Arene radical ions, 382
Arenes, synthesis from conjugated diynes, 189
Arenesulphonic acids, conversion to polysulphides, 345
Artemisia alcohol, 410
Arylaldehydes, reductive etherification, 490
Arylalkenes, 75
Arylalkoxycarbonylmercury halides, 427
Arylalkylselenides, 534
Arylalkynes, 174
–, hydration, 175
Arylalkynoic acids, 157
Arylcalcium halides, carboxylation, 408
Arylcarbenes, 109
–, geometric isomerism, 89
–, reactivity, 85
–, rearrangement, 87
4-Aryl-2,2-dimethyl-5-(4-nitrophenyl)-dioxoles, 95
Aryldisulphides, 33
Aryl halides, reactions with anylselenide anions, 534
Arylhalocarbenes, reaction with alkenes, 88, 89
Arylmercuric halides, carbonylation, 429
Arylmethylselenides, 534
–, demethylation, 534
Aryl migrations, 455
Aryl organomercurials, 420
Aryloxyphosphoranes, 493
Arylphosphonates, synthesis, 282
Arylrhodium complexes, 553
Arylselenides, 534
Arylselenoalkynes, 161
Arylselenols, reactions with enals, enones and enoic acids, 534
Arylselenonitriles, 526
Arylstrontium halides, carboxylation, 408
Aryl sulphones, as nucleophiles, 372

"Ate" complexes, 446–448
1*H*-Azepines, synthesis from nitrosobenzenes/triethylphosphite, 334
Azides, reactions with Grignard reagents, 395
Azidoalkanes, 417
10-Azidodihydrophenanthrene-9-ol, reaction with triethylphosphite, 334
Aziridine *N*-oxides, 81
Aziridines, 96
–, synthesis, 412
–, – from hydroxyamines, 512
Azobenzene, 103
Azobisisobutyronitrile, 520
Azomethanes, photolysis, 9

"Banana bonds", 34
Barbier reaction, 391
Bayer–Villiger reaction, 320
Benzannulated products, 111
Benzdiazoles, 301
Benzene, alkylation, 30
Benzenethiol, addition to conjugated allenic ketones, 352
Benzoazepines, 202
Benzofurans, 197
Benzoisothiazoles, 202
Benzophenone oxime, acyl derivatives, 288
1,5-Benzothiazepines, 203
Benzothiazoles, 202, 203
Benzoxathioles, 301
Benzyl cyanide, alkylation, 298
N-Benzyl-1,4-dihydronicotinamide, 453
Benzyl fluoride, synthesis from benzyl bromide, 222
Benzyl halides, stannylation, 480
Benzylic Grignard reagents, reductive coupling, 391
Benzylic ylides, 500
Benzyllithiums, 383
Benzylmethyl ketone, alkylation, 243
Benzylsulphinyl cyanides, 367
Benzylsulphonium salts, synthesis from benzyl silyl ethers, 359
Benzyltin compounds, 471
Benzyltrimethylammonium tetrabromooxomolybdate, 268, 325
Beryllium compounds, 389
Bialkyls, 252
Bicyclo[3.1.0]lactams, 113
Binuclear chromium clusters, reduction of conjugated double bond, 543
Biphenylenes, 209
Bis (acetylacetoneto) nickel(II), 403
(*Z*,*Z*)- Bis(3-acrylate) sulphides, 351
Bis(alkylthio) butenynes, 353
Bis (cyclopentadienyl) titanium dichloride, polystyrene bound, 48
Bis (cyclopentadienyl) zirconium(IV) complex, 547
Bis (2,2-debromocyclopropyl)ethene, 106
Bis (dialkylamino) alkynes, 161
Bis (dibenzylideneacetone) palladium, 135
Bis (diethylaluminium) sulphate, 479
Bis (1,2-dimethylpropyl) borane, 172
1,2-Bis (diphenylphosphino) ethane, 135
1,1-Bis (diphenylphosphino) ferrocene, 51
Bis (methoxycarbonyl) carbene, 110
2,4-Bis(4-methoxyphenyl)-1,3,2,4-dithiadiphosphetane-2,4-disulphide, 314
[Bis(salicylidene-γ-iminopropyl) methylamino]cobalt(II), 71
Bis(trialkylstannyl) ethenes, 468
Bis(tributyltin) sulphide, 348
Bis(trifluoromethyl) cadmium, 267
–, reaction with acid chlorides, 267
1,3-Bis(trimethylsilyl)cyclopropanone, reaction with diazomethane, 106
Bis(trimethylsilyl) esters, synthesis from diethylenolphosphate esters, 315
Bis(trimethylsilyl) peroxide, 338, 339
Bis(trimethylstannyl) methane, 477
Bisylids, synthesis, 335
9-Borabicyclo[3.3.1] nonane, 38, 51, 431
Borane/methylsulphide, 38
Borane/tetrahydrofuran, 315
Boranes, use in synthesis, 435
Boron, oxidative replacement, 435
–, replacement with halogen, 436
–, – with hydrogen, 437
–, – with mercury, 437
–, – with nitrogen, 437
Boronic acids, 430, 434

Boronic esters, 430, 434, 437
Boron sulphide, 366
Bovine serum albumin, 363
Bromal, 76
Bromine monofluoride, 247
Bromine/phosphorus trichloride, 230
Bromine/triphenylphosphine, bromination reagent, 230
1-Bromoacetophenone, 331
Bromoalkanes, oxidation, 263
–, reactions with mercury(II) perchlorate, 225
–, – with methyldimesyldorane, 224
–, synthesis from alcohols, 305, 360
–, – from alkyl nitrates, 328
Bromoalkenes, 259
2-Bromoalkenes, 260
β-Bromoalkyl peroxides, 79
–, synthesis, 318
Bromoalkynes, 38
1-Bromoalkynes, 158
α-Bromoamides, 409
Bromoboranes, 435
1-Bromo-1-chloroalk-1-enes, 259
1-Bromo-2-chlorobut-3-ene, 248
2-Bromo-1-chlorobut-3-ene, 248
1-Bromo-1,3-dienes, 172
Bromodiiodomethane, 251
(*E*)-2-Bromoethenyldibromoborane, 39
2-Bromoheptane, 230
1-Bromohept-1-yne, 264
α-Bromoketones synthesis, 79
2-Bromooctane, 230
1-Bromooctene, addition to norbornene, 263
2-Bromooctyne, 259
1-Bromo-2-phenylthioethene, cross coupling, 352
(*Z*)-1-Bromo-2-phenylthioethene, reaction with butylmagnesium bromide, 39
11-Bromoundecyl tosylate, 238
Brook rearrangement, 458, 464
Bunte salts, synthesis from amino thiols, 358
Buta-2,3-dichlorobuta-1,3-diene, 258
Butadiene, addition of hydrogen cyanide, 62
Butadienyl sulphonium salts, annulation reactions, 361
Butane, 17
–, oxidation, 25
(*R*,*R*)-Butane-2,3-diol, 443
But-1,3-diynes, 212
Butene, hydration, 80
2-Butene, addition of hydrogen fluoride, 220
(*E*)-2-Butene, *cis*-chlorotelluration-*trans*-dechlorotelluration, 49
Butene-2-yl radical, 48
2-Buten-1-ol, 260
Butenolides, 428, 551
But-2-enoylphosphonate, 330
Butylamine, hydrogenolysis, 44
–, manufacture, 63
Butylcyanoborohydride, 4
Butyl hydroperoxide, 57
Butyl ketone, reductive coupling, 47
Butylsulphinic acid, 366
Butyne-1,4-diol, monoacetylation, 142
Butyne iron complex, 183
But-3-ynoic acids, 198
But-3-ynols, 198

Caesium fluoroxysulphate, additions to alkenes, 77
Calcium halides, promotion of reactions between alcohols and hydrogen halides, 230
Calicheamicins, 205
$\Delta^{9(12)}$-Capnellene, 484
Carbanion formation, 503
Carbanions, phosphorus stabilised, 506
–, synthesis from thioaetals, 348
Carbazole, 236
Carbene, insertion reactions into C–C bonds, 99
–, reactions mediated by metal ions, 109
–, – by organometallic catalysts, 109
Carbene–alkene complexes, 89
Carbenes, 84–115, 384
–, cycloaddition to alkenes, 88
–, formation by laser flash photolysis, 88
–, insertion reactions, 98, 267, 385
–, MNDO calculations, 86
–, molecular orbital calculations, 84
–, reactions with dienes, 97
–, – with imines and enamines, 96

–, singlet–triplet interconversion, 84
–, structure and reactivity, 84
–, Wolff rearrangement, 110
Carbenoid, sigma-bond insertion, 98
Carboalumination, 390, 445
Carbocations, 16
Carbocupration, alkynes, 547
Carbocyclic rings, formation, 418
Carbohydrates, 448
Carbon–carbon bond formation, 383, 437, 478
Carbon dioxide, reduction, 47
Carbon–hydrogen bond formation, 383, 470
Carbon monoxide, absorption, 439
–, insertion, 548
–, use in industrial processes, 542
Carbon–nitrogen bond formation, 388
Carbon–oxygen bond formation, 471
Carbon–sulphur bond cleavage, 365
Carbon–sulphur bond formation, 388
Carbon tetrabromide, 250
Carbon tetraiodide, 250
Carbonylation, dienes, 566
–, organomercuric acetates, 429
–, reactions, 439, 551
Carbonylcarbenes, geometric isomerism, 89
Carbonyl compounds, allylation, 484
–, methenation, 558
–, reduction, 469
–, selenylation, 524
–, synthesis from nitroalkanes, 322
Carbonyl derivatives of allenes, 152
Carbonyl fluoride, 333
Carbonyl-olefination reactions, 509
Carbonyl-olefin complexes, 558
Carboxylate salts, reactions with Grignard reagents, 395
Carboxylic acids, 210
–, conversion into acyl chlorides, 511
–, esterification with trimethylphosphate, 342
–, oxidative decarboxylation, 43
–, reduction, 348, 448
–, – with borane/dimethyl sulphide, 272
–, – with catecholborane, 272
–, synthesis, 428
–, – from alkenes, 74
–, – from chloroformates, 506
–, – from enolate anion equivalents, 387
–, – *via* carbon dioxide insertion, 387
Carboxylic esters, nitration, 328
–, reactions with Grignard reagents, 394
–, synthesis, 428
Carotenoids, synthesis, 505
Catalysis for cracking of alkanes, 15, 16
Catalysts for "Wacker" process, 69–71
Catalytic [3,3]-sigmatropic rearrangements, 572
Catecholborane, hydroboration of alkenes, 434
Cationic iron complexes, 576, 577
Cecropia juvenile hormone, 475
Cerium(IV) salts, 365
Cetylpyridinium chloride, 255
Cetyltrimethylammonium permanganate, 73
Cetyltrimethylammonium salts, 227
Chalcones, oxidation, 455
Chelating alkene complexes, 572
Chelating alkenes, complexes with iron(0), 573
Chiral boronates, 437
Chiral hydroborating agents, 443
Chiral hydrogenations, 539
Chiral phosphorus ligands, 539
Chiral sigmatropic rearrangements, 443
Chiral synthesis, 443
Chiral thiols, 347
CHIRAPHOS, 540
Chloral, 76
Chloramine T, 351
Chlorformic esters, 227
Chlorine chlorosulphate, addition to alkenes, 78
Chlorine monofluoride, 248
Chloroalkanes, reactions with sodium formate, 224
1-Chloro-1-alkenes, hydroboronation, 259
1-Chloro-1-alken-1-ylboranes, 259
1-Chloro-1-alkenylsilanes, 259
β-Chloroalkyl chlorosulphates, 78
β-Chloroalkyltellurium, 60
N-Chloroamides, 219
2-Chloro-1,3,2-benzodioxaphosphole, 365
1-Chlorobenzotriazole, 260

α-Chloroboronic esters, 443
2-Chlorobuta-1,3-diene, 258
2-Chlorobutane, 57
4-Chlorobutyramides, reactions with triethyl phosphites, 331
1-Chlorodecan-2-ol, 58
2-Chlorodecanol, 58
α-Chlorodialkyl sulphoxides, 363
Chlorodifluoromethane, deprotonation, 95
N-Chlorodimethylamine, reactions with alkenes, 26, 59
Chloroepoxides, reactions with trialkylphosphites, 332
1-Chloroethyl chloroformate, 233
2-Chloroethyl chloroformate, 231
β-Chloroethylphosphites, transesterification, 329
Chlorofluorocarbene, 176
–, addition to alkenes, 89
1-Chloro-1-fluorocyclopropanes, 89
Chlorofluorohydrocarbons, 215
Chloroformamidines, synthesis from ureas, 511
2-Chloro-2-methylpropene, 57
2-Chloro-1-methylpyridinium iodide, 367
2-Chloromethyltetrahydrofuran, 166
Chloromethyltriphenylphosphonium salts, deprotonation, 261
3-Chloro-3-(4-nitrophenyl)diazirine, 1,3-dipolar addition reaction with benzaldehydes, 94
Chloronium ion intermediates, 57
(*S*)-2-Chlorooctane, 229
3-Chloroperbenzoic acid, 313
–, oxidative replacement of boron, 436
Chloroperoxidase, 249, 362
Chlorophenylcarbene, reaction with diethyl maleate, 92
N-Chlorophthalimide, 31
1-Chloropropene, 258
3-Chloroquinoline, 97
Chloro-substituted cyclopropanes, 412
N-Chlorosuccinimide, 365
N-Chlorosulphenyl compounds, synthesis from sulphur dichloride, 355
β-Chlorosulphonic esters, synthesis from methyl chlorosulphonate and 1-alkenes, 364
Chlorovinyl ketones, 146
–, reactions with phosphines, 455, 510
Chromium carbene complexes, reactions with morpholine, or pyrrolidine, 112
Chromium porphyrins, 69
Cinnamic acid/cyclodextrin complex, 248
(*E*)-Cinnamyltriphenyl tin, 487
Citral, epoxidation, 324
Claisen rearrangement, 316
Clemmenson reduction, 6
Coal, hydrocarbon source, 2
Cobalt(II) acetylacetonate/triethylaluminium catalyst, 46
Cobalt boride, 3
Cobalt carbonyl anion, 570
Cobalt(II) chloride–sodium borohydride, 53
Cobalt complexes, reduction of alkenes, 544
Cobalt compounds, 539
Cobalt η^5-dienyl complexes, 577
Cobalt hexacarbonyl, 168
Cobalt-nitrocomplexes, 70
Cobalt(II) phthalocyanines, 357
Compounds with metal bonded to four or more carbon atoms, 572
–, – to three carbon atoms, 565
–, – to two carbon atoms, 558
(*R*)-(–)-Coniine, 151
Conjugated dienes, dimerisation, 565
–, protection, 573
–, selective hydrogenation, 540
Conjugated diynes, 162
–, reduction by hydroalumination, 172
Conjugated enamines, reactions with Grignard reagents, 395
Conjugated enones, reactions with Grignard reagents, 394
–, – with Normant copper reagents, 400
–, – with organoseleniums, 516
Conjugated enynes, 165
Conjugated ketones, organic reactions with cubrates, 545
Conjugated triynes, 159
Conjugated ynones, 164, 169
Cope rearrangement, 133, 374, 572
Copper(II) acetate/iron(II) sulphate, 321
Copper alkynides, 165

Copper bonded complexes, 544
Copper perchlorate, 25
Coumarins, synthesis from aryl-acetylenes, 549
Coupling reactions, catalysed by transition metals, 406
Crabtree's iridium catalyst, 540
Cross coupling, promotion by transition metal catalysts, 442
Crotonaldehyde, 248
Crotyl chloride, 263
Crotyltributyl tin, 486
Crown thioethers, 351
Cumulenes, 260
Cuprates, reactions with aldehydes, 544, 546
–, – with enones, 544, 546
–, – with halides, 544, 546
–, – with ketones, 544, 546
–, – with tosylates, 544, 546
Cyanamides, 65
Cyanidation, 439
Cyanoacetylene dimer, 204
Cyanoallenes, reactions with hydrazines, 153
–, – with hydroxylamine, 153
Cyanosulphonium methylides, 360
Cyclic dienes, synthesis from cyclic 2-ethoxycarbonyl sulphones, 318
Cyclic enynes, 190
Cyclic ethers, synthesis, 530
Cyclic phosphine oxides, 504
Cyclic polysulphoxides, phase transfer reagents for alkylation reactions, 362
Cycloadditions, catalysis by organoaluminiums, 450
Cycloalkyl halides, 223
Cyclobutanes, 173
Cyclobutano[*d,e*]naphthalene, 87
Cyclobutanones, 106
Cyclobutenes, 106, 208
Cyclobutenones, 204, 205
Cyclobutylidene, intermolecular additions with alkenes, 105
–, synthesis 1,1-dibromocyclobutane, 105
β-Cyclodextrin/epichlorohydrin copolymer, 225
Cyclododecyl fluoride, 226
Cycloheptadiene, 133
1,2-Cyclohexadiene, Diels–Alder reactions, 115
–, INDO–MO calculations, 115
Cyclohexadiene iron tricarbonyl complexes, 574
Cyclohexadienes, 575
Cyclohexa-1,4-dienes, 204, 208
Cyclohexa-2-enones, base induced fragmentation, 160
Cyclohexane, hydroxylation, 22
–, reaction with tertiarybutylhypo-chlorite, 220
1,4-Cyclohexanediodides, 231
1,4-Cyclohexanediols, 231
Cyclohexanols, oxidative cleavage, 471
Cyclohexanone, oxidation, 320
Cyclohexene, epoxidation, 324
Cyclohexenes, 206
Cyclooctatetrenes, 211, 212
Cyclopentadiene, 195
Cyclopentadienimines, 190
Cyclopentadienyl acyliron complex, 554
Cyclopentadienyl complexes, 572
η^5-Cyclopentadienyl(isobutylene)iron dicarbonyl, 561
Cyclopentadienyl ketones, 142
Cyclopentanes, 249
Cyclopentanones, 111
–, synthesis, 575
Cyclopentanonyl esters, 142
Cyclopentenes, 210
Cyclopentenol, 152
Cyclopentenones, 192
3-Cyclopentenylidene, 105
Cyclophanes, 208
Cyclopropanation reactions, 265
Cyclopropanes, 88, 103, 147
–, oxidation, 421
–, oxymercuration–demercuration, 418
–, solvomercuration, 419
–, synthesis, 361, 384, 410, 423, 478, 492, 527, 556, 568
Cyclopropanoallenes, 90
Cyclopropanones, 144
Cyclopropene, planar singlet carbene, 86
Cyclopropene carbene, aromatic stabilisation, 86
Cyclopropenes, 384, 425
Cyclopropenones, 176, 206, 425

–, thermolysis, 161
Cyclopropylcarbinyl halides, 492
Cyclopropylcarbinylsilanes, 464
Cyclopropylium cations, 188
Cyclopropylmethylmagnesium bromide, 390
Cyclopropylsilanes, 464
Cyclotrimerisation of alkynes, 565
Cytochrome P-450, 22
–, models, 68

Danishefsky's diene, selenylated version, 171
Decan-2-one, 27
Decan-3-one, 27
Decarbonylation, 553
1-Decene, 55
(*Z*)-5-Decene, 39
Dehydration; of hydroxy compounds, 512
Deoxyribonucleosides, 167
1,3-Deoxystannylation, 492
Desulphurization, with phosphite esters, 333
Dethallation reactions, 453
Deuterioalkanes, 7
Dialdehydes, cyclisation, 543
Dialkoxyboranes, disproportionation, 434
Dialkylallylamines, 260
Dialkylaluminium amides, 448
Dialkylaroyl phosphonates, reactions with trialkyl phosphites, 281, 336
Dialkylcadium reagents, 413
Dialkylcadiums, reactions with acid halides, 414
–, – with carbonyl compounds, 414
Dialkyl disulphides, oxidation, 356
S,*S*-Dialkyl dithiocarbonates, 349
–, sources of thiolate ions, 352
Dialkyl ethers, reactions, 314, 315
–, synthesis, 310–314
–, – from alcohols and dialkylsulphates, 310
–, – from alkanals, 313
–, – from alkanols and haloalkanes, 310–312
–, – from alkanols and methyl iodide, 310
–, – from alkanols and the methylsulphinyl carbanion, 310
–, – from alkanones, 313
–, – from diazoalkanes and rhodium(II) acetate, 312
–, – from esters, 313
–, – from tertiarybutyl cation, 313
–, – from thallium(I) alkoxides, 312
–, – from thionoesters, 314
Dialkylhaloboranes, hydridisation, 37
Dialkyl hydrogenphosphites/alcohols, anodic oxidation, 341
Dialkyl ketones, 54
–, synthesis from alkenes, 53
Dialkylmercurials, 428
–, protonolysis, 416
Dialkylperoxides, oxidation, 319
–, synthesis from alkyl tosylates/potassium superoxide/crown ethers, 317
–, – from alkyl trifluoromethane sulphonates, 317
–, – from bromoalkanes/hydrogen peroxide/silver trifluoroacetate, 318
–, – from bromoalkanes/potassium superoxide/crown ethers, 317
–, – from iodoalkanes/hydrogen peroxide/silver trifluoroacetate, 318
–, – from trifluoromethane sulphonates, 317
Dialkyl phosphates, alkylation, 337
–, synthesis from pyridinium phosphate betaines, 339
Dialkyl polysulphides, synthesis from lithium alkyls, 357
Dialkyl sulphides, 348
–, oxidation by fungi, 362
–, – by sodium iodate(-)(*R*)-menthol, 362
Dialkyl sulphones, use in synthesis, 368
Dialkyl sulphoxides (alkylsulphinylalkanes), 362
–, reaction with diethylaminosulphur trifluoride, 364
–, synthesis from alkylaryl sulphoxides, 363
–, – from dialkyl sulphides, 362
Dialkylzincs, reactions with acid chlorides, 413
–, – with aryl halides, 413
Dialkyne polymers, 211
Dialkynylberylliums, 389

Dialkynylmercurials, 418
9,9-Dianthrylcarbene, 85
Diarylalkynes, 162
Diarylcarbenes, 85
–, addition to alkenes, 88
2,3-Diarylindan-1-ones, 193
Diarylmercurials, 428
Diaryltellurium bis(trichloroacetate), 267
1,8-Diazafluorenylidene, 85
1,2-Diazides, synthesis from alkenes, 65
Diazirines, as carbene precursors, 89
–, laser flashphotolysis, 93
Diazocarbonyl compounds, decomposition, 333
Diazo compounds, thermal decomposition, 363
α-Diazo-β-ketoalkylphosphonates, 110
Diazolidine, synthesis from diazo compounds, 106
Diazomalonates, 69
Diazomethane, reactions with alkenes, 83
–, – with zinc iodide, 412
Dibenzo-18-crown-6, 243
1,4-Dibenzoquinones, 205
Dibenzoylethene, 336
Diborane-dimethyl sulphide complex, 432
Dibromocarbene, 90
Dibromodichlorocyclopropanes, 92
1,2-Dibromo-1,1-diphenylethane, 258
1,6-Dibromohexane, 245
Dibromoiodomethane, 250
2,5-Dibromopyridine, alkynation, 166
1,1-Dibutyl-1-stannacyclohexa-2,5-diene, 473
Dibutyltin oxide, 325
Dibutyltin oxyperoxide, 325
1,1-Dichloroacetophenone, 331
1,1-Dichloroalkenes, 258
Dichlorobis(cyclopentadienyl) titanium(IV), 393
Dichloroborane, 366
Dichlorobromide anion, 248
Dichlorocarbene, addition to 1,2,2,-trimethylbicyclo[1.1.0]butane, 100
–, reactions with dichloroalkenes, 89
α,α-Dichlorocyclobutanones, 74
1,1-Dichloro-2,2-difluoroethene, 157
1,2-Dichloroethane, oxychlorination, 262
(Z)-1,2-Dichloroethene, alkynation, 166
Dichloroketene, 204
–, addition to alkenes, 74
3-Dichloromethyl-2,3-dimethylindole, 96
1,2-Dichloropropane, 258
2,2-Dichlorothioacetals, 353
2,5-Dichlorothiophenium-1-bis(methoxylcarbonyl)methylide, fragmentation, 110
Dicobaltoctacarbonyl, 429
–, complexes with alkynes, 563
Dicyanoacetic ester alkali metals salts, reactions, 352
Dicyanoanthracene, 75
Dicyanomethane, reaction with triethylphosphite, 332
Dicyclohexylborane, 420
–, addition to alkenes, 220
Dicyclohexylcarbodiimide, 368
N,N'-Dicyclohexylisourea, catalytic reduction, 6
1,1-Dicyclopropylallene, 131
1,3-Dideuterio-1,1,3,3-tetraphenylpropane, 128
Diels–Alder adducts, 111
Diels–Alder reactions, 56, 76, 82, 171, 198, 204, 574
–, catalysts, 450
–, with inverse electron demand, 137
2,4-Dienals, protection, 573
Dienamines, 207
(η^4-1,3-Diene)iron tricarbonyls, 574
(η^4-Diene)metal complexes, 572
Dienes, reduction, 46, 47
–, synthesis, 426, 427, 505, 534
–, – from sulphonium ylides, 362
1,3-Dienes, 135, 172, 189, 191, 204, 207, 208
–, reactions with iron carbonyls, 573
1,5-Dienes, cyclisation, 530
2E,4Z-Dienes, 153
Dienoic acids, synthesis, 373
(η^5-Dienyl)iron tricarbonyl complexes, 576
(η^5-Dienyl)metal complexes, 576
Dienynes, reactions with Normant copper reagents, 400

Diethoxyphosphonocyclopentan-2-ones, 110
Diethylacetylene, 188
Diethylaluminium chloride, 209
Diethylaluminium-2,2,6,6-tetramethylpiperide, 448
Diethylamine–chlorotrifluoroethene adduct, 256
N,N-Diethylaminoprop-1-yne, 177, 206
Diethyl azodiacarboxylate/triphenylphosphine, reagent for dehydration of alcohols, 512
Diethylenolphosphate esters, reactions with bromotrimethylsilane, 343
Diethyl ketenedicarboxylates, 200
N,N-Diethyl-β-ketoamides, 194
Diethylmethyloxonium ion, 45
Diethyl 2-phenethynylphosphonite, 329
Diethyl phosphorochloridodithionate/dimethylamine, 340
Diethyl phosphorocyanidate, synthesis from cyanogen bromide and triethylphosphite, 334
Diethyl phosphoroiodate, 333
–, structure correction, 333
Diethyl tartrates, 325
Diethyltertiarybutylphosphite, 329
(*E*)-3,4-Diethyl-2,2,5,5-tetramethylhex-3-ene, 47
Diethylzinc, 411
1,1-Difluoroalkanes, 245
Difluorocarbene, 95
–, addition to perfluoro-3-methylindene, 91
–, reaction with imines, 96
–, – with 2-methoxy-7,7-dimethylnorbornadiene, 100
–, – with quadricyclane, 99
cis-6,7-Difluorododecane, 258
Difluorodomethylene alkenes, 95
Difluoromethane disulphonyl fluoride, 376
α,α-Difluoromonoalkylamines, 256
7,8-Difluorotetradecane, 58
meso-7,8-Difluorotetradecane, 58
1,1-Dihalides, carbonyl equivalents, 347
1,2-Dihalides, synthesis, 246
1,1-Dihaloalkenes, 157
Dihaloborane–methylsulphide complexes, 52
Dihalocarbenes, 1,2-addition to 1,3-dienes, 97
–, 1,4-addition to 1,3-dienes, 97
Dihalocarbenes precursors, 490
Dihalogenoalkanes, reactions, 249
–, synthesis, 245
Dihalomethylhalogeno ketones, 76
Dihydroarenes, 361
Dihydrofurans, 142
4,5-Dihydrooxazoles, 66
1,5-Dihydropentalene, 106
Dihydropyrrolo[2,3-*b*]pyridines, 199
2,7-Dihydrothiepin-1,1-dioxides, 371
1,1-Diiodides, synthesis from 1,1-dibromides, 246
α,ω-Diiodides, 249
1,1-Diiodoalkenes, 262
Diiodotriethoxylphosphorane, correction of structure, 333
Diisobutylaluminium hydride, 172, 403, 447
1,2-Diketones, 210
–, synthesis from alkynes, 320
1,3-Diketones, synthesis, 319
Dimesitylborane, 433
4,5-Dimethoxy-1,2-benzoquinones, 165
Dimethoxycarbene, addition to 1,2,2,-trimethylbicyclo[1.1.0]butane, 100
Dimethoxycyclopentenetrione, 190
1,1-Dimethylalkenes, 162, 163
1,1-Dimethylallene, 132
4-(Dimethylamino)pyridine *N*-oxide, 345
Dimethylborane, 52
N,N-Dimethylbromoacetamide, 409
Dimethylbromosulphonium bromide, brominating reagent for alcohols, 360
2,3-Dimethylbutane, 27
–, cyanation, 30
2,3-Dimethylbutene, 90
2,3-Dimethylbut-2-ene, 57, 109
–, reaction with monochloroborane-methylsulphide, 51
3,3-Dimethylbutene, 57, 67
Dimethyldioxirane, 27, 68
Dimethyldithio *S,S*-dioxide acetals, 348
Dimethylenecycloalkanes, 323
Dimethylether, synthesis from methanol, 44

2,3-Dimethylindole, reaction with phenyl(trichloromethyl)mercury(II)–sodium iodide, 96
2,2-Dimethyl-1-iodo-5-hexene, reaction with lithium isopropoxide, 235
2,2-Dimethylpropane, bromination, 219
1,1-Dimethylsilaindene, 202
Dimethyl sulphide/boranes, 348
Dimethylsulphonium methide, reactions with enol ethers of β-diketones, 360
Dimethyl trifluoromethylmalonate, 297
Dimethylvinylcarbene, ambiphilic reactivity, 89
Dinitrogentetraoxide/1,5-diazabicyclo[5.4.0]undec-5-ene, 327
Diorganoberyllium compounds, 389
1,3,2-Dioxaborinanes, synthesis, 284
2,3-Dioxopent-3-ynoates, 175
DIPAMP, 540
1,5-Dipentalene, 106
Diphenylacetylene, 113, 193
Diphenylcarbene, 84
Diphenyl(chlorophenylmethylene)-iminium chloride, reactions with alcohols, 227
1,1-Diphenylethene, 261
2,5-Diphenylfuran, synthesis from dibenzoylethene, 336
Diphenyliodinium iodide, reactions with trialkylphosphites/copper salts, 337
Diphenylmethylene, 84
Diphenylphosphono cyclopentan-2-ones, 110
1,3-Dipolar cycloadditions, 508
Dipotassium allenides, 128
Disodium allenides, 128
Disodium tetracarbonylferrate, reactions with alkyl and acyl halides, 552
Displacement reactions with Grignard reagents, 396
2,4-Disubstituted furans, 360
Disulphides, anodic oxidation, 357
–, cleavage with organolithium derivatives, 388
–, reactions, 357
–, – with carbanions, 357
–, reduction to thiols, 357
–, synthesis, 355
–, – from alkylthiosulphates, 355
–, – from thionitrites, 355
–, – from trisulphides, 357
–, thiolate displacement reactions, 358
Ditertiaryalkyl ethers, synthesis, 311
Ditertiarybutoxyacetylene, 190
2,6-Ditertiarybutyl-4-cresol, 67
Ditertiarybutyl-*N*,*N*-diethylphosphoramidate, 338
Ditertiarybutyl-*N*,*N*-diethylphosphoramidate/1-tetrazole, 338
Ditertiarybutyltrioxide, synthesis, 319
Ditertiarydiperoxymonocarbonate, photolysis, 319
1,3,2-Dithiaborolane, 434
Dithioacetals, elimination reactions, 351
Divinylbenzene polymers, 224
Divinyl sulphides, 351
Diynes, 189
–, hydration, 175
1,3-Diynes, reduction, 172
1,4-Diynes, 209
1,3-Diyn-5-ols, reduction, 172
1-Dodecene, addition of fluorine, 58
Double Wittig reactions, 509
Dupont process, conversion of 1,3-butadiene to adiponitrile, 542
"Durham route", 212

Electron deficient allenes, reactions with *C*-phenyl-*N*-alkylnitrones, 138
Electrophilic carbene complexes, 555
Electrophilic carbenes, 555
β-Elimination of boron, 442
Elimination reactions, 41
Enals, conjugated, 497, 504
–, reactions with anylselenols, 534
–, reduction by formic acid/tris(triphenylphosphine) ruthenium dichloride, 299
–, – by triphenyltin formate, 300
–, synthesis, 505
–, – from aldehydes, 474
Enamines, hydroboration, 38
–, synthesis, 460, 505
β-Enaminesulphoxides, 364
Enaminoketones, 113
Endo-6-(2,2-difluorovinyl)bicyclo[3.1.0]-hex-2-ene, 99
Ene, alkylation, 489

Ene reactions, catalysis by organoaluminiums, 450
Enoate esters, selenenylation, 524
Enoic acid derivatives, reactions with anylselenols, 534
Enoic acids, synthesis, 505
Enol acetates, 526
–, synthesis, 422
γ,δ-Enolate esters, synthesis from allylic alcohols, 524, 529
Enolates, β-alkylation, 524
Enol ethers, synthesis, 316
Enol phosphates, cleavage by trialkylalanes, 343
Enol phosphonates, 159
Enolsilyl ethers, 471, 526
Enol triflates, 159
Enones, addition reactions, 475
–, conjugated, 504
–, –, synthesis, 506
–, protection, 471
–, reactions by arylselenols, 534
–, reduction, 544
–, – by sodium hydride/magnesium bromide, 299
–, – by sodium hydride/nickel(II) acetate, 299
–, – by triisobutylalane, 299
–, – by triisobutylane, 299
–, synthesis, 505
–, – by oxidative deselenation, 467, 522
γ,γ-Enones, isomerisation, 482
Enynes, 191
1,3-Enynes, 168, 172
Epoxidation catalysts, 67
Epoxides, deoxygenation, 69
–, reduction by lithium selectride, 279
–, ring-opening, 300, 301, 448
–, – by 1,5-diazabicyclo[5.4.0]undecene-5, 301
–, – by diethylaluminium-2,2,6,6-tetramethylpiperidine, 300
–, – by organoboranes, 300
–, – by organolithiums/lanthanoid derivatives, 301
–, – by trimethylsilyltriflate/2,5-lutidine, 301
–, synthesis, 448
–, – from carbonyl compounds, 385
α,β-Epoxysilanes, 460
Erythro acetylenic diols, 164
Esperamicins, 205
Esters, reaction with 1-chlorobenzotriazole, 315
–, reduction, 273
–, – by dimethylsulphide, 348
–, – by trichlorosilane/ditertiarybutyl peroxide, 313
–, selenenylation, 524
–, of inorganic acids, 310
Ethene, acetoxylation, 71
–, biosynthesis, 35
–, homologation, 50
–, oxidation, 70
Ethers, cleavage, 314, 315
–, – by sodium phenylselenide/borane, 478, 533
–, – with hydrogen bromide, 284
–, – with trimethylsilyl bromide, 285
α-Ethoxyaminophosphonates, synthesis, 335
Ethoxyethyl ethers, 73
2-(Ethoxymethyl)tetrahydrothiophene, synthesis, 358
Ethylaluminium dichloride, 209
Ethylamine, manufacture, 63
3-Ethylcyclohexanone, synthesis, 520
Ethylenebis(salicylideniminato)cobalt, 238
Ethylene glycol monoperfluoroalkanoates, 254
2-Ethyl-2-oxazoline, 328
3-Ethyl-2-pentene, chlorohydroxylation, 58
Ethyl sulphones, 370
P-O-Ethyl-*P*-thyrophosphate-tris(trialkylammonium) salts, 340
Ethyl trimethylsilylacetate, reactions with Grignard reagents, 40
Ethynyloxiranes, reactions with organoboranes, 300
Ethynylsulphones, Michael reactions, 369

Ferrocenes, 195
Finkelstein reaction, 221, 235, 242
–, alternatives, 234
Fischer carbene complexes, 111, 555

–, reactions with nitriles, 114
Fischer–Tropsch synthesis, 2, 46, 50, 542
Flash vacuum pyrolysis, 49
Fluorene anion, 236
Fluorenone, synthesis from 9-hydroxyfluorene, 319
Fluorenylidene, 84
1-Fluoroalkenes, 261
1*H*-1-Fluoro-1-alkenes, 95
2-Fluorobutane, 220
6, Fluoro-6-dodecenes, 258
α-Fluoroenamines, 256
1-Fluoroindene, 234
1-Fluoro-2-iodoindane, 248
1-Fluoro-2-iodo-1-phenylalkanes, 248
5-Fluorononane, 258
7-Fluoronortricyclene, 261
N-Fluorosulphonamides, 234
Fluorosulphonates, reactions with potassium iodide, 252
α-Fluoro thioethers, 364
Fluorotrichloromethane, 89
Formamides, dehydration, 512
Formic acid, enophile, 450
–, reactions with Grignard reagents, 395
2-Formylalkylphosphonates, 335
Fourier transform ion cyclotron resonance mass spectrometry, 45
Friedel Crafts alkylation, 55
Fumaric acid derivatives, 175
Functionalised allenes, 140
β-Functionalised organosilanes, 463
Furan-2-ones, 198
Furan-3-ones, 198
Furans, 153, 195–197, 361
–, synthesis, 509

Gas chromatography of group IIIA organometallics, 429
Gem dihalides, reactions with mercuric oxide/pyridine-hydrogen fluoride, 246
Geraniol, epoxidation, 325
Geranyl acetate, 531
Germanium peroxide, 317
Germanyl ethers, Wittig rearrangement, 314
Gif system, 24
Grignard reagents, 390
–, additions to alkenes, 391
–, – to alkynes, 391
–, catalyses using transition metals, 393
–, catalytic addition to conjugated enones, 405
–, coupling of alcohols, 407
–, – of ethers, 407
–, – of selenides, 407
–, – of silyl ethers, 407
–, – of thioethers, 407
–, – of tosylates, 407
–, cross coupling, 550
Guaianolides, 529

Hafnium compounds, 539
α-Haloacetals, elimination, 159
Haloacetophenones, reactions with neopentylphosphites, 331
–, – with trimethylphosphite, 331
Haloalkenes, synthesis, 425
α-Haloalkenylboranes, 438
2-Haloalkylmethyl sulphides, 348
α-Haloalkynborates, rearrangement, 437
Haloborane–dimethyl sulphide complexes, 435
Haloboranes, 434, 435
α-Haloborate complexes, preparation, 439
Halocarbenes, reactions with cyclopropanation, 89
α-Halocarbonyl compounds, 485
Halodemetallation, 470
α-Halo esters, 409
Halo-etherification, 530
α-Halo ethers, 316
Halofluoroalkanes, 59
Halogenation-deborohalogenation, 436
Halogen exchange, 221
Halogenoalkanes, fragmentation, 266
Halogenoalkenes, 5, 257, 259
–, ozonolysis, 262
–, synthesis from alcohols and alkenyl halides, 260
–, – from carbonyl compounds, 261
–, – from organoseleniums, 260
Halogenoalkynes, reactions with halogenocarbenes, 264
–, – with organometallics, 264
Halogenocarbenes, addition to alkenes, 267

–, cyclopropanation reactions, 265
–, physical properties, 264
–, reactions, 267
–, synthesis from organometallics, 267
–, – from organosilicons, 267
–, – from trihalogenoethanoic acid derivatives, 266
–, – from trihalogenomethanes, 265
–, toxicity, 265
Halogenochromates/tertiary amines, 286
α-Haloketones, transylidation, 506
Halolactonisation, 531
α-Halomethylmercury compounds, 423
Halonium ions, 45
α-Halotrialkylboranes, 438
Hemiacetals, reactions with Grignard reagents, 396
–, synthesis, 454
Hemithioacetals, synthesis, 347
Heptafluoropropene, 246
Heptane, 16
2-Heptanol, bromation, 230
Heptatriene, planar singlet carbene, 86
1-Heptyne, 264
1-Heptynedimethylalane, 129
Heterocycles, synthesis from alkenes, 82
Heterocyclic organomercurials, 420
Heterocyclic syntheses, 195
Heteropolyoxometalates, 21
Heteropolytungstic acid, 19
Hexachloroacetone/triphenylphosphine, 260
Hexachlorobutadiene, reactions with triols, 353
Hexachloroethane-phosphine complexes, 511
(*E,E*)-Hexadecan-5,7-diene, 263
Hexa-2,4-dienoate esters, reduction, 540
Hexafluorobut-2-yne, 211
1,1,2,3,3,3-Hexafluoropropyldiethylamine, 226, 256
1,1,2,2,3,3-Hexamethyl-4,5-bis-(methylene)cyclopentane reaction with dihalocarbenes, 97
Hexamethyldimethylene cyclopentane, 268
Hexamethylsiloxane, reactions with diphosphorus pentoxide, 340
Hexane, aromatization, 17
–, carbonylation, 30
Hexane-1,6-diol, 245
Hex-1-ene, 220
–, addition of carbon tetrachloride, 251
–, reaction with methylborane, 52
1-Hexyne, 111
Higher order cuprates, 545
Higher order mixed organocuprates, 546
Homoallyl alcohols, 485
–, synthesis, 302
Homoallyl sulphides, 349
Homoendate anion, equivalents, 361
Homoenolates, 386
Homogeneous catalysis, 2, 539, 542
Homolytic alkylation of ketones, 79
Homopropargylic acids, 129
Horner–Wittig reactions, 316
Housane, 140
Hunsdiecker reaction, 233
Hydrazones, reduction with bis(benzyloxy)borane, 7
–, – with catecholborane, 7
Hydride reagents, 4
Hydroalumination, alkenes, 445
–, alkynes, 446
Hydroboration, 431
–, alkenes, 37
–, amines, 348
–, kinetics, 52
–, reagents, 51
–, regioselectivity, 432
–, stereoselectivity, 433
Hydrocarbons, methylation, 8
Hydrocyanation, 62
–, alkenes and alkynes, 541
Hydroformylation catalysts, 54
Hydrogenation catalysts, nickel–boron alloy, 49
–, nickel–phosphorus alloy, 49
–, ruthenium complexes, 49
–, triphenylphosphine meta-trisulphonate, 50
–, uranium(III) chloride Wilkinsons catalyst, 49
Hydrogen fluoride/melamine, fluorinating, 226
Hydrogen fluoride–tantalum pentafluoride, 55
Hydrogen iodide, addition to alkenes, 57

Hydrogen peroxide, oxidative replacement of boron, 435
Hydrogen transfer reactions, 541
Hydrosilation, 541
2-Hydroxyalkanoic acids, oxidative decarboxylation, 47
3-Hydroxyalkanoic acids, deoxygenation, 47
β-Hydroxyalkylphosphorus compounds, 504
β-Hydroxyalkynes, 170
γ-Hydroxyalkynes, 164
Hydroxyallenes, 142
β-Hydroxyamides, 409
Hydroxyamines, dehydration, 512
3-Hydroxy-5-aryl-1-pentynes, 203
2-Hydroxycyclohexanone, synthesis, 320
2-Hydroxyethanesulphonyl chloride, 376
9-Hydroxyfluorene, oxidation, 319
2-Hydroxy ketones, 175, 321
–, synthesis from alkenes, 321
Hydroxylamine sulphonic acid, 437
Hydroxyphenyliodoso mesylate, 77
α-Hydroxyphosphonates, synthesis, 331
2-Hydroxypyridine esters, 243
N-Hydroxypyridine-2-thione esters, reductive decarbonylation, 7
3-Hydroxypyrroles, synthesis, 509
β-Hydroxyselenides, reductive elimination, 519
Hydroxysulphonium salts, synthesis from cyclic ethers, 359
–, vicinal, 355
β-Hydroxysulphoxides, 74, 75
–, pyrolysis, 364
1-Hydroxytriazole, 512

Imidazoles, 201
–, trifluoromethylation, 254
Imidazoline, 64
Imines, 557
–, oxidation, 322
–, reactions with carbenes, 411, 412
–, – with Grignard reagents, 395
Iminium salts, 192
Iminophosphoranes, synthesis, 512
β-Iminosulphones, synthesis from sulphonyl carbanions and nitriles, 373
Inactivated dienes, reactions with Normant copper reagents, 401
Incipient halide ion sources, 232
Indane, oxidation, 319
Indanes, 193
Indanone, synthesis from indane, 319
Indenes, 112
–, synthesis, 555
Indoles, 197, 202
Insertion reactions, alkylcarbenes, 101
–, alkylidenecarbenes, 101
Interhalogens, 173
Internal alkynes, 170
Internal hydroxyl nucleophiles, 559
Intramolecular aminations, 560
Intramolecular olefin insertion, 552
Iodine, 173
Iodine monofluoride, 248
Iodine tris(trifluoroacetate), 9, 26
Iodoalkanes, oxidation, 263
–, – by *m*-chloroperbenzoic acid, 225
Iodoalkenes, 173, 259
Iodoalkynes, 163, 167, 202
1-Iodobut-1-ene, 263
Iodobutyltributyl tin, 475
4-Iodocyclohexene, 231
Iododestannylation, 470
2-Iodo-2,3-dimethylbutane, 57
"Iodofluoride" adducts, 248
1-Iodohept-1-yne, reaction with butyllithium, 264
1-Iodohexane, 220, 261
1-Iodooctane, 232
2-Iodooctane, synthesis from 1-octene, 57
Iodosobenzene, 23, 69, 471
Iodosulphonylation, 369
Iridium compounds, 539
Iron carbene complexes, 556
Iron-oxo porphyrins, 25
Iron pentacarbonyl, 251
Iron perchlorate, 25
Iron–sodium hydride complexes, reduction of aldehydes, 544
–, – of ketones, 544
Iron tetraphenylporphyrin chloride, 22
Isoalkanes, removal, 9
Isobutane, oxidation, 23
Isobutene, acylation with acetylfluoroborate, 77

Isocyanates, cobalt complexes, 564
–, reactions with phosphocumulene ylides, 508
Isopinocamphenyl boranes, 443
Isoprene, addition of trichlorotrifluoroethane, 306
Isopropylethenes, NMR spectra, 35
Isopropylidene, synthesis from silylvinyl triflate, 103
–, – from vinyl triflate, 103
Isoquinolinium salts, 201
Isothiocyanates, reactions with phosphocumulene ylides, 508
Isoxazoles, 135
Isoxazolidines, 138, 163
Isoxazolines, synthesis from alkenes, 83

Karplus equation, 14
Ketals, synthesis, 421
Ketene thioacetals, synthesis from ortho thioesters, 351
α-Ketoaldehydes, 210
β-Ketoalkynes, 168, 169
α-Ketoamides, 210
α-Ketoesters, 210
β-Ketoesters, 169
–, α-alkylation, 543
–, synthesis, 319
Ketones, α-alkylation, 543
–, also see alkanones
–, catalysed addition reactions, 571
–, chlorination, 511
–, methylation, 8
–, oxidation by thallium salts, 455, 456
–, reactions with Grignard reagents, 393
–, reduction, 544
–, reductive amination, 543
–, selenylation, 522
–, synthesis, 298, 299, 421
–, – from acyl anion equivalents, 385
–, – from alkenes, 70, 78
–, – from alkenylmercuric chlorides, 429
–, – from boranes, 439
–, – from carboxylic acid derivatives, 385
Ketophosphonates, 331
2-Ketophosphonic esters, 332
Ketophosphonium salts, 506
β-Ketosilanes, 458
β-Ketosulphonic acids, synthesis from ketones, 375
β-Ketosulphoxides, as anion sources in alkylation and acylation reactions, 364
Kolbe reaction, 8

β-Lactams, synthesis, 558, 561
Lactones, 144
–, ring opening by sodium phenylselenide, 533
–, selenenylation, 524
–, synthesis from cyclic ketones, 320
Lanthanum modified alumina, 48
Lead alkynides, 169
Leukotrenes, synthesis, 505
Lewis acid complexes, 546
Lindlar catalyst, 171
γ-Lithioalkoxyallenes, as homoenolate equivalents, 147
Lithio(cyclopropyl)silanes, addition to aldehydes, 103
α-Lithioselenides, 515–517
α-Lithioselenoxides, 515
Lithium alkynides, 163–165
Lithium aluminium hydride, 293
Lithium butyldi-isobutylhydroaluminate, 4
Lithium copper hydrides, 4, 294
Lithium di-isobutylbutylhydroaluminate, 4
Lithium 1-(dimethylamino)-naphthalenide (LDMAN), 103
Lithium dimethylcopper, 74
Lithium/ethylamine, 6
Lithium halides, promotion of reactions between alcohols and hydrogen halide, 285
Lithium *N*-isopropylcyclohexylamide, 365
Lithium/selenium exchange, 515
Lithium triethylborohydride, 4, 293

Magnesium–anthracene systems, 391
Magnesium–ene reaction, 390
Magnesium enolates, 394
Maleic acid derivatives, 175
Maleic anhydride, 25
–, reaction with triethylphosphite, 336
Malonic esters, alkylation, 242
Manganese porphyrins, 69

Manganese tetraphenylporphyrin chloride, 22, 23
Marcus theory, 221
Meerwein–Ponndorf–Verley reaction, 273, 300
Menschutkin reaction, 232
Mercaptals, alkylation by alkyl halides, 346
Mercaptals (thioacetals), synthesis, 347
2-Mercapto-1,3-benzoxazole, 349
2-Mercaptocarboxylic acids, chiral syntheses, 346
Mercaptoethanol, 278
α-Mercurated carbonyl compounds, 422, 425
Mercuric alkenes, demercuration, 469
Mercuric perchlorate, 312
Mercuric thiocyanate, alkylation, 295
Mercuric trifluoroacetate, 319
Metal acyl"ate" complexes, 553
Metal acyl enolates, 554
Metalation, reactions, 393
Metal borides, 3
Metal carbene complexes, 187, 197
Metal-catalyses cyclooligomerisation, 564
Metal enolates, 386
Metal–halogen exchange, 392
Metal hydrides, 380
Metallaoxetanes, 68
Metallocycles, 187
Metallocyclobutane mechanism, 46
Metallocyclobutenones, rearrangement, 112
Methane, conversion, 2
–, dimerisation, 1
–, halogenation, 250
–, hydroxylation, 22
–, oligomerization, 1
–, oxidation, 20, 21
–, oxidative coupling, 18
Methanedisulphonyl fluoride, electrochemical fluorination, 376
Methanesulphonic acid, dealkylation of alkylboranes, 437
Methane thiol, synthesis from methanol and hydrogen sulphide, 345
Methanol, deoxygenation, 44
Methionine, 35
Methoxyallene, reaction with 4-methylphenylsulphenylchloride, 133
Methoxyallenes, 137
–, as acyl anion equivalents, 145
–, lithiation, 145
(2-Methoxybutadiene)iron tricarbonyl, 574
Methoxycarbonylallene, reaction with 2,3-dimethylbutene, 133
Methoxychlorocarbene, reactions with electron rich alkenes, 89
4-Methoxycyclopentene, 105
2-Methoxy-1,3-dioxolanes, 44
Methoxyethoxymethyl ethers, synthesis, 298
3-Methoxylphenylcarbene, generation from 3-methoxy-3-phenyldiazirine, 109
Methoxymethyl ethers, synthesis, 297
α-Methoxymethyltins, 474
1-Methoxyperfluoro-2-heptene, 256
2-Methoxypyridines, 201
Methoxythiomethyl ethers, synthesis, 298
Methylation reactions, 8
Methylborane, 52
2-Methylbutane, 25, 30
2-Methyl-3-butyn-2-ol, 168
Methyl dec-8-enoate, oxidation, 421
Methylenecyclohexanes, 470
Methylenecyclopentannulation, 186
Methylenecyclopropane, 105, 106
Methyleneoxazolines, 137
Methyl esters, synthesis of alkenes, 77
Methyl formate, addition to alkenes, 77
8-Methylhexadecane, 11
2-Methylhexane, 341
3-Methylhexane, 341
Methyl ketones, 364
–, synthesis from alkenes, 320
Methylmercury cyanate, 173
(*Z*)-3-Methyl-2-pentene, 37
Methyl phenylselenides, as nucleophiles, 532
2-Methylpropene, hydrochlorination, 57
Methyl propiolate, 207, 209
N-Methylpyrrolidone, 261
Methylseleno ketones, 523
Methylthiochloromethane, 298
Methyltitanium triisopropoxide, 275
7-Methyl-7-tridecanol, 52

Methyltrifluoromethyl sulphoxide, 363
Methyltri(octyl)ammonium chloride, 279
Michael addition, 349, 352, 368, 369
–, enones and enals, 504, 505
–, reverse, 510
Michaels Arbusov reaction, 502
Milbemycins, 183
Mixed cuprates, 544, 546
–, reduction of acid chlorides, 544
–, – of aldehydes, 544
–, – of alkylhalides, 544
–, – of enones, 544
–, – of ketones, 544
–, – of tosylates, 544
Molybdenum catalysed reactions with allylacetates, 570
Molybdenum catalysts, 321, 322
Molybdenum compounds, 539
Molybdenum hexacarbonyl, 6
Molybdenum-pyridine-HMPA, oxidative replacement of boron, 436
Monoalkynylalanes, 169
Monohalogenoalkanes, reactions, 234
–, synthesis from alkanes, 219
–, – from alkenes, 220
–, – from alkylhalides, 221
Monohydric alcohols, 269–309
Monsanto process for acetic acid synthesis, 542
Mortierelle isabellina NRRL1757, 362
Myrcene, 573

1-Naphthols, 187
Naphthyldiazomethanes, flash vacuum pyrolyses, 87
2-Naphthyltellurium trichloride, 60
Navenone A, synthesis, 514
Neopentyl alcohols, 282
Nickel boride, 3
Nickel catalysts, chiral induction, 406
Nickel complexes, reduction of organic halides, 544
–, – of tosylates, 544
Nickel compounds, 539
Nickel phosphine complexes, 550
Nickel salts, catalysts for cross coupling of organohalides and Grignard reagents, 406
Nickel tetracarbonyl, 425, 429
Nickel(II) phthalocyanine, epoxidation catalyst, 324
Nickeloles, 208
Nitric acid esters, reaction with nitriles, 328
Nitrile oxides, reactions with alkenes, 83
Nitriles, reduction, 348
–, synthesis from cyanates, 387
–, – from cyanogen halides, 387
Nitroalkanes, 417
–, oxidation, 322
Nitroalkenes, 417
4′-Nitrobenzenesulphenanilide, 177
N-Nitrocollidinium tetrafluoroborate, 326
Nitro compounds, reduction, 469
–, reductive cyclisation, 334
2-(1-Nitroethyl)-2-oxazoline, synthesis, 328
β-Nitromercurials, 417
Nitroperchlorates, 81
Nitroperoxypropyl nitrate, 80
1-(4-Nitrophenyl)-1,3-propyldinitrate, 329
Nitrosoureas, 105
Nitrostyrenes, coupling, 162
β-Nitrosulphones, reductive eliminations, 372
Nonatetrene, planar singlet carbene, 86
Non-conjugated dialkynes, 168
Non-conjugated dienes, synthesis, 506
Non-stabilised alkylidenetriphenylphosphoranes, reaction with chlorodifluoromethane, 95
Non-therminal alkynes, synthesis from phosphinyldiazomethane, 160
Norbornadiene, 99
2-Norbornene, 261
Normant copper reagents, 398
–, addition reactions, 398
–, carbonylation, 403
–, carboxylation, 403
–, coupling reactions, 402
Nucleophilic carbene complexes, 558
Nucleophilic carbenes, 555
Nucleoside phosphates, synthesis, 339
–, – from phosphites, 322
Nucleoside phosphites, oxidation, 322, 339

1-Octadecene, 42
1-Octanol, 287
(R)-2-Octanol, 284
1,2,4,6,7-Octapentaenes, 106
1-Octene, 41
–, addition of fluorine, 58
2-Octenes, 41
Octylamine, 327
Octylnitrate, synthesis, 327
Octyne, 259
Olefination, PO activated, 502
Olefin metathesis, 557
Onium intermediates, 45
Onium phase-transfer catalysts, 221
Organic halides, alkylation, 545
–, cross coupling, 483
–, – in chiral synthesis, 383
–, reduction, 469, 544
Organic hydroperoxides, 67
Organic perchlorates, 77
Organic tosylates, alkylation, 545
–, reduction, 544
Organoalanes, reactions with carbonyl compounds, 447
Organoaluminiums, 444, 524
–, reactions with acids, 448
–, – with aldehydes, 447
–, – with arylhalides, 450
–, – with epoxides, 447
–, – with esters, 449
–, – with ketones, 447
–, – with lactones, 449
–, – with sulphonates, 449
–, – with vinylhalides, 450
–, – with vinyl phosphates, 449
Organoantimonates, 513, 514
Organoarsenicals, 493, 513, 514
Organo azides, 388
Organobarium halides, reactions with carbonyl compounds, 408
–, – with conjugated enynes, 408
Organoberyllium compounds, toxicity, 389
Organobismuths, 513, 514
Organoboranes, oxidation, 283
–, – by amine-N-oxides, 280
Organoboron compounds, 430
–, NMR spectra, 430
–, photochemistry, 430
–, preparation, 430
Organoborons, 429
Organocadmium compounds, preparation, 414
–, use in synthesis, 415
Organocadmiums, 414
Organocalcium halides, reactions with carbonyl compounds, 408
–, – with conjugated enynes, 408
Organocopper coupling reagents, 8
Organocuprates, 522, 524, 545
–, oxymercuration, 425
Organogalliums, 444, 451
Organogermaniums, 457, 465, 477
Organoindiums, 451
Organolanthanides, 539
Organoleads, 457, 477
Organolithiums, 382
–, NMR spectroscopy, 383
–, reactions promoted by sonification, 303
–, use in organic synthesis, 383
Organomagnesium compounds, 390
Organomagnesium halides, 390
–, reactions with transition metals, 398
Organomercurials, 415
–, alkylation, 425
–, carbene-transfer reactions, 423
–, carbon–carbon bond formation, 422
–, carbon–halogen bond formation, 419
–, carbon–hydrogen bond formation, 416
–, carbon–nitrogen bond formation, 422
–, carbon–oxygen bond formation, 420
–, carbon–phosphorus bond formation, 422
–, carbon–sulphur bond formation, 422
–, carbonylation, 428
–, coupling of alkenyl groups, 426
–, deuteriolysis, 416
–, displacement reactions, 416, 422
–, NMR spectroscopy, 415
–, photolysis, 202
–, preparations, 415
–, protonolysis, 416
–, reactions with alkenes, 423, 424
–, – with alkynes, 423
–, – with diethyl diazocarboxylate, 424
–, reductive displacement of mercury, 416
–, structure, 415

–, synthesis, 446
–, toxicity, 415
–, transition-metal catalysed reactions, 425
–, use in synthesis, 416
Organometallic compounds, analysis, 380
–, chromatography, 381
–, electron spin resonance, 380
–, free radicals, 381
–, gas chromatography, 381, 389
–, general properties and reactions, 381
–, high energy processes, 381
–, high-field Fourier-transform multinuclear NMR, 380
–, infrared and Raman spectra, 380
–, methods of preparation, 378
–, Mössbauer spectroscopy, 380
–, NMR spectroscopy, 380
–, photochemistry, 381
–, photoelectron spectroscopy, 380
–, soft ionisation methods in mass spectrometry, 380
–, spectroscopy, 380
–, X-ray crystallography, 381
Organopalladium compounds, reactions with alkoxycarbonylmercury chlorides, 427
Organopoloniums, 514, 536
Organoseleniums, 344–377, 514–537
Organosilicons, 457–465, 477, 478, 481
Organosilyl anions, 463
Organostannyl carbenes, 478
Organostrontium halides, reactions with conjugated enynes, 408
Organosulphurs, 344–377
Organotelluriums, 344–377, 514, 536, 537
Organothalliums, 451
–, uses in synthesis, 452
Organotin compounds, addition reactions, 467
–, biocides, 466
–, displacement of halides, 466
–, lithiation, 472
–, NMR spectra, 466
–, nucleophilic displacement, 466
–, reducing agents, 469
–, substitution reactions, 478
–, toxicity, 466
–, use in synthesis, 469
Organotin epoxides, 471
Organotins, 457, 466, 477
Organotin thiolates, 369
Organotoranes, carbonylation, 282, 283
Organozinc bromides, 409
Organozincs, 409
–, reactions promoted by sonification, 303
Organozirconiums, 524
Orthoesters, reactions with Grignard reagents, 396
3-Oxacarbacyclin precursors, 503
Oxalate hydroxamic esters, 6
Oxaphosphetane ring formation, 36
1,2-Oxaphospholenes, 135
Oxaphospholes, 330
1,4-Oxathiane-borane, 432
1,3-Oxathiolanones, synthesis from chiral hydroxyacids, 346
Oxathioles, 127
β-Oxaylides, 498
6*H*-1,2-Oxazines, 137
Oxaziridines, synthesis from imines, 322
1,3-Oxazoles, 198
2-Oxazolidinones, 66
Oxazolines, substrates for carboxylic acid synthesis, 386
2-Oxazolines, nitration, 328
Oxidative addition-insertion, 550
Oxidative addition-transmetalation, 550
N-Oxides, synthesis from tertiaryamines, 321
Oximes, reactions with tri-alkylphosphites, 335
Oxiranes, 169
–, catalyses ring opening, 406
–, cleavage, 230, 533
–, deoxygenation, 533
–, reactions with Grignard reagents, 396
–, – with Normant copper reagents, 401
–, – with ylides, 507
Oxo acids of sulphur, 348
3-Oxoalkenylphosphonium salts, 509, 510
5-Oxohexanoic acids, 194
Oxo-reaction, 542
Oxychlorination, 262
Oxymercuration, 416

Oxythallation, allenes, alkenes, alkynes, 452
Ozonides, structure, 66
Ozonolysis, organomercurials, 420

Palladium acetate–triphenyl phosphine catalysts, 566
Palladium carbonylation catalysts, 428
Palladium complexes, coupling of alkenylmercury halides, 426
Palladium compounds, 539
Palladium diene complexes, 572
Palladium(0) complexes, reduction of allyl ethers, 544
–, – of sulphones, 544
–, – of thioethers, 544
–, – of trialkylsilyl ethers, 544
Palladium(0) tinhydride, reduction of acid chlorides, 544
Palladium(II)-nitrocomplexes, 70
Palladium(II) triphenylphosphine, 54
[2.2]Paracyclophane, 204, 333
Paraformaldehyde, homologation with Grignard reagents, 396
Paraquinones, 187
Pentacarbonyl iron, 403
Pentadiene, planar singlet carbene, 86
1,3-Pentadienes, reactions with iron carbonyls, 576
2,4-Pentadienol complexes, 576
Pentadienyl cations, 105
Pentadienyl–$Fe(CO)_3$ compounds, 576
1,1,2,3,3-Pentafluoropropyl iodide, addition to ethene, 253
Pentamethylcyclodienyl titanium(II) chloride/sodium naphthenate, 49
Pentane, oxidation, 25
–, photochemical carbonylation, 29
2-Pentanone, asymmetric reduction, 278
E-Pent-2-ene, 541
Z-2-Pentene, isomerisation, 541
1-Pentyne, 114
Pent-4-yne-1-ol, 166
Peptide synthesis, 602
Perfluoroalkanoic acids, 255
Perfluoroalkanoyl chlorides, 262
Perfluoroalkenes, addition reactions, 246, 256
–, nucleophilic reactions, 256
–, reactions with sulphur dioxide, 375
Perfluoroalkenylhalides, 262
Perfluoroalkylalkynes, reactions with ethyl chloroformate, 262
Perfluoroalkylbromides, 252
Perfluoroalkyliodides, addition reactions, 253
–, coupling with alkyl halides, 255
–, enzymic reactions, 255
–, homolysis, 253, 254
–, hydrolysis, oxidation, reductive dimerisation, 252
–, reactions with electrophiles, 254
–, – with nucleophiles, 253
Perfluoroalkylsulphides, 346
Perfluoro-3,4-dimethylhex-3-ene, reactions with alkyl lithiums, 262
Perfluoro-1-heptene, 256
Perfluoropropene, reactions with dialkylamines, 256
Perfluorosulphonic esters, 374
Perfluorosulphonyl chlorides, 376
Perfluorosulphonyl iodides, 376
Peroxodisulphate ion/copper(II) chloride, 232
Peroxymercuration, 263, 318
Perylene, reactions with lithium alkoxides, 235
Peterson elimination, 460
Peterson olefination, 369, 463, 477
Phenothiazine, 236
2-Phenoxy-1,3,2-benzodioxaphosphole, 366
Phenylacetylene, 158
1-Phenylalkenes, iodofluorine adducts, 248
1-Phenylalkynes, 452
Phenylallenes, photochemistry, 130
Phenylchlorocarbene, reactions with electron rich alkenes, 89
Phenylchromium carbene complexes, 111
Phenylcyclopropane, peroxymercuration/demercuration, 319
Phenyldichlorophosphate, 231
Phenyldimethylsilyl cuprates, reactions with α,β-unsaturated carbonyl compounds, 132
Phenylethynol, 163
Phenyliodonium acetate, 160

2-Phenyl-3-isopropylindazole, 103
Phenylpropenones, 145
Phenylselenides, 147
–, reductive deselenation, 6
α-Phenylselenoaldehydes, synthesis from enol ethers, 529
Phenylselenoalkenyl sulphones, 185
Phenylselenocarbocyclisation, 531
Phenylseleno derivatives, 7
Phenylselenoetherification, 530
α-Phenylselenoketones, 524, 526
–, alkylation, 521
Phenylselenol, reactions with alcohols, 534
Phenylselenolactonisation, 531
Phenylselenotrimethylmethane, formyl anion equivalent, 517
Phenylselenotrimethylsilane, 534
β-(Phenylseleno)vinylsulphones, 153
Phenylselenylbromide, addition to alkenes, 260
Phenylsulphides, 101
N-Phenylsulphimides, 350
Phenylsulphinylcyanides, 367
Phenylsulphonylallene, reaction with diazopropane, 138
Phenyltellurium tribromide, 61
Phenylthiocarbene, 101
Phenyltrimethylsilylselenide, 365
Pheromones, synthesis, 504
Phosphate esters, hydrolysis, 342
Phosphate monoesters, 339
–, synthesis, 339, 340
–, – from alkanols and trimethylsilylpolyphosphate, 340
Phosphinates, 502
Phosphine oxides, 110
Phosphines, reactions with arylselenonitriles, 526
–, synthesis, 493
Phosphinoruthenium hydrides, 306
Phosphite/metal complexes, 334
Phosphocumulene ylides, 508
Phosphonamides, 502
Phosphonate anions, 506
Phosphonic esters, synthesis from phosphate triesters, 343
Phosphonoacetate esters, 110
Phosphonothionates, 502
Phosphonylsulphide intermediates, oxidation, 363
α-Phosphonylsulphoxides, synthesis, 363
Phosphoramidates, deprotonation, 513
Phosphoranes, synthesis uses, 494–503
Phosphoric acid esters, phosphorylation reagents, 340
Phosphoric acids, conversion into phosphoric acid chlorides, 511
Phosphoric esters, synthesis from phophate triesters, 343
Phosphoric triesters, synthesis from phosphorous triesters, 338
Phosphorous acid β-chloroethyl esters, 329
Phosphorus, organochemistry, 493
–, reductive removal, 506
Phosphorus heterocycles, 199
Phosphorus–oxygen bond cleavages, 397
Phosphorus pentasulphide, 365
Phosphorus stabilised carbanions, reactions with electrophiles rather than carbonyl compounds, 506, 507
Phosphorus–sulphur ester bond cleavages, 397
Phosphorus triesters, oxidation, 338
Phosphorus trihalides, polymer bound, 230
Phosphorus triiodide, 365
Photochemical dearylation, 339
Phthalic anhydride, 25
Phthalimidomethyl alkylsulphoxides, 366
Pinacol-binacolone rearrangement, 450
Pinacol thiocarbonate, reaction with triethylphosphite, 333
α-Pinene, hydroboration, 283
Piperidines, 151
Polyacetylenes, 211, 212
Poly(diacetylenes), 212, 213
Polyethylene glycols, 222
–, dehydrochlorination promoters, 257
Polyfluoroalkenes, 255
–, synthesis from trifluorochloroethene, 255
Polyfluoroalkyl halides, reactions, 252
–, synthesis, 252
Polyfluoroalkyl iodides, synthesis, 252
Polyhalogenoalkanes, 250

–, environmental toxicity, 214
Polyhalogenoalkenes, copper catalyses additions to alkenes, 55
Polyhalogenoalkylsilanes, carbene precursors, 267
Polymerisation, alkenes and alkynes, 542
Polystyrene anion exchange resins, 241
Polysubstituted alkynes, 176
Polysulphide mediated reactions, 357
Polysulphides, 357, 359
–, lubricant additives, 358
–, reduction to thiols, 345
–, synthesis from disulphides, 358
–, – from nitrososulphides, 358
–, – from thiols, 358
Polyvinylpyridine, 224, 228
Potassium/18-crown-6/tertiarybutylamine, 6
Potassium phosphinotrihydroborate, 4
Potassium superoxide, reactions with alkyl tosylates, 317
–, – with bromoalkanes, 317
Potassium tertiarybutylhydroperoxide, 317
Primary amines, synthesis, 334
Propadienylcyclopentadienes, 106
Propargyl acetate, 168
Propargyl alcohols, 194, 198
Propargylallyl alcohols, synthesis, 303
Propargylamines, 165
Propargyl bromide, 184
Propargyl bromides, reactions with aluminium, and trimethyl borate, 155
Propargyl chloride, metallation, 164
Propargyl ethers, reactions with tertiarybutyl lithium, 147
Propargyl halides, 165
Propargyl selenoxides, 517
Propargyl toxylates, 165
N-Propargyl-2-vinylpyrrolidines, 201
Propargyl ylides, 501
Propene, epoxidation, 69, 324
–, ozonolysis, 66
Propene oxide, 324
Propenylallenes, photochemistry, 140
Propylene glycerol-1,2-dinitrate, 80
Propylsulphinic acid, 366
Prostaglandins, synthesis, 440, 454, 470, 472, 505
Protodesilylation, 459
Pseudohalides, 522
Pseudomonas oleovorens, 362
Pyranones, 206
Pyran-2-ones, 197
Pyrans, 145
Pyrazoles, 155, 195, 196
Pyrazolines, conversion into pyrazoles, 138
Pyrazolinones, 160
Pyridine-*N*-oxide, synthesis, 321
Pyridines, 189
Pyridinium chlorochromate, oxidation of borate esters, 436
–, – of primary alkylboranes, 436
Pyridinium phosphate betaines, 339
(Pyridin-2-yl)selenyloctadecane, reaction with hydrogen peroxide, 42
Pyridones, synthesis, 564
Pyrid-2-ones, 200, 201
Pyrones, synthesis, 555
2-Pyrones, 189
Pyrophosphoric acid, 339
Pyrroles, synthesis, 509
Pyrrolidines, 151
(*E*)-4-Pyrrolidinyl-3-heptene, hydroboration, 38
Pyrrolidone anion, generated electrochemically, 242
Pyrrolizines, synthesis, 509

Quadricyclane, 268
Quasiylides, synthesis, 336
Quinoline chlorochromate, 287
Quinolines, synthesis, 509
1,4-Quinones, synthesis from malonyl- and phthaloyl-cobalt complexes, 564
Quinoxaline, acylation, 322

Radical alkylations, 442
Ramberg–Bäcklund reaction, 42, 371, 373
Raney nickel, 6, 58
Reactions of alkanes, 15
– of transition-metal organic compounds, 539
Reactive ylides, 497
1,2-Rearrangement reactions, 458
1,4-Rearrangement reactions, 458

1,3-Rearrangements, 458
Reductive deselenization of selenocompounds, 6
Reductive desulphurization of thiocompounds, 6
Reductive metalation, phenyl trioethers, 382
Reformatsky reaction, 409
Retro Diels–Alder reactions, 162
Rhenium compounds, 539
Rhodium complexes, 54
–, coupling of alkenylmercury halides, 426
Rhodium compounds, 539
Rhodium η^5-dienyl complexes, 577
Rhodium phosphine complexes, 54
Ribose derivatives selenylation, 527
Ring opening polymerisation of cycloolefins, 556
Ritter reaction, 328
Ruthenium catalysts, 24
Ruthenium compounds, 539

Samarium diiodide, 58, 243, 343
Schiff bases, synthesis from 2-alkenals, 335
Schlosser–Wittig reaction, 498
"Schrock" type carbene complexes, 555
Secondary alcohols, deoxygenation, 5
Selenamides, as nucleophiles, 524
Selenate esters, synthesis, 534
Selenenamines, 524
Selenides, synthesis from alcohols, 535
Seleninic anhydrides, 536
Selenium, oxidative removal, 517
–, reactions with Grignard reagents, 398
–, reductive removal, 520
Selenium(IV) chloride, reactions with alcohols, 228
Selenium dioxide, oxidations, 535
Selenium electrophiles, 522
Selenium reagents, use in Wittig reaction, 36
– in synthesis, 515
Selenium stabilised carbanions, 515
Selenoacetals, 515
–, hydrolysis, 519
Selenoalkynes, 169, 171, 471, 526
α-Selenocarbonyl compounds, deselenylation, 520
Seleno compounds, reduction, 469
1,2,3-Selenodiazoles, decomposition, 161
Selenoketals, 515
α-Selenoketones, 536
–, synthesis, 526
Selenonium salts, synthesis, 519
Seleno-orthoesters, 519
Selenophiles, 460, 515
Selenoxides, 534
Selenyl electrophiles, 530
Selenyl halides, reactions with pseudohalides, 467, 522
Sharpless epoxidation, 306, 325
[3,3]-Sigmatropic rearrangements, 561
Silanes, deoxygenation of alcohols, 5
Sila Pummerer rearrangement, 458
Silicon β-effect, 459, 461
Silicon sulphide, 365
α-Siloxyaldehydes, 488
Silver trifluoroacetate, 318
Silver wool, 224
Silylalkynes, 167
Silylallenes, 140
C-Silylallenes, 146
Silyl cuprates, reactions with enones, 463
Silyl-1,5-diyn-3-enes, 168
β-Silylenolates, synthesis, 463
Silylenol ethers, synthesis from β-ketosilanes, 458
Silyl-1,3-enynes, 168
Silyl ethers, Wittig rearrangement, 314
N-Silylimines, reactions with Grignard reagents, 395
Silylosyallenes, 142
β-Silylvinyl lithiums, 474
Silylylides, 362
Simmons–Smith reaction, 410
Skell–Woodworth rules, 88
Snoopy reaction, 498
Sodium amalgam, 6
Sodium bis(2-methoxy ethoxy)aluminium hydride, 365
Sodium iodiide/chlorotrimethylsilane, reversal of phenylselenoesterification, 531
Sodium metaperiodate/ruthenium dioxide, 74
Sodium naphthalene, 239

Sodium perborate, oxidative replacement of boron, 436
Sodium-potassium alloy, 6
Sodium *p*-toluenesulphinate, alkylation, 242
Sodium *p*-toluenesulphinate/ 1,2-dimethoxyethane, 241
Solvomercuration, 415
Solvomercuration–palladation, 421
Soybean oil, 2
Spherical hydrocarbons, 99
Spiro compounds, 207
Spiroketals, 183
Stable sulphonium ylides, 361
Stable ylides, 496
β-Stannylalcohols, deoxystannylation, 491
Stannyl anions, 468
β-Stannylpropanal, 492
N-Stannyltetrazoles, 201
Steroidal nitrites, oxidation, 327
Steroids, synthesis, 190, 247
Stilbenes, synthesis, 505
Strained alkenes, 34
Substituted cumulenes, 173
Succinic esters, 427
Sulphenamides, 357
Sulphenic acids, reduction, 353
–, synthesis from butyl sulphoxides, 353
Sulphenic acids and their derivatives, 353, 354
Sulphenic esters, hydrolysis, 354
–, synthesis from sulphenyl chlorides, 354
–, – from sulphenyl iodides, 354
β-Sulphenylacrylic acids, 352
Sulphenyl chlorides, synthesis from thiolacetates, 354
Sulphenyl compounds, synthesis from alkenes, 81
Sulphenyl halides, reactions, 354
Sulphides, oxidation, 322, 369
–, synthesis, 350
– from thioiminium salts, 350
Sulphinate esters, synthesis from butylhydroxyalkyl sulphoxides, 364
Sulphinates, rearrangement to sulphones, 371
Sulphinic acid derivatives, iododesulphonylation, 376
Sulphinic acids, reactions with dinitrogen tetraoxide, 367
–, reduction, 353
Sulphonamides, 65
–, deprotonation, 513
Sulphonate esters, reduction, 376
Sulphonates, reactions with selenides, 534
Sulphone carbanions, 372
Sulphones, desulphonylation, 371
–, nitration, 328
–, reduction, 544
–, synthesis, 242
–, – from sulphides, 267
Sulphonic acids, esterification with orthoformates, 375
–, protection as 2,2,2-trifluroethyl esters, 375
–, reactions, 376
–, synthesis from disulphides, 374
–, – from thiols, 374
Sulphonic esters, nitration, 328
Sulphonium methide, 360
Sulphonium salts, 358
–, reactions with sulphides, 360
–, synthesis from alkenes, 360
–, – from Pummerer rearrangement of sulphoxides, 360
Sulphonomercuration, alkynes, 369
Sulphonyl diazomethanes, substitutes for diazomethane, 368
Sulphonyl fluorides, 370
Sulphonyl nitrates, 367
Sulphoxides, 353
–, deoxygenation to sulphides, 365
–, oxidation, 368
–, synthesis, 322, 353
–, – from sulphides, 322
Sulphur, reactions with Grignard reagents, 398
Sulphur dioxide, promoter of dehydrogenation, 33
–, reactions with butadienes, 371
–, – with 1,6-trienes, 371
Sulphuryl chloride, 173
Sulphuryl chloride/silica, 365
β-Sultones, NMR studies, 375
Symmetric organocuprates, 545

–, alkylation reactions, 545
Synthesis gas, 1, 542

Tantalum alkylidene complexes, 558
Tebbe's reagent, 558
Tellurium, reactions with Grignard reagents, 398
Tellurium(IV) chloride, reactions with alcohols, 283
Telluro compounds, reduction, 469
Terminal acetylenes, synthesis from ketones, 159
Terminal alkenes, oxidation, 320
–, peroxymercuration, 318
–, solvomercuration, 420
–, synthesis from carboxylic acid chlorides, 42
Terminal alkynes, 170
–, alkylation, 166
–, dimerization, 172
Terminal allenes, 237
–, synthesis, 116
Tertiaryalkyl halides, 223
–, synthesis from alcohols, 228
Tertiary amines, oxidation to *N*-oxides, 321
Tertiaryamylhydroperoxide, 321, 322
Tertiarybutylallyl ether, alkylation, 316
Tertiarybutylhydroperoxide, 254, 318, 319, 322, 324, 325, 536
–, reactions, 323
Tertiarybutylhydroperoxide/iron(II) sulphate, 322
Tertiarybutylhydroperoxide/osmium tetraoxide, 321
Tertiarybutyl nitrate, synthesis, 327
Tertiarybutyl nitrite, oxidation, 327
Tertiarybutylperoxide, 290, 305
Tertiarybutylperoxide/molybdenum hexacarbonyl, 288
2-Tertiarybutylperoxyalkylmercury(II) acetates, 318
Tertiaryhydroperoxides, 321
Tervalent phosphorus-polyhalogenoalkane reagents, 511
Tetraalkyltin reagents, 430
Tetrabutylammonium iodate, 365
Tetrabutylammonium *p*-toluenesulphinate, 296
Tetrabutyloctahydrotriborate, 273
Tetrachlorocyclopropanes,1,2-dechlorination, 89
Tetrachloromethane/phosphine complexes, 511
Tetrachloromethane/phosphinic polymers, halogenating reagents, 229
Tetrachloromethane/triphenyl-phosphine, halogenating reagent, 229
Tetraenes, 130
Tetrafluoroethene, 246
2,3,3,3-Tetrafluoropropionates, 256
Tetrahalogenomethanes, synthesis, 250
Tetrahydrothiophene, explosive oxidation, 362
Tetrakis(triphenylphosphine)palladium, 145, 428
Tetrakis(triphenylphosphine)palladium/dimethylformamide, 288
Tetramethylallene, reaction with diethyl acetylenedicarboxylate, 139
–, – with sulphur dioxide, 139
Tetramethylammonium ditertiary-butylphosphate, reactions with alkylhalides, 338
Tetramethylbutane, 32
Tetramethylene sulphoxide, 362
N,*N*,*N'*,*N'*-Tetramethylethylene-diamine, 283
1,1,3,3-Tetraphenylallene, lithiation, 128
1,1,3,3-Tetraphenylallyl potassium, 128
Tetraphenylporphyrin, 22
Tetrapropylammonium perruthenate, 289
Tetrasulphides, synthesis from methoxycarbonylhydrazones, 357
Tetratertiarybutylethene, attempted synthesis, 35
1,1,2,2-Tetra(tertiarybutyl)ethane, 9
Tetra[triphenylphosphinyl]-ruthenium(II)hydride, 77
Tetra-(2,4,6-triphenylphenyl)porphyrin, 22
Thallium, replacement by acetoxy groups, 453
–, – by alkoxy groups, 453
–, – by cyanide ion, 453

–, – by halides, 453
–, – by hydrogen, 453
–, – by selenocyanates, 453
–, – by thiocyanates, 453
Thallium(II)carboxylate dibromides, 233
Thallium triacetate, 451, 452
Thallium trifluoroacetate, 451
Thallium trinitrate, 366, 454
Thexylborane, 433
Thexylchloroborane/methyl sulphide complex, 365
Thiacrown ethers, 348
Thia heterocydes, 199
Thiazetidines, reduction, 345
Thiazoline esters, 558
Thiiranes, reduction by chiral boranes, 347
–, synthesis from alkenes, 82
Thiiranium species, 177
Thioacetals, 360
Thioalkylcarbenes, reactions, 348
Thioanisoles, 197
Thiocarboxylic acids, 347
Thio compounds, reduction, 469
Thiocyanates, oxidation, 367
Thioether carbene complexes, 188
Thioethers, 349
–, reactions with Grignard reagents, 396
Thiol–alkene cooxidation, 74
Thiols (mercaptansor hydrosulphides), 344
–, chlorination, 511
–, cyclisation, 531
–, esterification, 347
–, industrial uses, 344
–, occurrence, 344
–, oxidation with *m*-chloroperoxybenzoic acid, 366
–, physical properties, 344
–, preparations, 344
–, reactions, 344, 347
–, synthesis, 240
–, – from diaza compounds, 346
–, thermodynamic properties, 344
–, toxicities, 344
Thione to thiol rearrangement, 345
Thionoesters, synthesis from carboxylic esters and 2,4-bis(4-methoxyphenyl)-1,3,2,4-dithiadiphosphetane-2,4-disulphide, 314
Thiophenes, 195, 196
–, synthesis, 509
Thiophosphoryl bromide, 365
Thiosilanes, 349
Thiosulphate salts, 358
Thiosulphenyl chlorides (chlorodisulphides), 356
Thiosulphinates, oxidation, 377
Thiosulphinyl chlorides, conversion into thiosulphinate esters, 368
–, synthesis from thiols, 367, 368
Thiosulphonates, 345
β-Thiosulphonium salts, 355
Thiyl radical, 48
Threo-β-hydroxyalkylacetylenes, 141
Threo-vicinal disulphones, 373
Thymine, alkylation with trialkylphosphates, 341
Tin hydrides, addition to alkenes and alkynes, 469
Tin peroxide, 317
Titanium alkynides, 169
Titanium catalysts, 322
Titanium trichlorides, 365
Titanocene hydrides, 393
Titanylallenes, 141
Toluenes, nitration, 328
Tosylhydrazones, reduction with hydride reagents, 7
N-Tosylsulphilimines, 350
Transition-metal alkyne complexes, 563
Transition-metal carbenes, 454
Transition-metal carbonyls, 551
Transition-metal catalysed decomposition of diazo compounds, 557
Transition-metal hydrides, reduction of conjugated double bonds, 542
Transition-metal organometallics, 539
Transmetalations, electrophilic metals, 549
Trialkylaluminiums/tetrakis(triphenylphosphine)palladium, 343
Trialkyl boranes, 275, 431
–, free radical bromination by *N*-bromosuccinimide, 438
–, iodination, 220
–, oxidation, 318
–, reactions with formaldehyde, 282

–, – with oxygen, 318
Trialkylphosphates, 337, 340
–, dealkylation with *p*-roluenesulphonic acid, 341
–, solvents for arene electrophilic substition reactions, 342
–, – for nitration of arenes, 342
–, synthesis from dialkylhydrogen phosphites, 341
–, – from dialkylphosphates, 342
–, transesterification, 342
Trialkylphosphites, 331, 502
–, catalysts of thiol synthesis, 240
–, dealkylation, 337
–, fluorination, 333
–, reaction with dialkylaroylphosphonates, 336
–, – with diphenyliodinium iodide, 337
–, synthesis from alcohols, 329
–, – from alcohols by oxidative phosphorylation, 338
Trialkylphosphites/copper(I) halides, 330
Trialkylsilylethers, reduction, 544
γ-Trialkylstannyl compounds, 471
Trialkynylalane, 165
1,2,3-Triazoles, synthesis, 508, 510
Tributyldiisophosphorane, 241
Tributylstannyllithium, 348
Tributyltin hydride, 5, 275, 366
Trichlorocyclopropanes, 158
1,1,2-Trichlorostyrene, 158
1,1,2-Trichlorotrifluoroethane, 254
1,2,3-Trideuterio-1,1,3,3-tetraphenylpropane, 128
Triethylborane, 5
Triethylphosphite, desulphurisation reactions, 333
–, reactions with alkylazides, 334
–, – with *trans*-10-azidodihydrophenanthrene-9-ol, 334
–, – with cyanogen bromide, 334
–, – with dicyanomethane, 332
–, – with iodine, 333
–, – with maleic anhydride, 335
–, – with nitrosobenzene and benzene, 334
–, – with oximes, 335
–, – with pinacolthiocarbonate, 333
–, – with Schiff bases, 280, 335
Triethylphosphite/carbon tetrachloride, 330
Triethylsilane, 3, 5
3,3,3-Trifluoro-1-arylpropynes, 161
Trifluorochloroethene, lithiation/silylation, 255
2,2,2-Trifluorodiazoethane, 375
α-Trifluorodimethylsulphone, 42
Trifluoromethanesulphenyl chloride, reactions with orthoesters, 354
Trifluoromethane sulphonates, 262
Trifluoromethane sulphonic acid, 82
Trifluoromethyl copper complexes, 255
1,1,2-Trihalogenoalkenes, 157
Trihalogeno esters reactions with Grignard reagents, 394
Trihalogenomethylalkenols, 76
Triiron dodecacarbonyl/inorganic oxide catalysts, 46
Triisobutylalane, 299
Triisopropylphosphite, 331
Triisopropylsilyl ethers, synthesis, 298
Trimethylaluminium addition to alkynes, 548
Trimethylamine-*N*-oxide, 280
–, oxidative replacement of boron, 436
Trimethylborate, 273
Trimethylboroxine-pyridine complexes, 430
Trimethylbromosilane dimethyl sulphoxide, 348
Trimethylchlorosilaneb dimethyl sulphoxide, 348
2,3,3-Trimethylpenta-1,4-dienes, 100
Trimethylphosphate, 342
–, reaction with polyphosphoric acid, 341
Trimethylphosphite, 330, 332, 333
–, reaction with bibenzoylethene, 281, 336
Trimethylphosphite/copper(I) iodide, 333
Trimethylsiloxy thiols, alkylation, 352
Trimethylsilylalkynes, 140, 459
1-Trimethylsilylalkynes, 180
Trimethylsilylallenes, reactions with aldehydes and ketones, 140
Trimethylsilyl bromide, 230
Trimethylsilyl chloride, 5
Trimethylsilyl chloride/zinc, 365
Trimethylsilylcyanide, 174

Trimethylsilyldiazomethane, 106
β-Trimethylsilylethanol, protection reagent for "active" hydrogen atoms, 463
Trimethylsilyl ethers, synthesis, 298
Trimethylsilyl iodide, 231
1-Trimethylsilyl-1-methoxyallene, 147
Trimethylsilyloxyenones, flash vacuum pyrolysis, 152
Trimethylsilyl pyrophosphate, 231
Trimethylsilyl triflate/triethylamine, 339
1-Trimethylsilyl-2-ynes, reduction, 172
Trimethylsulphonium chloride, 359
Triorganostannyl trichloroacetates, 490
Triphenylarsine ligands, 484
Triphenyldiiodophosphorane, 296
Triphenylphosphine, 297
Triphenylphosphine/carbon tetrabromide, 295
Triphenylphosphine dibromide, polymer supported, 295
Triphenylpropargyltin, 237
Triphenylpyrilium nitrate, 327
2,4,6-Triphenylpyrylium salts, 232
Triphenylsilane, 5
Triphenylstannylallenes, 171
Triphenyltin formate, 300
Tris (3,6-dioxaheptyl) amine, 6
Trisisopropylsilyl ethers, synthesis, 298
1,3,5-Tris(tertiarybutyl)benzene, 342
Tris (trimethylsilyl) butanone, 106
Tris(triphenylphosphine)rhodium chloride, 316
Trisubstituted alkenes, oxidation, 321
–, synthesis, 39
Trisulphides, conversion to disulphides, 357
–, synthesis from cyclic thiosulphonates, 357
–, – from thiols and diimidazol-l-yl sulphide, 357
Tritertiarybutylphosphite, 329
Trityl perchlorate, 313
Trityl tetrafluoroborate, 292, 315
Tungstate catalysts, 346
Tungsten carbene complexes, 555
Tungsten catalysts, 322
Tungsten compounds, 539
Tungstic acid, 74

Ultrasound, 52
–, dispersion of sodium or potassium, 382
–, effect on organometallic compounds, 381
Undec-5-yne, 264
Unsaturated alcohols, epoxidation, 324, 325
Unsaturated aldehydes; see also enals, 176, 177
Unsaturated aluminium compounds, rearrangement reactions, 429
Unsaturated boron compounds, rearrangement reactions, 429
α,β-Unsaturated carboxylic acids, 176
α,β-Unsaturated esters, oxidation, 319
α,β-Unsaturated ketones, 171, 181
–, oxidation, 319
–, reaction with titanium alkoxy "ate" complexes, 304
α,β-Unsaturated nitriles, 174, 182
Unsaturated nitriles, synthesis from bromostyrenes, 263
γ,δ-Unsaturated sulphones, 370
α,β-Unsaturated sulphonic acids, synthesis by Wittig–Horner reactions, 375
Uracil, alkylation, 341
Uranium(III) chloride Wilkinsons catalyst, 49
Ureas, 65
–, chlorination, 511
Urethanes, 65
–, cyclisation, 531

(–) Valeranone, 410
Vanadium catalysts, 267
– in epoxidations, 324
Vanadium oxydiacetonide, 324
Vicinal diacetates, synthesis from alkenes, 72
Vicinal diamines, synthesis from alkenes, 64, 65
Vicinal dicarboxylic acids, synthesis from alkenes, 74
Vicinal dichloroalkanes, 59
Vicinal diesters, synthesis from alkenes, 78
Vicinal dihalides, dehalogenation, 45, 46
–, dehydrohalogenation, 157

Vicinal diols, deoxygenation, 44
–, oxydative cleavage, 322, 323
Vicinal ditriflates, synthesis from alkenes, 78
Vicinal fluoroalkyl sulphates, 77
Vinyl acetates, 180, 418
Vinylalanes, 448, 547
Vinylallenes, sigmatropic rearrangements, 130
Vinyl anions, 472
Vinylboranes, 431
Vinylboronates, reactions with iodine, 259
Vinyl bromides, halogen exchange, 261
–, synthesis, 260
–, – from trimethylvinylsilanes, 259
Vinylcarbenes, geometric isomerism, 89
Vinyl chlorides, 261
–, dehydrochlorination, 264
Vinylcopper intermediates, 545
Vinylcoppers, coupling with 1-halogeno-1-alkenes, 263
Vinylcuprates, 470
–, addition to α,β-sulphones, 39
–, reactions, 547
2-Vinylcyclopropanediazonium ion, 105
Vinylcyclopropylidenes, 105
Vinyl ethers, 135
–, lithiation, 474
–, metalation, 316
–, rearrangement, 317
–, synthesis from α-haloethers, 316
Vinyl fluorides, 262
Vinyl halides, 257, 470
–, reactions with arylselenide anions, 534
–, synthesis, 420, 505
Vinylketimines, reaction with acetylenic esters, 207
Vinylketones, reactions with phosphocumulene ylides, 508
Vinyllithiums, 470
Vinylmagnesium halides, 397
Vinylmercurials, 420
–, photolysis, 422
Vinylmercuric salts, decomposition to vinyl-esters and -ethers, 420
Vinyloxiranes, reactions with organoboranes, 300
Vinyl phosphates, 332
Vinyl phosphonates, synthesis, 330, 332
–, – from vinyl halides, 332
Vinyl phosphonium salts, 509, 554
Vinylselenides, 134, 178, 525, 526, 536
–, hydrolysis, 519
Vinylselenoxides, 515
Vinylsilanes, 130, 179, 180, 370, 459
–, epoxidation, 460
–, protiodesilylation, 459
–, synthesis from vinylsulphones, 491
Vinylstannanes, 178, 467
Vinyl sulphides, 351, 389
–, Michael additions, 373
–, reduction to alkyl sulphides, 349, 350
(*E*)-Vinyl sulphides, 177, 352
Vinyl sulphones, 370, 491
–, cross coupling, 39
–, nucleophilic displacement reactions, 372
–, reduction with sodium dithionite, 39
Vinyl sulphonium salts, 361
Vinyl sulphoxides, 397
–, optically pure, 362
Vinylsulphur compounds, synthesis, 505
Vinyltetrahydrofurans, 145
Vinylthioethers, reactions with Grignard reagents, 396
Vinyltin compounds, deuterolysis, 470
–, halodestannylation, 470
–, polymeric, 473
–, protonolysis, 470
Vinyltins, 480
Vinyltrimethylsilane, 433
Vitamin E side-chain, synthesis, 193, 570
Vitride, 345

"Wacker" process, 69
Wadworth–Horner–Emmons reaction, 338, 384, 386, 502–505
Water gas reaction, 542
Wilkinson's catalyst, 49, 429, 539, 553
Wittig reaction, 3, 9, 36, 95, 494–503
–, use of selenium compounds, 37
Wittig reagents, 558
Wittig rearrangement, 314, 477
Wolff–Kishner reduction, 6

X-ray structural analysis data of Group IA organometal derivatives, 382

Xanthates, 346

Ylide anions, reactions with aldehydes, 499
Ylides, acylation, 506
–, alkylation, 506
–, benzylic, 500
–, propargylic, 501
–, reactions with episulphites, 37
–, – with selenium compounds, 37
–, – with sulphur, 37
–, synthesis, 494
–, – by desilylation, 493
Ynoic acids, 198
Ynones, photocyclisation, 203
–, reduction, 172

Zeolite-carbenoid species, 44
Zeolite catalysts, 15–17
Ziegler–Natta polymerisation, 544
Zinc, 5
Zinc alkyls, use in synthesis, 409
Zinc alkynides, 167
Zirconium alkenyls, 548
Zirconium alkylidene complexes, 558
Zirconium alkyls, transmetalation, 548
Zirconium catalysed organo-aluminium addition to acetylenes, 431
Zirconium complexes, 445
Zirconium compounds, 539
Zirconium hydrides, silica bound, 48
Zirconocene halides, 63

Ref. 547 R61

Second supplements [sic] to
the 2nd edition of Rodd's